KT-117-534

THE  TIMES REFERENCE
ATLAS OF THE WORLD

TIMES BOOKS

LONDON

CONTENTS

© Collins Bartholomew Ltd

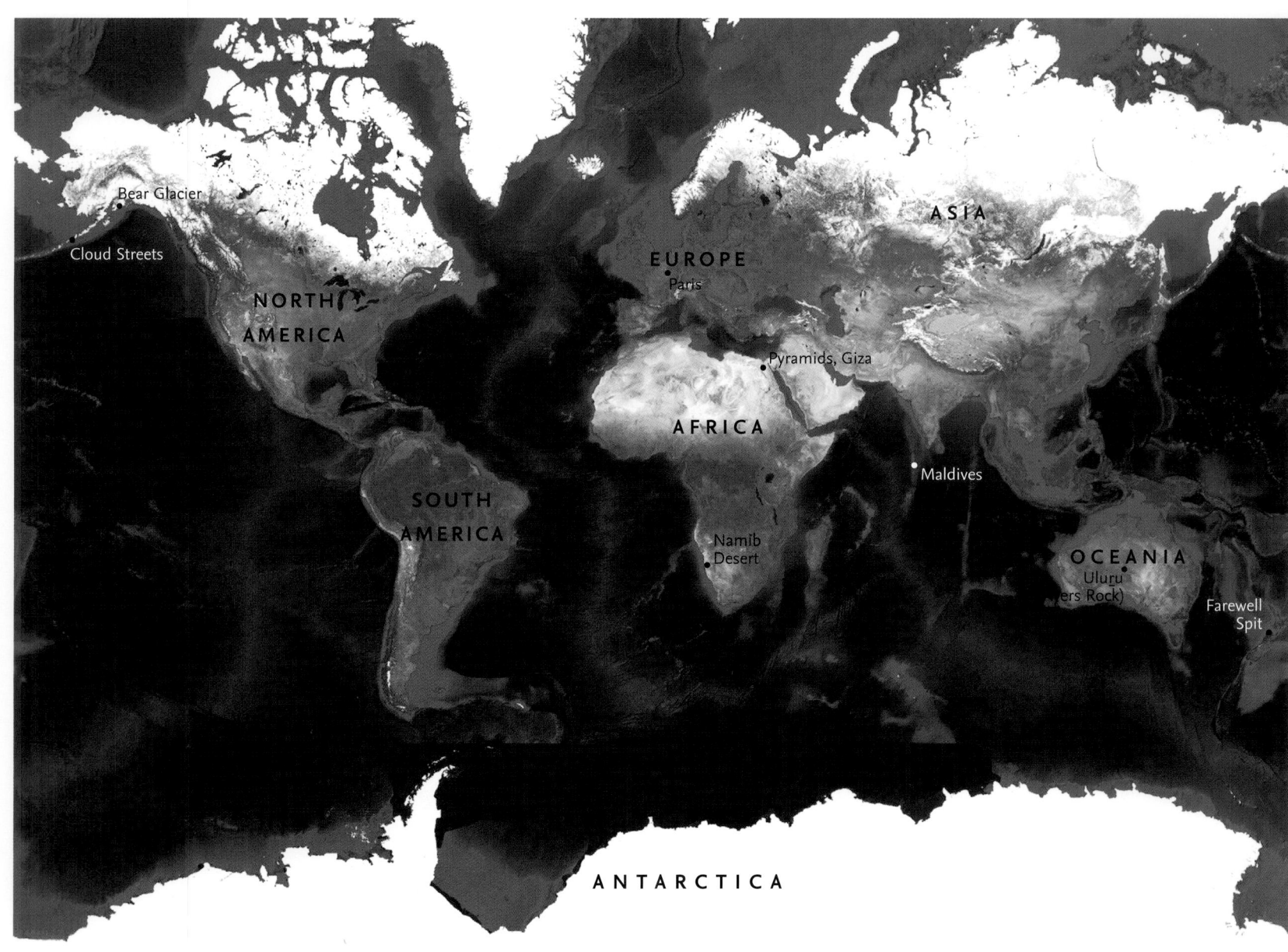

6–7 OCEANIA

The continent of Oceania comprises Australia, New Guinea, New Zealand and the islands of the Pacific Ocean. The main Pacific island groups of Melanesia, Micronesia and Polynesia sit amongst the complex of ridges and troughs which make up the Pacific seafloor. Notable among these, and visible extending northwards from New Zealand, are the Kermadec and Tonga trenches – the latter reaching a depth of 10 800 m at Horizon Deep. Australia itself appears largely dry and barren, its vast interior consisting of several deserts, with brighter salt lakes in the low artesian basin of the east central area. The east coast of Australia, separated from the interior by the Great Dividing Range – the source of the continent's longest rivers the Murray and the Darling – is more densely vegetated. New Guinea is covered by dense tropical forest, while New Zealand displays a great variety of land cover types, most prominent being the snow-capped Southern Alps on South Island.

8–9 ASIA

This vast continent – the world's largest – covers an enormous area and contains a great variety of landscapes, evident on this image. It stretches from the Mediterranean Sea in the west to the far east of the Russian Federation and Japan, and from arctic Siberia in the north to the tropical islands of Indonesia. The Caspian Sea – the world's largest lake – is prominent in the west. The snow-capped Caucasus mountains stretching from the Caspian Sea to the Black Sea clearly mark the divide between Asia and Europe. Just east of the Caspian Sea lies the complex shape of the Aral Sea. This was once the world's fourth largest lake, but is now drastically reduced in size because of climate change and the extraction of water for irrigation. In the centre of the image, the long arc of the mountain ranges of the Himalaya, Karakoram, Hindu Kush and Tien Shan circle the featureless Tarim Pendi basin and the lake-riddled Plateau of Tibet.

10–11 EUROPE

The generally densely vegetated continent of Europe contains some dramatic geographical features. Its northern and western limits are marked by the complex coastlines of Iceland, Scandinavia and north western Russian Federation, while the British Isles sit on the flat, wide continental shelf. Europe's mountain ranges divide the continent – in the southwest, the Pyrenees separate France from the drier Iberian Peninsula, the wide arc of the Alps separates Italy from the rest of western Europe, the Carpathian Mountains, appearing as a dark curve between the Alps and the Black Sea, mark the edge of the vast European plains, and the Caucasus, stretching between the Black Sea and the Caspian Sea, create a prominent barrier between Europe and Asia. Two of Europe's greatest rivers are also clearly visible on this image – the Volga, Europe's longest river, flowing south from the Ural Mountains into the Caspian Sea and the Dnieper flowing across the plains into the northern Black Sea.

12–13 AFRICA

This image of Africa clearly shows the change in vegetation through the equatorial regions from the vast, dry Sahara desert covering much of the north of the continent, through the rich forests of the Congo basin – the second largest drainage basin in the world – to the high plateau of southern Africa. Lake Victoria dominates central east Africa and the Nile and its delta create a distinctive feature in the desert in the northeast. The path of the Great Rift Valley can be traced by the pattern of linear lakes in east Africa, to Ethiopia, and along the Red Sea. The dark fan-shaped feature in central southern Africa is the Okavango Delta in Botswana – one of the world's most ecologically sensitive areas. To the east of the continent lies Madagascar, and in the Indian Ocean northeast of this is the Mascarene Ridge sea feature stretching from the Seychelles in the north to Mauritius and Réunion in the south.

14–15 NORTH AMERICA

Many well-known geographical features are identifiable on this image of North America, which also illustrates the contrasts in landscapes across the continent. Greenland, the world's largest island, sits off the northeast coast while the dramatic chain of the Aleutian Islands in the northwest stretches from Alaska across the Bering Sea to the Kamchatka Peninsula in the Russian Federation. Further south in the Pacific Ocean, at the far left of the image, lie the Hawai'ian Islands and their very distinctive ocean ridge. There is a strong west-east contrast across the continent. The west is dominated by the Rocky Mountains, which give way to the Great Plains. In the east, the Great Lakes, the largest of which, Lake Superior, is second in size only to the Caspian Sea, the valley of the Mississippi and the Coastal Plain are prominent. In the southeast the complex floor of the Caribbean Sea is visible, particularly the dramatic Cayman Trench, stretching from the Gulf of Honduras to southern Cuba.

16–17 SOUTH AMERICA

The Andes mountains stretch along the whole length of the west coast of South America, widening into the high plains of the Altiplano in Bolivia and Peru in the centre of the continent. Lake Titicaca, the world's highest large navigable lake, lies on the Altiplano, straddling the Bolivia–Peru border. Running parallel to the Andes, just off the west coast, is the Peru–Chile Trench which marks the active boundary between the Nazca and South American tectonic plates. Movement between these plates gives rise to numerous volcanoes in the Andes. The Amazon river runs across almost the whole width of the continent in the north, meeting the Atlantic Ocean in its wide delta on the northeast coast. The vast Amazon basin is one of the most ecologically diverse areas of the Earth. In the south, the wide continental shelf stretches eastward from the tip of the continent to the Falkland Islands and South Georgia on the bottom edge of the image.

18–19 ANTARCTICA

Protected from commercial exploitation and from the implementation of territorial claims by the Antarctic Treaty implemented in 1959, Antarctica is perhaps the world's greatest unspoilt, and relatively unexplored, wilderness. This image combines bathymetric data (incomplete in some, black, areas) with satellite images to show the extent of the continental ice sheet in an austral summer. Floating sea ice is not shown. The Antarctic Peninsula – home to numerous scientific research stations – in the top left of the image reaching towards South America, the huge Ronne and Ross ice shelves, and the Transantarctic Mountains – dividing the continent into West and East Antarctica – are the dominant physical features.

20–21 FAREWELL SPIT, NEW ZEALAND

Farewell Spit on the northern tip of New Zealand's South Island stretches 30 km eastwards from Cape Farewell into the Tasman Sea. It has formed as sand, eroded from the sandstone cliffs on nearby Cape Farewell, has been carried eastward by sea currents and deposited in a great arc. The sandy dunes on the north side of the spit are steep and unstable, appearing on this image as a white band. Dunes on the south side of the spit, facing Golden Bay, appear green as they are more stable and covered with vegetation. In Golden Bay, eighty square kilometres of mudflats are exposed at low tide; they provide a rich feeding ground for birds and more than eighty species of wetland birds have been recorded here, including many migratory waders. Detail of the topography and drainage of the flats appears on the image in blue and pink. The spit is predicted to grow by 2 km in the next five years.

22–23 CLOUD STREETS, ALASKA, USA

Extending for approximately 800 km to the southwest of mainland Alaska and ending at the Aleutian Islands, the Alaskan Peninsula separates the Pacific Ocean and the Bering Sea. Neat rows of clouds can be seen over the land and surrounding sea and sea ice on this image. These patterns are called 'cloud streets' and consist of cumulus clouds which form when cold air from the ice meets and chills the warmer, moist air over the ocean. As the air temperature drops, the water freezes into small clouds which are arranged in neat rows aligned with the prevailing wind direction. Although some cloud streets can be seen over the land of the peninsula and over the cracking sea ice near the middle of the western edge of the image, most are over the open water.

24–25 MALDIVES, INDIAN OCEAN

The Maldives are a chain of 1192 small coral islands arranged in twenty-six clusters called atolls. Each atoll is made up of several dozen coral reefs. They lie along a line over 800 km long on the north–south Maldive Ridge in the Indian Ocean, 700 km southwest of Sri Lanka. Only 250 of the islands in the Maldives are inhabited. This image shows North and South Maalhosmadulu Atolls. Rubble and sediment, mainly from dead coral, have built up on some of these reefs to form low, flat islands. These show as green islands surrounded by white beaches. Atolls are formed as oceanic volcanic islands with fringing coral reefs, sink below the ocean surface as the rocks cool. The volcanic peaks eventually disappear altogether, leaving a ring of coral with a central lagoon; this process may take as long as 30 million years. With a highest point only 2.4 m above sea level, the Maldives are especially vulnerable to rising sea levels caused by global warming.

26–27 NAMIB DESERT, NAMIBIA

The Namib is believed to be the planet's oldest desert, having endured arid or semi-arid conditions for at least 55 million years. It extends inland from the Atlantic coast to the foot of the Namib Escarpment and stretches 1287 km north to south. The Namib is one of the driest places on earth with sparse, and highly unpredictable, rainfall. Coastal fog, which can extend 50 km inland, is the life-blood of the Namib as it creates enough moisture for many species to survive and contributes to a highly diverse animal life. Bright yellow areas on this image are part of the vast dune sea; the linear and crescent-shaped dunes are up to 300 m high, making them some of the highest in the world. The intermittent Tsauchab River can be seen running through the centre of this image and terminating in large mud flat – the Sossos Vlei – which fills with water during the infrequent rains.

28–29 BEAR GLACIER, ALASKA, USA

Bear Glacier is one of forty glaciers which flow off the vast Harding Icefield, 200 km south of Anchorage in Alaska. This IKONOS satellite image offers a detailed view of the ice surface near the tip of the glacier. Bear Glacier ends in a small glacial lake called Strohn Lake, the colour of which is a typical blue-green colour. This colour results from highly reflective glacial flour – fine grained sediment eroded by the ice, which is suspended in the water. Pieces of ice which have broken off the tip of the glacier can be seen floating across the lake. The stripes running down the centre of the glacier are medial moraines, which formed when two glaciers merged farther up the valley. The intricate pattern of crevasses – cracks which form on the ice surface as it moves over rough terrain – is beautifully displayed on this image. Like many of the glaciers in Alaska, Bear Glacier has shrunk significantly in the last century.

30–31 PYRAMIDS, GIZA, EGYPT

Among the largest structures ever built, the pyramids at Giza in Egypt are over 4 500 years old. They were built as tombs and monuments for the Pharaohs, the rulers of Egypt. The largest, and oldest, of the three main pyramids at this site is the Great Pyramid, built by the Pharaoh Khufu. The medium sized Khafre pyramid was built by his son, and the smallest, Menkaure pyramid, was the last to be built. The complex also includes many tombs and temples for queens, other members of royal families and royal attendants. The white line on the image, running east from the Khafre pyramid leads to the Valley Temple and the Great Sphinx. This half lion - half human statue is the largest monolith statue in the world, the head of which is thought to represent that of Khafre. The Great Pyramid was listed as one of the Seven Wonders of the Ancient World and is the only one still in existence. The pyramids were also designated a world heritage site by UNESCO in 1979.

32–33 ULURU (AYERS ROCK), AUSTRALIA

Uluru, or Ayers Rock, in Australia's Northern Territory is one of Australia's most recognizable natural features. It stands 348 m above the surrounding plain as a monumental geological feature called an inselberg. The sandstone it consists of is more resistant to erosion than the softer rocks, now eroded away, which used to surround it. The strata of the sandstone are almost vertical and features on the surface of the rock, created by erosion, are clearly shown on this image. The Uluru-Kata Tjuta National Park was inscribed on the World Heritage List in 1987, one of the few sites to be listed for both its natural and cultural values – there are paintings and carvings on the surface of Uluru which were made thousands of years ago by Aborigines who believed the rock to be a sacred place.

34–35 PARIS, FRANCE

Paris is the capital of, and largest city in France. It is located on the river Seine and contains many historic monuments and gardens. This image covers part of central Paris. South of the river, just on the bend is the distinctive shape of the Eiffel Tower facing the Jardin du Trocadero and the curving wings of the Palais de Chaillot just across the river. To the north east of the Palais is the wheel shaped Place Charles de Gaulle with the Arc de Triomphe in the centre. The Avenue des Champs-Elysées runs from the Arc de Triomphe south-east to the Jardin des Tuileries and the Louvre museum with the Pyramid at its entrance.

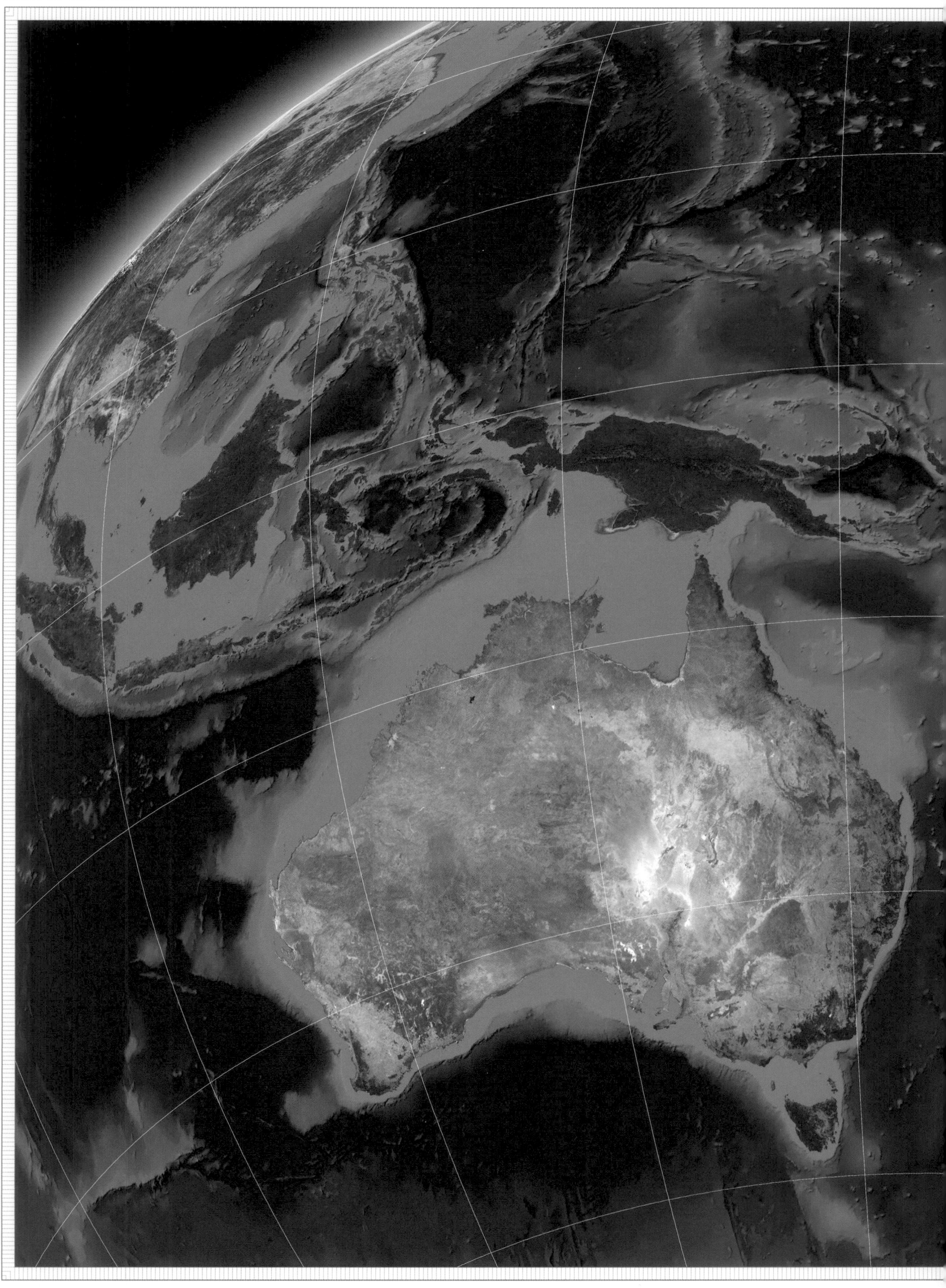

© Collins Bartholomew Ltd

© Collins Bartholomew Ltd

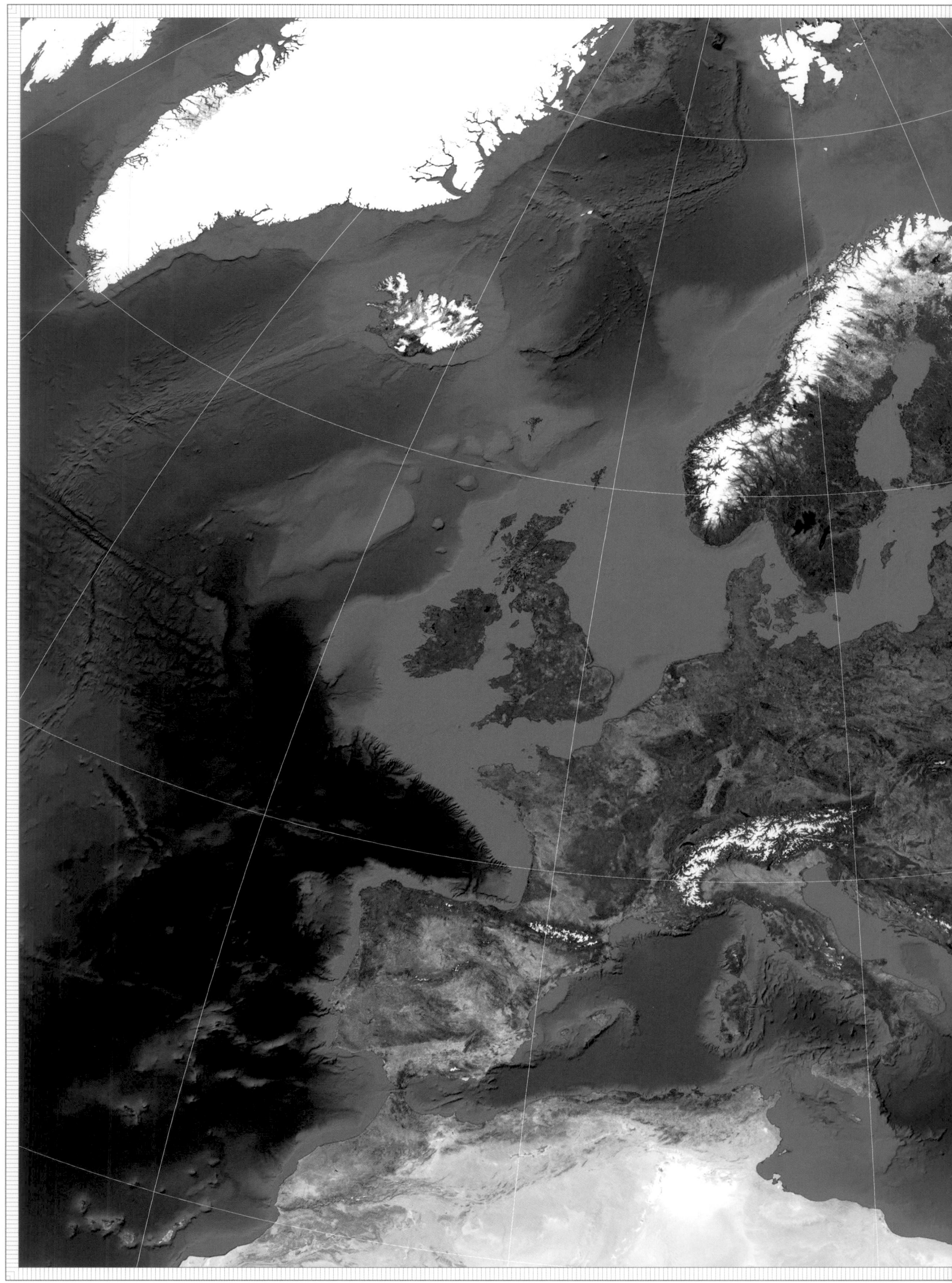

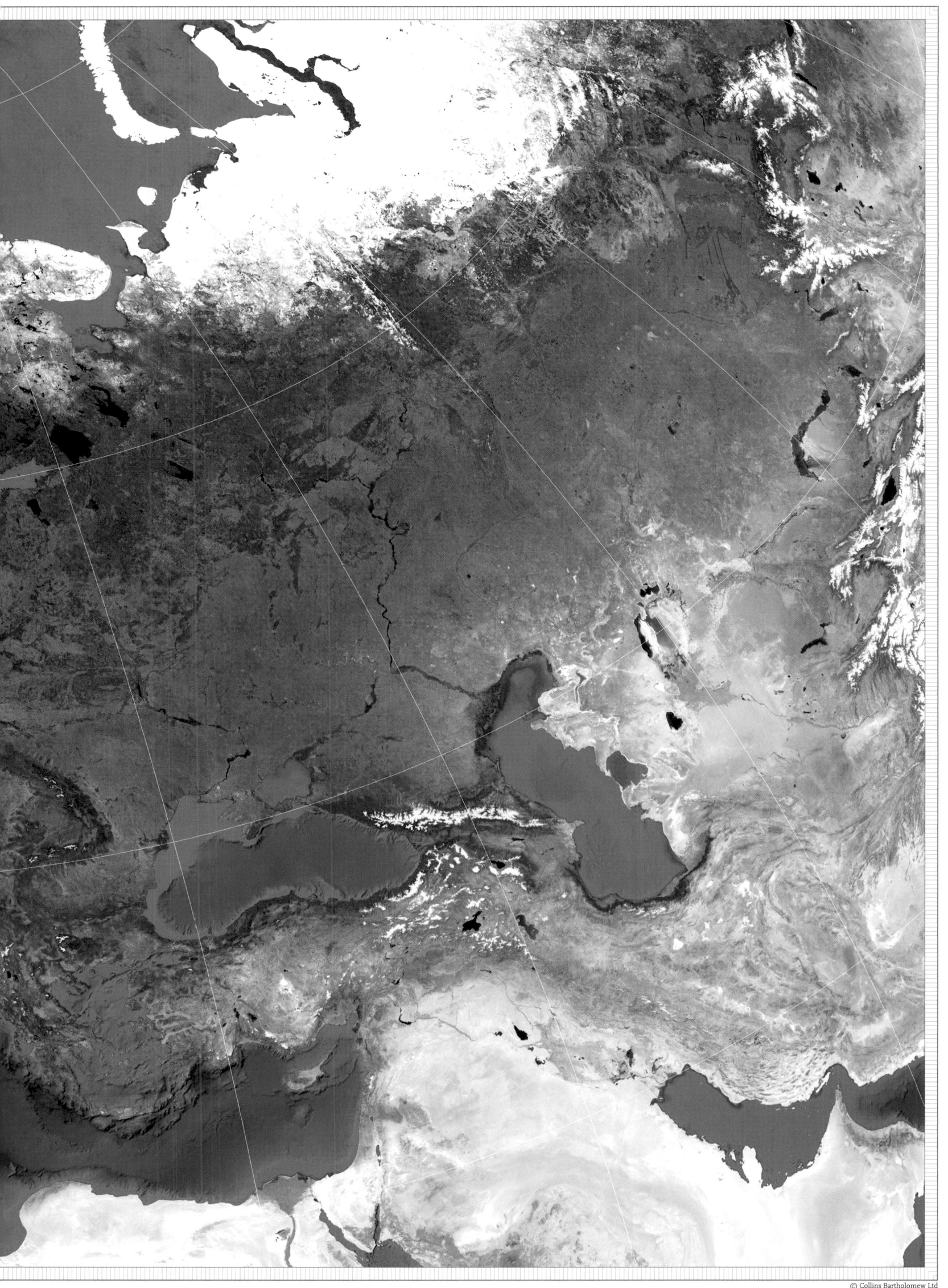

© Collins Bartholomew Ltd

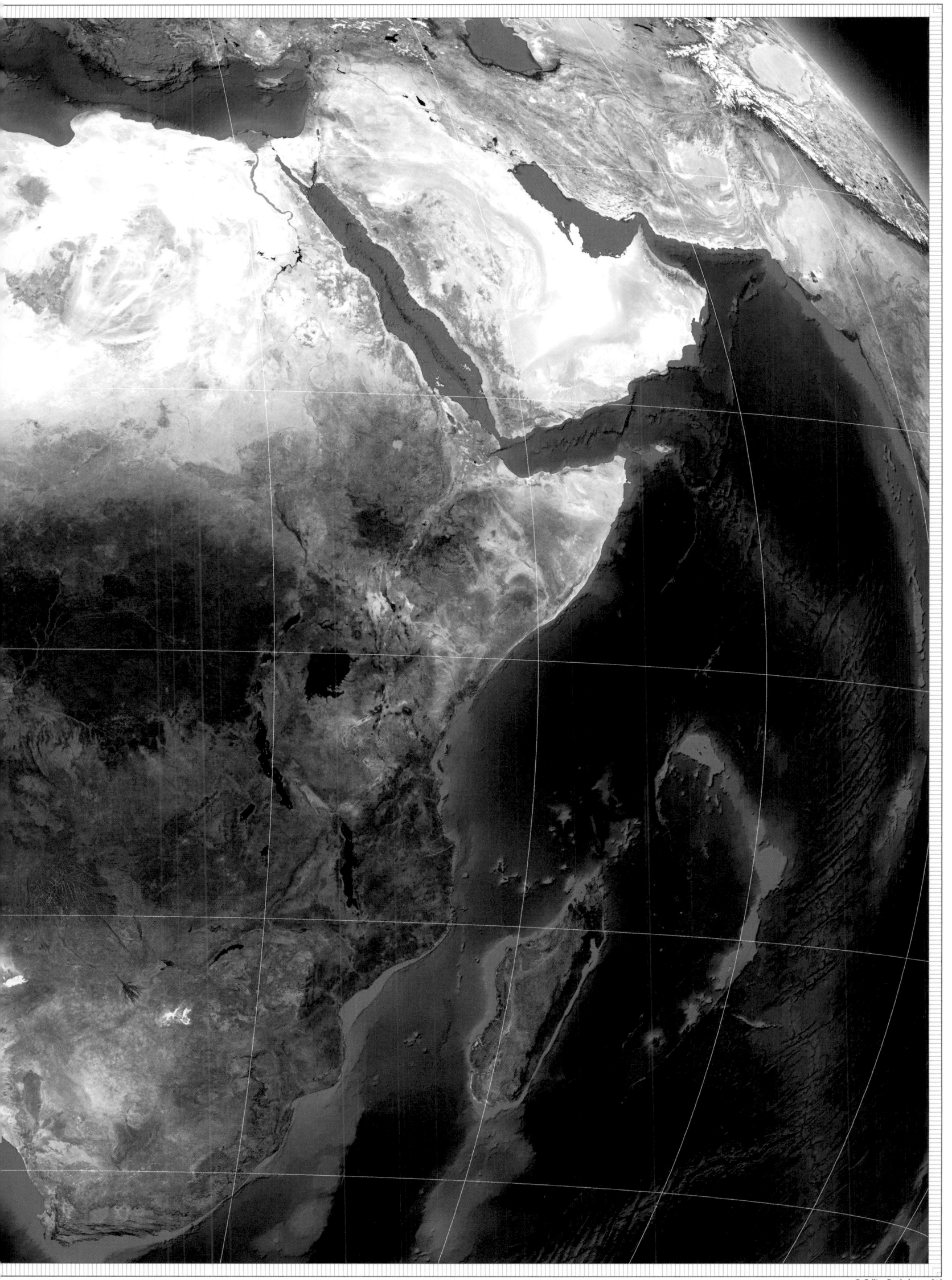

© Collins Bartholomew Ltd

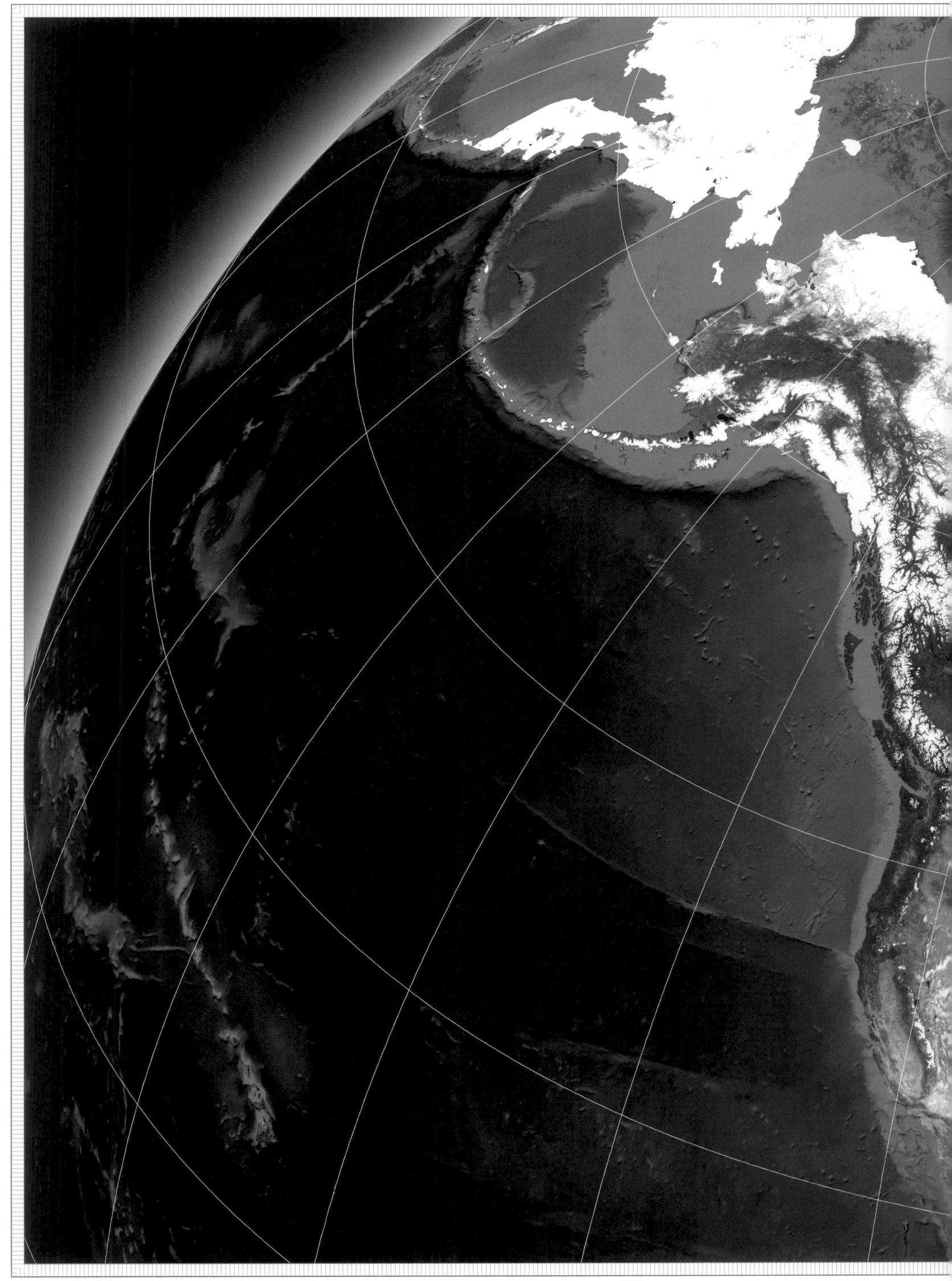

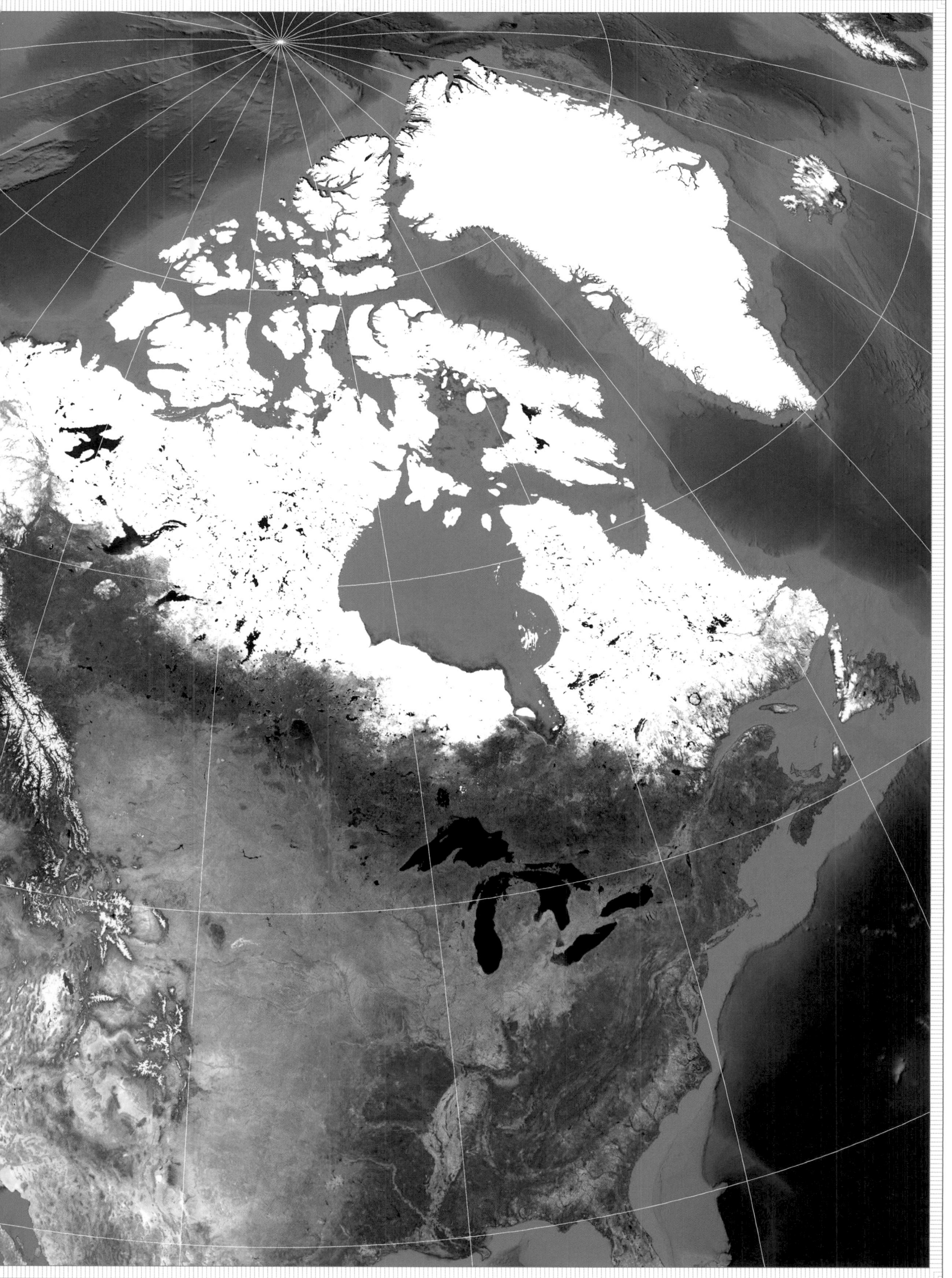

© Collins Bartholomew Ltd

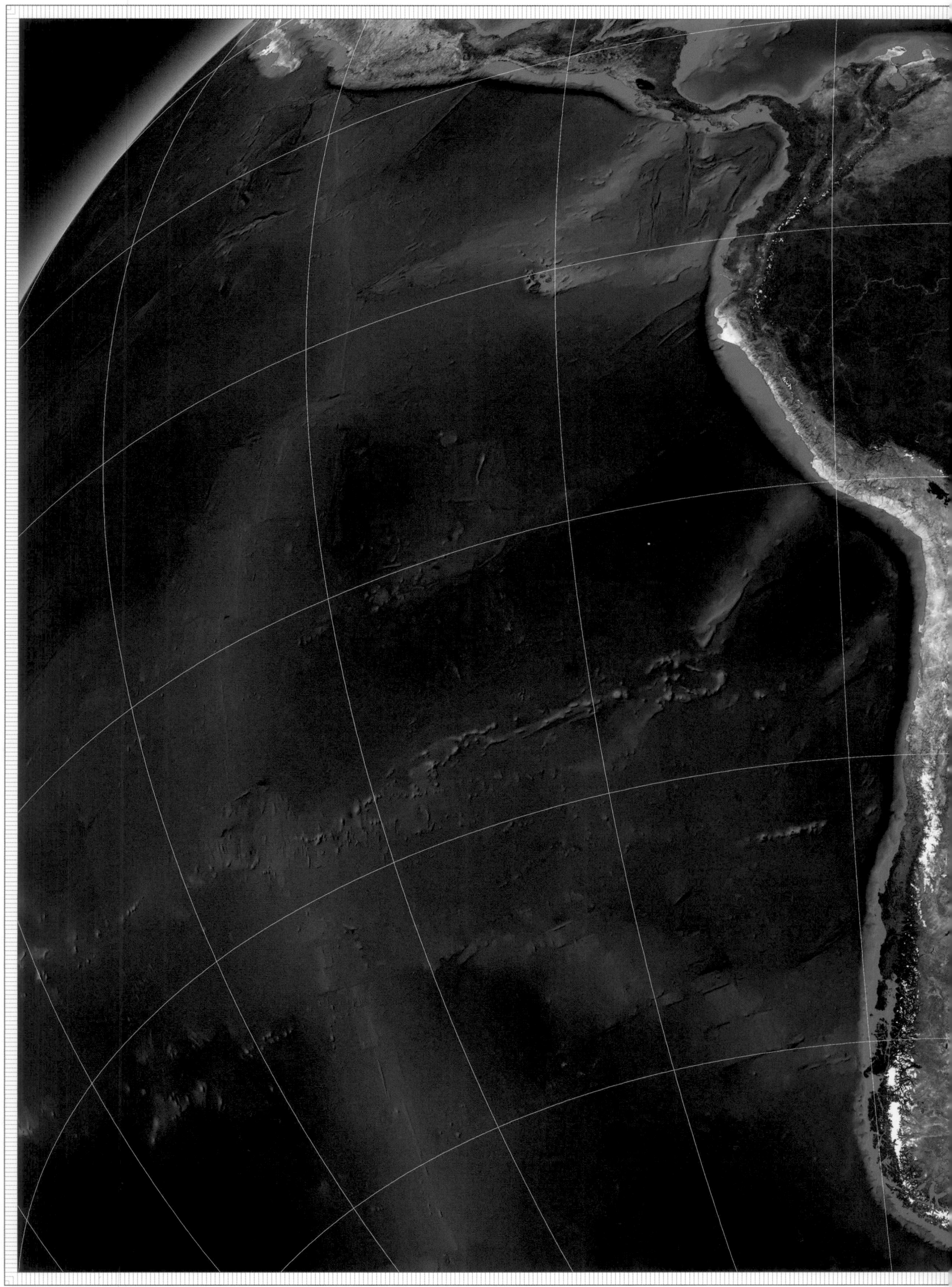

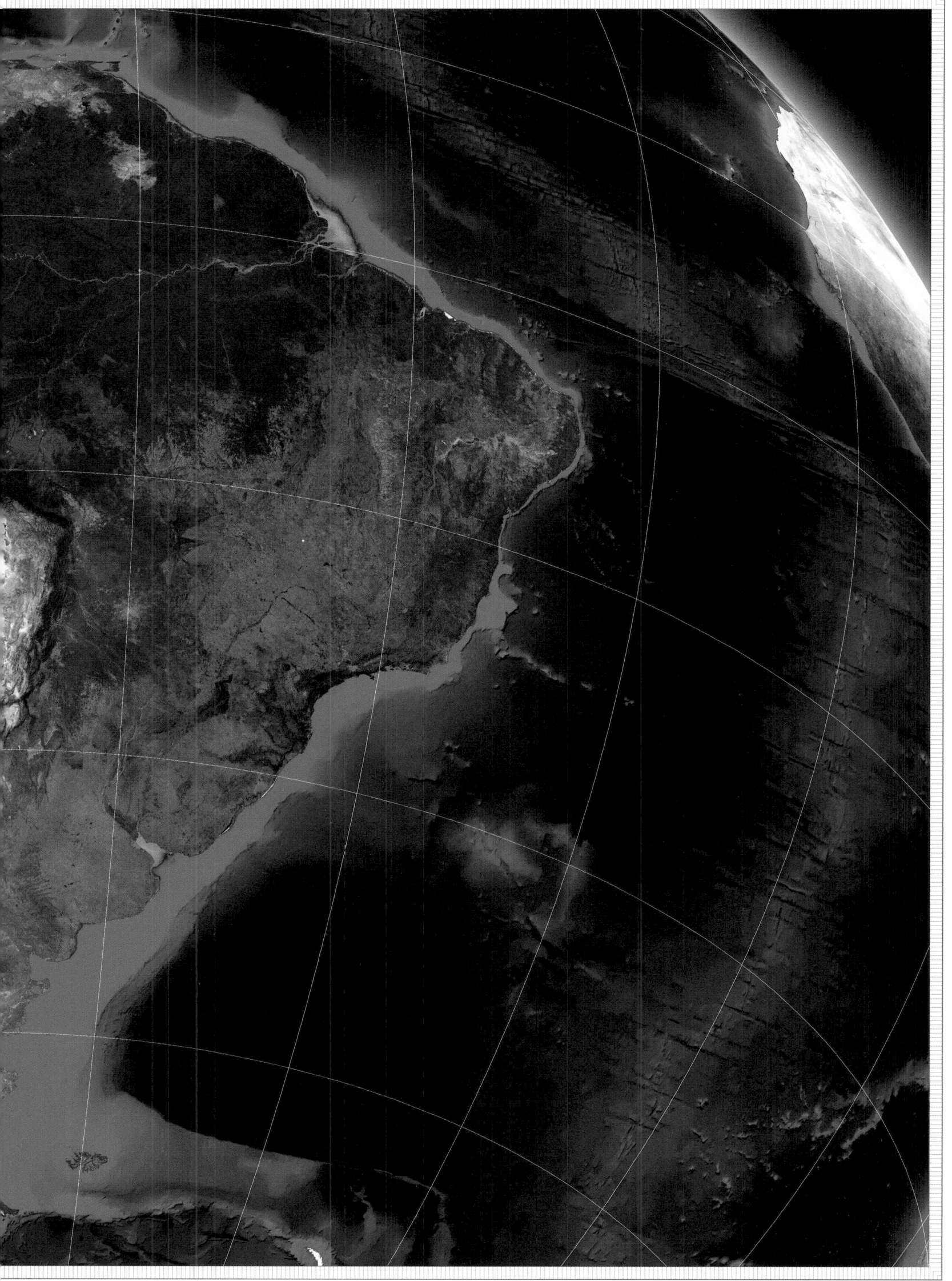

© Collins Bartholomew Ltd

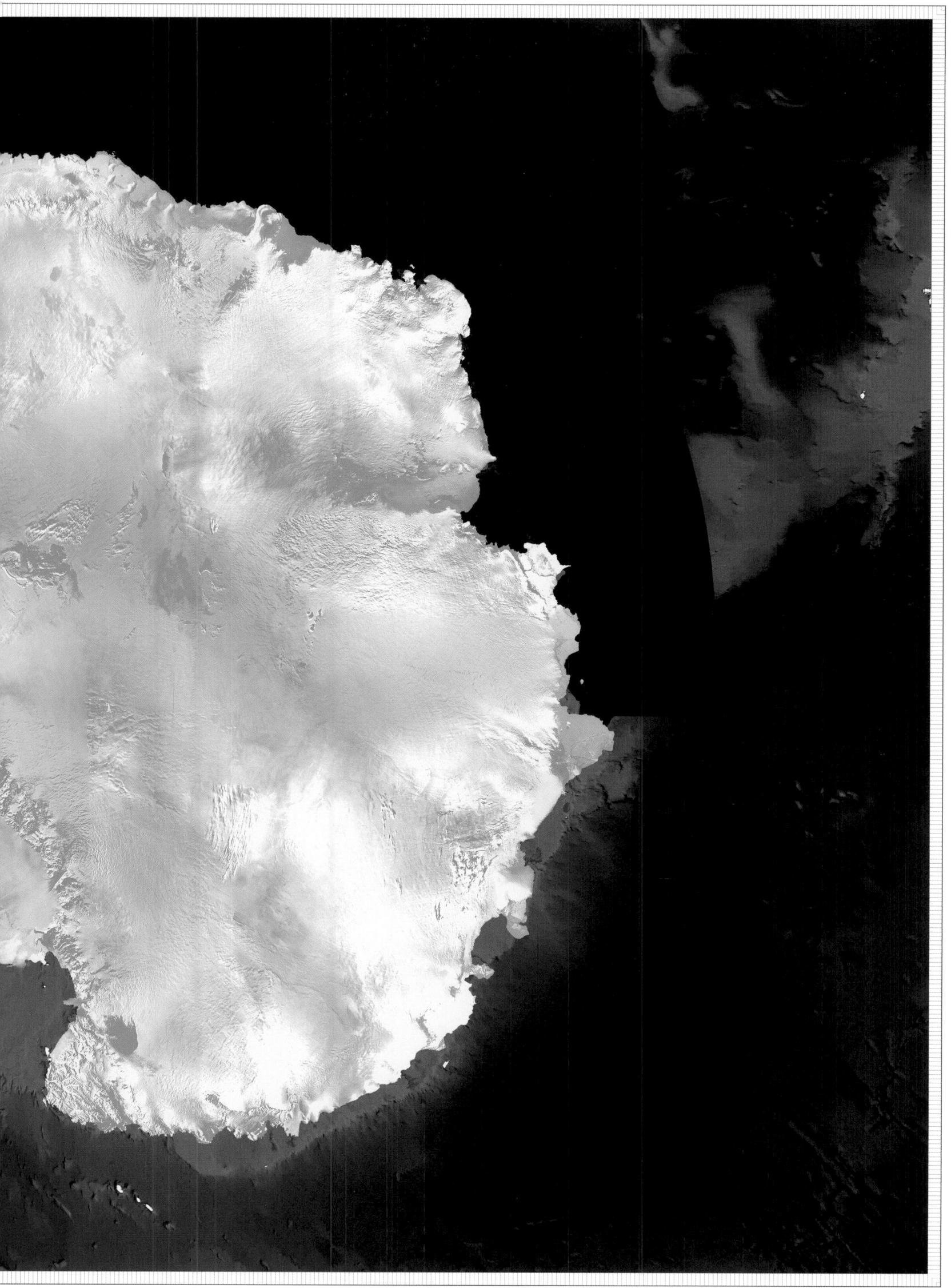

© Collins Bartholomew Ltd

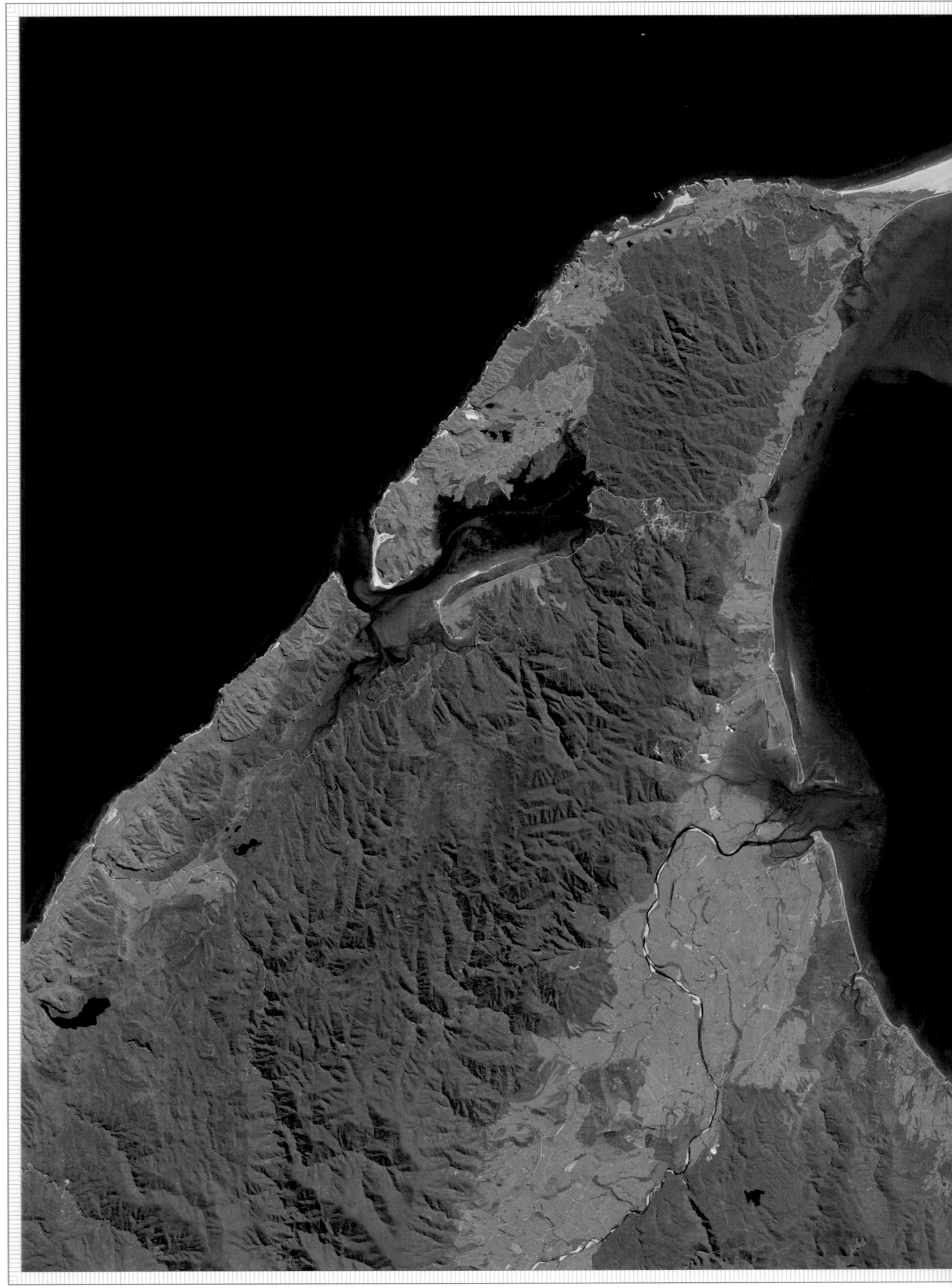

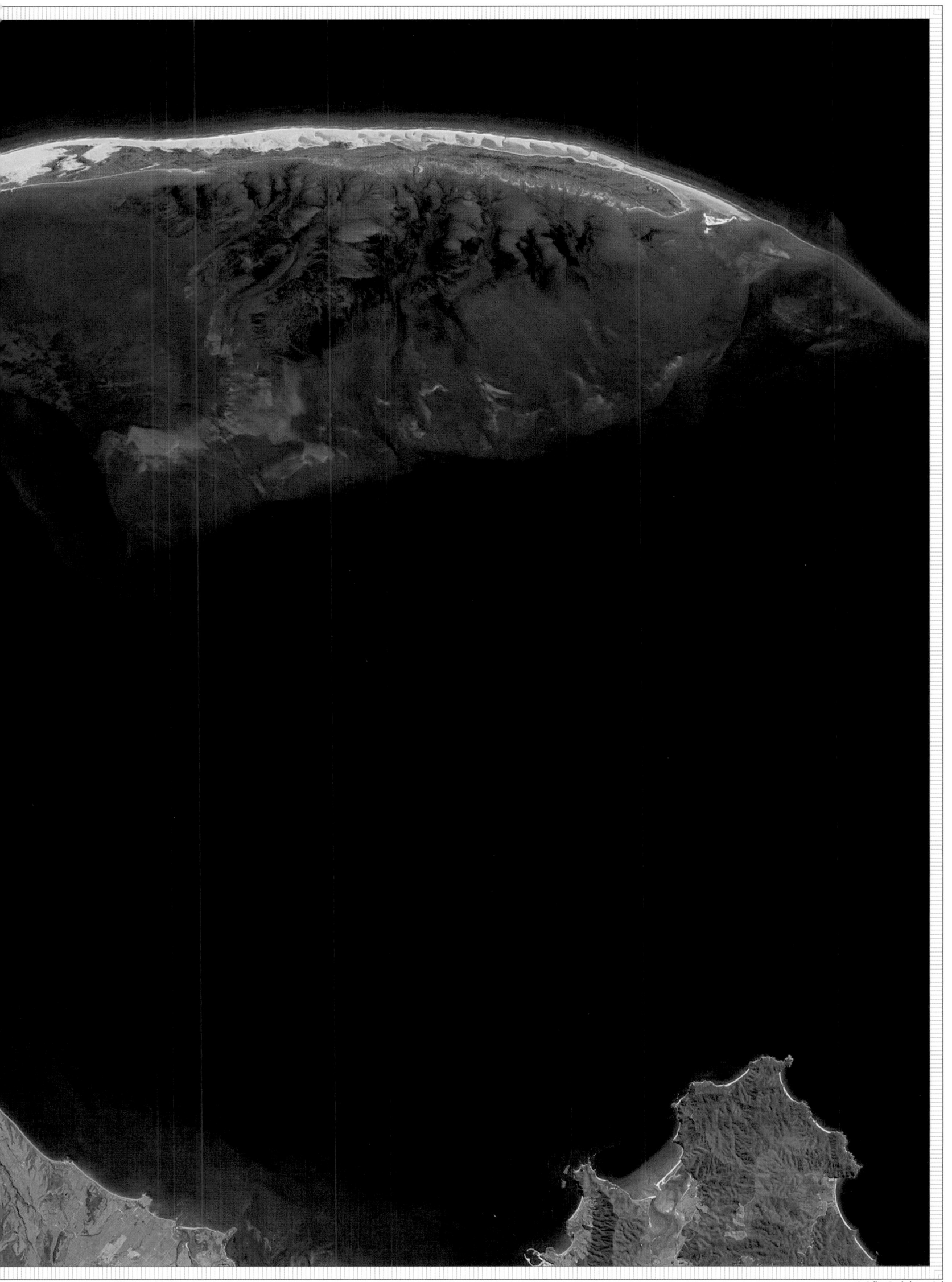

© Collins Bartholomew Ltd

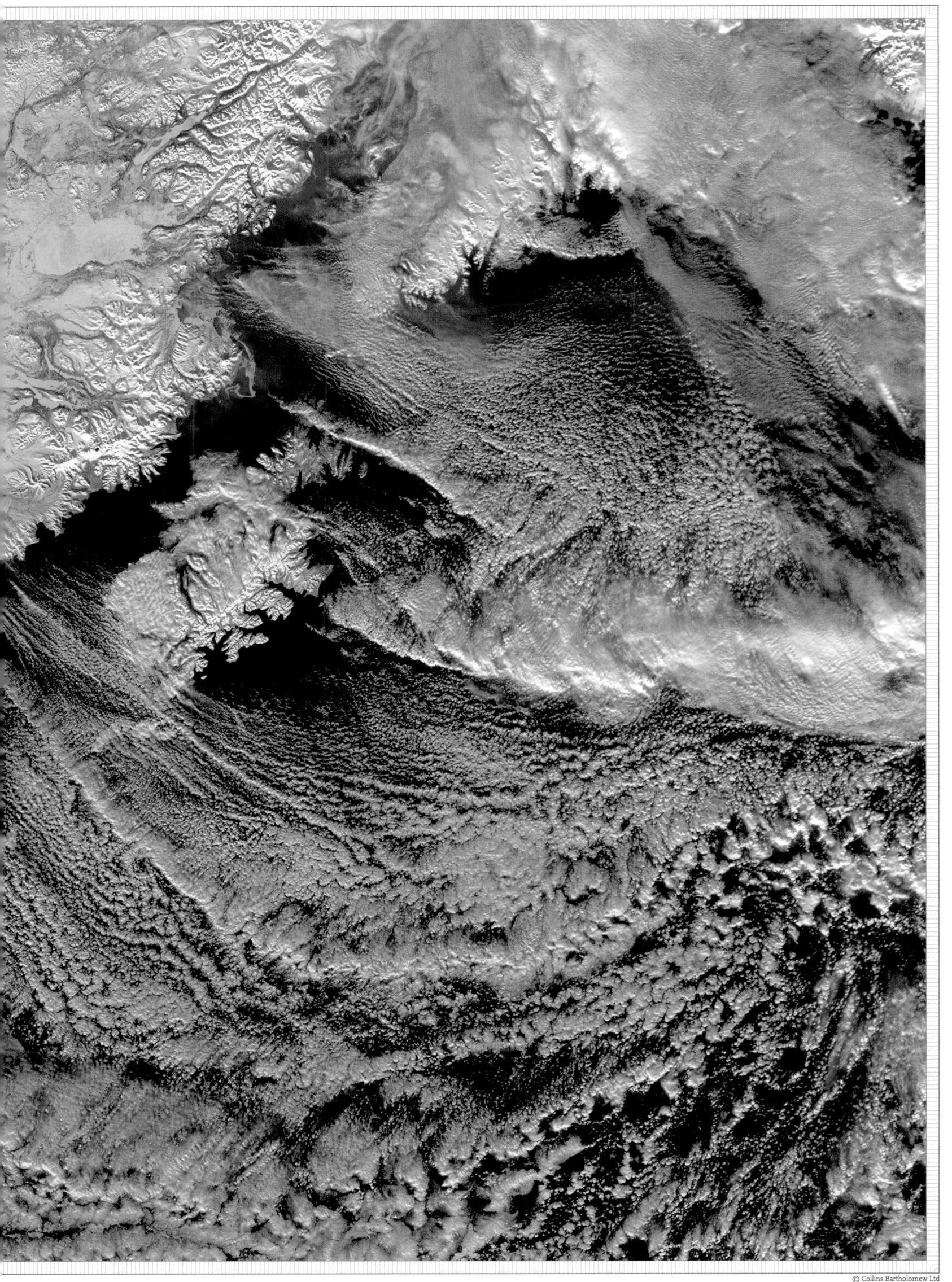

© Collins Bartholomew Ltd

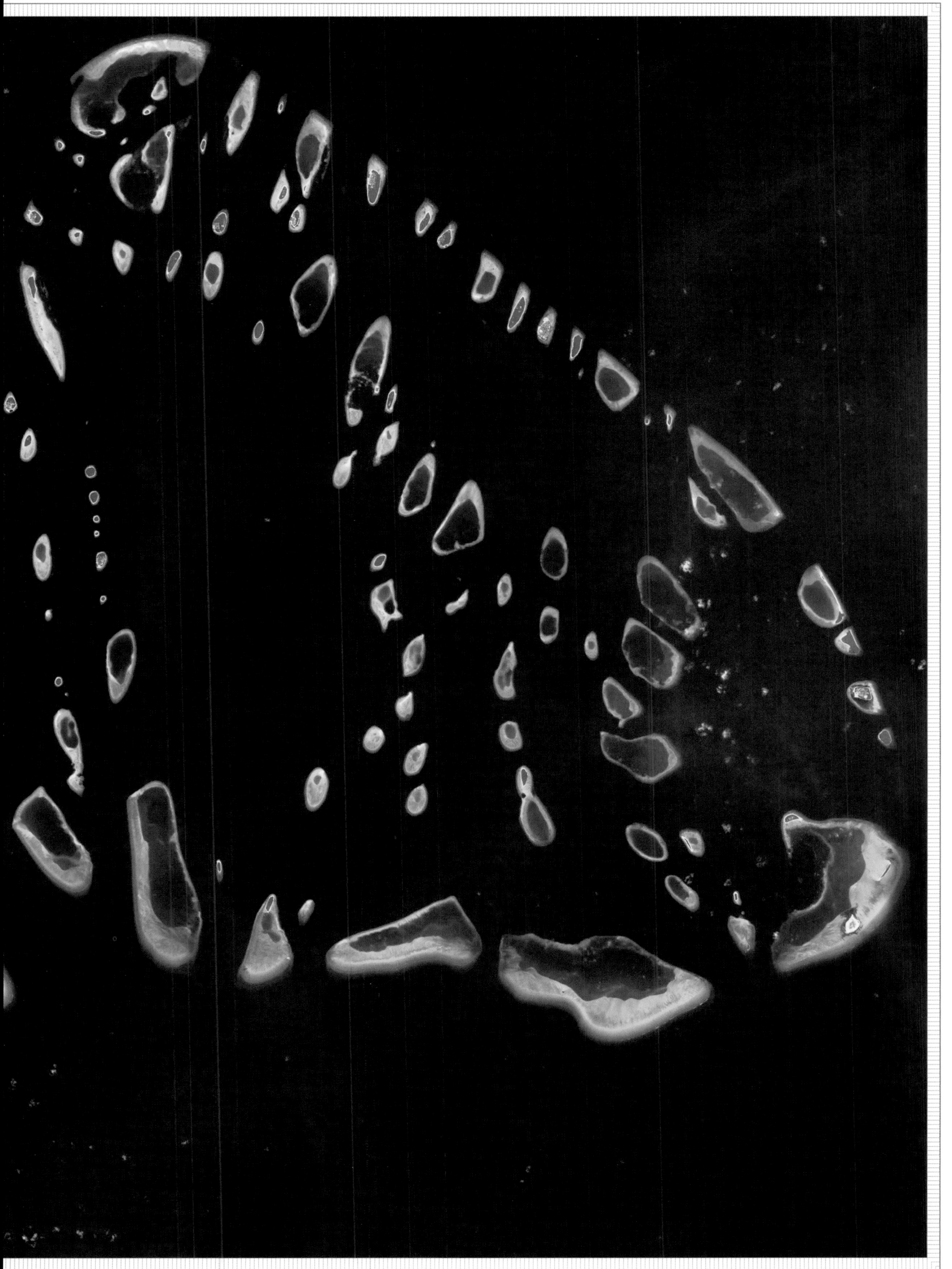

© Collins Bartholomew Ltd

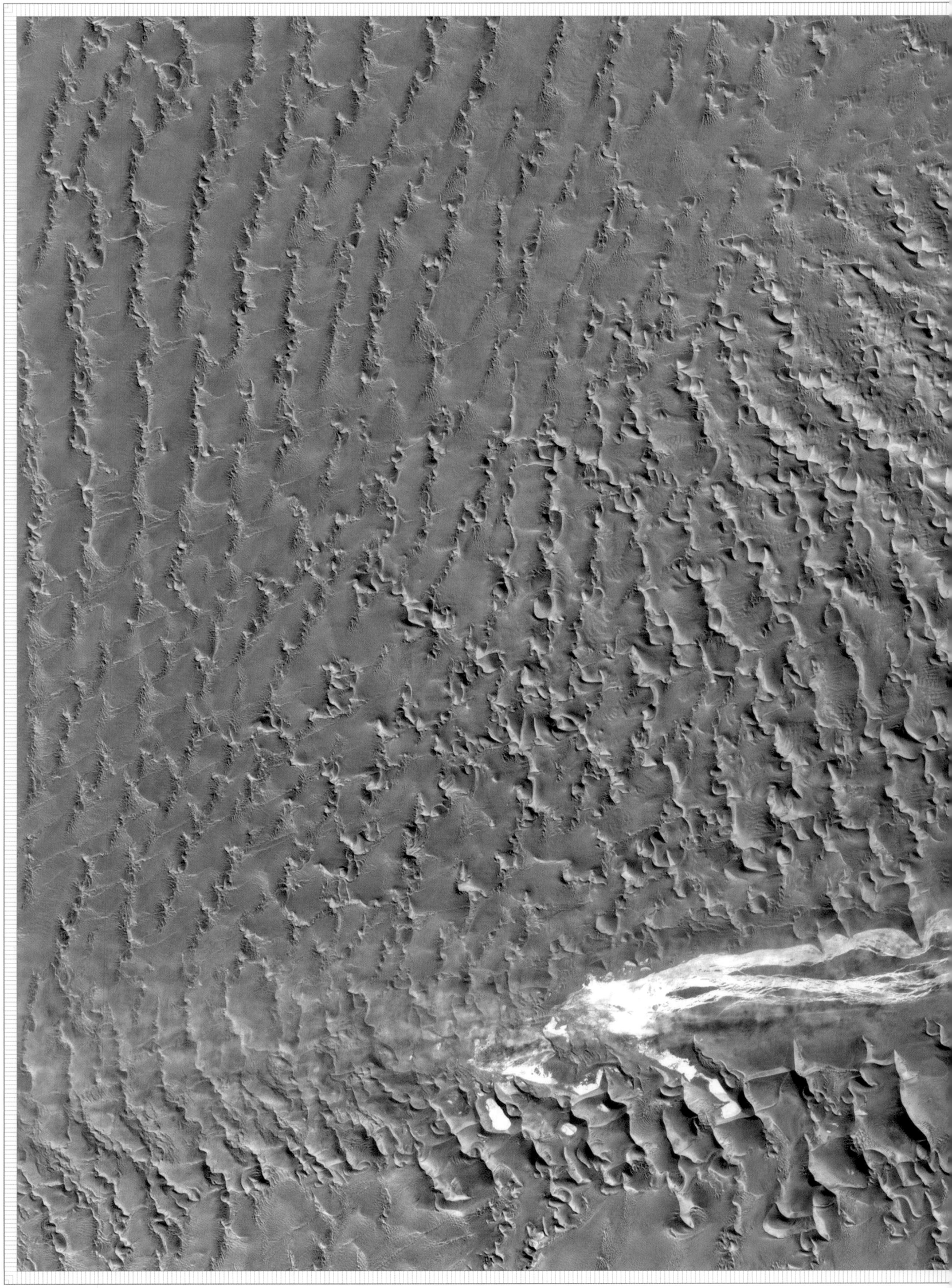

© Collins Bartholomew Ltd

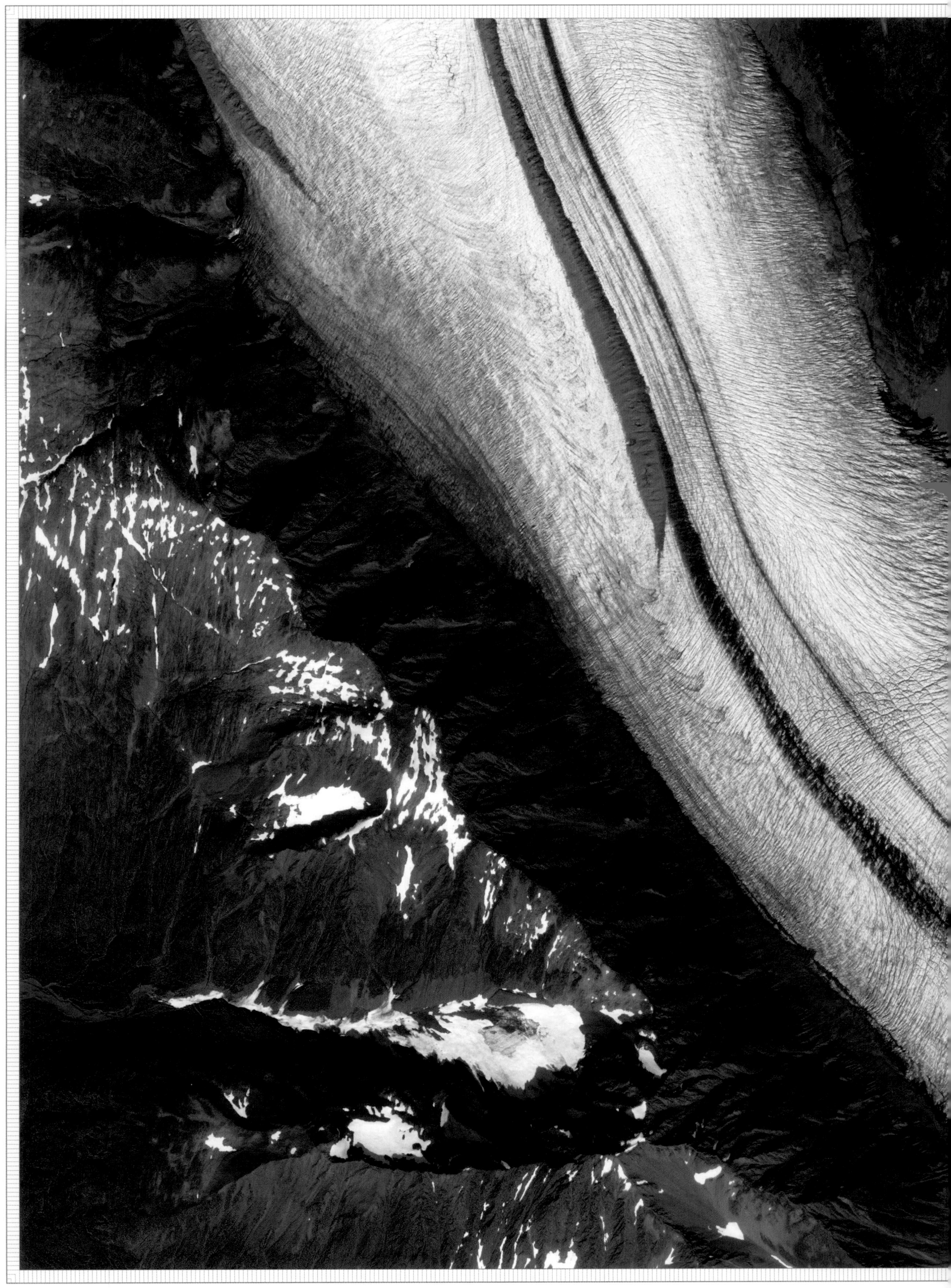

© Collins Bartholomew Ltd

© Collins Bartholomew Ltd

© Collins Bartholomew Ltd

© Collins Bartholomew Ltd

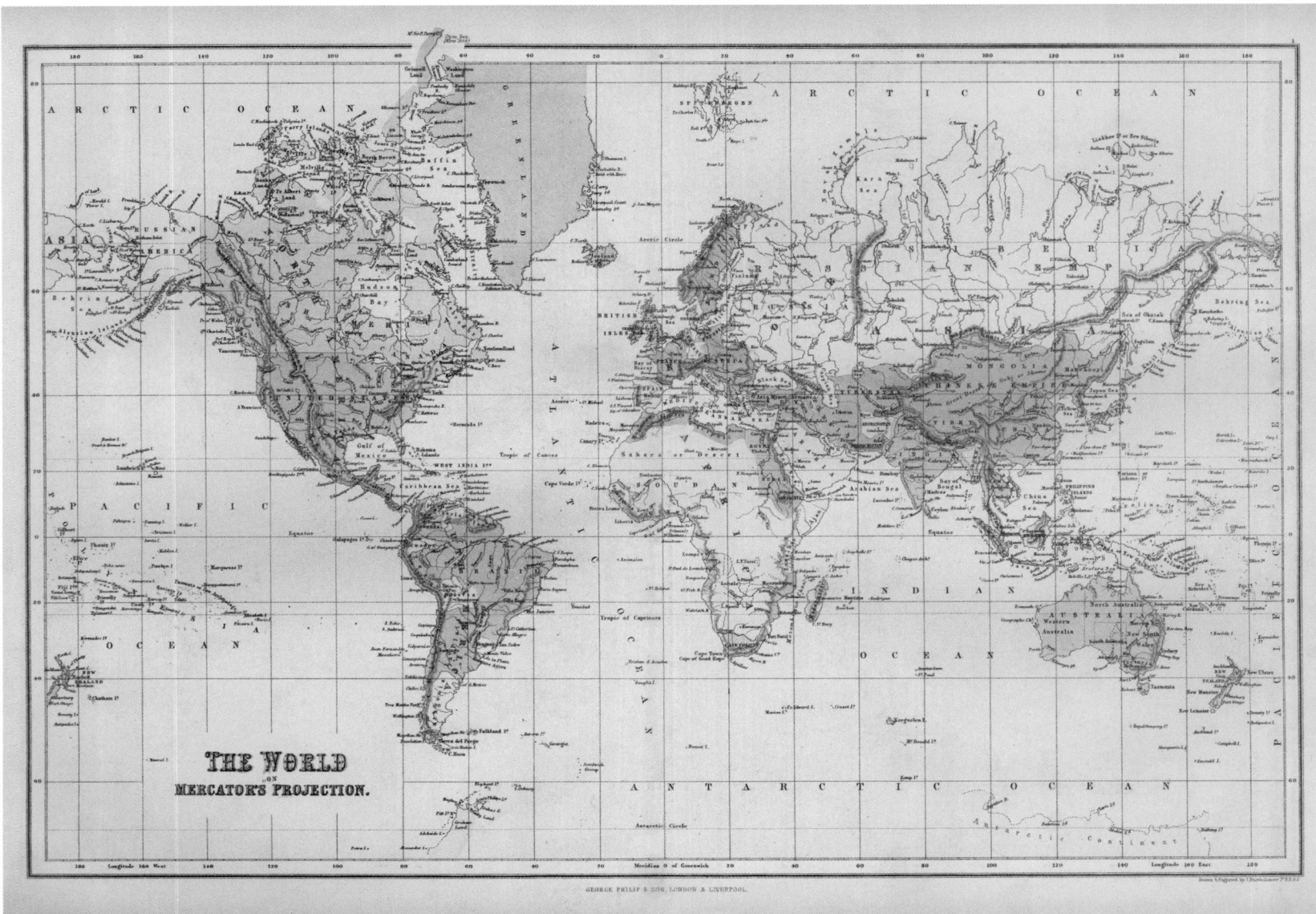

THE WORLD
ON
MERCATOR'S PROJECTION.

GEORGE PHILIP & SON, LONDON & LIVERPOOL.

38–39 THE WORLD ON MERCATOR'S PROJECTION 1858

From the *Family Atlas of Physical, General and Classical Geography*. Drawn and engraved by J. Bartholomew Jr F.R.G.S.

The nineteenth century was known as the 'Age of Empire', when all the major European powers harboured imperial ambitions and used their commercial and military might to extend their influence. In the first half of the century, the process had been gradual. Britain had emerged as the pre-eminent overseas power, extending the boundaries of her established colonial possessions in North America, India and Australia. In the second half of the century, the pace of imperial expansion increased markedly and the world depicted here was on a cusp of a dramatic change.

40–41 POLITICAL DIVISIONS OF THE WORLD 1914

From the *International Reference Atlas of the World*. Cartography by J. G. Bartholomew LL.D., F.R.G.S., Cartographer to the King.

This map shows the imperial divisions of the world at the onset of the First World War in 1914. European colonial empires had grown rapidly over the past century and by now the Great Powers of Europe had engrossed nine-tenths of Africa and much of Asia. Prior to 1914, Europe had been run on balance-of-power politics, where a status quo was maintained between the major powers, often with unofficial agreements and alliances.

42–43 WORLD POLITICAL DIVISIONS 1936

From the *Advanced Atlas, Fifth Edition*. Cartography by John Bartholomew M.C., M.A., F.R.S.E., F.R.G.S. Cartographer to the King.

The political situation of the world three years before the outbreak of the Second World War can be seen from this map. The power of empires had waned significantly after the First World War, and a number of treaties and pacts were signed between countries to safeguard against military attacks. Growing political and social conflict was leading to nationalist uprisings, while both communism and fascism were on the rise in Europe.

44–45 WORLD ROUTES OF COMMERCE 1950

From the *Advanced Atlas of Modern Geography*, cartography by John Bartholomew, M.C., Director, the Geographical Institute, Edinburgh.

Just as the First World War had acted as a catalyst for massive change to the existing world order, so too the fallout from the Second World War brought significant political, territorial and economic upheaval across the globe. The most significant development in world politics post-1945 was the emergence of the USA and USSR as hostile superpowers, and the ideological alignment of other nations with each respective camp. The armed stand-off which emerged between the two power blocs became known as the Cold War and lasted until the fall of Soviet communism in the early 1990s.

46–47 WORLD POLITICAL CHART 1963

From the *Edinburgh World Atlas, Fifth Edition*. Cartography by John Bartholomew C.B.E., M.C., LL.D., F.R.S.E., F.R.G.S.

Almost twenty years on from the end of the Second World War, the 'Age of Empire' was close to its end. International politics had instead become dominated by two superpowers – the United States and the Union of Soviet Socialist Republics (USSR) – who were opposed to each other during the lengthy Cold War. Significant changes affected French and British possessions worldwide at this time. After the war, European powers no longer had the military strength to defend against nationalist movements, nor the economic strength to enforce their rule. Decolonisation in Africa increased.

48–49 STATES OF THE WORLD 1982

From the *Bartholomew World Atlas, Twelfth Edition*. Cartography by John Bartholomew, M.A., F.R.S.E., Director, the Geographical Institute, Edinburgh.

This map represents a transition between one extensive series of changes and another – beforehand, the decolonisation which had gone on, especially in Africa; and afterwards, the collapse of communist regimes in the 1990s. One of the inset maps plots the many changes of sovereignty that had occurred since 1939. Britain sought to maintain association with its former colonies through the Commonwealth. The changes that came after this map was published mostly resulted from political changes in the Soviet Union and its Warsaw Pact allies.

50–51 NORTH POLAR REGIONS AND SOUTH POLAR REGIONS 1898

From the *Citizen's Atlas*, cartography by J. G. Bartholomew, F.R.G.S.

These maps of the polar regions allow a variety of interesting comparisons to be drawn about these extremes of the Earth and how they were being explored.

Far from being homogeneous ice masses, the northern and southern polar regions are physically very different. The North Pole is at the centre of the Arctic Ocean and is an almost landlocked body of water largely composed of drifting pack ice; the South Pole, by contrast, lies on a continental land mass.

The motivations of those who ventured into these unexplored regions were different. For many northern polar explorers, the intention was to find a navigable passage through the ice to open up a trade route to link the Atlantic and Pacific Oceans – the so-called Northwest Passage. Unlike the northern polar region, the physical boundaries of the southern polar region were very poorly understood, as is evident from the map. It was not until the 1820s that Antarctica was first actually sighted, probably by the Russian explorer, Bellingshausen. From the 1830s a series of national expeditions embarked for Antarctica.

52–53 AUSTRALIA 1898

From the *Citizen's Atlas*, cartography by J. G. Bartholomew, F.R.G.S.

This map records the geopolitical make-up of Australia immediately prior to the ending of British colonial rule. Three years after it was drawn, the federal Commonwealth of Australia came into being, holding Dominion status under the British Crown. The political boundaries depicted here had been established during the course of the nineteenth century, and with minor exceptions have remained unchanged to the present day. A particularly striking feature of the map is the pattern of settlement. The temperate and subtropical climates of the southwestern and eastern coastal areas attracted the original European settlers and 60 per cent of the Australian population still lives there. The drive inland was accelerated by a series of goldrushes beginning in the 1850s.

54–55 TURKEY IN EUROPE, GREECE &c. 1898

From the *Citizen's Atlas*, cartography by J. G. Bartholomew, F.R.G.S.

By the end of the nineteenth century, the Ottoman or Turkish Empire had existed for 600 years. It was an Islamic successor to both the Roman and Holy Roman Empires, and at the peak of its powers in the seventeenth century had stretched from Gibraltar to the Caspian Sea, and from Vienna to the mouth of the Persian Gulf.

By 1898, however, the 'Sick Man of Europe' (as Tsar Nicholas I had called it) was in terminal decline, financially bankrupt and losing territory to foreign incursions and nationalist unrest in its fringe provinces. The Great Powers – Britain, France, Germany, Austria-Hungary and Russia – were keen to exploit local tensions and nationalist uprisings in the Balkans to destabilise the Ottoman Empire.

56–57 AFRICA 1898

From the *Citizen's Atlas*, Cartography by J. G. Bartholomew, F.R.G.S.

A comparison between the maps of Africa on p.38–39 [World 1858] and this one dramatically illustrates the speed with which the continent was parcelled up between the competing European powers in the latter half of the nineteenth century. In 1880, the 'Scramble for Africa' began in earnest with Britain in many ways the chief beneficiary. A primary motivation was to secure communication channels with India, the keystone of her empire. It was for this reason that Egypt was effectively annexed in 1882 to protect the strategically vital Suez Canal.

58–59 DOMINION OF CANADA 1898

From the *Citizen's Atlas*, cartography by J. G. Bartholomew, F.R.G.S.

This map provides a fascinating snapshot of the development of modern Canada barely thirty years after the British colonies of North America were united and granted legislative autonomy under the Crown. This confederation was the first stage in the development of the independent nation. Many of the territorial names on the map such as Quebec, Ontario, Nova Scotia and Manitoba echo those of the country's modern provinces. Others – Assiniboia, Athabasca and Keewatin amongst them – are much less familiar.

60–61 TURKEY IN ASIA, PERSIA, ARABIA, &c. 1914

From the *International Reference Atlas of the World*, cartography by J. G. Bartholomew LL.D., F.R.G.S., Cartographer to the King.

The late nineteenth and early twentieth centuries had seen a dramatic contraction of the Ottoman Empire, particularly in the Balkans but also in North Africa, where Algeria and Tunisia had been ceded to France, and Libya to Italy. Although never formally invaded, Persia was economically dependent on Europe, and as a result, Britain and Russia effectively divided the country between them from 1907 into two spheres of economic interest in which each power could exert its influence.

In October 1914, Turkey entered the First World War on Germany's side, setting in motion the train of events that would bring about the final dissolution of the Ottoman Empire and the ultimate creation of the modern Turkish republic.

62–63 INDIAN EMPIRE 1914

From the *International Reference Atlas of the World*. Cartography by J. G. Bartholomew LL.D., F.R.G.S., Cartographer to the King.

Dating from immediately before the outbreak of the First World War, this map illustrates the reach of British imperial power on the Indian subcontinent. Extending far beyond the political borders of modern India, it also included the present-day states of Pakistan, Bangladesh and Myanmar (Burma). Although not officially part of British India, the Crown colony of Ceylon (now Sri Lanka) and the northern kingdoms of Nepal and Bhutan also fell under its influence. The map also shows the mixture of direct rule and local autonomy that was vital to the smooth administration of such a vast and diverse territory.

64–65 EUROPE POLITICAL 1922

From *The Times Survey Atlas of the World*, Prepared at "The Edinburgh Geographical Institute" under the direction of J. G. Bartholomew, LL.D., F.R.S.E., F.R.G.S., Cartographer to the King.

The aftermath of the First World War and the Treaty of Versailles in 1919 redrew the world map and brought an end to the centuries of dynastic power in central and eastern Europe. The new separate states of Austria, Hungary, Czechoslovakia and Yugoslavia emerged after the demise of the Habsburg dynasty. At this time, ethnic nationalism was threatening European colonial empires with ideas of democracy and social reform.

Following the Russian Revolution in 1917, the Russian Empire lost much of its western frontier – the new Baltic states of Estonia, Latvia and Lithuania successfully fought independence and were recognised as independent countries in 1920. The map, however, shows these states during only a brief spell of independence.

66–67 CHINA 1922

From *The Times Survey Atlas of the World*, prepared at "The Edinburgh Geographical Institute" under the direction of J.G. Bartholomew, LL.D. F.R.S.E., F.R.G.S., Cartographer to the King.

The map shows how English-speakers were used to seeing Chinese place names long before modern 'Pinyin' spellings, which are characterised by frequent occurrences of q, x, y and z. Most obviously, some main names are in English – eg Inner Mongolia – and some are partly English – eg Gulf of Liao-tung. Hyphens are used in the old Wade-Giles spelling system to show sounds that come from separate characters.

The inset map from 1958 shows how the cartographer's view of Chinese names had evolved in the meantime: the names look somewhat more familiar to the modern eye, the hyphens have mostly gone, and some spellings are simpler. The most noticeable difference between the two maps is that the important places are in much bigger type, aiding clarity and legibility.

68–69 WORLD POWERS 1957

From *The Times Atlas of the World, Mid-Century Edition 1958*. Cartography by John Bartholomew, M.C., LL.D.

The most striking feature of this map is its unusual viewpoint (or projection). Devised in 1948 by John Bartholomew, the Atlantis Projection abandons the common atlas convention of showing the Arctic at the top and the Antarctic at the bottom. Here the projection is tilted to focus on the Atlantic Ocean. In this instance it is particularly effective in conveying the combative nature of relations between the United States and the Soviet Union, the two 'superpowers' which emerged to dominate the new world order following the Second World War.

Within a few short years of this particular map being drawn, significant colour changes would be required for a number of countries: Alaska would become a full member state of the USA (1959); Fidel Castro would establish a Marxist government on America's doorstep in Cuba (1959), and the process by which many African nations would shake off the last remnants of European colonialism would begin in earnest.

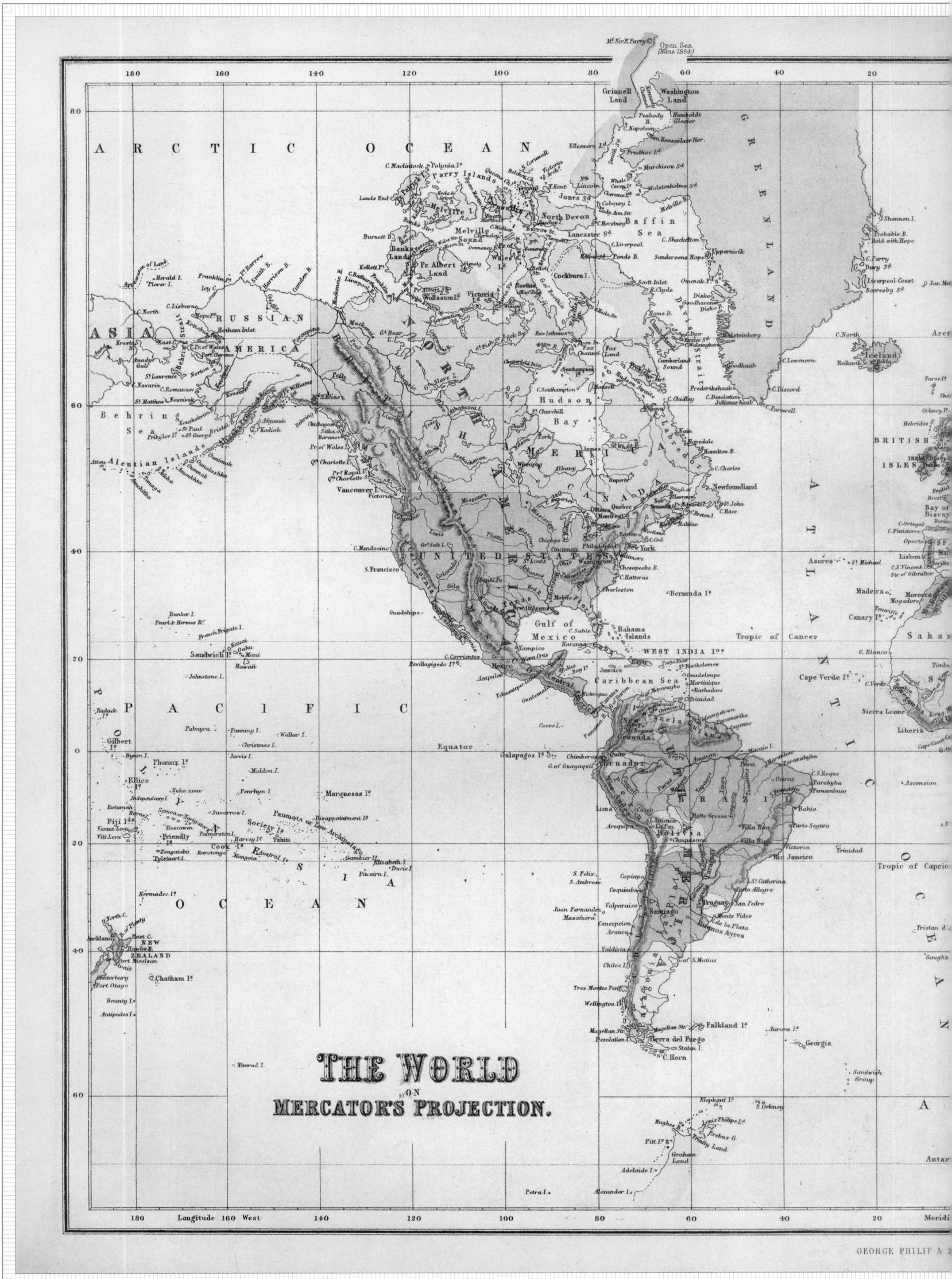

THE WORLD
ON
MERCATOR'S PROJECTION.

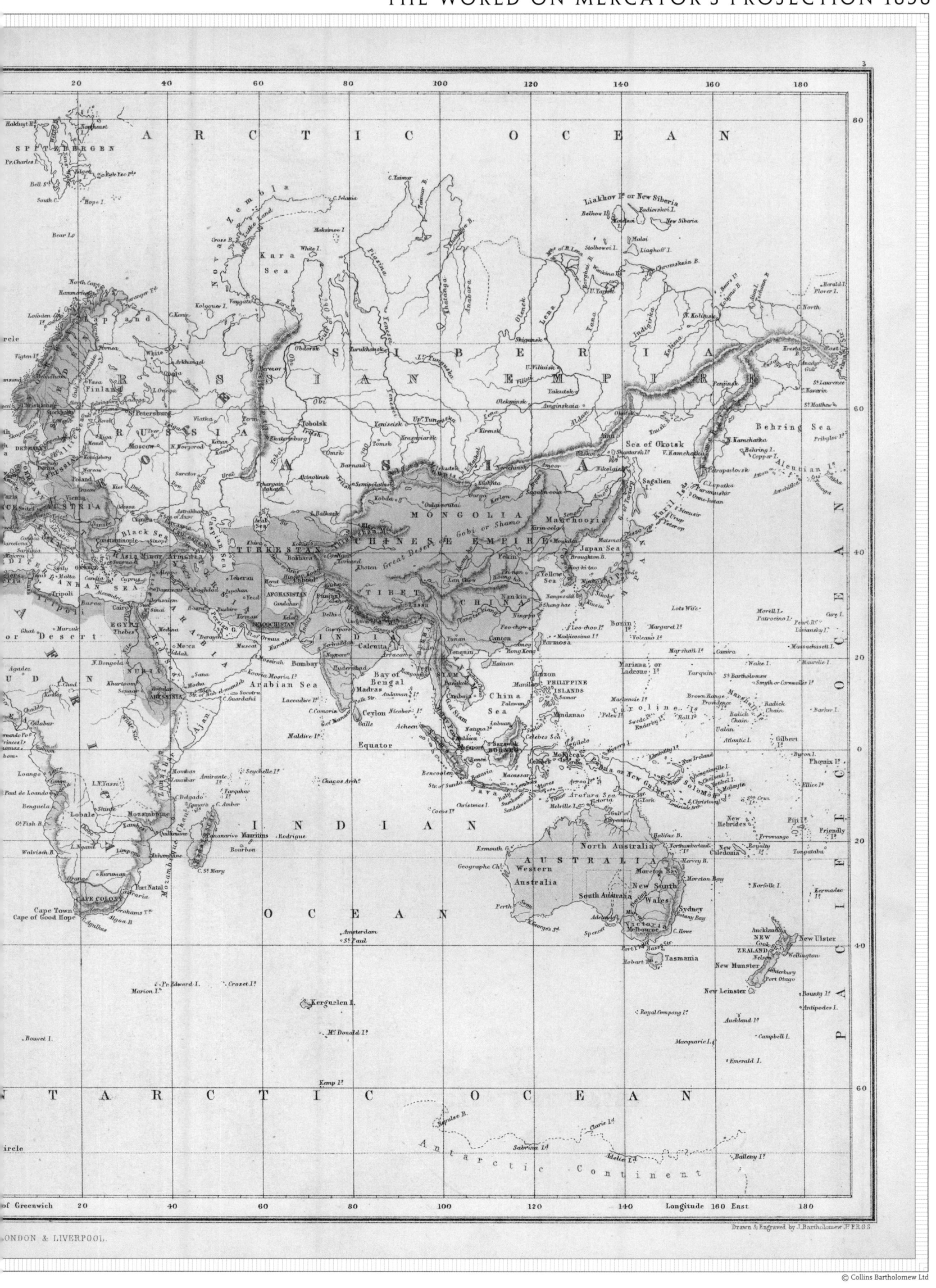

© Collins Bartholomew Ltd

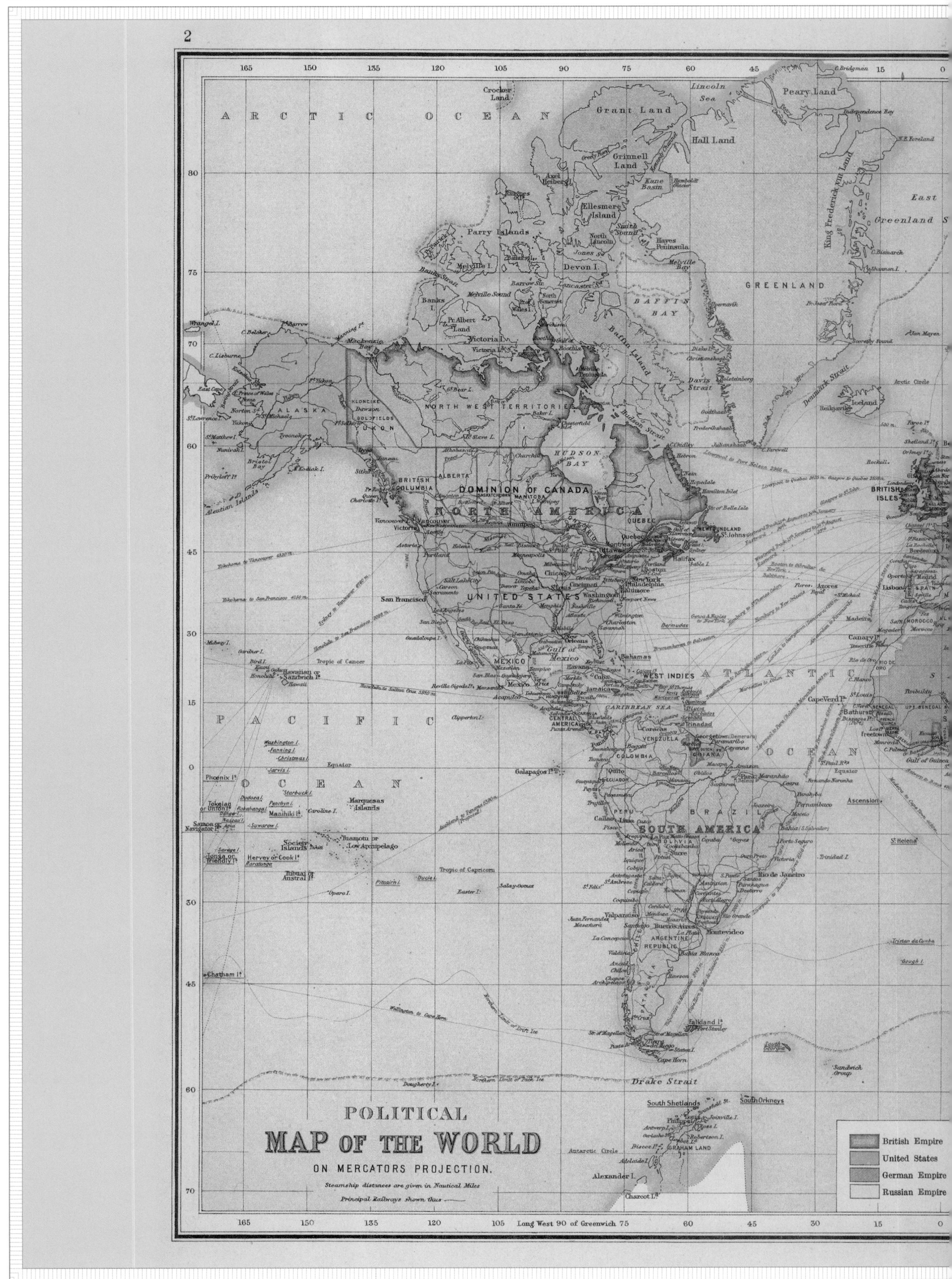

POLITICAL
MAP OF THE WORLD
ON MERCATORS PROJECTION.

Steamship distances are given in Nautical Miles

Principal Railways shown thus ————

	British Empire
	United States
	German Empire
	Russian Empire

3

	French Possessions
	Portuguese Possessions
	Dutch Possessions
	Chinese Possessions

John Bartholomew & Co. Edinr.

© Collins Bartholomew Ltd

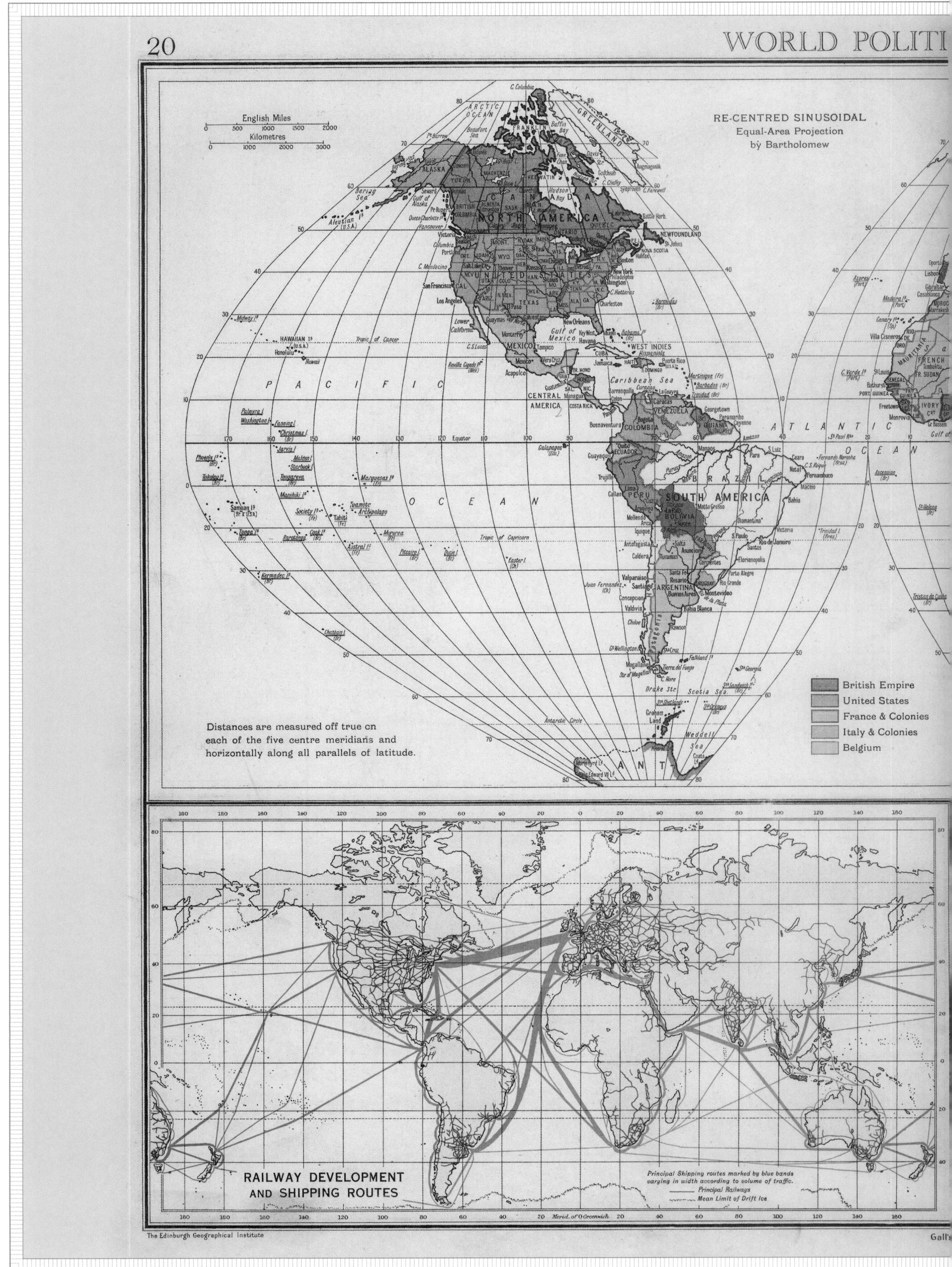

20

WORLD POLITI

RE-CENTRED SINUSOIDAL
Equal-Area Projection
by Bartholomew

English Miles

0   500   1000   1500   2000
Kilometres
0   1000   2000   3000

Distances are measured off true on
each of the five centre meridians and
horizontally along all parallels of latitude.

| British Empire |
| United States |
| France & Colonies |
| Italy & Colonies |
| Belgium |

RAILWAY DEVELOPMENT
AND SHIPPING ROUTES

Principal Shipping routes marked by blue bands
varying in width according to volume of traffic.
——— Principal Railways
- - - Mean Limit of Drift Ice

The Edinburgh Geographical Institute

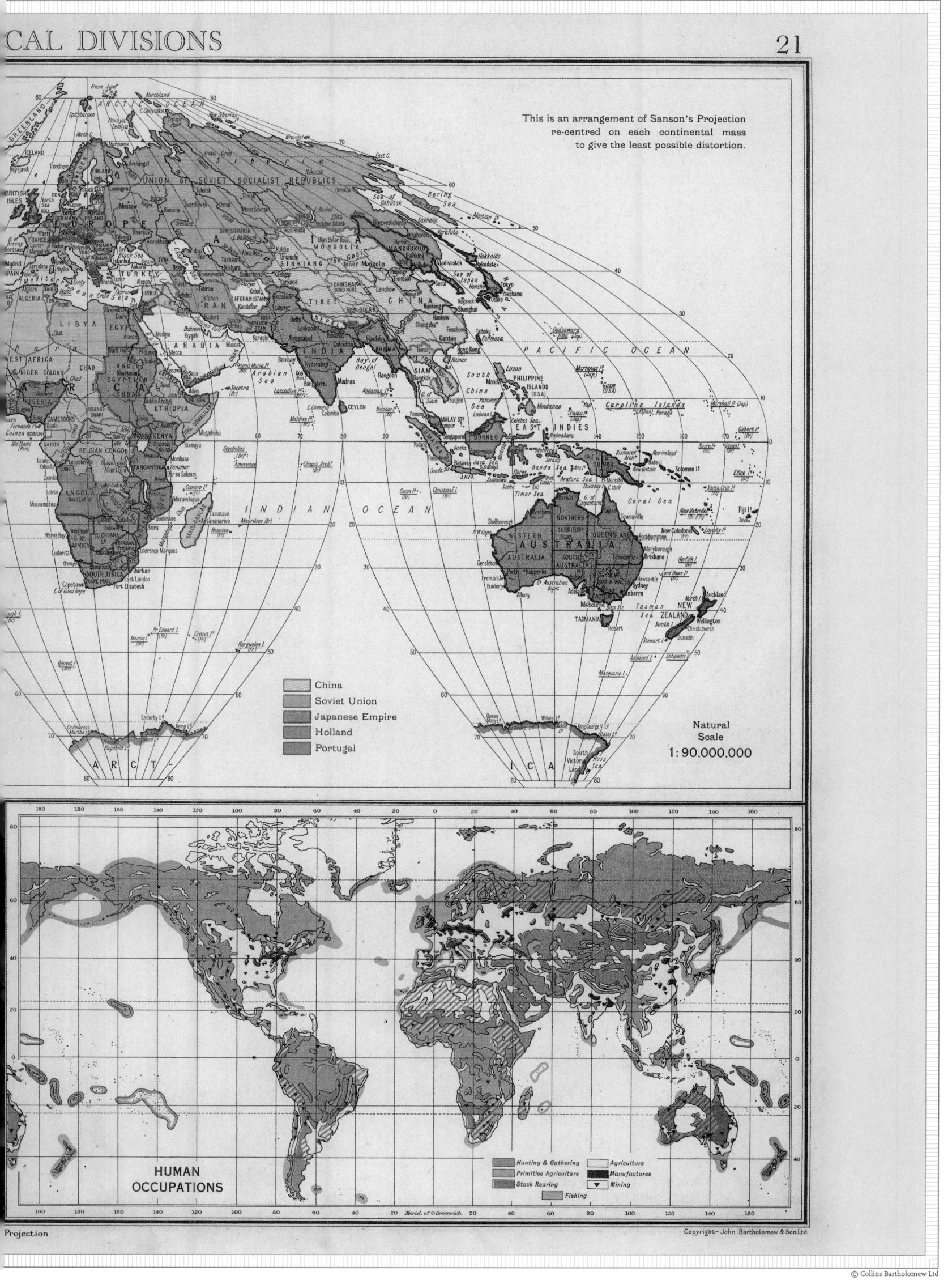

This is an arrangement of Sanson's Projection re-centred on each continental mass to give the least possible distortion.

Legend:
- China
- Soviet Union
- Japanese Empire
- Holland
- Portugal

Natural Scale
1:90,000,000

HUMAN OCCUPATIONS

- Hunting & Gathering
- Primitive Agriculture
- Stock Rearing
- Fishing
- Agriculture
- Manufactures
- Mining

Projection

Copyright- John Bartholomew & Son.Ltd.

© Collins Bartholomew Ltd

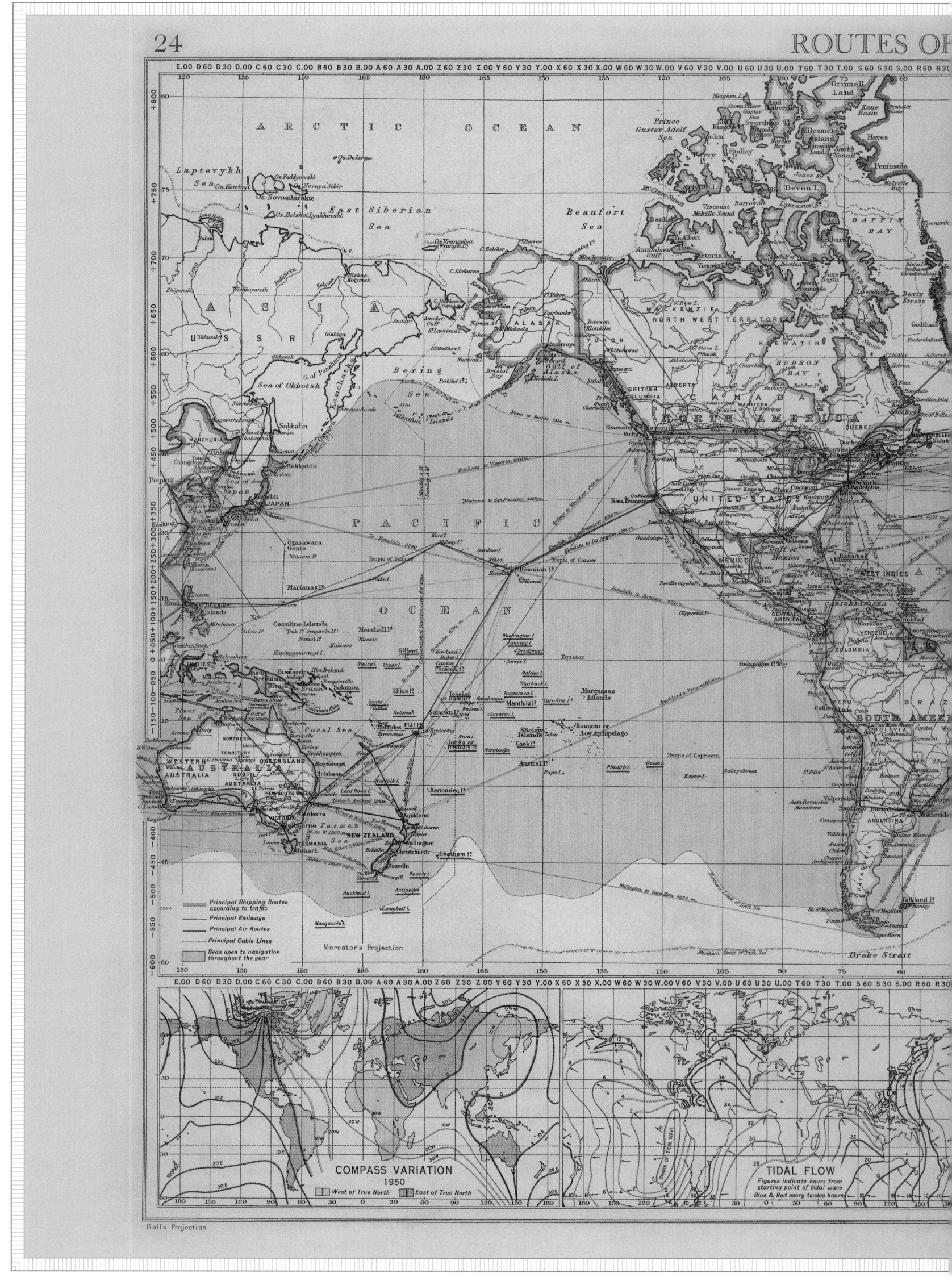

STANDARD TIME
Red and Blue areas are based on standard differences in hours from G.M.T. Yellow areas on half-hourly differences, e.g. India is 5½ hours fast on Greenwich.

LANGUAGES OF COMMERCE

English	Slavonic
French	Mongolian
Spanish	Other Languages
Portuguese	
Other European	
Arabic Group	

British Commonwealth — Latin American States
United States of America — Middle East States
French Territories — Soviet Russian Group
Netherlands Territories — Far Eastern Group
W. European States — African States

Copyright- John Bartholomew & Son Ltd. Edinburgh

© Collins Bartholomew Ltd

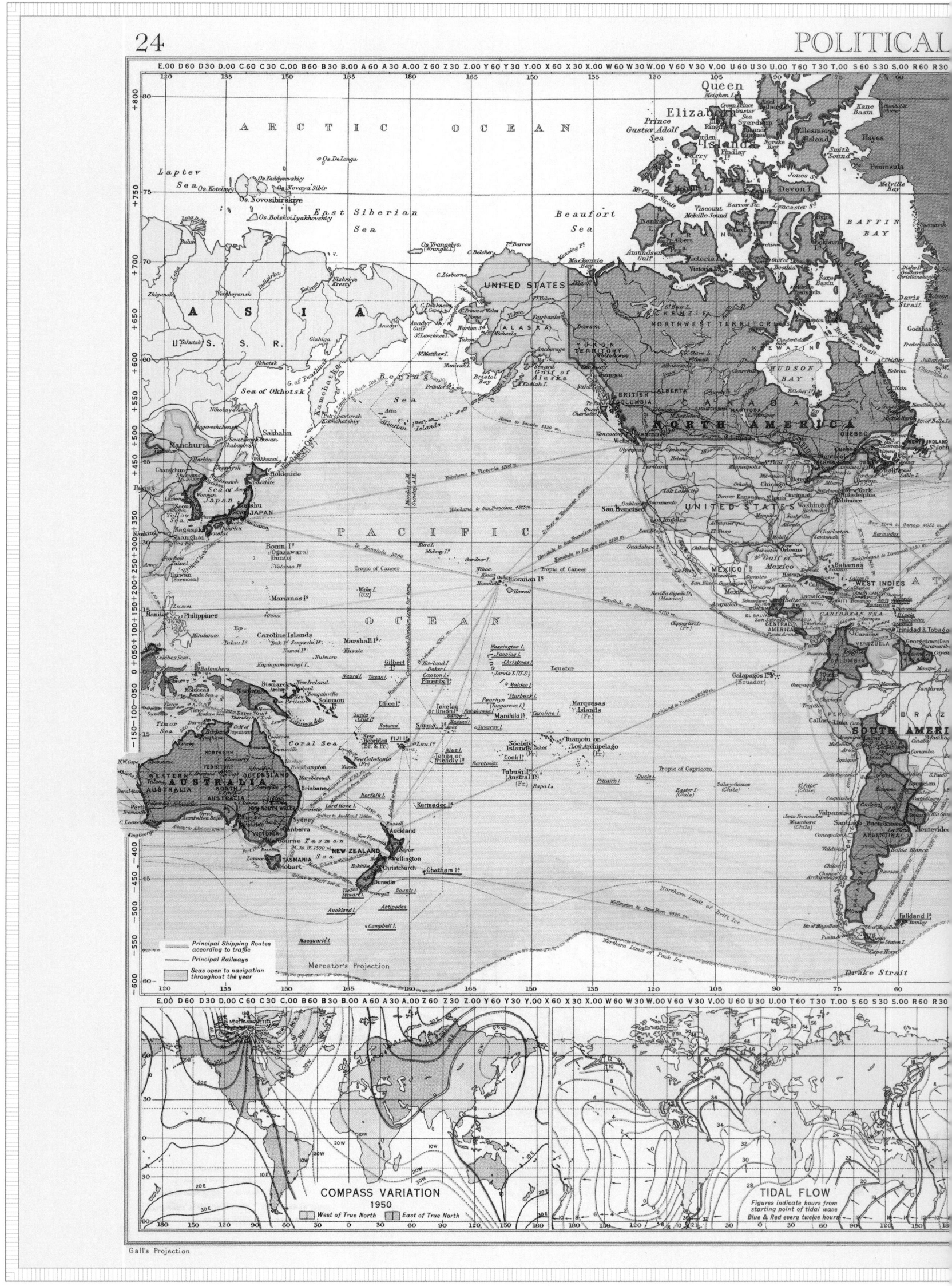

POLITICAL

ARCTIC OCEAN

Laptev Sea

East Siberian Sea

Beaufort Sea

BAFFIN BAY

ASIA
U.S.S.R.

UNITED STATES

NORTHWEST TERRITORIES

CANADA

NORTH AMERICA

Sea of Okhotsk

HUDSON BAY

Manchuria

Sea of Japan

JAPAN

PACIFIC

OCEAN

UNITED STATES

Tropic of Cancer

Tropic of Cancer

MEXICO

Gulf of Mexico

WEST INDIES

Philippines

Caroline Islands

Marshall Is.

Gilbert

CARIBBEAN SEA

CENTRAL AMERICA

VENEZUELA

COLOMBIA

Galapagos Is. (Ecuador)

Equator

Marquesas Islands

PERU

BRAZIL

Coral Sea

New Hebrides (Br. & Fr.)

FIJI Is.

Society Islands (Fr.)

Tropic of Capricorn

SOUTH AMERICA

WESTERN AUSTRALIA

NORTHERN TERRITORY

QUEENSLAND

SOUTH AUSTRALIA

AUSTRALIA

NEW SOUTH WALES

VICTORIA

TASMANIA

Tasman Sea

NEW ZEALAND

Wellington

Christchurch

Dunedin

ARGENTINA

Northern Limit of Drift Ice

Mercator's Projection

Wellington to Cape Horn

Northern Limit of Pack Ice

Drake Strait

	Principal Shipping Routes according to traffic
	Principal Railways
	Seas open to navigation throughout the year

COMPASS VARIATION
1950

West of True North   East of True North

TIDAL FLOW
Figures indicate hours from starting point of tidal wave
Blue & Red every twelve hours

STANDARD TIME

Red and Blue areas are based on standard differences in hours from G.M.T., Yellow areas on half-hourly differences, e.g. India is 5½ hours fast on Greenwich.

LANGUAGES OF COMMERCE

■ English	Portuguese	Slavonic
French	Other European	Mongolian
Spanish	Arabic Group	Other Languages

ⓒ — John Bartholomew & Son, Ltd., Edinburgh

© Collins Bartholomew Ltd

24

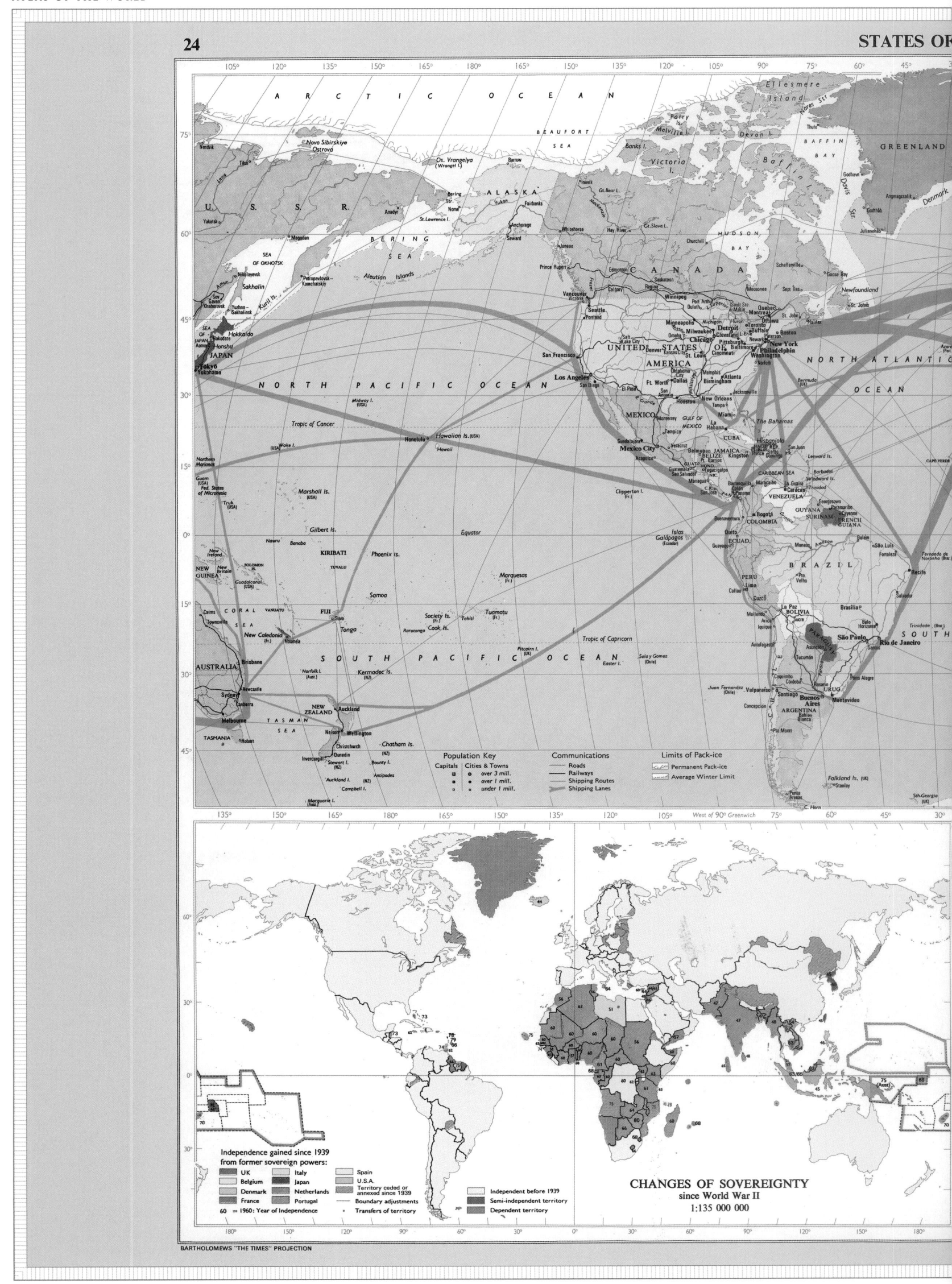

ARCTIC OCEAN

U.S.S.R.

ALASKA

CANADA

GREENLAND

NORTH PACIFIC OCEAN

UNITED STATES OF AMERICA

NORTH ATLANTIC OCEAN

JAPAN

MEXICO

GULF OF MEXICO

CUBA

Los Angeles

Tropic of Cancer

Hawaiian Is. (USA)

Mexico City

VENEZUELA

GUYANA
SURINAM
FRENCH GUIANA

COLOMBIA

ECUAD.

Equator

KIRIBATI

TUVALU

Marquesas (Fr.)

BRAZIL

PERU

BOLIVIA

NEW GUINEA

FIJI

Samoa

Society Is. (Fr.) Tuamotu (Fr.)

Tonga   Rarotonga Cook Is.   Tahiti

Tropic of Capricorn

São Paulo

Rio de Janeiro

AUSTRALIA

SOUTH PACIFIC OCEAN

ARGENTINA

URUG.

Buenos Aires

Montevideo

TASMANIA

NEW ZEALAND

TASMAN SEA

Wellington

Falkland Is. (UK)

Population Key

Capitals | Cities & Towns
over 3 mill.
over 1 mill.
under 1 mill.

Communications
Roads
Railways
Shipping Routes
Shipping Lanes

Limits of Pack-ice
Permanent Pack-ice
Average Winter Limit

West of 90° Greenwich

CHANGES OF SOVEREIGNTY
since World War II
1:135 000 000

Independence gained since 1939
from former sovereign powers:

UK
Belgium
Denmark
France
60 = 1960: Year of Independence

Italy
Japan
Netherlands
Portugal
Spain
U.S.A.
Territory ceded or annexed since 1939
Boundary adjustments
Transfers of territory

Independent before 1939
Semi-independent territory
Dependent territory

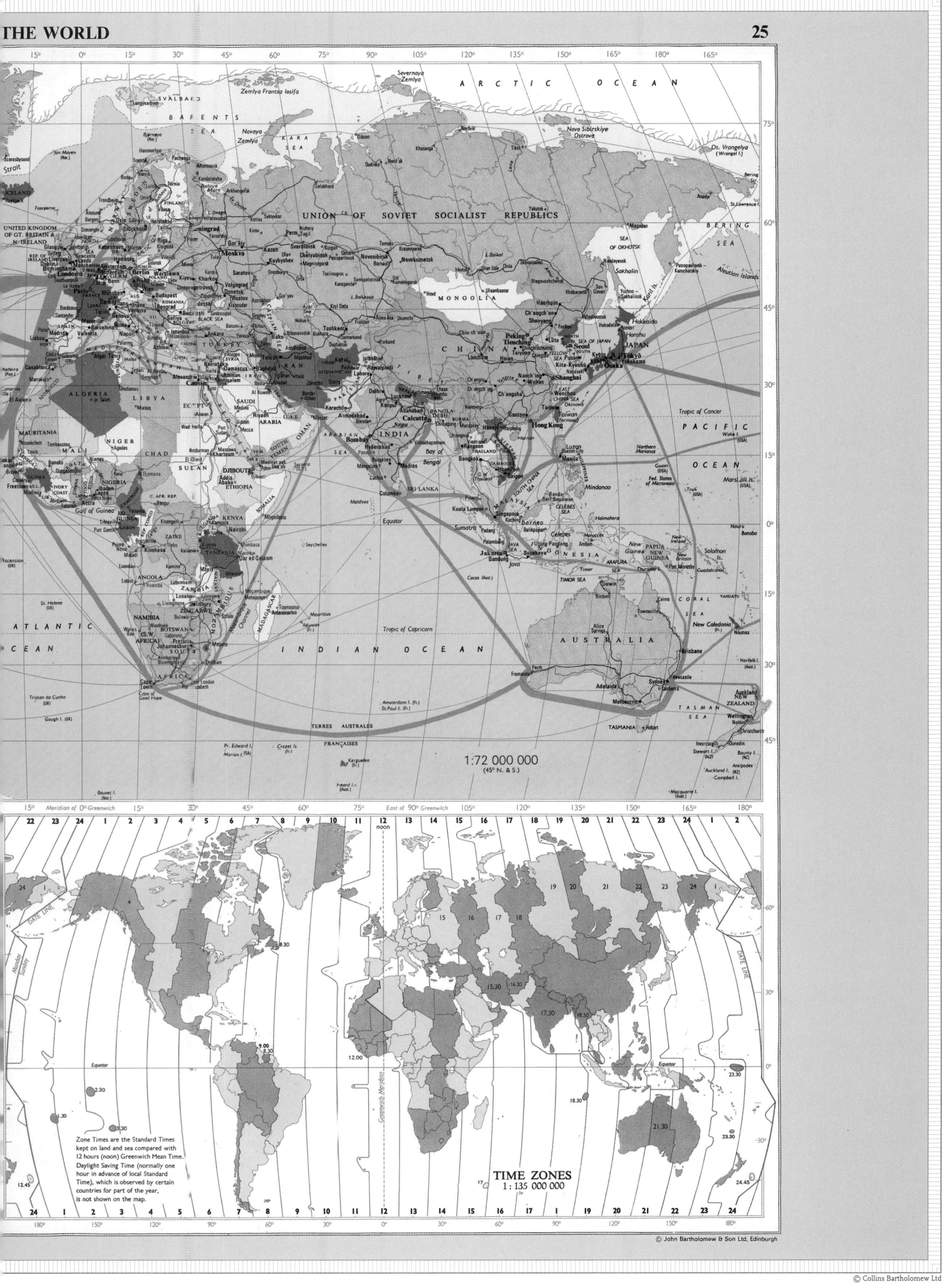

1:72 000 000
(45° N. & S.)

Zone Times are the Standard Times
kept on land and sea compared with
12 hours (noon) Greenwich Mean Time.
Daylight Saving Time (normally one
hour in advance of local Standard
Time), which is observed by certain
countries for part of the year,
is not shown on the map.

TIME ZONES
1 : 135 000 000

© John Bartholomew & Son Ltd, Edinburgh

© Collins Bartholomew Ltd

NORTH POLAR REGIONS

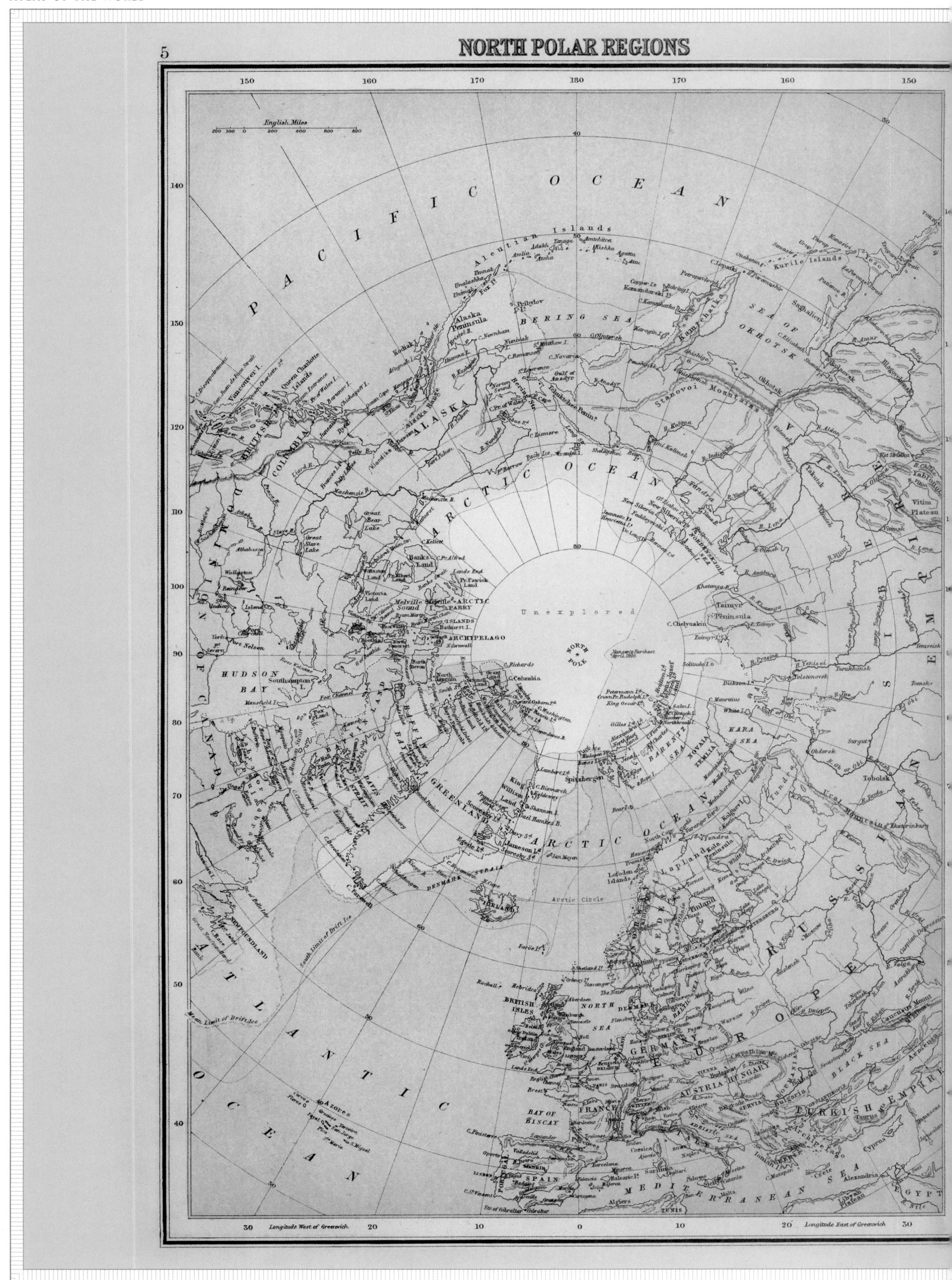

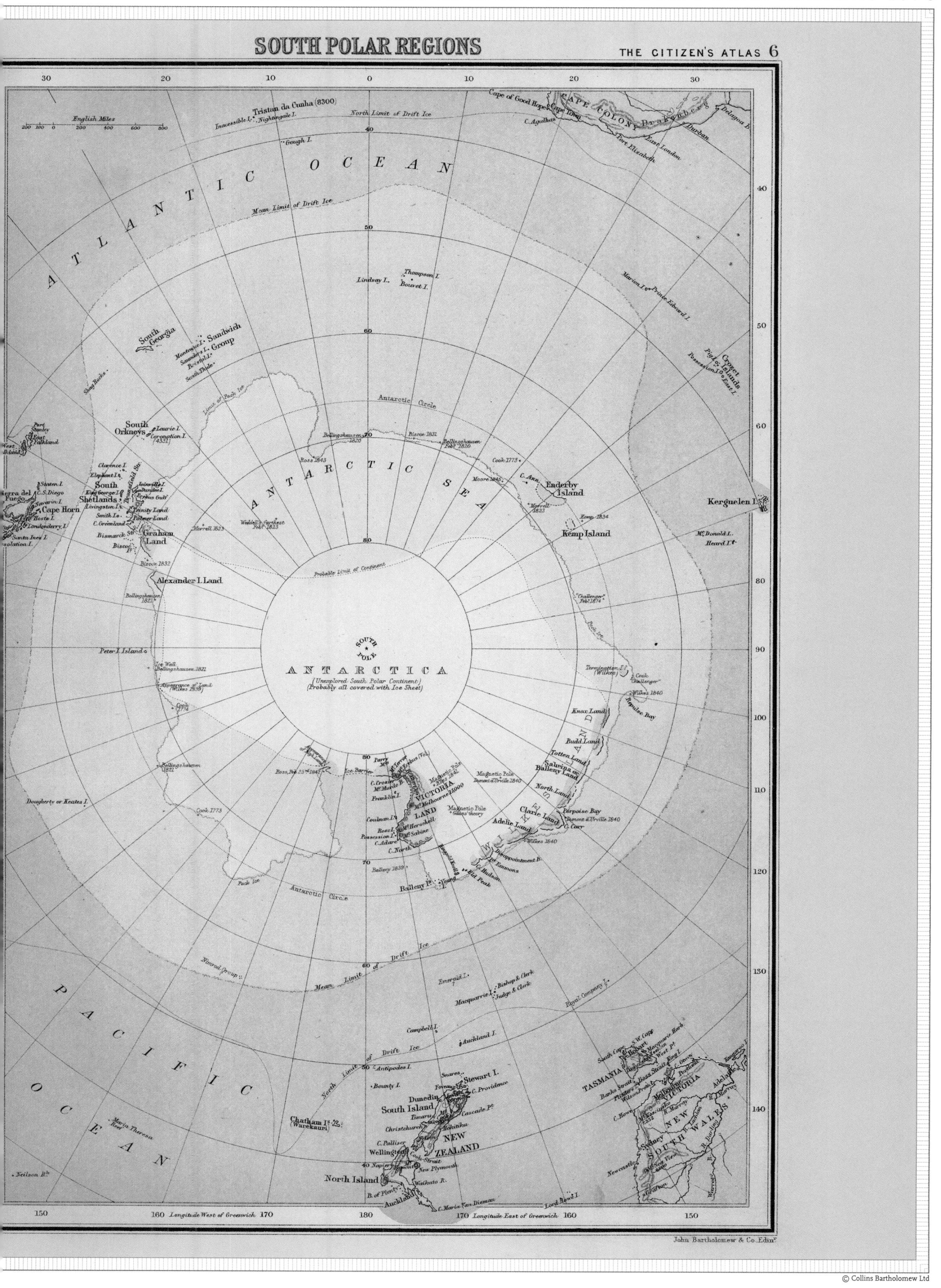

SOUTH POLAR REGIONS

John Bartholomew & Co. Edin.

© Collins Bartholomew Ltd

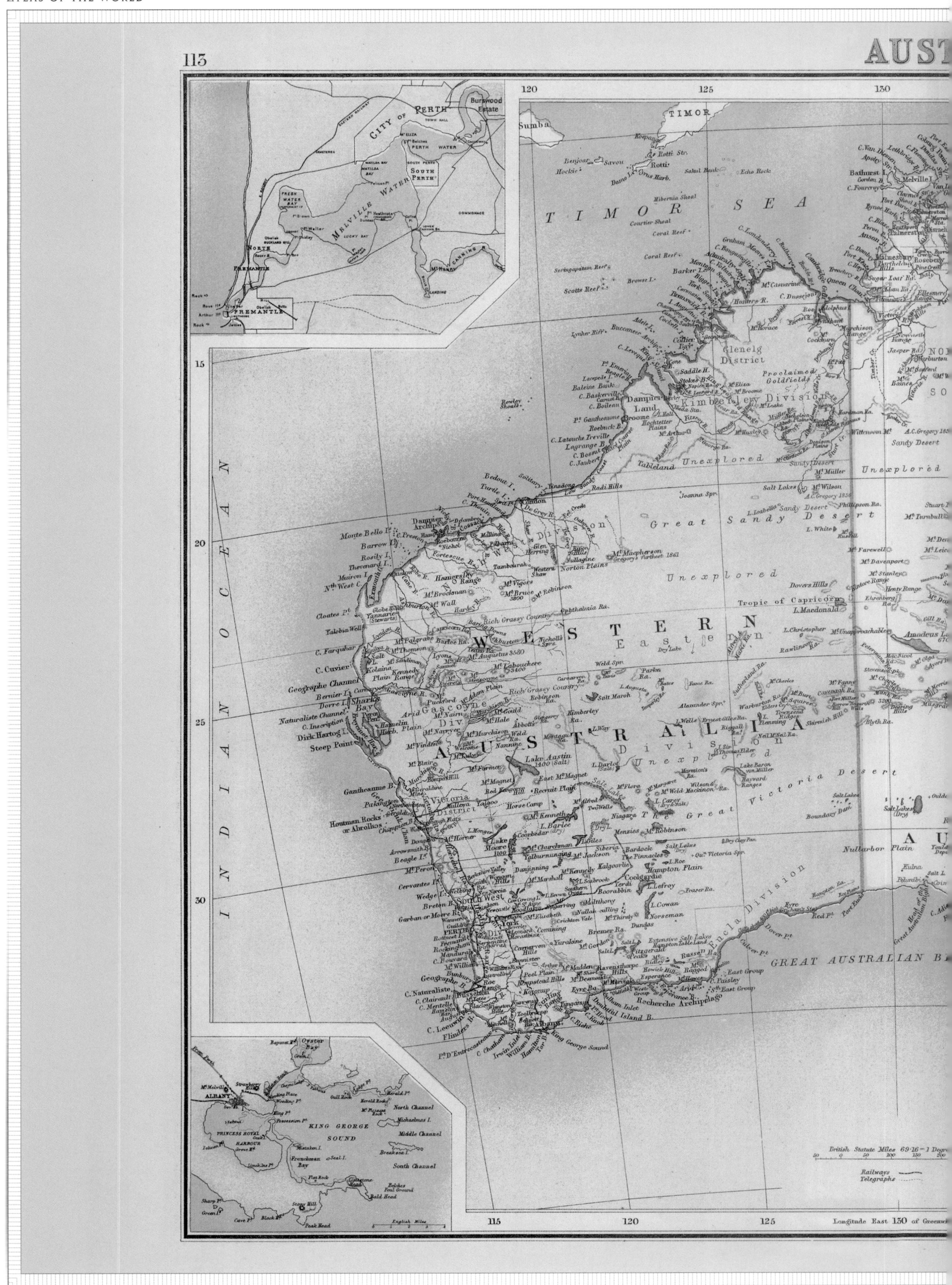

John Bartholomew & Co. Edin.

© Collins Bartholomew Ltd

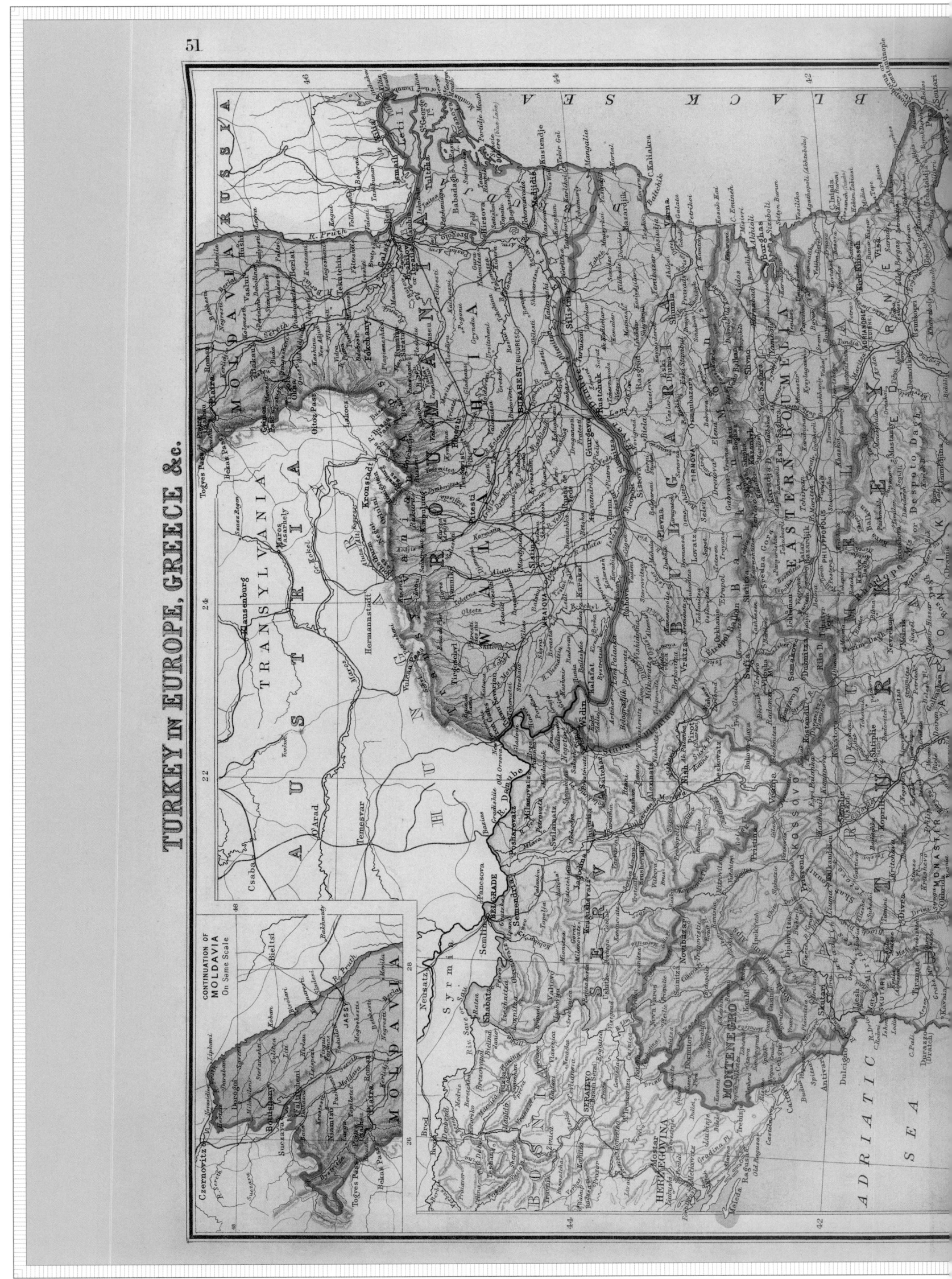

TURKEY IN EUROPE, GREECE &c.

CONTINUATION OF
MOLDAVIA
On Same Scale

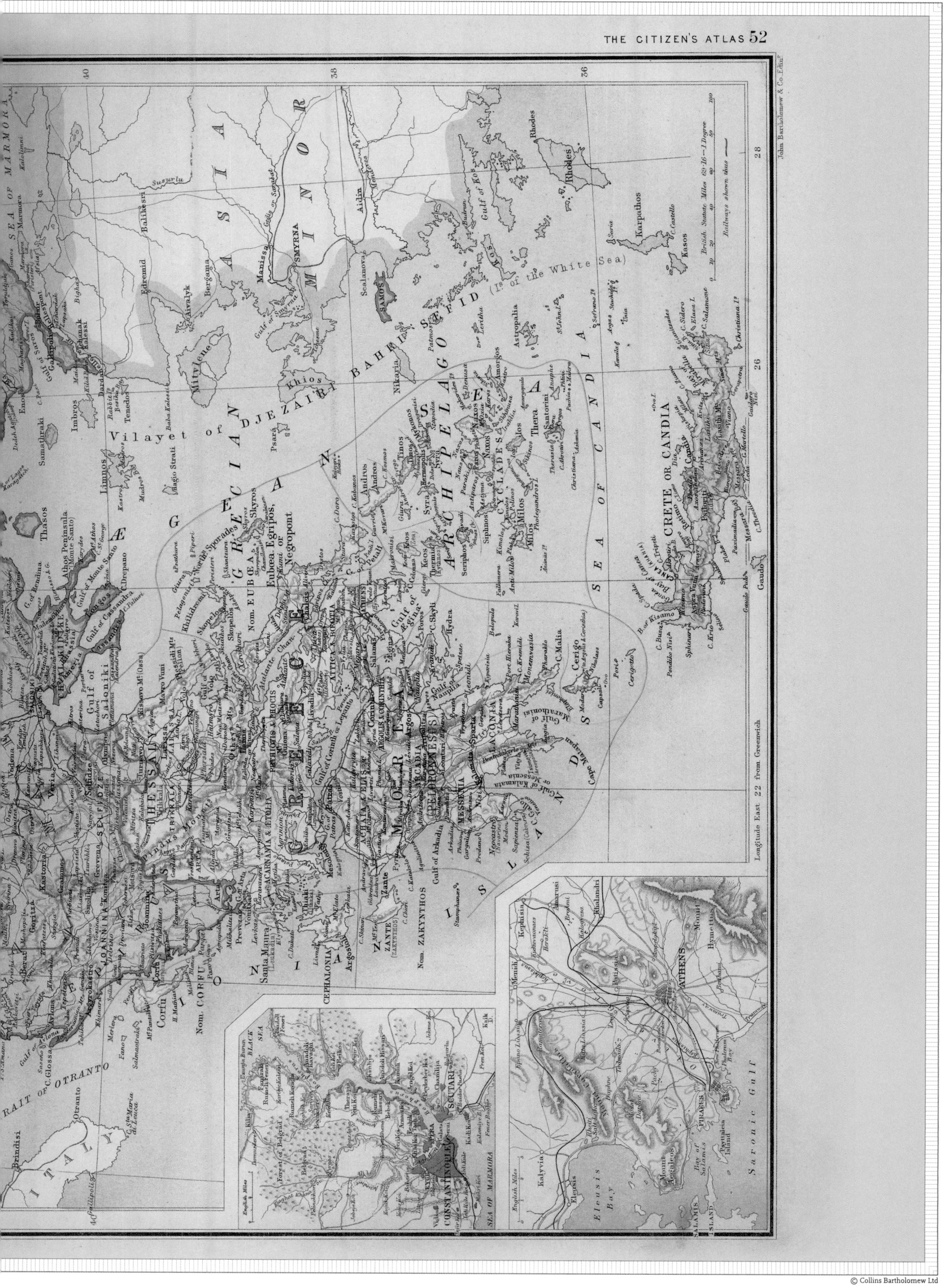

John Bartholomew & Co. Edinr

© Collins Bartholomew Ltd

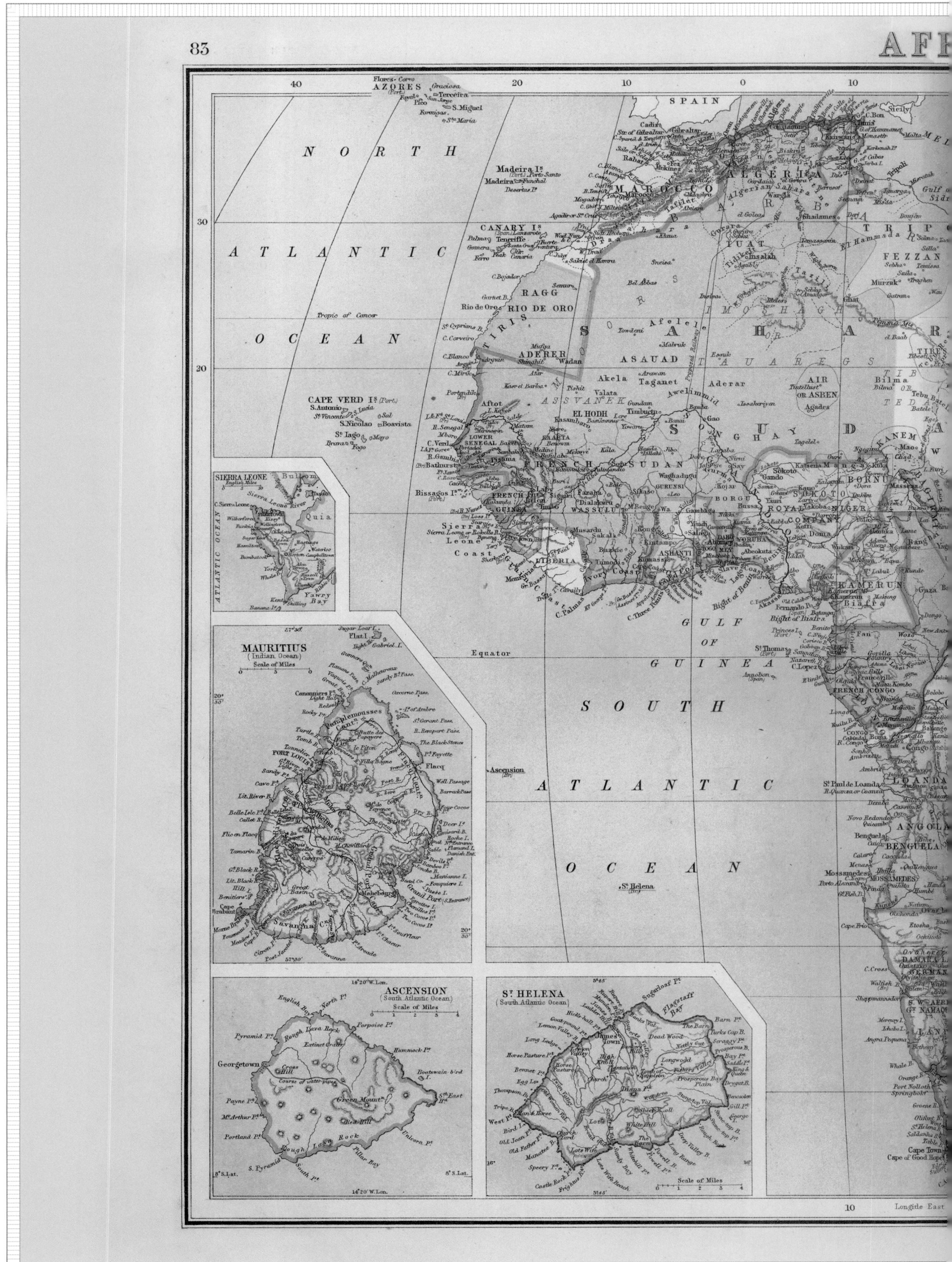

AF

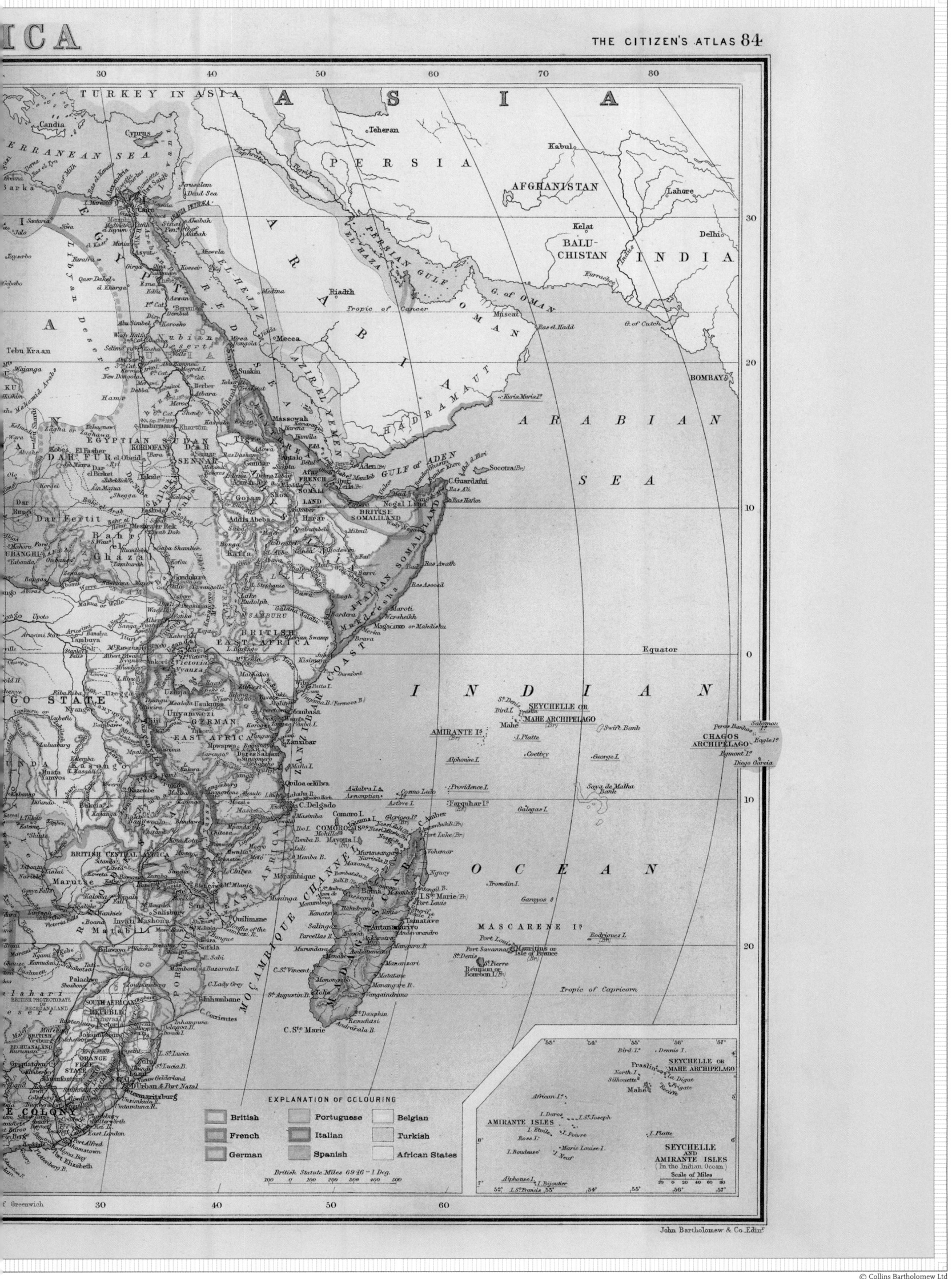

EXPLANATION OF COLOURING

British   Portuguese   Belgian

French   Italian   Turkish

German   Spanish   African States

British Statute Miles 69·16 = 1 Deg.
200   0   100   200   300   400   500

SEYCHELLE OR
MAHE ARCHIPELAGO

SEYCHELLE
AND
AMIRANTE ISLES
(In the Indian Ocean)
Scale of Miles

AMIRANTE ISLES

John Bartholomew & Co. Edin.

© Collins Bartholomew Ltd

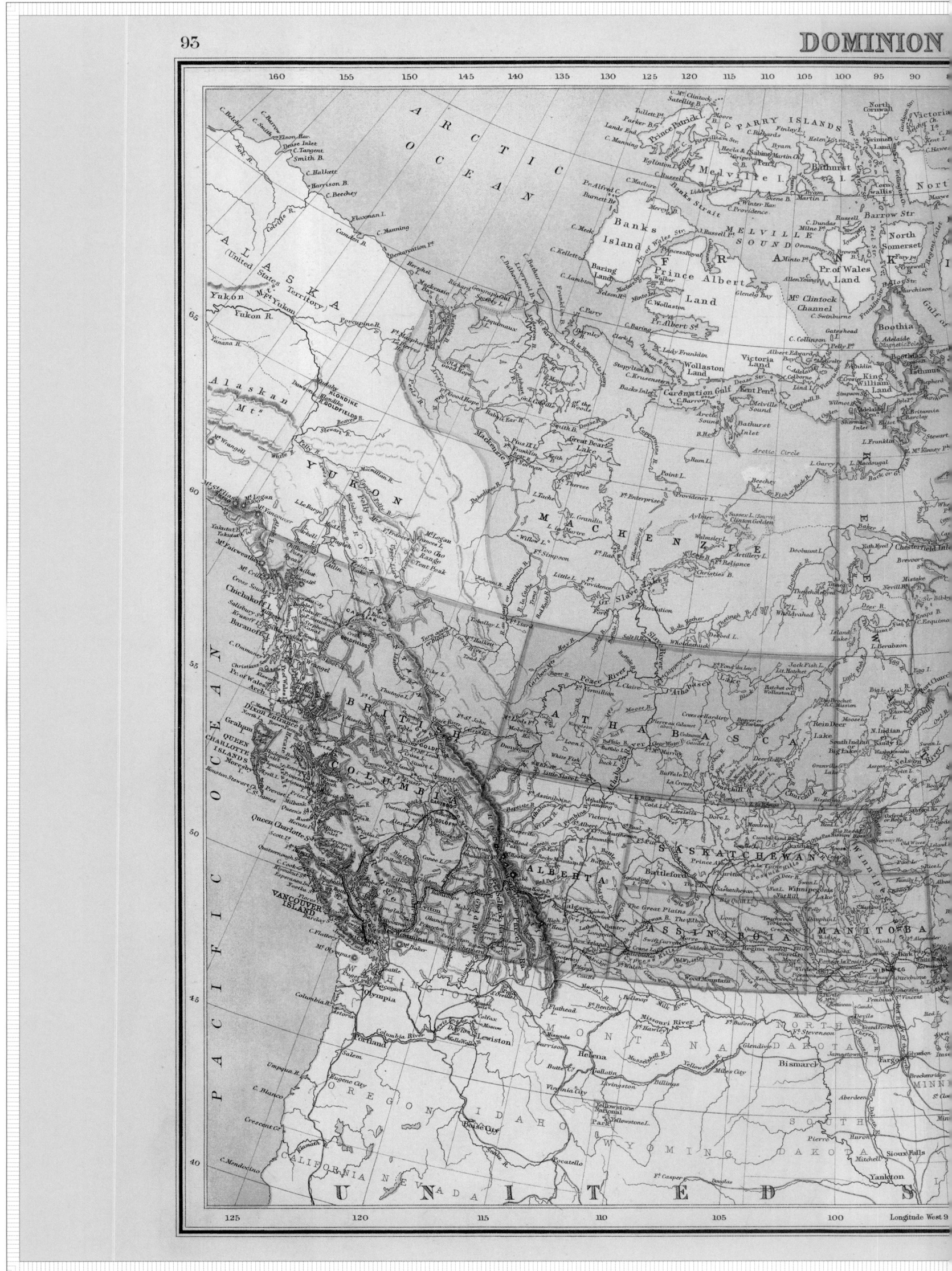

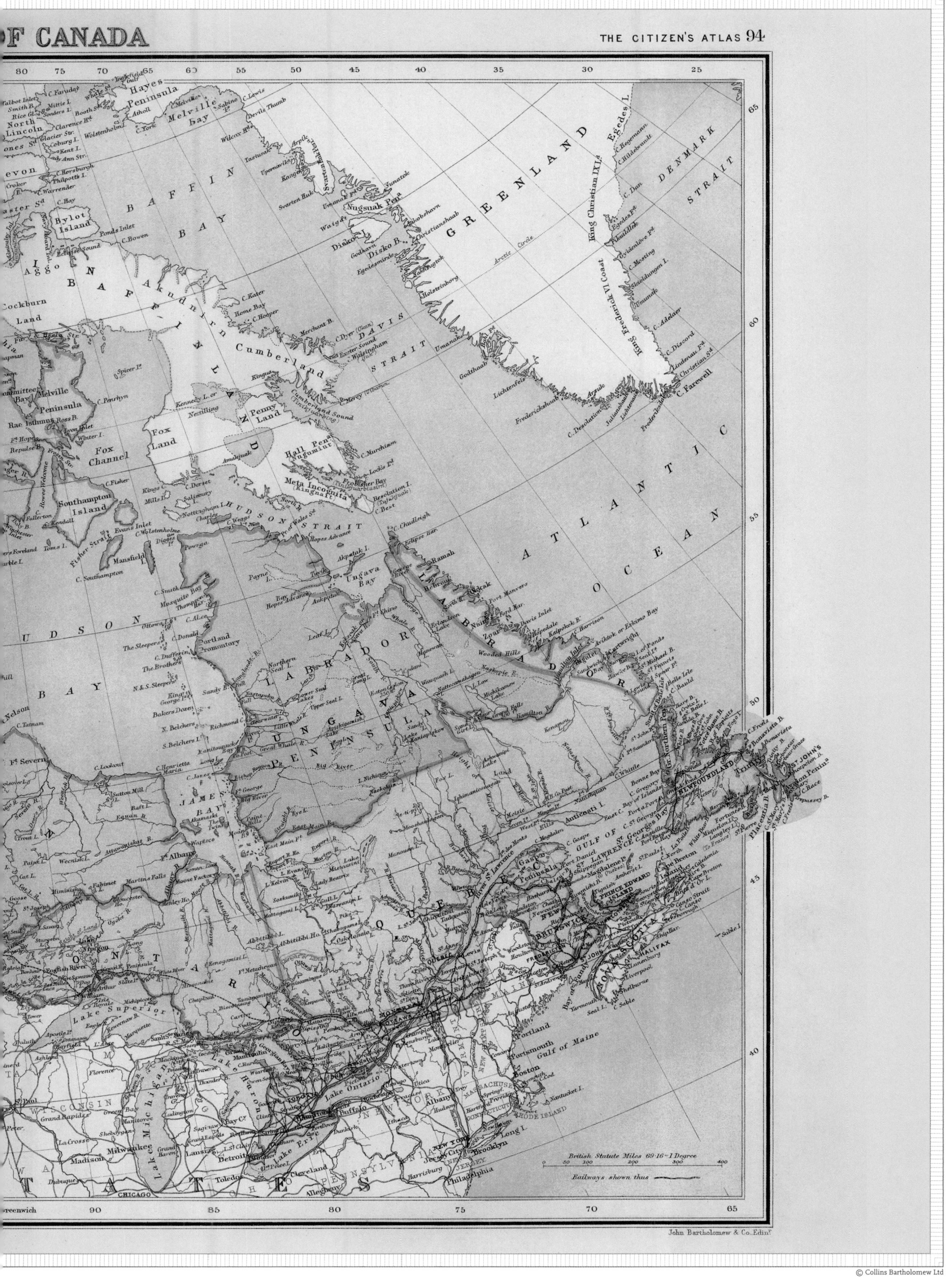

OF CANADA

John Bartholomew & Co., Edin^r

© Collins Bartholomew Ltd

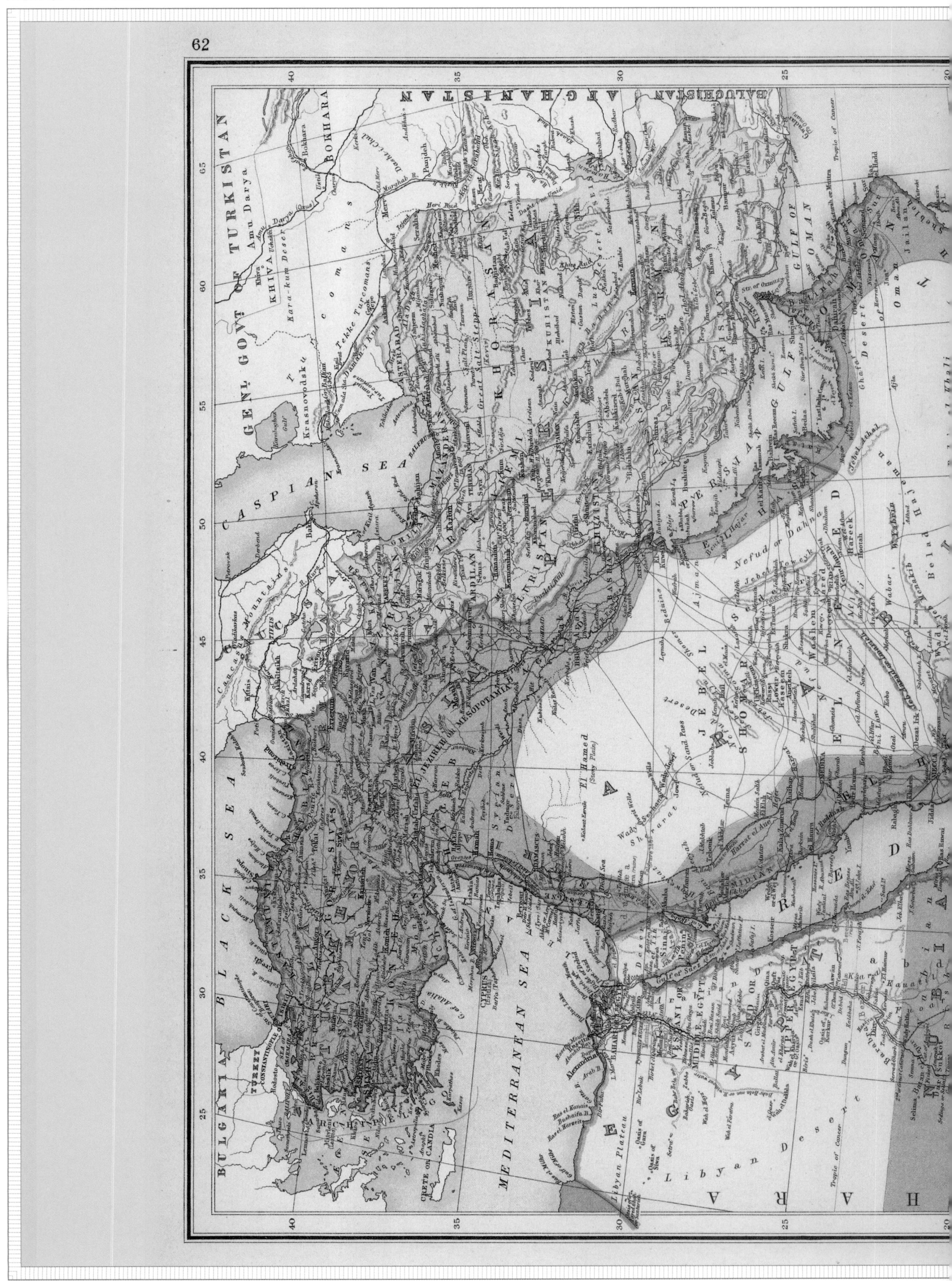

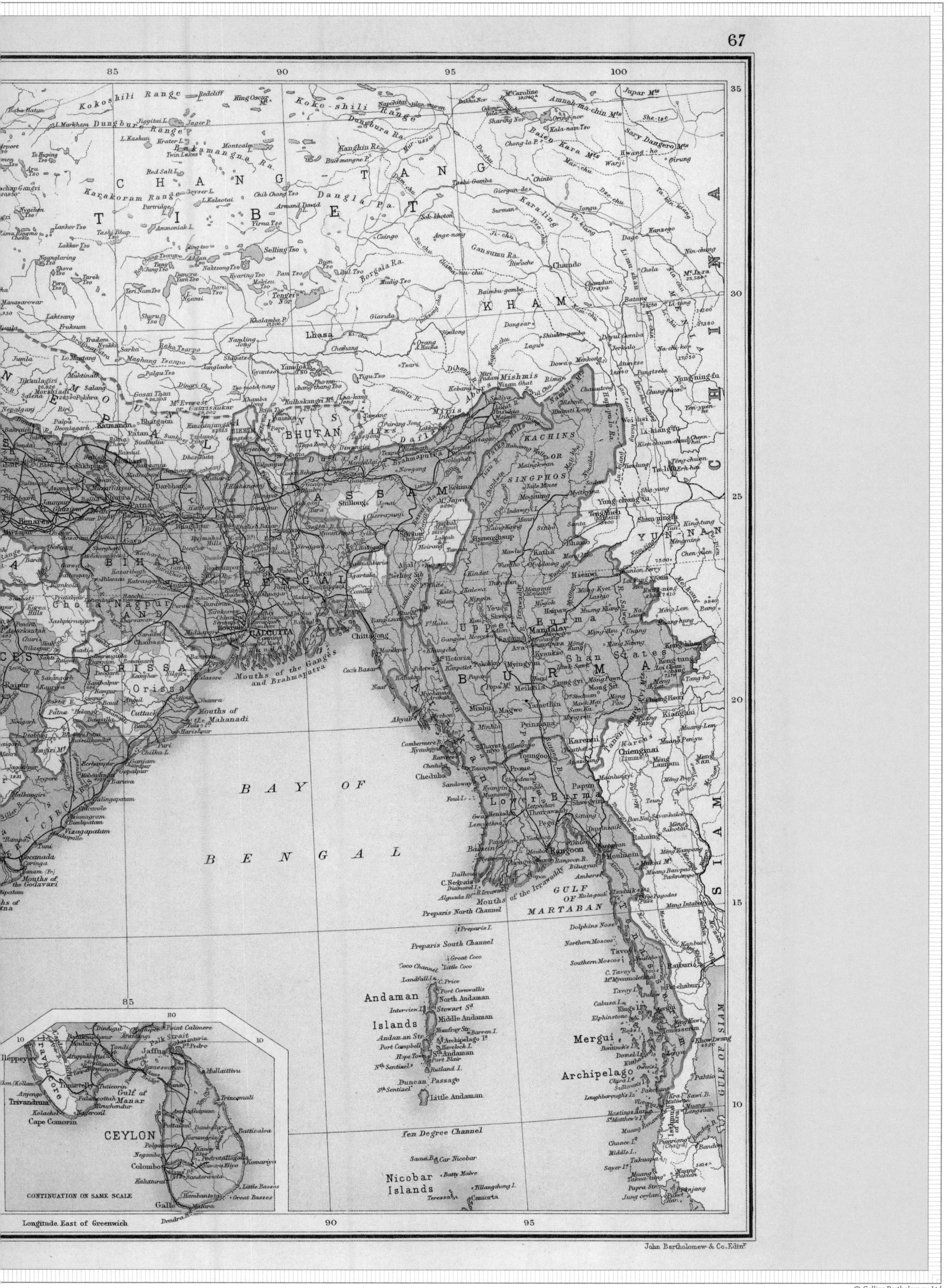

© Collins Bartholomew Ltd

POLITICAL

PLATE 10

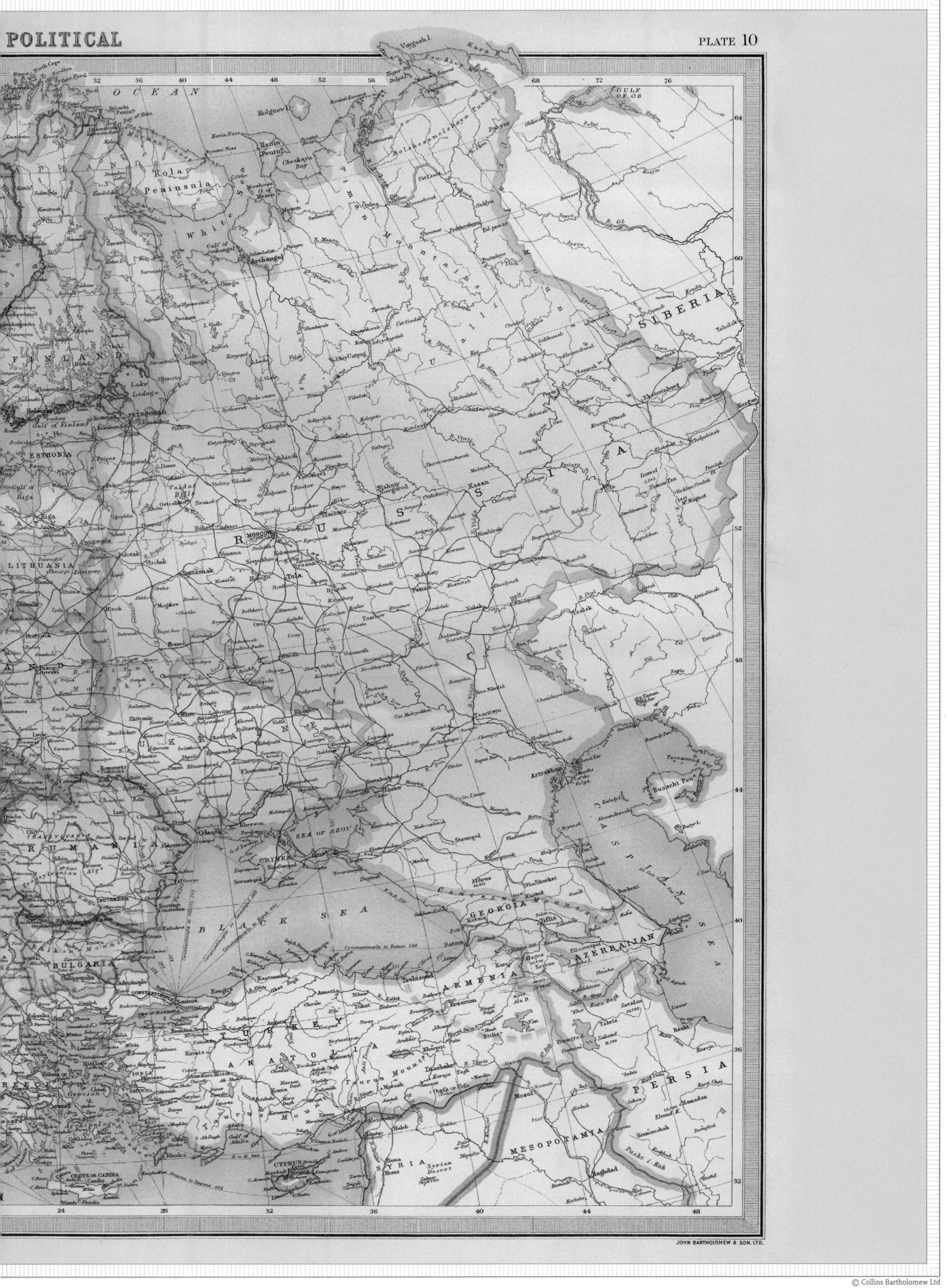

JOHN BARTHOLOMEW & SON, LTD.

© Collins Bartholomew Ltd

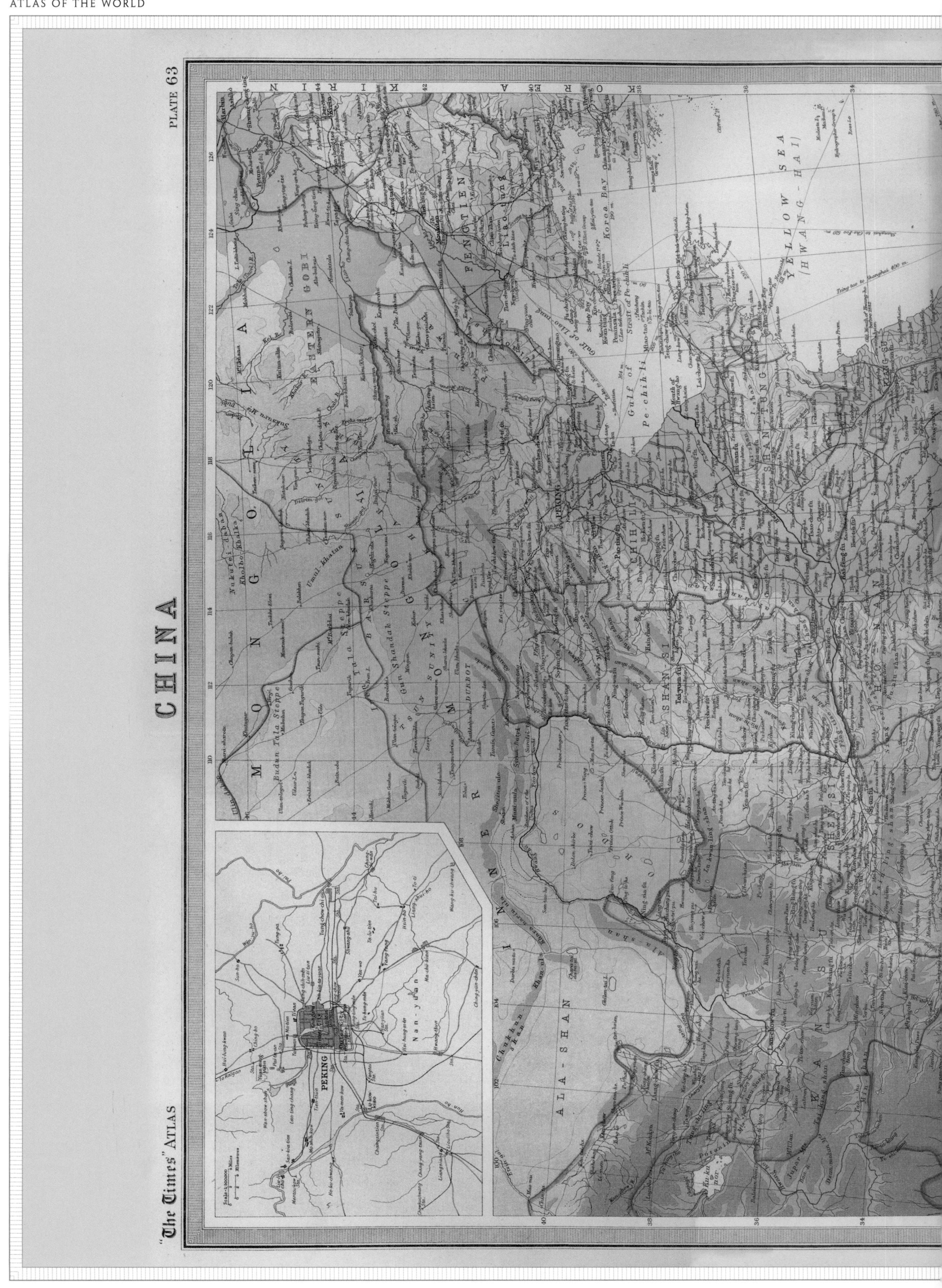

PLATE 63

CHINA

"The Times" ATLAS

PLATE 63

© Collins Bartholomew Ltd

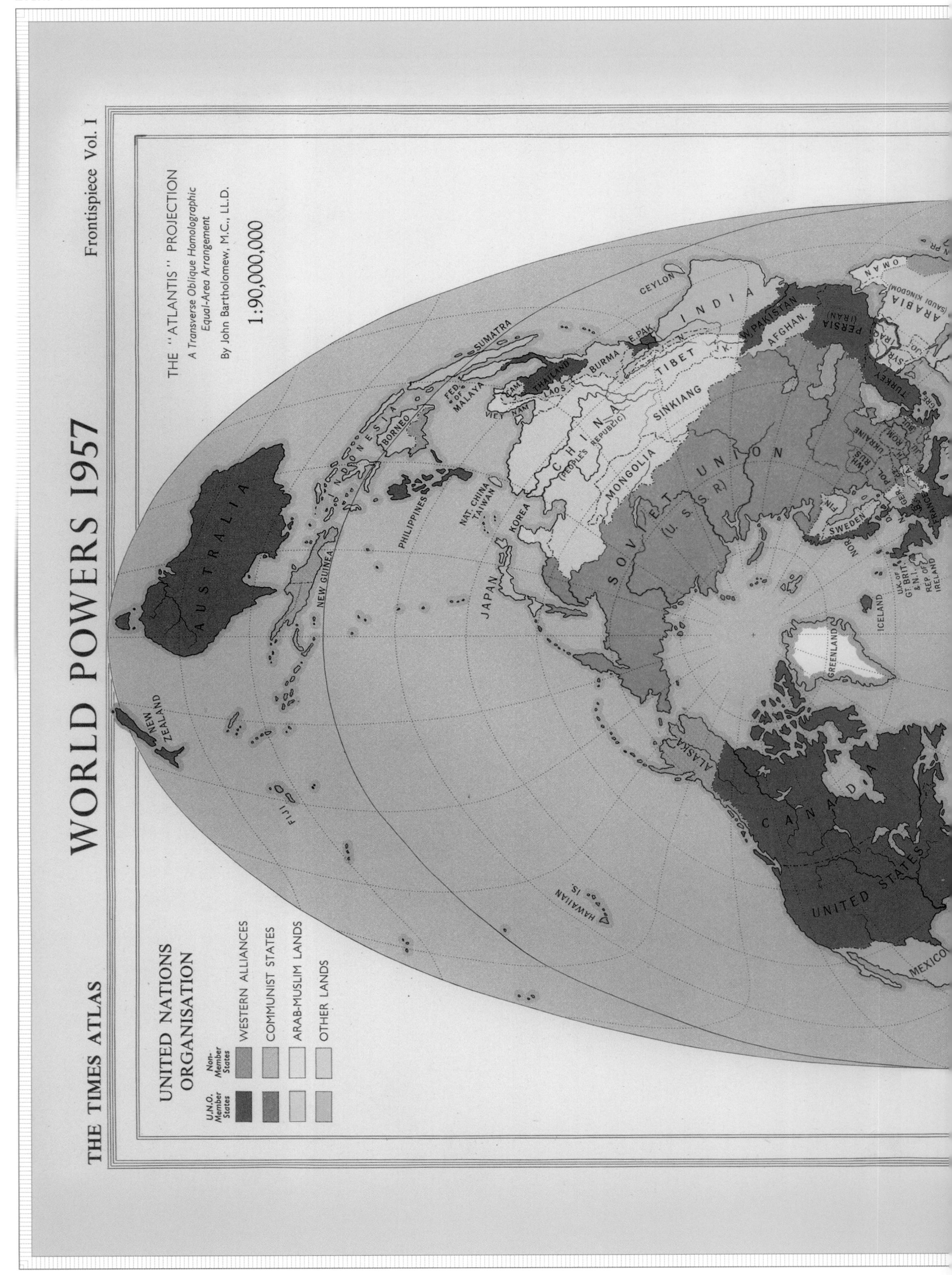

THE TIMES ATLAS

WORLD POWERS 1957

Frontispiece Vol. I

THE "ATLANTIS" PROJECTION
A Transverse Oblique Homolographic
Equal-Area Arrangement
By John Bartholomew, M.C., LL.D.

1:90,000,000

UNITED NATIONS
ORGANISATION

	WESTERN ALLIANCES
	COMMUNIST STATES
	ARAB-MUSLIM LANDS
	OTHER LANDS

U.N.O.
Member
States

Non-
Member
States

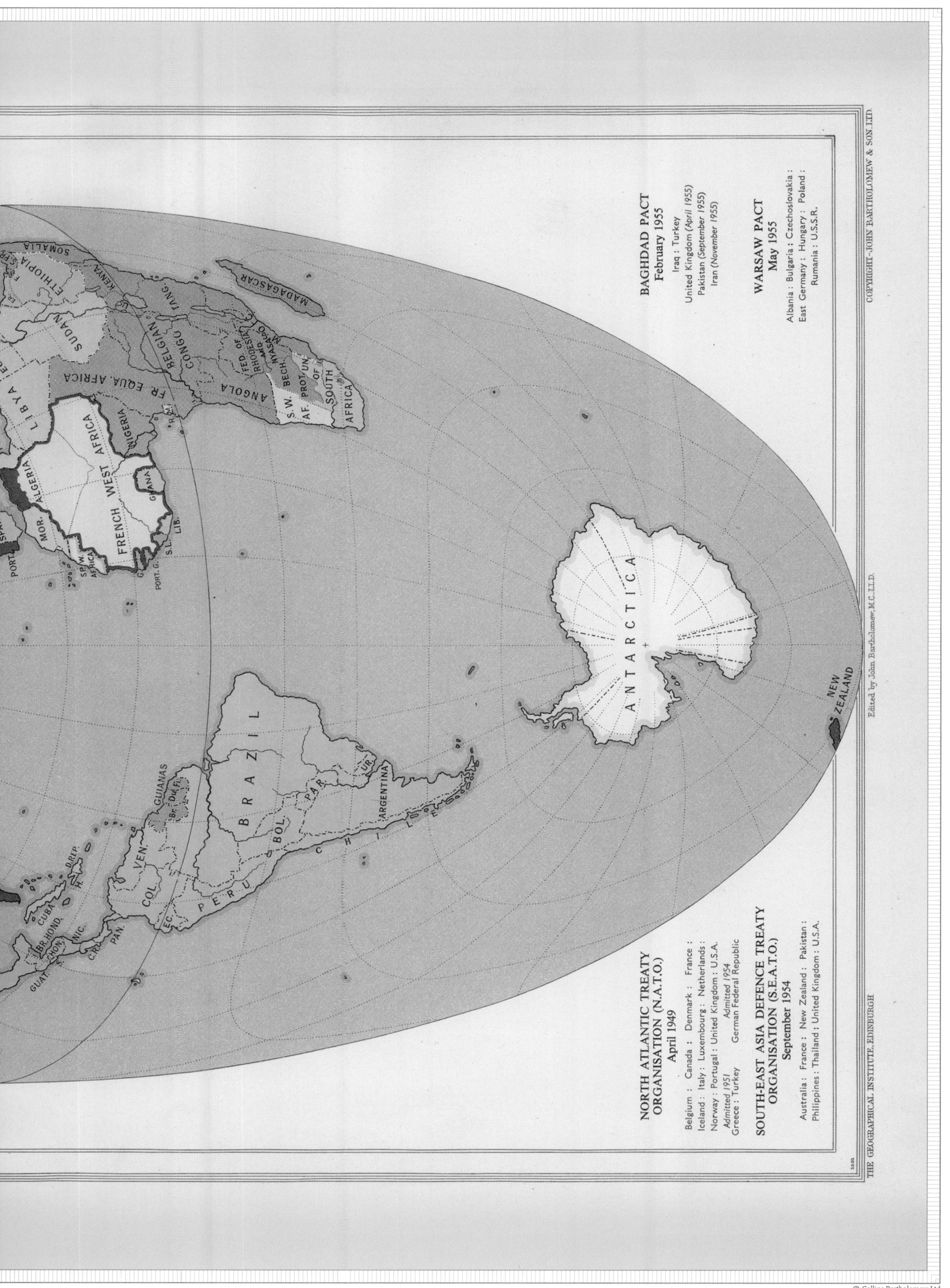

BAGHDAD PACT
February 1955

Iraq : Turkey
United Kingdom (April 1955)
Pakistan (September 1955)
Iran (November 1955)

WARSAW PACT
May 1955

Albania : Bulgaria : Czechoslovakia :
East Germany : Hungary : Poland :
Rumania : U.S.S.R.

NORTH ATLANTIC TREATY
ORGANISATION (N.A.T.O.)
April 1949

Belgium : Canada : Denmark : France :
Iceland : Italy : Luxembourg : Netherlands :
Norway : Portugal : United Kingdom : U.S.A.
Admitted 1951   Admitted 1954
Greece : Turkey   German Federal Republic

SOUTH-EAST ASIA DEFENCE TREATY
ORGANISATION (S.E.A.T.O.)
September 1954

Australia : France : New Zealand : Pakistan :
Philippines : Thailand : United Kingdom : U.S.A.

THE GEOGRAPHICAL INSTITUTE, EDINBURGH

Edited by John Bartholomew, M.C. I.I.D.

COPYRIGHT—JOHN BARTHOLOMEW & SON LTD.

© Collins Bartholomew Ltd

STATES AND TERRITORIES

All 195 independent countries and all populated dependent and disputed territories are included in this list of the states and territories of the world; the list is arranged in alphabetical order by the conventional name form. For independent states, the full name is given below the conventional name, if this is different; for territories, the status is given. The capital city name is given in conventional English form with selected alternative, usually local, form in brackets.

Area and population statistics are the latest available and include estimates. The information on languages and religions is based on the latest information on 'de facto' speakers of the language or 'de facto' adherents of the religion. This varies greatly from country to country because some countries include questions in censuses while others do not, in which case best estimates are used. The order of the languages and religions reflects their relative importance within the country; generally, languages or religions are included when more than one per cent of the population are estimated to be speakers or adherents.

ABBREVIATIONS

CURRENCIES

CFA	Communauté Financière Africaine
CFP	Comptoirs Français du Pacifique

Membership of selected international organizations is shown by the abbreviations below; dependent territories do not normally have separate memberships of these organizations.

ORGANIZATIONS

APEC	Asia-Pacific Economic Cooperation
ASEAN	Association of Southeast Asian Nations
CARICOM	Caribbean Community
CIS	Commonwealth of Independent States
Comm.	The Commonwealth
EU	European Union
GCC	Gulf Cooperation Council
NATO	North Atlantic Treaty Organization
OECD	Organisation for Economic Co-operation and Development
OPEC	Organization of the Petroleum Exporting Countries
SADC	Southern African Development Community
UN	United Nations

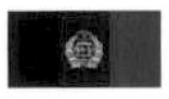

AFGHANISTAN
Islamic Republic of Afghanistan

Area Sq Km	652 225	**Languages**	Dari, Pushtu, Uzbek, Turkmen
Area Sq Miles	251 825	**Religions**	Sunni Muslim, Shi'a Muslim
Population	28 150 000	**Currency**	Afghani
Capital	Kābul	**Organizations**	UN

Map page 141

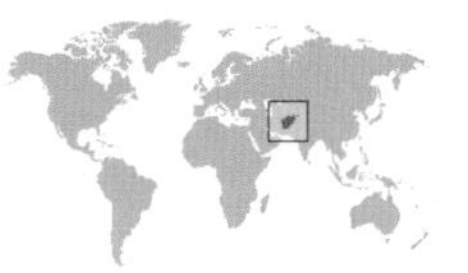

A landlocked country in central Asia with central highlands bordered by plains in the north and southwest, and by the mountains of the Hindu Kush in the northeast. The climate is dry continental. Over the last thirty years war has disrupted the economy, which is highly dependent on farming and livestock rearing. Most trade is with the former USSR, Pakistan and Iran.

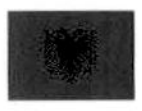

ALBANIA
Republic of Albania

Area Sq Km	28 748	**Languages**	Albanian, Greek
Area Sq Miles	11 100	**Religions**	Sunni Muslim, Albanian Orthodox, Roman Catholic
Population	3 155 000	**Currency**	Lek
Capital	Tirana (Tiranë)	**Organizations**	NATO, UN

Map page 171

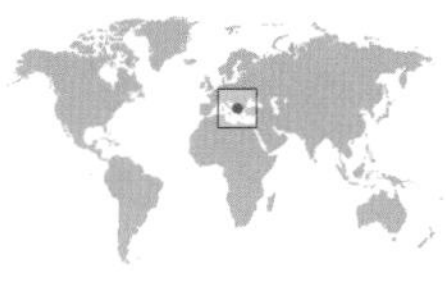

Albania lies in the western Balkan Mountains in southeastern Europe, bordering the Adriatic Sea. It is mountainous, with coastal plains where half the population lives. The economy is based on agriculture and mining. Albania is one of the poorest countries in Europe and relies heavily on foreign aid.

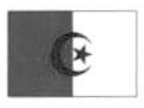

ALGERIA
People's Democratic Republic of Algeria

Area Sq Km	2 381 741	**Languages**	Arabic, French, Berber
Area Sq Miles	919 595	**Religions**	Sunni Muslim
Population	34 895 000	**Currency**	Algerian dinar
Capital	Algiers (Alger)	**Organizations**	OPEC, UN

Map page 176

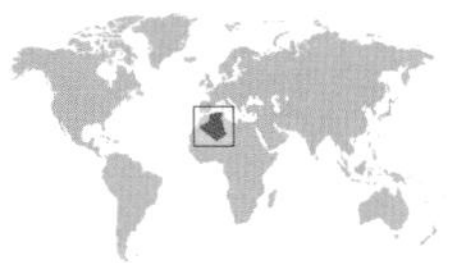

Algeria, the second largest country in Africa, lies on the Mediterranean coast of northwest Africa and extends southwards to the Atlas Mountains and the dry sandstone plateau and desert of the Sahara. The climate ranges from Mediterranean on the coast to semi-arid and arid inland. The most populated areas are the coastal plains and the fertile northern slopes of the Atlas Mountains. Oil, natural gas and related products account for over ninety-five per cent of export earnings. Agriculture employs about a quarter of the workforce, producing mainly food crops. Algeria's main trading partners are Italy, France and the USA.

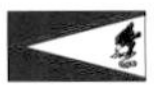

American Samoa
United States Unincorporated Territory

Area Sq Km	197	**Languages**	Samoan, English
Area Sq Miles	76	**Religions**	Protestant, Roman Catholic
Population	67 000	**Currency**	United States dollar
Capital	Fagatogo		

Map page 123

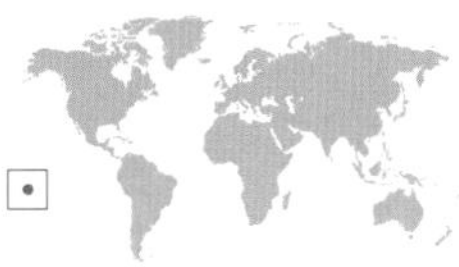

Lying in the south Pacific Ocean, American Samoa consists of five main islands and two coral atolls. The largest island is Tutuila. Tuna and tuna products are the main exports, and the main trading partner is the USA.

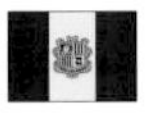

ANDORRA
Principality of Andorra

Area Sq Km	465	**Languages**	Spanish, Catalan, French
Area Sq Miles	180	**Religions**	Roman Catholic
Population	86 000	**Currency**	Euro
Capital	Andorra la Vella	**Organizations**	UN

Map page 167

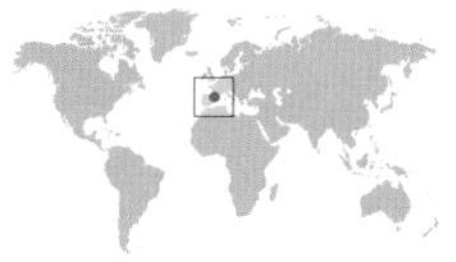

A landlocked state in southwest Europe, Andorra lies in the Pyrenees mountain range between France and Spain. It consists of deep valleys and gorges, surrounded by mountains. Tourism, encouraged by the development of ski resorts, is the mainstay of the economy. Banking is also an important economic activity.

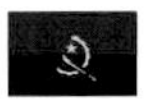

ANGOLA
Republic of Angola

Area Sq Km	1 246 700	**Languages**	Portuguese, Bantu, local languages
Area Sq Miles	481 354	**Religions**	Roman Catholic, Protestant, traditional beliefs
Population	18 498 000	**Currency**	Kwanza
Capital	Luanda	**Organizations**	OPEC, SADC, UN

Map page 176–177

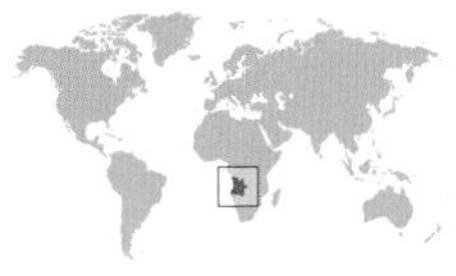

Angola lies on the Atlantic coast of south central Africa. Its small northern province, Cabinda, is separated from the rest of the country by part of the Democratic Republic of the Congo. Much of Angola is high plateau. In the west is a narrow coastal plain and in the southwest is desert. The climate is equatorial in the north but desert in the south. Over eighty per cent of the population relies on subsistence agriculture. Angola is rich in minerals (particularly diamonds), and oil accounts for approximately ninety per cent of export earnings. The USA, South Korea and Portugal are its main trading partners.

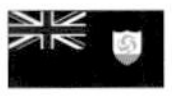

Anguilla
United Kingdom Overseas Territory

Area Sq Km	155	**Languages**	English
Area Sq Miles	60	**Religions**	Protestant, Roman Catholic
Population	15 000	**Currency**	East Caribbean dollar
Capital	The Valley		

Map page 205 Anguilla lies at the northern end of the Leeward Islands in the eastern Caribbean. Tourism and fishing form the basis of the economy.

ANTIGUA AND BARBUDA

Area Sq Km	442	**Languages**	English, Creole
Area Sq Miles	171	**Religions**	Protestant, Roman Catholic
Population	88 000	**Currency**	East Caribbean dollar
Capital	St John's	**Organizations**	CARICOM, Comm., UN

Map page 205

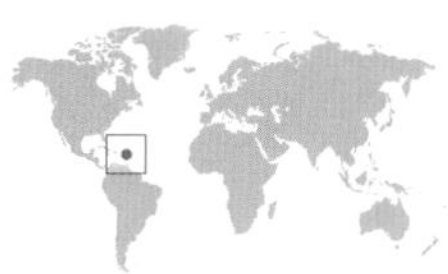

The state comprises the islands of Antigua, Barbuda and the tiny rocky outcrop of Redonda, in the Leeward Islands in the eastern Caribbean. Antigua, the largest and most populous island, is mainly hilly scrubland, with many beaches. The climate is tropical, and the economy relies heavily on tourism. Most trade is with other eastern Caribbean states and the USA.

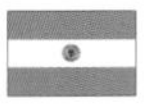

ARGENTINA
Argentine Republic

Area Sq Km	2 766 889	**Languages**	Spanish, Italian, Amerindian languages
Area Sq Miles	1 068 302	**Religions**	Roman Catholic, Protestant
Population	40 276 000	**Currency**	Argentinian peso
Capital	Buenos Aires	**Organizations**	UN

Map page 212

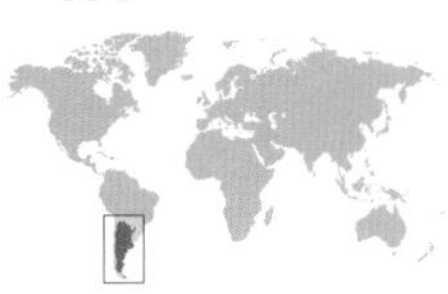

Argentina, the second largest state in South America, extends from Bolivia to Cape Horn and from the Andes mountains to the Atlantic Ocean. It has four geographical regions: subtropical forests and swampland in the northeast; temperate fertile plains or Pampas in the centre; the wooded foothills and valleys of the Andes in the west; and the cold, semi-arid plateaus of Patagonia in the south. The highest mountain in South America, Cerro Aconcagua, is in Argentina. Nearly ninety per cent of the population lives in towns and cities. The country is rich in natural resources including petroleum, natural gas, ores and precious metals. Agricultural products dominate exports, which also include motor vehicles and crude oil. Most trade is with Brazil and the USA.

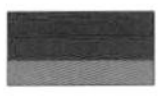

ARMENIA
Republic of Armenia

Area Sq Km	29 800	**Languages**	Armenian, Azeri
Area Sq Miles	11 506	**Religions**	Armenian Orthodox
Population	3 083 000	**Currency**	Dram
Capital	Yerevan (Erevan)	**Organizations**	CIS, UN

Map page 137

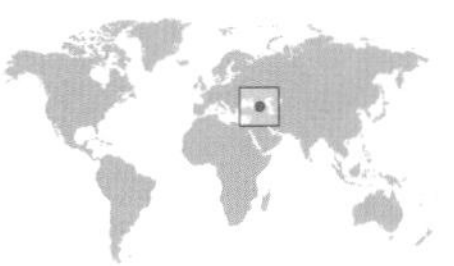

A landlocked state in southwest Asia, Armenia lies in the south of the Lesser Caucasus mountains. It is a mountainous country with a continental climate. One-third of the population lives in the capital, Yerevan. Exports include diamonds, scrap metal and machinery. Many Armenians depend on remittances from abroad.

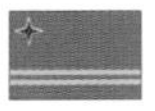

Aruba
Self-governing Netherlands Territory

Area Sq Km	193	**Languages**	Papiamento, Dutch, English
Area Sq Miles	75	**Religions**	Roman Catholic, Protestant
Population	107 000	**Currency**	Aruban florin
Capital	Oranjestad		

Map page 213 The most southwesterly of the islands in the Lesser Antilles in the Caribbean, Aruba lies just off the coast of Venezuela. Tourism, offshore finance and oil refining are the most important sectors of the economy. The USA is the main trading partner.

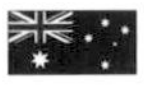

AUSTRALIA
Commonwealth of Australia

Area Sq Km	7 692 024	**Languages**	English, Italian, Greek
Area Sq Miles	2 969 907	**Religions**	Protestant, Roman Catholic, Orthodox
Population	21 293 000	**Currency**	Australian dollar
Capital	Canberra	**Organizations**	APEC, Comm., OECD, UN

Map page 124

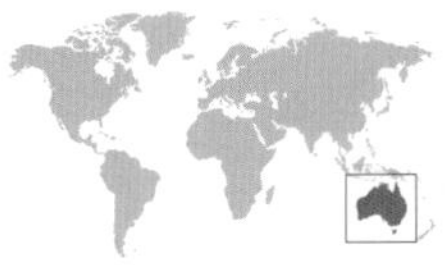

Australia, the world's sixth largest country, occupies the smallest, flattest and driest continent. The western half of the country is mostly arid plateaus, ridges and vast deserts. The central eastern area comprises the lowlands of river systems draining into Lake Eyre, while to the east is the Great Dividing Range, a belt of ridges and plateaus running from Queensland to Tasmania. Climatically, more than two-thirds of the country is arid or semi-arid. The north is tropical monsoon, the east subtropical, and the southwest and southeast temperate. The majority of Australia's highly urbanized population lives along the east, southeast and southwest coasts.

Australia has vast mineral deposits and various sources of energy. It is among the world's leading producers of iron ore, bauxite, nickel, copper and uranium. It is a major producer of coal, and oil and natural gas are also being exploited. Although accounting for only five per cent of the workforce, agriculture continues to be an important sector of the economy, with food and agricultural raw materials making up most of Australia's export earnings. Fuel, ores and metals, and manufactured goods, account for the remainder of exports. Japan and the USA are Australia's main trading partners.

Australian Capital Territory (Federal Territory)

Area Sq Km (Sq Miles) 2 358 (910)　　Population 346 400　　Capital Canberra

Jervis Bay Territory (Territory)

Area Sq Km (Sq Miles) 73 (28)　　Population 611

New South Wales (State)

Area Sq Km (Sq Miles) 800 642 (309 130)　　Population 7 017 100　　Capital Sydney

Northern Territory (Territory)

Area Sq Km (Sq Miles) 1 349 129 (520 902)　　Population 221 100　　Capital Darwin

Queensland (State)

Area Sq Km (Sq Miles) 1 730 648 (668 207)　　Population 4 320 100　　Capital Brisbane

South Australia (State)

Area Sq Km (Sq Miles) 983 482 (379 725)　　Population 1 607 700　　Capital Adelaide

Tasmania (State)

Area Sq Km (Sq Miles) 68 401 (26 410)　　Population 498 900　　Capital Hobart

Victoria (State)

Area Sq Km (Sq Miles) 227 416 (87 806)　　Population 5 340 300　　Capital Melbourne

Western Australia (State)

Area Sq Km (Sq Miles) 2 529 875 (976 790)　　Population 2 188 500　　Capital Perth

AUSTRIA

Republic of Austria

Area Sq Km	83 855	Languages	German, Croatian, Turkish
Area Sq Miles	32 377	Religions	Roman Catholic, Protestant
Population	8 364 000	Currency	Euro
Capital	Vienna (Wien)	Organizations	EU, OECD, UN

Map page 168

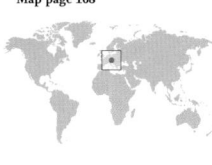

Two-thirds of Austria, a landlocked state in central Europe, lies within the Alps, with lower mountains to the north. The only lowlands are in the east. The Danube river valley in the northeast contains almost all the agricultural land and most of the population. Although the climate varies with altitude, in general summers are warm and winters cold with heavy snowfalls. Manufacturing industry and tourism are the most important sectors of the economy. Exports are dominated by manufactured goods. Germany is Austria's main trading partner.

AZERBAIJAN

Republic of Azerbaijan

Area Sq Km	86 600	Languages	Azeri, Armenian, Russian, Lezgian
Area Sq Miles	33 436	Religions	Shi'a Muslim, Sunni Muslim, Russian and Armenian Orthodox
Population	8 832 000		
Capital	Baku	Currency	Azerbaijani manat
		Organizations	CIS, UN

Map page 137

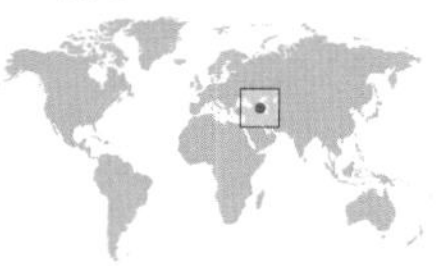

Azerbaijan lies to the southeast of the Caucasus mountains, on the Caspian Sea. Its region of Naxçıvan is separated from the rest of the country by part of Armenia. It has mountains in the northeast and west, valleys in the centre, and a low coastal plain. The climate is continental. It is rich in energy and mineral resources. Oil production, onshore and offshore, is the main industry and the basis of heavy industries. Agriculture is important, with cotton and tobacco the main cash crops.

THE BAHAMAS

Commonwealth of the Bahamas

Area Sq Km	13 939	Languages	English, Creole
Area Sq Miles	5 382	Religions	Protestant, Roman Catholic
Population	342 000	Currency	Bahamian dollar
Capital	Nassau	Organizations	CARICOM, Comm., UN

Map page 205

The Bahamas, an archipelago made up of approximately seven hundred islands and over two thousand cays, lies to the northeast of Cuba and east of the Florida coast of the USA. Twenty-two islands are

inhabited, and two-thirds of the population lives on the main island of New Providence. The climate is warm for much of the year, with heavy rainfall in the summer. Tourism is the islands' main industry. Offshore banking, insurance and ship registration are also major foreign exchange earners.

BAHRAIN

Kingdom of Bahrain

Area Sq Km	691	Languages	Arabic, English
Area Sq Miles	267	Religions	Shi'a Muslim, Sunni Muslim, Christian
Population	791 000		
Capital	Manama (Al Manāmah)	Currency	Bahraini dinar
		Organizations	GCC, UN

Map page 140

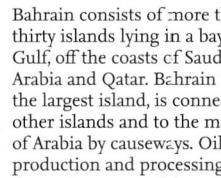

Bahrain consists of more than thirty islands lying in a bay in The Gulf, off the coasts of Saudi Arabia and Qatar. Bahrain Island, the largest island, is connected to other islands and to the mainland of Arabia by causeways. Oil production and processing are the main sectors of the economy.

BANGLADESH

People's Republic of Bangladesh

Area Sq Km	143 998	Languages	Bengali, English
Area Sq Miles	55 598	Religions	Sunni Muslim, Hindu
Population	162 221 000	Currency	Taka
Capital	Dhaka (Dacca)	Organizations	Comm., UN

Map page 145

The south Asian state of Bangladesh is in the northeast of the Indian subcontinent, on the Bay of Bengal. It consists almost entirely of the low-lying alluvial plains and deltas of the Ganges and Brahmaputra rivers. The southwest is swampy, with mangrove forests in the delta area. The north, northeast and southeast have low forested hills. Bangladesh is one of the world's most densely populated and least developed countries. The economy is based on agriculture, though the garment industry is the main export sector. Storms during the summer monsoon season often cause devastating flooding and crop destruction. The country relies on large-scale foreign aid and remittances from workers abroad.

BARBADOS

Area Sq Km	430	Languages	English, Creole
Area Sq Miles	166	Religions	Protestant, Roman Catholic
Population	256 000	Currency	Barbados dollar
Capital	Bridgetown	Organizations	CARICOM, Comm., UN

Map page 205

The most easterly of the Caribbean islands, Barbados is small and densely populated. It has a tropical climate and is subject to hurricanes. The economy is based on tourism, financial services, light industries and sugar production.

BELARUS

Republic of Belarus

Area Sq Km	207 600	Languages	Belorussian, Russian
Area Sq Miles	80 155	Religions	Belorussian Orthodox, Roman Catholic
Population	9 634 000		
Capital	Minsk	Currency	Belarus rouble
		Organizations	CIS, UN

Map page 169

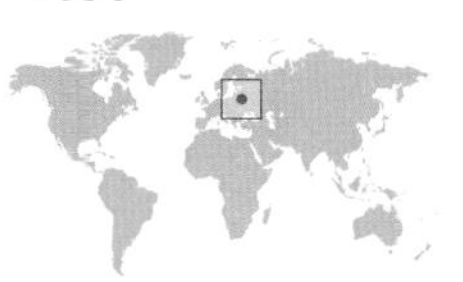

Belarus, a landlocked state in eastern Europe, consists of low hills and plains, with many lakes, rivers and, in the south, extensive marshes. Forests cover approximately one-third of the country. It has a continental climate. Agriculture contributes one-third of national income, with beef cattle and grains as the major products. Manufacturing industries produce a range of items, from construction equipment to textiles. The Russian Federation and Ukraine are the main trading partners.

BELGIUM

Kingdom of Belgium

Area Sq Km	30 520	Languages	Dutch (Flemish), French (Walloon), German
Area Sq Miles	11 784		
Population	10 647 000	Religions	Roman Catholic, Protestant
Capital	Brussels (Bruxelles)	Currency	Euro
		Organizations	EU, NATO, OECD, UN

Map page 164

Belgium lies on the North Sea coast of western Europe. Beyond low sand dunes and a narrow belt of reclaimed land, fertile plains extend to the Sambre-Meuse river valley. The land rises to the forested Ardennes plateau in the

southeast. Belgium has mild winters and cool summers. It is densely populated and has a highly urbanized population. With few mineral resources, Belgium imports raw materials for processing and manufacture. The agricultural sector is small, but provides for most food needs. A large services sector reflects Belgium's position as the home base for over eight hundred international institutions. The headquarters of the European Union are in the capital, Brussels.

BELIZE

Area Sq Km	22 965	Languages	English, Spanish, Mayan, Creole
Area Sq Miles	8 867	Religions	Roman Catholic, Protestant
Population	307 000	Currency	Belize dollar
Capital	Belmopan	Organizations	CARICOM, Comm., UN

Map page 207

Belize lies on the Caribbean coast of central America and includes numerous cays and a large barrier reef offshore. The coastal areas are flat and swampy. To the southwest are the Maya Mountains. Tropical jungle covers much of the country and the climate is humid tropical, but tempered by sea breezes. A third of the population lives in the capital. The economy is based primarily on agriculture, forestry and fishing, and exports include raw sugar, orange concentrate and bananas.

BENIN

Republic of Benin

Area Sq Km	112 620	Languages	French, Fon, Yoruba, Adja, local languages
Area Sq Miles	43 483		
Population	8 935 000	Religions	Traditional beliefs, Roman Catholic, Sunni Muslim
Capital	Porto-Novo	Currency	CFA franc
		Organizations	UN

Map page 176

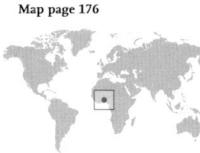

Benin is in west Africa, on the Gulf of Guinea. The climate is tropical in the north, equatorial in the south. The economy is based mainly on agriculture and transit trade. Agricultural products account for two-thirds of export earnings. Oil, produced offshore, is also a major export.

Bermuda
United Kingdom Overseas Territory

Area Sq Km	54	Languages	English
Area Sq Miles	21	Religions	Protestant, Roman Catholic
Population	65 000	Currency	Bermuda dollar
Capital	Hamilton		

Map page 205 In the Atlantic Ocean to the east of the USA, Bermuda comprises a group of small islands with a warm and humid climate. The economy is based on international business and tourism.

BHUTAN

Kingdom of Bhutan

Area Sq Km	46 620	Languages	Dzongkha, Nepali, Assamese
Area Sq Miles	18 000	Religions	Buddhist, Hindu
Population	697 000	Currency	Ngultrum, Indian rupee
Capital	Thimphu	Organizations	UN

Map page 145

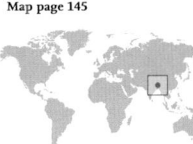

Bhutan lies in the eastern Himalaya mountains, between China and India. It is mountainous in the north, with fertile valleys. The climate ranges between permanently cold in the far north and subtropical in the south. Most of the population is involved in livestock rearing and subsistence farming. Bhutan is the world's largest producer of cardamom. Tourism is an increasingly important foreign currency earner.

BOLIVIA
Republic of Bolivia

Area Sq Km	1 098 581	Languages	Spanish, Quechua, Aymara
Area Sq Miles	424 164	Religions	Roman Catholic, Protestant, Baha'i
Population	9 863 000	Currency	Boliviano
Capital	La Paz/Sucre	Organizations	UN

Map page 210

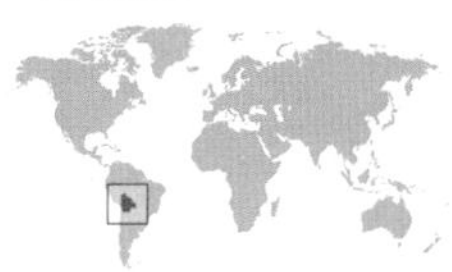

Bolivia is a landlocked state in central South America. Most Bolivians live on the high plateau within the Andes mountains. The lowlands range between dense rainforest in the northeast and semi-arid grasslands in the southeast. Bolivia is rich in minerals (zinc, tin and gold), and sales generate approximately half of export income. Natural gas, timber and soya beans are also exported. The USA is the main trading partner.

© Collins Bartholomew Ltd

BOSNIA-HERZEGOVINA
Republic of Bosnia and Herzegovina

Area Sq Km	51 130	**Languages**	Bosnian, Serbian, Croatian
Area Sq Miles	19 741	**Religions**	Sunni Muslim, Serbian Orthodox, Roman Catholic, Protestant
Population	3 767 000	**Currency**	Marka
Capital	Sarajevo	**Organizations**	UN

Map page 170–171

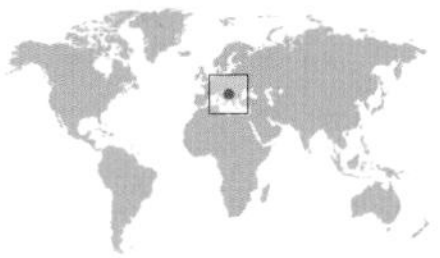

Bosnia-Herzegovina lies in the western Balkan Mountains of southern Europe, on the Adriatic Sea. It is mountainous, with ridges running northwest- southeast. The main lowlands are around the Sava valley in the north. Summers are warm, but winters can be very cold. The economy relies heavily on overseas aid.

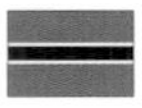

BOTSWANA
Republic of Botswana

Area Sq Km	581 370	**Languages**	English, Setswana, Shona, local languages
Area Sq Miles	224 468	**Religions**	Traditional beliefs, Protestant, Roman Catholic
Population	1 950 000	**Currency**	Pula
Capital	Gaborone	**Organizations**	Comm., SADC, UN

Map page 179

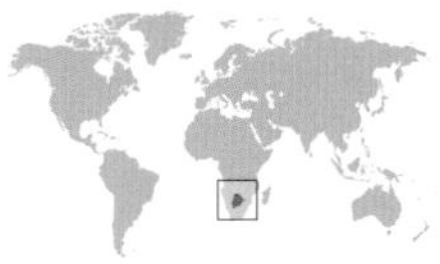

Botswana is a landlocked state in southern Africa. Over half of the country lies within the Kalahari Desert, with swamps to the north and salt-pans to the northeast. Most of the population lives near the eastern border. The climate is subtropical, but drought-prone. The economy was founded on cattle rearing, and although beef remains an important export, the economy is now based on mining. Diamonds account for seventy per cent of export earnings. Copper-nickel matte is also exported. Most trade is with other members of the Southern African Customs Union.

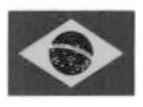

BRAZIL
Federative Republic of Brazil

Area Sq Km	8 514 879	**Languages**	Portuguese
Area Sq Miles	3 287 613	**Religions**	Roman Catholic, Protestant
Population	193 734 000	**Currency**	Real
Capital	Brasília	**Organizations**	UN

Map page 210–211

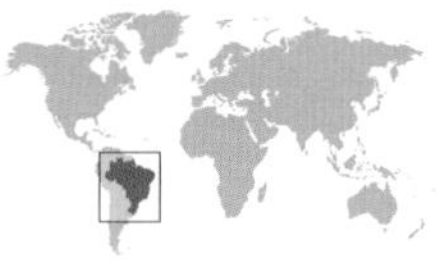

Brazil, in eastern South America, covers almost half of the continent, and is the world's fifth largest country. The northwest contains the vast basin of the Amazon, while the centre-west is largely a vast plateau of savanna and rock escarpments. The northeast is mostly semi-arid plateaus, while to the east and south are rugged mountains, fertile valleys and narrow, fertile coastal plains. The Amazon basin is hot, humid and wet; the rest of the country is cooler and drier, with seasonal variations. The northeast is drought-prone. Most Brazilians live in urban areas along the coast and on the central plateau. Brazil has well-developed agricultural, mining and service sectors, and the economy is larger than that of all other South American countries combined. Brazil is the world's biggest producer of coffee, and other agricultural crops include grains and sugar cane. Mineral production includes iron, aluminium and gold. Manufactured goods include food products, transport equipment, machinery and industrial chemicals. The main trading partners are the USA and Argentina. Despite its natural wealth, Brazil has a large external debt and a growing poverty gap.

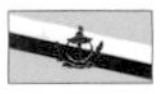

BRUNEI
Brunei Darussalam

Area Sq Km	5 765	**Languages**	Malay, English, Chinese
Area Sq Miles	2 226	**Religions**	Sunni Muslim, Buddhist, Christian
Population	400 000	**Currency**	Brunei dollar
Capital	Bandar Seri Begawan	**Organizations**	APEC, ASEAN, Comm., UN

Map page 155

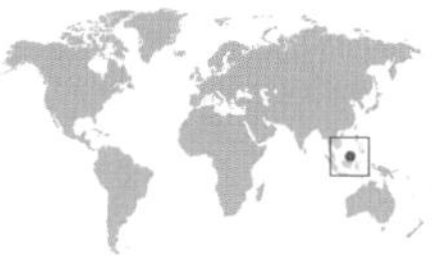

The southeast Asian oil-rich state of Brunei lies on the northwest coast of the island of Borneo, on the South China Sea. Its two enclaves are surrounded by the Malaysian state of Sarawak. Tropical rainforest covers over two-thirds of the country. The economy is dominated by the oil and gas industries.

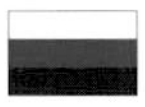

BULGARIA
Republic of Bulgaria

Area Sq Km	110 994	**Languages**	Bulgarian, Turkish, Romany, Macedonian
Area Sq Miles	42 855	**Religions**	Bulgarian Orthodox, Sunni Muslim
Population	7 545 000	**Currency**	Lev
Capital	Sofia (Sofiya)	**Organizations**	EU, NATO, UN

Map page 171

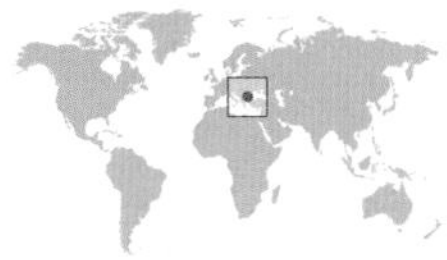

Bulgaria, in southern Europe, borders the western shore of the Black Sea. The Balkan Mountains separate the Danube plains in the north from the Rhodope Mountains and the lowlands in the south. The economy has a strong agricultural base. Manufacturing industries include machinery, consumer goods, chemicals and metals. Most trade is with the Russian Federation, Italy and Germany.

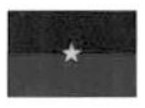

BURKINA
Democratic Republic of Burkina Faso

Area Sq Km	274 200	**Languages**	French, Moore (Mossi), Fulani, local languages
Area Sq Miles	105 869	**Religions**	Sunni Muslim, traditional beliefs, Roman Catholic
Population	15 757 000	**Currency**	CFA franc
Capital	Ouagadougou	**Organizations**	UN

Map page 176

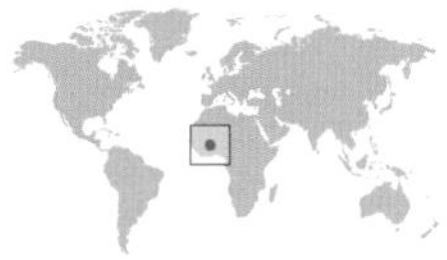

Burkina, a landlocked country in west Africa, lies within the Sahara desert to the north and semi-arid savanna to the south. Rainfall is erratic, and droughts are common. Livestock rearing and farming are the main activities, and cotton, livestock, groundnuts and some minerals are exported. Burkina relies heavily on foreign aid, and is one of the poorest and least developed countries in the world.

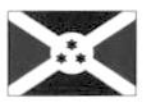

BURUNDI
Republic of Burundi

Area Sq Km	27 835	**Languages**	Kirundi (Hutu, Tutsi), French
Area Sq Miles	10 747	**Religions**	Roman Catholic, traditional beliefs, Protestant
Population	8 303 000	**Currency**	Burundian franc
Capital	Bujumbura	**Organizations**	UN

Map page 178

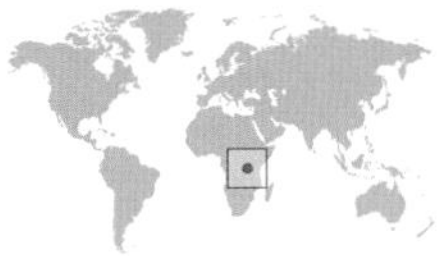

The densely populated east African state of Burundi consists of high plateaus rising from the shores of Lake Tanganyika in the southwest. It has a tropical climate and depends on subsistence farming. Coffee is its main export, and its main trading partners are Germany and Belgium. The country has been badly affected by internal conflict since the early 1990s.

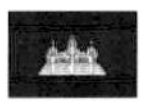

CAMBODIA
Kingdom of Cambodia

Area Sq Km	181 035	**Languages**	Khmer, Vietnamese
Area Sq Miles	69 884	**Religions**	Buddhist, Roman Catholic, Sunni Muslim
Population	14 805 000	**Currency**	Riel
Capital	Phnom Penh (Phnom Pénh)	**Organizations**	ASEAN, UN

Map page 154

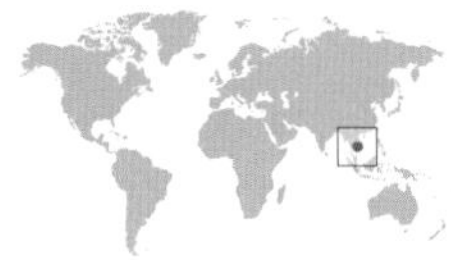

Cambodia lies in southeast Asia on the Gulf of Thailand, and occupies the Mekong river basin, with the Tônlé Sap (Great Lake) at its centre. The climate is tropical monsoon. Forests cover half the country. Most of the population lives on the plains and is engaged in farming (chiefly rice growing), fishing and forestry. The economy is recovering slowly following the devastation of civil war in the 1970s.

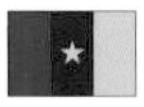

CAMEROON
Republic of Cameroon

Area Sq Km	475 442	**Languages**	French, English, Fang, Bamileke, local languages
Area Sq Miles	183 569	**Religions**	Roman Catholic, traditional beliefs, Sunni Muslim, Protestant
Population	19 522 000	**Currency**	CFA franc
Capital	Yaoundé	**Organizations**	Comm., UN

Map page 176–177

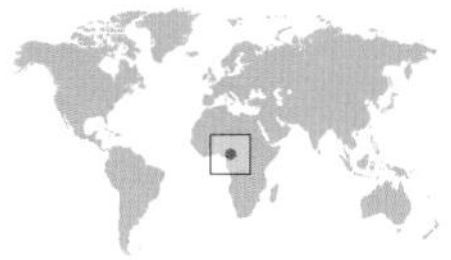

Cameroon is in west Africa, on the Gulf of Guinea. The coastal plains and southern and central plateaus are covered with tropical forest. Despite oil resources and favourable agricultural conditions Cameroon still faces problems of underdevelopment. Oil, timber and cocoa are the main exports. France is the main trading partner.

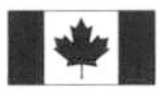

CANADA

Area Sq Km	9 984 670	**Languages**	English, French
Area Sq Miles	3 855 103	**Religions**	Roman Catholic, Protestant, Eastern Orthodox, Jewish
Population	33 573 000	**Currency**	Canadian dollar
Capital	Ottawa	**Organizations**	APEC, Comm., NATO, OECD, UN

Map page 184–185

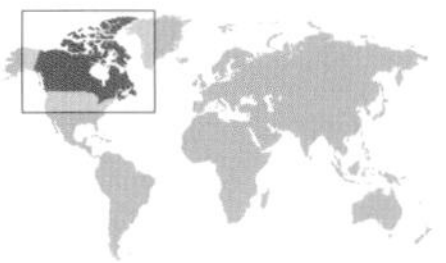

The world's second largest country, Canada covers the northern two-fifths of North America and has coastlines on the Atlantic, Arctic and Pacific Oceans. In the west are the Coast Mountains, the Rocky Mountains and interior plateaus. In the centre lie the fertile Prairies. Further east, covering about half the total land area, is the Canadian Shield, a relatively flat area of infertile lowlands around Hudson Bay, extending to Labrador on the east coast. The Shield is bordered to the south by the fertile Great Lakes-St Lawrence lowlands. In the far north climatic conditions are polar, while the rest has a continental climate. Most Canadians live in the urban areas of the Great Lakes-St Lawrence basin. Canada is rich in mineral and energy resources. Only five per cent of land is arable. Canada is among the world's leading producers of wheat, of wood from its vast coniferous forests, and of fish and seafood from its Atlantic and Pacific fishing grounds. It is a major producer of nickel, uranium, copper, iron ore, zinc and other minerals, as well as oil and natural gas. Its abundant raw materials are the basis for many manufacturing industries. Main exports are machinery, motor vehicles, oil, timber, newsprint and paper, wood pulp and wheat. Since the 1989 free trade agreement with the USA and the 1994 North America Free Trade Agreement, trade with the USA has grown and now accounts for around seventy-five per cent of imports and around eighty-five per cent of exports.

	Area Sq Km (Sq Miles)	Population	Capital
Alberta (Province)	661 848 (255 541)	3 610 782	Edmonton
British Columbia (Province)	944 735 (364 764)	4 405 534	Victoria
Manitoba (Province)	647 797 (250 116)	1 210 547	Winnipeg
New Brunswick (Province)	72 908 (28 150)	747 790	Fredericton
Newfoundland and Labrador (Province)	405 212 (156 453)	508 944	St John's
Northwest Territories (Territory)	1 346 106 (519 734)	43 151	Yellowknife
Nova Scotia (Province)	55 284 (21 345)	939 125	Halifax
Nunavut (Territory)	2 093 190 (808 185)	31 522	Iqaluit (Frobisher Bay)
Ontario (Province)	1 076 395 (415 598)	12 977 059	Toronto
Prince Edward Island (Province)	5 660 (2 185)	140 750	Charlottetown
Québec (Province)	1 542 056 (595 391)	7 771 854	Québec
Saskatchewan (Province)	651 036 (251 366)	1 020 847	Regina
Yukon (Territory)	482 443 (186 272)	33 372	Whitehorse

CAPE VERDE
Republic of Cape Verde

Area Sq Km	4 033	**Languages**	Portuguese, Creole
Area Sq Miles	1 557	**Religions**	Roman Catholic, Protestant
Population	506 000	**Currency**	Cape Verde escudo
Capital	Praia	**Organizations**	UN

Map page 176

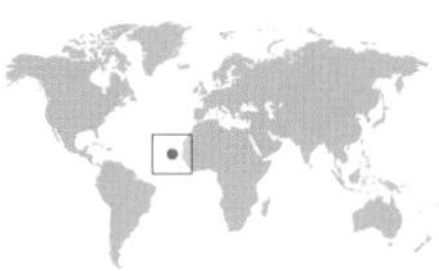

Cape Verde is a group of semi-arid volcanic islands lying off the coast of west Africa. The economy is based on fishing and subsistence farming but relies on emigrant workers' remittances and foreign aid.

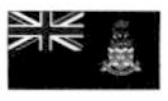

Cayman Islands
United Kingdom Overseas Territory

Area Sq Km 259
Area Sq Miles 100
Population 56 000
Capital George Town
Languages English
Religions Protestant, Roman Catholic
Currency Cayman Islands dollar

Map page 205 A group of islands in the Caribbean, northwest of Jamaica. There are three main islands: Grand Cayman, Little Cayman and Cayman Brac. The Cayman Islands are one of the world's major offshore financial centres. Tourism is also important to the economy.

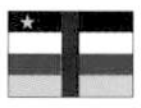

CENTRAL AFRICAN REPUBLIC

Area Sq Km 622 436
Area Sq Miles 240 324
Population 4 422 000
Capital Bangui
Languages French, Sango, Banda, Baya, local languages
Religions Protestant, Roman Catholic, traditional beliefs, Sunni Muslim
Currency CFA franc
Organizations UN

Map page 177

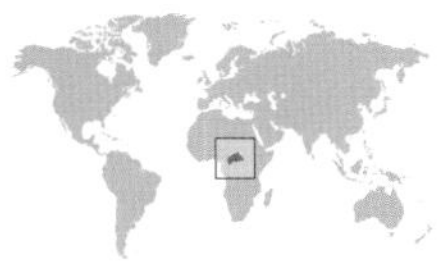

A landlocked country in central Africa, the Central African Republic is mainly savanna plateau, drained by the Ubangi and Chari river systems, with mountains to the east and west. The climate is tropical, with high rainfall. Most of the population lives in the south and west, and a majority of the workforce is involved in subsistence farming. Some cotton, coffee, tobacco and timber are exported, but diamonds account for around half of export earnings.

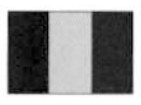

CHAD
Republic of Chad

Area Sq Km 1 284 000
Area Sq Miles 495 755
Population 11 206 000
Capital Ndjamena
Languages Arabic, French, Sara, local languages
Religions Sunni Muslim, Roman Catholic, Protestant, traditional beliefs
Currency CFA franc
Organizations UN

Map page 177

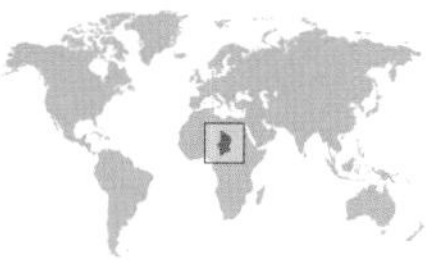

Chad is a landlocked state of north-central Africa. It consists of plateaus, the Tibesti mountains in the north and the Lake Chad basin in the west. Climatic conditions range between desert in the north and tropical forest in the southwest. With few natural resources, Chad relies on subsistence farming, exports of raw cotton, and foreign aid. The main trading partners are France, Portugal and Cameroon.

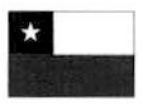

CHILE
Republic of Chile

Area Sq Km 756 945
Area Sq Miles 292 258
Population 16 970 000
Capital Santiago
Languages Spanish, Amerindian languages
Religions Roman Catholic, Protestant
Currency Chilean peso
Organizations APEC, OECD, UN

Map page 212

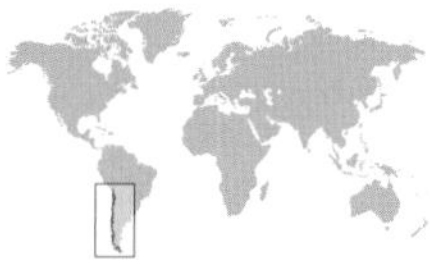

Chile lies along the Pacific coast of the southern half of South America. Between the Andes in the east and the lower coastal ranges is a central valley, with a mild climate, where most Chileans live. To the north is the arid Atacama Desert and to the south is cold, wet forested grassland. Chile has considerable mineral resources and is the world's leading exporter of copper. Nitrates, molybdenum, gold and iron ore are also mined. Agriculture (particularly viticulture), forestry and fishing are also important to the economy.

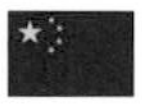

CHINA
People's Republic of China

Area Sq Km 9 584 492
Area Sq Miles 3 700 593
Population 1 330 265 000
Capital Beijing (Peking)
Languages Mandarin, Wu, Cantonese, Hsiang, regional languages
Religions Confucian, Taoist, Buddhist, Christian, Sunni Muslim
Currency Yuan, Hong Kong dollar, Macao pataca
Organizations APEC, UN

Map page 146

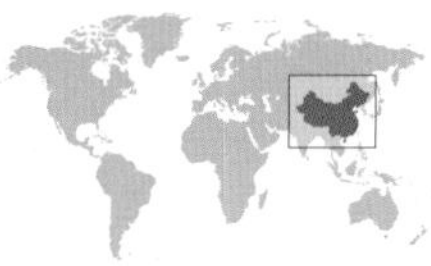

China, the world's most populous and fourth largest country, occupies a large part of east Asia, borders fourteen states and has coastlines on the Yellow, East China and South China Seas. It has a huge variety of landscapes. The southwest contains the high Plateau of Tibet, flanked by the Himalaya and Kunlun Shan mountains. The north is mountainous with arid basins and extends from the Tien Shan and Altai Mountains and the vast Taklimakan Desert in the west to the plateau and Gobi Desert in the centre-east. Eastern China is predominantly lowland and is divided broadly into the basins of the Yellow River (Huang He) in the north, the Yangtze (Chang Jiang) in the centre and the Pearl River (Xi Jiang) in the southeast. Climatic conditions and vegetation are as diverse as the topography: much of the country experiences temperate conditions, while the southwest has an extreme mountain climate and the southeast enjoys a moist, warm subtropical climate. Nearly seventy per cent of China's huge population lives in rural areas, and agriculture employs around half of the working population. The main crops are rice, wheat, soya beans, peanuts, cotton, tobacco and hemp. China is rich in coal, oil and natural gas and has the world's largest potential in hydroelectric power. It is a major world producer of iron ore, molybdenum, copper, asbestos and gold. Economic reforms from the early 1980's led to an explosion in manufacturing development concentrated on the 'coastal economic open region'. The main exports are machinery, textiles, footwear, toys and sports goods. Japan and the USA are China's main trading partners.

Anhui (Province)
Area Sq Km (Sq Miles) 139 000 (53 668) **Population** 61 180 000 **Capital** Hefei

Beijing (Municipality)
Area Sq Km (Sq Miles) 16 800 (6 487) **Population** 16 330 000 **Capital** Beijing (Peking)

Chongqing (Municipality)
Area Sq Km (Sq Miles) 23 000 (8 880) **Population** 28 160 000 **Capital** Chongqing

Fujian (Province)
Area Sq Km (Sq Miles) 121 400 (46 873) **Population** 35 810 000 **Capital** Fuzhou

Gansu (Province)
Area Sq Km (Sq Miles) 453 700 (175 175) **Population** 26 170 000 **Capital** Lanzhou

Guangdong (Province)
Area Sq Km (Sq Miles) 178 000 (68 726) **Population** 94 490 000 **Capital** Guangzhou (Canton)

Guangxi Zhuangzu Zizhiqu (Autonomous Region)
Area Sq Km (Sq Miles) 236 000 (91 120) **Population** 47 680 000 **Capital** Nanning

Guizhou (Province)
Area Sq Km (Sq Miles) 176 000 (67 954) **Population** 37 620 000 **Capital** Guiyang

Hainan (Province)
Area Sq Km (Sq Miles) 34 000 (13 127) **Population** 8 450 000 **Capital** Haikou

Hebei (Province)
Area Sq Km (Sq Miles) 187 700 (72 471) **Population** 69 430 000 **Capital** Shijiazhuang

Heilongjiang (Province)
Area Sq Km (Sq Miles) 454 600 (175 522) **Population** 38 240 000 **Capital** Harbin

Henan (Province)
Area Sq Km (Sq Miles) 167 000 (64 479) **Population** 93 600 000 **Capital** Zhengzhou

Hong Kong (Special Administrative Region)
Area Sq Km (Sq Miles) 1 075 (415) **Population** 6 926 000 **Capital** Hong Kong

Hubei (Province)
Area Sq Km (Sq Miles) 185 900 (71 776) **Population** 56 990 000 **Capital** Wuhan

Hunan (Province)
Area Sq Km (Sq Miles) 210 000 (81 081) **Population** 63 550 000 **Capital** Changsha

Jiangsu (Province)
Area Sq Km (Sq Miles) 102 600 (39 614) **Population** 76 250 000 **Capital** Nanjing

Jiangxi (Province)
Area Sq Km (Sq Miles) 166 900 (64 440) **Population** 43 680 000 **Capital** Nanchang

Jilin (Province)
Area Sq Km (Sq Miles) 187 000 (72 201) **Population** 27 300 000 **Capital** Changchun

Liaoning (Province)
Area Sq Km (Sq Miles) 147 400 (56 911) **Population** 42 980 000 **Capital** Shenyang

Macao (Special Administrative Region)
Area Sq Km (Sq Miles) 17 (7) **Population** 526 000 **Capital** Macao

Nei Mongol Zizhiqu Inner Mongolia (Autonomous Region)
Area Sq Km (Sq Miles) 1 183 000 (456 759) **Population** 24 050 000 **Capital** Hohhot

Ningxia Huizu Zizhiqu (Autonomous Region)
Area Sq Km (Sq Miles) 66 400 (25 637) **Population** 6 100 000 **Capital** Yinchuan

Qinghai (Province)
Area Sq Km (Sq Miles) 721 000 (278 380) **Population** 5 520 000 **Capital** Xining

Shaanxi (Province)
Area Sq Km (Sq Miles) 205 600 (79 383) **Population** 37 480 000 **Capital** Xi'an

Shandong (Province)
Area Sq Km (Sq Miles) 153 300 (59 189) **Population** 93 670 000 **Capital** Jinan

Shanghai (Municipality)
Area Sq Km (Sq Miles) 6 300 (2 432) **Population** 18 580 000 **Capital** Shanghai

Shanxi (Province)
Area Sq Km (Sq Miles) 156 300 (60 348) **Population** 33 930 000 **Capital** Taiyuan

Sichuan (Province)
Area Sq Km (Sq Miles) 569 000 (219 692) **Population** 81 270 000 **Capital** Chengdu

Tianjin (Municipality)
Area Sq Km (Sq Miles) 11 300 (4 363) **Population** 11 150 000 **Capital** Tianjin

Xinjiang Uygur Zizhiqu Sinkiang (Autonomous Region)
Area Sq Km (Sq Miles) 1 600 000 (617 763) **Population** 20 950 000 **Capital** Ürümqi

Xizang Zizhiqu Tibet (Autonomous Region)
Area Sq Km (Sq Miles) 1 228 400 (474 288) **Population** 2 840 000 **Capital** Lhasa

Yunnan (Province)
Area Sq Km (Sq Miles) 394 000 (152 124) **Population** 45 140 000 **Capital** Kunming

Zhejiang (Province)
Area Sq Km (Sq Miles) 101 800 (39 305) **Population** 50 600 000 **Capital** Hangzhou

Taiwan: The People's Republic of China claims Taiwan as its 23rd Province

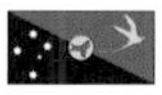

Christmas Island
Australian External Territory

Area Sq Km 135
Area Sq Miles 52
Population 1 351
Capital The Settlement (Flying Fish Cove)
Languages English
Religions Buddhist, Sunni Muslim, Protestant, Roman Catholic
Currency Australian dollar

Map page 147 The island is situated in the east of the Indian Ocean, to the south of Indonesia. The economy was formerly based on phosphate extraction, although reserves are now nearly depleted. Tourism is developing and is a major employer.

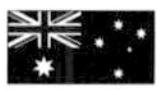

Cocos Islands (Keeling Islands)
Australian External Territory

Area Sq Km 14
Area Sq Miles 5
Population 621
Capital West Island
Languages English
Religions Sunni Muslim, Christian
Currency Australian dollar

Map page 147 The Cocos Islands consist of numerous islands on two coral atolls in the eastern Indian Ocean between Sri Lanka and Australia. Most of the population lives on West Island or Home Island. Coconuts are the only cash crop, and the main export.

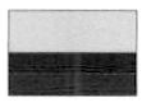

COLOMBIA
Republic of Colombia

Area Sq Km 1 141 748
Area Sq Miles 440 831
Population 45 660 000
Capital Bogotá
Languages Spanish, Amerindian languages
Religions Roman Catholic, Protestant
Currency Colombian peso
Organizations UN

Map page 213

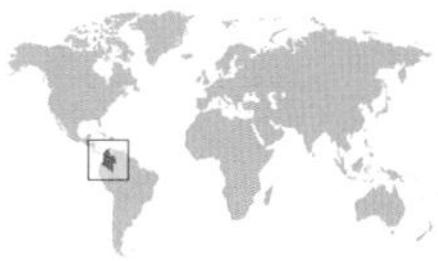

A state in northwest South America, Colombia has coastlines on the Pacific Ocean and the Caribbean Sea. Behind coastal plains lie three ranges of the Andes mountains, separated by high valleys and plateaus where most Colombians live. To the southeast are grasslands and the forests of the Amazon. The climate is tropical, although temperatures vary with altitude. Only five per cent of land is cultivable. Coffee (Colombia is the world's second largest producer), sugar, bananas, cotton and flowers are exported. Coal, nickel, gold, silver, platinum and emeralds (Colombia is the world's largest producer) are mined. Oil and its products are the main export. Industries include the processing of minerals and crops. The main trade partner is the USA. Internal violence – both politically motivated and relating to Colombia's leading role in the international trade in illegal drugs – continues to hinder development.

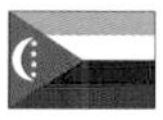

COMOROS
Union of the Comoros

Area Sq Km 1 862
Area Sq Miles 719
Population 676 000
Capital Moroni
Languages Comorian, French, Arabic
Religions Sunni Muslim, Roman Catholic
Currency Comoros franc
Organizations UN

Map page 179

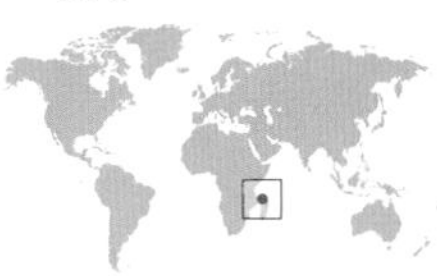

This state, in the Indian Ocean off the east African coast, comprises three volcanic islands of Ngazidja (Grande Comore), Nzwani (Anjouan) and Mwali (Mohéli), and some coral atolls. These tropical islands are mountainous, with poor soil and few natural resources. Subsistence farming predominates. Vanilla, cloves and ylang-ylang (an essential oil) are exported, and the economy relies heavily on workers' remittances from abroad.

© Collins Bartholomew Ltd

CONGO
Republic of the Congo

Area Sq Km	342 000	**Languages**	French, Kongo, Monokutuba, local languages
Area Sq Miles	132 047	**Religions**	Roman Catholic, Protestant, traditional beliefs, Sunni Muslim
Population	3 683 000	**Currency**	CFA franc
Capital	Brazzaville	**Organizations**	UN

Map page 178

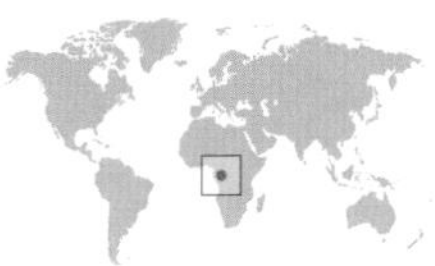

Congo, in central Africa, is mostly a forest or savanna-covered plateau drained by the Ubangi-Congo river systems. Sand dunes and lagoons line the short Atlantic coast. The climate is hot and tropical. Most Congolese live in the southern third of the country. Half of the workforce are farmers, growing food and cash crops including sugar, coffee, cocoa and oil palms. Oil and timber are the mainstays of the economy, and oil generates over fifty per cent of the country's export revenues.

CONGO, DEMOCRATIC REPUBLIC OF THE

Area Sq Km	2 345 410	**Languages**	French, Lingala, Swahili, Kongo, local languages
Area Sq Miles	905 568	**Religions**	Christian, Sunni Muslim
Population	66 020 000	**Currency**	Congolese franc
Capital	Kinshasa	**Organizations**	SADC, UN

Map page 178–179

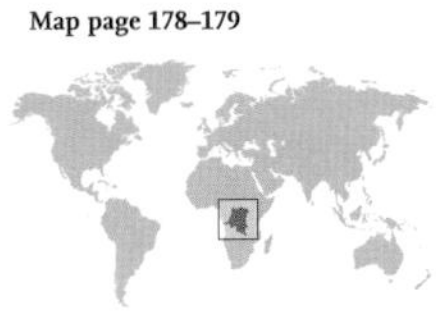

This central African state, formerly Zaire, consists of the basin of the Congo river flanked by plateaus, with high mountain ranges to the east and a short Atlantic coastline to the west. The climate is tropical, with rainforest close to the Equator and savanna to the north and south. Fertile land allows a range of food and cash crops to be grown, chiefly coffee. The country has vast mineral resources, with copper, cobalt and diamonds being the most important.

Cook Islands
New Zealand Overseas Territory

Area Sq Km	293	**Languages**	English, Maori
Area Sq Miles	113	**Religions**	Protestant, Roman Catholic
Population	20 000	**Currency**	New Zealand dollar
Capital	Avarua		

Map page 123 These consist of groups of coral atolls and volcanic islands in the southwest Pacific Ocean. The main island is Rarotonga. Distance from foreign markets and restricted natural resources hinder development.

COSTA RICA
Republic of Costa Rica

Area Sq Km	51 100	**Languages**	Spanish
Area Sq Miles	19 730	**Religions**	Roman Catholic, Protestant
Population	4 579 000	**Currency**	Costa Rican colón
Capital	San José	**Organizations**	UN

Map page 206

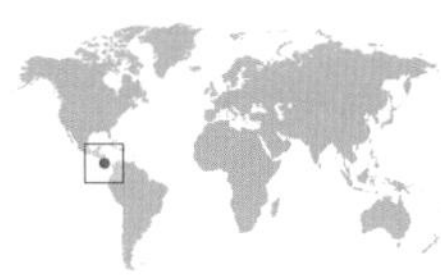

Costa Rica, in central America, has coastlines on the Caribbean Sea and Pacific Ocean. From tropical coastal plains, the land rises to mountains and a temperate central plateau, where most of the population lives. The economy depends on agriculture and tourism, with ecotourism becoming increasingly important. Main exports are textiles, coffee and bananas, and almost half of all trade is with the USA.

CÔTE D'IVOIRE (Ivory Coast)
Republic of Côte d'Ivoire

Area Sq Km	322 463	**Languages**	French, Creole, Akan, local languages
Area Sq Miles	124 504	**Religions**	Sunni Muslim, Roman Catholic, traditional beliefs, Protestant
Population	21 075 000	**Currency**	CFA franc
Capital	Yamoussoukro	**Organizations**	UN

Map page 176

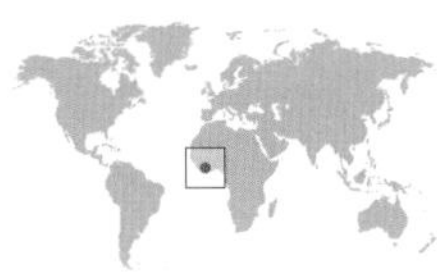

Côte d'Ivoire (Ivory Coast) is in west Africa, on the Gulf of Guinea. In the north are plateaus and savanna; in the south are low undulating plains and rainforest, with sand-bars and lagoons on the coast. Temperatures are warm, and rainfall is heavier in the south. Most of the workforce is engaged in farming. Côte d'Ivoire is a major producer of cocoa and coffee, and agricultural products (also including cotton and timber) are the main exports. Oil and gas have begun to be exploited.

CROATIA
Republic of Croatia

Area Sq Km	56 538	**Languages**	Croatian, Serbian
Area Sq Miles	21 829	**Religions**	Roman Catholic, Serbian Orthodox, Sunni Muslim
Population	4 416 000	**Currency**	Kuna
Capital	Zagreb	**Organizations**	NATO, UN

Map page 170

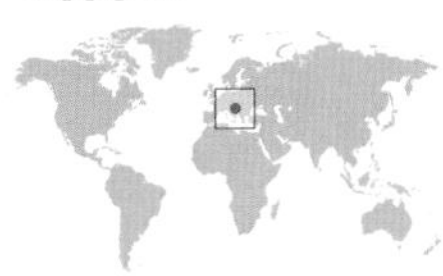

The southern European state of Croatia has a long coastline on the Adriatic Sea, with many offshore islands. Coastal areas have a Mediterranean climate; inland is cooler and wetter. Croatia was once strong agriculturally and industrially, but conflict in the early 1990s, and associated loss of markets and a fall in tourist revenue, caused economic difficulties from which recovery has been slow.

CUBA
Republic of Cuba

Area Sq Km	110 860	**Languages**	Spanish
Area Sq Miles	42 803	**Religions**	Roman Catholic, Protestant
Population	11 204 000	**Currency**	Cuban peso
Capital	Havana (La Habana)	**Organizations**	UN

Map page 205

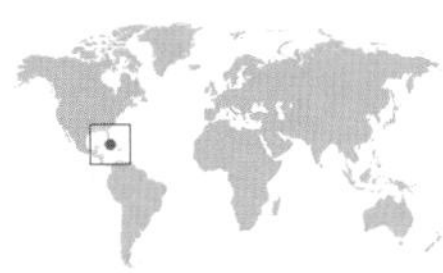

The country comprises the island of Cuba (the largest island in the Caribbean), and many islets and cays. A fifth of Cubans live in and around Havana. Cuba is slowly recovering from the withdrawal of aid and subsidies from the former USSR. Sugar remains the basis of the economy, although tourism is developing and is, together with remittances from workers abroad, an important source of revenue.

CYPRUS
Republic of Cyprus

Area Sq Km	9 251	**Languages**	Greek, Turkish, English
Area Sq Miles	3 572	**Religions**	Greek Orthodox, Sunni Muslim
Population	871 000	**Currency**	Euro
Capital	Nicosia (Lefkosia)	**Organizations**	Comm., EU, UN

Map page 136

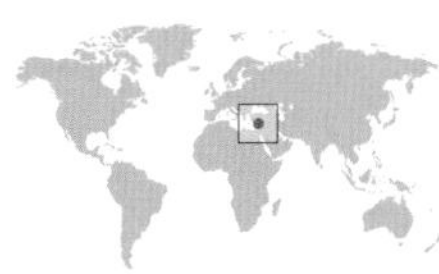

The eastern Mediterranean island of Cyprus has effectively been divided into two since 1974. The economy of the Greek-speaking south is based mainly on specialist agriculture and tourism, with shipping and offshore banking. The ethnically Turkish north depends on agriculture, tourism and aid from Turkey. The island has hot dry summers and mild winters. Cyprus joined the European Union in May 2004.

CZECH REPUBLIC

Area Sq Km	78 864	**Languages**	Czech, Moravian, Slovakian
Area Sq Miles	30 450	**Religions**	Roman Catholic, Protestant
Population	10 369 000	**Currency**	Koruna
Capital	Prague (Praha)	**Organizations**	EU, NATO, OECD, UN

Map page 168

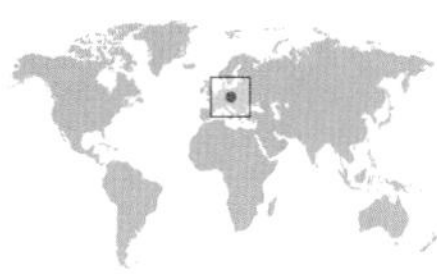

The landlocked Czech Republic in central Europe consists of rolling countryside, wooded hills and fertile valleys. The climate is continental. The country has substantial reserves of coal and lignite, timber and some minerals, chiefly iron ore. It is highly industrialized, and major manufactured goods include industrial machinery, consumer goods, cars, iron and steel, chemicals and glass. Germany is the main trading partner. The Czech Republic joined the European Union in May 2004.

DENMARK
Kingdom of Denmark

Area Sq Km	43 075	**Languages**	Danish
Area Sq Miles	16 631	**Religions**	Protestant
Population	5 470 000	**Currency**	Danish krone
Capital	Copenhagen (København)	**Organizations**	EU, NATO, OECD, UN

Map page 159

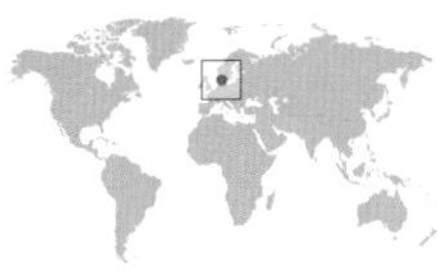

In northern Europe, Denmark occupies the Jutland (Jylland) peninsula and nearly five hundred islands in and between the North and Baltic Seas. The country is low-lying, with long, indented coastlines. The climate is cool and temperate, with rainfall throughout the year. A fifth of the population lives in and around the capital, Copenhagen (København), on the largest of the islands, Zealand (Sjælland). The country's main natural resource is its agricultural potential: two-thirds of the total area is fertile farmland or pasture. Agriculture is high-tech, and with forestry and fishing employs only around six per cent of the workforce. Denmark is self-sufficient in oil and natural gas, produced from fields in the North Sea. Manufacturing, largely based on imported raw materials, accounts for over half of all exports, which include machinery, food, furniture and pharmaceuticals. The main trading partners are Germany and Sweden.

DJIBOUTI
Republic of Djibouti

Area Sq Km	23 200	**Languages**	Somali, Afar, French, Arabic
Area Sq Miles	8 958	**Religions**	Sunni Muslim, Christian
Population	864 000	**Currency**	Djibouti franc
Capital	Djibouti	**Organizations**	UN

Map page 178

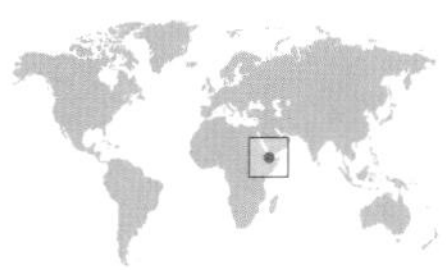

Djibouti lies in northeast Africa, on the Gulf of Aden at the entrance to the Red Sea. Most of the country is semi-arid desert with high temperatures and low rainfall. More than two-thirds of the population live in the capital. There is some camel, sheep and goat herding, but with few natural resources the economy is based on services and trade. Djibouti serves as a free trade zone for northern Africa, and the capital's port is a major transhipment and refuelling destination. It is linked by rail to Addis Ababa in Ethiopia.

DOMINICA
Commonwealth of Dominica

Area Sq Km	750	**Languages**	English, Creole
Area Sq Miles	290	**Religions**	Roman Catholic, Protestant
Population	67 000	**Currency**	East Caribbean dollar
Capital	Roseau	**Organizations**	CARICOM, Comm., UN

Map page 205

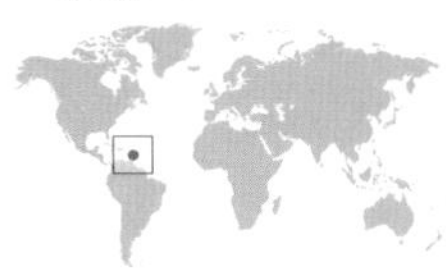

Dominica is the most northerly of the Windward Islands, in the eastern Caribbean. It is very mountainous and forested, with a coastline of steep cliffs. The climate is tropical and rainfall is abundant. Approximately a quarter of Dominicans live in the capital. The economy is based on agriculture, with bananas (the major export), coconuts and citrus fruits the most important crops. Tourism is a developing industry.

DOMINICAN REPUBLIC

Area Sq Km	48 442	**Languages**	Spanish, Creole
Area Sq Miles	18 704	**Religions**	Roman Catholic, Protestant
Population	10 090 000	**Currency**	Dominican peso
Capital	Santo Domingo	**Organizations**	UN

Map page 205

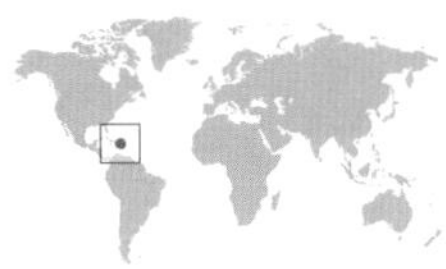

The state occupies the eastern two-thirds of the Caribbean island of Hispaniola (the western third is Haiti). It has a series of mountain ranges, fertile valleys and a large coastal plain in the east. The climate is hot tropical, with heavy rainfall. Sugar, coffee and cocoa are the main cash crops. Nickel (the main export), and gold are mined, and there is some light industry. The USA is the main trading partner. Tourism is the main foreign exchange earner.

EAST TIMOR
Democratic Republic of Timor-Leste

Area Sq Km	14 874	**Languages**	Portuguese, Tetun, English
Area Sq Miles	5 743	**Religions**	Roman Catholic
Population	1 134 000	**Currency**	United States dollar
Capital	Dili	**Organizations**	UN

Map page 147

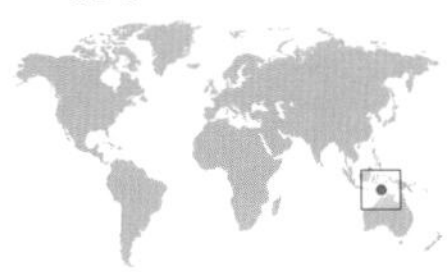

The island of Timor is part of the Indonesian archipelago, to the north of western Australia. East Timor occupies the eastern section of the island, and a small coastal enclave (Ocussi) to the west. A referendum in 1999 ended Indonesia's occupation, after which the country was under UN transitional administration until full independence was achieved in 2002. The economy is in a poor state and East Timor is heavily dependent on foreign aid.

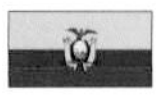

ECUADOR

Republic of Ecuador

Area Sq Km	272 045	**Languages**	Spanish, Quechua, and other Amerindian languages
Area Sq Miles	105 037		
Population	13 625 000	**Religions**	Roman Catholic
Capital	Quito	**Currency**	United States dollar
		Organizations	OPEC, UN

Map page 210

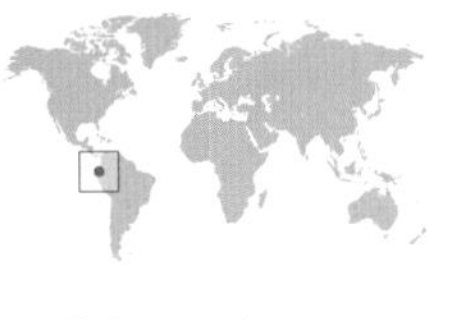

Ecuador is in northwest South America, on the Pacific coast. It consists of a broad coastal plain, high mountain ranges in the Andes, and part of the forested upper Amazon basin to the east. The climate is tropical, moderated by altitude. Most people live on the coast or in the mountain valleys. Ecuador is one of South America's main oil producers, and mineral reserves include gold. Most of the workforce depends on agriculture. Petroleum, bananas, shrimps, coffee and cocoa are exported. The USA is the main trading partner.

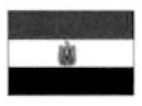

EGYPT

Arab Republic of Egypt

Area Sq Km	1 000 250	**Languages**	Arabic
Area Sq Miles	386 199	**Religions**	Sunni Muslim, Coptic Christian
Population	82 999 000	**Currency**	Egyptian pound
Capital	Cairo (Al Qāhirah)	**Organizations**	UN

Map page 177

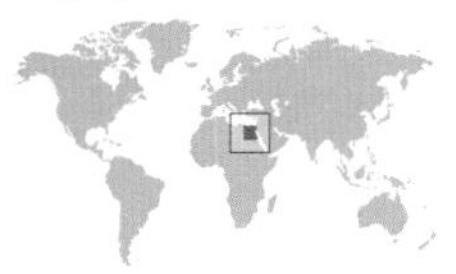

Egypt, on the eastern Mediterranean coast of north Africa, is low-lying, with areas below sea level in the Qattara depression. It is a land of desert and semi-desert, except for the Nile valley, where ninety-nine per cent of Egyptians live. The Sinai peninsula in the northeast of the country forms the only land bridge between Africa and Asia. The summers are hot, the winters mild and rainfall is negligible. Less than four per cent of land (chiefly around the Nile floodplain and delta) is cultivated. Farming employs about one-third of the workforce; cotton is the main cash crop. Egypt imports over half its food needs. There are oil and natural gas reserves, although nearly a quarter of electricity comes from hydroelectric power. Main exports are oil and oil products, cotton, textiles and clothing.

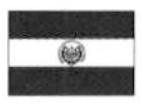

EL SALVADOR

Republic of El Salvador

Area Sq Km	21 041	**Languages**	Spanish
Area Sq Miles	8 124	**Religions**	Roman Catholic, Protestant
Population	6 163 000	**Currency**	El Salvador colón, United States dollar
Capital	San Salvador	**Organizations**	UN

Map page 207

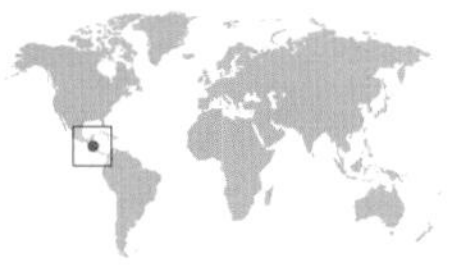

Located on the Pacific coast of central America, El Salvador consists of a coastal plain and volcanic mountain ranges which enclose a densely populated plateau area. The coast is hot, with heavy summer rainfall; the highlands are cooler. Coffee (the chief export), sugar and cotton are the main cash crops. The main trading partners are the USA and Guatemala.

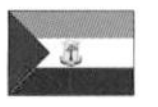

EQUATORIAL GUINEA

Republic of Equatorial Guinea

Area Sq Km	28 051	**Languages**	Spanish, French, Fang
Area Sq Miles	10 831	**Religions**	Roman Catholic, traditional beliefs
Population	676 000	**Currency**	CFA franc
Capital	Malabo	**Organizations**	UN

Map page 176

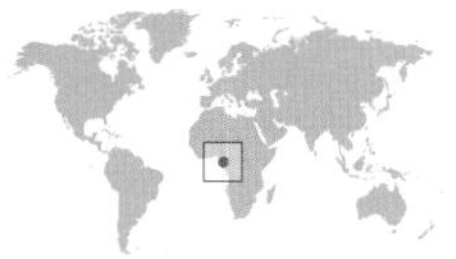

The state consists of Rio Muni, an enclave on the Atlantic coast of central Africa, and the islands of Bioko, Annobón and the Corisco group. Most of the population lives on the coastal plain and upland plateau of Rio Muni. The capital city, Malabo, is on the fertile volcanic island of Bioco. The climate is hot, humid and wet. Oil production started in 1992, and oil is now the main export, along with timber. The economy depends heavily on foreign aid.

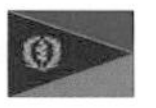

ERITREA

State of Eritrea

Area Sq Km	117 400	**Languages**	Tigrinya, Tigre
Area Sq Miles	45 328	**Religions**	Sunni Muslim, Coptic Christian
Population	5 073 000	**Currency**	Nakfa
Capital	Asmara	**Organizations**	UN

Map page 178

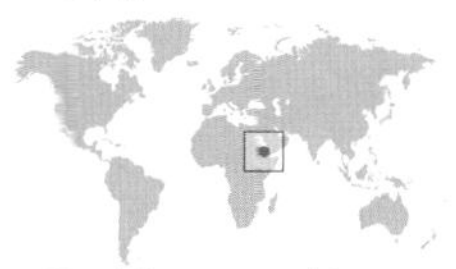

Eritrea, on the Red Sea coast of northeast Africa, consists of a high plateau in the north with a coastal plain which widens to the south. The coast is hot; inland is cooler. Rainfall is unreliable. The agriculture-based economy has suffered from over thirty years of war and occasional poor rains. Eritrea is one of the least developed countries in the world.

ESTONIA

Republic of Estonia

Area Sq Km	45 200	**Languages**	Estonian, Russian
Area Sq Miles	17 452	**Religions**	Protestant, Estonian and Russian Orthodox
Population	1 340 000	**Currency**	Kroon
Capital	Tallinn	**Organizations**	EU, NATO, OECD, UN

Map page 159

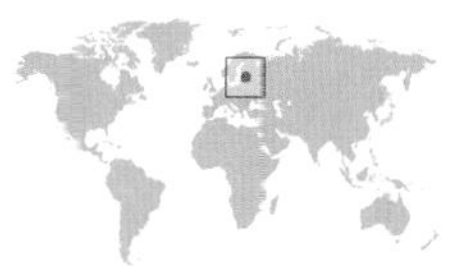

Estonia is in northern Europe, on the Gulf of Finland and the Baltic Sea. The land, over one-third of which is forested, is generally low-lying with many lakes. Approximately one-third of Estonians live in the capital, Tallinn. Exported goods include machinery, wood products, textiles and food products. The main trading partners are the Russian Federation, Finland and Sweden. Estonia joined the European Union in May 2004.

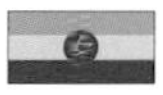

ETHIOPIA

Federal Democratic Republic of Ethiopia

Area Sq Km	1 133 880	**Languages**	Oromo, Amharic, Tigrinya, local languages
Area Sq Miles	437 794	**Religions**	Ethiopian Orthodox, Sunni Muslim, traditional beliefs
Population	82 825 000	**Currency**	Birr
Capital	Addis Ababa (Ādīs Ābeba)	**Organizations**	UN

Map page 178

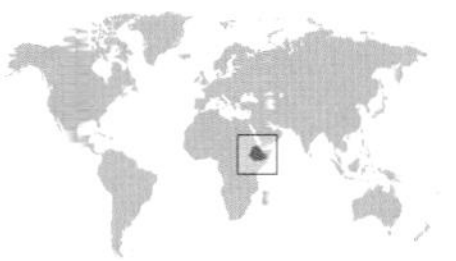

A landlocked country in northeast Africa, Ethiopia comprises a mountainous region in the west which is traversed by the Great Rift Valley. The east is mostly arid plateau land. The highlands are warm with summer rainfall. Most people live in the central–northern area. In recent years civil war, conflict with Eritrea and poor infrastructure have hampered economic development. Subsistence farming is the main activity, although droughts have led to frequent famines. Coffee is the main export and there is some light industry. Ethiopia is one of the least developed countries in the world.

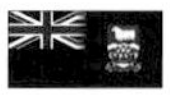

Falkland Islands

United Kingdom Overseas Territory

Area Sq Km	12 170	**Languages**	English
Area Sq Miles	4 699	**Religions**	Protestant, Roman Catholic
Population	2 955	**Currency**	Falkland Islands pound
Capital	Stanley		

Map page 212 Lying in the southwest Atlantic Ocean, northeast of Cape Horn, two main islands, West Falkland and East Falkland and many smaller islands, form the territory of the Falkland Islands. The economy is based on sheep farming and the sale of fishing licences.

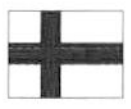

Faroe Islands

Self-governing Danish Territory

Area Sq Km	1 399	**Languages**	Faroese, Danish
Area Sq Miles	540	**Religions**	Protestant
Population	50 000	**Currency**	Danish krone
Capital	Thorshavn (Tórshavn)		

Map page 158 A self-governing territory, the Faroe Islands lie in the north Atlantic Ocean between the UK and Iceland. The islands benefit from the North Atlantic Drift ocean current, which has a moderating effect on the climate. The economy is based on deep-sea fishing.

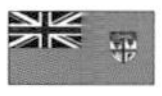

FIJI

Sovereign Democratic Republic of Fiji

Area Sq Km	18 330	**Languages**	English, Fijian, Hindi
Area Sq Miles	7 077	**Religions**	Christian, Hindu, Sunni Muslim
Population	849 000	**Currency**	Fiji dollar
Capital	Suva	**Organizations**	UN

Map page 125

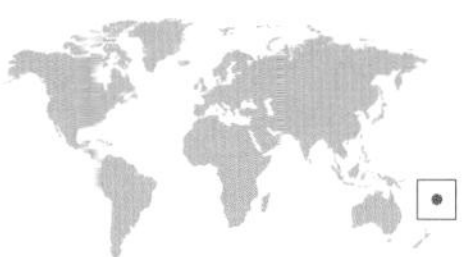

The southwest Pacific republic of Fiji comprises two mountainous and volcanic islands, Vanua Levu and Viti Levu, and over three hundred smaller islands. The climate is tropical and the economy is based on agriculture (chiefly sugar, the main export), fishing, forestry, gold mining and tourism.

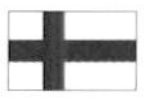

FINLAND

Republic of Finland

Area Sq Km	338 145	**Languages**	Finnish, Swedish
Area Sq Miles	130 559	**Religions**	Protestant, Greek Orthodox
Population	5 326 000	**Currency**	Euro
Capital	Helsinki (Helsingfors)	**Organizations**	EU, OECD, UN

Map page 158–159

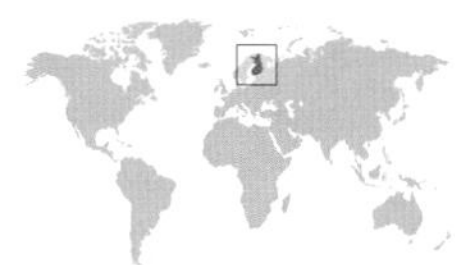

Finland is in northern Europe, and nearly one-third of the country lies north of the Arctic Circle. Forests cover over seventy per cent of the land area, and ten per cent is covered by lakes. Summers are short and warm, and winters are long and severe, particularly in the north. Most of the population lives in the southern third of the country, along the coast or near the lakes. Timber is a major resource and there are important minerals, chiefly chromium. Main industries include metal working, electronics, paper and paper products, and chemicals. The main trading partners are Germany, Sweden and the UK.

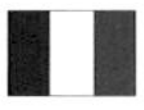

FRANCE

French Republic

Area Sq Km	543 965	**Languages**	French, Arabic
Area Sq Miles	210 026	**Religions**	Roman Catholic, Protestant, Sunni Muslim
Population	62 343 000	**Currency**	Euro
Capital	Paris	**Organizations**	EU, NATO, OECD, UN

Map page 166

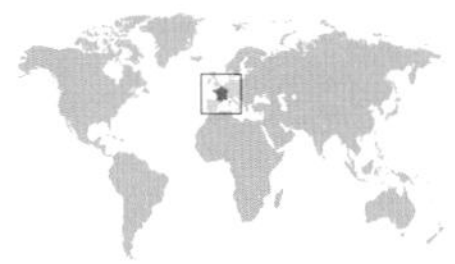

France lies in western Europe and has coastlines on the Atlantic Ocean and the Mediterranean Sea. It includes the Mediterranean island of Corsica. Northern and western regions consist mostly of flat or rolling countryside, and include the major lowlands of the Paris basin, the Loire valley and the Aquitaine basin, drained by the Seine, Loire and Garonne river systems respectively. The centre-south is dominated by the hill region of the Massif Central. To the east are the Vosges and Jura mountains and the Alps. In the southwest, the Pyrenees form a natural border with Spain. The climate is temperate with warm summers and cool winters, although the Mediterranean coast has hot, dry summers and mild winters. Over seventy per cent of the population lives in towns, with almost a sixth of the population living in the Paris area. The French economy has a substantial and varied agricultural base. It is a major producer of both fresh and processed food. There are relatively few mineral resources; it has coal reserves, and some oil and natural gas, but it relies heavily on nuclear and hydroelectric power and imported fuels. France is one of the world's major industrial countries. Main industries include food processing, iron, steel and aluminium production, chemicals, cars, electronics and oil refining. The main exports are transport equipment, plastics and chemicals. Tourism is a major source of revenue and employment. Trade is predominantly with other European Union countries.

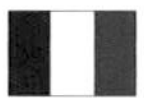

French Guiana

French Overseas Department

Area Sq Km	90 000	**Languages**	French, Creole
Area Sq Miles	34 749	**Religions**	Roman Catholic
Population	226 000	**Currency**	Euro
Capital	Cayenne		

Map page 211 French Guiana, on the north coast of South America, is densely forested. The climate is tropical, with high rainfall. Most people live in the coastal strip, and agriculture is mostly subsistence farming. Forestry and fishing are important, but mineral resources are largely unexploited and industry is limited. French Guiana depends on French aid. The main trading partners are France and the USA.

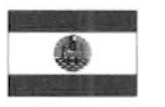

French Polynesia

French Overseas Country

Area Sq Km	3 265	**Languages**	French, Tahitian, Polynesian languages
Area Sq Miles	1 261	**Religions**	Protestant, Roman Catholic
Population	269 000	**Currency**	CFP franc
Capital	Papeete		

Map page 123 Extending over a vast area of the southeast Pacific Ocean, French Polynesia comprises more than one hundred and thirty islands and coral atolls. The main island groups are the Marquesas Islands, the Tuamotu Archipelago and the Society Islands. The capital, Papeete, is on Tahiti in the Society Islands. The climate is subtropical, and the economy is based on tourism. The main export is cultured pearls.

© Collins Bartholomew Ltd

GABON
Gabonese Republic

Area Sq Km	267 667	Languages	French, Fang, local languages
Area Sq Miles	103 347	Religions	Roman Catholic, Protestant, traditional beliefs
Population	1 475 000		
Capital	Libreville	Currency	CFA franc
		Organizations	UN

Map page 178

Gabon, on the Atlantic coast of central Africa, consists of low plateaus and a coastal plain lined by lagoons and mangrove swamps. The climate is tropical and rainforests cover over three-quarters of the land area. Over seventy per cent of the population lives in towns. The economy is heavily dependent on oil, which accounts for around seventy-five per cent of exports; manganese, uranium and timber are the other main exports. Agriculture is mainly at subsistence level.

THE GAMBIA
Republic of the Gambia

Area Sq Km	11 295	Languages	English, Malinke, Fulani, Wolof
Area Sq Miles	4 361	Religions	Sunni Muslim, Protestant
Population	1 705 000	Currency	Dalasi
Capital	Banjul	Organizations	Comm., UN

Map page 176

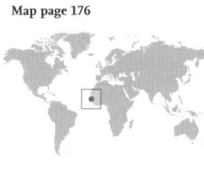

The Gambia, on the coast of west Africa, occupies a strip of land along the lower Gambia river. Sandy beaches are backed by mangrove swamps, beyond which is savanna. The climate is tropical, with most rainfall in the summer. Over seventy per cent of Gambians are farmers, growing chiefly groundnuts (the main export), cotton, oil palms and food crops. Livestock rearing and fishing are important, while manufacturing is limited. Re-exports, mainly from Senegal, and tourism are major sources of income.

Gaza
Semi-autonomous region

Area Sq Km	363	Languages	Arabic
Area Sq Miles	140	Religions	Sunni Muslim, Shi'a Muslim
Population	1 486 816	Currency	Israeli shekel
Capital	Gaza		

Map page 136   Gaza is a narrow strip of land on the southeast corner of the Mediterranean Sea, between Egypt and Israel. This Palestinian territory has limited autonomy from Israel, but hostilities between Israel and the indigenous Arab population continue to restrict its economic development.

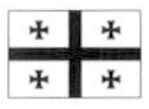

GEORGIA
Republic of Georgia

Area Sq Km	69 700	Languages	Georgian, Russian, Armenian, Azeri, Ossetian, Abkhaz
Area Sq Miles	26 911		
Population	4 260 000	Religions	Georgian Orthodox, Russian Orthodox, Sunni Muslim
Capital	T'bilisi		
		Currency	Lari
		Organizations	CIS, UN

Map page 173

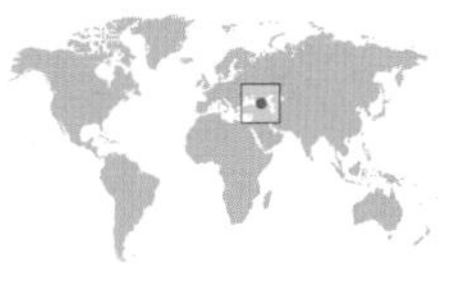

Georgia is in the northwest Caucasus area of southwest Asia, on the eastern coast of the Black Sea. Mountain ranges in the north and south flank the Kura and Rioni valleys. The climate is generally mild, and along the coast it is subtropical. Agriculture is important, with tea, grapes, and citrus fruits the main crops. Mineral resources include manganese ore and oil, and the main industries are steel, oil refining and machine building. The main trading partners are the Russian Federation and Turkey.

GERMANY
Federal Republic of Germany

Area Sq Km	357 022	Languages	German, Turkish
Area Sq Miles	137 849	Religions	Protestant, Roman Catholic
Population	82 167 000	Currency	Euro
Capital	Berlin	Organizations	EU, NATO, OECD, UN

Map page 168

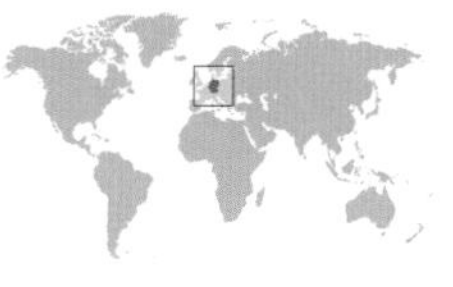

The central European state of Germany borders nine countries and has coastlines on the North and Baltic Seas. Behind the indented coastline, and covering about one-third of the country, is the north German plain, a region of fertile farmland and sandy heaths drained by the country's major rivers. The central highlands are a belt of forested hills and plateaus which stretch from the Eifel region in the west to the Erzgebirge mountains along the border with the Czech Republic. Farther south the land rises to the Swabian Alps (Schwäbische Alb),

with the high rugged and forested Black Forest (Schwarzwald) in the southwest. In the far south the Bavarian Alps form the border with Austria. The climate is temperate, with continental conditions in eastern areas. The population is highly urbanized, with over eighty-five per cent living in cities and towns. With the exception of coal, lignite, potash and baryte, Germany lacks minerals and other industrial raw materials. It has a small agricultural base, although a few products (chiefly wines and beers) enjoy an international reputation. Germany is the world's third ranking economy after the USA and Japan. Its industries are amongst the world's most technologically advanced. Exports include machinery, vehicles and chemicals. The majority of trade is with other countries in the European Union, the USA and Japan.

Baden-Württemberg (State)

Area Sq Km (Sq Miles) 35 752 (13 804)   Population 10 751 000   Capital Stuttgart

Bayern (State)

Area Sq Km (Sq Miles) 70 550 (27 240)   Population 12 521 000   Capital Munich (München)

Berlin (State)

Area Sq Km (Sq Miles) 892 (344)   Population 3 426 000   Capital Berlin

Brandenburg (State)

Area Sq Km (Sq Miles) 29 476 (11 381)   Population 2 528 000   Capital Potsdan

Bremen (State)

Area Sq Km (Sq Miles) 404 (156)   Population 661 000   Capital Bremen

Hamburg (State)

Area Sq Km (Sq Miles) 755 (292)   Population 1 772 000   Capital Hamburg

Hessen (State)

Area Sq Km (Sq Miles) 21 114 (8 152)   Population 6 072 000   Capital Wiesbaden

Mecklenburg-Vorpommern (State)

Area Sq Km (Sq Miles) 23 173 (8 947)   Population 1 670 000   Capital Schwerin

Niedersachsen (State)

Area Sq Km (Sq Miles) 47 616 (18 385)   Population 7 959 000   Capital Hannover

Nordrhein-Westfalen (State)

Area Sq Km (Sq Miles) 34 082 (13 159)   Population 17 964 000   Capital Düsseldorf

Rheinland-Pfalz (State)

Area Sq Km (Sq Miles) 19 847 (7 663)   Population 4 037 000   Capital Mainz

Saarland (State)

Area Sq Km (Sq Miles) 2 568 (992)   Population 1 033 000   Capital Saarbrücken

Sachsen (State)

Area Sq Km (Sq Miles) 18 413 (7 109)   Population 4 200 000   Capital Dresden

Sachsen-Anhalt (State)

Area Sq Km (Sq Miles) 20 447 (7 895)   Population 2 393 000   Capital Magdeburg

Schleswig-Holstein (State)

Area Sq Km (Sq Miles) 15 761 (6 085)   Population 2 836 000   Capital Kiel

Thüringen (State)

Area Sq Km (Sq Miles) 16 172 (6 244)   Population 2 274 000   Capital Erfurt

GHANA
Republic of Ghana

Area Sq Km	238 537	Languages	English, Hausa, Akan, local languages
Area Sq Miles	92 100	Religions	Christian, Sunni Muslim, traditional beliefs
Population	23 837 000		
Capital	Accra	Currency	Cedi
		Organizations	Comm., UN

Map page 176

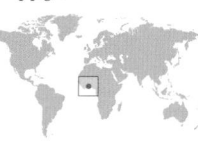

A west African state on the Gulf of Guinea, Ghana is a land of plains and low plateaus covered with savanna and rainforest. In the east is the Volta basin and Lake Volta. The climate is tropical, with the highest rainfall in the south, where most of the population lives. Agriculture employs around sixty per cent of the workforce. Main exports are gold, timber, cocoa, bauxite and manganese ore.

Gibraltar
United Kingdom Overseas Territory

Area Sq Km	7	Languages	English, Spanish
Area Sq Miles	3	Religions	Roman Catholic, Protestant, Sunni Muslim
Population	31 000		
Capital	Gibraltar	Currency	Gibraltar pound

Map page 167   Gibraltar lies on the south coast of Spain at the western entrance to the Mediterranean Sea. The economy depends on tourism, offshore banking and shipping services.

GREECE
Hellenic Republic

Area Sq Km	131 957	Languages	Greek
Area Sq Miles	50 949	Religions	Greek Orthodox, Sunni Muslim
Population	11 161 000	Currency	Euro
Capital	Athens (Athina)	Organizations	EU, NATO, OECD, UN

Map page 171

Greece comprises a mountainous peninsula in the Balkan region of southeastern Europe and many islands in the Ionian, Aegean and Mediterranean Seas. The islands make up over one-fifth of its area. Mountains and hills cover much of the country. The main lowland areas are the plains of Thessaly in the centre and around Thessaloniki in the northeast. Summers are hot and dry while winters are mild and wet, but colder in the north with heavy snowfalls in the mountains. One-third of Greeks live in the Athens area. Employment in agriculture accounts for approximately twenty per cent of the workforce, and exports include citrus fruits, raisins, wine, olives and olive oil. Aluminium and nickel are mined and a wide range of manufactures are produced, including food products and tobacco, textiles, clothing, and chemicals. Tourism is an important industry and there is a large services sector. Most trade is with other European Union countries.

Greenland
Self-governing Danish Territory

Area Sq Km	2 175 600	Languages	Greenlandic, Danish
Area Sq Miles	840 004	Religions	Protestant
Population	57 000	Currency	Danish krone
Capital	Nuuk (Godthåb)		

Map page 185

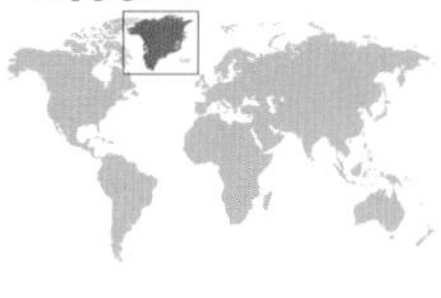

Situated to the northeast of North America between the Atlantic and Arctic Oceans, Greenland is the largest island in the world. It has a polar climate and over eighty per cent of the land area is covered by permanent ice cap. The economy is based on fishing and fish processing.

GRENADA

Area Sq Km	378	Languages	English, Creole
Area Sq Miles	146	Religions	Roman Catholic, Protestant
Population	104 000	Currency	East Caribbean dollar
Capital	St George's	Organizations	CARICOM, Comm., UN

Map page 213

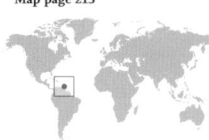

The Caribbean state comprises Grenada, the most southerly of the Windward Islands, and the southern islands of the Grenadines. Grenada has wooded hills, with beaches in the southwest. The climate is warm and wet. Agriculture is the main activity, with bananas, nutmeg and cocoa the main exports. Tourism is the main foreign exchange earner.

Guadeloupe
French Overseas Department

Area Sq Km	1 780	Languages	French, Creole
Area Sq Miles	687	Religions	Roman Catholic
Population	465 000	Currency	Euro
Capital	Basse-Terre		

Map page 205   Guadeloupe, in the Leeward Islands in the Caribbean, consists of two main islands (Basse-Terre and Grande-Terre, connected by a bridge), Marie-Galante, and a few outer islands. The climate is tropical, but moderated by trade winds. Bananas, sugar and rum are the main exports and tourism is a major source of income.

Guam
United States Unincorporated Territory

Area Sq Km	541	Languages	Chamorro, English, Tagalog
Area Sq Miles	209	Religions	Roman Catholic
Population	178 000	Currency	United States dollar
Capital	Hagåtña		

Map page 147   Lying at the south end of the Northern Mariana Islands in the western Pacific Ocean, Guam has a humid tropical climate. The island has a large US military base and the economy relies on that and on tourism.

GUATEMALA
Republic of Guatemala

Area Sq Km	108 890	Languages	Spanish, Mayan languages
Area Sq Miles	42 043	Religions	Roman Catholic, Protestant
Population	14 027 000	Currency	Quetzal, United States dollar
Capital	Guatemala City	Organizations	UN

Map page 207

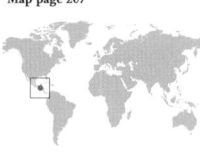

The most populous country in Central America after Mexico, Guatemala has long Pacific and short Caribbean coasts separated by a mountain chain which includes several active volcanoes. The climate is hot tropical in the lowlands and cooler in the highlands, where most of the population lives. Farming is the main activity and coffee, sugar and bananas are the main exports. There is some manufacturing of clothing and textiles. The main trading partner is the USA.

Guernsey
United Kingdom Crown Dependency

Area Sq Km	78	Languages	English, French
Area Sq Miles	30	Religions	Protestant, Roman Catholic
Population	64 801	Currency	Pound sterling
Capital	St Peter Port		

Map page 166   Guernsey is one of the Channel Islands, lying off northern France. The dependency also includes the nearby islands of Alderney, Sark and Herm. Financial services are an important part of the island's economy.

GUINEA
Republic of Guinea

Area Sq Km	245 857	Languages	French, Fulani, Malinke, local languages
Area Sq Miles	94 926		
Population	10 069 000	Religions	Sunni Muslim, traditional beliefs, Christian
Capital	Conakry	Currency	Guinea franc
		Organizations	UN

Map page 176

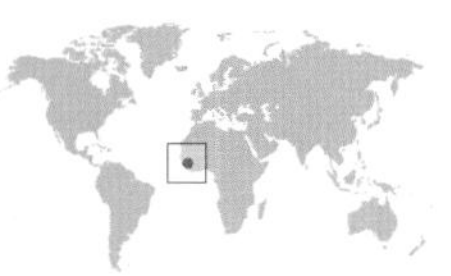

Guinea is in west Africa, on the Atlantic Ocean. There are mangrove swamps along the coast, while inland are lowlands and the Fouta Djallon mountains and plateaus. To the east are savanna plains drained by the upper Niger river system. The southeast is hilly. The climate is tropical, with high coastal rainfall. Agriculture is the main activity, employing nearly eighty per cent of the workforce, with coffee, bananas and pineapples the chief cash crops. There are huge reserves of bauxite, which accounts for more than seventy per cent of exports. Other exports include aluminium oxide, gold, coffee and diamonds.

GUINEA-BISSAU
Republic of Guinea-Bissau

Area Sq Km	36 125	Languages	Portuguese, Crioulo, local languages
Area Sq Miles	13 948	Religions	Traditional beliefs, Sunni Muslim, Christian
Population	1 611 000		
Capital	Bissau	Currency	CFA franc
		Organizations	UN

Map page 176

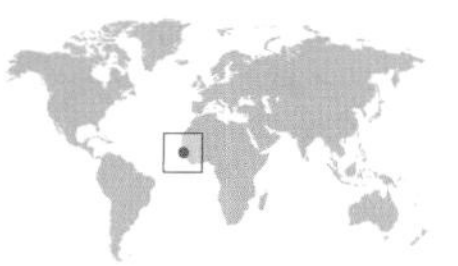

Guinea-Bissau is on the Atlantic coast of west Africa. The mainland coast is swampy and contains many estuaries. Inland are forested plains, and to the east are savanna plateaus. The climate is tropical. The economy is based mainly on subsistence farming. There is little industry, and timber and mineral resources are largely unexploited. Cashews account for seventy per cent of exports. Guinea-Bissau is one of the least developed countries in the world.

GUYANA
Co-operative Republic of Guyana

Area Sq Km	214 969	Languages	English, Creole, Amerindian languages
Area Sq Miles	83 000	Religions	Protestant, Hindu, Roman Catholic, Sunni Muslim
Population	762 000		
Capital	Georgetown	Currency	Guyana dollar
		Organizations	CARICOM, Comm., UN

Map page 210–211

Guyana, on the northeast coast of South America, consists of highlands in the west and savanna uplands in the southwest. Most of the country is densely forested. A lowland coastal belt supports crops and most of the population. The generally hot, humid and wet conditions are modified along the coast by sea breezes. The economy is based on agriculture, bauxite, and forestry. Sugar, bauxite, gold, rice and timber are the main exports.

HAITI
Republic of Haiti

Area Sq Km	27 750	Languages	French, Creole
Area Sq Miles	10 714	Religions	Roman Catholic, Protestant, Voodoo
Population	10 033 000	Currency	Gourde
Capital	Port-au-Prince	Organizations	CARICOM, UN

Map page 205

Haiti, occupying the western third of the Caribbean island of Hispaniola, is a mountainous state with small coastal plains and a central valley. The Dominican Republic occupies the rest of the island. The climate is tropical, and is hottest in coastal areas. Haiti has few natural resources, is densely populated and relies on exports of local crafts and coffee, and remittances from workers abroad.

HONDURAS
Republic of Honduras

Area Sq Km	112 088	Languages	Spanish, Amerindian languages
Area Sq Miles	43 277	Religions	Roman Catholic, Protestant
Population	7 466 000	Currency	Lempira
Capital	Tegucigalpa	Organizations	UN

Map page 206

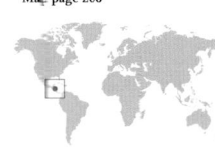

Honduras, in central America, is a mountainous and forested country with lowland areas along its long Caribbean and short Pacific coasts. Coastal areas are hot and humid with heavy summer rainfall; inland is cooler and drier. Most of the population lives in the central valleys. Coffee and bananas are the main exports, along with shellfish and zinc. Industry involves mainly agricultural processing.

HUNGARY
Republic of Hungary

Area Sq Km	93 030	Languages	Hungarian
Area Sq Miles	35 919	Religions	Roman Catholic, Protestant
Population	9 993 000	Currency	Forint
Capital	Budapest	Organizations	EU, NATO, OECD, UN

Map page 168–169

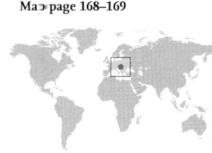

The Danube river flows north-south through central Hungary, a landlocked country in eastern Europe. In the east lies a great plain, flanked by highlands in the north. In the west low mountains and Lake Balaton separate a smaller plain and southern uplands. The climate is continental. Sixty per cent of the population lives in urban areas, and one-fifth lives in the capital, Budapest. Some minerals and energy resources are exploited, chiefly bauxite, coal and natural gas. Hungary has an industrial economy based on metals, machinery, transport equipment, chemicals and food products. The main trading partners are Germany and Austria. Hungary joined the European Union in May 2004.

ICELAND
Republic of Iceland

Area Sq Km	102 820	Languages	Icelandic
Area Sq Miles	39 699	Religions	Protestant
Population	323 000	Currency	Icelandic króna
Capital	Reykjavik	Organizations	NATO, OECD, UN

Map page 158

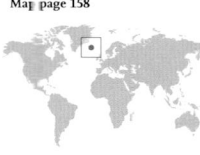

Iceland lies in the north Atlantic Ocean near the Arctic Circle, to the northwest of Scandinavia. The landscape is volcanic, with numerous hot springs, geysers, and approximately two hundred volcanoes. One-tenth of the country is covered by ice caps. Only coastal lowlands are cultivated and settled, and over half the population lives in the Reykjavik area. The climate is mild, moderated by the North Atlantic Drift ocean current and by southwesterly winds. The mainstays of the economy are fishing and fish processing, which account for seventy per cent of exports. Agriculture involves mainly sheep and dairy farming. Hydroelectric and geothermal energy resources are considerable. The main industries produce aluminium, ferro-silicon and fertilizers. Tourism, including ecotourism, is growing in importance.

INDIA
Republic of India

Area Sq Km	3 064 898	Languages	Hindi, English, many regional languages
Area Sq Miles	1 183 364		
Population	1 198 003 000	Religions	Hindu, Sunni Muslim, Shi'a Muslim, Sikh, Christian
Capital	New Delhi		
		Currency	Indian rupee
		Organizations	Comm., UN

Map page 134–135

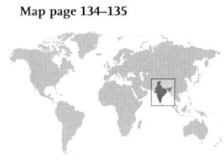

The south Asian country of India occupies a peninsula that juts out into the Indian Ocean between the Arabian Sea and Bay of Bengal. The heart of the peninsula is the Deccan plateau, bordered on either side by ranges of hills, the western Ghats and the lower eastern Ghats, which fall away to narrow coastal plains. To the north is a broad plain, drained by the Indus, Ganges and Brahmaputra rivers and their tributaries. The plain is intensively farmed and is the most populous region. In the west is the Thar Desert. The mountains of the Himalaya form India's northern border, together with parts of the Karakoram and Hindu Kush ranges in the northwest. The climate shows marked seasonal variation: a hot season from March to June; a monsoon season from June to October; and a cold season from November to February. Rainfall ranges between very high in the northeast Assam region to negligible in the Thar Desert. Temperatures range from very cold in the Himalaya to tropical heat over much of the south. Over seventy per cent of the huge population – the second largest in the world – is rural, although Delhi, Mumbai (Bombay) and Kolkata (Calcutta) all rank among the ten largest cities in the world. Agriculture, forestry and fishing account for a quarter of national output and two-thirds of employment. Much of the farming is on a subsistence basis and involves mainly rice and wheat. India is a major world producer of tea, sugar, jute, cotton and tobacco. Livestock is reared mainly for dairy products and hides. There are major reserves of coal, reserves of oil and natural gas, and many minerals, including iron, manganese, bauxite, diamonds and gold. The manufacturing sector is large and diverse – mainly chemicals and chemical products, textiles, iron and steel, food products, electrical goods and transport equipment; software and pharmaceuticals are also important. All the main manufactured products are exported, together with diamonds and jewellery. The USA, Germany, Japan and the UK are the main trading partners.

INDONESIA
Republic of Indonesia

Area Sq Km	1 919 445	Languages	Indonesian, local languages
Area Sq Miles	741 102	Religions	Sunni Muslim, Protestant, Roman Catholic, Hindu, Buddhist
Population	229 965 000		
Capital	Jakarta	Currency	Rupiah
		Organizations	APEC, ASEAN, UN

Map page 147

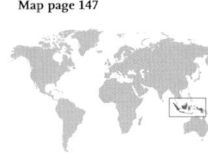

Indonesia, the largest and most populous country in southeast Asia, consists of over thirteen thousand islands extending between the Pacific and Indian Oceans. Sumatra, Java, Sulawesi (Celebes), Kalimantan (two-thirds of Borneo) and Papua (formerly Irian Jaya, western New Guinea) make up ninety per cent of the land area. Most of Indonesia is mountainous and covered with rainforest or mangrove swamps, and there are over three hundred volcanoes, many active. Two-thirds of the population lives in the lowland areas of the islands of Java and Madura. The climate is tropical monsoon. Agriculture is the largest sector of the economy and Indonesia is among the world's top producers of rice, palm oil, tea, coffee, rubber and tobacco. Many goods are produced, including textiles, clothing, cement, tin, fertilizers and vehicles. Main exports are oil, natural gas, timber products and clothing. Main trading partners are Japan, the USA and Singapore. Indonesia is a relatively poor country, and ethnic tensions and civil unrest often hinder economic development.

IRAN
Islamic Republic of Iran

Area Sq Km	1 648 000	Languages	Farsi, Azeri, Kurdish, regional languages
Area Sq Miles	636 296		
Population	74 196 000	Religions	Shi'a Muslim, Sunni Muslim
Capital	Tehrān	Currency	Iranian rial
		Organizations	OPEC, UN

Map page 140–141

Iran is in southwest Asia, and has coasts on The Gulf, the Caspian Sea and the Gulf of Oman. In the east is a high plateau, with large salt pans and a vast sand desert. In the west the Zagros Mountains form a series of ridges, and to the north lie the Elburz Mountains. Most farming and settlement is on the narrow plain along the Caspian Sea and in the foothills of the north and west. The climate is one of extremes, with hot summers and very cold winters. Most of the light rainfall is in the winter months. Agriculture involves approximately one-third of the workforce. Wheat is the main crop, but fruit (especially dates) and pistachio nuts are grown for export. Petroleum (the main export) and natural gas are Iran's leading natural resources. Manufactured goods include carpets, clothing, food products and construction materials.

© Collins Bartholomew Ltd

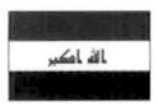

IRAQ

Republic of Iraq

Area Sq Km	438 317	**Languages**	Arabic, Kurdish, Turkmen
Area Sq Miles	169 235	**Religions**	Shi'a Muslim, Sunni Muslim, Christian
Population	30 747 000	**Currency**	Iraqi dinar
Capital	Baghdād	**Organizations**	OPEC, UN

Map page 137

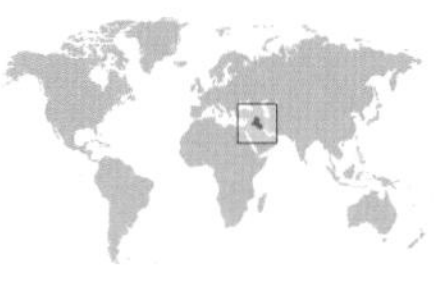

Iraq, in southwest Asia, has at its heart the lowland valley of the Tigris and Euphrates rivers. In the southeast, where the two rivers join, are the Mesopotamian marshes and the Shaṭṭ al 'Arab waterway leading to The Gulf. The north is hilly, while the west is mostly desert. Summers are hot and dry, and winters are mild with light, unreliable rainfall. The Tigris-Euphrates valley contains most of the country's arable land. One in five of the population lives in the capital, Baghdad. The economy has suffered following the 1991 Gulf War and the invasion of US-led coalition forces in 2005. The latter resulted in the overthrow of the dictator Saddam Hussein, but there is continuing internal instability. Oil is normally the main export.

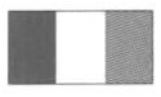

IRELAND

Republic of Ireland

Area Sq Km	70 282	**Languages**	English, Irish
Area Sq Miles	27 136	**Religions**	Roman Catholic, Protestant
Population	4 515 000	**Currency**	Euro
Capital	Dublin (Baile Átha Cliath)	**Organizations**	EU, OECD, UN

Map page 163

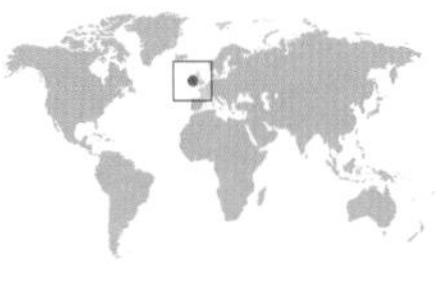

The Irish Republic occupies some eighty per cent of the island of Ireland, in northwest Europe. It is a lowland country of wide valleys, lakes and peat bogs, with isolated mountain ranges around the coast. The west coast is rugged and indented with many bays. The climate is mild due to the modifying effect of the North Atlantic Drift ocean current and rainfall is plentiful, although highest in the west. Nearly sixty per cent of the population lives in urban areas, Dublin and Cork being the main cities. Resources include natural gas, peat, lead and zinc. Agriculture, the traditional mainstay, now employs less than ten per cent of the workforce, while industry employs nearly thirty per cent. The main industries are electronics, pharmaceuticals and engineering as well as food processing, brewing and textiles. Service industries are expanding, with tourism a major earner. The UK is the main trading partner.

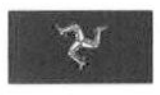

Isle of Man

United Kingdom Crown Dependency

Area Sq Km	572	**Languages**	English
Area Sq Miles	221	**Religions**	Protestant, Roman Catholic
Population	80 000	**Currency**	Pound sterling
Capital	Douglas		

Map page 160

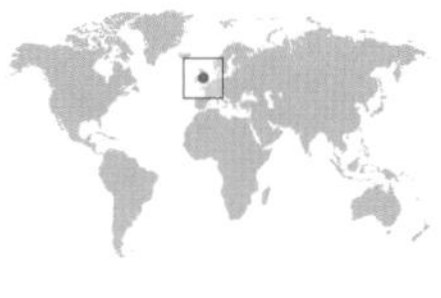

The Isle of Man lies in the Irish Sea between England and Northern Ireland. The island is self-governing, although the UK is responsible for its defence and foreign affairs. It is not part of the European Union, but has a special relationship with the EU which allows for free trade. Eighty per cent of the economy is based on the service sector, particularly financial services.

ISRAEL

State of Israel

Area Sq Km	20 770	**Languages**	Hebrew, Arabic
Area Sq Miles	8 019	**Religions**	Jewish, Sunni Muslim, Christian, Druze
Population	7 170 000	**Currency**	Shekel
Capital	Jerusalem (Yerushalayim) (El Quds) De facto capital. Disputed.	**Organizations**	OECD, UN

Map page 136

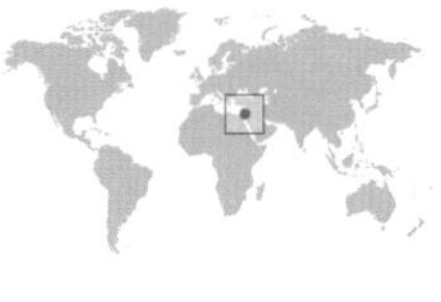

Israel lies on the Mediterranean coast of southwest Asia. Beyond the coastal Plain of Sharon are the hills and valleys of Samaria, with the Galilee highlands to the north. In the east is a rift valley, which extends from Lake Tiberias (Sea of Galilee) to the Gulf of Aqaba and contains the Jordan river and the Dead Sea. In the south is the Negev, a triangular semi-desert plateau. Most of the population lives on the coastal plain or in northern and central areas. Much of Israel has warm summers and mild, wet winters. The south is hot and dry. Agricultural production was boosted by the occupation of the West Bank in 1967. Manufacturing makes the largest contribution to the economy, and tourism is also important. Israel's main exports are machinery and transport equipment, software, diamonds, clothing, fruit and vegetables. The country relies heavily on foreign aid. Security issues relating to territorial disputes over the West Bank and Gaza have still to be resolved.

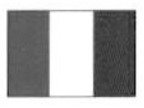

ITALY

Italian Republic

Area Sq Km	301 245	**Languages**	Italian
Area Sq Miles	116 311	**Religions**	Roman Catholic
Population	59 870 000	**Currency**	Euro
Capital	Rome (Roma)	**Organizations**	EU, NATO, OECD, UN

Map page 170–171

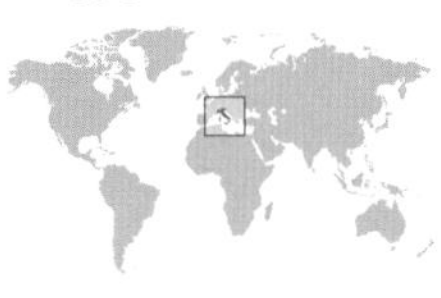

Most of the southern European state of Italy occupies a peninsula that juts out into the Mediterranean Sea. It includes the islands of Sicily and Sardinia and approximately seventy much smaller islands in the surrounding seas. Italy is mountainous, dominated by the Alps, which form its northern border, and the various ranges of the Apennines, which run almost the full length of the peninsula. Many of Italy's mountains are of volcanic origin, and its active volcanoes are Vesuvius, near Naples, Etna and Stromboli. The main lowland area, the Po river valley in the northeast, is the main agricultural and industrial area and is the most populous region. Italy has a Mediterranean climate, although the north experiences colder, wetter winters, with heavy snow in the Alps. Natural resources are limited, and only about twenty per cent of the land is suitable for cultivation. The economy is fairly diversified. Some oil, natural gas and coal are produced, but most fuels and minerals used by industry are imported. Agriculture is important, with cereals, vines, fruit and vegetables the main crops. Italy is the world's largest wine producer. The north is the centre of Italian industry, especially around Turin, Milan and Genoa. Leading manufactures include industrial and office equipment, domestic appliances, cars, textiles, clothing, leather goods, chemicals and metal products. There is a strong service sector, and with over twenty-five million visitors a year, tourism is a major employer and accounts for five per cent of the national income. Finance and banking are also important. Most trade is with other European Union countries.

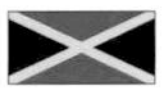

JAMAICA

Area Sq Km	10 991	**Languages**	English, Creole
Area Sq Miles	4 244	**Religions**	Protestant, Roman Catholic
Population	2 719 000	**Currency**	Jamaican dollar
Capital	Kingston	**Organizations**	CARICOM, Comm., UN

Map page 205

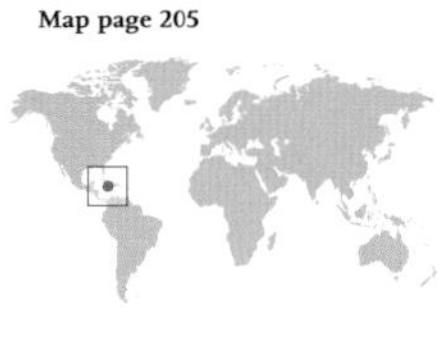

Jamaica, the third largest Caribbean island, has beaches and densely populated coastal plains traversed by hills and plateaus rising to the forested Blue Mountains in the east. The climate is tropical, but cooler and wetter on high ground. The economy is based on tourism, agriculture, mining and light manufacturing. Bauxite, aluminium oxide, sugar and bananas are the main exports. The USA is the main trading partner. Foreign aid is also significant.

Jammu and Kashmir

Disputed territory (India/Pakistan/China)

Area Sq Km	222 236	**Population**	13 000 000
Area Sq Miles	85 806	**Capital**	Srinagar

Map page 144 A disputed region in the north of the Indian subcontinent, to the west of the Karakoram and Himalaya mountains. The 'Line of Control' separates the northwestern, Pakistani-controlled area and the southeastern, Indian-controlled area. China occupies the Himalayan section known as the Aksai Chin, which is also claimed by India.

JAPAN

Area Sq Km	377 727	**Languages**	Japanese
Area Sq Miles	145 841	**Religions**	Shintoist, Buddhist, Christian
Population	127 156 000	**Currency**	Yen
Capital	Tōkyō	**Organizations**	APEC, OECD, UN

Map page 150–151

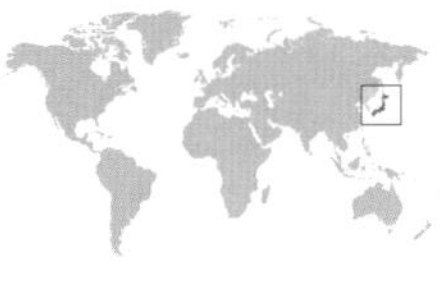

Japan lies in the Pacific Ocean off the coast of eastern Asia and consists of four main islands – Hokkaidō, Honshū, Shikoku and Kyūshū – and more than three thousand smaller islands in the surrounding Sea of Japan, East China Sea and Pacific Ocean. The central island of Honshū accounts for sixty per cent of the total land area and contains eighty per cent of the population. Behind the long and deeply indented coastline, nearly three-quarters of the country is mountainous and heavily forested. Japan has over sixty active volcanoes, and is subject to frequent earthquakes and typhoons. The climate is generally temperate maritime, with warm summers and mild winters, except in western Hokkaidō and northwest Honshū, where the winters are very cold with heavy snow. Only fourteen per cent of the land area is suitable for cultivation, and its few raw materials (coal, oil, natural gas, lead, zinc and copper) are insufficient for its industry. Most materials must be imported, including about ninety per cent of energy requirements. Yet Japan has the world's second largest industrial economy, with a range of modern heavy and light industries centred mainly around the major ports of Yokohama, Ōsaka and Tōkyō. It is the world's largest manufacturer of cars, motorcycles and merchant ships, and a major producer of steel, textiles, chemicals and cement. It is also a leading producer of many consumer durables, such as washing machines, and electronic equipment, chiefly office equipment and computers. Japan has a strong service sector, banking and finance being particularly important, and Tōkyō has one of the world's major stock exchanges. Owing to intensive agricultural production, Japan is seventy per cent self-sufficient in food. The main food crops are rice, barley, fruit, wheat and soya beans. Livestock rearing (chiefly cattle, pigs and chickens) and fishing are also important, and Japan has one of the largest fishing fleets in the world. A major trading nation, Japan has trade links with many countries in southeast Asia and in Europe, although its main trading partner is the USA.

Jersey

United Kingdom Crown Dependency

Area Sq Km	116	**Languages**	English, French
Area Sq Miles	45	**Religions**	Protestant, Roman Catholic
Population	90 800	**Currency**	Pound sterling
Capital	St Helier		

Map page 166 One of the Channel Islands lying off the west coast of the Cherbourg peninsula in northern France. Financial services are the most important part of the economy.

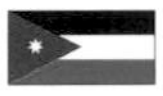

JORDAN

Hashemite Kingdom of Jordan

Area Sq Km	89 206	**Languages**	Arabic
Area Sq Miles	34 443	**Religions**	Sunni Muslim, Christian
Population	6 316 000	**Currency**	Jordanian dinar
Capital	'Ammān	**Organizations**	UN

Map page 136–137

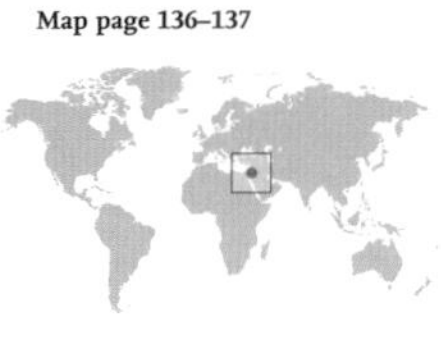

Jordan, in southwest Asia, is landlocked apart from a short coastline on the Gulf of Aqaba. Much of the country is rocky desert plateau. To the west of the mountains, the land falls below sea level to the Dead Sea and the Jordan river. The climate is hot and dry. Most people live in the northwest. Phosphates, potash, pharmaceuticals, fruit and vegetables are the main exports. The tourist industry is important, and the economy relies on workers' remittances from abroad and foreign aid.

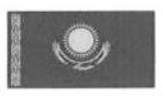

KAZAKHSTAN

Republic of Kazakhstan

Area Sq Km	2 717 300	**Languages**	Kazakh, Russian, Ukrainian, German, Uzbek, Tatar
Area Sq Miles	1 049 155	**Religions**	Sunni Muslim, Russian Orthodox, Protestant
Population	15 637 000	**Currency**	Tenge
Capital	Astana (Akmola)	**Organizations**	CIS, UN

Map page 138–139

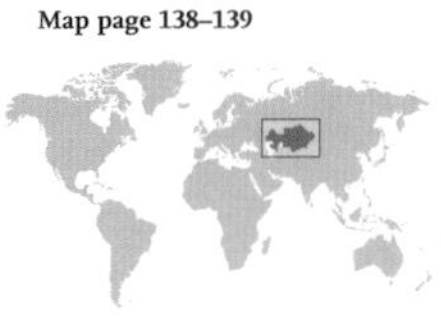

Stretching across central Asia, Kazakhstan covers a vast area of steppe land and semi-desert. The land is flat in the west, with large lowlands around the Caspian Sea, rising to mountains in the southeast. The climate is continental. Agriculture and livestock rearing are important, and cotton and tobacco are the main cash crops. Kazakhstan is very rich in minerals, including coal, chromium, gold, molybdenum, lead and zinc, and has substantial reserves of oil and gas. Mining, metallurgy, machine building and food processing are major industries. Oil, gas and minerals are the main exports, and the Russian Federation is the dominant trading partner.

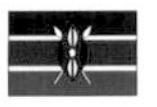

KENYA

Republic of Kenya

Area Sq Km	582 646	**Languages**	Swahili, English, local languages
Area Sq Miles	224 961	**Religions**	Christian, traditional beliefs
Population	39 802 000	**Currency**	Kenyan shilling
Capital	Nairobi	**Organizations**	Comm., UN

Map page 178

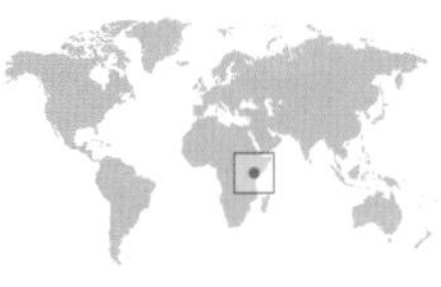

Kenya is in east Africa, on the Indian Ocean. Inland beyond the coastal plains the land rises to plateaus interrupted by volcanic mountains. The Great Rift Valley runs north-south to the west of the capital, Nairobi. Most of the population lives in the central area. Conditions are tropical on the coast, semi-desert in the north and savanna in the south. Hydroelectric power from the Upper Tana river provides most of the country's electricity. Agricultural products, mainly tea, coffee, fruit and vegetables, are the main exports. Light industry is important, and tourism, oil refining and re-exports for landlocked neighbours are major foreign exchange earners.

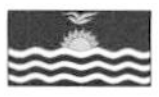

KIRIBATI

Republic of Kiribati

Area Sq Km	717	**Languages**	Gilbertese, English
Area Sq Miles	277	**Religions**	Roman Catholic, Protestant
Population	98 000	**Currency**	Australian dollar
Capital	Bairiki	**Organizations**	Comm., UN

Map page 123

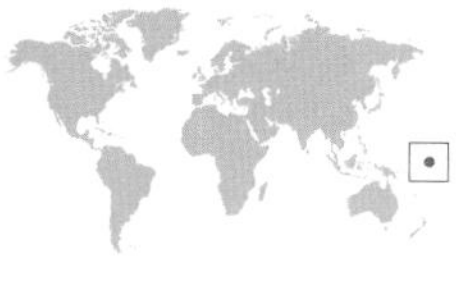

Kiribati, in the Pacific Ocean, straddles the Equator and comprises coral islands in the Gilbert, Phoenix and Line Island groups and the volcanic island of Banaba. Most people live on the Gilbert Islands, and the capital, Bairiki, is on Tarawa island in this group. The climate is hot, and wetter in the north. Copra and fish are exported. Kiribati relies on remittances from workers abroad and foreign aid.

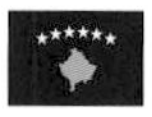

KOSOVO

Republic of Kosovo

Area Sq Km	10 908	**Languages**	Albanian, Serbian
Area Sq Miles	4 212	**Religions**	Sunni Muslim, Serbian Orthodox
Population	2 153 139	**Currency**	Euro
Capital	Prishtinë (Priština)		

Map page 171

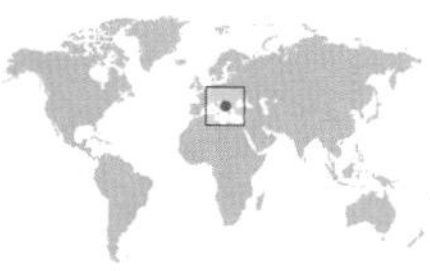

Kosovo, traditionally an autonomous southern province of Serbia, was the focus of ethnic conflict between Serbs and the majority ethnic Albanians in the 1990s until international intervention in 1999, after which it was administered by the UN. Kosovo declared its independence from Serbia in February 2008. The landscape is largely hilly or mountainous, especially along the southern and western borders.

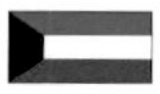

KUWAIT

State of Kuwait

Area Sq Km	17 818	**Languages**	Arabic
Area Sq Miles	6 880	**Religions**	Sunni Muslim, Shi'a Muslim, Christian, Hindu
Population	2 985 000	**Currency**	Kuwaiti dinar
Capital	Kuwait (Al Kuwayt)	**Organizations**	GCC, OPEC, UN

Map page 137

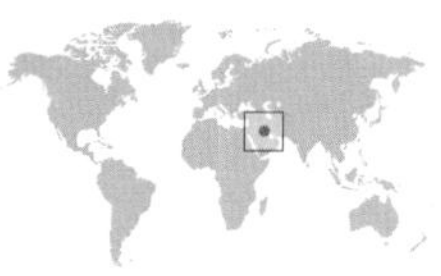

Kuwait lies on the northwest shores of The Gulf in southwest Asia. It is mainly low-lying desert, with irrigated areas along the bay, Kuwait Jun, where most people live. Summers are hot and dry, and winters are cool with some rainfall. The oil industry, which accounts for eighty per cent of exports, has largely recovered from the damage caused by the Gulf War in 1991. Income is also derived from extensive overseas investments. Japan and the USA are the main trading partners.

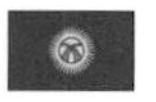

KYRGYZSTAN

Kyrgyz Republic

Area Sq Km	198 500	**Languages**	Kyrgyz, Russian, Uzbek
Area Sq Miles	76 641	**Religions**	Sunni Muslim, Russian Orthodox
Population	5 482 000	**Currency**	Kyrgyz som
Capital	Bishkek (Frunze)	**Organizations**	CIS, UN

Map page 139

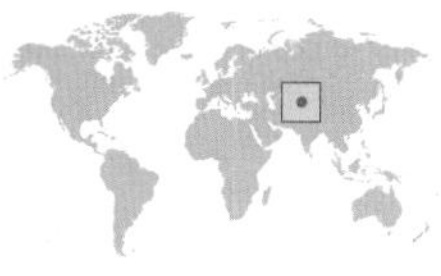

A landlocked central Asian state, Kyrgyzstan is rugged and mountainous, lying to the west of the Tien Shan mountain range. Most of the population lives in the valleys of the north and west. Summers are hot and winters are cold. Agriculture (chiefly livestock farming) is the main activity. Some oil and gas, coal, gold, antimony and mercury are produced. Manufactured goods include machinery, metals and metal products, which are the main exports. Most trade is with Germany, the Russian Federation, Kazakhstan and Uzbekistan.

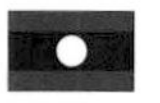

LAOS

Lao People's Democratic Republic

Area Sq Km	236 800	**Languages**	Lao, local languages
Area Sq Miles	91 429	**Religions**	Buddhist, traditional beliefs
Population	6 320 000	**Currency**	Kip
Capital	Vientiane (Viangchan)	**Organizations**	ASEAN, UN

Map page 147

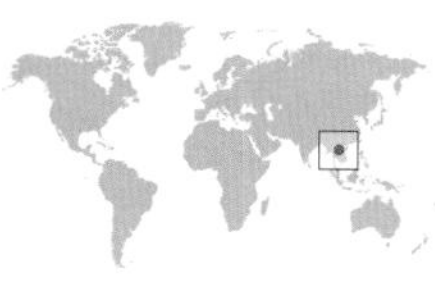

A landlocked country in southeast Asia, Laos is a land of mostly forested mountains and plateaus. The climate is tropical monsoon. Most of the population lives in the Mekong valley and the low plateau in the south, where food crops, chiefly rice, are grown. Hydroelectricity from a plant on the Mekong river, timber, coffee and tin are exported. Laos relies heavily on foreign aid.

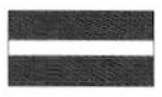

LATVIA

Republic of Latvia

Area Sq Km	64 589	**Languages**	Latvian, Russian
Area Sq Miles	24 938	**Religions**	Protestant, Roman Catholic, Russian Orthodox
Population	2 249 000	**Currency**	Lats
Capital	Rīga	**Organizations**	EU, NATO, UN

Map page 159

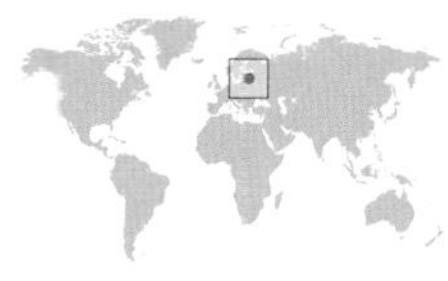

Latvia is in northern Europe, on the Baltic Sea and the Gulf of Riga. The land is flat near the coast but hilly with woods and lakes inland. The country has a modified continental climate. One-third of the people live in the capital, Rīga. Crop and livestock farming are important. There are few natural resources. Industries and main exports include food products, transport equipment, wood and wood products and textiles. The main trading partners are the Russian Federation and Germany. Latvia joined the European Union in May 2004.

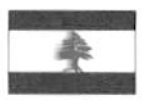

LEBANON

Republic of Lebanon

Area Sq Km	10 452	**Languages**	Arabic, Armenian, French
Area Sq Miles	4 036	**Religions**	Shi'a Muslim, Sunni Muslim, Christian
Population	4 224 000	**Currency**	Lebanese pound
Capital	Beirut (Beyrouth)	**Organizations**	UN

Map page 136

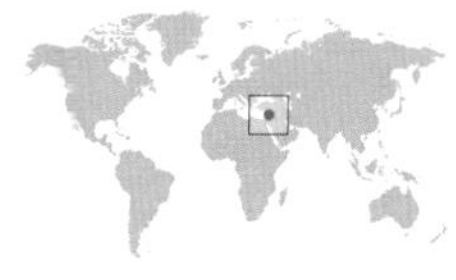

Lebanon lies on the Mediterranean coast of southwest Asia. Beyond the coastal strip, where most of the population lives, are two parallel mountain ranges, separated by the Bekaa Valley (El Beq'a). The economy and infrastructure have been recovering since the 1975–1991 civil war crippled the traditional sectors of financial services and tourism. Italy, France and the UAE are the main trading partners.

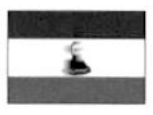

LESOTHO

Kingdom of Lesotho

Area Sq Km	30 355	**Languages**	Sesotho, English, Zulu
Area Sq Miles	11 720	**Religions**	Christian, traditional beliefs
Population	2 067 000	**Currency**	Loti, South African rand
Capital	Maseru	**Organizations**	Comm., SADC, UN

Map page 181

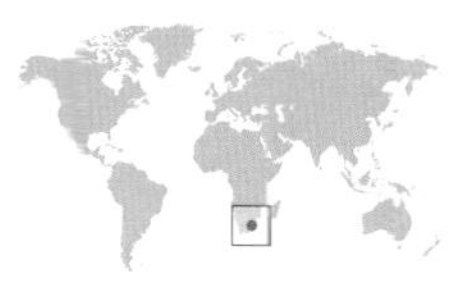

Lesotho is a landlocked state surrounded by the Republic of South Africa. It is a mountainous country lying within the Drakensberg mountain range. Farming and herding are the main activities. The economy depends heavily on South Africa for transport links and employment. A major hydroelectric plant completed in 1998 allows the sale of water to South Africa. Exports include manufactured goods (mainly clothing and road vehicles), food, live animals, wool and mohair.

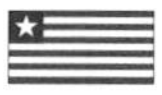

LIBERIA

Republic of Liberia

Area Sq Km	111 369	**Languages**	English, Creole, local languages
Area Sq Miles	43 000	**Religions**	Traditional beliefs, Christian, Sunni Muslim
Population	3 955 000	**Currency**	Liberian dollar
Capital	Monrovia	**Organizations**	UN

Map page 176

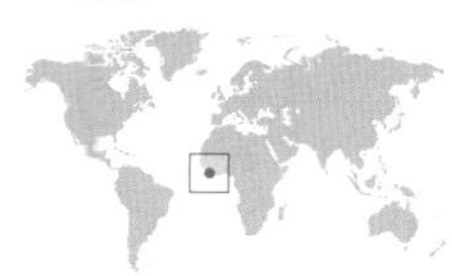

Liberia is on the Atlantic coast of west Africa. Beyond the coastal belt of sandy beaches and mangrove swamps the land rises to a forested plateau and highlands along the Guinea border. A quarter of the population lives along the coast. The climate is hot with heavy rainfall. Liberia is rich in mineral resources and forests. The economy is based on the production and export of basic products. Exports include diamonds, iron ore, rubber and timber. Liberia has a huge international debt and relies heavily on foreign aid.

LIBYA

Great Socialist People's Libyan Arab Jamahiriya

Area Sq Km	1 759 540	**Languages**	Arabic, Berber
Area Sq Miles	679 362	**Religions**	Sunni Muslim
Population	6 420 000	**Currency**	Libyan dinar
Capital	Tripoli (Tarābulus)	**Organizations**	OPEC, UN

Map page 176–177

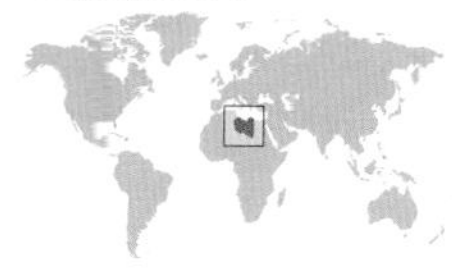

Libya lies on the Mediterranean coast of north Africa. The desert plains and hills of the Sahara dominate the landscape and the climate is hot and dry. Most of the population lives in cities near the coast, where the climate is cooler with moderate rainfall. Farming and herding, chiefly in the northwest, are important but the main industry is oil. Libya is a major producer, and oil accounts for virtually all of its export earnings. Italy and Germany are the main trading partners.

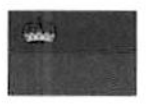

LIECHTENSTEIN

Principality of Liechtenstein

Area Sq Km	160	**Languages**	German
Area Sq Miles	62	**Religions**	Roman Catholic, Protestant
Population	36 000	**Currency**	Swiss franc
Capital	Vaduz	**Organizations**	UN

Map page 166

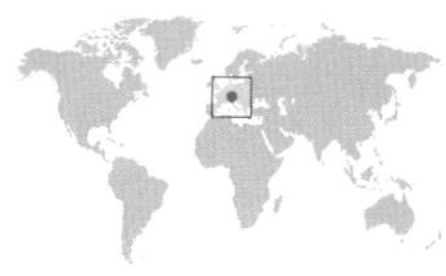

A landlocked state between Switzerland and Austria, Liechtenstein has an industrialized, free-enterprise economy. Low business taxes have attracted companies to establish offices which provide approximately one-third of state revenues. Banking is also important. Major products include precision instruments, ceramics and textiles.

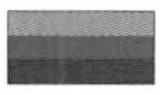

LITHUANIA

Republic of Lithuania

Area Sq Km	65 200	**Languages**	Lithuanian, Russian, Polish
Area Sq Miles	25 174	**Religions**	Roman Catholic, Protestant, Russian Orthodox
Population	3 287 000	**Currency**	Litas
Capital	Vilnius	**Organizations**	EU, NATO, UN

Map page 159

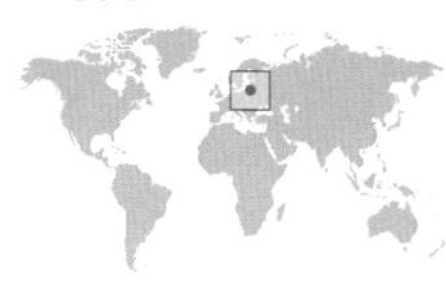

Lithuania is in northern Europe on the eastern shores of the Baltic Sea. It is mainly lowland with many lakes, rivers and marshes. Agriculture, fishing and forestry are important, but manufacturing dominates the economy. The main exports are machinery, mineral products and chemicals. The Russian Federation and Germany are the main trading partners. Lithuania joined the European Union in May 2004.

LUXEMBOURG

Grand Duchy of Luxembourg

Area Sq Km	2 586	**Languages**	Letzeburgish, German, French
Area Sq Miles	998	**Religions**	Roman Catholic
Population	486 000	**Currency**	Euro
Capital	Luxembourg	**Organizations**	EU, NATO, OECD, UN

Map page 164

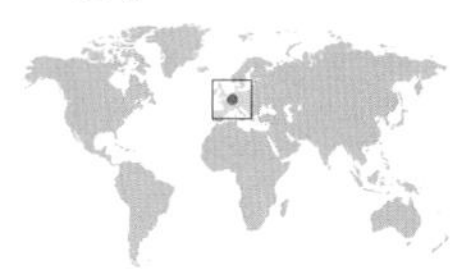

Luxembourg, a small landlocked country in western Europe, borders Belgium, France and Germany. The hills and forests of the Ardennes dominate the north, with rolling pasture to the south, where the main towns, farms and industries are found. The iron and steel industry is still important, but light industries (including textiles, chemicals and food products) are growing. Luxembourg is a major banking centre. Main trading partners are Belgium, Germany and France.

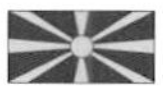

MACEDONIA (F.Y.R.O.M.)

Republic of Macedonia

Area Sq Km	25 713	**Languages**	Macedonian, Albanian, Turkish
Area Sq Miles	9 928	**Religions**	Macedonian Orthodox, Sunni Muslim
Population	2 042 000	**Currency**	Macedonian denar
Capital	Skopje	**Organizations**	NATO, UN

Map page 171

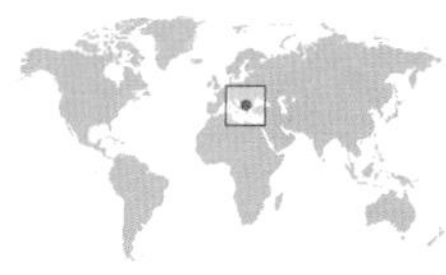

The Former Yugoslav Republic of Macedonia is a landlocked state in southern Europe. Lying within the southern Balkan Mountains, it is traversed northwest-southeast by the Vardar valley. The climate is continental. The economy is based on industry, mining and agriculture, but conflicts in the region have reduced trade and caused economic difficulties. Foreign aid and loans are now assisting in modernization and development of the country.

© Collins Bartholomew Ltd

MADAGASCAR
Republic of Madagascar

Area Sq Km	587 041	Languages	Malagasy, French
Area Sq Miles	226 658	Religions	Traditional beliefs, Christian, Sunni Muslim
Population	19 625 000		
Capital	Antananarivo	Currency	Malagasy franc
		Organizations	SADC, UN

Map page 179

Madagascar lies off the east coast of southern Africa. The world's fourth largest island, it is mainly a high plateau, with a coastal strip to the east and scrubby plain to the west. The climate is tropical, with heavy rainfall in the north and east. Most of the population lives on the plateau. Although the amount of arable land is limited, the economy is based on agriculture. The main industries are agricultural processing, textile manufacturing and oil refining. Foreign aid is important. Exports include coffee, vanilla, cotton cloth, sugar and shrimps. France is the main trading partner.

MALAWI
Republic of Malawi

Area Sq Km	118 484	Languages	Chichewa, English, local languages
Area Sq Miles	45 747	Religions	Christian, traditional beliefs, Sunni Muslim
Population	15 263 000		
Capital	Lilongwe	Currency	Malawian kwacha
		Organizations	Comm., SADC, UN

Map page 179

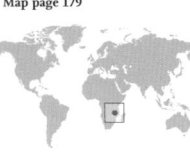

Landlocked Malawi in central Africa is a narrow hilly country at the southern end of the Great Rift Valley. One-fifth is covered by Lake Nyasa. Most of the population lives in rural areas in the southern regions. The climate is mainly subtropical, with varying rainfall. The economy is predominantly agricultural, with tobacco, tea and sugar the main exports. Malawi is one of the world's least developed countries and relies heavily on foreign aid. South Africa is the main trading partner.

MALAYSIA
Federation of Malaysia

Area Sq Km	332 965	Languages	Malay, English, Chinese, Tamil, local languages
Area Sq Miles	128 559		
Population	27 468 000	Religions	Sunni Muslim, Buddhist, Hindu, Christian, traditional beliefs
Capital	Kuala Lumpur/ Putrajaya		
		Currency	Ringgit
		Organizations	APEC, ASEAN, Comm., UN

Map page 155

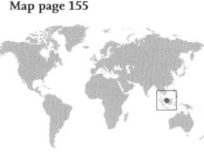

Malaysia, in southeast Asia, comprises two regions, separated by the South China Sea. The western region occupies the southern Malay Peninsula, which has a chain of mountains dividing the eastern coastal strip from wider plains to the west. East Malaysia, consisting of the states of Sabah and Sarawak in the north of the island of Borneo, is mainly rainforest-covered hills and mountains with mangrove swamps along the coast. Both regions have a tropical climate with heavy rainfall. About eighty per cent of the population lives in Peninsular Malaysia. The country is rich in natural resources and has reserves of minerals and fuels. It is an important producer of tin, oil, natural gas and tropical hardwoods. Agriculture remains a substantial part of the economy, but industry is the most important sector. The main exports are transport and electronic equipment, oil, chemicals, palm oil, wood and rubber. The main trading partners are Japan, the USA and Singapore.

MALDIVES
Republic of the Maldives

Area Sq Km	298	Languages	Divehi (Maldivian)
Area Sq Miles	115	Religions	Sunni Muslim
Population	309 000	Currency	Rufiyaa
Capital	Male	Organizations	Comm., UN

Map page 130

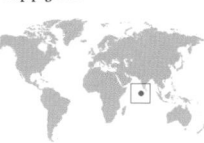

The Maldive archipelago comprises over a thousand coral atolls (around two hundred of which are inhabited) in the Indian Ocean, southwest of India. Over eighty per cent of the land area is less than one metre above sea level. The main atolls are North and South Male and Addu. The climate is hot, humid and monsoonal. There is little cultivation and almost all food is imported. Tourism has expanded rapidly and is the most important sector of the economy.

MALI
Republic of Mali

Area Sq Km	1 240 140	Languages	French, Bambara, local languages
Area Sq Miles	478 821	Religions	Sunni Muslim, traditional beliefs, Christian
Population	13 010 000		
Capital	Bamako	Currency	CFA franc
		Organizations	UN

Map page 176

A landlocked state in west Africa, Mali is low-lying, with a few rugged hills in the northeast. Northern regions lie within the Sahara desert. To the south, around the Niger river, are marshes and savanna grassland. Rainfall is unreliable. Most of the population lives along the Niger and Falémé rivers. Exports include cotton, livestock and gold. Mali is one of the least developed countries in the world and relies heavily on foreign aid.

MALTA
Republic of Malta

Area Sq Km	316	Languages	Maltese, English
Area Sq Miles	122	Religions	Roman Catholic
Population	409 000	Currency	Euro
Capital	Valletta	Organizations	Comm., EU, UN

Map page 170

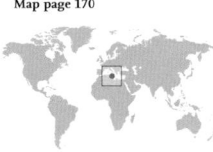

The islands of Malta and Gozo lie in the Mediterranean Sea, off the coast of southern Italy. The islands have hot, dry summers and mild winters. The economy depends on foreign trade, tourism and the manufacture of electronics and textiles. Main trading partners are the USA, France and Italy. Malta joined the European Union in May 2004.

MARSHALL ISLANDS
Republic of the Marshall Islands

Area Sq Km	181	Languages	English, Marshallese
Area Sq Miles	70	Religions	Protestant, Roman Catholic
Population	62 000	Currency	United States dollar
Capital	Delap-Uliga-Djarrit	Organizations	UN

Map page 123

The Marshall Islands consist of over a thousand atolls, islands and islets, within two chains in the north Pacific Ocean. The main atolls are Majuro (home to half the population), Kwajalein, Jaluit, Enewetak and Bikini. The climate is tropical, with heavy autumn rainfall. About half the workforce is employed in farming or fishing. Tourism is a small source of foreign exchange and the islands depend heavily on aid from the USA.

Martinique
French Overseas Department

Area Sq Km	1 079	Languages	French, Creole
Area Sq Miles	417	Religions	Roman Catholic, traditional beliefs
Population	405 000	Currency	Euro
Capital	Fort-de-France		

Map page 205 — Martinique, one of the Caribbean Windward Islands, has volcanic peaks in the north, a populous central plain, and hills and beaches in the south. Tourism is a major source of foreign exchange, and substantial aid is received from France. The main trading partners are France and Guadeloupe.

MAURITANIA
Islamic Arab and African Republic of Mauritania

Area Sq Km	1 030 700	Languages	Arabic, French, local languages
Area Sq Miles	397 955	Religions	Sunni Muslim
Population	3 291 000	Currency	Ouguiya
Capital	Nouakchott	Organizations	UN

Map page 176

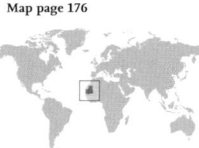

Mauritania is on the Atlantic coast of northwest Africa and lies almost entirely within the Sahara desert. Oases and a fertile strip along the Senegal river to the south are the only areas suitable for cultivation. The climate is generally hot and dry. About a quarter of Mauritanians live in the capital, Nouakchott. Most of the workforce depends on livestock rearing and subsistence farming. There are large deposits of iron ore which account for more than half of total exports. Mauritania's coastal waters are among the richest fishing grounds in the world. The main trading partners are France, Japan and Italy.

MAURITIUS
Republic of Mauritius

Area Sq Km	2 040	Languages	English, Creole, Hindi, Bhojpurī, French
Area Sq Miles	788		
Population	1 288 000	Religions	Hindu, Roman Catholic, Sunni Muslim
Capital	Port Louis		
		Currency	Mauritius rupee
		Organizations	Comm., SADC, UN

Map page 175

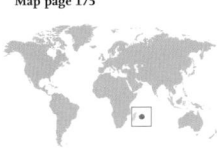

The state comprises Mauritius, Rodrigues and some twenty small islands in the Indian Ocean, east of Madagascar. The main island of Mauritius is volcanic in origin and has a coral coast, rising to a central plateau. Most of the population lives on the north and west sides of the island. The climate is warm and humid. The economy is based on sugar production, light manufacturing (chiefly clothing) and tourism.

Mayotte
French Departmental Collectivity

Area Sq Km	373	Languages	French, Mahorian
Area Sq Miles	144	Religions	Sunni Muslim, Christian
Population	194 000	Currency	Euro
Capital	Dzaoudzi		

Map page 179

Lying in the Indian Ocean off the east coast of central Africa, Mayotte is geographically part of the Comoro archipelago. The economy is based on agriculture, but Mayotte depends heavily on aid from France.

MEXICO
United Mexican States

Area Sq Km	1 972 545	Languages	Spanish, Amerindian languages
Area Sq Miles	761 604	Religions	Roman Catholic, Protestant
Population	109 610 000	Currency	Mexican peso
Capital	Mexico City (México)	Organizations	APEC, OECD, UN

Map page 206–207

The largest country in Central America, Mexico extends south from the USA to Guatemala and Belize, and from the Pacific Ocean to the Gulf of Mexico. The greater part of the country is high plateau flanked by the western and eastern ranges of the Sierra Madre mountains. The principal lowland is the Yucatán peninsula in the southeast. The climate varies with latitude and altitude: hot and humid in the lowlands, warm on the plateau and cool with cold winters in the mountains. The north is arid, while the far south has heavy rainfall. Mexico City is the second largest conurbation in the world and the country's centre of trade and industry. Agriculture involves a fifth of the workforce; crops include grains, coffee, cotton and vegetables. Mexico is rich in minerals, including copper, zinc, lead, tin, sulphur, and silver. It is one of the world's largest producers of oil, from vast reserves in the Gulf of Mexico. The oil and petrochemical industries still dominate the economy, but a variety of manufactured goods are produced, including iron and steel, motor vehicles, textiles, chemicals and food and tobacco products. Tourism is growing in importance. Over three-quarters of all trade is with the USA.

MICRONESIA, FEDERATED STATES OF

Area Sq Km	701	Languages	English, Chuukese, Pohnpeian, local languages
Area Sq Miles	271		
Population	111 000	Religions	Roman Catholic, Protestant
Capital	Palikir	Currency	United States dollar
		Organizations	UN

Map page 122–123

Micronesia comprises over six hundred atolls and islands of the Caroline Islands in the north Pacific Ocean. A third of the population lives on Pohnpei. The climate is tropical, with heavy rainfall. Fishing and subsistence farming are the main activities. Fish, garments and bananas are the main exports. Income is also derived from tourism and the licensing of foreign fishing fleets. The islands depend heavily on aid from the USA.

MOLDOVA
Republic of Moldova

Area Sq Km	33 700	Languages	Romanian, Ukrainian, Gagauz, Russian
Area Sq Miles	13 012	Religions	Romanian Orthodox, Russian Orthodox
Population	3 604 000	Currency	Moldovan leu
Capital	Chişinău (Kishinev)	Organizations	CIS, UN

Map page 173

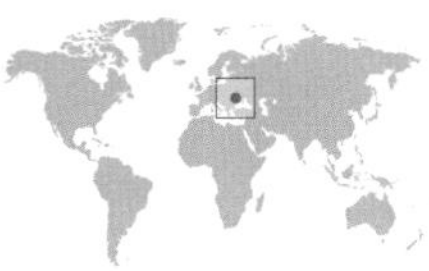

Moldova lies between Romania and Ukraine in eastern Europe. It consists of hilly steppe land, drained by the Prut and Dniester rivers. Moldova has no mineral resources, and the economy is mainly agricultural, with sugar beet, tobacco, wine and fruit the chief products. Food processing, machinery and textiles are the main industries. The Russian Federation is the main trading partner.

MONACO
Principality of Monaco

Area Sq Km	2	Languages	French, Monégasque, Italian
Area Sq Miles	1	Religions	Roman Catholic
Population	33 000	Currency	Euro
Capital	Monaco-Ville	Organizations	UN

Map page 166

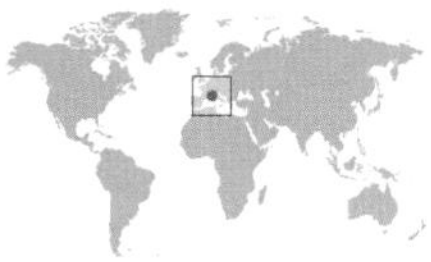

The principality occupies a rocky peninsula and a strip of land on France's Mediterranean coast. Monaco's economy depends on service industries (chiefly tourism, banking and finance) and light industry.

MONGOLIA

Area Sq Km	1 565 000	Languages	Khalka (Mongolian), Kazakh, local languages
Area Sq Miles	604 250	Religions	Buddhist, Sunni Muslim
Population	2 671 000	Currency	Tugrik (tögrög)
Capital	Ulan Bator (Ulaanbaatar)	Organizations	UN

Map page 146

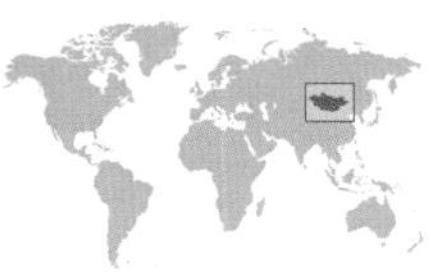

Mongolia is a landlocked country in eastern Asia between the Russian Federation and China. Much of it is high steppe land, with mountains and lakes in the west and north. In the south is the Gobi Desert. Mongolia has long, cold winters and short, mild summers. A quarter of the population lives in the capital, Ulaanbaatar. Livestock breeding and agricultural processing are important. There are substantial mineral resources. Copper and textiles are the main exports. China and the Russian Federation are the main trading partners.

MONTENEGRO
Republic of Montenegro

Area Sq Km	13 812	Languages	Serbian (Montenegrin), Albanian
Area Sq Miles	5 333	Religions	Montenegrin Orthodox, Sunni Muslim
Population	624 000	Currency	Euro
Capital	Podgorica	Organizations	UN

Map page 171

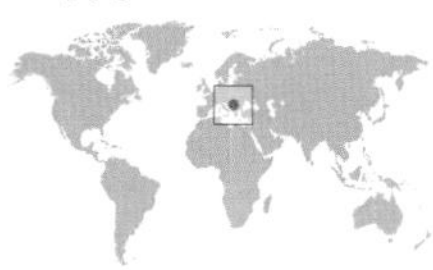

Montenegro was the last constituent republic of the former Yugoslavia to become an independent nation, in June 2006. At that time it opted to split from the state union of Serbia and Montenegro. Montenegro separates the much larger Serbia from the Adriatic coast. The landscape is rugged and mountainous, and the climate Mediterranean.

Montserrat
United Kingdom Overseas Territory

Area Sq Km	100	Languages	English
Area Sq Miles	39	Religions	Protestant, Roman Catholic
Population	4 655	Currency	East Caribbean dollar
Capital	Brades (temporary capital)	Organizations	CARICOM

Map page 205

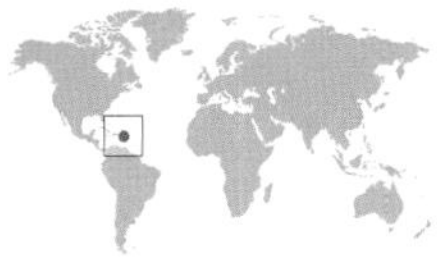

An island in the Leeward Islands group in the Lesser Antilles, in the Caribbean. From 1995 to 1997 the volcanoes in the Soufrière Hills erupted for the first time since 1630. Over sixty per cent of the island was covered in volcanic ash and Plymouth, the capital was, virtually destroyed. Many people emigrated, and the remaining population moved to the north of the island. Brades has replaced Plymouth as the temporary capital. Reconstruction is being funded by aid from the UK.

MOROCCO
Kingdom of Morocco

Area Sq Km	446 550	Languages	Arabic, Berber, French
Area Sq Miles	172 414	Religions	Sunni Muslim
Population	31 993 000	Currency	Moroccan dirham
Capital	Rabat	Organizations	UN

Map page 176

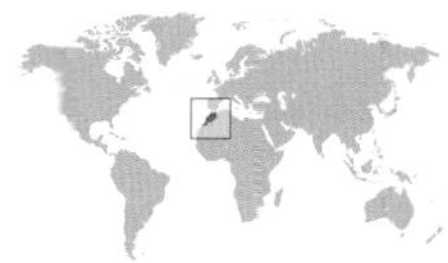

Lying in the northwest of Africa, Morocco has both Atlantic and Mediterranean coasts. The Atlas Mountains separate the arid south and disputed region of western Sahara from the fertile west and north, which have a milder climate. Most Moroccans live on the Atlantic coastal plain. The economy is based on agriculture, phosphate mining and tourism; the most important industries are food processing, textiles and chemicals.

MOZAMBIQUE
Republic of Mozambique

Area Sq Km	799 380	Languages	Portuguese, Makua, Tsonga, local languages
Area Sq Miles	308 642	Religions	Traditional beliefs, Roman Catholic, Sunni Muslim
Population	22 894 000	Currency	Metical
Capital	Maputo	Organizations	Comm., SADC, UN

Map page 179

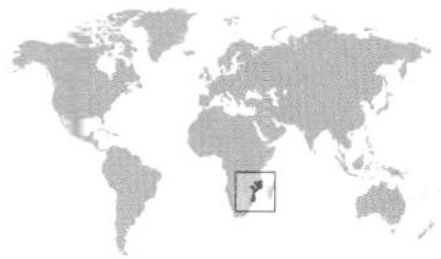

Mozambique lies on the east coast of southern Africa. The land is mainly a savanna plateau drained by the Zambezi and Limpopo rivers, with highlands to the north. Most of the population lives on the coast or in the river valleys. In general the climate is tropical with winter rainfall, but droughts occur. The economy is based on subsistence agriculture. Exports include shrimps, cashews, cotton and sugar, but Mozambique relies heavily on aid, and remains one of the least developed countries in the world.

MYANMAR (Burma)
Union of Myanmar

Area Sq Km	676 577	Languages	Burmese, Shan, Karen, local languages
Area Sq Miles	261 228	Religions	Buddhist, Christian, Sunni Muslim
Population	50 020 000	Currency	Kyat
Capital	Rangoon (Yangôn)/ Nay Pyi Taw	Organizations	ASEAN, UN

Map page 147

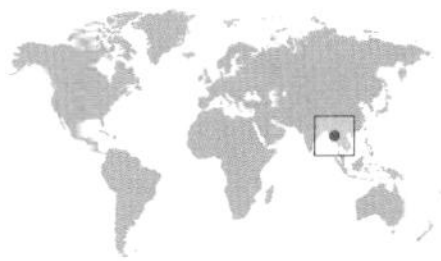

Myanmar (Burma) is in southeast Asia, bordering the Bay of Bengal and the Andaman Sea. Most of the population lives in the valley and delta of the Irrawaddy river, which is flanked by mountains and high plateaus. The climate is hot and monsoonal, and rainforest covers much of the land. Most of the workforce is employed in agriculture. Myanmar is rich in minerals, including zinc, lead, copper and silver. Political and social unrest and lack of foreign investment have affected economic development.

NAMIBIA
Republic of Namibia

Area Sq Km	824 292	Languages	English, Afrikaans, German, Ovambo, local languages
Area Sq Miles	318 261	Religions	Protestant, Roman Catholic
Population	2 171 000	Currency	Namibian dollar
Capital	Windhoek	Organizations	Comm., SADC, UN

Map page 179

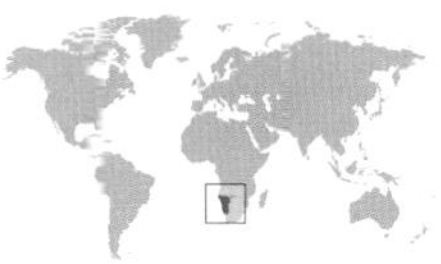

Namibia lies on the southern Atlantic coast of Africa. Mountain ranges separate the coastal Namib Desert from the interior plateau, bordered to the south and east by the Kalahari Desert. The country is hot and dry, but some summer rain in the north supports crops and livestock. Employment is in agriculture and fishing, although the economy is based on mineral extraction – diamonds, uranium, lead, zinc and silver. The economy is closely linked to the Republic of South Africa.

NAURU
Republic of Nauru

Area Sq Km	21	Languages	Nauruan, English
Area Sq Miles	8	Religions	Protestant, Roman Catholic
Population	10 000	Currency	Australian dollar
Capital	Yaren	Organizations	Comm., UN

Map page 125

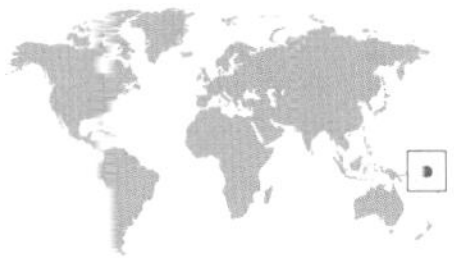

Nauru is a coral island near the Equator in the Pacific Ocean. It has a fertile coastal strip and a barren central plateau. The climate is tropical. The economy is based on phosphate mining, but reserves are near exhaustion and replacement of this income is a serious long-term problem.

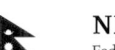

NEPAL
Federal Democratic Republic of Nepal

Area Sq Km	147 181	Languages	Nepali, Maithili, Bhojpuri, English, local languages
Area Sq Miles	56 827	Religions	Hindu, Buddhist, Sunni Muslim
Population	29 331 000	Currency	Nepalese rupee
Capital	Kathmandu	Organizations	UN

Map page 144–145

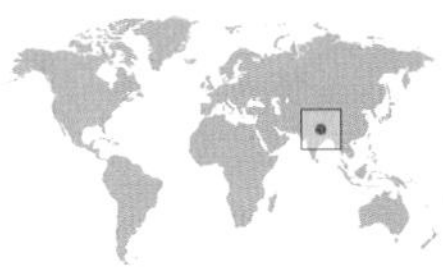

Nepal lies in the eastern Himalaya mountains between India and China. High mountains (including Everest) dominate the north. Most people live in the temperate central valleys and subtropical southern plains. The economy is based largely on agriculture and forestry. There is some manufacturing, chiefly of textiles and carpets, and tourism is important. Nepal relies heavily on foreign aid.

NETHERLANDS
Kingdom of the Netherlands

Area Sq Km	41 526	Languages	Dutch, Frisian
Area Sq Miles	16 033	Religions	Roman Catholic, Protestant, Sunni Muslim
Population	16 592 000	Currency	Euro
Capital	Amsterdam/ The Hague ('s-Gravenhage)	Organizations	EU, NATO, OECD, UN

Map page 164

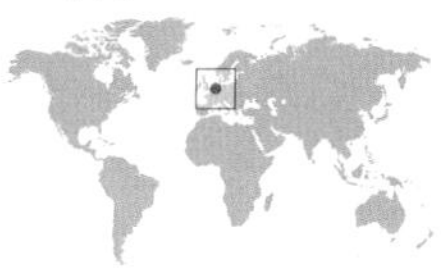

The Netherlands lies on the North Sea coast of western Europe. Apart from low hills in the far southeast, the land is flat and low-lying, much of it below sea level. The coastal region includes the delta of five rivers and polders (reclaimed land), protected by sand dunes, dykes and canals. The climate is temperate, with cool summers and mild winters. Rainfall is spread evenly throughout the year. The Netherlands is a densely populated and highly urbanized country, with the majority of the population living in the cities of Amsterdam, Rotterdam and The Hague. Horticulture and dairy farming are important activities, although they employ less than four per cent of the workforce. The Netherlands ranks as the world's third agricultural exporter, and is a leading producer and exporter of natural gas from reserves in the North Sea. The economy is based mainly on international trade and manufacturing industry. The main industries produce food products, chemicals, machinery, electrical and electronic goods and transport equipment. Germany is the main trading partner, followed by other European Union countries.

Netherlands Antilles
Self-governing Netherlands Territory

Area Sq Km	800	Languages	Dutch, Papiamento, English
Area Sq Miles	309	Religions	Roman Catholic, Protestant
Population	198 000	Currency	Netherlands Antilles guilder
Capital	Willemstad		

Map page 213

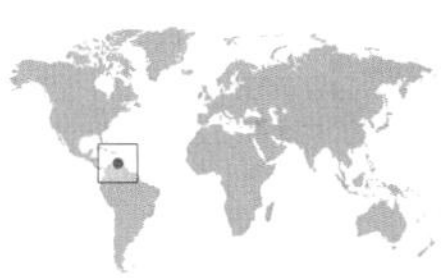

The territory comprises two separate island groups: Curaçao and Bonaire off the northern coast of Venezuela, and Saba, Sint Eustatius and the southern part of St Martin (Sint Maarten) in the Lesser Antilles. Tourism, oil refining and offshore finance are the mainstays of the economy. The main trading partners are the USA, Venezuela and Mexico.

New Caledonia
French Overseas Collectivity

Area Sq Km	19 058	Languages	French, local languages
Area Sq Miles	7 358	Religions	Roman Catholic, Protestant, Sunni Muslim
Population	250 000	Currency	CFP franc
Capital	Nouméa		

Map page 125

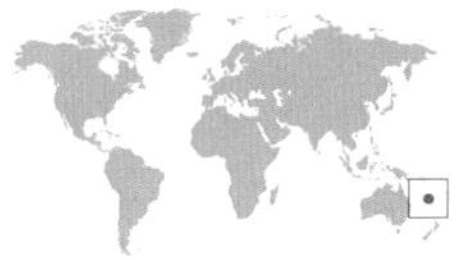

An island group lying in the southwest Pacific, with a sub-tropical climate. New Caledonia has over one-fifth of the world's nickel reserves, and the main economic activity is metal mining. Tourism is also important. New Caledonia relies on aid from France.

© Collins Bartholomew Ltd

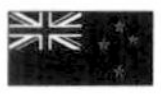

NEW ZEALAND

Area Sq Km	270 534	Languages	English, Maori
Area Sq Miles	104 454	Religions	Protestant, Roman Catholic
Population	4 266 000	Currency	New Zealand dollar
Capital	Wellington	Organizations	APEC, Comm., OECD, UN

Map page 128

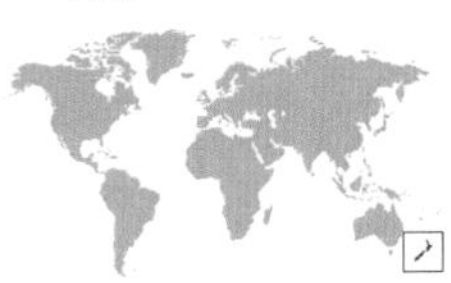

New Zealand comprises two main islands separated by the narrow Cook Strait, and a number of smaller islands. North Island, where three-quarters of the population lives, has mountain ranges, broad fertile valleys and a central plateau with hot springs and active volcanoes. South Island is also mountainous, with the Southern Alps running its entire length. The only major lowland area is the Canterbury Plains in the centre-east. The climate is generally temperate, although South Island has colder winters. Farming is the mainstay of the economy. New Zealand is one of the world's leading producers of meat (beef, lamb and mutton), wool and dairy products; fruit and fish are also important. Hydroelectric and geothermal power provide much of the country's energy needs. Other industries produce timber, wood pulp, iron, aluminium, machinery and chemicals. Tourism is the fastest growing sector of the economy. The main trading partners are Australia, the USA and Japan.

NICARAGUA

Republic of Nicaragua

Area Sq Km	130 000	Languages	Spanish, Amerindian languages
Area Sq Miles	50 193	Religions	Roman Catholic, Protestant
Population	5 743 000	Currency	Córdoba
Capital	Managua	Organizations	UN

Map page 206

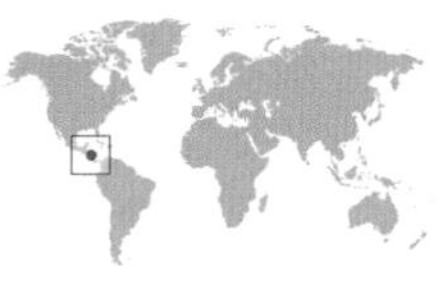

Nicaragua lies at the heart of Central America, with both Pacific and Caribbean coasts. Mountain ranges separate the east, which is largely rainforest, from the more developed western regions, which include Lake Nicaragua and some active volcanoes. The highest land is in the north. The climate is tropical. Nicaragua is one of the western hemisphere's poorest countries, and the economy is largely agricultural. Exports include coffee, seafood, cotton and bananas. The USA is the main trading partner. Nicaragua has a huge national debt, and relies heavily on foreign aid.

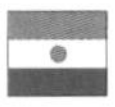

NIGER

Republic of Niger

Area Sq Km	1 267 000	Languages	French, Hausa, Fulani, local languages
Area Sq Miles	489 191	Religions	Sunni Muslim, traditional beliefs
Population	15 290 000	Currency	CFA franc
Capital	Niamey	Organizations	UN

Map page 176–177

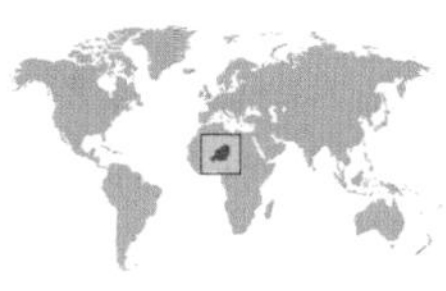

A landlocked state of west Africa, Niger lies mostly within the Sahara desert, but with savanna in the south and in the Niger valley area. The mountains of the Massif de l'Aïr dominate central regions. Much of the country is hot and dry. The south has some summer rainfall, although droughts occur. The economy depends on subsistence farming and herding, and uranium exports, but Niger is one of the world's least developed countries and relies heavily on foreign aid. France is the main trading partner.

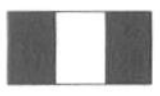

NIGERIA

Federal Republic of Nigeria

Area Sq Km	923 768	Languages	English, Hausa, Yoruba, Ibo, Fulani, local languages
Area Sq Miles	356 669	Religions	Sunni Muslim, Christian, traditional beliefs
Population	154 729 000	Currency	Naira
Capital	Abuja	Organizations	Comm., OPEC, UN

Map page 176–177

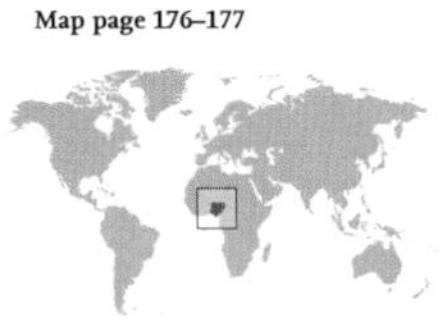

Nigeria is in west Africa, on the Gulf of Guinea, and is the most populous country in Africa. The Niger delta dominates coastal areas, fringed with sandy beaches, mangrove swamps and lagoons. Inland is a belt of rainforest which gives way to woodland or savanna on high plateaus. The far north is the semi-desert edge of the Sahara. The climate is tropical, with heavy summer rainfall in the south but low rainfall in the north. Most of the population lives in the coastal lowlands or in the west. About half the workforce is involved in agriculture, mainly growing subsistence crops. Agricultural production, however, has failed to keep up with demand, and Nigeria is now a net importer of food. Cocoa and rubber are the only significant export crops. The economy is heavily dependent on vast oil resources in the Niger delta and in shallow offshore waters, and oil accounts for over ninety per cent of export earnings. Nigeria also has natural gas reserves and some mineral deposits, but these are largely undeveloped. Industry involves mainly oil refining, chemicals (chiefly fertilizers), agricultural processing, textiles, steel manufacture and vehicle assembly. Political instability in the past has left Nigeria with heavy debts, poverty and unemployment.

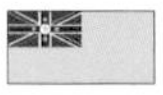

Niue

Self-governing New Zealand Territory

Area Sq Km	258	Languages	English, Nivean
Area Sq Miles	100	Religions	Christian
Population	1 625	Currency	New Zealand dollar
Capital	Alofi		

Map page 125

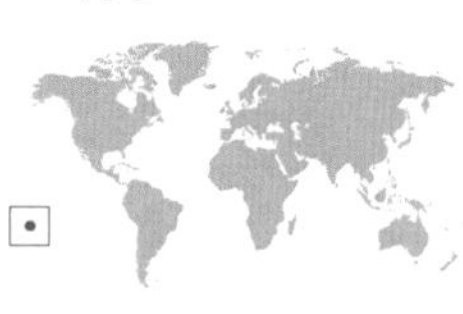

Niue, one of the largest coral islands in the world, lies in the south Pacific Ocean about 500 kilometres (300 miles) east of Tonga. The economy depends on aid and remittances from New Zealand. The population is declining because of migration to New Zealand.

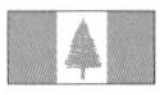

Norfolk Island

Australian External Territory

Area Sq Km	35	Languages	English
Area Sq Miles	14	Religions	Protestant, Roman Catholic
Population	2 523	Currency	Australian dollar
Capital	Kingston		

Map page 125

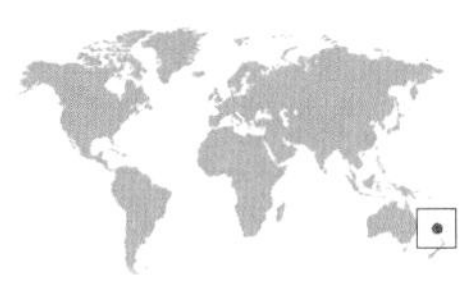

In the south Pacific Ocean, Norfolk Island lies between Vanuatu and New Zealand. Tourism has increased steadily and is the mainstay of the economy and provides revenues for agricultural development.

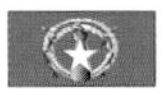

Northern Mariana Islands

United States Commonwealth

Area Sq Km	477	Languages	English, Chamorro, local languages
Area Sq Miles	184	Religions	Roman Catholic
Population	87 000	Currency	United States dollar
Capital	Capitol Hill		

Map page 147

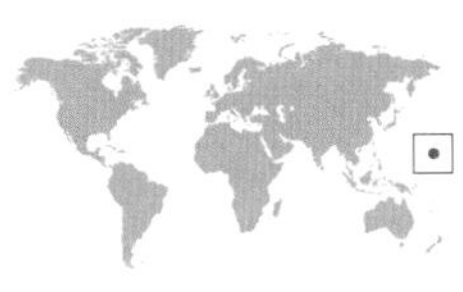

A chain of islands in the northwest Pacific Ocean, extending over 550 kilometres (350 miles) north to south. The main island is Saipan. Tourism is a major industry, employing approximately half the workforce.

NORTH KOREA

Democratic People's Republic of Korea

Area Sq Km	120 538	Languages	Korean
Area Sq Miles	46 540	Religions	Traditional beliefs, Chondoist, Buddhist
Population	23 906 000	Currency	North Korean won
Capital	P'yŏngyang	Organizations	UN

Map page 152

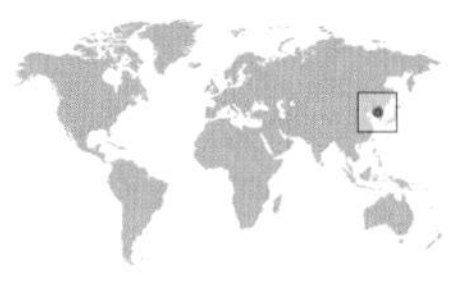

Occupying the northern half of the Korean peninsula in eastern Asia, North Korea is a rugged and mountainous country. The principal lowlands and the main agricultural areas are the plains in the southwest. More than half the population lives in urban areas, mainly on the coastal plains. North Korea has a continental climate, with cold, dry winters and hot, wet summers. Approximately one-third of the workforce is involved in agriculture, mainly growing food crops on cooperative farms. Various minerals, notably iron ore, are mined and are the basis of the country's heavy industries. Exports include minerals (lead, magnesite and zinc) and metal products (chiefly iron and steel). The economy declined after 1991, when ties to the former USSR and eastern bloc collapsed, and there have been serious food shortages.

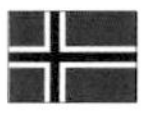

NORWAY

Kingdom of Norway

Area Sq Km	323 878	Languages	Norwegian
Area Sq Miles	125 050	Religions	Protestant, Roman Catholic
Population	4 812 000	Currency	Norwegian krone
Capital	Oslo	Organizations	NATO, OECD, UN

Map page 158–159

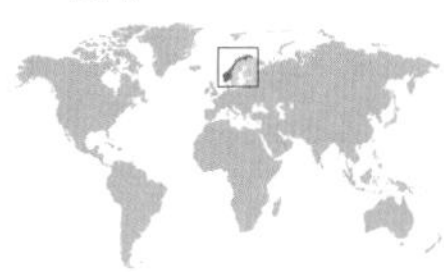

Norway stretches along the north and west coasts of Scandinavia, from the Arctic Ocean to the North Sea. Its extensive coastline is indented with fjords and fringed with many islands. Inland, the terrain is mountainous, with coniferous forests and lakes in the south. The only major lowland areas are along the southern North Sea and Skagerrak coasts, where most of the population lives. The climate is modified by the effect of the North Atlantic Drift ocean current. Norway has vast petroleum and natural gas resources in the North Sea. It is one of western Europe's leading producers of oil and gas, and exports of oil account for approximately half of total export earnings. Related industries include engineering (oil and gas platforms) and petrochemicals. More traditional industries process local raw materials, particularly fish, timber and minerals. Agriculture is limited, but fishing and fish farming are important. Norway is the world's leading exporter of farmed salmon. Merchant shipping and tourism are major sources of foreign exchange.

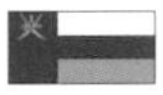

OMAN

Sultanate of Oman

Area Sq Km	309 500	Languages	Arabic, Baluchi, Indian languages
Area Sq Miles	119 499	Religions	Ibadhi Muslim, Sunni Muslim
Population	2 845 000	Currency	Omani riyal
Capital	Muscat (Masqaṭ)	Organizations	GCC, UN

Map page 142

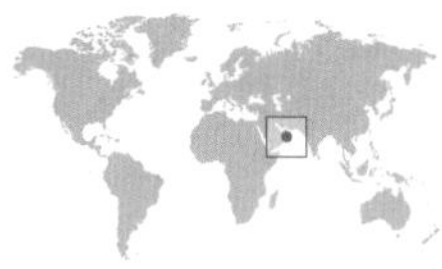

In southwest Asia, Oman occupies the east and southeast coasts of the Arabian Peninsula and an enclave north of the United Arab Emirates. Most of the land is desert, with mountains in the north and south. The climate is hot and mainly dry. Most of the population lives on the coastal strip on the Gulf of Oman. The majority depend on farming and fishing, but the oil and gas industries dominate the economy with around eighty per cent of export revenues coming from oil.

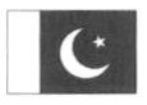

PAKISTAN

Islamic Republic of Pakistan

Area Sq Km	803 940	Languages	Urdu, Punjabi, Sindhi, Pushtu, English
Area Sq Miles	310 403	Religions	Sunni Muslim, Shi'a Muslim, Christian, Hindu
Population	180 808 000	Currency	Pakistani rupee
Capital	Islamabad	Organizations	Comm., UN

Map page 141

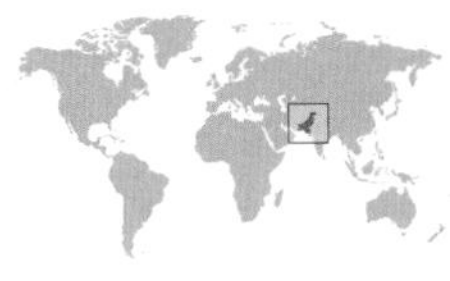

Pakistan is in the northwest part of the Indian subcontinent in south Asia, on the Arabian Sea. The east and south are dominated by the great basin of the Indus river system. This is the main agricultural area and contains most of the predominantly rural population. To the north the land rises to the mountains of the Karakoram, Hindu Kush and Himalaya mountains. The west is semi-desert plateaus and mountain ranges. The climate ranges between dry desert, and arctic tundra on the mountain tops. Temperatures are generally warm and rainfall is monsoonal. Agriculture is the main sector of the economy, employing approximately half of the workforce, and is based on extensive irrigation schemes. Pakistan is one of the world's leading producers of cotton and a major exporter of rice. Pakistan produces natural gas and has a variety of mineral deposits including coal and gold, but they are little developed. The main industries are textiles and clothing manufacture and food processing, with fabrics and ready-made clothing the leading exports. Pakistan also produces leather goods, fertilizers, chemicals, paper and precision instruments. The country depends heavily on foreign aid and remittances from workers abroad.

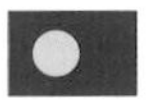

PALAU

Republic of Palau

Area Sq Km	497	Languages	Palauan, English
Area Sq Miles	192	Religions	Roman Catholic, Protestant, traditional beliefs
Population	20 000	Currency	United States dollar
Capital	Melekeok	Organizations	UN

Map page 147

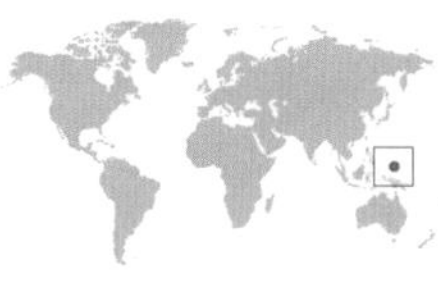

Palau comprises over three hundred islands in the western Caroline Islands, in the west Pacific Ocean. The climate is tropical. The economy is based on farming, fishing and tourism, but Palau is heavily dependent on aid from the USA.

PANAMA

Republic of Panama

Area Sq Km	77 082	**Languages**	Spanish, English, Amerindian languages
Area Sq Miles	29 762	**Religions**	Roman Catholic, Protestant, Sunni Muslim
Population	3 454 000	**Currency**	Balboa
Capital	Panama City (Panamá)	**Organizations**	UN

Map page 206

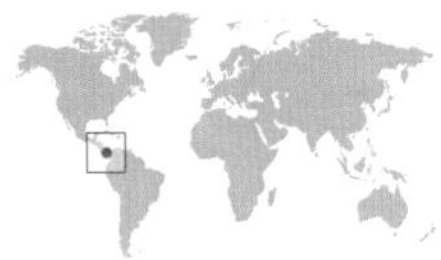

Panama is the most southerly state in central America and has Pacific and Caribbean coasts. It is hilly, with mountains in the west and jungle near the Colombian border. The climate is tropical. Most of the population lives on the drier Pacific side. The economy is based mainly on services related to the Panama Canal: shipping, banking and tourism. Exports include bananas, shrimps, coffee, clothing and fish products. The USA is the main trading partner.

PAPUA NEW GUINEA

Independent State of Papua New Guinea

Area Sq Km	462 840	**Languages**	English, Tok Pisin (Creole), local languages
Area Sq Miles	178 704	**Religions**	Protestant, Roman Catholic, traditional beliefs
Population	6 732 000	**Currency**	Kina
Capital	Port Moresby	**Organizations**	APEC, Comm., UN

Map page 124

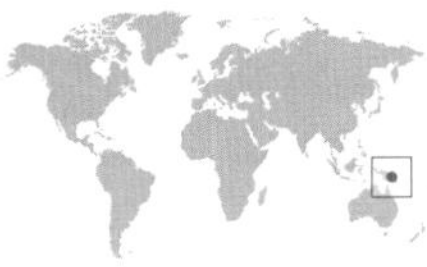

Papua New Guinea occupies the eastern half of the island of New Guinea and includes many island groups. It has a forested and mountainous interior, bordered by swampy plains, and a tropical monsoon climate. Most of the workforce are farmers. Timber, copra, coffee and cocoa are important, but exports are dominated by minerals, chiefly gold and copper. The country depends on foreign aid. Australia, Japan and Singapore are the main trading partners.

PARAGUAY

Republic of Paraguay

Area Sq Km	406 752	**Languages**	Spanish, Guaraní
Area Sq Miles	157 048	**Religions**	Roman Catholic, Protestant
Population	6 349 000	**Currency**	Guaraní
Capital	Asunción	**Organizations**	UN

Map page 212

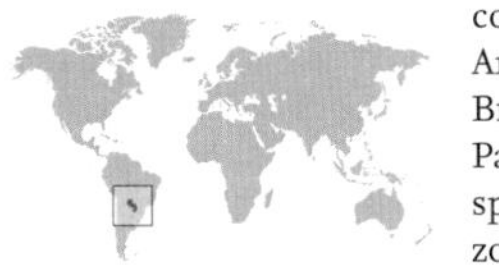

Paraguay is a landlocked country in central South America, bordering Bolivia, Brazil and Argentina. The Paraguay river separates a sparsely populated western zone of marsh and flat alluvial plains from a more developed, hilly and forested region to the east and south. The climate is subtropical. Virtually all electricity is produced by hydroelectric plants, and surplus power is exported to Brazil and Argentina. The hydroelectric dam at Itaipú is one of the largest in the world. The mainstay of the economy is agriculture and related industries. Exports include cotton, soya bean and edible oil products, timber and meat. Brazil and Argentina are the main trading partners.

PERU

Republic of Peru

Area Sq Km	1 285 216	**Languages**	Spanish, Quechua, Aymara
Area Sq Miles	496 225	**Religions**	Roman Catholic, Protestant
Population	29 165 000	**Currency**	Sol
Capital	Lima	**Organizations**	APEC, UN

Map page 210

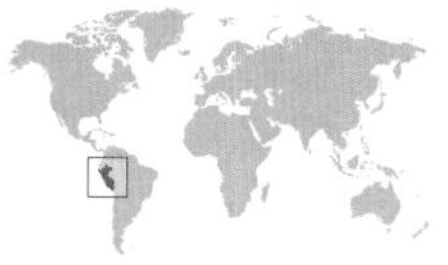

Peru lies on the Pacific coast of South America. Most Peruvians live on the coastal strip and on the plateaus of the high Andes mountains. East of the Andes is the Amazon rainforest. The coast is temperate with low rainfall while the east is hot, humid and wet. Agriculture involves one-third of the workforce and fishing is also important. Agriculture and fishing have both been disrupted by the El Niño climatic effect in recent years. Sugar, cotton, coffee and, illegally, coca are the main cash crops. Copper and copper products, fishmeal, zinc products, coffee, petroleum and its products, and textiles are the main exports. The USA and the European Union are the main trading partners.

PHILIPPINES

Republic of the Philippines

Area Sq Km	300 000	**Languages**	English, Filipino, Tagalog, Cebuano, local languages
Area Sq Miles	115 831	**Religions**	Roman Catholic, Protestant, Sunni Muslim, Aglipayan
Population	91 983 000	**Currency**	Philippine peso
Capital	Manila	**Organizations**	APEC, ASEAN, UN

Map page 153

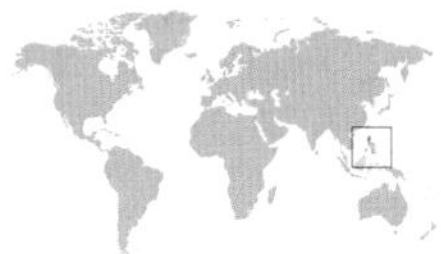

The Philippines, in southeast Asia, consists of over seven thousand islands and atolls lying between the South China Sea and the Pacific Ocean. The islands of Luzon and Mindanao account for two-thirds of the land area. They and nine other fairly large islands are mountainous and forested. There are active volcanoes, and earthquakes and tropical storms are common. Most of the population lives in the plains on the larger islands or on the coastal strips. The climate is hot and humid with heavy monsoonal rainfall. Rice, coconuts, sugar cane, pineapples and bananas are the main agricultural crops, and fishing is also important. Main exports are electronic equipment, machinery and transport equipment, garments and coconut products. Foreign aid and remittances from workers abroad are important to the economy, which faces problems of high population growth rate and high unemployment. The USA and Japan are the main trading partners.

Pitcairn Islands

United Kingdom Overseas Territory

Area Sq Km	45	**Languages**	English
Area Sq Miles	17	**Religions**	Protestant
Population	66	**Currency**	New Zealand dollar
Capital	Adamstown		

Map page 123

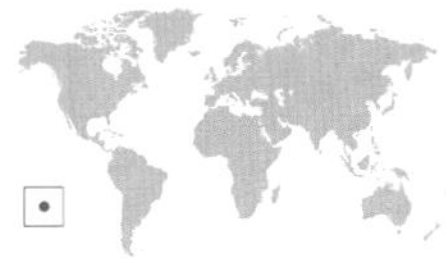

An island group in the southeast Pacific Ocean consisting of Pitcairn Island and three uninhabited islands. It was originally settled by mutineers from HMS *Bounty* in 1790.

POLAND

Polish Republic

Area Sq Km	312 683	**Languages**	Polish, German
Area Sq Miles	120 728	**Religions**	Roman Catholic, Polish Orthodox
Population	38 074 000	**Currency**	Złoty
Capital	Warsaw (Warszawa)	**Organizations**	EU, NATO, OECD, UN

Map page 168–169

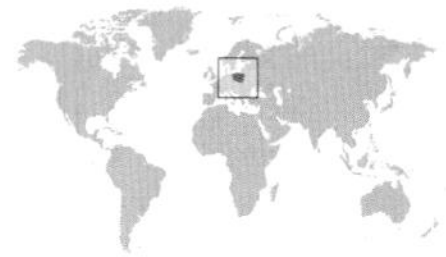

Poland lies on the Baltic coast of eastern Europe. The Oder (Odra) and Vistula (Wisła) river deltas dominate the coast. Inland, much of the country is low-lying, with woods and lakes. In the south the land rises to the Sudeten Mountains and the western part of the Carpathian Mountains, which form the borders with the Czech Republic and Slovakia respectively. The climate is continental. Around a quarter of the workforce is involved in agriculture, and exports include livestock products and sugar. The economy is heavily industrialized, with mining and manufacturing accounting for forty per cent of national income. Poland is one of the world's major producers of coal, and also produces copper, zinc, lead, sulphur and natural gas. The main industries are machinery and transport equipment, shipbuilding, and metal and chemical production. Exports include machinery and transport equipment, manufactured goods, food and live animals. Germany is the main trading partner. Poland joined the European Union in May 2004.

PORTUGAL

Portuguese Republic

Area Sq Km	88 940	**Languages**	Portuguese
Area Sq Miles	34 340	**Religions**	Roman Catholic, Protestant
Population	10 707 000	**Currency**	Euro
Capital	Lisbon (Lisboa)	**Organizations**	EU, NATO, OECD, UN

Map page 167

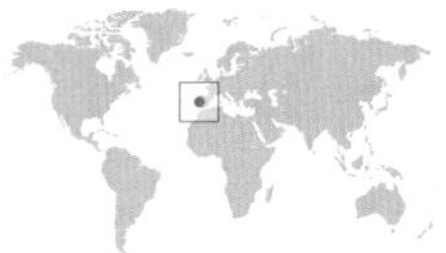

Portugal lies in the western part of the Iberian peninsula in southwest Europe, has an Atlantic coastline and is bordered by Spain to the north and east. The island groups of the Azores and Madeira are parts of Portugal. On the mainland, the land north of the river Tagus (Tejo) is mostly highland, with extensive forests of pine and cork. South of the river is undulating lowland. The climate in the north is cool and moist; the south is warmer, with dry, mild winters. Most Portuguese live near the coast, and more than one-third of the total population lives around the capital, Lisbon (Lisboa). Agriculture, fishing and forestry involve approximately ten per cent of the workforce. Mining and manufacturing are the main sectors of the economy. Portugal produces kaolin, copper, tin, zinc, tungsten and salt. Exports include textiles, clothing and footwear, electrical machinery and transport equipment, cork and wood products, and chemicals. Service industries, chiefly tourism and banking, are important to the economy, as are remittances from workers abroad. Most trade is with other European Union countries.

Puerto Rico

United States Commonwealth

Area Sq Km	9 104	**Languages**	Spanish, English
Area Sq Miles	3 515	**Religions**	Roman Catholic, Protestant
Population	3 982 000	**Currency**	United States dollar
Capital	San Juan		

Map page 205

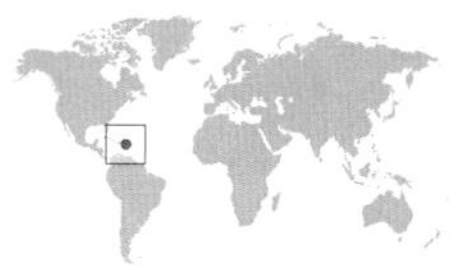

The Caribbean island of Puerto Rico has a forested, hilly interior, coastal plains and a tropical climate. Half of the population lives in the San Juan area. The economy is based on manufacturing (chiefly chemicals, electronics and food), tourism and agriculture. The USA is the main trading partner.

QATAR

State of Qatar

Area Sq Km	11 437	**Languages**	Arabic
Area Sq Miles	4 416	**Religions**	Sunni Muslim
Population	1 409 000	**Currency**	Qatari riyal
Capital	Doha (Ad Dawḩah)	**Organizations**	GCC, OPEC, UN

Map page 140

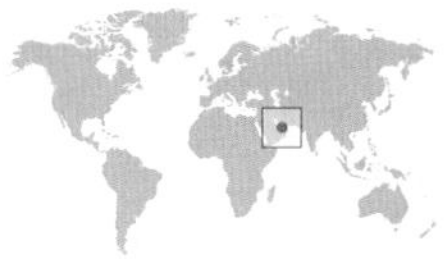

Qatar occupies a peninsula in southwest Asia that extends northwards from east-central Saudi Arabia into The Gulf. The land is flat and barren with sand dunes and salt pans. The climate is hot and mainly dry. Most people live in the area of the capital, Doha. The economy is heavily dependent on oil and natural gas production and the oil-refining industry. Income also comes from overseas investment. Japan is the largest trading partner.

Réunion

French Overseas Department

Area Sq Km	2 551	**Languages**	French, Creole
Area Sq Miles	985	**Religions**	Roman Catholic
Population	827 000	**Currency**	Euro
Capital	St-Denis		

Map page 175

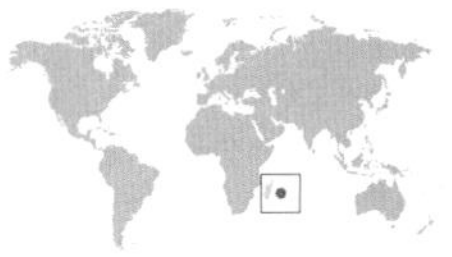

The Indian Ocean island of Réunion is mountainous, with coastal lowlands and a warm climate. The economy depends on tourism, French aid, and exports of sugar. In 2005 France transferred the administration of various small uninhabited islands in the seas around Madagascar from Réunion to the French Southern and Antarctic Lands.

ROMANIA

Area Sq Km	237 500	**Languages**	Romanian, Hungarian
Area Sq Miles	91 699	**Religions**	Romanian Orthodox, Protestant, Roman Catholic
Population	21 275 000	**Currency**	Romanian leu
Capital	Bucharest (București)	**Organizations**	EU, NATO, UN

Map page 171

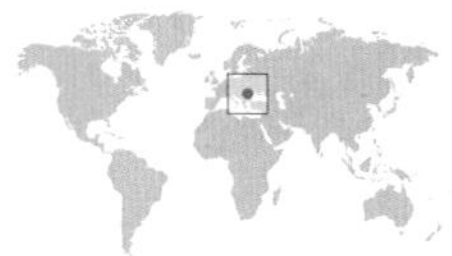

Romania lies in eastern Europe, on the northwest coast of the Black Sea. Mountains separate the Transylvanian Basin in the centre of the country from the populous plains of the east and south and from the Danube delta. The climate is continental. Romania has mineral resources (zinc, lead, silver and gold) and oil and natural gas reserves. Economic development has been slow and sporadic, but measures to accelerate change were introduced in 1999. Agriculture employs over one-third of the workforce. The main exports are textiles, mineral products, chemicals, machinery and footwear. The main trading partners are Germany and Italy.

© Collins Bartholomew Ltd

RUSSIAN FEDERATION

Area Sq Km	17 075 400	**Languages**	Russian, Tatar, Ukrainian, local languages
Area Sq Miles	6 592 849	**Religions**	Russian Orthodox, Sunni Muslim, Protestant
Population	140 874 000	**Currency**	Russian rouble
Capital	Moscow (Moskva)	**Organizations**	APEC, CIS, UN

Map page 132–133

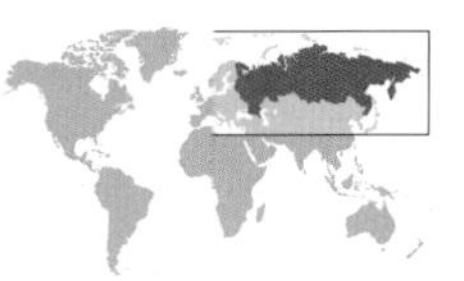

The Russian Federation occupies much of eastern Europe and all of northern Asia, and is the world's largest country. It borders fourteen countries to the west and south and has long coastlines on the Arctic and Pacific Oceans to the north and east. European Russia lies west of the Ural Mountains. To the south the land rises to uplands and the Caucasus mountains on the border with Georgia and Azerbaijan. East of the Urals lies the flat West Siberian Plain and the Central Siberian Plateau. In the south-east is Lake Baikal, the world's deepest lake, and the Sayan ranges on the border with Kazakhstan and Mongolia. Eastern Siberia is rugged and mountainous, with many active volcanoes in the Kamchatka Peninsula. The country's major rivers are the Volga in the west and the Ob', Irtysh, Yenisey, Lena and Amur in Siberia. The climate and vegetation range between arctic tundra in the north and semi-arid steppe towards the Black and Caspian Sea coasts in the south. In general, the climate is continental with extreme temperatures. The majority of the population (the eighth largest in the world), and industry and agriculture are concentrated in European Russia. The economy is dependent on exploitation of raw materials and on heavy industry. Russia has a wealth of mineral resources, although they are often difficult to exploit because of climate and remote locations. It is one of the world's leading producers of petroleum, natural gas and coal as well as iron ore, nickel, copper, bauxite, and many precious and rare metals. Forests cover over forty per cent of the land area and supply an important timber, paper and pulp industry. Approximately eight per cent of the land is suitable for cultivation, but farming is generally inefficient and food, especially grains, must be imported. Fishing is important and Russia has a large fleet operating around the world. The transition to a market economy has been slow and difficult, with considerable underemployment. As well as mining and extractive industries there is a wide range of manufacturing industry, from steel mills to aircraft and space vehicles, shipbuilding, synthetic fabrics, plastics, cotton fabrics, consumer durables, chemicals and fertilizers. Exports include fuels, metals, machinery, chemicals and forest products. The most important trading partners include Germany, the USA and Belarus.

RWANDA

Republic of Rwanda

Area Sq Km	26 338	**Languages**	Kinyarwanda, French, English
Area Sq Miles	10 169	**Religions**	Roman Catholic, traditional beliefs, Protestant
Population	9 998 000	**Currency**	Rwandan franc
Capital	Kigali	**Organizations**	Comm., UN

Map page 178

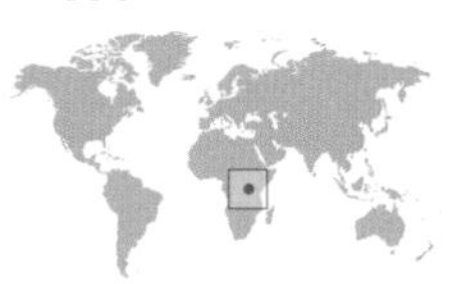

Rwanda, the most densely populated country in Africa, is situated in the mountains and plateaus to the east of the western branch of the Great Rift Valley in east Africa. The climate is warm with a summer dry season. Rwanda depends on subsistence farming, coffee and tea exports, light industry and foreign aid. The country is slowly recovering from serious internal conflict which caused devastation in the early 1990s.

St-Barthélemy

French Overseas Collectivity

Area Sq Km	21	**Languages**	French
Area Sq Miles	8	**Religions**	Roman Catholic
Population	8 450	**Currency**	Euro
Capital	Gustavia		

Map page 205

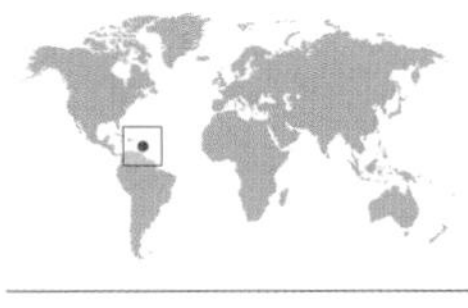

An island in the Leeward Islands in the Lesser Antilles, in the Caribbean south of St-Martin. It was separated from Guadeloupe politically in 2007. Tourism is the main economic activity.

St Helena, Ascension and Tristan da Cunha

United Kingdom Overseas Territory

Area Sq Km	410	**Languages**	English
Area Sq Miles	158	**Religions**	Protestant, Roman Catholic
Population	5 619	**Currency**	St Helena pound, Pound sterling
Capital	Jamestown		

Map page 174

St Helena and its dependencies Ascension and Tristan da Cunha are isolated island groups lying in the south Atlantic Ocean. St Helena is a rugged island of volcanic origin. The main activity is fishing, but the economy relies on financial aid from the UK. Main trading partners are the UK and South Africa.

ST KITTS AND NEVIS

Federation of St Kitts and Nevis

Area Sq Km	261	**Languages**	English, Creole
Area Sq Miles	101	**Religions**	Protestant, Roman Catholic
Population	52 000	**Currency**	East Caribbean dollar
Capital	Basseterre	**Organizations**	CARICOM, Comm., UN

Map page 205

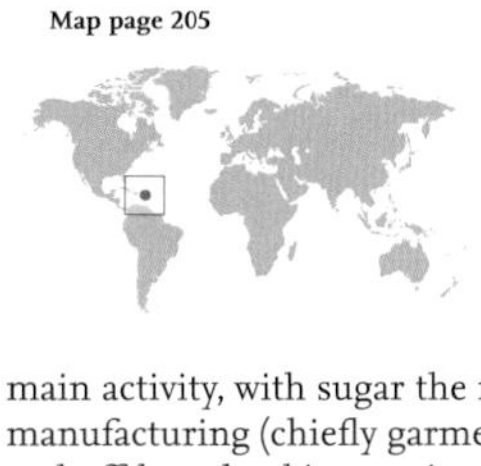

St Kitts and Nevis are in the Leeward Islands, in the Caribbean. Both volcanic islands are mountainous and forested, with sandy beaches and a warm, wet climate. About three-quarters of the population lives on St Kitts. Agriculture is the main activity, with sugar the main product. Tourism and manufacturing (chiefly garments and electronic components) and offshore banking are important activities.

ST LUCIA

Area Sq Km	616	**Languages**	English, Creole
Area Sq Miles	238	**Religions**	Roman Catholic, Protestant
Population	172 000	**Currency**	East Caribbean dollar
Capital	Castries	**Organizations**	CARICOM, Comm., UN

Map page 205

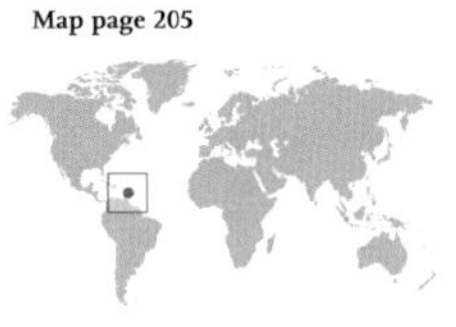

St Lucia, one of the Windward Islands in the Caribbean Sea, is a volcanic island with forested mountains, hot springs, sandy beaches and a wet tropical climate. Agriculture is the main activity, with bananas accounting for approximately forty per cent of export earnings. Tourism, agricultural processing and light manufacturing are increasingly important.

St-Martin

French Overseas Collectivity

Area Sq Km	54	**Languages**	French
Area Sq Miles	21	**Religions**	Roman Catholic
Population	35 692	**Currency**	Euro
Capital	Marigot		

Map page 205

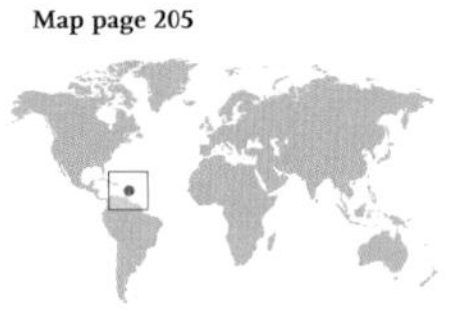

The northern part of St-Martin, one of the Leeward Islands, in the Caribbean. The other part of the island is part of the Netherlands Antilles (Sint Maarten). It was separated from Guadeloupe politically in 2007. Tourism is the main source of income.

St Pierre and Miquelon

French Territorial Collectivity

Area Sq Km	242	**Languages**	French
Area Sq Miles	93	**Religions**	Roman Catholic
Population	6 125	**Currency**	Euro
Capital	St-Pierre		

Map page 185

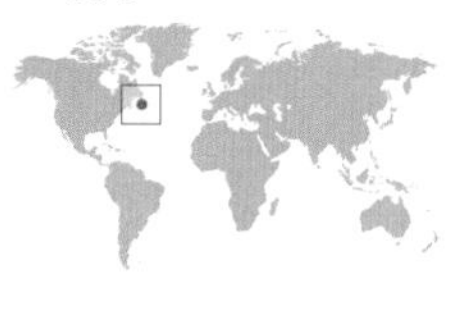

A group of islands off the south coast of Newfoundland in eastern Canada. The islands are largely unsuitable for agriculture, and fishing and fish processing are the most important activities. The islands rely heavily on financial assistance from France.

ST VINCENT AND THE GRENADINES

Area Sq Km	389	**Languages**	English, Creole
Area Sq Miles	150	**Religions**	Protestant, Roman Catholic
Population	109 000	**Currency**	East Caribbean dollar
Capital	Kingstown	**Organizations**	CARICOM, Comm., UN

Map page 205

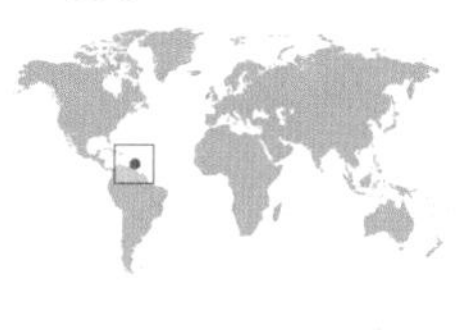

St Vincent, whose territory includes islets and cays in the Grenadines, is in the Windward Islands, in the Caribbean. St Vincent itself is forested and mountainous, with an active volcano, Soufrière. The climate is tropical and wet. The economy is based mainly on agriculture and tourism. Bananas account for approximately one-third of export earnings and arrowroot is also important. Most trade is with the USA and other CARICOM countries.

SAMOA

Independent State of Samoa

Area Sq Km	2 831	**Languages**	Samoan, English
Area Sq Miles	1 093	**Religions**	Protestant, Roman Catholic
Population	179 000	**Currency**	Tala
Capital	Apia	**Organizations**	Comm., UN

Map page 125

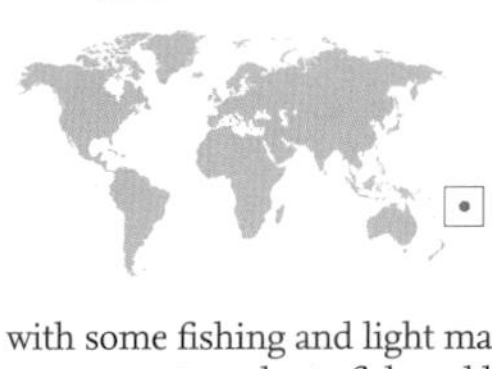

Samoa consists of two larger mountainous and forested islands, Savai'i and Upolu, and seven smaller islands, in the south Pacific Ocean. Over half the population lives on Upolu. The climate is tropical. The economy is based on agriculture, with some fishing and light manufacturing. Traditional exports are coconut products, fish and beer. Tourism is increasing, but the islands depend on workers' remittances and foreign aid.

SAN MARINO

Republic of San Marino

Area Sq Km	61	**Languages**	Italian
Area Sq Miles	24	**Religions**	Roman Catholic
Population	31 000	**Currency**	Euro
Capital	San Marino	**Organizations**	UN

Map page 170

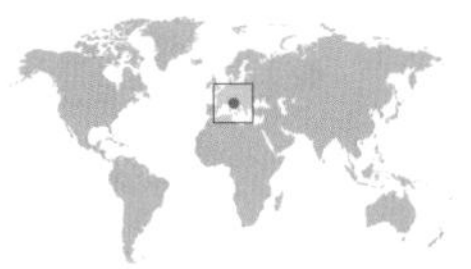

Landlocked San Marino lies in northeast Italy. A third of the people live in the capital. There is some agriculture and light industry, but most income comes from tourism. Italy is the main trading partner.

SÃO TOMÉ AND PRÍNCIPE

Democratic Republic of São Tomé and Príncipe

Area Sq Km	964	**Languages**	Portuguese, Creole
Area Sq Miles	372	**Religions**	Roman Catholic, Protestant
Population	163 000	**Currency**	Dobra
Capital	São Tomé	**Organizations**	UN

Map page 176

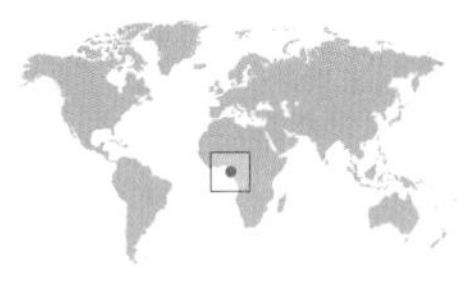

The two main islands and adjacent islets lie off the coast of west Africa in the Gulf of Guinea. São Tomé is the larger island, with over ninety per cent of the population. Both São Tomé and Príncipe are mountainous and tree-covered, and have a hot and humid climate. The economy is heavily dependent on cocoa, which accounts for around ninety per cent of export earnings.

SAUDI ARABIA

Kingdom of Saudi Arabia

Area Sq Km	2 200 000	**Languages**	Arabic
Area Sq Miles	849 425	**Religions**	Sunni Muslim, Shi'a Muslim
Population	25 721 000	**Currency**	Saudi Arabian riyal
Capital	Riyadh (Ar Riyāḍ)	**Organizations**	GCC, OPEC, UN

Map page 142

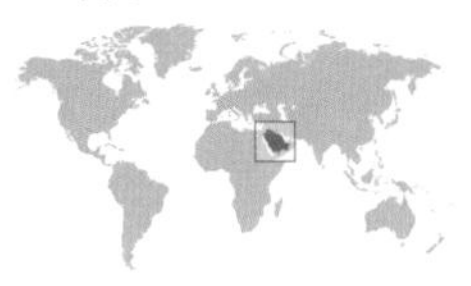

Saudi Arabia occupies most of the Arabian Peninsula in southwest Asia. The terrain is desert or semi-desert plateaus, which rise to mountains running parallel to the Red Sea in the west and slope down to plains in the southeast and along The Gulf in the east. Over eighty per cent of the population lives in urban areas. There are around four million foreign workers in Saudi Arabia, employed mainly in the oil and service industries. Summers are hot, winters are warm and rainfall is low. Saudi Arabia has the world's largest reserves of oil and significant natural gas reserves, both onshore and in The Gulf. Crude oil and refined products account for over ninety per cent of export earnings. Other industries and irrigated agriculture are being encouraged, but most food and raw materials are imported. Saudi Arabia has important banking and commercial interests. Japan and the USA are the main trading partners.

SENEGAL

Republic of Senegal

Area Sq Km	196 720	**Languages**	French, Wolof, Fulani, local languages
Area Sq Miles	75 954	**Religions**	Sunni Muslim, Roman Catholic, traditional beliefs
Population	12 534 000	**Currency**	CFA franc
Capital	Dakar	**Organizations**	UN

Map page 176

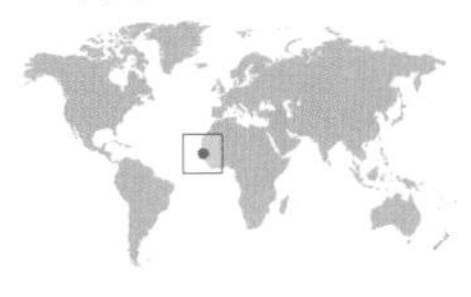

Senegal lies on the Atlantic coast of west Africa. The north is arid semi-desert, while the south is mainly fertile savanna bushland. The climate is tropical with summer rains, although droughts occur. One-fifth of the population lives in and around Dakar, the capital and main port. Fish, groundnuts and phosphates are the main exports. France is the main trading partner.

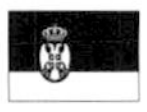

SERBIA

Republic of Serbia

Area Sq Km	77 453	**Languages**	Serbian, Hungarian
Area Sq Miles	29 904	**Religions**	Serbian Orthodox, Roman Catholic, Sunni Muslim
Population	9 850 000	**Currency**	Serbian dinar
Capital	Beograd (Belgrade)	**Organizations**	UN

Map page 171

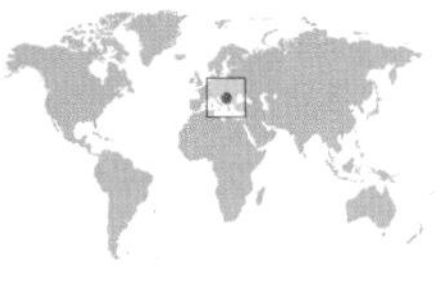

Following ethnic conflict and the break-up of Yugoslavia through the 1990s, the state union of Serbia and Montenegro retained the name Yugoslavia until 2003. The two then became separate independent countries in 2006. The southern Serbian province of Kosovo declared its independence from Serbia in February 2008. The landscape is rugged, mountainous and forested in the south, while the north is low-lying and drained by the Danube river system.

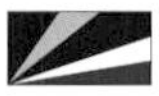

SEYCHELLES

Republic of Seychelles

Area Sq Km	455	**Languages**	English, French, Creole
Area Sq Miles	176	**Religions**	Roman Catholic, Protestant
Population	84 000	**Currency**	Seychelles rupee
Capital	Victoria	**Organizations**	Comm., SADC, UN

Map page 175

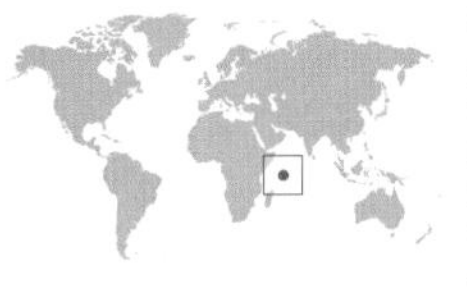

The Seychelles comprises an archipelago of over one hundred granitic and coral islands in the western Indian Ocean. Over ninety per cent of the population lives on the main island, Mahé. The climate is hot and humid with heavy rainfall. The economy is based mainly on tourism, fishing and light manufacturing.

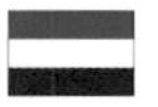

SIERRA LEONE

Republic of Sierra Leone

Area Sq Km	71 740	**Languages**	English, Creole, Mende, Temne, local languages
Area Sq Miles	27 699	**Religions**	Sunni Muslim, traditional beliefs
Population	5 696 000	**Currency**	Leone
Capital	Freetown	**Organizations**	Comm., UN

Map page 176

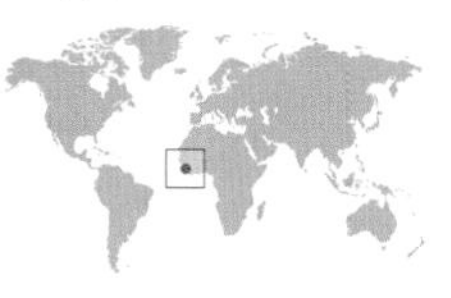

Sierra Leone lies on the Atlantic coast of west Africa. Its coastline is heavily indented and is lined with mangrove swamps. Inland is a forested area rising to savanna plateaus, with mountains to the northeast. The climate is tropical and rainfall is heavy. Most of the workforce is involved in subsistence farming. Cocoa and coffee are the main cash crops. Diamonds and rutile (titanium ore) are the main exports. Sierra Leone is one of the world's poorest countries, and the economy relies on substantial foreign aid.

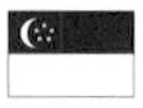

SINGAPORE

Republic of Singapore

Area Sq Km	639	**Languages**	Chinese, English, Malay, Tamil
Area Sq Miles	247	**Religions**	Buddhist, Taoist, Sunni Muslim, Christian, Hindu
Population	4 737 000	**Currency**	Singapore dollar
Capital	Singapore	**Organizations**	APEC, ASEAN, Comm., UN

Map page 150

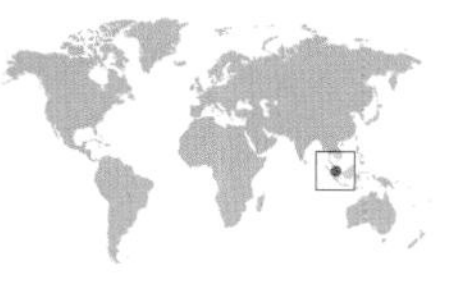

The state comprises the main island of Singapore and over fifty other islands, lying off the southern tip of the Malay Peninsula in southeast Asia. Singapore is generally low-lying and includes land reclaimed from swamps and the sea. It is hot and humid, with heavy rainfall throughout the year. There are fish farms and vegetable gardens in the north and east of the island, but most food is imported. Singapore also lacks mineral and energy resources. Manufacturing industries and services are the main sectors of the economy. Their rapid development has fuelled the nation's impressive economic growth during recent decades. Main industries include electronics, oil refining, chemicals, pharmaceuticals, ship repair, food processing and textiles. Singapore is also a major financial centre. Its port is one of the world's largest and busiest and acts as an entrepôt for neighbouring states. Tourism is also important. Japan, the USA and Malaysia are the main trading partners.

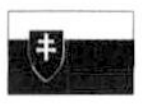

SLOVAKIA

Slovak Republic

Area Sq Km	49 035	**Languages**	Slovak, Hungarian, Czech
Area Sq Miles	18 933	**Religions**	Roman Catholic, Protestant, Orthodox
Population	5 406 000	**Currency**	Euro
Capital	Bratislava	**Organizations**	EU, NATO, OECD, UN

Map page 168–169

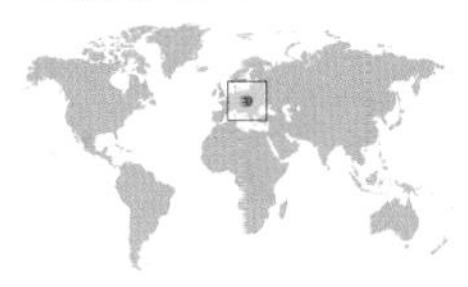

A landlocked country in central Europe, Slovakia is mountainous in the north, but low-lying in the southwest. The climate is continental. There is a range of manufacturing industries, and the main exports are machinery and transport equipment, but in recent years there have been economic difficulties and growth has been slow. Slovakia joined the European Union in May 2004. Most trade is with other EU countries, especially the Czech Republic.

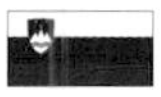

SLOVENIA

Republic of Slovenia

Area Sq Km	20 251	**Languages**	Slovene, Croatian, Serbian
Area Sq Miles	7 819	**Religions**	Roman Catholic, Protestant
Population	2 020 000	**Currency**	Euro
Capital	Ljubljana	**Organizations**	EU, NATO, OECD, UN

Map page 170

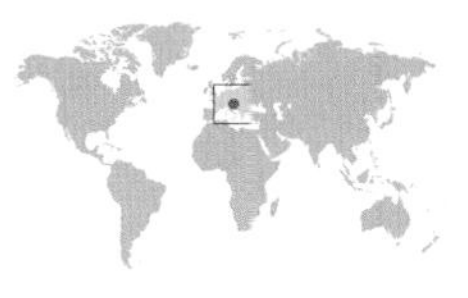

Slovenia lies in the northwest Balkan Mountains of southern Europe and has a short coastline on the Adriatic Sea. It is mountainous and hilly, with lowlands on the coast and in the Sava and Drava river valleys. The climate is generally continental inland and Mediterranean nearer the coast. The main agricultural products are potatoes, grain and sugar beet; the main industries include metal processing, electronics and consumer goods. Trade has been re-orientated towards western markets and the main trading partners are Germany and Italy. Slovenia joined the European Union in May 2004.

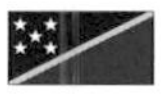

SOLOMON ISLANDS

Area Sq Km	28 370	**Languages**	English, Creole, local languages
Area Sq Miles	10 954	**Religions**	Protestant, Roman Catholic
Population	523 000	**Currency**	Solomon Islands dollar
Capital	Honiara	**Organizations**	Comm., UN

Map page 125

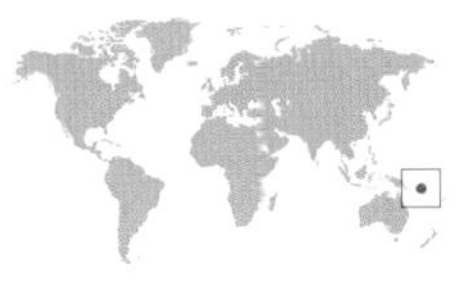

The state consists of the Solomon, Santa Cruz and Shortland Islands in the southwest Pacific Ocean. The six main islands are volcanic, mountainous and forested, although Guadalcanal, the most populous, has a large lowland area. The climate is generally hot and humid. Subsistence farming, forestry and fishing predominate. Exports include timber products, fish, copra and palm oil. The islands depend on foreign aid.

SOMALIA

Somali Republic

Area Sq Km	637 657	**Languages**	Somali, Arabic
Area Sq Miles	246 201	**Religions**	Sunni Muslim
Population	9 133 000	**Currency**	Somali shilling
Capital	Mogadishu (Muqdisho)	**Organizations**	UN

Map page 178

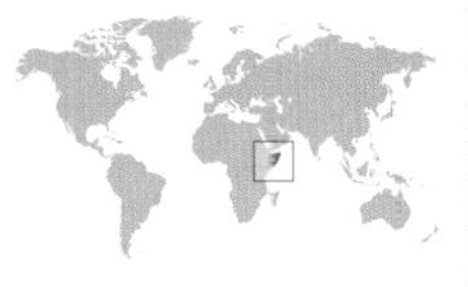

Somalia is in northeast Africa, on the Gulf of Aden and Indian Ocean. It consists of a dry scrubby plateau, rising to highlands in the north. The climate is hot and dry, but coastal areas and the Jubba and Webi Shabeelle river valleys support crops and most of the population. Subsistence farming and livestock rearing are the main activities. Exports include livestock and bananas. Frequent drought and civil war have prevented economic development. Somalia is one of the poorest, most unstable and least developed countries in the world.

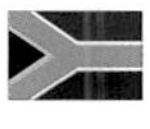

SOUTH AFRICA, REPUBLIC OF

Area Sq Km	1 219 090	**Languages**	Afrikaans, English, nine other official languages
Area Sq Miles	470 693	**Religions**	Protestant, Roman Catholic, Sunni Muslim, Hindu
Population	50 110 000	**Currency**	Rand
Capital	Pretoria (Tshwane)/ Cape Town	**Organizations**	Comm., SADC, UN

Map page 180–181

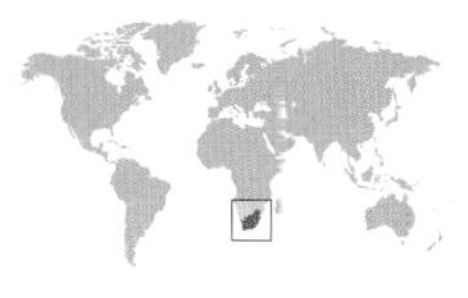

The Republic of South Africa occupies most of the southern part of Africa. It surrounds Lesotho and has a long coastline on the Atlantic and Indian Oceans. Much of the land is a vast plateau, covered with grassland or bush and drained by the Orange and Limpopo river systems. A fertile coastal plain rises to mountain ridges in the south and east, including Table Mountain near Cape Town and the Drakensberg range in the east. Gauteng is the most populous province, with Johannesburg and Pretoria its main cities. South Africa has warm summers and mild winters. Most of the country has the majority of its rainfall in summer, but the coast around Cape Town has winter rains. South Africa has the largest economy in Africa, although wealth is unevenly distributed and unemployment is very high. Agriculture employs approximately one-third of the workforce, and produce includes fruit, wine, wool and maize. The country is the world's leading producer of gold and chromium and an important producer of diamonds. Many other minerals are also mined. The main industries are mineral and food processing, chemicals, electrical equipment, textiles and motor vehicles. Financial services are also important.

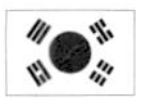

SOUTH KOREA

Republic of Korea

Area Sq Km	99 274	**Languages**	Korean
Area Sq Miles	38 330	**Religions**	Buddhist, Protestant, Roman Catholic
Population	48 333 000	**Currency**	South Korean won
Capital	Seoul (Sŏul)	**Organizations**	APEC, OECD, UN

Map page 152

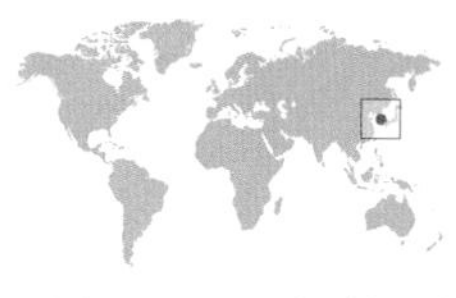

The state consists of the southern half of the Korean Peninsula in eastern Asia and many islands lying off the western and southern coasts in the Yellow Sea. The terrain is mountainous, although less rugged than that of North Korea. Population density is high and the country is highly urbanized; most of the population lives on the western coastal plains and in the river basins of the Han-gang in the northwest and the Naktong-gang in the southeast. The climate is continental, with hot, wet summers and dry, cold winters. Arable land is limited by the mountainous terrain, but because of intensive farming South Korea is nearly self-sufficient in food. Sericulture (silk) is important, as is fishing, which contributes to exports. South Korea has few mineral resources, except for coal and tungsten. It has achieved high economic growth based mainly on export manufacturing. The main manufactured goods are cars, electronic and electrical goods, ships, steel, chemicals and toys, as well as textiles, clothing, footwear and food products. The USA and Japan are the main trading partners.

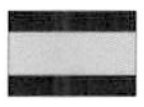

SPAIN

Kingdom of Spain

Area Sq Km	504 782	**Languages**	Spanish, Castilian, Catalan, Galician, Basque
Area Sq Miles	194 897	**Religions**	Roman Catholic
Population	44 904 000	**Currency**	Euro
Capital	Madrid	**Organizations**	EU, NATO, OECD, UN

Map page 167

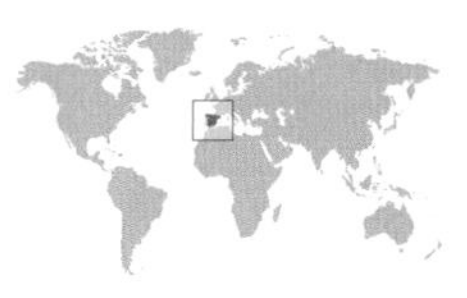

Spain occupies the greater part of the Iberian peninsula in southwest Europe, with coastlines on the Atlantic Ocean and Mediterranean Sea. It includes the Balearic Islands in the Mediterranean, the Canary Islands in the Atlantic, and two enclaves in north Africa (Ceuta and Melilla). Much of the mainland is a high plateau drained by the Douro (Duero), Tagus (Tajo) and Guadiana rivers. The plateau is interrupted by a low mountain range and bounded to the east and north also by mountains, including the Pyrenees, which form the border with France and Andorra. The main lowland areas are the Ebro basin in the northeast, the eastern coastal plains and the Guadalquivir basin in the southwest. Over three-quarters of the population lives in urban areas. The plateau experiences hot summers and cold winters. Conditions are cooler and wetter to the north, and warmer and drier to the south. Agriculture involves about ten per cent of the workforce, and fruit, vegetables and wine are exported. Fishing is an important industry, and Spain has a large fishing fleet. Mineral resources include lead, copper, mercury and fluorspar. Some oil is produced, but Spain has to import most energy needs. The economy is based mainly on manufacturing and services. The principal products are machinery, transport equipment, motor vehicles and food products, with a wide variety of other manufactured goods. With approximately fifty million visitors a year, tourism is a major industry. Banking and commerce are also important. Approximately seventy per cent of trade is with other European Union countries.

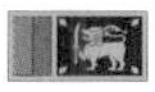

SRI LANKA

Democratic Socialist Republic of Sri Lanka

Area Sq Km	65 610	**Languages**	Sinhalese, Tamil, English
Area Sq Miles	25 332	**Religions**	Buddhist, Hindu, Sunni Muslim, Roman Catholic
Population	20 238 000	**Currency**	Sri Lankan rupee
Capital	Sri Jayewardenepura Kotte	**Organizations**	Comm., UN

Map page 143

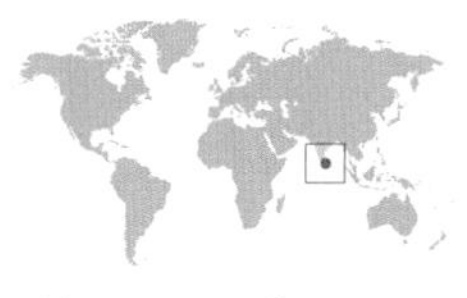

Sri Lanka lies in the Indian Ocean off the southeast coast of India in south Asia. It has rolling coastal plains, with mountains in the centre-south. The climate is hot and monsoonal. Most people live on the west coast. Manufactures (chiefly textiles and clothing), tea, rubber, copra and gems are exported. The economy relies on foreign aid and workers' remittances. The USA and the UK are the main trading partners.

© Collins Bartholomew Ltd

SUDAN
Republic of the Sudan

Area Sq Km	2 505 813	Languages	Arabic, Dinka, Nubian, Beja, Nuer, local languages
Area Sq Miles	967 500		
Population	42 272 000	Religions	Sunni Muslim, traditional beliefs, Christian
Capital	Khartoum		
		Currency	Sudanese pound (Sudani)
		Organizations	UN

Map page 177

Africa's largest country, the Sudan is in the northeast of the continent, on the Red Sea. It lies within the upper Nile basin, much of which is arid plain but with swamps to the south. Mountains lie to the northeast, west and south. The climate is hot and arid with light summer rainfall, and droughts occur. Most people live along the Nile and are farmers and herders. Cotton, gum arabic, livestock and other agricultural products are exported. The government is working with foreign investors to develop oil resources, but civil war in the south and ethnic cleansing in Darfur continue to restrict the growth of the economy. Main trading partners are Saudi Arabia, China and Libya.

SURINAME
Republic of Suriname

Area Sq Km	153 820	Languages	Dutch, Surinamese, English, Hindi
Area Sq Miles	63 251	Religions	Hindu, Roman Catholic, Protestant, Sunni Muslim
Population	520 000		
Capital	Paramaribo	Currency	Suriname guilder
		Organizations	CARICOM, UN

Map page 211

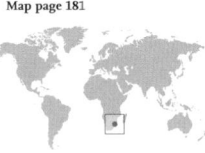

Suriname, on the Atlantic coast of northern South America, consists of a swampy coastal plain (where most of the population lives), central plateaus, and highlands in the south. The climate is tropical, and rainforest covers much of the land. Bauxite mining is the main industry, and alumina and aluminium are the chief exports, with shrimps, rice, bananas and timber also exported. The main trading partners are the Netherlands, Norway and the USA.

SWAZILAND
Kingdom of Swaziland

Area Sq Km	17 364	Languages	Swazi, English
Area Sq Miles	6 704	Religions	Christian, traditional beliefs
Population	1 185 000	Currency	Emalangeni, South African rand
Capital	Mbabane	Organizations	Comm., SADC, UN

Map page 181

Landlocked Swaziland in southern Africa lies between Mozambique and the Republic of South Africa. Savanna plateaus descend from mountains in the west towards hill country in the east. The climate is subtropical, but temperate in the mountains. Subsistence farming predominates. Asbestos and diamonds are mined. Exports include sugar, fruit and wood pulp. Tourism and workers' remittances are important to the economy. Most trade is with South Africa.

SWEDEN
Kingdom of Sweden

Area Sq Km	449 964	Languages	Swedish
Area Sq Miles	173 732	Religions	Protestant, Roman Catholic
Population	9 249 000	Currency	Swedish krona
Capital	Stockholm	Organizations	EU, OECD, UN

Map page 158–159

Sweden occupies the eastern part of the Scandinavian peninsula in northern Europe and borders the Baltic Sea, the Gulf of Bothnia, and the Kattegat and Skagerrak, connecting with the North Sea. Forested mountains cover the northern half, part of which lies within the Arctic Circle. The southern part of the country is a lowland lake region where most of the population lives. Sweden has warm summers and cold winters, which are more severe in the north. Natural resources include coniferous forests, mineral deposits and water resources. Some dairy products, meat, cereals and vegetables are produced in the south. The forests supply timber for export and for the important pulp, paper and furniture industries. Sweden is an important producer of iron ore and copper. Zinc, lead, silver and gold are also mined. Machinery and transport equipment, chemicals, pulp and wood, and telecommunications equipment are the main exports. The majority of trade is with other European Union countries.

SWITZERLAND
Swiss Confederation

Area Sq Km	41 293	Languages	German, French, Italian, Romansch
Area Sq Miles	15 943	Religions	Roman Catholic, Protestant
Population	7 568 000	Currency	Swiss franc
Capital	Bern	Organizations	OECD, UN

Map page 166

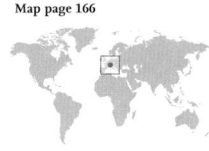

Switzerland is a mountainous landlocked country in west central Europe. The southern regions lie within the Alps, while the northwest is dominated by the Jura mountains. The rest of the land is a high plateau, where most of the population lives. The climate varies greatly, depending on altitude and relief, but in general summers are mild and winters are cold with heavy snowfalls. Switzerland has one of the highest standards of living in the world, yet it has few mineral resources, and most food and industrial raw materials are imported. Manufacturing makes the largest contribution to the economy. Engineering is the most important industry, producing precision instruments and heavy machinery. Other important industries are chemicals and pharmaceuticals. Banking and financial services are very important, and Zürich is one of the world's leading banking cities. Tourism, and international organizations based in Switzerland, are also major foreign currency earners. Germany is the main trading partner.

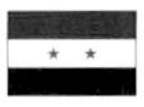

SYRIA
Syrian Arab Republic

Area Sq Km	185 180	Languages	Arabic, Kurdish, Armenian
Area Sq Miles	71 498	Religions	Sunni Muslim, Shi'a Muslim, Christian
Population	21 906 000	Currency	Syrian pound
Capital	Damascus (Dimashq)	Organizations	UN

Map page 136–137

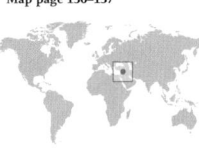

Syria is in southwest Asia, has a short coastline on the Mediterranean Sea, and stretches inland to a plateau traversed northwest-southeast by the Euphrates river. Mountains flank the southwest borders with Lebanon and Israel. The climate is Mediterranean in coastal regions, hotter and drier inland. Most Syrians live on the coast or in the river valleys. Cotton, cereals and fruit are important products, but the main exports are petroleum and related products, and textiles.

TAIWAN
Republic of China

Area Sq Km	36 179	Languages	Mandarin, Min, Hakka, local languages
Area Sq Miles	13 969	Religions	Buddhist, Taoist, Confucian, Christian
Population	23 046 000	Currency	Taiwan dollar
Capital	T'aipei	Organizations	APEC

Map page 149

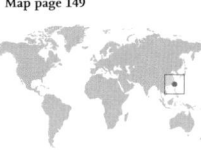

The east Asian state consists of the island of Taiwan, separated from mainland China by the Taiwan Strait, and several much smaller islands. Much of Taiwan is mountainous and forested. Densely populated coastal plains in the west contain the bulk of the population and most economic activity. Taiwan has a tropical monsoon climate, with warm, wet summers and mild winters. Agriculture is highly productive. The country is virtually self-sufficient in food and exports some products. Coal, oil and natural gas are produced and a few minerals are mined, but none of them are of great significance to the economy. Taiwan depends heavily on imports of raw materials and exports of manufactured goods. The main manufactures are electrical and electronic goods, including television sets, personal computers and calculators, textiles, fertilizers, clothing, footwear and toys. The main trading partners are the USA, Japan and Germany. The People's Republic of China claims Taiwan as its 23rd Province.

TAJIKISTAN
Republic of Tajikistan

Area Sq Km	143 100	Languages	Tajik, Uzbek, Russian
Area Sq Miles	55 251	Religions	Sunni Muslim
Population	6 952 000	Currency	Somoni
Capital	Dushanbe	Organizations	CIS, UN

Map page 139

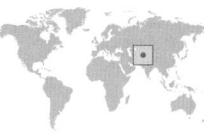

Landlocked Tajikistan in central Asia is a mountainous country, dominated by the mountains of the Alai Range and the Pamir. In the less mountainous western areas summers are warm, although winters are cold. Agriculture is the main sector of the economy, chiefly cotton growing and cattle breeding. Mineral deposits include lead, zinc, and uranium. Processed metals, textiles and clothing are the main manufactured goods; the main exports are aluminium and cotton. Uzbekistan, Kazakhstan and the Russian Federation are the main trading partners.

TANZANIA
United Republic of Tanzania

Area Sq Km	945 087	Languages	Swahili, English, Nyamwezi, local languages
Area Sq Miles	364 900		
Population	43 739 000	Religions	Shi'a Muslim, Sunni Muslim, traditional beliefs, Christian
Capital	Dodoma		
		Organizations	Comm., SADC, UN

Map page 178–179

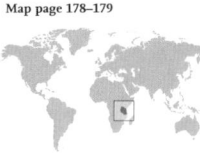

Tanzania lies on the coast of east Africa and includes the island of Zanzibar in the Indian Ocean. Most of the mainland is a savanna plateau lying east of the Great Rift Valley. In the north, near the border with Kenya, is Kilimanjaro, the highest mountain in Africa. The climate is tropical. The economy is predominantly based on agriculture, which employs an estimated ninety per cent of the workforce. Agricultural processing and gold and diamond mining are the main industries, although tourism is growing. Coffee, cotton, cashew nuts and tobacco are the main exports, with cloves from Zanzibar. Most export trade is with India and the UK. Tanzania depends heavily on foreign aid.

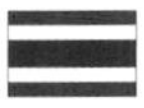

THAILAND
Kingdom of Thailand

Area Sq Km	513 115	Languages	Thai, Lao, Chinese, Malay, Mon-Khmer languages
Area Sq Miles	198 115		
Population	67 764 000	Religions	Buddhist, Sunni Muslim
Capital	Bangkok (Krung Thep)	Currency	Baht
		Organizations	APEC, ASEAN, UN

Map page 154

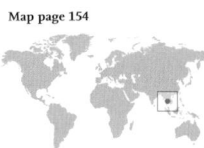

The largest country in the Indo-China peninsula, Thailand has coastlines on the Gulf of Thailand and Andaman Sea. Central Thailand is dominated by the Chao Phraya river basin, which contains Bangkok, the capital city and centre of most economic activity. To the east is a dry plateau drained by tributaries of the Mekong river, while to the north, west and south, extending down most of the Malay peninsula, are forested hills and mountains. Many small islands line the coast. The climate is hot, humid and monsoonal. About half the workforce is involved in agriculture. Fishing and fish processing are important. Thailand produces natural gas, some oil and lignite, minerals (chiefly tin, tungsten and baryte) and gemstones. Manufacturing is the largest contributor to national income, with electronics, textiles, clothing and footwear, and food processing the main industries. With around seven million visitors a year, tourism is the major source of foreign exchange. Thailand is one of the world's leading exporters of rice and rubber, and a major exporter of maize and tapioca. Japan and the USA are the main trading partners.

TOGO
Republic of Togo

Area Sq Km	56 785	Languages	French, Ewe, Kabre, local languages
Area Sq Miles	21 925	Religions	Traditional beliefs, Christian, Sunni Muslim
Population	6 619 000		
Capital	Lomé	Currency	CFA franc
		Organizations	UN

Map page 176

Togo is a long narrow country in west Africa with a short coastline on the Gulf of Guinea. The interior consists of plateaus rising to mountainous areas. The climate is tropical, and is drier inland. Agriculture is the mainstay of the economy. Phosphate mining and food processing are the main industries. Cotton, phosphates, coffee and cocoa are the main exports. Lomé, the capital, is an entrepôt trade centre.

Tokelau
New Zealand Overseas Territory

Area Sq Km	10	Languages	English, Tokelauan
Area Sq Miles	4	Religions	Christian
Population	1 466	Currency	New Zealand dollar

Map page 125 Tokelau consists of three atolls, Atafu, Nukunonu and Fakaofa, lying in the Pacific Ocean north of Samoa. Subsistence agriculture is the main activity, and the islands rely on aid from New Zealand and remittances from workers overseas.

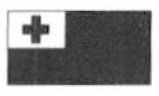

TONGA

Kingdom of Tonga

Area Sq Km	748	**Languages**	Tongan, English
Area Sq Miles	289	**Religions**	Protestant, Roman Catholic
Population	104 000	**Currency**	Pa'anga
Capital	Nuku'alofa	**Organizations**	Comm., UN

Map page 125

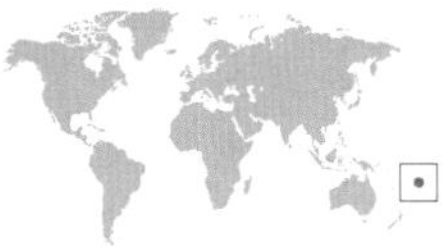

Tonga comprises some one hundred and seventy islands in the south Pacific Ocean, northeast of New Zealand. The three main groups are Tongatapu (where sixty per cent of Tongans live), Ha'apai and Vava'u. The climate is warm and wet, and the economy relies heavily on agriculture. Tourism and light industry are also important to the economy. Exports include squash, fish, vanilla beans and root crops. Most trade is with New Zealand, Japan and Australia.

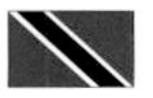

TRINIDAD AND TOBAGO

Republic of Trinidad and Tobago

Area Sq Km	5 130	**Languages**	English, Creole, Hindi
Area Sq Miles	1 981	**Religions**	Roman Catholic, Hindu, Protestant, Sunni Muslim
Population	1 339 000		
Capital	Port of Spain	**Currency**	Trinidad and Tobago dollar
		Organizations	CARICOM, Comm., UN

Map page 213

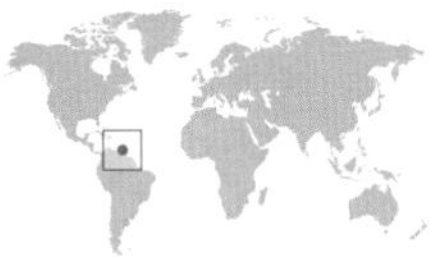

Trinidad, the most southerly Caribbean island, lies off the Venezuelan coast. It is hilly in the north, with a central plain. Tobago, to the northeast, is smaller, more mountainous and less developed. The climate is tropical. The main crops are cocoa, sugar cane, coffee, fruit and vegetables. Oil and petrochemical industries dominate the economy. Tourism is also important. The USA is the main trading partner.

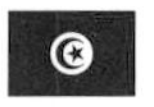

TUNISIA

Tunisian Republic

Area Sq Km	164 150	**Languages**	Arabic, French
Area Sq Miles	63 379	**Religions**	Sunni Muslim
Population	10 272 000	**Currency**	Tunisian dinar
Capital	Tunis	**Organizations**	UN

Map page 176

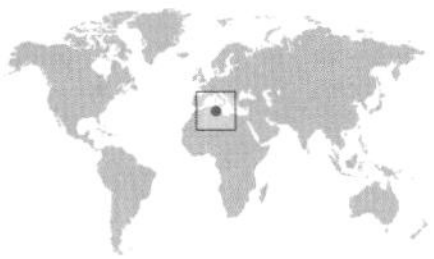

Tunisia is on the Mediterranean coast of north Africa. The north is mountainous with valleys and coastal plains, has a Mediterranean climate and is the most populous area. The south is hot and arid. Oil and phosphates are the main resources, and the main crops are olives and citrus fruit. Tourism is an important industry. Exports include petroleum products, textiles, fruit and phosphorus. Most trade is with European Union countries.

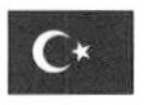

TURKEY

Republic of Turkey

Area Sq Km	779 452	**Languages**	Turkish, Kurdish
Area Sq Miles	300 948	**Religions**	Sunni Muslim, Shi'a Muslim
Population	74 816 000	**Currency**	Lira
Capital	Ankara	**Organizations**	NATO, OECD, UN

Map page 136–137

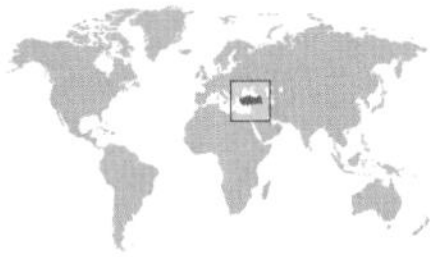

Turkey occupies a large peninsula of southwest Asia and has coastlines on the Black, Mediterranean and Aegean Seas. It includes eastern Thrace, which is in southeastern Europe and is separated from the rest of the country by the Bosporus, the Sea of Marmara and the Dardanelles. The Asian mainland consists of the semi-arid Anatolian plateau, flanked to the north, south and east by mountains. Over forty per cent of Turks live in central Anatolia and on the Marmara and Aegean coastal plains. The coast has a Mediterranean climate, but inland conditions are more extreme with hot, dry summers and cold, snowy winters. Agriculture involves about forty per cent of the workforce, and products include cotton, grain, tobacco, fruit, nuts and livestock. Turkey is a leading producer of chromium, iron ore, lead, tin, borate, and baryte while coal is also mined. The main manufactured goods are clothing, textiles, food products, steel and vehicles. Tourism is a major industry, with nine million visitors a year. Germany and the USA are the main trading partners. Remittances from workers abroad are important to the economy.

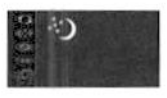

TURKMENISTAN

Republic of Turkmenistan

Area Sq Km	488 100	**Languages**	Turkmen, Uzbek, Russian
Area Sq Miles	188 456	**Religions**	Sunni Muslim, Russian Orthodox
Population	5 110 000	**Currency**	Turkmen manat
Capital	Aşgabat (Ashkhabad)	**Organizations**	UN

Map page 138

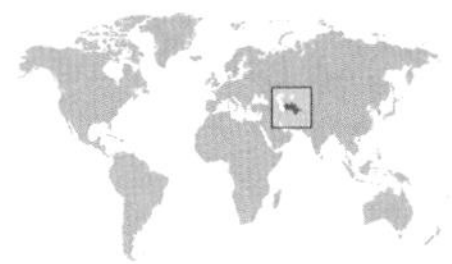

Turkmenistan, in central Asia, comprises the plains of the Karakum Desert, the foothills of the Kopet Dag mountains in the south, the Amudar'ya valley in the north and the Caspian Sea plains in the west. The climate is dry, with extreme temperatures. The economy is based mainly on irrigated agriculture (chiefly cotton growing), and natural gas and oil. Main exports are natural gas, oil and cotton fibre. Ukraine, Iran, Turkey and the Russian Federation are the main trading partners.

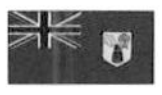

Turks and Caicos Islands

United Kingdom Overseas Territory

Area Sq Km	430	**Languages**	English
Area Sq Miles	166	**Religions**	Protestant
Population	33 000	**Currency**	United States dollar
Capital	Grand Turk (Cockburn Town)		

Map page 205

The state consists of over forty low-lying islands and cays in the northern Caribbean. Only eight islands are inhabited, and two-fifths of the people live on Grand Turk and Salt Cay. The climate is tropical, and the economy is based on tourism, fishing and offshore banking.

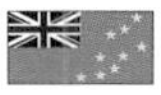

TUVALU

Area Sq Km	25	**Languages**	Tuvaluan, English
Area Sq Miles	10	**Religions**	Protestant
Population	10 000	**Currency**	Australian dollar
Capital	Vaiaku	**Organizations**	Comm., UN

Map page 125

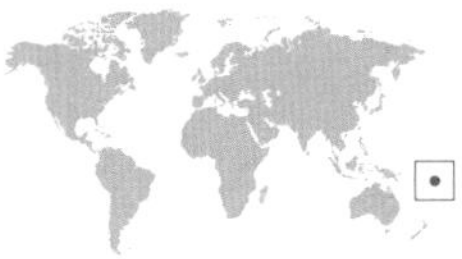

Tuvalu comprises nine low-lying coral atolls in the south Pacific Ocean. One-third of the population lives on Funafuti, and most people depend on subsistence farming and fishing. The islands export copra, stamps and clothing, but rely heavily on foreign aid. Most trade is with Fiji, Australia and New Zealand.

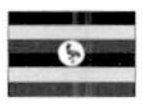

UGANDA

Republic of Uganda

Area Sq Km	241 038	**Languages**	English, Swahili, Luganda, local languages
Area Sq Miles	93 065		
Population	32 710 000	**Religions**	Roman Catholic, Protestant, Sunni Muslim, traditional beliefs
Capital	Kampala		
		Currency	Ugandan shilling
		Organizations	Comm., UN

Map page 178

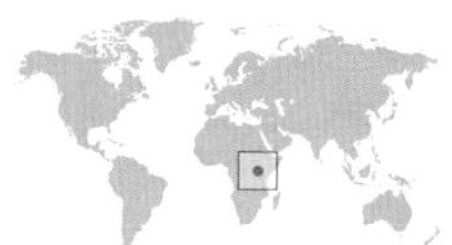

A landlocked country in east Africa, Uganda consists of a savanna plateau with mountains and lakes. The climate is warm and wet. Most people live in the southern half of the country. Agriculture employs around eighty per cent of the workforce and dominates the economy. Coffee, tea, fish and fish products are the main exports. Uganda relies heavily on aid.

UKRAINE

Area Sq Km	603 700	**Languages**	Ukrainian, Russian
Area Sq Miles	233 090	**Religions**	Ukrainian Orthodox, Ukrainian Catholic, Roman Catholic
Population	45 708 000		
Capital	Kiev (Kyiv)	**Currency**	Hryvnia
		Organizations	CIS, UN

Map page 173

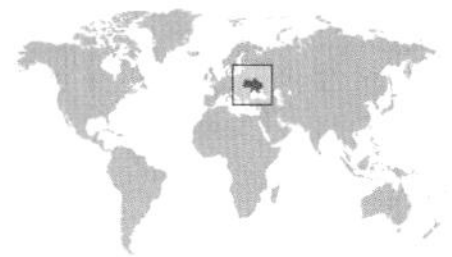

The country lies on the Black Sea coast of eastern Europe. Much of the land is steppe, generally flat and treeless, but with rich black soil, and it is drained by the river Dnieper. Along the border with Belarus are forested, marshy plains. The only uplands are the Carpathian Mountains in the west and smaller ranges on the Crimean peninsula. Summers are warm and winters are cold, with milder conditions in the Crimea. About a quarter of the population lives in the mainly industrial areas around Donets'k, Kiev and Dnipropetrovs'k. The Ukraine is rich in natural resources: fertile soil, substantial mineral and natural gas deposits, and forests. Agriculture and livestock rearing are important, but mining and manufacturing are the dominant sectors of the economy. Coal, iron and manganese mining, steel and metal production, machinery, chemicals and food processing are the main industries. The Russian Federation is the main trading partner.

UNITED ARAB EMIRATES

Federation of Emirates

Area Sq Km	77 700	**Languages**	Arabic, English
Area Sq Miles	30 000	**Religions**	Sunni Muslim, Shi'a Muslim
Population	4 599 000	**Currency**	United Arab Emirates dirham
Capital	Abu Dhabi (Abū Ẓabī)	**Organizations**	GCC, OPEC, UN

Map page 140

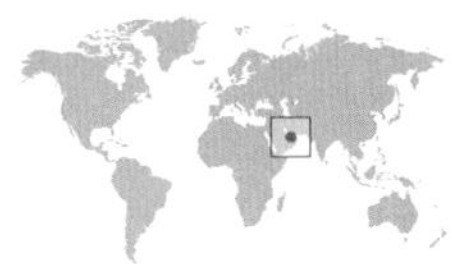

The UAE lies on the Gulf coast of the Arabian Peninsula. Six emirates are on The Gulf, while the seventh, Fujairah, is on the Gulf of Oman. Most of the land is flat desert with sand dunes and salt pans. The only hilly area is in the northeast. Over eighty per cent of the population lives in three of the emirates - Abu Dhabi, Dubai and Sharjah. Summers are hot and winters are mild, with occasional rainfall in coastal areas. Fruit and vegetables are grown in oases and irrigated areas, but the Emirates' wealth is based on hydrocarbons found in Abu Dhabi, Dubai, Sharjah and Ras al Khaimah. The UAE is one of the major oil producers in the Middle East. Dubai is an important entrepôt trade centre The main trading partner is Japan.

Abu Dhabi (Emirate)

Area Sq Km (Sq Miles)	67 340 (26 000)	**Population**	1 559 000	**Capital**	Abu Dhabi (Abū Ẓabī)

Ajman (Emirate)

Area Sq Km (Sq Miles)	259 (100)	**Population**	237 000	**Capital**	'Ajman

Dubai (Emirate)

Area Sq Km (Sq Miles)	3 885 (1 500)	**Population**	1 596 000	**Capital**	Dubai (Dubayy)

Fujairah (Emirate)

Area Sq Km (Sq Miles)	1 165 (450)	**Population**	143 000	**Capital**	Fujairah

Ra's al Khaymah (Emirate)

Area Sq Km (Sq Miles)	1 684 (650)	**Population**	231 000	**Capital**	Ra's al Khaymah

Sharjah (Emirate)

Area Sq Km (Sq Miles)	2 590 (1 000)	**Population**	946 000	**Capital**	Sharjah (Ash Shāriqan)

Umm al Qaywayn (Emirate)

Area Sq Km (Sq Miles)	777 (300)	**Population**	53 000	**Capital**	Umm al Qaywayn

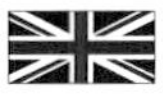

UNITED KINGDOM

United Kingdom of Great Britain and Northern Ireland

Area Sq Km	243 609	**Languages**	English, Welsh, Gaelic
Area Sq Miles	94 058	**Religions**	Protestant, Roman Catholic, Muslim
Population	61 565 000	**Currency**	Pound sterling
Capital	London	**Organizations**	Comm., EU, NATO, OECD, UN

Map page 160–163

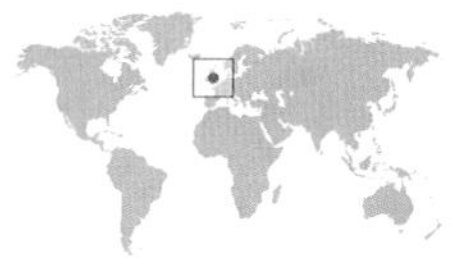

The United Kingdom, in northwest Europe, occupies the island of Great Britain, part of Ireland, and many small adjacent islands. Great Britain comprises England, Scotland and Wales. England covers over half the land area and supports over four-fifths of the population, at its densest in the southeast. The English landscape is flat or rolling with some uplands, notably the Cheviot Hills on the Scottish border, the Pennines in the centre-north, and the hills of the Lake District in the northwest. Scotland consists of southern uplands, central lowlands, the Highlands (which include the UK's highest peak) and many islands. Wales is a land of hills, mountains and river valleys. Northern Ireland contains uplands, plains and the UK's largest lake, Lough Neagh. The climate of the UK is mild, wet and variable. There are few mineral deposits, but important energy resources. Agricultural activities involve sheep and cattle rearing, dairy farming, and crop and fruit growing in the east and southeast. Productivity is high, but approximately one-third of food is imported. The UK produces petroleum and natural gas from reserves in the North Sea and is self-sufficient in energy in net terms. Major manufactures are food and drinks, motor vehicles and parts, aerospace equipment, machinery, electronic and electrical equipment, and chemicals and chemical products. However, the economy is dominated by service industries, including banking, insurance, finance and business services. London, the capital, is one of the world's major financial centres. Tourism is also a major industry, with approximately twenty-five million visitors a year. International trade is also important, equivalent to one-third of national income. Over half of the UK's trade is with other European Union countries.

© Collins Bartholomew Ltd

England (Constituent country)
Area Sq Km (Sq Miles) 130 433 (50 360) **Population** 51 446 000
Capital London

Northern Ireland (Province)
Area Sq Km (Sq Miles) 13 576 (5 242) **Population** 1 775 000
Capital Belfast

Scotland (Constituent country)
Area Sq Km (Sq Miles) 78 822 (30 433) **Population** 5 169 000
Capital Edinburgh

Wales (Principality)
Area Sq Km (Sq Miles) 20 778 (8 022) **Population** 2 993 000
Capital Cardiff

UNITED STATES OF AMERICA
Federal Republic

Area Sq Km	9 826 635	**Languages**	English, Spanish
Area Sq Miles	3 794 085	**Religions**	Protestant, Roman Catholic, Sunni Muslim, Jewish
Population	314 659 000		
Capital	Washington D.C.	**Currency**	United States dollar
		Organizations	APEC, NATO, OECD, UN

Map page 192–193

The USA comprises forty-eight contiguous states in North America, bounded by Canada and Mexico, plus the states of Alaska, to the northwest of Canada, and Hawaii, in the north Pacific Ocean. The populous eastern states cover the Atlantic coastal plain (which includes the Florida peninsula and the Gulf of Mexico coast) and the Appalachian Mountains. The central states occupy a vast interior plain drained by the Mississippi-Missouri river system. To the west lie the Rocky Mountains, separated from the Pacific coastal ranges by intermontane plateaus. The Pacific coastal zone is also mountainous, and prone to earthquakes. Hawaii is a group of some twenty volcanic islands. Climatic conditions range between arctic in Alaska to desert in the intermontane plateaus. Most of the USA has a temperate climate, although the interior has continental conditions. There are abundant natural resources, including major reserves of minerals and energy resources. The USA has the largest and most technologically advanced economy in the world, based on manufacturing and services. Although agriculture accounts for approximately two per cent of national income, productivity is high and the USA is a net exporter of food, chiefly grains and fruit. Cotton is the major industrial crop. The USA produces iron ore, copper, lead, zinc, and many other minerals. It is a major producer of coal, petroleum and natural gas, although being the world's biggest energy user it imports significant quantities of petroleum and its products. Manufacturing is diverse. The main industries are petroleum, steel, motor vehicles, aerospace, telecommunications, electronics, food processing, chemicals and consumer goods. Tourism is a major foreign currency earner, with approximately forty-five million visitors a year. Other important service industries are banking and finance, Wall Street in New York being one of the world's major stock exchanges. Canada and Mexico are the main trading partners.

Alabama (State)
Area Sq Km (Sq Miles) 135 765 (52 419) **Population** 4 661 900
Capital Montgomery

Alaska (State)
Area Sq Km (Sq Miles) 1 717 854 (663 267) **Population** 686 293
Capital Juneau

Arizona (State)
Area Sq Km (Sq Miles) 295 253 (113 998) **Population** 6 500 180
Capital Phoenix

Arkansas (State)
Area Sq Km (Sq Miles) 137 733 (53 179) **Population** 2 855 390
Capital Little Rock

California (State)
Area Sq Km (Sq Miles) 423 971 (163 696) **Population** 36 756 666
Capital Sacramento

Colorado (State)
Area Sq Km (Sq Miles) 269 602 (104 094) **Population** 4 939 456
Capital Denver

Connecticut (State)
Area Sq Km (Sq Miles) 14 356 (5 543) **Population** 3 501 252
Capital Hartford

Delaware (State)
Area Sq Km (Sq Miles) 6 446 (2 489) **Population** 873 092
Capital Dover

District of Columbia (District)
Area Sq Km (Sq Miles) 176 (68) **Population** 591 833
Capital Washington

Florida (State)
Area Sq Km (Sq Miles) 170 305 (65 755) **Population** 18 328 340
Capital Tallahassee

Georgia (State)
Area Sq Km (Sq Miles) 153 910 (59 425) **Population** 9 685 744
Capital Atlanta

Hawaii (State)
Area Sq Km (Sq Miles) 28 311 (10 931) **Population** 1 288 198
Capital Honolulu

Idaho (State)
Area Sq Km (Sq Miles) 216 445 (83 570) **Population** 1 523 816
Capital Boise

Illinois (State)
Area Sq Km (Sq Miles) 149 997 (57 914) **Population** 12 901 563
Capital Springfield

Indiana (State)
Area Sq Km (Sq Miles) 94 322 (36 418) **Population** 6 376 792
Capital Indianapolis

Iowa (State)
Area Sq Km (Sq Miles) 145 744 (56 272) **Population** 3 002 555
Capital Des Moines

Kansas (State)
Area Sq Km (Sq Miles) 213 096 (82 277) **Population** 2 802 134
Capital Topeka

Kentucky (State)
Area Sq Km (Sq Miles) 104 659 (40 409) **Population** 4 269 245
Capital Frankfort

Louisiana (State)
Area Sq Km (Sq Miles) 134 265 (51 840) **Population** 4 410 796
Capital Baton Rouge

Maine (State)
Area Sq Km (Sq Miles) 91 647 (35 385) **Population** 1 316 456
Capital Augusta

Maryland (State)
Area Sq Km (Sq Miles) 32 134 (12 407) **Population** 5 633 597
Capital Annapolis

Massachusetts (State)
Area Sq Km (Sq Miles) 27 337 (10 555) **Population** 6 497 967
Capital Boston

Michigan (State)
Area Sq Km (Sq Miles) 250 493 (96 716) **Population** 10 003 422
Capital Lansing

Minnesota (State)
Area Sq Km (Sq Miles) 225 171 (86 939) **Population** 5 220 393
Capital St Paul

Mississippi (State)
Area Sq Km (Sq Miles) 125 433 (48 430) **Population** 2 938 618
Capital Jackson

Missouri (State)
Area Sq Km (Sq Miles) 180 533 (69 704) **Population** 5 911 605
Capital Jefferson City

Montana (State)
Area Sq Km (Sq Miles) 380 837 (147 042) **Population** 967 440
Capital Helena

Nebraska (State)
Area Sq Km (Sq Miles) 200 346 (77 354) **Population** 1 783 432
Capital Lincoln

Nevada (State)
Area Sq Km (Sq Miles) 286 352 (110 561) **Population** 2 600 167
Capital Carson City

New Hampshire (State)
Area Sq Km (Sq Miles) 24 216 (9 350) **Population** 1 315 809
Capital Concord

New Jersey (State)
Area Sq Km (Sq Miles) 22 587 (8 721) **Population** 8 682 661
Capital Trenton

New Mexico (State)
Area Sq Km (Sq Miles) 314 914 (121 589) **Population** 1 984 356
Capital Santa Fe

New York (State)
Area Sq Km (Sq Miles) 141 299 (54 556) **Population** 19 490 297
Capital Albany

North Carolina (State)
Area Sq Km (Sq Miles) 139 391 (53 819) **Population** 9 222 414
Capital Raleigh

North Dakota (State)
Area Sq Km (Sq Miles) 183 112 (70 700) **Population** 641 481
Capital Bismarck

Ohio (State)
Area Sq Km (Sq Miles) 116 096 (44 825) **Population** 11 485 910
Capital Columbus

Oklahoma (State)
Area Sq Km (Sq Miles) 181 035 (69 898) **Population** 3 642 361
Capital Oklahoma City

Oregon (State)
Area Sq Km (Sq Miles) 254 806 (98 381) **Population** 3 790 060
Capital Salem

Pennsylvania (State)
Area Sq Km (Sq Miles) 119 282 (46 055) **Population** 12 448 279
Capital Harrisburg

Rhode Island (State)
Area Sq Km (Sq Miles) 4 002 (1 545) **Population** 1 050 788
Capital Providence

South Carolina (State)
Area Sq Km (Sq Miles) 82 931 (32 020) **Population** 4 479 800
Capital Columbia

South Dakota (State)
Area Sq Km (Sq Miles) 199 730 (77 116) **Population** 804 194
Capital Pierre

Tennessee (State)
Area Sq Km (Sq Miles) 109 150 (42 143) **Population** 6 214 888
Capital Nashville

Texas (State)
Area Sq Km (Sq Miles) 695 622 (268 581) **Population** 24 326 974
Capital Austin

Utah (State)
Area Sq Km (Sq Miles) 219 887 (84 899) **Population** 2 736 424
Capital Salt Lake City

Vermont (State)
Area Sq Km (Sq Miles) 24 900 (9 614) **Population** 621 270
Capital Montpelier

Virginia (State)
Area Sq Km (Sq Miles) 110 784 (42 774) **Population** 7 769 089
Capital Richmond

Washington (State)
Area Sq Km (Sq Miles) 184 666 (71 300) **Population** 6 549 224
Capital Olympia

West Virginia (State)
Area Sq Km (Sq Miles) 62 755 (24 230) **Population** 1 814 468
Capital Charleston

Wisconsin (State)
Area Sq Km (Sq Miles) 169 639 (65 498) **Population** 5 627 967
Capital Madison

Wyoming (State)
Area Sq Km (Sq Miles) 253 337 (97 814) **Population** 532 668
Capital Cheyenne

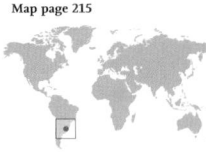

URUGUAY
Oriental Republic of Uruguay

Area Sq Km	176 215	**Languages**	Spanish
Area Sq Miles	68 037	**Religions**	Roman Catholic, Protestant, Jewish
Population	3 361 000	**Currency**	Uruguayan peso
Capital	Montevideo	**Organizations**	UN

Map page 215

Uruguay, on the Atlantic coast of central South America, is a low-lying land of prairies. The coast and the River Plate estuary in the south are fringed with lagoons and sand dunes. Almost half the population lives in the capital, Montevideo. Uruguay has warm summers and mild winters. The economy is based on cattle and sheep ranching, and the main industries produce food products, textiles, and petroleum products. Meat, wool, hides, textiles and agricultural products are the main exports. Brazil and Argentina are the main trading partners.

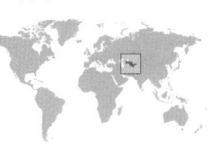

UZBEKISTAN
Republic of Uzbekistan

Area Sq Km	447 400	**Languages**	Uzbek, Russian, Tajik, Kazakh
Area Sq Miles	172 742	**Religions**	Sunni Muslim, Russian Orthodox
Population	27 488 000	**Currency**	Uzbek som
Capital	Tashkent (Toshkent)	**Organizations**	CIS, UN

Map page 138–139

A landlocked country of central Asia, Uzbekistan consists mainly of the flat Kyzylkum Desert. High mountains and valleys are found towards the southeast borders with Kyrgyzstan and Tajikistan. Most settlement is in the Fergana basin. The climate is hot and dry. The economy is based mainly on irrigated agriculture, chiefly cotton production. Uzbekistan is rich in minerals, including gold, copper, lead, zinc and uranium, and it has one of the largest gold mines in the world. Industry specializes in fertilizers and machinery for cotton harvesting and textile manufacture. The Russian Federation is the main trading partner.

VANUATU
Republic of Vanuatu

Area Sq Km	12 190	Languages	English, Bislama (Creole), French
Area Sq Miles	4 707	Religions	Protestant, Roman Catholic, traditional beliefs
Population	240 000		
Capital	Port Vila	Currency	Vatu
		Organizations	Comm., UN

Map page 125

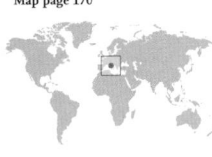

Vanuatu occupies an archipelago of approximately eighty islands in the southwest Pacific. Many of the islands are mountainous, of volcanic origin and densely forested. The climate is tropical, with heavy rainfall. Half of the population lives on the main islands of Éfaté and Espíritu Santo, and the majority of people are employed in agriculture. Copra, beef, timber, vegetables, and cocoa are the main exports. Tourism is becoming important to the economy. Australia, Japan and Germany are the main trading partners.

VATICAN CITY
Vatican City State or Holy See

Area Sq Km	0.5	Languages	Italian
Area Sq Miles	0.2	Religions	Roman Catholic
Population	557	Currency	Euro
Capital	Vatican City (Città del Vaticano)		

Map page 170

The world's smallest sovereign state, the Vatican City occupies a hill to the west of the river Tiber within the Italian capital, Rome. It is the headquarters of the Roman Catholic church, and income comes from investments, voluntary contributions and tourism.

VENEZUELA
Republic of Venezuela

Area Sq Km	912 050	Languages	Spanish, Amerindian languages
Area Sq Miles	352 144	Religions	Roman Catholic, Protestant
Population	28 583 000	Currency	Bolívar fuérte
Capital	Caracas	Organizations	OPEC, UN

Map page 213

Venezuela is in northern South America, on the Caribbean. Its coast is much indented, with the oil-rich area of Lake Maracaibo at the western end, and the swampy Orinoco Delta to the east. Mountain ranges run parallel to the coast, and turn southwestwards to form a northern extension of the Andes. Central Venezuela is an area of lowland grasslands drained by the Orinoco river system. To the south are the Guiana Highlands, which contain the Angel Falls, the world's highest waterfall. Almost ninety per cent of the population lives in towns, mostly in the coastal mountain areas. The climate is tropical, with most rainfall in summer. Farming is important, particularly cattle ranching and dairy farming; coffee, maize, rice and sugar cane are the main crops. Venezuela is a major oil producer, and oil accounts for about seventy-five per cent of export earnings. Aluminium, iron ore, copper and gold are also mined, and manufactures include petrochemicals, aluminium, steel, textiles and food products. The USA and Puerto Rico are the main trading partners.

VIETNAM
Socialist Republic of Vietnam

Area Sq Km	329 565	Languages	Vietnamese, Thai, Khmer, Chinese, local languages
Area Sq Miles	127 246	Religions	Buddhist, Taoist, Roman Catholic, Cao Dai, Hoa Hao
Population	88 069 000		
Capital	Ha Nôi (Hanoi)	Currency	Dong
		Organizations	APEC, ASEAN, UN

Map page 147

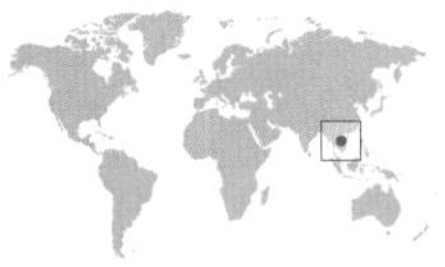

Vietnam lies in southeast Asia on the west coast of the South China Sea. The Red River delta lowlands in the north are separated from the huge Mekong delta in the south by long, narrow coastal plains backed by the mountainous and forested terrain of the Annam Highlands. Most of the population lives in the river deltas. The climate is tropical, with summer monsoon rains. Over three-quarters of the workforce is involved in agriculture, forestry and fishing. Coffee, tea and rubber are important cash crops, but Vietnam is the world's second largest rice exporter. Oil, coal and copper are produced, and other main industries are food processing, clothing and footwear, cement and fertilizers. Exports include oil, coffee, rice, clothing, fish and fish products. Japan and Singapore are the main trading partners.

Virgin Islands (U.K.)
United Kingdom Overseas Territory

Area Sq Km	153	Languages	English
Area Sq Miles	59	Religions	Protestant, Roman Catholic
Population	23 000	Currency	United States dollar
Capital	Road Town		

Map page 205

The Caribbean territory comprises four main islands and over thirty islets at the eastern end of the Virgin Islands group. Apart from the flat coral atoll of Anegada, the islands are volcanic in origin and hilly. The climate is subtropical, and tourism is the main industry.

Virgin Islands (U.S.A.)
United States Unincorporated Territory

Area Sq Km	352	Languages	English, Spanish
Area Sq Miles	136	Religions	Protestant, Roman Catholic
Population	111 000	Currency	United States dollar
Capital	Charlotte Amalie		

Map page 205

The territory consists of three main islands and over fifty islets in the Caribbean's western Virgin Islands. The islands are hilly, of volcanic origin, and the climate is subtropical. The economy is based on tourism, with some manufacturing, including a major oil refinery on St Croix.

Wallis and Futuna Islands
French Overseas Collectivity

Area Sq Km	274	Languages	French, Wallisian, Futunian
Area Sq Miles	106	Religions	Roman Catholic
Population	15 000	Currency	CFP franc
Capital	Matā'utu		

Map page 125

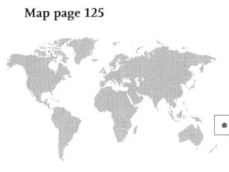

The south Pacific territory comprises the volcanic islands of the Wallis archipelago and the Hoorn Islands. The climate is tropical. The islands depend on subsistence farming, the sale of licences to foreign fishing fleets, workers' remittances from abroad and French aid.

West Bank
Disputed territory

Area Sq Km	5 860	Languages	Arabic, Hebrew
Area Sq Miles	2 263	Religions	Sunni Muslim, Jewish, Shi'a Muslim, Christian
Population	2 448 433		
		Currency	Jordanian dinar, Israeli shekel

Map page 136

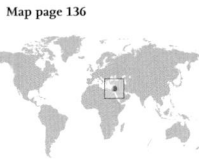

The territory consists of the west bank of the river Jordan and parts of Judea and Samaria. The land was annexed by Israel in 1967, but some areas have been granted autonomy under agreements between Israel and the Palestinian Authority. Conflict between the Israelis and the Palestinians continues to restrict economic development.

Western Sahara
Disputed territory

Area Sq Km	266 000	Languages	Arabic
Area Sq Miles	102 703	Religions	Sunni Muslim
Population	513 000	Currency	Moroccan dirham
Capital	Laâyoune		

Map page 176

Situated on the northwest coast of Africa, the territory of the Western Sahara is now effectively controlled by Morocco. The land is low, flat desert with higher land in the northeast. There is little cultivation and only about twenty per cent of the land is pasture. Livestock herding, fishing and phosphate mining are the main activities. All trade is controlled by Morocco.

YEMEN
Republic of Yemen

Area Sq Km	527 968	Languages	Arabic
Area Sq Miles	203 850	Religions	Sunni Muslim, Shi'a Muslim
Population	23 580 000	Currency	Yemeni riyal
Capital	Şan'ā'	Organizations	UN

Map page 142

Yemen occupies the southwestern part of the Arabian Peninsula, on the Red Sea and the Gulf of Aden. Beyond the Red Sea coastal plain the land rises to a mountain range and then descends to desert plateaus. Much of the country is hot and arid, but there is more rainfall in the west, where most of the population lives. Farming and fishing are the main activities, with cotton the main cash crop. The main exports are crude oil, fish, coffee and dried fruit. Despite some oil resources Yemen is one of the poorest countries in the Arab world. Main trading partners are Thailand, China, South Korea and Saudi Arabia.

ZAMBIA
Republic of Zambia

Area Sq Km	752 614	Languages	English, Bemba, Nyanja, Tonga, local languages
Area Sq Miles	290 586	Religions	Christian, traditional beliefs
Population	12 935 000	Currency	Zambian kwacha
Capital	Lusaka	Organizations	Comm., SADC, UN

Map page 179

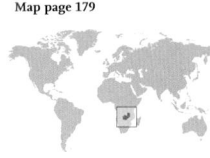

A landlocked state in south central Africa, Zambia consists principally of high savanna plateaus and is bordered by the Zambezi river in the south. Most people live in the Copperbelt area in the centre-north. The climate is tropical, with a rainy season from November to May. Agriculture employs approximately eighty per cent of the workforce, but is mainly at subsistence level. Copper mining is the mainstay of the economy, although reserves are declining. Copper and cobalt are the main exports. Most trade is with South Africa.

ZIMBABWE
Republic of Zimbabwe

Area Sq Km	390 759	Languages	English, Shona, Ndebele
Area Sq Miles	150 873	Religions	Christian, traditional beliefs
Population	12 523 000	Currency	Zimbabwean dollar (suspended)
Capital	Harare	Organizations	SADC, UN

Map page 179

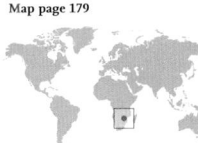

Zimbabwe, a landlocked state in south-central Africa, consists of high plateaus flanked by the Zambezi river valley and Lake Kariba in the north and the Limpopo river in the south. Most of the population lives in the centre of the country. There are significant mineral resources, including gold, nickel, copper, asbestos, platinum and chromium. Agriculture is a major sector of the economy, with crops including tobacco, maize, sugar cane and cotton. Beef cattle are also important. Exports include tobacco, gold, ferroalloys, nickel and cotton. South Africa is the main trading partner. The economy has suffered recently through significant political unrest and instability.

© Collins Bartholomew Ltd

EARTHQUAKES AND VOLCANOES

DISTRIBUTION OF EARTHQUAKES AND VOLCANOES

- ● Deadliest earthquake
- ● Earthquake of magnitude >=7.5
- ∘ Earthquake of magnitude 5.5 – 7.5
- ▲ Major volcano
- ▲ Other volcano

DEADLIEST EARTHQUAKES 1900–2010

Year	Location	Deaths
1905	**Kangra**, India	19 000
1907	west of **Dushanbe**, Tajikistan	12 000
1908	**Messina**, Italy	110 000
1915	**Abruzzo**, Italy	35 000
1917	**Bali**, Indonesia	15 000
1920	**Ningxia Province**, China	200 000
1923	**Tōkyō**, Japan	142 807
1927	**Qinghai Province**, China	200 000
1932	**Gansu Province**, China	70 000
1933	**Sichuan Province**, China	10 000
1934	**Nepal/India**	10 700
1935	**Quetta**, Pakistan	30 000
1939	**Chillán**, Chile	28 000
1939	**Erzincan**, Turkey	32 700
1948	**Aşgabat**, Turkmenistan	19 800
1962	northwest **Iran**	12 225
1970	**Huánuco Province**, Peru	66 794
1974	**Yunnan** and **Sichuan Provinces**, China	20 000
1975	**Liaoning Province**, China	10 000
1976	central **Guatemala**	22 778
1976	**Tangshan**, Hebei Province, China	255 000
1978	**Khorāsān Province**, Iran	20 000
1980	**Chlef**, Algeria	11 000
1988	**Spitak**, Armenia	25 000
1990	**Manjil**, Iran	50 000
1999	**İzmit (Kocaeli)**, Turkey	17 000
2001	**Gujarat**, India	20 000
2003	**Bam**, Iran	26 271
2004	**Sumatra**, Indonesia/Indian Ocean	>225 000
2005	northwest **Pakistan**	74 648
2008	**Sichuan Province**, China	>60 000
2010	**Léogâne**, Haiti	222 570

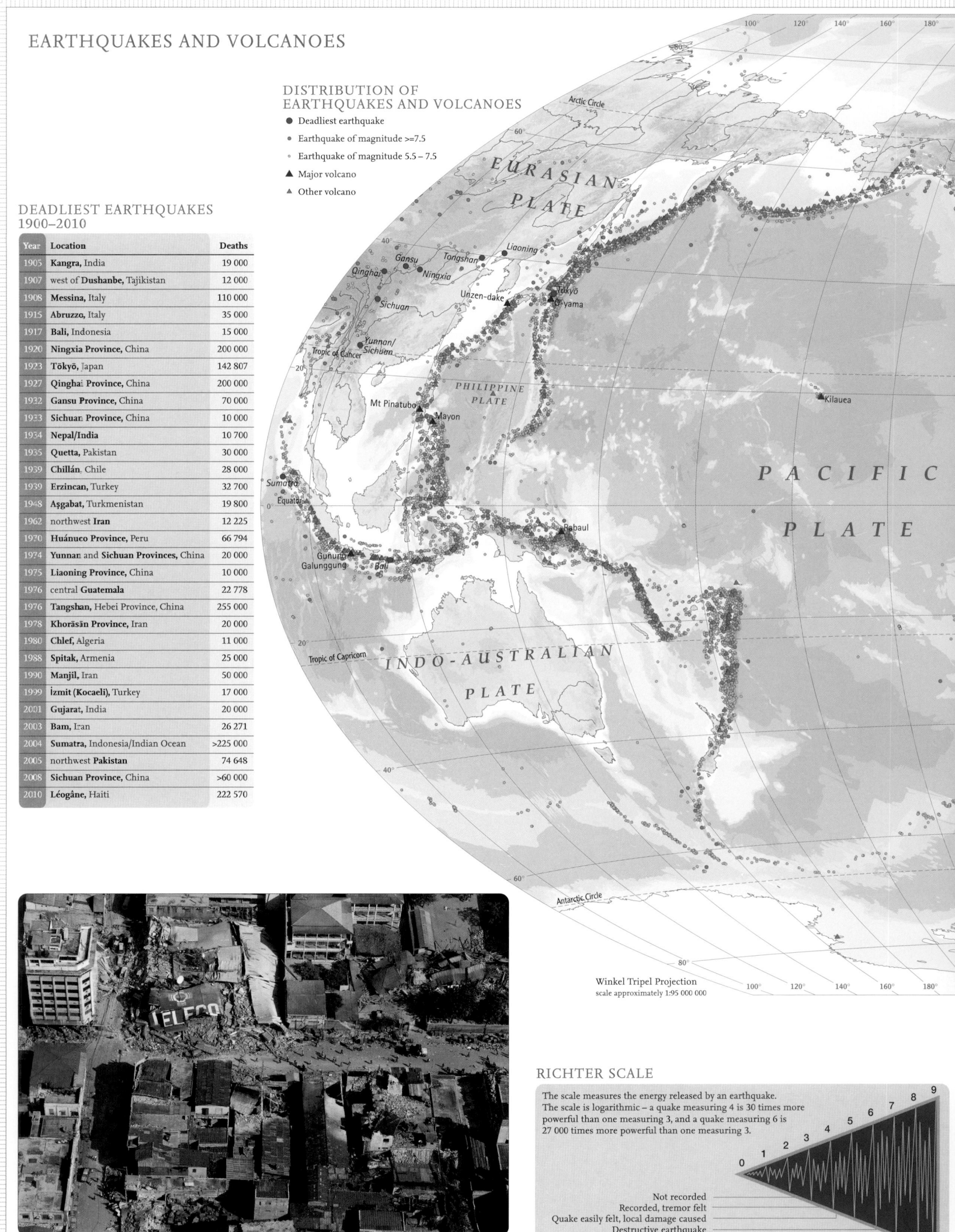

Winkel Tripel Projection
scale approximately 1:95 000 000

The capital city of Port-au-Prince in Haiti was destroyed by a massive 7.0 magnitude earthquake in January 2010. It was estimated that 222 570 people died and more than one million people lost their homes.

RICHTER SCALE

The scale measures the energy released by an earthquake. The scale is logarithmic – a quake measuring 4 is 30 times more powerful than one measuring 3, and a quake measuring 6 is 27 000 times more powerful than one measuring 3.

Not recorded
Recorded, tremor felt
Quake easily felt, local damage caused
Destructive earthquake
Major earthquake
Most powerful earthquake recorded – 8.9

MAJOR VOLCANIC ERUPTIONS 1980–2010

Year	Volcano	Country
1980	Mt St Helens	USA
1982	El Chichónal	Mexico
1982	Gunung Galunggung	Indonesia
1983	Kilauea	Hawaii, USA
1983	Ō-yama	Japan
1985	Nevado del Ruiz	Colombia
1991	Mt Pinatubo	Philippines
1991	Unzen-dake	Japan
1993	Mayon	Philippines
1993	Volcán Galeras	Colombia
1994	Volcán Llaima	Chile
1994	Rabaul	Papua New Guinea
1997	Soufrière Hills	Montserrat
2000	Hekla	Iceland
2001	Mount Etna	Italy
2002	Nyiragongo	Democratic Republic of the Congo

Mount Bromo, Java, Indonesia, one of the many active volcanoes that have formed around the edge of the Pacific Ocean.

© Collins Bartholomew Ltd

CLIMATE I

MAJOR CLIMATIC REGIONS AND SUB-TYPES 2006

Köppen classification system
Winkel Tripel Projection
scale 1:110 000 000

● Climate graph location
○ Weather extreme location

Polar
| EF | Ice cap |
| ET | Tundra |

Cooler humid
Dc Dd	Subarctic
Db	Continental cool summer
Da	Continental warm summer

Warmer humid
Cb Cc	Temperate
Ca	Humid subtropical
Cs	Mediterranean

Dry
| BS | Steppe |
| BW | Desert |

Tropical humid
| Aw As | Savanna |
| Af Am | Rain forest |

A Rainy climate with no winter: coolest month above 18°C (64.4°F).

B Dry climates; limits are defined by formulae based on rainfall effectiveness:
BS Steppe or semi-arid climate.
BW Desert or arid climate.

*C Rainy climates with mild winters: coolest month above 0°C (32°F), but below 18°C (64.4°F); warmest month above 10°C (50°F).

*D Rainy climates with severe winters: coldest month below 0°C (32°F); warmest month above 10°C (50°F).

E Polar climates with no warm season: warmest month below 10°C (50°F).
ET Tundra climate: warmest month below 10°C (50°F) but above 0°C (32°F).
EF Perpetual frost: all months below 0°C (32°F).

a Warmest month above 22°C (71.6°F).

b Warmest month below 22°C (71.6°F).

c Less than four months over 10°C (50°F).

d As 'c', but with severe cold: coldest month below -38°C (-36.4°F).

f Constantly moist rainfall throughout the year.

*h Warmer dry: all months above 0°C (32°F).

*k Cooler dry: at least one month below 0°C (32°F).

m Monsoon rain: short dry season, but is compensated by heavy rains during rest of the year.

n Frequent fog.

s Dry season in summer.

w Dry season in winter.

* Modification of Köppen definition

WORLD WEATHER EXTREMES

	Location
Highest shade temperature	57.8°C / 136°F Al 'Azīzīyah, Libya (13th September 1922)
Hottest place – Annual mean	34.4°C / 93.9°F Dalol, Ethiopia
Driest place – Annual mean	0.1 mm / 0.004 inches Atacama Desert, Chile
Most sunshine – Annual mean	90% Yuma, Arizona, USA (over 4 000 hours)
Lowest screen temperature	-89.2°C / -128.6°F Vostok Station, Antarctica (21st July 1983)
Coldest place – Annual mean	-56.6°C / -69.9°F Plateau Station, Antarctica
Wettest place – Annual mean	11 873 mm / 467.4 inches Meghalaya, India
Most rainy days	Up to 350 per year Mount Waialeale, Hawaii, USA
Windiest place	322 km per hour / 200 miles per hour in gales, Commonwealth Bay, Antarctica
Highest surface wind speed	512 km per hour / 318 miles per hour in a tornado, Oklahoma City, Oklahoma, USA (3rd May 1999)
Greatest snowfall	31 102 mm / 1 224.5 inches Mount Rainier, Washington, USA (19th February 1971 – 18th February 1972)
Highest barometric pressure	1 083.8 mb Agata, Siberia, Russian Federation (31st December 1968)
Lowest barometric pressure	870 mb 483 km / 300 miles west of Guam, Pacific Ocean (12th October 1979)

Tropical Cyclone Nargis, reached Category 4 status with winds of 210 km per hr (130 miles per hr), while crossing the Bay of Bengal. By the time this image was taken on 3 May 2008, it had weakened to tropical storm strength, but the path of the cyclone took it over the coastal plains of Myanmar and almost directly over the city of Rangoon. There was extensive flooding and many thousands of people were killed.

TRACKS OF TROPICAL STORMS

(wind speeds often over 160 km per hour)
scale 1:247 000 000

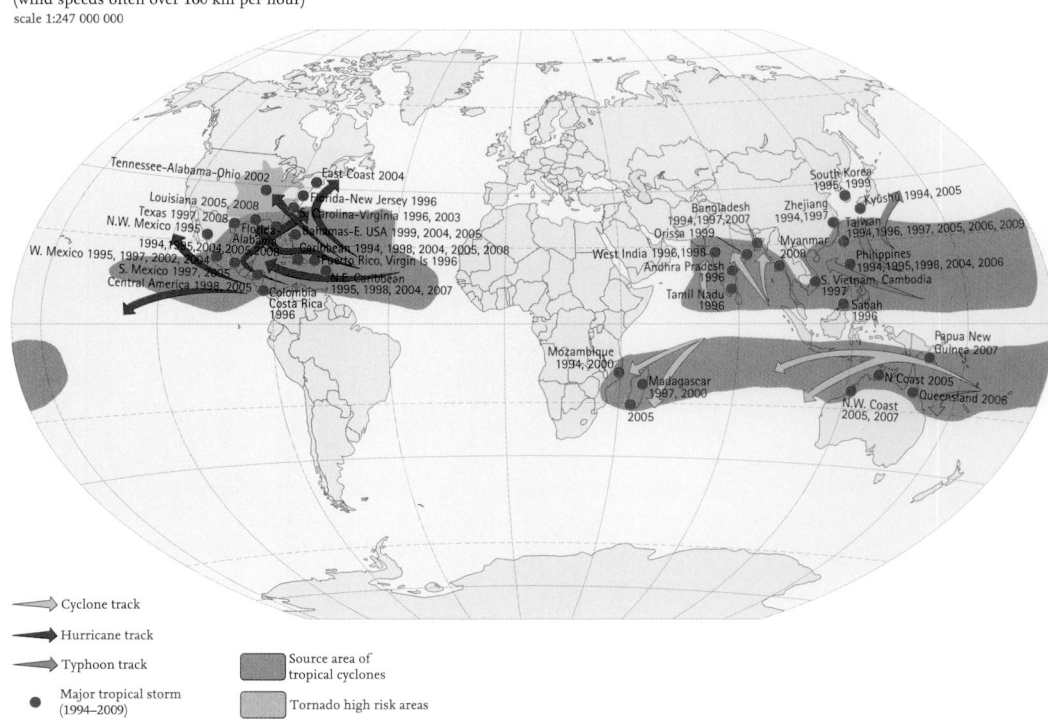

⇨ Cyclone track
➡ Hurricane track
⇨ Typhoon track
● Major tropical storm (1994–2009)

▮ Source area of tropical cyclones
▮ Tornado high risk areas

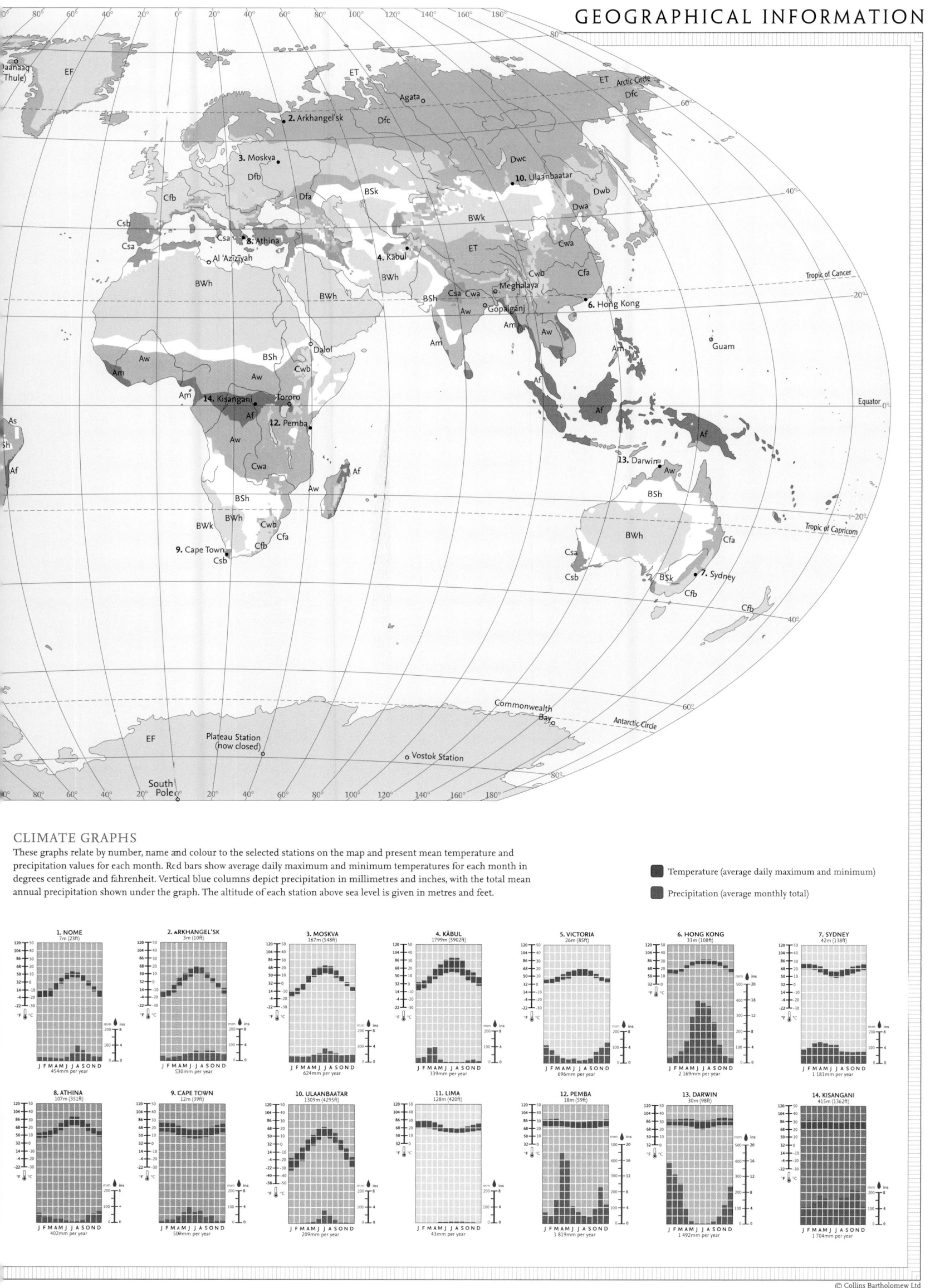

CLIMATE GRAPHS

These graphs relate by number, name and colour to the selected stations on the map and present mean temperature and precipitation values for each month. Red bars show average daily maximum and minimum temperatures for each month in degrees centigrade and fahrenheit. Vertical blue columns depict precipitation in millimetres and inches, with the total mean annual precipitation shown under the graph. The altitude of each station above sea level is given in metres and feet.

■ Temperature (average daily maximum and minimum)

■ Precipitation (average monthly total)

1. NOME 7m (23ft) — 454mm per year

2. ARKHANGEL'SK 3m (10ft) — 530mm per year

3. MOSKVA 167m (548ft) — 624mm per year

4. KĀBUL 1799m (5902ft) — 339mm per year

5. VICTORIA 26m (85ft) — 696mm per year

6. HONG KONG 33m (108ft) — 2 169mm per year

7. SYDNEY 42m (138ft) — 1 181mm per year

8. ATHINA 107m (351ft) — 402mm per year

9. CAPE TOWN 12m (39ft) — 509mm per year

10. ULAANBAATAR 1309m (4295ft) — 209mm per year

11. LIMA 128m (420ft) — 43mm per year

12. PEMBA 18m (59ft) — 1 819mm per year

13. DARWIN 30m (98ft) — 1 492mm per year

14. KISANGANI 415m (1362ft) — 1 704mm per year

© Collins Bartholomew Ltd

CLIMATE II

CLIMATE CHANGE

In 2008 the global mean temperature was over 0.7°C higher than that at the end of the nineteenth century. Most of this warming is caused by human activities which result in a build-up of greenhouse gases, mainly carbon dioxide, allowing heat to be trapped within the atmosphere. Carbon dioxide emissions have increased since the beginning of the industrial revolution due to burning of fossil fuels, increased urbanization, population growth, deforestation and industrial pollution.

Annual climate indicators such as number of frost-free days, length of growing season, heat wave frequency, number of wet days, length of dry spells and frequency of weather extremes are used to monitor climate change. The map below shows how future changes in temperature will not be spread evenly around the world. Some regions will warm faster than the global average, while others will warm more slowly.

The McCarty Glacier in the Kenai Peninsula in Alaska is a tidewater glacier which has retreated around 16 km between 1909 (top) and 2004 (bottom).

PROJECTION OF GLOBAL TEMPERATURES 2090–2099

Based on IPCC scenario A1B. Change relative to 1980–1999.

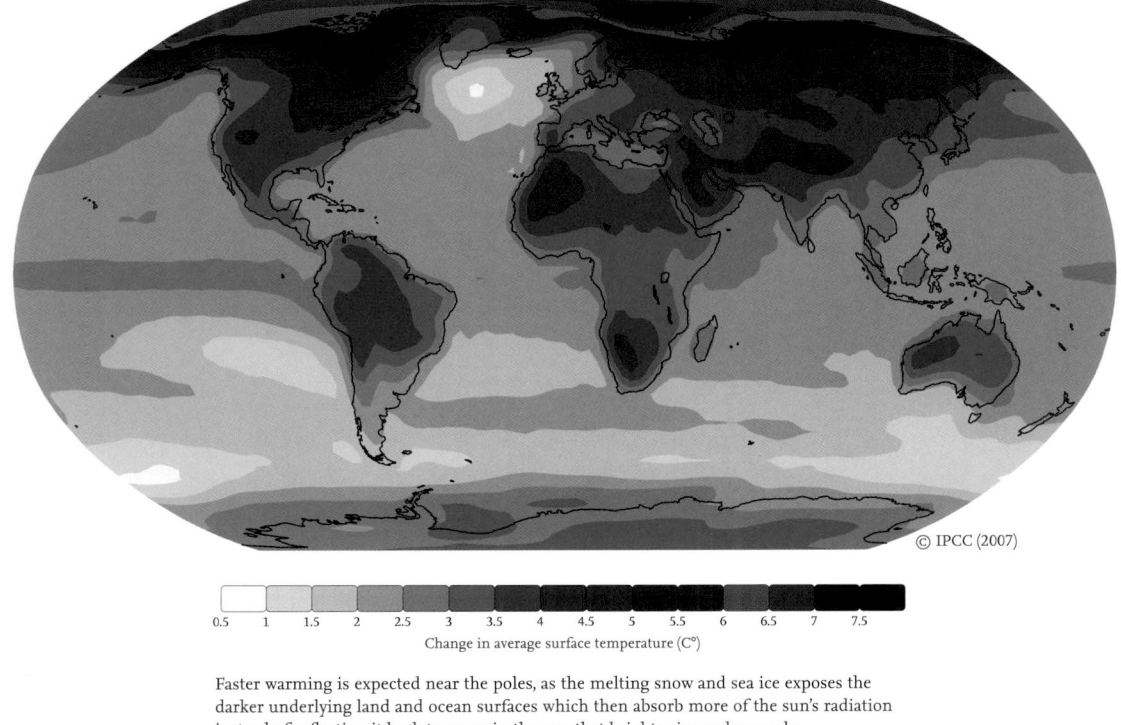

© IPCC (2007)

| 0.5 | 1 | 1.5 | 2 | 2.5 | 3 | 3.5 | 4 | 4.5 | 5 | 5.5 | 6 | 6.5 | 7 | 7.5 |

Change in average surface temperature (C°)

Faster warming is expected near the poles, as the melting snow and sea ice exposes the darker underlying land and ocean surfaces which then absorb more of the sun's radiation instead of reflecting it back to space in the way that brighter ice and snow do.

THREAT OF RISING SEA LEVEL

It has been suggested that further global warming of between 1.0 and 6.4 C° may occur by the end of the 21st century. Sea level is projected to rise by between 28 cm and 58 cm, threatening a number of coastal cities, low-lying deltas and small islands. Larger rises are predicted in some locations than others.

AREAS AT RISK OF SUBMERSION

○ Major cities

▭ Coastal areas at greatest risk

◣ Islands and archipelagos

◣ Areas of low-lying islands

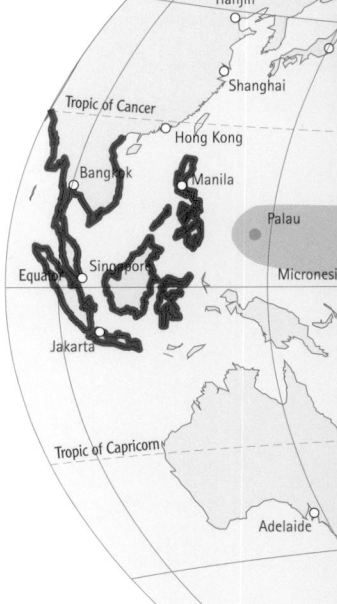

LOWEST PACIFIC ISLANDS

Location	Maximum height above sea level	Land area sq km	sq miles	Population
Kingman Reef	1 m (3 ft)	1	0.4	0
Palmyra Atoll	2 m (7 ft)	12	5	0
Ashmore and Cartier Islands	3 m (10 ft)	5	2	0
Howland Island	3 m (10 ft)	2	1	0
Johnston Atoll	5 m (16 ft)	3	1	0
Tokelau	5 m (16 ft)	10	4	1 000
Tuvalu	5 m (16 ft)	25	10	10 000
Coral Sea Islands Territory	6 m (20 ft)	22	8	0
Wake Island	6 m (20 ft)	7	3	0
Jarvis Island	7 m (23 ft)	5	2	0

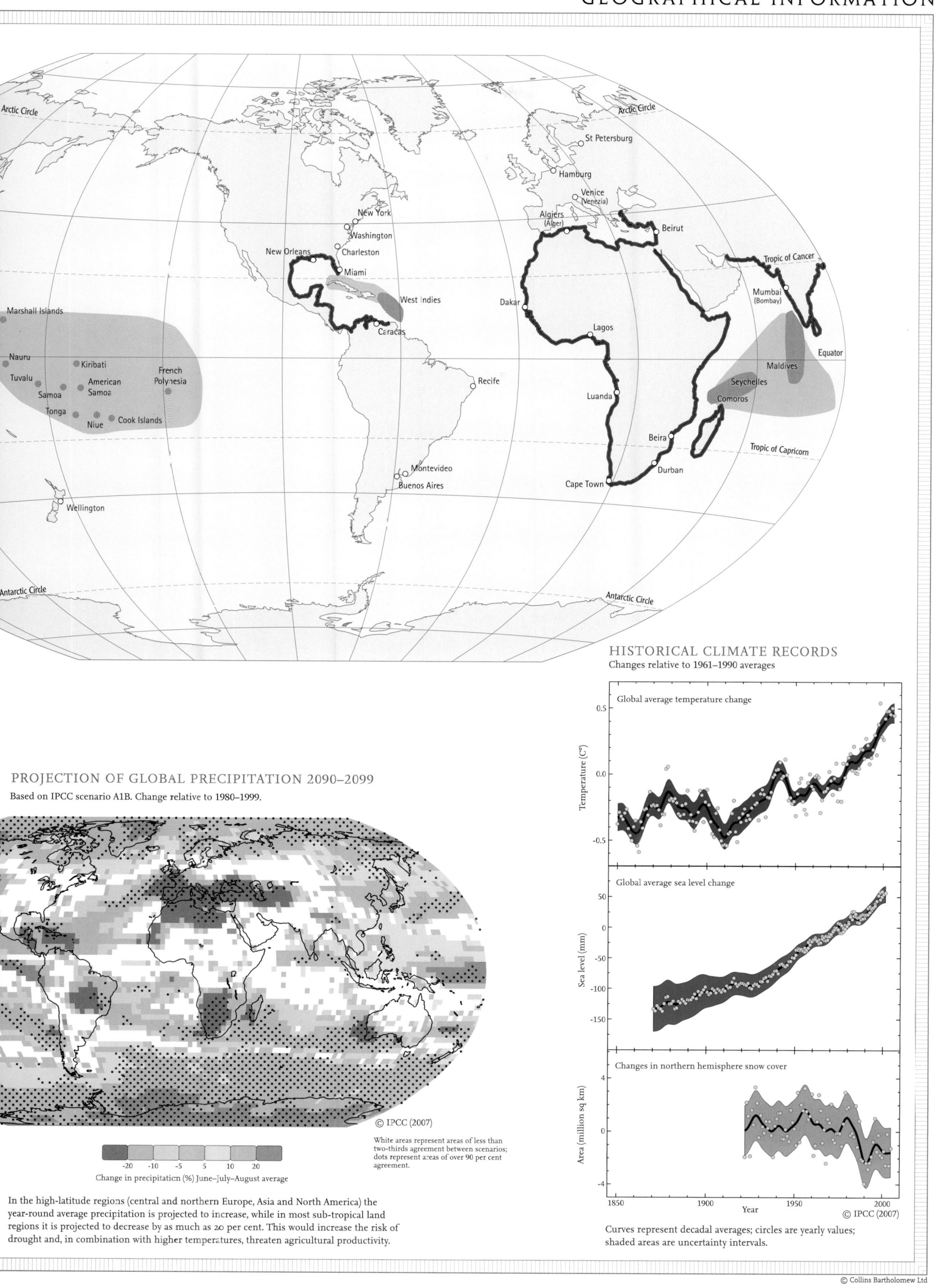

© IPCC (2007)

HISTORICAL CLIMATE RECORDS
Changes relative to 1961–1990 averages

Global average temperature change

Global average sea level change

Changes in northern hemisphere snow cover

© IPCC (2007)

Curves represent decadal averages; circles are yearly values;
shaded areas are uncertainty intervals.

PROJECTION OF GLOBAL PRECIPITATION 2090–2099
Based on IPCC scenario A1B. Change relative to 1980–1999.

-20   -10   -5   5   10   20
Change in precipitation (%) June–July–August average

White areas represent areas of less than
two-thirds agreement between scenarios;
dots represent areas of over 90 per cent
agreement.

In the high-latitude regions (central and northern Europe, Asia and North America) the
year-round average precipitation is projected to increase, while in most sub-tropical land
regions it is projected to decrease by as much as 20 per cent. This would increase the risk of
drought and, in combination with higher temperatures, threaten agricultural productivity.

© Collins Bartholomew Ltd

POPULATION

TOP TWENTY COUNTRIES BY POPULATION AND POPULATION DENSITY 2009

Total population	Country	Rank	Country*	Inhabitants per sq mile	Inhabitants per sq km
1 330 265 000	China	1	Bangladesh	2 918	1 127
1 198 003 000	India	2	Taiwan	1 650	637
314 659 000	United States of America	3	South Korea	1 261	487
229 965 000	Indonesia	4	Netherlands	1 035	400
193 734 000	Brazil	5	India	1 012	391
180 808 000	Pakistan	6	Haiti	936	362
162 221 000	Bangladesh	7	Belgium	904	349
154 729 000	Nigeria	8	Japan	872	337
140 874 000	Russian Federation	9	Sri Lanka	799	308
127 156 000	Japan	10	Philippines	794	307
109 610 000	Mexico	11	Vietnam	692	267
91 983 000	Philippines	12	United Kingdom	655	253
88 069 000	Vietnam	13	Germany	596	230
82 999 000	Egypt	14	Pakistan	583	225
82 825 000	Ethiopia	15	Dominican Republic	539	208
82 167 000	Germany	16	Nepal	516	199
74 816 000	Turkey	17	Italy	515	199
74 196 000	Iran	18	North Korea	514	198
67 764 000	Thailand	19	Nigeria	434	168
66 020 000	Democratic Republic of the Congo	20	China	360	139

*Only countries with a population of over 10 million are considered.

AGE PYRAMIDS

World population by five-year age group and sex.

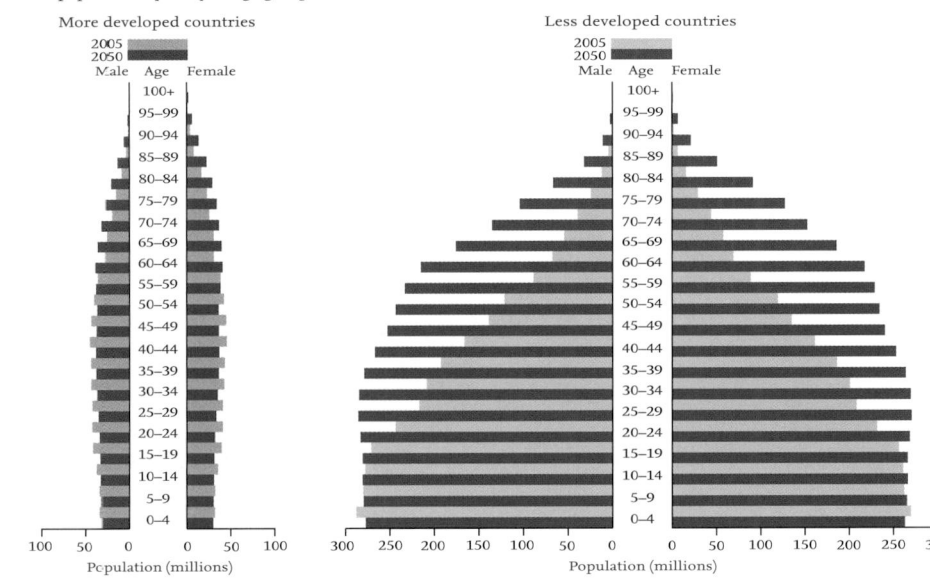

KEY POPULATION STATISTICS FOR MAJOR REGIONS

	Population 2009 (millions)	Growth (per cent)	Infant mortality rate	Total fertility rate	Life expectancy (years)	% aged 60 and over 2010	% aged 60 and over 2050
World	6 829	1.2	47	2.6	68	11	22
More developed regions[1]	1 223	0.1	6	1.6	77	22	33
Less developed regions[2]	5 596	1.4	52	2.7	66	9	20
Africa	1 010	2.4	83	4.6	54	5	11
Asia	4 121	1.2	42	2.4	69	10	24
Europe[3]	732	-0.1	7	1.5	75	22	34
Latin America and the Caribbean[4]	582	1.3	22	2.3	73	10	26
North America	348	0.6	6	2.0	79	18	28
Oceania	35	1.0	23	2.4	76	15	24

Except for population and % aged 60 and over figures, the data are annual averages projected for the period 2005–2010.

1. Europe, North America, Australia, New Zealand and Japan.
2. Africa, Asia (excluding Japan), Latin America and the Caribbean, and Oceania (excluding Australia and New Zealand).
3. Includes Russian Federation.
4. South America, Central America (including Mexico) and all Caribbean Islands.

WORLD POPULATION DISTRIBUTION

Winkel Tripel Projection
scale approximately 1:109 000 000

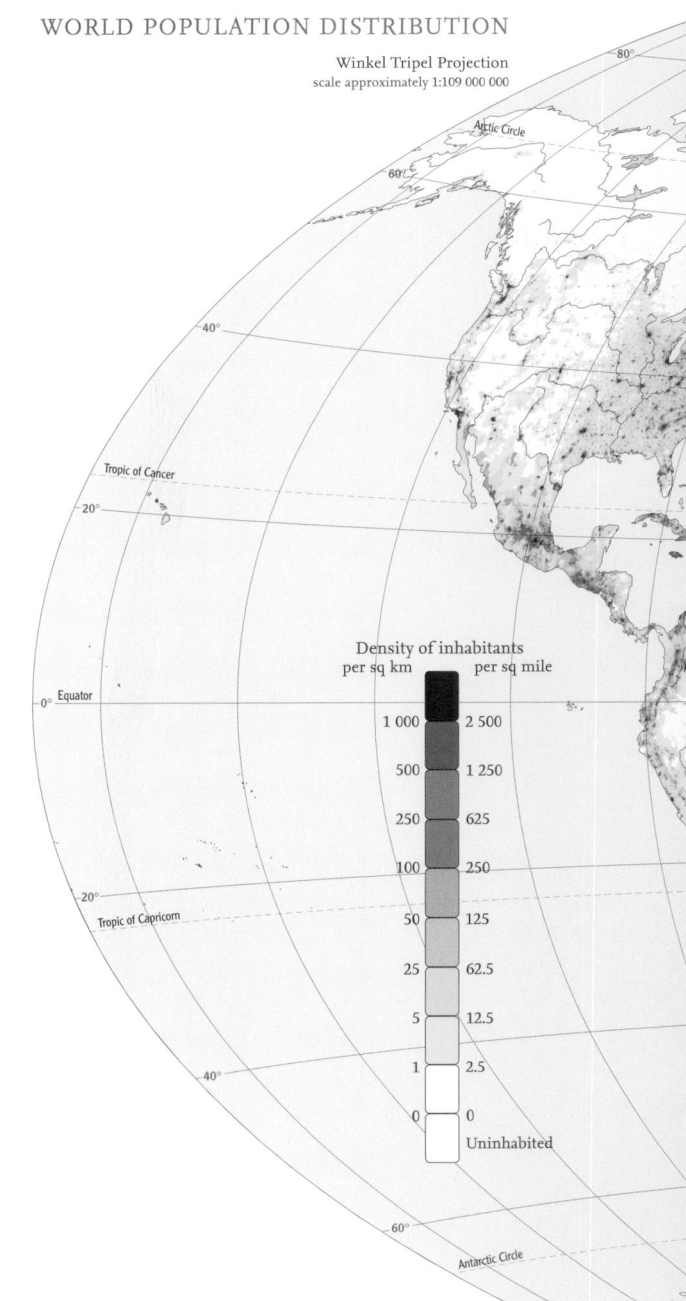

Density of inhabitants

per sq km	per sq mile
1 000	2 500
500	1 250
250	625
100	250
50	125
25	62.5
5	12.5
1	2.5
0	0
Uninhabited	

Population growth in the 20th century was rapid and continued growth could carry the world's population past seven billion by 2015.

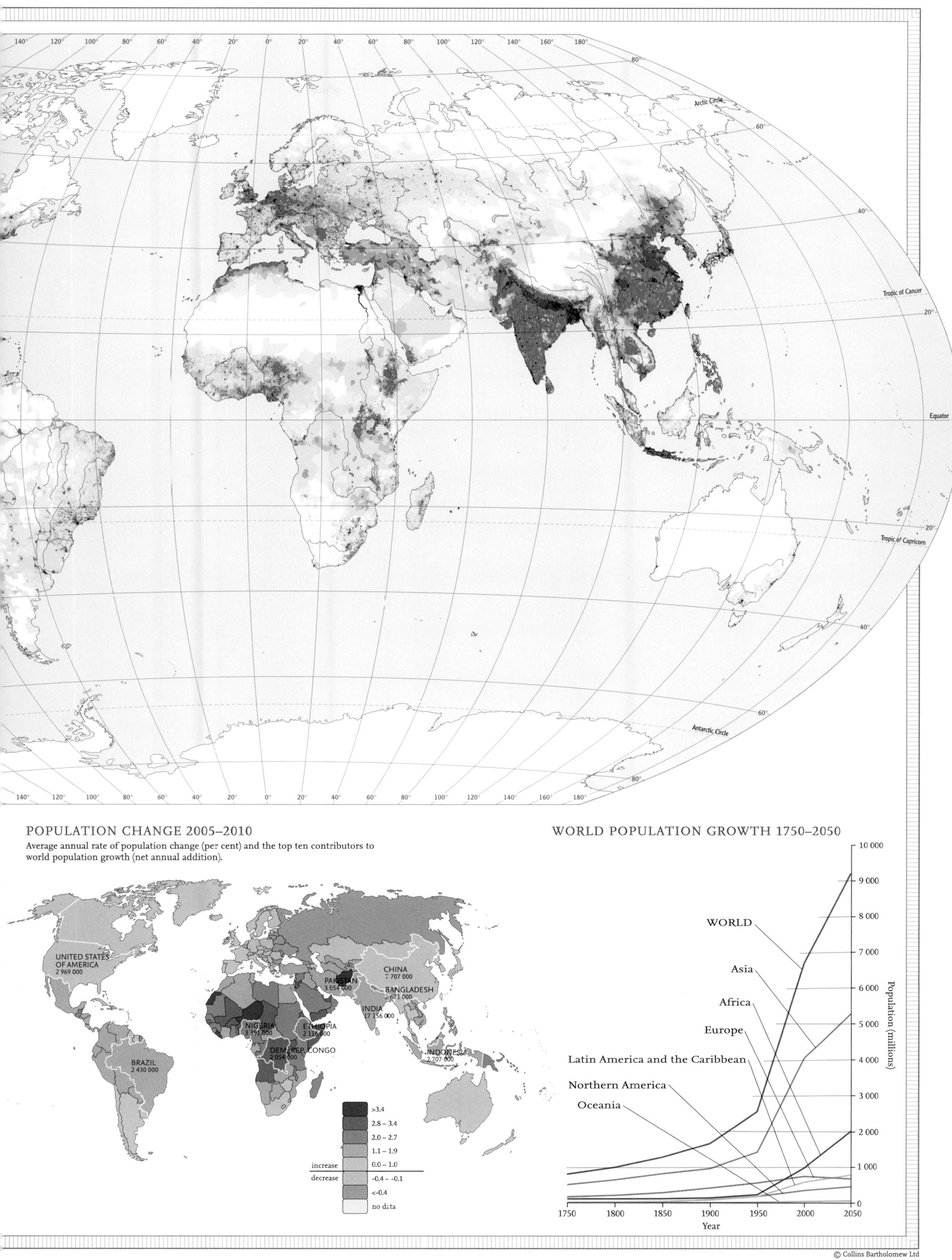

POPULATION CHANGE 2005–2010

Average annual rate of population change (per cent) and the top ten contributors to world population growth (net annual addition).

UNITED STATES
OF AMERICA
2 969 000

CHINA
7 707 000

PAKISTAN
3 054 000

BANGLADESH
2 671 000

INDIA
17 156 000

NIGERIA
3 391 000

ETHIOPIA
2 116 000

DEM. REP. CONGO
2 054 000

INDONESIA
2 707 000

BRAZIL
2 430 000

> 3.4
2.8 – 3.4
2.0 – 2.7
1.1 – 1.9
increase 0.0 – 1.0
decrease -0.4 – -0.1
< -0.4
no data

WORLD POPULATION GROWTH 1750–2050

WORLD

Asia

Africa

Europe

Latin America and the Caribbean

Northern America

Oceania

Population (millions)

10 000
9 000
8 000
7 000
6 000
5 000
4 000
3 000
2 000
1 000
0

1750   1800   1850   1900   1950   2000   2050

Year

© Collins Bartholomew Ltd

COMMUNICATIONS

WORLD COMMUNICATIONS EQUIPMENT 1993–2007

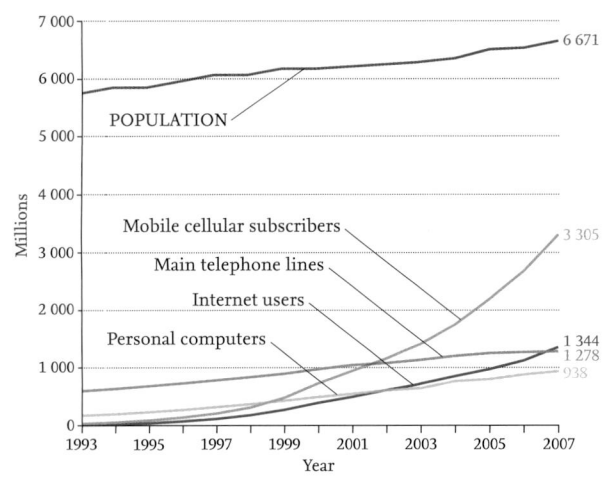

TOP BROADBAND ECONOMIES 2008

Countries with the highest broadband penetration rate – subscribers per 100 inhabitants

	Top Economies	Rate
1	Sweden	37.3
2	Denmark	36.8
3	Netherlands	35.0
4	Norway	34.0
5	Switzerland	33.0
6	Iceland	32.9
7	South Korea	32.0
8	Finland	30.6
9	Luxembourg	30.3
10	Canada	29.0
11	France	28.6
12	United Kingdom	28.3
13	Belgium	28.3
14	Germany	27.4
15	Hong Kong, China	26.8
16	USA	25.6
17	Macao, China	25.1
18	Australia	24.5
19	Malta	24.2
20	Estonia	23.9

INTERNATIONAL TELECOMMUNICATIONS INDICATORS BY REGION 2007

Telecommunications indicators by region 2007

- Africa
- North America
- Latin America and the Caribbean*
- Europe
- Asia
- Oceania

*Includes Mexico

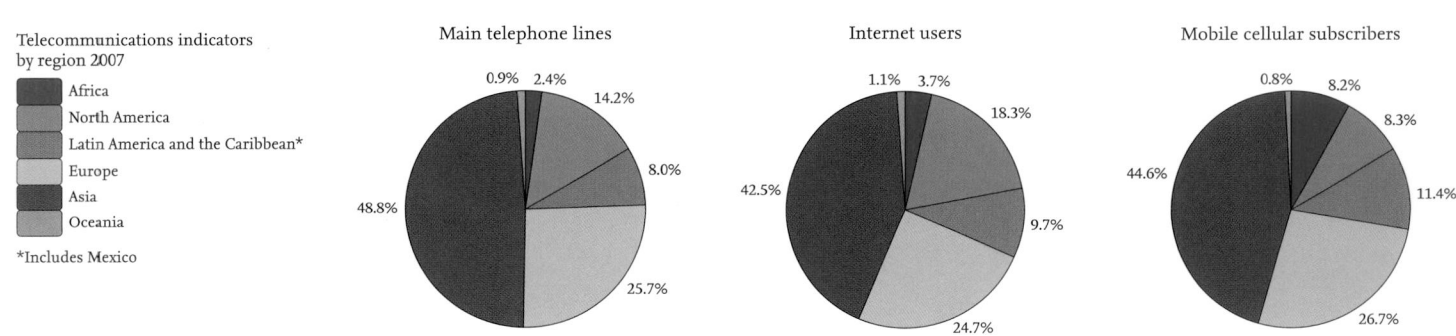

INTERNATIONAL TELECOMMUNICATIONS TRAFFIC 2008

Winkel Tripel Projection
1:116 000 000

Telephone lines per 100 inhabitants 2008

- over 50.0
- 35.0 – 50.0
- 15.0 – 34.9
- 10.0 – 14.9
- 5.0 – 9.9
- 1.0 – 4.9
- 0 – 0.9
- no data

Total telephone lines 2008

Europe
Total telephone lines
318 558 000

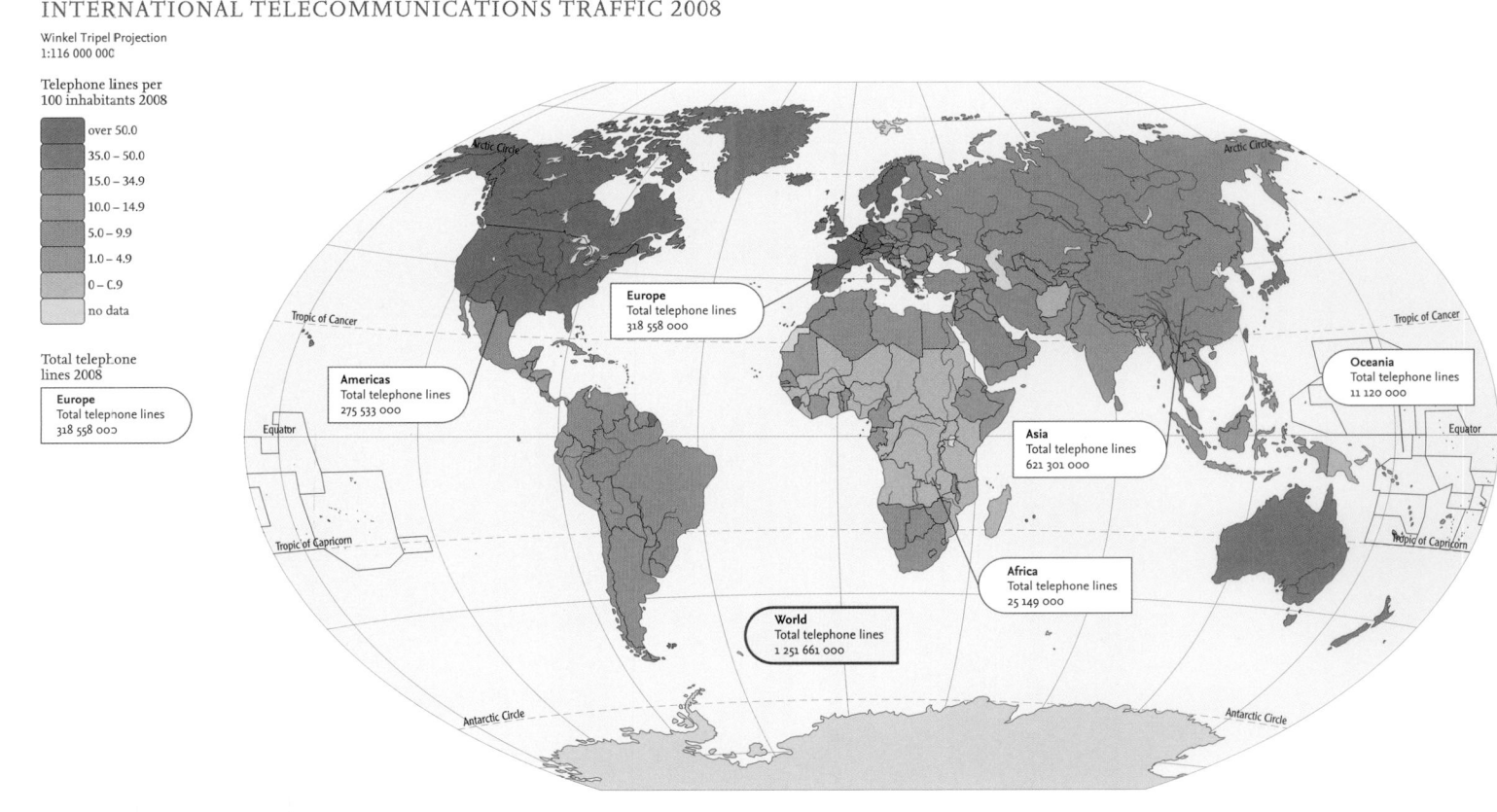

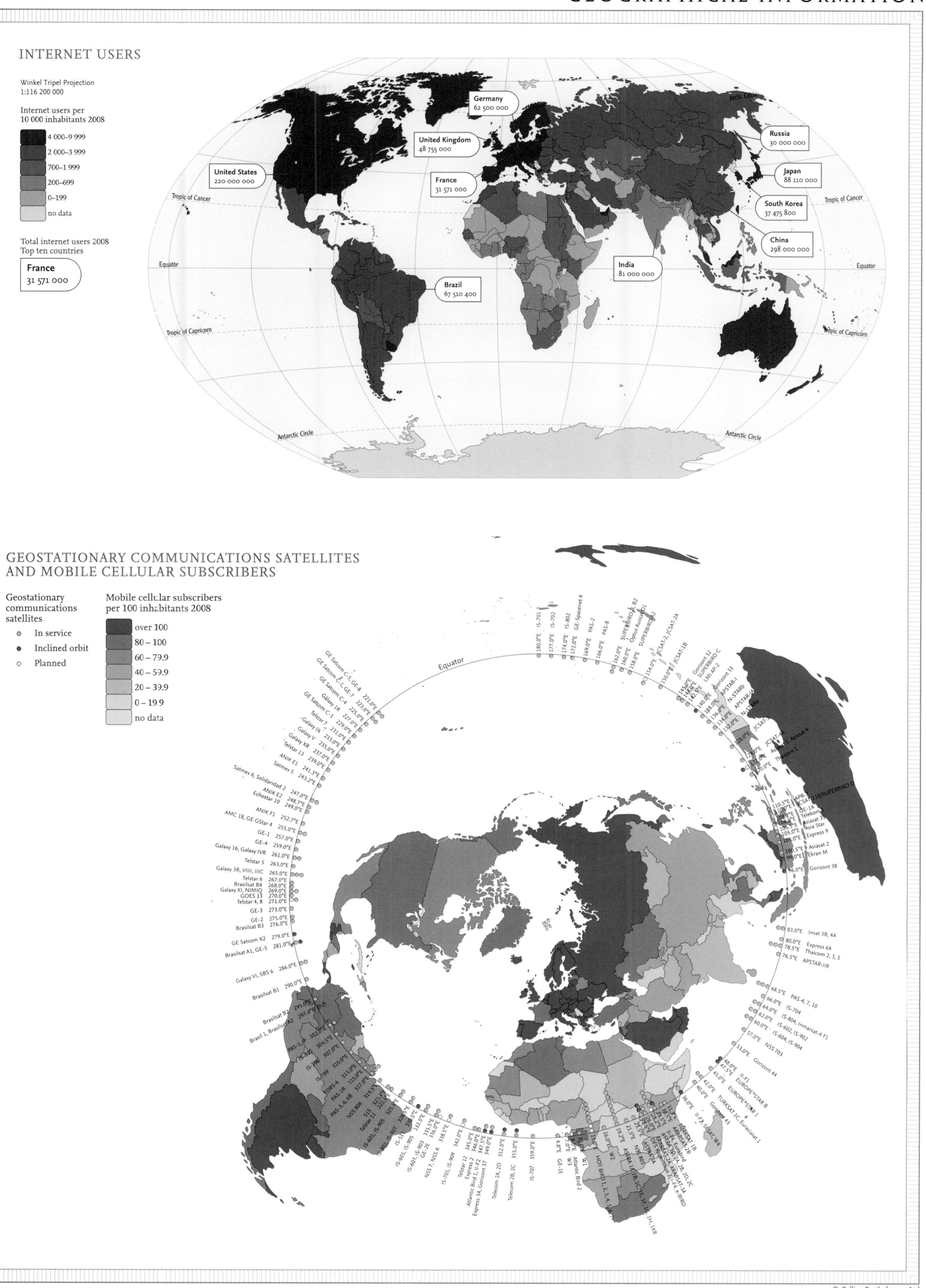

INTERNET USERS

Winkel Tripel Projection
1:116 200 000

Internet users per
10 000 inhabitants 2008

- 4 000–9 999
- 2 000–3 999
- 700–1 999
- 200–699
- 0–199
- no data

Total internet users 2008
Top ten countries

France
31 571 000

Germany
62 500 000

United Kingdom
48 755 000

United States
220 000 000

France
31 571 000

Russia
30 000 000

Japan
88 110 000

South Korea
37 475 800

China
298 000 000

India
81 000 000

Brazil
67 510 400

Arctic Circle
Tropic of Cancer
Equator
Tropic of Capricorn
Antarctic Circle

GEOSTATIONARY COMMUNICATIONS SATELLITES AND MOBILE CELLULAR SUBSCRIBERS

Geostationary
communications
satellites

- In service
- Inclined orbit
- Planned

Mobile cellular subscribers
per 100 inhabitants 2008

- over 100
- 80 – 100
- 60 – 79.9
- 40 – 59.9
- 20 – 39.9
- 0 – 19.9
- no data

© Collins Bartholomew Ltd

PHYSICAL FEATURES

The images below illustrate some of the major physical features of the world.

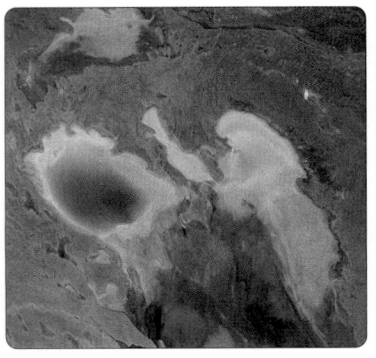

Lake Eyre, South Australia

Mississippi-Missouri, United States of America

The Caspian Sea

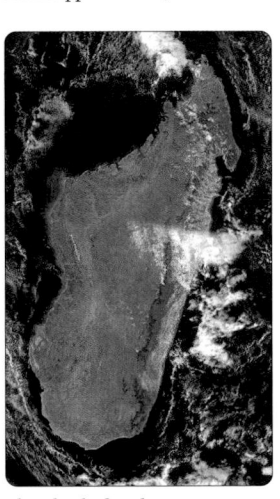

The island of Madagasgar

OCEANIA Total Land Area 8 844 516 sq km / 3 414 887 sq miles

HIGHEST MOUNTAINS	metres	feet
Puncak Jaya, Indonesia	5 030	16 502
Puncak Trikora, Indonesia	4 730	15 518
Puncak Mandala, Indonesia	4 700	15 420
Puncak Yamin, Indonesia	4 595	15 075
Mt Wilhelm, Papua New Guinea	4 509	14 793
Mt Kubor, Papua New Guinea	4 359	14 301

LONGEST RIVERS	km	miles
Murray-Darling	3 672	2 282
Darling	2 844	1 767
Murray	2 375	1 476
Murrumbidgee	1 690	1 050
Lachlan	1 480	920
Macquarie	950	590

LARGEST ISLANDS	sq km	sq miles
New Guinea	808 510	312 167
South Island, New Zealand	151 215	58 384
North Island, New Zealand	115 777	44 701
Tasmania	67 800	26 178

LARGEST LAKES	sq km	sq miles
Lake Eyre	0–8 900	0–3 436
Lake Torrens	0–5 780	0–2 232

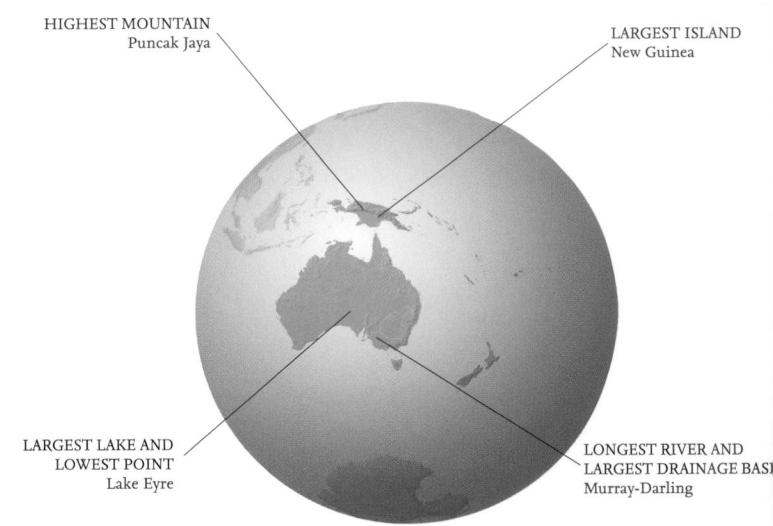

HIGHEST MOUNTAIN
Puncak Jaya

LARGEST ISLAND
New Guinea

LARGEST LAKE AND
LOWEST POINT
Lake Eyre

LONGEST RIVER AND
LARGEST DRAINAGE BASIN
Murray-Darling

ANTARCTICA Total Land Area 12 093 000 sq km / 4 669 133 sq miles (excluding ice shelves)

HIGHEST MOUNTAINS	metres	feet
Vinson Massif	4 897	16 066
Mt Tyree	4 852	15 918
Mt Kirkpatrick	4 528	14 855
Mt Markham	4 351	14 275
Mt Jackson	4 190	13 747
Mt Sidley	4 181	13 717

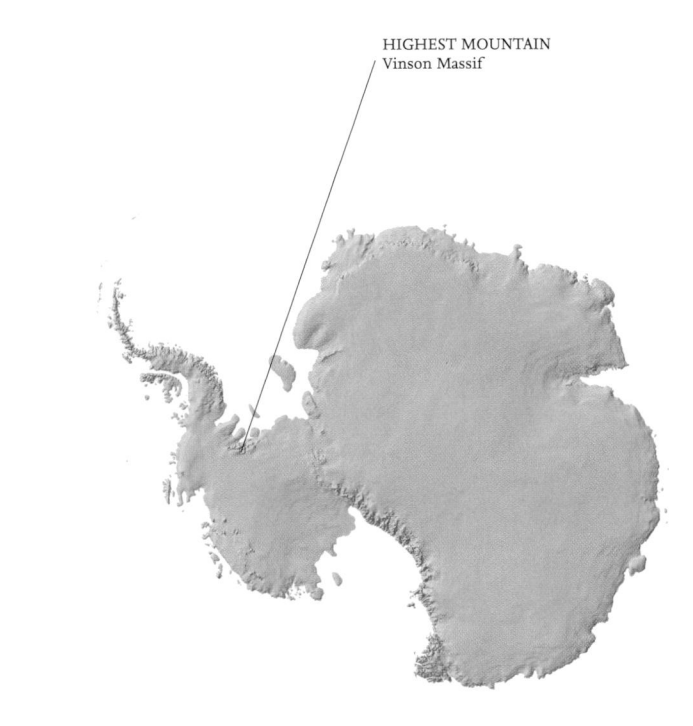

HIGHEST MOUNTAIN
Vinson Massif

ASIA Total Land Area 45 036 492 sq km / 17 388 686 sq miles

HIGHEST MOUNTAINS	metres	feet
Mt Everest (Sagarmatha/ Qomolangma Feng), China/Nepal	8 848	29 028
K2 (Qogir Feng), China/Pakistan	8 611	28 251
Kangchenjunga, India/Nepal	8 586	28 169
Lhotse, China/Nepal	8 516	27 939
Makalu, China/Nepal	8 463	27 765
Cho Oyu, China/Nepal	8 201	26 906

LONGEST RIVERS	km	miles
Yangtze (Chang Jiang)	6 380	3 965
Ob'-Irtysh	5 568	3 460
Yenisey-Angara-Selenga	5 550	3 449
Yellow (Huang He)	5 464	3 395
Irtysh	4 440	2 759
Mekong	4 425	2 750

LARGEST ISLANDS	sq km	sq miles
Borneo	745 561	287 861
Sumatra (Sumatera)	473 606	182 859
Honshū	227 414	87 805
Celebes (Sulawesi)	189 216	73 056
Java (Jawa)	132 188	51 038
Luzon	104 690	40 421

LARGEST LAKES	sq km	sq miles
Caspian Sea	371 000	143 243
Lake Baikal (Ozero Baykal)	30 500	11 776
Lake Balkhash (Ozero Balkhash)	17 400	6 718
Aral Sea (Aral'skoye More)	17 158	6 625
Ysyk-Köl	6 200	2 394

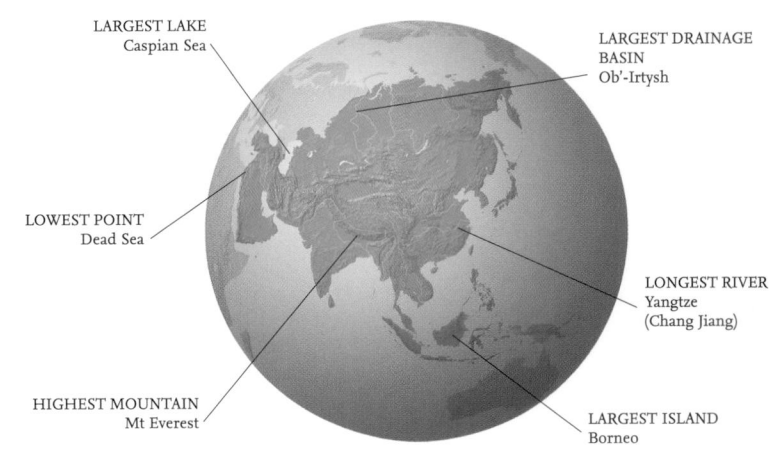

LARGEST LAKE
Caspian Sea

LARGEST DRAINAGE BASIN
Ob'-Irtysh

LOWEST POINT
Dead Sea

LONGEST RIVER
Yangtze (Chang Jiang)

HIGHEST MOUNTAIN
Mt Everest

LARGEST ISLAND
Borneo

EUROPE Total Land Area 9 908 599 sq km / 3 825 731 sq miles

HIGHEST MOUNTAINS	metres	feet
El'brus, Russian Federation	5 642	18 510
Gora Dykh-Tau, Russian Federation	5 204	17 073
Shkhara, Georgia/Russian Federation	5 201	17 063
Kazbek, Georgia/Russian Federation	5 047	16 558
Mont Blanc, France/Italy	4 808	15 774
Dufourspitze, Italy/Switzerland	4 634	15 203

LONGEST RIVERS	km	miles
Volga	3 688	2 292
Danube	2 850	1 771
Dnieper	2 285	1 420
Kama	2 028	1 260
Don	1 931	1 200
Pechora	1 802	1 120

LARGEST ISLANDS	sq km	sq miles
Great Britain	218 476	84 354
Iceland	102 820	39 699
Novaya Zemlya	90 650	35 000
Ireland	83 045	32 064
Spitsbergen	37 814	14 600
Sicily (Sicilia)	25 426	9 817

LARGEST LAKES	sq km	sq miles
Caspian Sea	371 000	143 243
Lake Ladoga (Ladozhskoye Ozero)	18 390	7 100
Lake Onega (Onezhskoye Ozero)	9 600	3 707
Vänern	5 585	2 156
Rybinskoye Vodokhranilishche	5 180	2 000

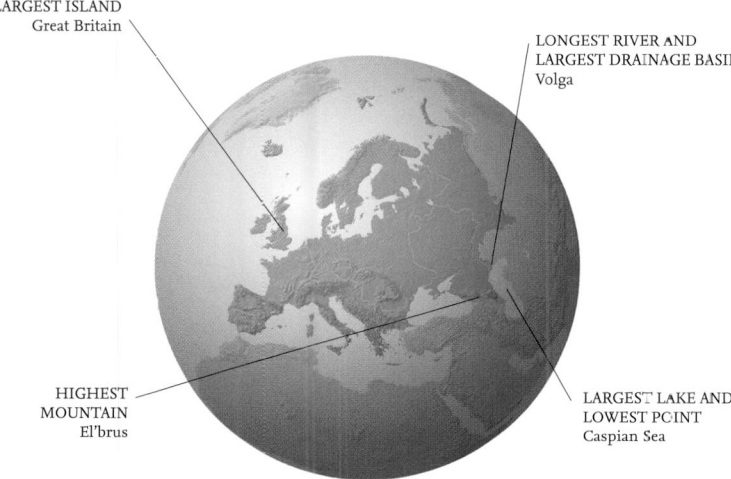

LARGEST ISLAND
Great Britain

LONGEST RIVER AND
LARGEST DRAINAGE BASIN
Volga

HIGHEST
MOUNTAIN
El'brus

LARGEST LAKE AND
LOWEST POINT
Caspian Sea

NORTH AMERICA Total Land Area 24 680 331 sq km / 9 529 129 sq miles

HIGHEST MOUNTAINS	metres	feet
Mt McKinley, USA	6 194	20 321
Mt Logan, Canada	5 959	19 550
Pico de Orizaba, Mexico	5 610	18 405
Mt St Elias, USA	5 489	18 008
Volcán Popocatépetl, Mexico	5 452	17 887
Mt Foraker, USA	5 303	17 398

LONGEST RIVERS	km	miles
Mississippi-Missouri	5 969	3 709
Mackenzie-Peace-Finlay	4 241	2 635
Missouri	4 086	2 539
Mississippi	3 765	2 340
Yukon	3 185	1 979
Rio Grande (Río Bravo del Norte)	3 057	1 900

LARGEST ISLANDS	sq km	sq miles
Greenland	2 175 600	839 999
Baffin Island	507 451	195 927
Victoria Island	217 291	83 896
Ellesmere Island	196 236	75 767
Cuba	110 860	42 803
Newfoundland	108 860	42 031
Hispaniola	76 192	29 418

LARGEST LAKES	sq km	sq miles
Lake Superior	82 100	31 699
Lake Huron	59 600	23 012
Lake Michigan	57 800	22 317
Great Bear Lake	31 328	12 096
Great Slave Lake	28 568	11 030
Lake Erie	25 700	9 923
Lake Winnipeg	24 387	9 416
Lake Ontario	18 960	7 320

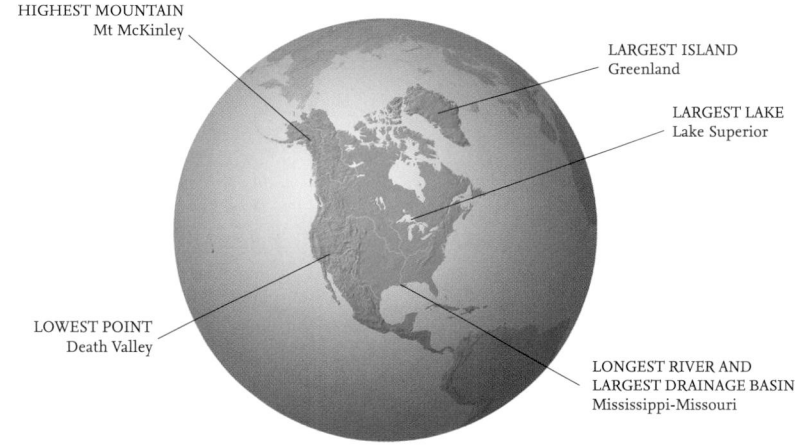

HIGHEST MOUNTAIN
Mt McKinley

LARGEST ISLAND
Greenland

LARGEST LAKE
Lake Superior

LOWEST POINT
Death Valley

LONGEST RIVER AND
LARGEST DRAINAGE BASIN
Mississippi-Missouri

AFRICA Total Land Area 30 343 578 sq km / 11 715 721 sq miles

HIGHEST MOUNTAINS	metres	feet
Kilimanjaro, Tanzania	5 892	19 330
Mt Kenya (Kirinyaga), Kenya	5 199	17 057
Margherita Peak, Democratic Republic of the Congo/Uganda	5 110	16 765
Meru, Tanzania	4 565	14 977
Ras Dejen, Ethiopia	4 533	14 872
Mt Karisimbi, Rwanda	4 510	14 796

LONGEST RIVERS	km	miles
Nile	6 695	4 160
Congo	4 667	2 900
Niger	4 184	2 600
Zambezi	2 736	1 700
Webi Shabeelle	2 490	1 547
Ubangi	2 250	1 398

LARGEST LAKES	sq km	sq miles
Lake Victoria	68 870	26 591
Lake Tanganyika	32 600	12 587
Lake Nyasa (Lake Malawi)	29 500	11 390
Lake Volta	8 482	3 275
Lake Turkana	6 500	2 510
Lake Albert	5 600	2 162

LARGEST ISLANDS	sq km	sq miles
Madagascar	587 040	226 656

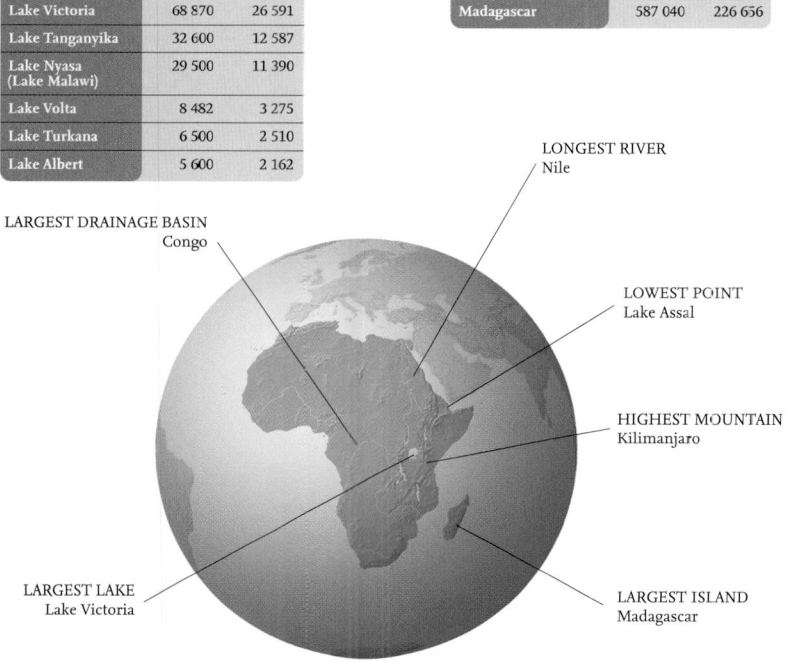

LARGEST DRAINAGE BASIN
Congo

LONGEST RIVER
Nile

LOWEST POINT
Lake Assal

HIGHEST MOUNTAIN
Kilimanjaro

LARGEST LAKE
Lake Victoria

LARGEST ISLAND
Madagascar

SOUTH AMERICA Total Land Area 17 815 420 sq km / 6 878 572 sq miles

HIGHEST MOUNTAINS	metres	feet
Cerro Aconcagua, Argentina	6 959	22 831
Nevado Ojos del Salado, Argentina/Chile	6 908	22 664
Cerro Bonete, Argentina	6 872	22 546
Cerro Pissis, Argentina	6 858	22 500
Cerro Tupungato, Argentina/Chile	6 800	22 309
Cerro Mercedario, Argentina	6 770	22 211

LONGEST RIVERS	km	miles
Amazon (Amazonas)	6 516	4 049
Río de la Plata-Paraná	4 500	2 796
Purus	3 218	2 000
Madeira	3 200	1 988
São Francisco	2 900	1 802
Tocantins	2 750	1 709

LARGEST ISLANDS	sq km	sq miles
Isla Grande de Tierra del Fuego	47 000	18 147
Isla de Chiloé	8 394	3 241
East Falkland	6 760	2 610
West Falkland	5 413	2 090

LARGEST LAKES	sq km	sq miles
Lake Titicaca	8 340	3 220

LONGEST RIVER AND
LARGEST DRAINAGE BASIN
Amazon (Amazonas)

LARGEST LAKE
Lake Titicaca

LOWEST POINT
Laguna del Carbón

HIGHEST MOUNTAIN
Cerro Aconcagua

LARGEST ISLAND
Isla Grande de Tierra del Fuego

© Collins Bartholomew Ltd

WORLD CITIES

KEY TO CITY PLANS

- Built-up area
- Park/Open space
- Cemetery
- Water
- Marsh
- River/Canal
- Road
- Railway
- Administrative boundary
- Airport
- General place of interest
- Place of worship
- Academic/Municipal building
- Transport location

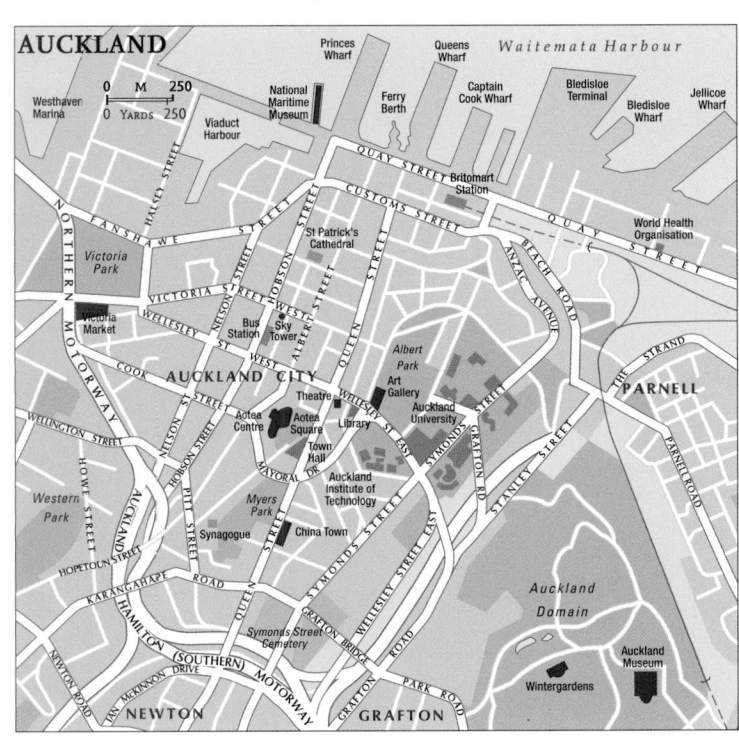

AUCKLAND

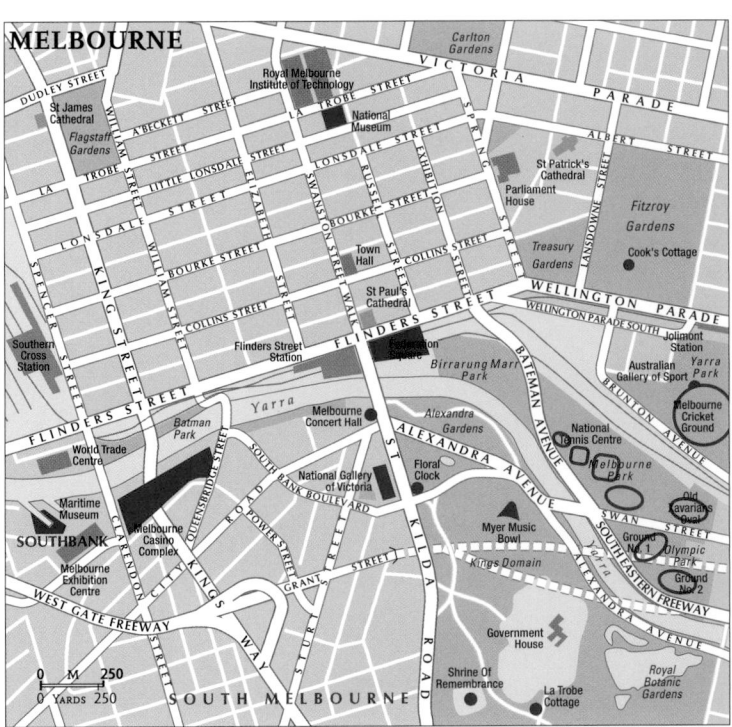

MELBOURNE

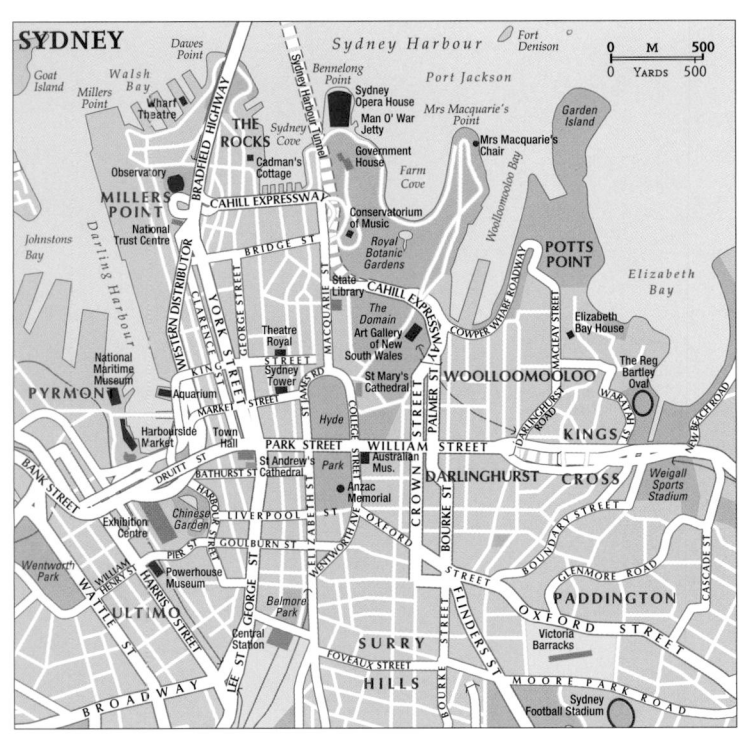

SYDNEY

JAKARTA

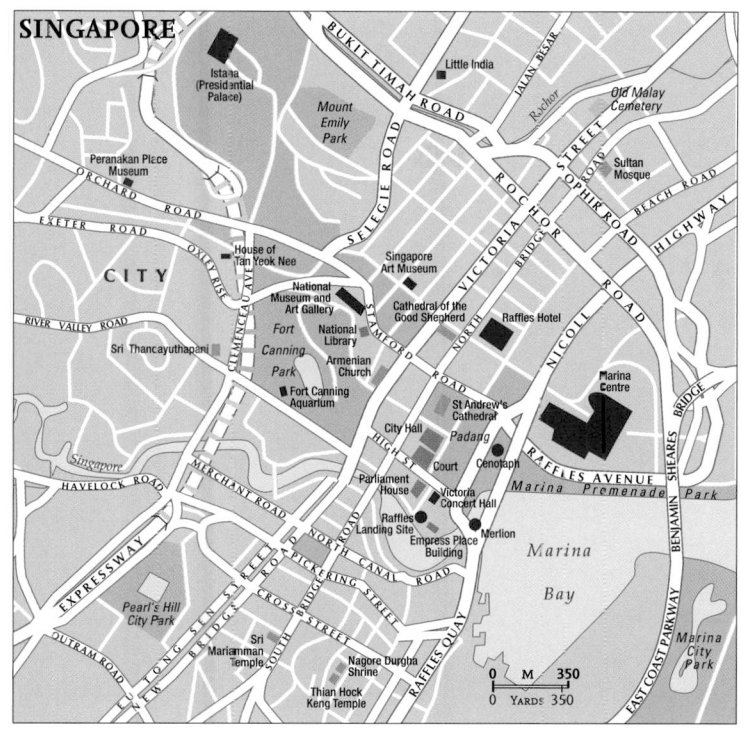

SINGAPORE

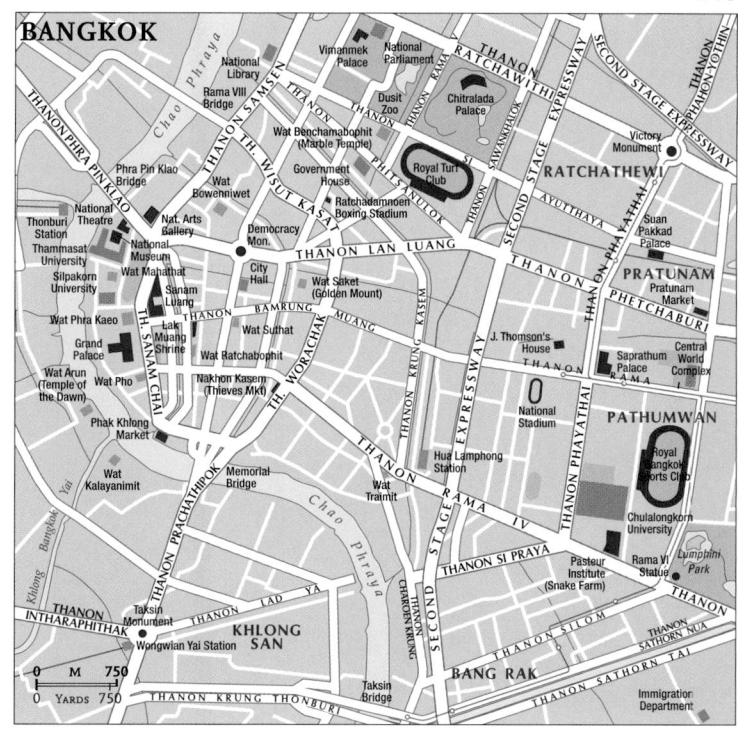

BANGKOK

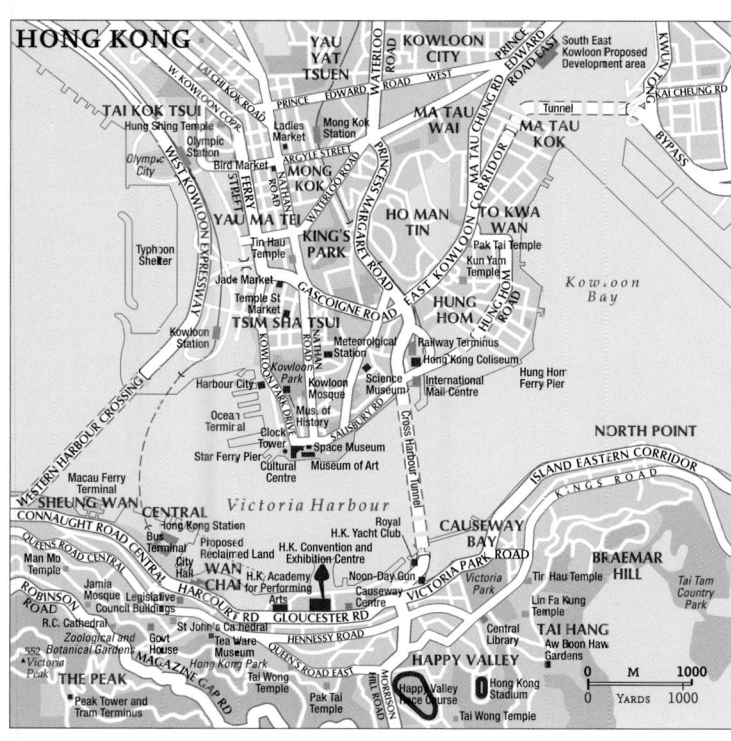

HONG KONG

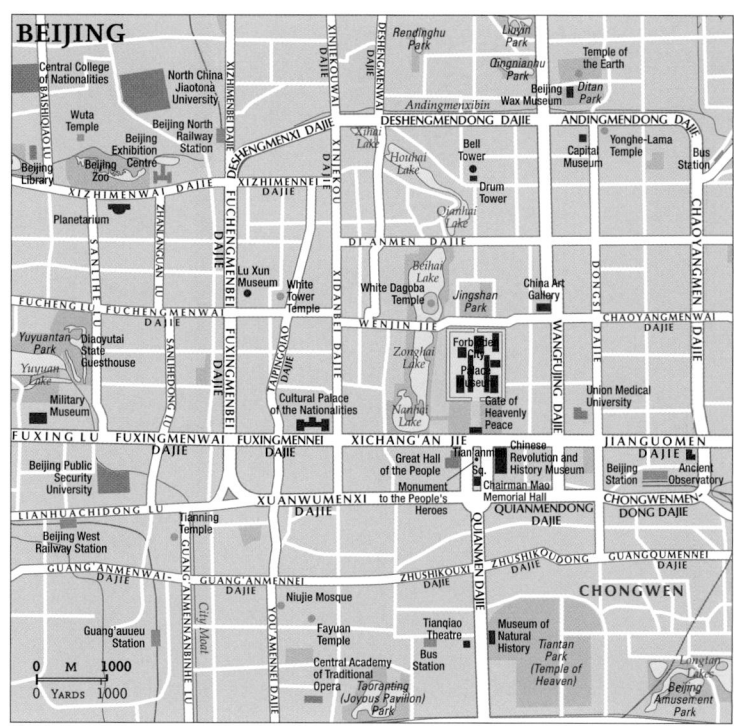

BEIJING

SHANGHAI

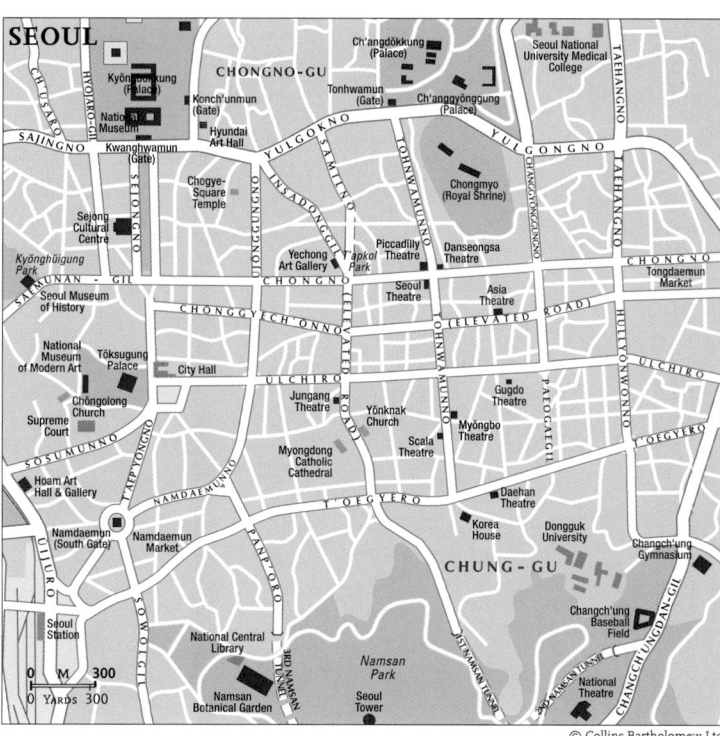

SEOUL

© Collins Bartholomew Ltd

TŌKYŌ

TOKOROZAWA-SHI

Kitano

Sakanoshita

NIIZA-SHI

ASAKA-SHI

SAITAMA

WAKŌ-SHI

Itabashi
Art Gallery

Tōkyō-
daibutsu
Temple

ITABASH

URAWA TOKOROZAWA BYPASS

Azuma-gawa

Yanase-gawa

Yanase-gawa

Seibukyujomae
Station

Seibuen
Park

Seibuen
Station

Tama-ko

Sayama
Park

Ōizumi
Central
Park

Hikarigaoka
Park

NERIMA-KU

HIGASHIYAMATO-SHI

Higashiyamato
Green Park

HIGASHIMURAYAMA-
SHI

KIYOSE-SHI

HIGASHIKURUME-SHI

Shikii-gawa

Kurabe-gawa

SHIKI-KAIDO

OME-KAIDO

SHIN-OME KAIDO

Higashimurayama
Central Park

Kodaira
Cemetery

Yanagikubo

HŌYA-SHI

Makino
Memorial
Garden

Nerima
Art Gallery

SHIN-OME KAIDO

Sanpoji
Temple

Chihiro-
Iwasaki
Memorial
Gallery

Nakano
Historical
Museum

Medicinal
Plant Garden

OME-KAIDO

OME-KAIDO

OME-KAIDO

Ogawa

TANASHI-SHI

SHIN-OME KAIDO

Myōshōji-gawa

Araiyakushi
Temple

Toy
Museum

KODAIRA-SHI

Koganei
Country
Club

Koganei
Park

ITSUKAICHI-KAIDO

MUSASHINO-SHI

Zenpukuji-gawa

Kichijōji
Station

TŌKYŌ

ITSUKAICHI-KAIDO

KOGANEI-SHI

Inokashira
Natural
Park

SUGINAMI-KU

NAKANO

KOKUBUNJI-SHI

Man-yo
Botanical Garden

TOHACHI-DORO

Wadabori
Park

Takachiho
University of Commerce

HONA

Tōkyō University
of Agriculture
and Engineering

Tama
Cemetery

Nogawa
Park

MITAKA-SHI

HITOMI-KAIDO

HITOMI-KAIDO

INOKASHIRA-DORI

KUNITACHI-SHI

Yaho-ten-mangu
Shrine

KOSHU-KAIDO

Okunitama-jinja
Shrine

National
Observatory

TOHACHI-DORO

Jindai
Botanical
Garden

Chōfu
Airfield

Jindaiji
Temple

SETAGAYA-
KU

Tama-gawa

FUCHŪ-SHI

CHUO EXPRESSWAY

Tōkyō
Racetrack

Koremasa
Station

CHŪŌ EXPRESSWAY

CHŌFU-SHI

KOSHU-KAIDO

EXPRESSWAY

Gotokuji
Temple

Shoin-jinja
Shrine

Tamagawa
Green Park

KOMAZU-DORI

Tōkyō
University of
Agriculture

Sakuragaoka
Country Club

Keio Hyakkaen
Garden

KOMAE-SHI

TAMA-SHI

Okari-gawa

U.S. Army Tama
Golf Course

Tama
Country Club

Sakuragaoka
Park

Tama-gawa

Misawa-gawa

TSURUKAWA-KAIDO

EXPRESSWAY

Kinuta Park

Setagaya
Art Gallery

Komazawa
Olympic Park

Central Tōkyō

Kitanomaru
Park

Science and
Technology
Museum

0   M   250

0   YARDS   250

INNER LOOP EXPRESSWAY

HONGO-DORI

Craft Gallery

National Museum
of Modern Art

CHIYODA-KU

East
Garden

Communications
Museum

Seikado
Library

Fukiage
Imperial
Residence

Cabinet
Library

Futako-tamagawa
Green Park
Playground

SHINJUKU-DORI

Imperial
Palace
Gardens

New
Imperial
Palace

EIFAI-DORI

Mukogaoka
Amusement
Park

Tama
University of Arts

National
Theatre

Tōkyō
Station

Gotō Art
Museum

Joshinji
Temple

Zushi

Supreme
Court

Sakurada-dori

Outer
Garden

Mizonokuchi

HIBIYA-DORI

FUCHU-KAIDO

Sakurada-dori Moat

National
Diet Library

Sukurada
Gate

TAKATSU-
KU

National
Diet Building

Parliamentary
Museum

High
Court

Imperial
Theatre

KAWASAKI-SHI

Negishi

CHIYODI-DORI

Kawasaki
City
Museum

Prime Minister's
Residence

Hibiya
Park

Yūrakuchō
Station

NAKAHARA-
KU

TOMEI EXPRESSWAY

DAISAN-KEIHIN-DORI

NAKAHARA-DORI

TAMA-TSURUMI-DORI

Maginu

Kizuki

EXPRESSWAY NO.1

Hibiya
Concert
Hall

Hibiya
Library

Nissei
Theatre

Central Art
Gallery

MIYAMAE-KU

SOTOBORI-DORI

Midori

TSUZUKI-KU

Sakura-gawa

Hibiya
Public Hall

HARUMI-DORI

CHUO-DORI

KANAGAWA

EXPRESSWAY NO.3

MINATO-KU

Kabukiza
Theatre

YOKOHAMA-SHI

Katsuda

Nakayama

Hiyoshi

Hara-Machida

MIDORI-KU

Kawawa

Tsunashima

CHIBA

TODA-SHI

KAWAGUCHI-SHI

MATSUDO-SHI

Toneri Park

Mizumoto Park

KITA-KU

ADACHI-KU

KANNANA-DORI

KATSUSHIKA-KU

ICHIKAWA-SHI

Keisei-kanamachi Station

Shibamata-taishakuten Temple

Johoku Central Park

Nihon Calligraphy Museum

Itabashi Childrens Zoo

ARAKAWA-KU

TOSHIMA-KU

Ikebukuro Station

Togenuki-jizo Temple

Kishibojin Shrine

Gokoku-ji Imperial Family Grave

Kisshoji Temple

Asakusa-Choaokan Gallery

Daimyo Clock Museum

Yanaka Cemetery

Ueno Park

Japan Toy Museum

SUMIDA-KU

EDOGAWA-KU

Koishikawa Botanical Garden

BUNKYŌ-KU

Metropolitan Art Gallery

National Museum

National Science Museum

St Mary's Cathedral

Tōkyō University

Ueno Zoo

Ueno Royal Museum

Ueno Station

Sensōji Temple

Asakusa Station

SHINJUKU-KU

Science University of Tōkyō

Tōkyō Dome

Kanda Myōjin Shrine

TAITŌ-KU

Kameido-tenmangu Shrine

Hōsenji Temple

Yasukuni-Jinja Shrine

Transportation Museum

Torigoe-jinja Shrine

SEE INSET

Science and Technology Museum

CHŪŌ-KU

Shinjuku Station

Budōkan (Judo Hall)

National Museum of Modern Art

Suijengu Shrine

Fukagawa Edo Museum

Metropolitan Government Offices

Historical Museum

Communications Museum

Tōkyō Station

Japanese Sword Museum

Geinin-Kan (State Guesthouse)

Shinjuku Gyoen Garden

New Imperial Palace

National Theatre

Tōkyō Stock Exchange

Fukagawa-Fudoson Temple

National Noh Theatre

CHIYODA-KU

Mullion

Meiji Jingu Shrine

National Jingu Stadium

Suntory Museum of Art

National Diet Building

Kabukiza Theatre

Tomioka-Hachimangu Shrine

KŌTŌ-KU

Edogawa Natural Zoo

URAYASU-SHI

Ohta Memorial Museum of Art

Yoyogi Park

Riccar Art Museum

Tsukiji-Honhanji Temple

Tōkyō University of Mercantile Marine

KASAIBASHI-DORI

Subway Museum

Metropolitan Modern Literature Museum

National Yoyogi Sports Centre

Shoto Museum of Art

Nezu Art Museum

Okura Shukokan Museum

NHK Broadcasting Museum

Hamarikyū Garden

The Furniture Museum

Kasairinkai Park

Aquarium

Aoyama Cemetery

Tōkyō Tower

Zōjō-ji Temple

World Trade Centre

Tōkyō Internat onal Trade Centre

Yumenoshima Park

Tōkyō Heliport

Tōkyō Disneyland

MINATO-KU

Riccar Art Gallery

Rainbow Bridge

National Park for Nature Study

Sengakuji Temple

Tōkyō International Trade Centre

Meguro Art Gallery

Hatakeyama Collection

Tōkyō University of Fisheries

TELEPORT TOWN

Tōkyō Port

Wakasu Golf Course

Daienji Temple

Shinagawa Station

Shiokaze Park

Meguro-Fudo Temple

Gotanda Station

Shinagawa-jinja Shrine

Museum of Maritime Science

MEGURO-KU

SHINAGAWA-KU

Tōkyō Institute of Technology

Oi Race Course

Oi Wharf Central Marine Park

Tōkyō-wan

Tomioka Art Museum

Ryushi Memorial Museum

Honmonji Temple

ŌTA-KU

Kamata Station

Tamagawa Green Park

Tōkyō International Airport (Haneda)

Yako

0   M   2000

0   YARDS   2000

© Collins Bartholomew Ltd

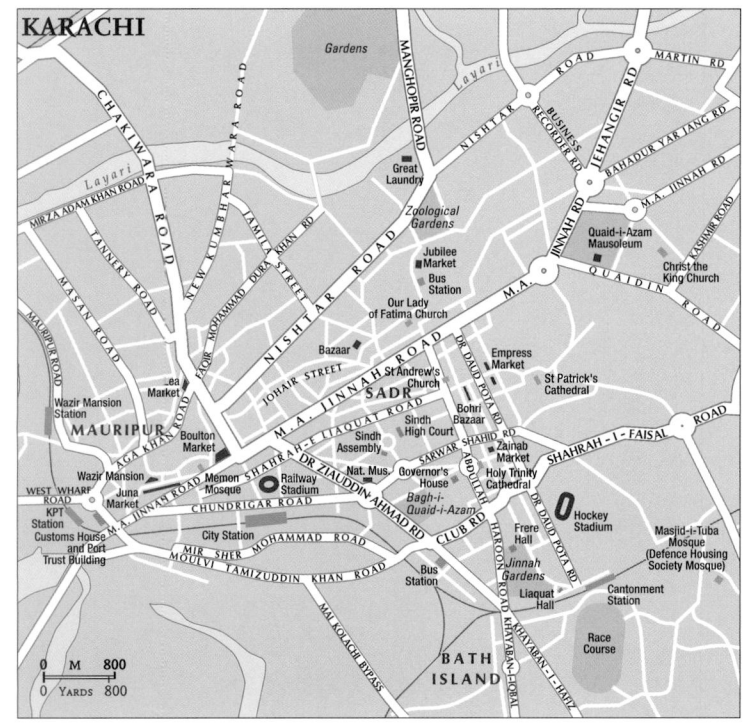

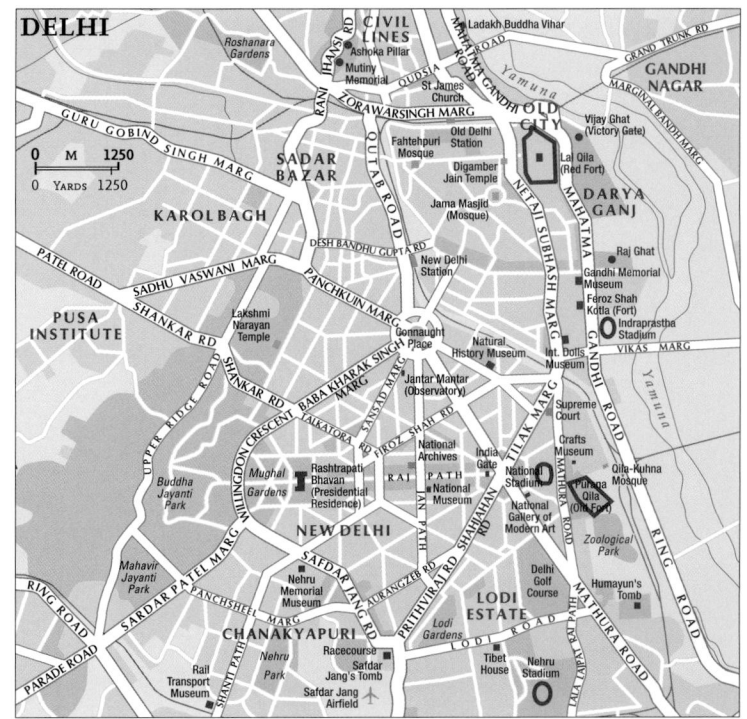

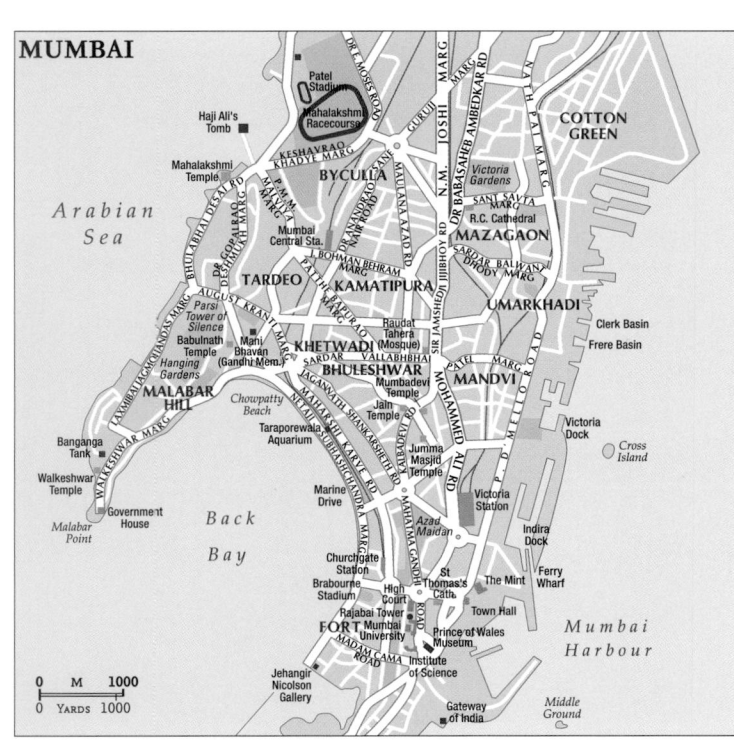

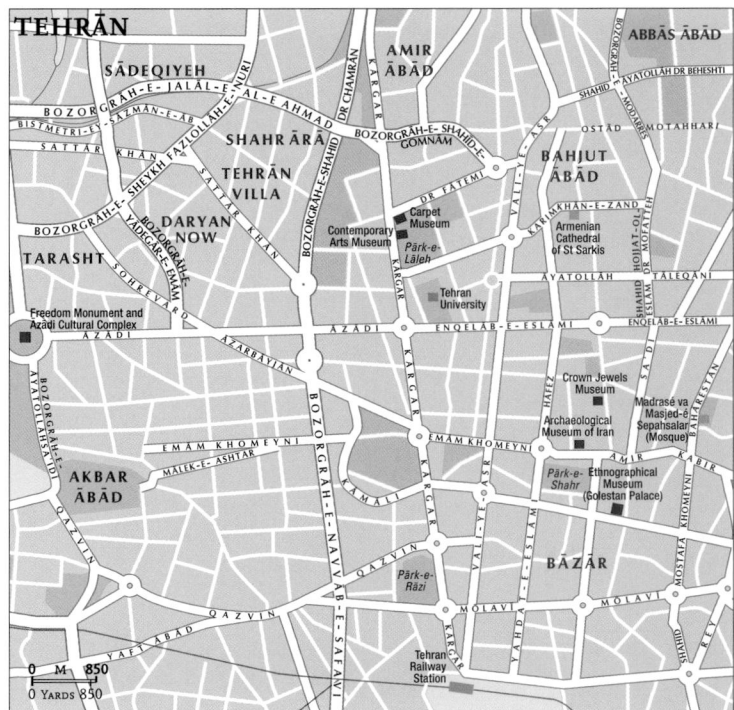

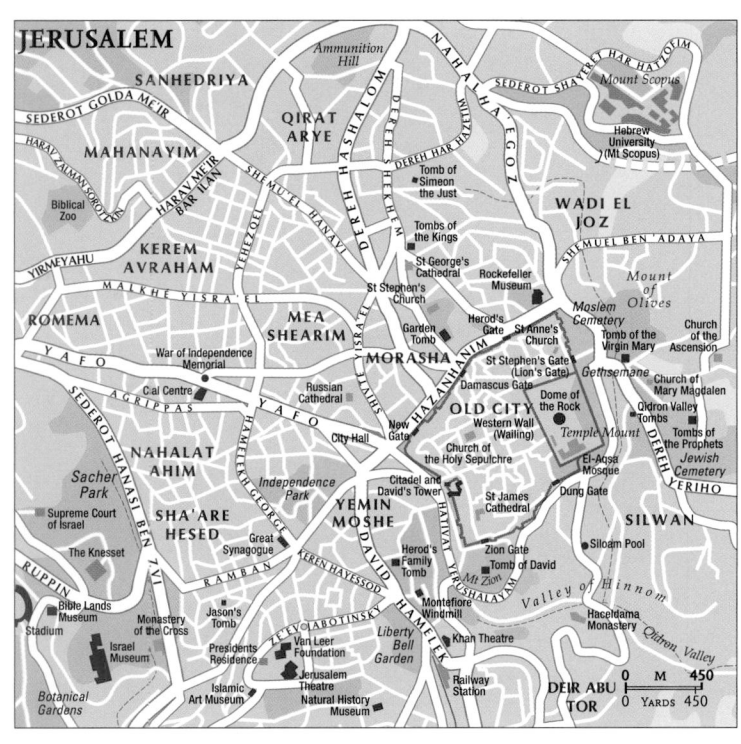

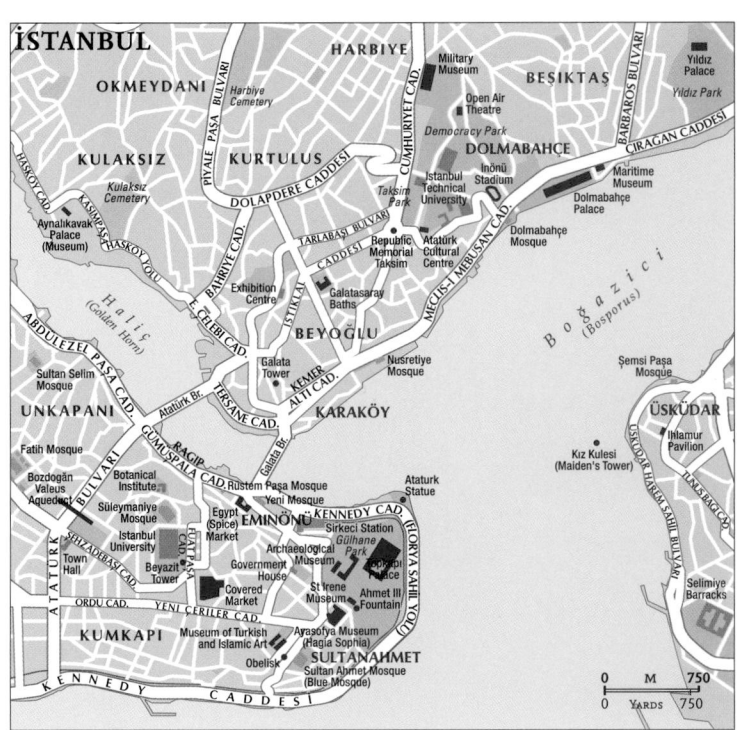

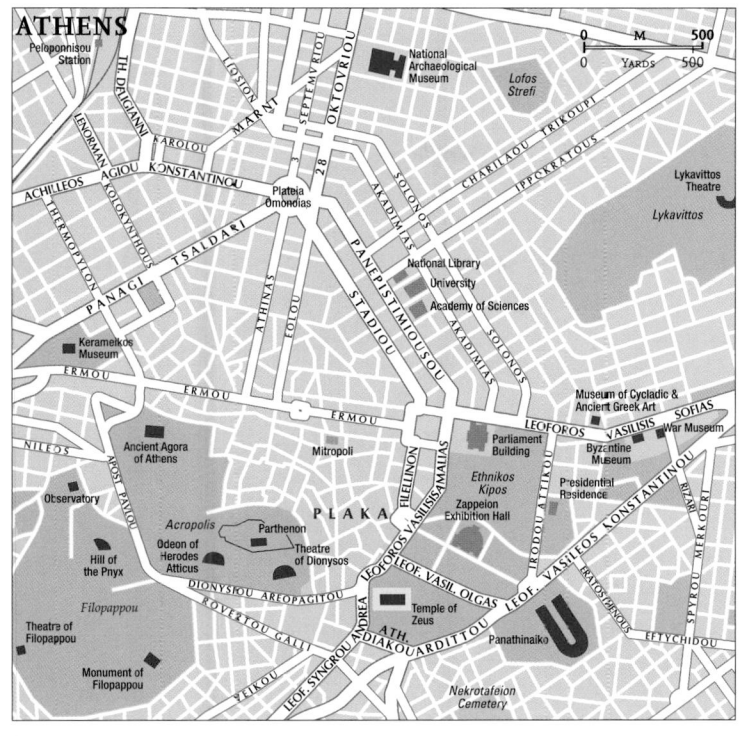

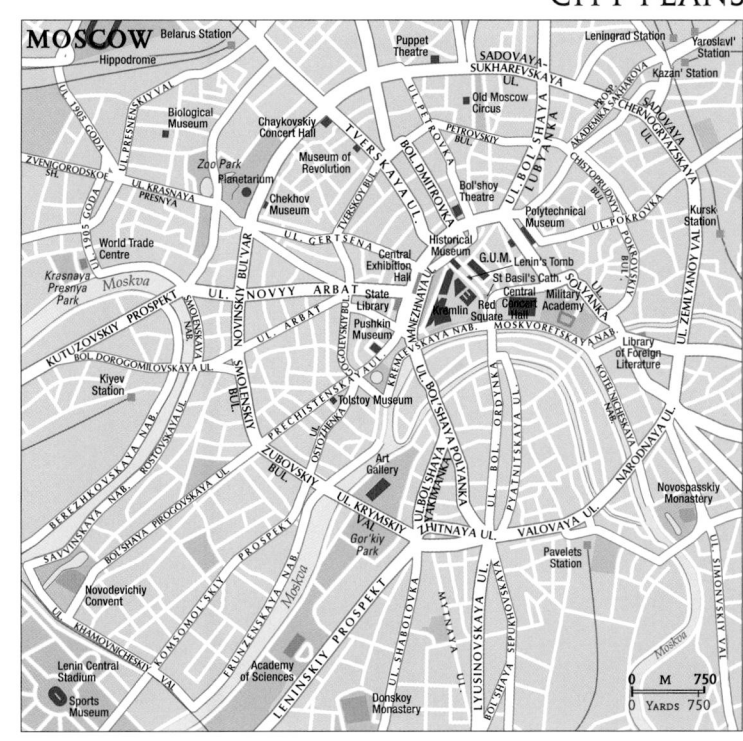

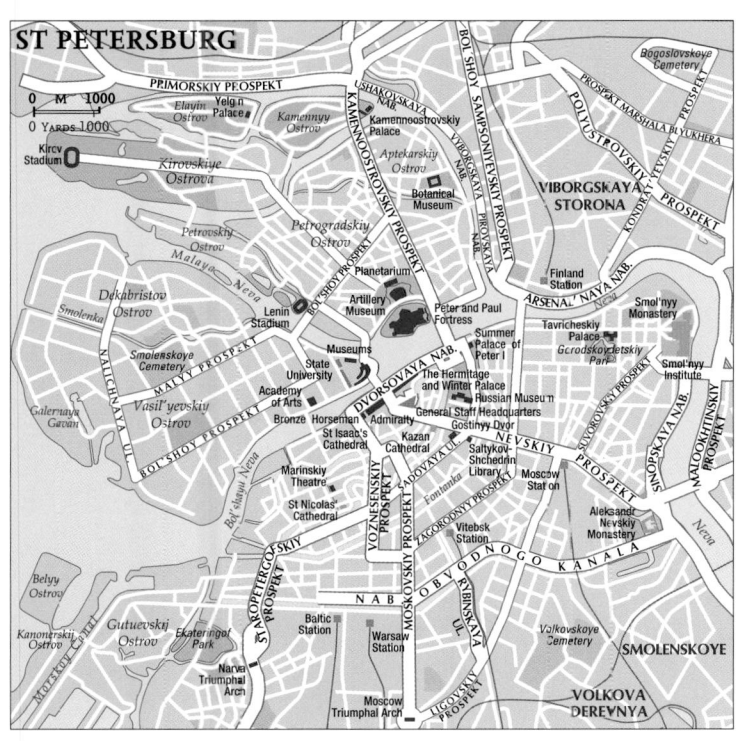

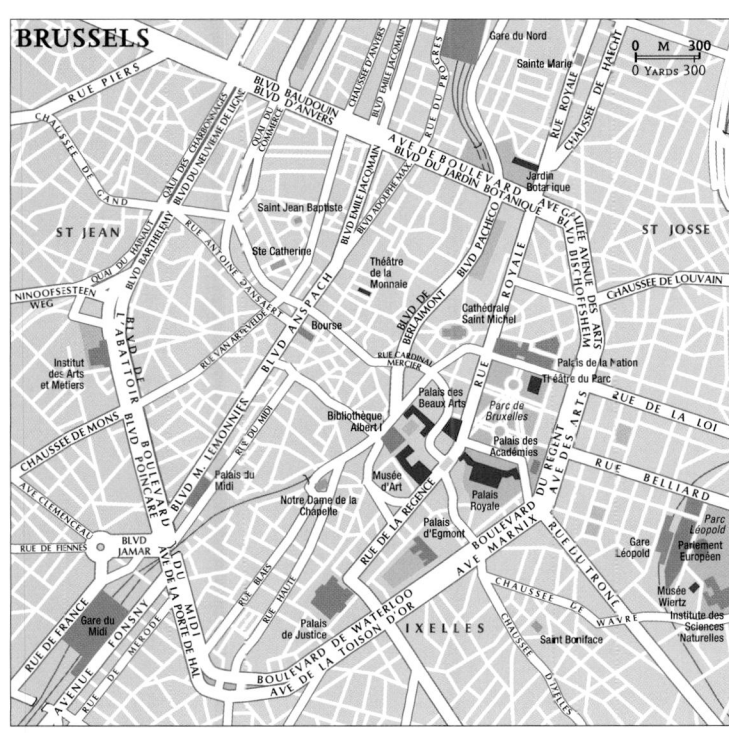

© Collins Bartholomew Ltd

LONDON

Central London

Belmont
Burnt Oak
RAF Museums
Holders Hill
A1
GREAT N. WAY
EDGWARE ROAD

Queensbury
Hendon
Golders Green

Kingsbury
Cricklewood
FINCHLEY
A41
Dollis Hill

Northwick Park
Fryent Country Park
Wembley Park
Gladstone Park
Willesden

BRENT
Wembley Stadium
Wembley
NORTH CIRCULAR ROAD
Willesden Green
Kilburn

Sunbury Golf Course
Alperton
Harlesden
HARROW ROAD

Perivale
Grand Union Canal
Park Royal
Paddington

Ealing Golf Course
WESTERN
North Acton

EALING
HANGER LANE
Acton
East Acton
WESTWAY A40
North Kensington
Padd

Ealing
AVENUE
Wormwood Scrubs
Notting Hill
A219

THE VALE
Shepherd's Bush
A402

Hayes End
Gunnersbury
HAMMERSMITH
Holland Park

Hayes
Hanwell
Olympia
AND FULHAM
A4

Southall
CHISWICK HIGH ROAD
Chiswick
Earls Court Exhibition Centre
Footb Stadi.

Yiewsley
Norwood Green
M4
Gunnersbury Park
Chiswick House
A316
Earls Court
Hammersmith Bridge

West Drayton
Grand Union Canal
Castelnau
Barn Elms Wildfowl Reserve

North Hyde
Osterley Park
Brentford
M4
Syon House
Royal Botanic Gardens Kew
Football Stadium
FULHAM RD
KING'S RD

Harlington
BATH ROAD A4
Heston
GREAT WEST ROAD
Syon Park
Barnes
Parson Green

Cranford
Osterley
KEW ROAD
Mortlake
Putney Bridge

Heathrow Airport (London)
A30
Hounslow West
Isleworth
Richmond
SOUTH CIRCULAR ROAD
Putney

Stanwell
Hounslow
ROEHAMPTON LANE
Putney Heath
Wan

East Bedfont
HOUNSLOW
Rugby Ground
WAND
Southfields

Hounslow Heath
A316
Richmond
Richmond Park

Ashford
Feltham
RICHMOND UPON
Twickenham
Thames
All England Lawn Tennis and Croquet Club

Hanworth
THAMES
Wimbledon Common
Wimbledon Park

A308
Crane
Teddington
KINGSTON HILL
Coombe Hill Golf Course
Wimbledon

Kempton Park Racecourse
Hampton
Bushy Park
Norbiton
COOMBE LANE

Queen Mary Reservoir
Sunbury
A308
New Malden
Bushy Mead

Molesey Reservoirs
Hampton Court Palace
Hampton Court Park
Kingston Upon Thames
West Barnes
Morden Park
Motspur Park

West Molesey
East Molesey
Thames Ditton
KINGSTON UPON THAMES
Surbiton
ME

Shepperton
Queen Elizabeth II Reservoir
Island Barn Reservoir
Mole
Long Ditton
Old Malden
A3

Walton-on-Thames

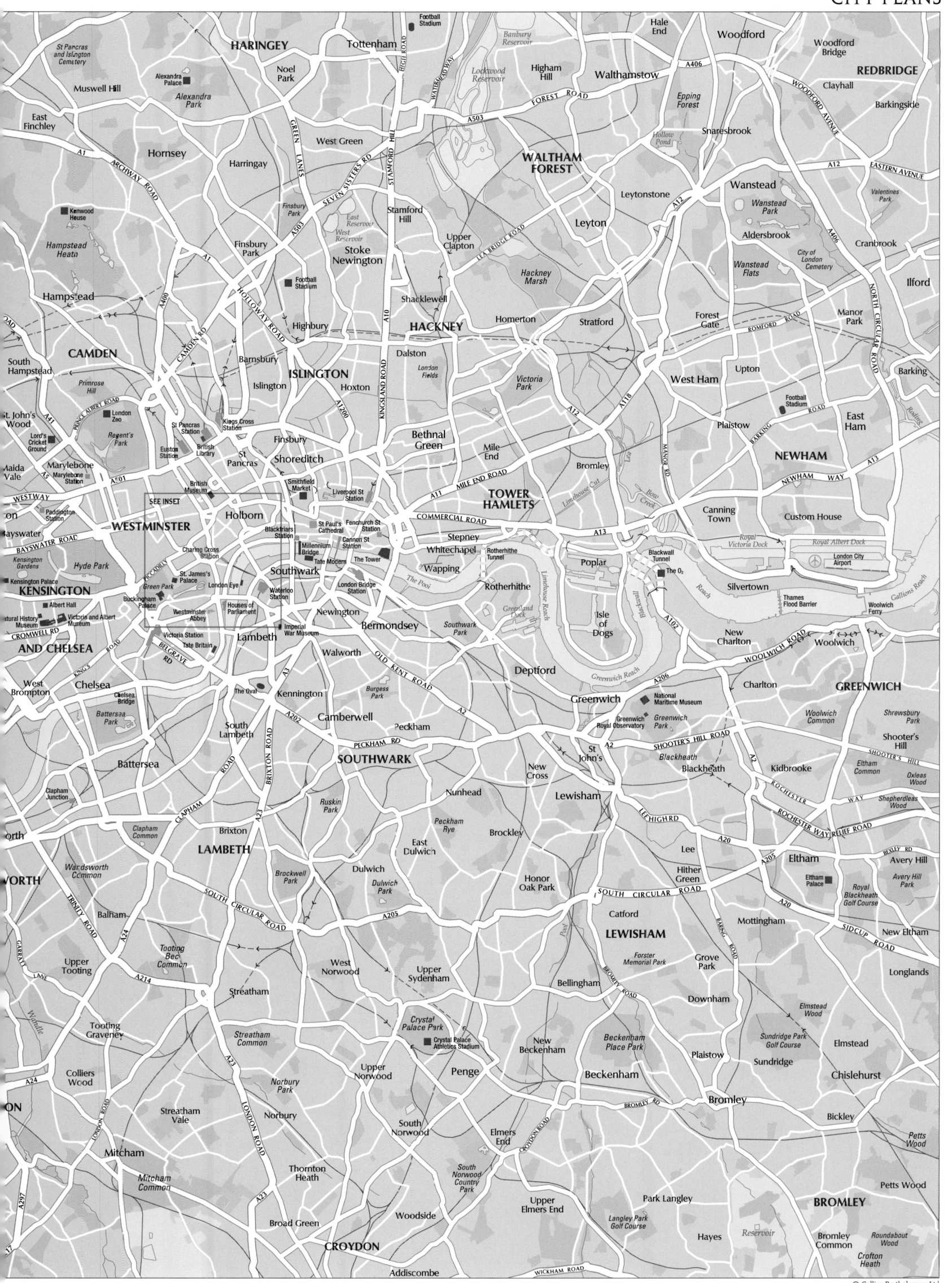

St Pancras and Islington Cemetery
HARINGEY
Tottenham
Football Stadium
Banbury Reservoir
Hale End
Woodford
Woodford Bridge
REDBRIDGE
Muswell Hill
Alexandra Palace
Noel Park
HIGH ROAD
Lockwood Reservoir
Higham Hill
A406
Walthamstow
Clayhall
Barkingside
East Finchley
Alexandra Park
A503
WATERLOO WAY
Epping Forest
WOODFORD AVENUE
A12
EASTERN AVENUE
Hornsey
West Green
SEVEN SISTERS RD
STAMFORD HILL
WALTHAM FOREST
Hollow Pond
Wanstead
FOREST ROAD
Kenwood House
ARCHWAY ROAD
Harringay
GREEN LANES
Finsbury Park
East Reservoir
A1
Leytonstone
A12
Wanstead Park
Valentines Park
Hampstead Heath
HOLLOWAY ROAD
Stamford Hill
West Reservoir
Leyton
Aldersbrook
A406
Cranbrook
Hampstead
A1
Stoke Newington
A503
Upper Clapton
LEA BRIDGE ROAD
Hackney Marsh
Wanstead Flats
City of London Cemetery
Ilford
CAMDEN
Highbury
Shacklewell
Homerton
Stratford
Forest Gate
ROMFORD ROAD
Manor Park
NORTH CIRCULAR ROAD
South Hampstead
ISLINGTON
Barnsbury
HACKNEY
Dalston
A118
West Ham
Upton
Barking
Primrose Hill
Islington
Hoxton
London Fields
Victoria Park
A12
Plaistow
BARKING ROAD
East Ham
St. John's Wood
London Zoo
A501
Finsbury
A1200
KINGSLAND ROAD
A10
Bethnal Green
A12
MANOR RD
Football Stadium
NEWHAM
NEWHAM WAY
Regent's Park
St Pancras Station
Kings Cross Station
St Pancras
Shoreditch
Mile End
Bromley
A13
Lord's Cricket Ground
Euston Station
British Library
MILE END ROAD
A11
Canning Town
Custom House
Maida Vale
Marylebone
Marylebone Station
British Museum
Smithfield Market
Liverpool St Station
TOWER HAMLETS
Stepney
A13
Limehouse Cut
Bow Creek
Royal Victoria Dock
Royal Albert Dock
London City Airport
WESTWAY
Paddington Station
A501
WESTMINSTER
Holborn
St Paul's Cathedral
Fenchurch St Station
COMMERCIAL ROAD
Whitechapel
Rotherhithe Tunnel
Poplar
Blackwall Tunnel
The O2
Silvertown
Thames Flood Barrier
Gallions Reach
BAYSWATER ROAD
Blackfriars Station
Millennium Bridge
Cannon St Station
A13
Wapping
A102
Woolwich Ferry
Kensington Gardens
Hyde Park
Charing Cross Station
Tate Modern
The Tower
London Bridge Station
The Pool
Rotherhithe
Greenland Dock
Isle of Dogs
Blackwall Reach
New Charlton
Woolwich
KENSINGTON
Green Park
St. James's Palace
London Eye
Southwark
Newington
Bermondsey
Southwark Park
Deptford
A206
WOOLWICH ROAD
Woolwich
GREENWICH
Buckingham Palace
Waterloo Station
Houses of Parliament
Imperial War Museum
Greenwich Reach
Greenwich
Charlton
Albert Hall
Westminster Abbey
Lambeth
Walworth
National Maritime Museum
Natural History Museum
Victoria Station
Tate Britain
OLD KENT ROAD
Burgess Park
Greenwich Royal Observatory
Greenwich Park
Woolwich Common
Shrewsbury Park
Victoria and Albert Museum
CROMWELL RD
AND CHELSEA
BELGRAVE RD
The Oval
Kennington
Camberwell
A2
St John's
A2
SHOOTER'S HILL ROAD
Blackheath
SHOOTER'S HILL ROAD
Shooter's Hill
West Brompton
KINGS ROAD
Chelsea
Chelsea Bridge
A202
Peckham
PECKHAM RD
New Cross
Blackheath
ROCHESTER WAY
Kidbrooke
Eltham Common
Oxleas Wood
Battersea Park
South Lambeth
Lewisham
Lee
A20
Eltham
Avery Hill
Battersea
A3
Nunhead
Peckham Rye
Brockley
Hither Green
A205
ROCHESTER WAY RELIEF ROAD
BEXLEY RD
Clapham Junction
Ruskin Park
SOUTHWARK
Honor Oak Park
SOUTH CIRCULAR ROAD
A20
Eltham Palace
Royal Blackheath Golf Course
Avery Hill Park
CLAPHAM ROAD
Brixton
East Dulwich
Lee
Catford
BARING ROAD
Mottingham
SIDCUP ROAD
New Eltham
Wandsworth Common
Clapham Common
LAMBETH
Dulwich
Dulwich Park
BROMLEY ROAD
LEWISHAM
Grove Park
Longlands
TRINITY ROAD
Balham
SOUTH CIRCULAR ROAD
Brockwell Park
A205
Forster Memorial Park
Bellingham
Downham
Elmstead Wood
GARRATT LANE
A24
Tooting Bec Common
A214
West Norwood
Upper Sydenham
A2212
Sundridge Park Golf Course
Elmstead
Upper Tooting
Streatham
Crystal Palace Park
New Beckenham
Beckenham Place Park
Plaistow
Sundridge
Chislehurst
Tooting Graveney
Streatham Common
Crystal Palace Athletics Stadium
Upper Norwood
Penge
Beckenham
BROMLEY RD
Bromley
Bickley
Colliers Wood
A23
Norbury Park
A24
A215
Norbury
South Norwood
CROYDON ROAD
Petts Wood
Streatham Vale
Thornton Heath
South Norwood Country Park
Elmers End
Park Langley
BROMLEY
Petts Wood
Mitcham
A297
Mitcham Common
Broad Green
Woodside
Upper Elmers End
Langley Park Golf Course
Hayes
Reservoir
Bromley Common
Roundabout Wood
Crofton Heath
A17
CROYDON
Addiscombe
WICKHAM ROAD

© Collins Bartholomew Ltd

PARIS

Central Paris

© Collins Bartholomew Ltd

ROME

MILAN

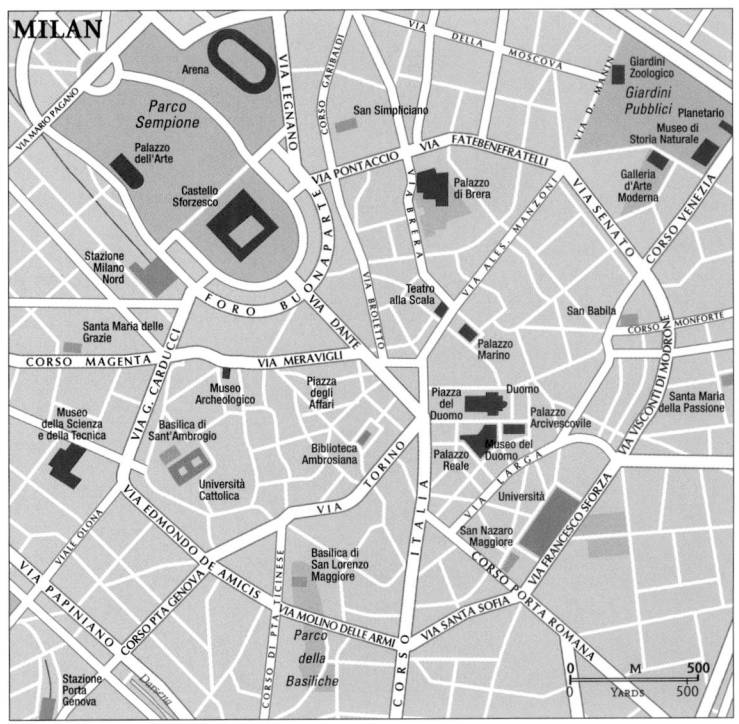

MADRID

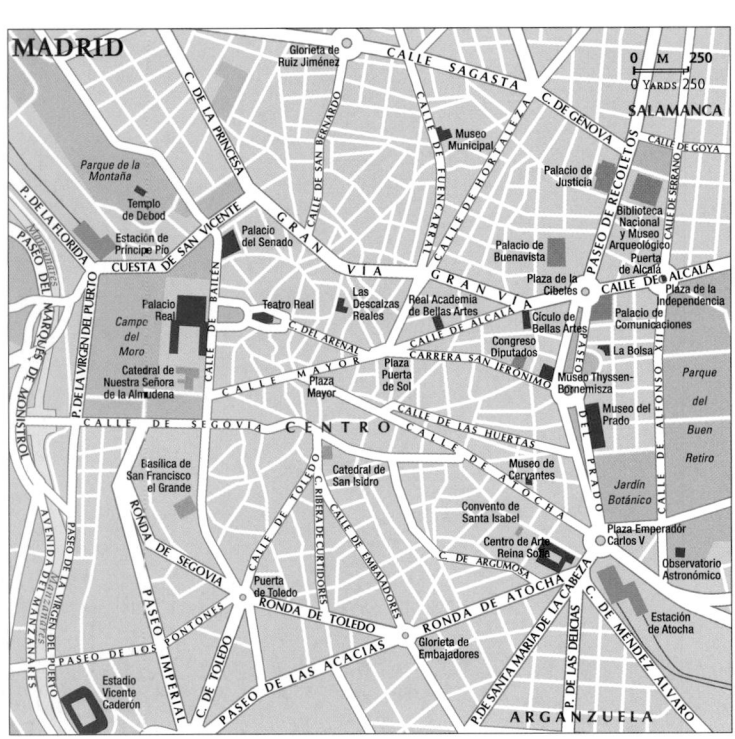

BARCELONA

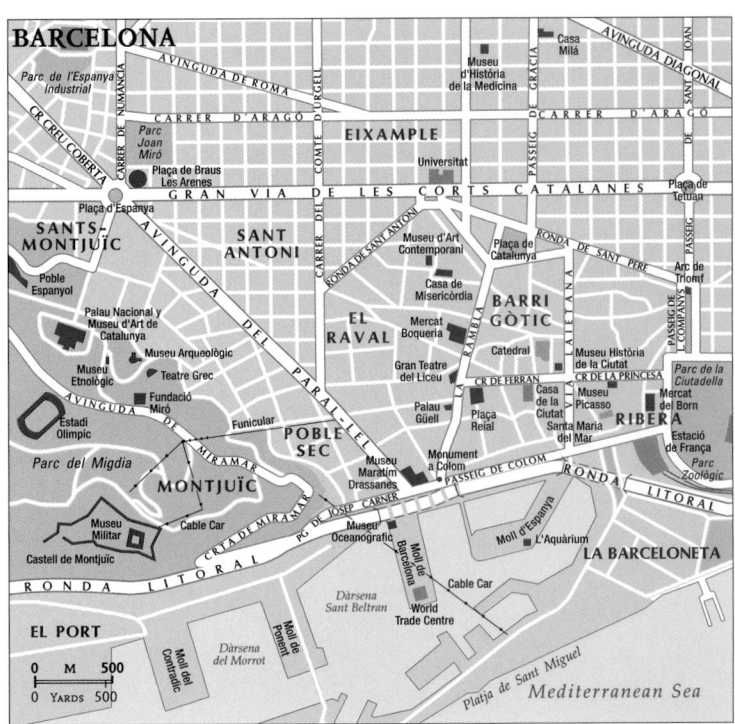

CAIRO

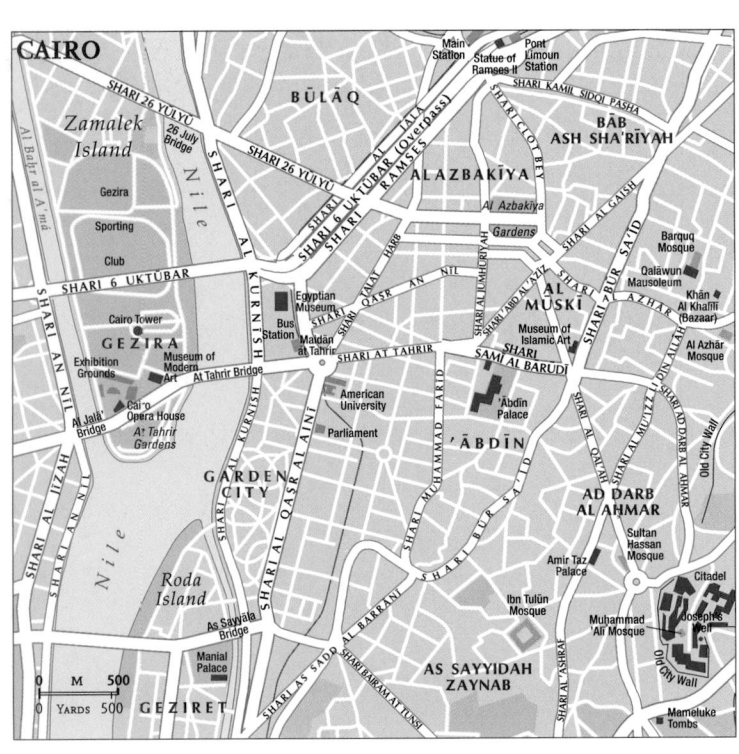

CAPE TOWN

MONTRÉAL
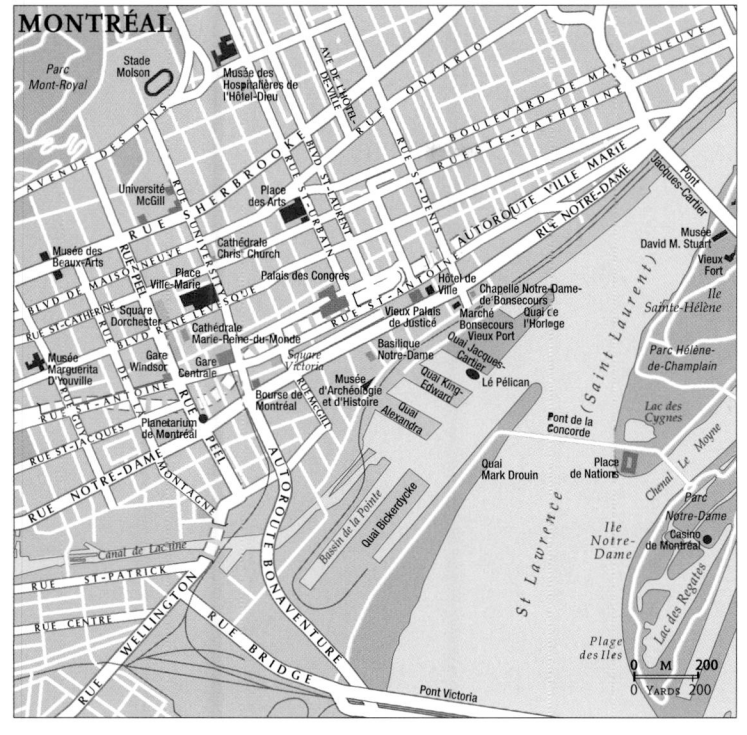

TORONTO
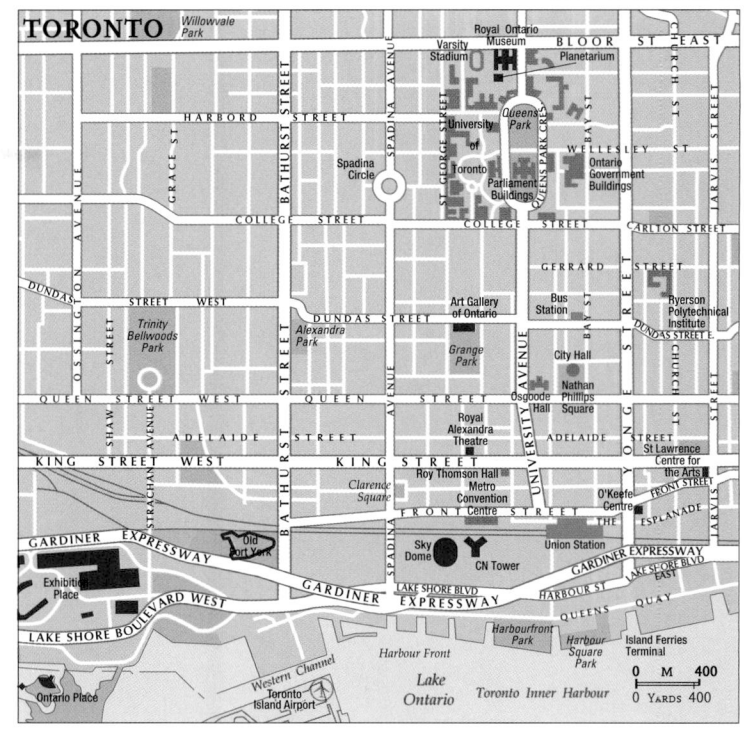

CHICAGO

WASHINGTON D.C.
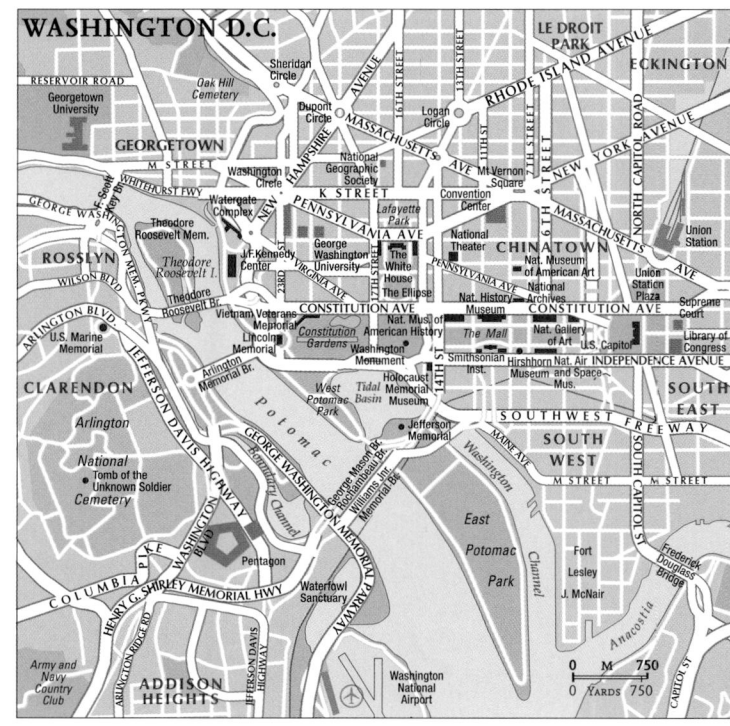

LOS ANGELES

SAN FRANCISCO

© Collins Bartholomew Ltd

NEW YORK

ATLANTIC OCEAN

© Collins Bartholomew Ltd

MEXICO CITY

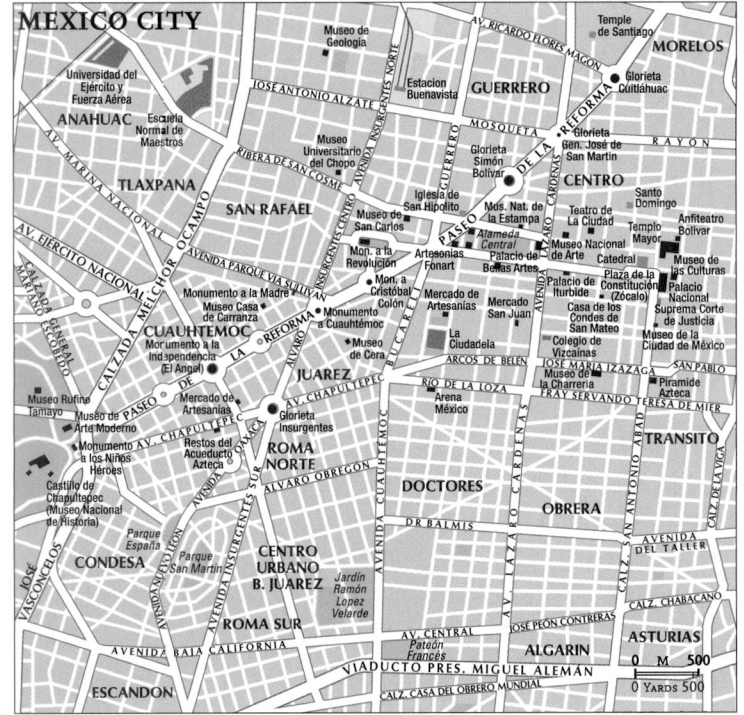

Universidad del Ejército y Fuerza Aérea · Escuela Normal de Maestros · ANAHUAC · TLAXPANA · SAN RAFAEL · Museo de Geología · Estación Buenavista · GUERRERO · Temple de Santiago · MORELOS · Glorieta Cuitláhuac · Museo Universitario del Chopo · CENTRO · Glorieta Simón Bolívar · Iglesia de San Hipólito · Mds. Nat. de la Estampa · Iglesia de San José de San Martín · Santo Domingo · Museo de San Carlos · Alameda Central · Museo Nacional de Arte · Templo Mayor · Catedral · Anfiteatro Bolívar · Museo de las Culturas · Mon. a la Revolución · Teatro de la Ciudad · Palacio de Bellas Artes · Artesanías Fonart · Mon. a Cristóbal Colón · Mercado de Artesanías · CUAUHTÉMOC · Museo Casa de Carranza · Monumento a la Madre · Palacio de Iturbide · Plaza de la Constitución (Zócalo) · Palacio Nacional · Suprema Corte de Justicia · Casa de los Condes de San Mateo de Vizcayas · Museo de la Ciudad de México · Monumento a Cuauhtémoc · Museo de Cera · Colegio de Belen · ARCOS DE BELEN · La Ciudadela · Museo de José María Izazaga · Pirámide Azteca · Monumento a la Independencia (El Ángel) · JUAREZ · Mercado de Artesanías · Museo Rufino Tamayo · Museo del Paseo · Arte Moderno · Glorieta Insurgentes · ROMA NORTE · Monumento a los Niños Héroes · Restos del Acueducto Azteca · Río de la Loza · Arena México · FRAY SERVANDO TERESA DE MIER · TRANSITO · Castillo de Chapultepec (Museo Nacional de Historia) · DOCTORES · OBRERA · DR BALMIS · AV. DEL TALLER · Parque España · CONDESA · Parque San Martín · CENTRO URBANO B. JUAREZ · Jardín Ramón López Velarde · ROMA SUR · ALGARIN · ASTURIAS · AV. CENTRAL Panteón Francés · JOSE PEON CONTRERAS · CALZ. CHABACANO · ESCANDON · AVENIDA BAJA CALIFORNIA · VIADUCTO PRES. MIGUEL ALEMÁN · CALZ. CASA DEL OBRERO MUNDIAL · 0 M 500 · 0 Yards 500

LIMA

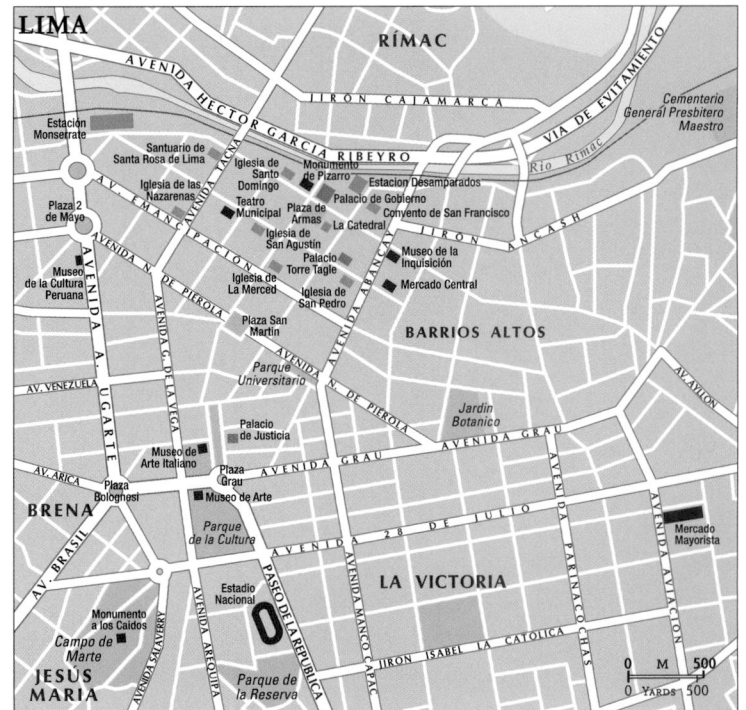

RÍMAC · AVENIDA HECTOR GARCIA RIBEYRO · IRON CAJAMARCA · VIA DE EVITAMIENTO · Estación Monserate · Cementerio General Presbitero Maestro · Santuario de Santa Rosa de Lima · Iglesia de Santo Domingo · Monumento a Pizarro · Estación Desamparados · Río Rímac · IRON ANCASH · Plaza 2 de Mayo · Iglesia de las Nazarenas · Teatro Municipal · Plaza de Armas · Palacio de Gobierno · Iglesia de San Agustín · La Catedral · Convento de San Francisco · Museo de la Cultura Peruana · Palacio Torre Tagle · Museo de la Inquisición · Iglesia de La Merced · Mercado Central · BARRIOS ALTOS · Plaza San Martín · Iglesia de San Pedro · Parque Universitario · Jardin Botanico · Palacio de Justicia · Museo de Arte Italiano · Plaza Bolognesi · Museo de Arte · BRENA · Parque de la Cultura · 28 DE JULIO · Mercado Mayorista · LA VICTORIA · Monumento a los Caidos · Estadio Nacional · Campo de Marte · Parque de la Reserva · JESÚS MARIA · JIRON ISABEL LA CATOLICA · 0 M 500 · 0 Yards 500

RIO DE JANEIRO

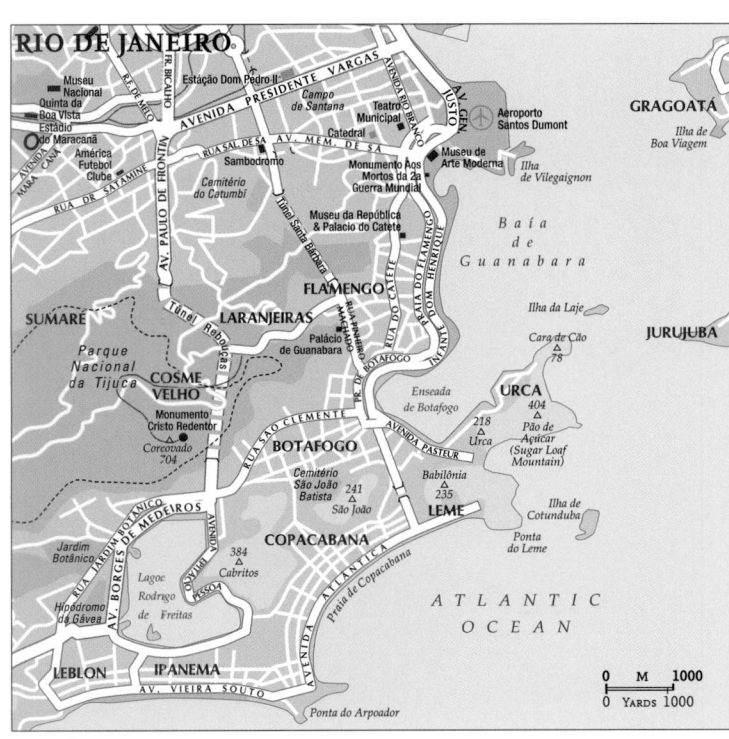

Museu Nacional · Quinta da Boa Vista · Estádio do Maracanã · América Futebol Clube · Estação Dom Pedro II · AVENIDA PRESIDENTE VARGAS · Campo de Santana · Teatro Municipal · Aeroporto Santos Dumont · GRAGOATÁ · Ilha de Boa Viagem · Catedral · Sambódromo · Museu de Arte Moderna · Ilha de Vilegaignon · Cemitério do Catumbi · Mausoléu Aos Mortos da 2a Guerra Mundial · Baía de Guanabara · Museu da República & Palácio do Catete · Ilha da Laje · SUMARÉ · FLAMENGO · LARANJEIRAS · Palácio de Guanabara · JURUJUBA · Cara de Cão 78 · Parque Nacional da Tijuca · COSME VELHO · Monumento Cristo Redentor Corcovado 704 · BOTAFOGO · Enseada de Botafogo · URCA 404 · 218 Urca · Pão de Açúcar (Sugar Loaf Mountain) · Babilônia 235 · LEME · Cemitério São João Batista 241 São João · Ilha de Cotunduba · COPACABANA · 384 Cabritos · Ponta do Leme · Jardim Botânico · Lagoa Rodrigo de Freitas · Hipódromo de Gávea · AVENIDA ATLANTICA · Praia de Copacabana · ATLANTIC OCEAN · LEBLON · IPANEMA · AV. VIEIRA SOUTO · Ponta do Arpoador · 0 M 1000 · 0 Yards 1000

SÃO PAULO

Aerodromo de Marte · AVENIDA ASSIS CHATEAUBRIAND · Rio Tietê · AVENIDA PRESIDENTE CASTELO BRANCO · BARRA FUNDA · BOM RETIRO · PARI · Estação do Barra Funda · AV. MARQUEZ DE TUCE · Parque da Luz · Estação da Luz · SANTA EFIGÉNIA · SANTA CECÍLIA · Igreja Santa Cecília · Mercado de Flores · Mercado Municipal · Basílica São Bento · Parque Dom Pedro II · Estação Roosevelt · Santa Casa de Misericórdia · Praça da República · Teatro Municipal · Igreja de Santo Antônio · Patio do Colégio · Praça Buenos Aires · Biblioteca Municipal · CONSOLAÇÃO · BRÁS · Cemitério da Consolação · Igreja NS da Consolação · Igreja São Francisco de Assis · Catedral Metropolitana · SÉ · Igreja São Gonçalo · MOÓCA · BELA VISTA · Igreja das Almas · Igreja NS Achiropita · Igreja São · CERQUEIRA CÉSAR · Museu Memórias do Bixiga · LIBERDADE · Parque 9 de Julho · AVENIDA 9 DE JULHO · 0 M 500 · 0 Yards 500

BUENOS AIRES

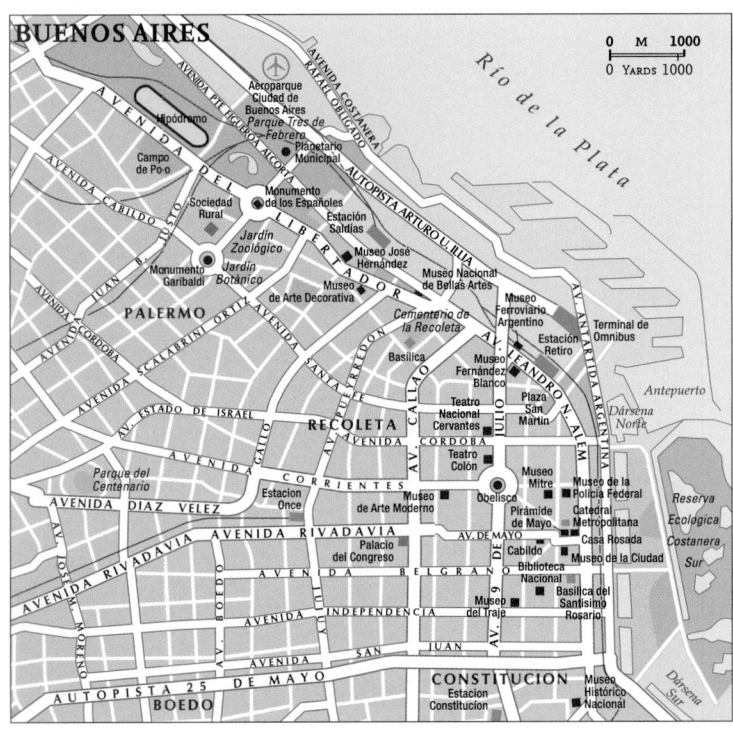

Aeroparque Ciudad de Buenos Aires · Hipódromo · Rio de la Plata · Parque Tres de Febrero · Planetario Municipal · Campo de Polo · Sociedad Rural · Monumento de los Españoles · Estación Saldías · Jardin Zoológico · Museo José Hernández · AUTOPISTA ARTURO ILLIA · Monumento Garibaldi · Jardín Botánico · Museo Nacional de Bellas Artes · Museo de Arte Decorativo · PALERMO · Cementerio de la Recoleta · Museo Ferroviario Argentino · Terminal de Omnibus · Basílica · Estación Retiro · RECOLETA · Museo Fernández Blanco · Plaza San Martín · Dársena Norte · Teatro Nacional Cervantes · Antepuerto · Teatro Colón · Museo de Arte Moderno · Museo Mitre · Obelisco · Reserva Ecológica Costanera Sur · Parque del Centenario · AVENIDA CORRIENTES · Pirámide de Mayo · Catedral Metropolitana · Museo de la Policía Federal · Casa Rosada · AVENIDA RIVADAVIA · Cabildo · Museo de la Ciudad · Palacio del Congreso · Basílica del Santísimo Rosario · AVENIDA BELGRANO · Museo del Traje · AVENIDA INDEPENDENCIA · CONSTITUCION · Estación Constitución · Museo Histórico Nacional · AUTOPISTA 25 DE MAYO · BOEDO · 0 M 1000 · 0 Yards 1000

CARACAS

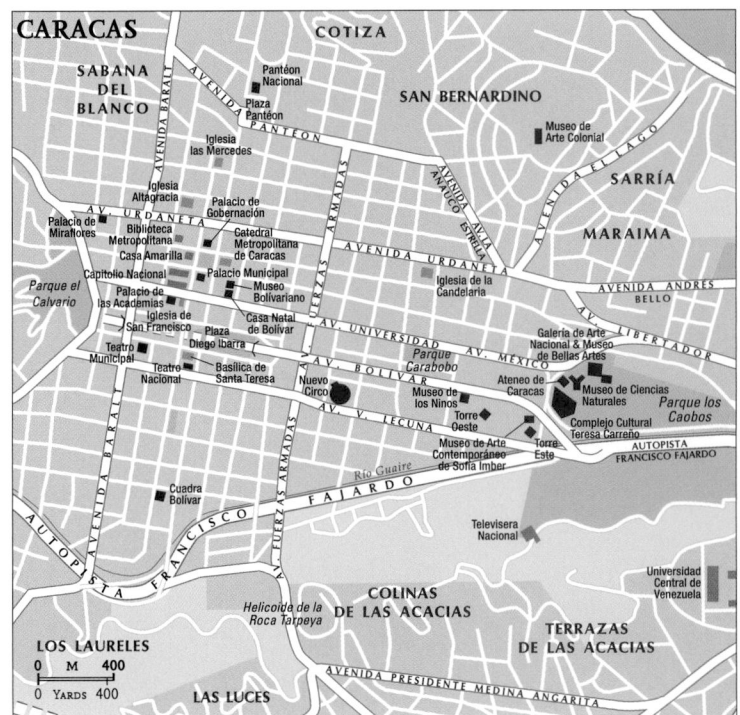

COTIZA · Panteón Nacional · SABANA DEL BLANCO · Plaza Panteón · SAN BERNARDINO · Iglesia las Mercedes · Museo de Arte Colonial · Iglesia Altagracia · SARRÍA · Palacio de Gobernación · MARAIMA · Palacio de Miraflores · Biblioteca Metropolitana · Catedral Metropolitana de Caracas · Casa Amarilla · Palacio Municipal · AVENIDA URDANETA · Iglesia de la Candelaria · Capitolio Nacional · Museo Bolivariano · AVENIDA ANDRES BELLO · Parque el Calvario · Palacio de las Academias · Casa Natal de Bolívar · Galería de Arte Nacional & Museo de Bellas Artes · Iglesia de San Francisco · Plaza Diego Ibarra · Teatro Municipal · Basílica de Santa Teresa · Ateneo de Caracas · Museo de Ciencias Naturales · Nuevo Circo · Museo de los Niños · Parque los Caobos · Teatro Nacional · Torre Oeste · Torre Este · Complejo Cultural Teresa Carreño · Cuadra Bolívar · Museo de Arte Contemporáneo de Sofía Imber · AUTOPISTA FRANCISCO FAJARDO · Rio Guaire · Televisora Nacional · COLINAS DE LAS ACACIAS · Helicoide de la Roca Tarpeya · Universidad Central de Venezuela · TERRAZAS DE LAS ACACIAS · AVENIDA PRESIDENTE MEDINA ANGARITA · LOS LAURELES · LAS LUCES · 0 M 400 · 0 Yards 400

RELIEF

Contour intervals used in layer-colouring for land height and sea depth

Reference maps   Ocean maps

METRES FEET		
6000 / 19686		4000 / 13124
5000 / 16404		3000 / 9843
4000 / 13124		2000 / 6562
3000 / 9843		1000 / 3281
2000 / 6562		500 / 1640
1000 / 3281		200 / 656
500 / 1640		0 / 0
200 / 656		200 / 656
0 / 0		2000 / 6562
LAND BELOW SEA LEVEL		3000 / 9843
		4000 / 13124
200 / 656		5000 / 16404
2000 / 6562		6000 / 19686
4000 / 13124		7000 / 22967
6000 / 19686		9000 / 29529
M FT		M FT

1234 △ Summit
Height in metres

123 Ocean deep
Depth in metres

LAND AND WATER FEATURES

——	River
- - - -	Impermanent river/Wadi
+++++	Canal
··········	Flood dyke
——	Coral reef
········· ··	Escarpment
I	Dam/Barrage
⊻ 123	Pass — Height in metres
1234 ▲	Volcano — Height in metres
‖	Waterfall
ˇ	Oasis

	Lake
	Salt lake/Lagoon
	Dry salt lake/Salt pan
	Impermanent lake
	Impermanent salt lake
	Marsh
	Sandy desert/Dunes
	Rocky desert
	Lava field
	Ice cap/Glacier

TRANSPORT

═══	Motorway — Shown on large-scale maps only
——	Main road
——	Other road
- - - -	Track
―•⋯•―	Road tunnel
——	Main railway
——	Other railway
―•⋯•―	Railway tunnel
✈	Main airport
✈	Regional airport

CITIES AND TOWNS

Population	National Capital	Administrative Capital (Shown for selected countries only)	Other City or Town
over 10 million	**Tōkyō** ▣	**Karachi** ⊙	**New York** ⊙
5 million to 10 million	**Santiago** ▣	**Tianjin** ⊙	**Philadelphia** ⊙
1 million to 5 million	**Damascus** ▣	**Douala** ⊙	**Barranquilla** ⊙
500 000 to 1 million	**Bangui** ▣	**Bulawayo** ◎	**El Paso** ◎
100 000 to 500 000	Wellington ▫	Mansa ○	Mobile ○
50 000 to 100 000	Port of Spain ▫	Lubango ○	Zaraza ○
10 000 to 50 000	Malabo ▫	Chinhoyi ○	El Tigre ○
under 10 000	Roseau ▫	Ati ○	Soledad ○

BOUNDARIES

▬▬▬	International boundary
·▬·▬·	Disputed international boundary/ alignment unconfirmed
•••••	Ceasefire line
▬▬▬	Administrative boundary

MISCELLANEOUS SYMBOLS

- - - -	National park
··········	Reserve
∿∿∿∿	Ancient wall
∴	Site of specific interest
	Built-up area

STYLES OF LETTERING

Cities and towns are explained above

Country	**FRANCE**	Island	*Gran Canaria*	
Overseas Territory/Dependency	**Guadeloupe**	Lake	*Lake Erie*	
Disputed Territory	AKSAI CHIN	Mountain	*Mont Blanc*	
Administrative name (Shown for selected countries only)	SCOTLAND	River	*Thames*	
Area name	PATAGONIA	Region	*LAPPLAND*	

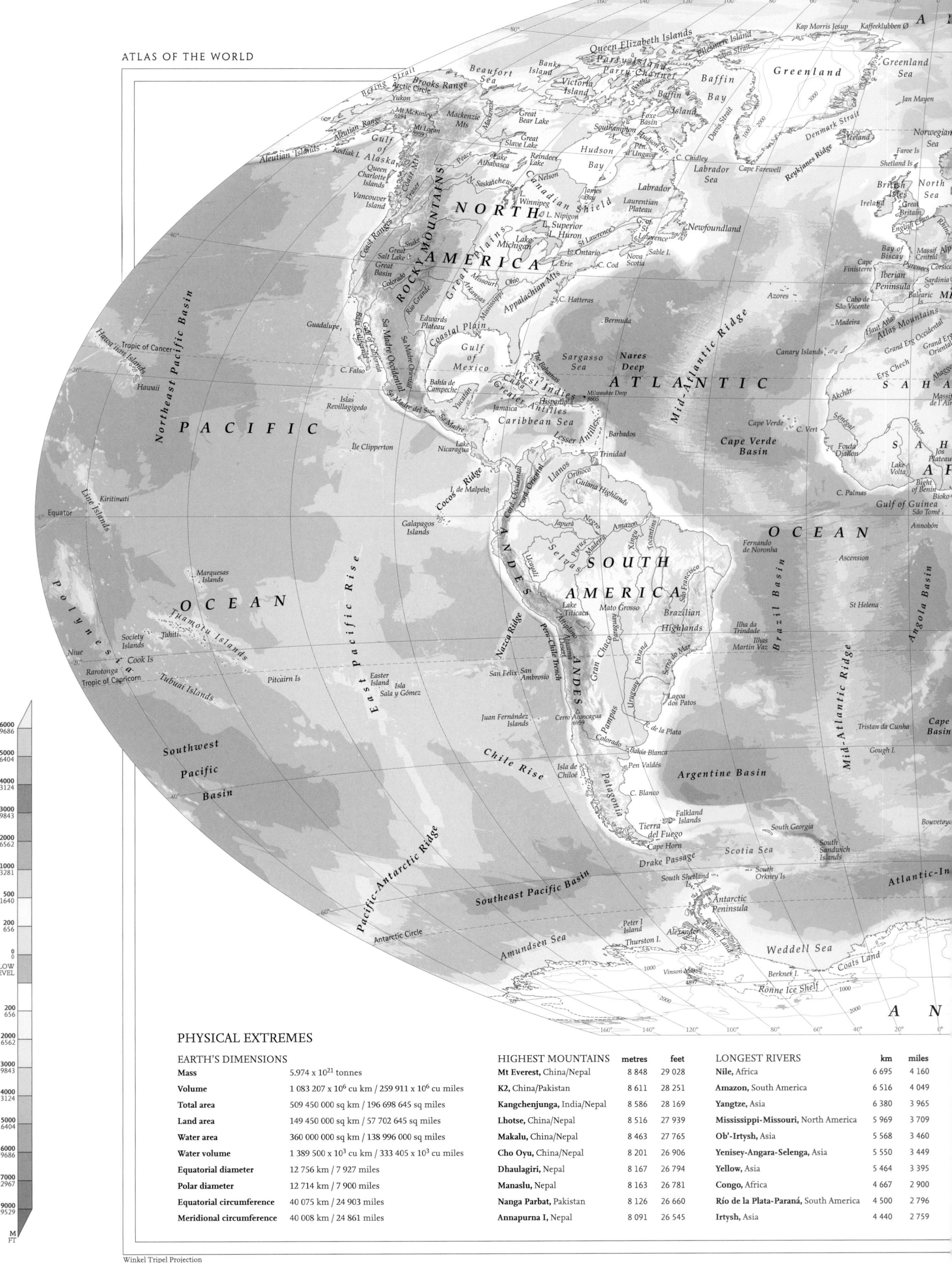

6000	19686
5000	16404
4000	13124
3000	9843
2000	6562
1000	3281
500	1640
200	656
0	0
LAND BELOW SEA LEVEL	
200	656
2000	6562
3000	9843
4000	13124
5000	16404
6000	19686
7000	22967
9000	29529
M	
FT	

PHYSICAL EXTREMES

EARTH'S DIMENSIONS

Mass	5.974 x 10²¹ tonnes
Volume	1 083 207 x 10⁶ cu km / 259 911 x 10⁶ cu miles
Total area	509 450 000 sq km / 196 698 645 sq miles
Land area	149 450 000 sq km / 57 702 645 sq miles
Water area	360 000 000 sq km / 138 996 000 sq miles
Water volume	1 389 500 x 10³ cu km / 333 405 x 10³ cu miles
Equatorial diameter	12 756 km / 7 927 miles
Polar diameter	12 714 km / 7 900 miles
Equatorial circumference	40 075 km / 24 903 miles
Meridional circumference	40 008 km / 24 861 miles

HIGHEST MOUNTAINS	metres	feet
Mt Everest, China/Nepal	8 848	29 028
K2, China/Pakistan	8 611	28 251
Kangchenjunga, India/Nepal	8 586	28 169
Lhotse, China/Nepal	8 516	27 939
Makalu, China/Nepal	8 463	27 765
Cho Oyu, China/Nepal	8 201	26 906
Dhaulagiri, Nepal	8 167	26 794
Manaslu, Nepal	8 163	26 781
Nanga Parbat, Pakistan	8 126	26 660
Annapurna I, Nepal	8 091	26 545

LONGEST RIVERS	km	miles
Nile, Africa	6 695	4 160
Amazon, South America	6 516	4 049
Yangtze, Asia	6 380	3 965
Mississippi-Missouri, North America	5 969	3 709
Ob'-Irtysh, Asia	5 568	3 460
Yenisey-Angara-Selenga, Asia	5 550	3 449
Yellow, Asia	5 464	3 395
Congo, Africa	4 667	2 900
Río de la Plata-Paraná, South America	4 500	2 796
Irtysh, Asia	4 440	2 759

Winkel Tripel Projection

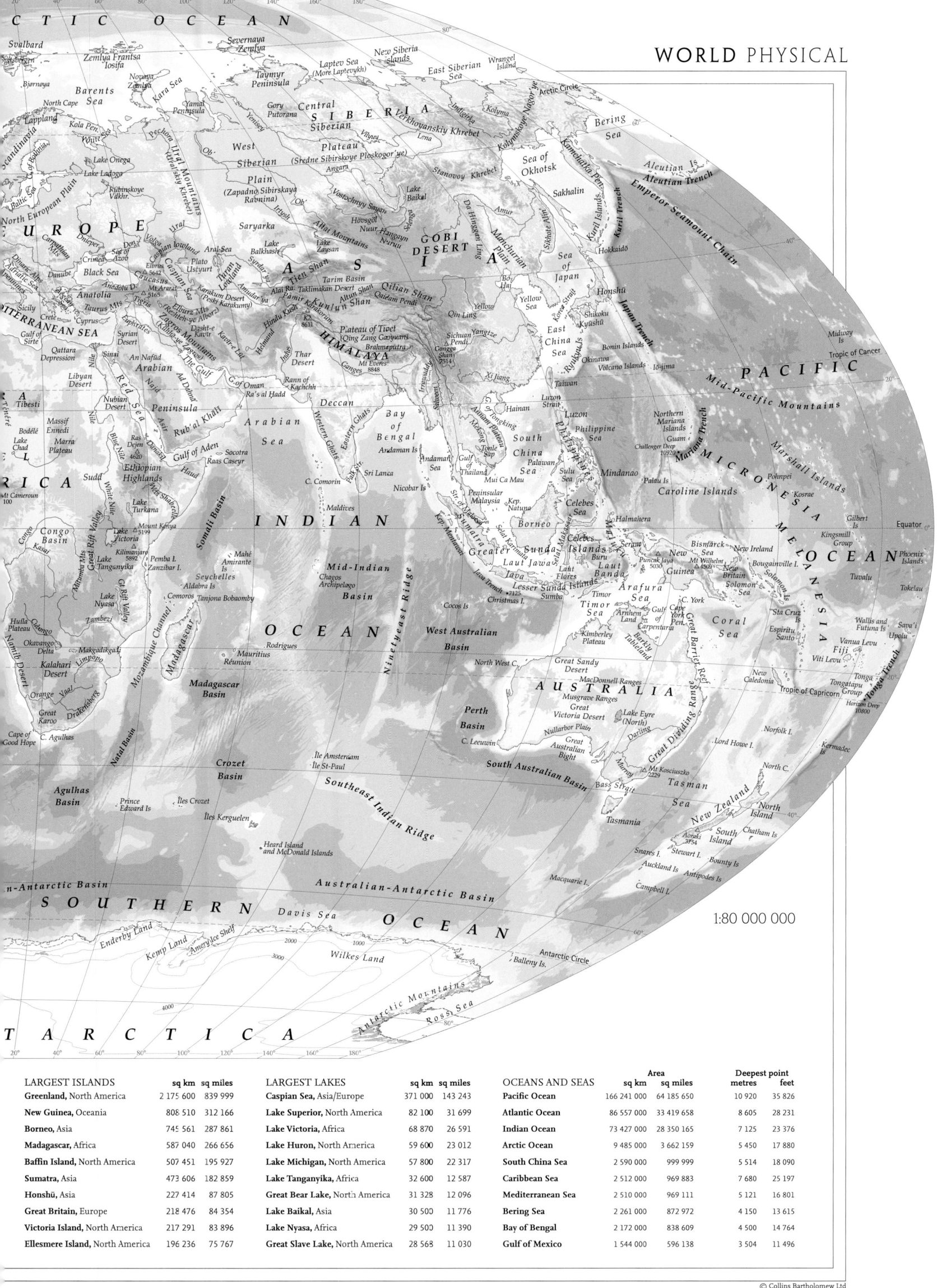

1:80 000 000

LARGEST ISLANDS	sq km	sq miles
Greenland, North America	2 175 600	839 999
New Guinea, Oceania	808 510	312 166
Borneo, Asia	745 561	287 861
Madagascar, Africa	587 040	266 656
Baffin Island, North America	507 451	195 927
Sumatra, Asia	473 606	182 859
Honshū, Asia	227 414	87 805
Great Britain, Europe	218 476	84 354
Victoria Island, North America	217 291	83 896
Ellesmere Island, North America	196 236	75 767

LARGEST LAKES	sq km	sq miles
Caspian Sea, Asia/Europe	371 000	143 243
Lake Superior, North America	82 100	31 699
Lake Victoria, Africa	68 870	26 591
Lake Huron, North America	59 600	23 012
Lake Michigan, North America	57 800	22 317
Lake Tanganyika, Africa	32 600	12 587
Great Bear Lake, North America	31 328	12 096
Lake Baikal, Asia	30 500	11 776
Lake Nyasa, Africa	29 500	11 390
Great Slave Lake, North America	28 568	11 030

OCEANS AND SEAS	Area sq km	Area sq miles	Deepest point metres	Deepest point feet
Pacific Ocean	166 241 000	64 185 650	10 920	35 826
Atlantic Ocean	86 557 000	33 419 658	8 605	28 231
Indian Ocean	73 427 000	28 350 165	7 125	23 376
Arctic Ocean	9 485 000	3 662 159	5 450	17 880
South China Sea	2 590 000	999 999	5 514	18 090
Caribbean Sea	2 512 000	969 883	7 680	25 197
Mediterranean Sea	2 510 000	969 111	5 121	16 801
Bering Sea	2 261 000	872 972	4 150	13 615
Bay of Bengal	2 172 000	838 609	4 500	14 764
Gulf of Mexico	1 544 000	596 138	3 504	11 496

© Collins Bartholomew Ltd

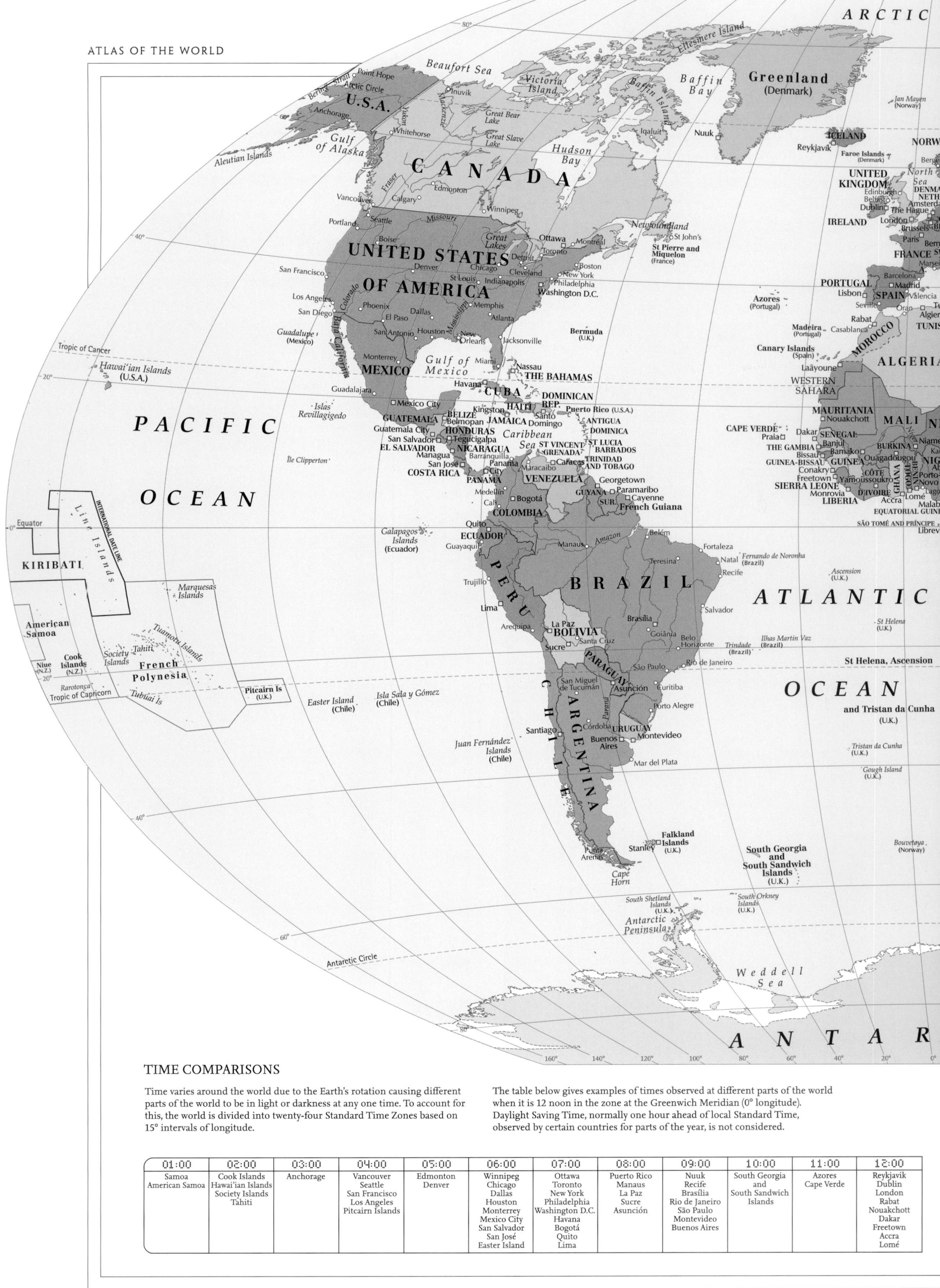

TIME COMPARISONS

Time varies around the world due to the Earth's rotation causing different parts of the world to be in light or darkness at any one time. To account for this, the world is divided into twenty-four Standard Time Zones based on 15° intervals of longitude.

The table below gives examples of times observed at different parts of the world when it is 12 noon in the zone at the Greenwich Meridian (0° longitude). Daylight Saving Time, normally one hour ahead of local Standard Time, observed by certain countries for parts of the year, is not considered.

01:00	02:00	03:00	04:00	05:00	06:00	07:00	08:00	09:00	10:00	11:00	12:00
Samoa American Samoa	Cook Islands Hawai'ian Islands Society Islands Tahiti	Anchorage	Vancouver Seattle San Francisco Los Angeles Pitcairn Islands	Edmonton Denver	Winnipeg Chicago Dallas Houston Monterrey Mexico City San Salvador San José Easter Island	Ottawa Toronto New York Philadelphia Washington D.C. Havana Bogotá Quito Lima	Puerto Rico Manaus La Paz Sucre Asunción	Nuuk Recife Brasília Rio de Janeiro São Paulo Montevideo Buenos Aires	South Georgia and South Sandwich Islands	Azores Cape Verde	Reykjavik Dublin London Rabat Nouakchott Dakar Freetown Accra Lomé

Winkel Tripel Projection

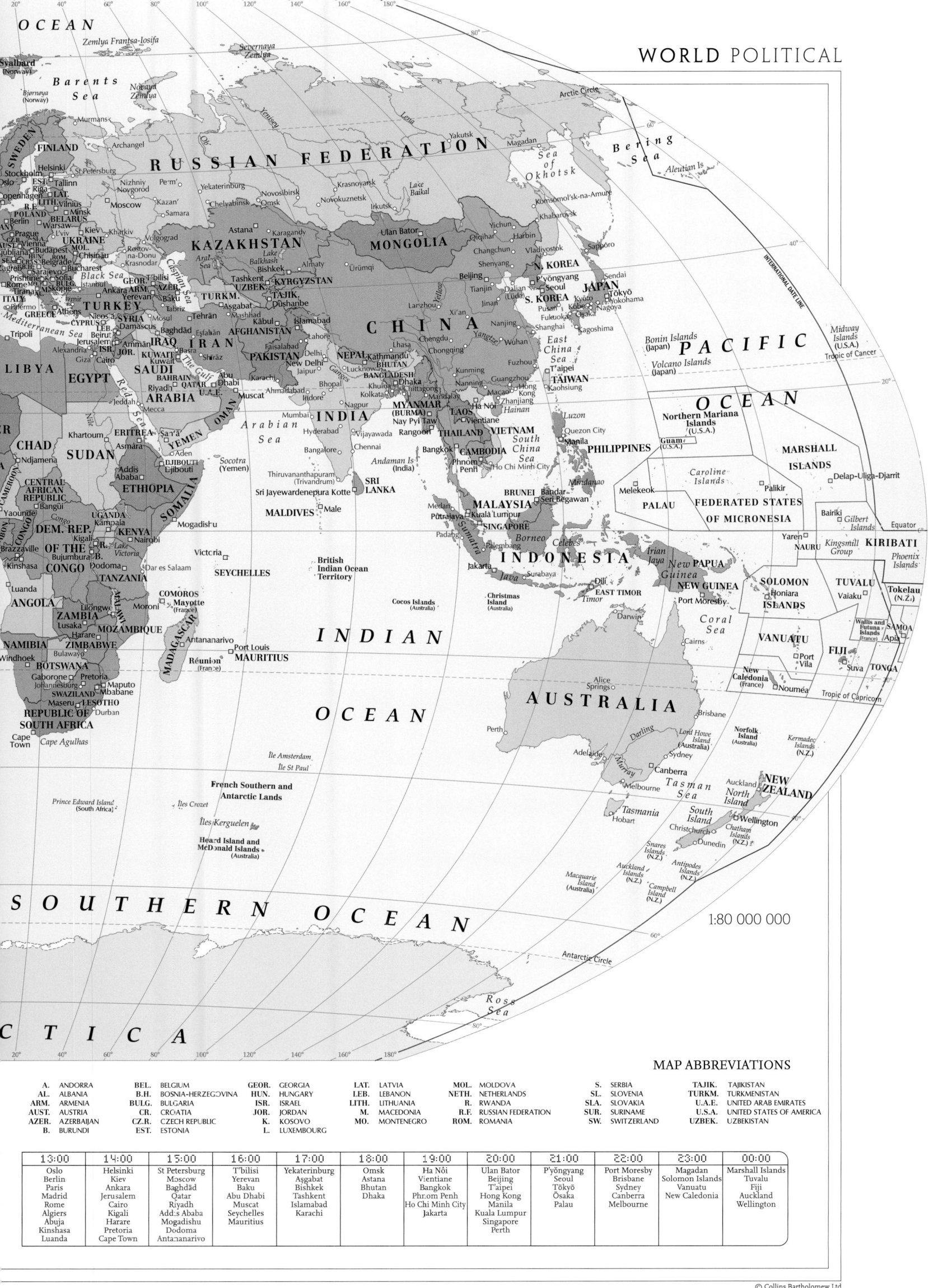

1:80 000 000

MAP ABBREVIATIONS

A.	ANDORRA	BEL.	BELGIUM	GEOR.	GEORGIA	LAT.	LATVIA	MOL.	MOLDOVA	S.	SERBIA	TAJIK.	TAJIKISTAN
AL.	ALBANIA	B.H.	BOSNIA-HERZEGOVINA	HUN.	HUNGARY	LEB.	LEBANON	NETH.	NETHERLANDS	SL.	SLOVENIA	TURKM.	TURKMENISTAN
ARM.	ARMENIA	BULG.	BULGARIA	ISR.	ISRAEL	LITH.	LITHUANIA	R.	RWANDA	SLA.	SLOVAKIA	U.A.E.	UNITED ARAB EMIRATES
AUST.	AUSTRIA	CR.	CROATIA	JOR.	JORDAN	M.	MACEDONIA	R.F.	RUSSIAN FEDERATION	SUR.	SURINAME	U.S.A.	UNITED STATES OF AMERICA
AZER.	AZERBAIJAN	CZ.R.	CZECH REPUBLIC	K.	KOSOVO	MO.	MONTENEGRO	ROM.	ROMANIA	SW.	SWITZERLAND	UZBEK.	UZBEKISTAN
B.	BURUNDI	EST.	ESTONIA	L.	LUXEMBOURG								

13:00	14:00	15:00	16:00	17:00	18:00	19:00	20:00	21:00	22:00	23:00	00:00
Oslo	Helsinki	St Petersburg	T'bilisi	Yekaterinburg	Omsk	Ha Nôi	Ulan Bator	P'yŏngyang	Port Moresby	Magadan	Marshall Islands
Berlin	Kiev	Moscow	Yerevan	Aşgabat	Astana	Vientiane	Beijing	Seoul	Brisbane	Solomon Islands	Tuvalu
Paris	Ankara	Baghdād	Baku	Bishkek	Bhutan	Bangkok	T'aipei	Tōkyō	Sydney	Vanuatu	Fiji
Madrid	Jerusalem	Qatar	Abu Dhabi	Tashkent	Dhaka	Phrom Penh	Hong Kong	Ōsaka	Canberra	New Caledonia	Auckland
Rome	Cairo	Riyadh	Muscat	Islamabad		Ho Chi Minh City	Manila	Palau	Melbourne		Wellington
Algiers	Kigali	Addis Ababa	Seychelles	Karachi		Jakarta	Kuala Lumpur				
Abuja	Harare	Mogadishu	Mauritius				Singapore				
Kinshasa	Pretoria	Dodoma					Perth				
Luanda	Cape Town	Antananarivo									

© Collins Bartholomew Ltd

INDIAN OCEAN

ASIA

AUSTRALIA

PAPUA NEW GUINEA

F G H I J

165° 180° 45° 165° 30° 150°

1

H a w a i i a n I s l a n d s

Kure Atoll

Midway Islands (U.S.A.)

Pearl and Hermes Atoll

Wake Island (U.S.A.)

Lisianski Island Laysan Island

Gardner Pinnacles

Necker Island

Tropic of Cancer

2

MARSHALL ISLANDS

Ralik Chain Ratak Chain

Kwajalein Maloelap

Palikir Kosrae

Pohnpei

Delap-Uliga-Djarrit

Mili

P A C I F I C

Johnston Atoll (U.S.A.)

Kaua'i O'ahu Maui

Hawai'i

30° 150°

F MICRONESIA

Gilbert Islands **Bairiki**

Tarawa Aranuka

Yaren **NAURU**

Banaba Nonouti Tabiteuea

Beru Nikunau Onotoa Nui Vaitupu

O C E A N

Howland Island (U.S.A.) Baker Island (U.S.A.)

Kingman Reef (U.S.A.)

Palmyra Atoll (U.S.A.)

Teraina

Tabuaeran

3

ougainville Island Nukumanu Islands

Choiseul Ontong Java Atoll

Santa Isabel

SOLOMON ISLANDS

New Georgia Malaita

Honiara San Cristobal

Guadalcanal

Rennell

Tamana Arorae

Kingsmill Group

TUVALU

Nanumea Niutao

Nanumanga

Nukufetau Funafuti

Nukulaelae

Vaiaku

Niulakita

Phoenix Islands

McKean Nikumaroro Orona

Kanton Rawaki Manra

K I R I B A T I

Jarvis Island (U.S.A.)

Kiritimati

15° 135°

L i n e I s l a n d s

Malden Island Starbuck Island

Caroline Island (Millennium Island)

Nuku Hiva

4

Banks Islands

Espíritu Santo Maéwo

VANUATU Malakula Ambrym

Epi **Efaté**

Port Vila Erromango

Tanna Anatom

Duff Islands

Santa Cruz Islands

Rotuma (Fiji)

Wallis and Futuna Islands (France) Iles Wallis

Îles de Hoorn

Matā'utu

Yasawa Group

Viti Levu Vanua Levu

Suva Kandavu Moala

FIJI

Nukunono Atafu

Fakaofo

Swains Island

SAMOA Savai'i

Apia Upolu

Niuafo'ou

Tafahi

Tokelau (New Zealand)

American Samoa

Tutuila Manu'a Is.

Fagatogo

Pukapuka Nassau

Manihiki

Rakahanga

Suwarrow

Penrhyn

Vostok Island

Flint Island

5

Îles Chesterfield (France)

New Caledonia (France)

Cato Island and Bank

Iles Loyauté (France)

Nouméa

Hunter I.

Île des Pins

Ceva-i-Ra

Ono-i-Lau

Vava'u Group

Tofua

TONGA

Nuku'alofa

Tongatapu Group

Ata

Alofi

Niue (New Zealand)

Palmerston

Aitutaki

Cook Islands (New Zealand)

Rarotonga Mangaia

Manuae Mauke

Maria

Motu One Rangiroa

Îles du Roi-Georges

Fakarava

Papeete **Tahiti**

Society Islands

Hereheretue

Îles du Duc de Gloucester

French Polynesia

Marquesas Islands

Hiva Oa

Îles du Désappointement

Pukapuka

Anaa

Hao

Rurutu

Tubuai

Raivavae

Tubuai Islands

Polynesia

Rapa

Marutea

Groupe Actéon

Îles Gambier

Marotiri

Equator 0°

15° 120°

6

Norfolk Island (Australia)

Lord Howe Island (Australia)

Raoul Island

Kermadec Islands (New Zealand)

Cape Maria van Diemen

Whangarei Great Barrier Island

North Island **Auckland** Manukau

NEW ZEALAND Hamilton

New Plymouth Gisborne

Napier

Nelson **Wellington** Palmerston North

Greymouth Blenheim

Aoraki

South Island

Cape Providence SOUTHERN ALPS Christchurch

Timaru Oamaru

Dunedin

Stewart Island **Invercargill**

Chatham Islands (New Zealand)

Pitt Island

7

Snares Islands (New Zealand)

Bounty Islands (New Zealand)

Antipodes Islands (New Zealand)

Auckland Islands (New Zealand)

Campbell Island (New Zealand)

Macquarie Island (Australia)

Adamstown

Pitcairn Islands (U.K.) Henderson I. Ducie I.

Pitcairn Island

Tropic of Capricorn

TASMAN SEA

INTERNATIONAL DATE LINE

165° 180° 165° 150° 45° 135° 120° 30°

F G H I J

MILES	KM
1200	2000
	1600
800	1200
	800
400	
	400
0	0

1:32 000 000

© Collins Bartholomew Ltd

123

147

147

Celebes Sea
Manado
Semenanjung Minahasa
Tondano
Tolitoli
Gorontalo
Sangkulirang
Moutong
Kepulauan
Togian
Celebes
(Sulawesi)
Donggala Palu
Teluk Tomini
Pangkalsiang
Ralu
Ujekuli
Bangai
Kolonedale
Kolaka
Kendari
Parepare
Watampone
Teluk Bone
Singkang
Buton
Makassar
(Ujung Pandang)
Bulukumba
Sinjai
Tukangbesi

INDONESIA

Halmahera Waigeo
Ternate
Selat Dampir Kwoka
Tidore
Sorong Jazirah Doberai
Salawati Ranski
Misool Inanwatan
Seram (Ceram Sea)
Laut Seram
Buru
Ambon
(Amboina)
Banda (Banda Sea)
Laut Banda
Kepulauan Kai
Kai Besar Kai
Kepulauan
Aru
Trangan

NEW GUINEA
IRIAN JAYA
Jayapura

PAPUA
NEW GUINEA

Bismarck Archipelago
Bismarck Sea

Gulf of Papua
Port Moresby

Torres Strait

INDIAN OCEAN

Timor Sea

Arafura Sea

Gulf of Carpentaria

Cape York Peninsula

Great Barrier Reef

Coral Sea Islands Territory (Aust.)

Darwin

Arnhem Land

NORTHERN TERRITORY

Tanami Desert

QUEENSLAND

Townsville

Cairns

GREAT SANDY DESERT

Alice Springs
Macdonnell Ranges

Simpson Desert

Uluru (Ayers Rock)

WESTERN AUSTRALIA

Gibson Desert

A U S T R A L I A

SOUTH AUSTRALIA

NEW SOUTH WALES

GREAT VICTORIA DESERT

Lake Eyre

Port Hedland

Perth
Fremantle

Nullarbor Plain

Great Australian Bight

Adelaide

Melbourne

VICTORIA

Sydney
Canberra
A.C.T.

TASMANIA

Hobart

Lambert Azimuthal Equal Area Projection

6000
19686
5000
16404
4000
13124
3000
9843
2000
6562
1000
3281
500
1640
200
656
0
0
LAND BELOW
SEA LEVEL
200
656
2000
6562
4000
13124
6000
19686
M
FT

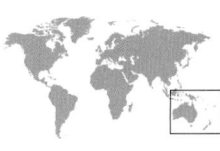

1:20 000 000

© Collins Bartholomew Ltd

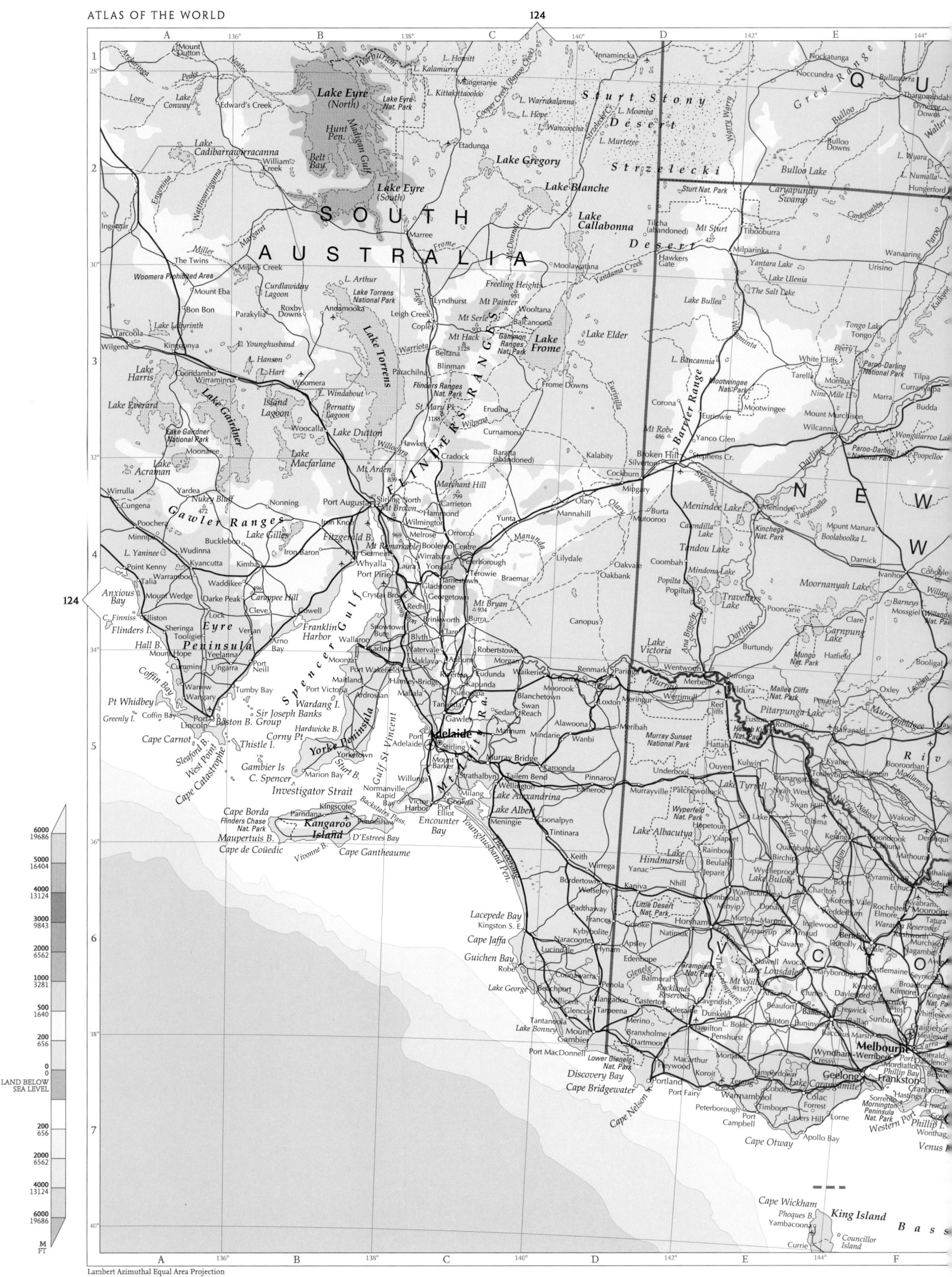

6000
19686

5000
16404

4000
13124

3000
9843

2000
6562

1000
3281

500
1640

0
0

LAND BELOW
SEA LEVEL

200
656

2000
6562

4000
13124

6000
19686

M
FT

Lambert Azimuthal Equal Area Projection

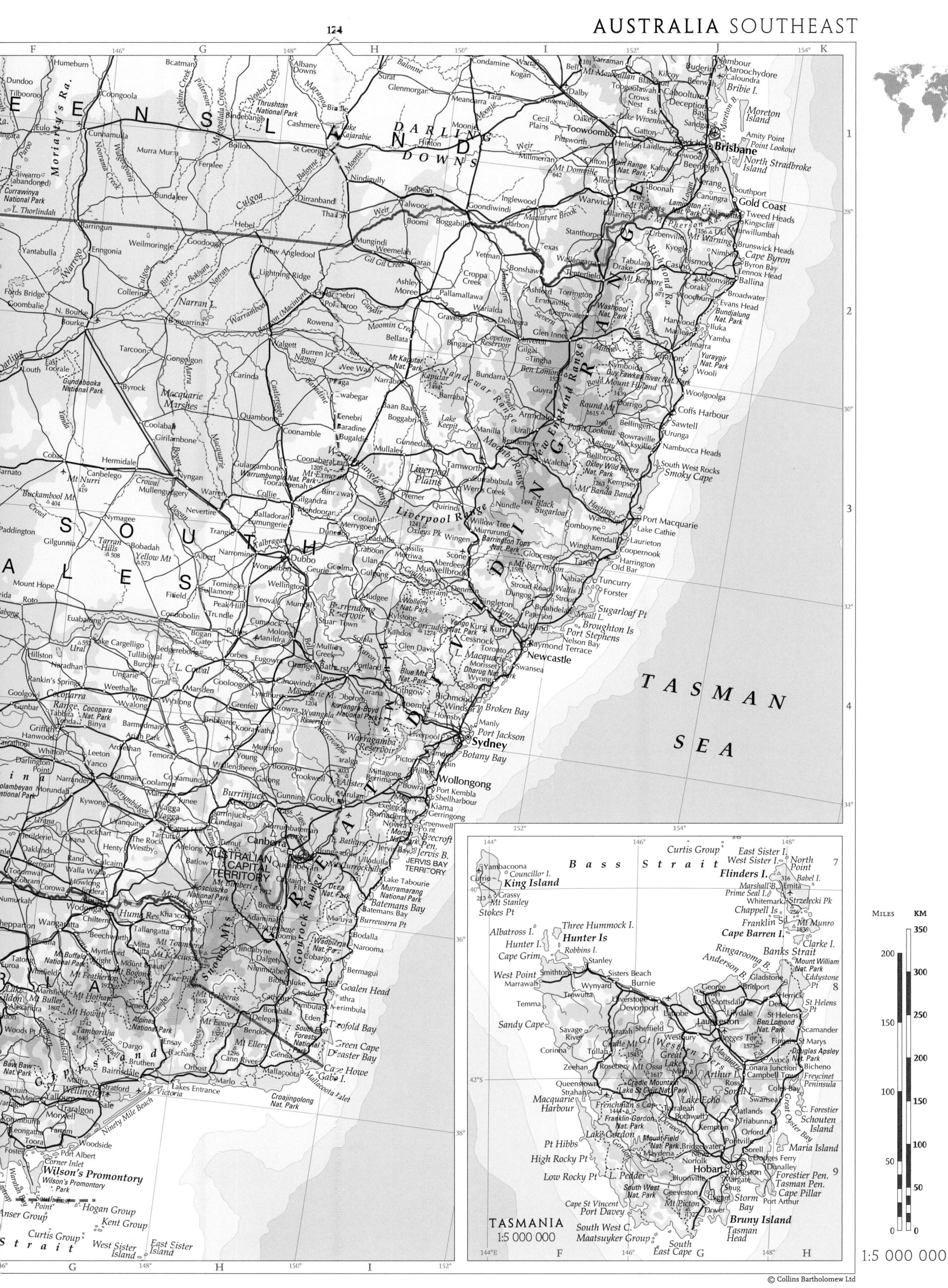

TASMAN

SEA

TASMANIA
1:5 000 000

© Collins Bartholomew Ltd

MILES   KM

1:5 000 000

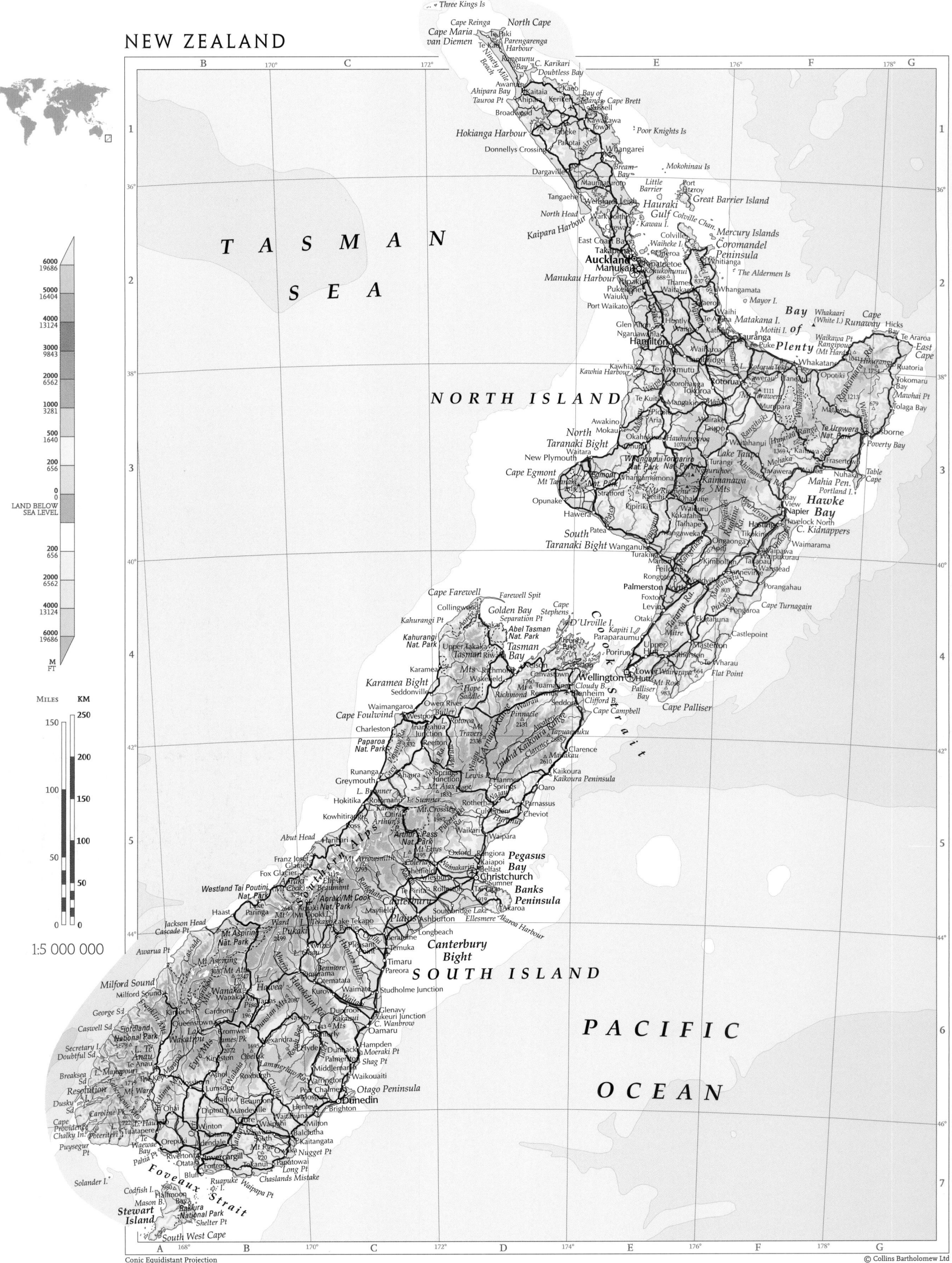

NEW ZEALAND

TASMAN SEA

PACIFIC

OCEAN

NORTH ISLAND

SOUTH ISLAND

1:5 000 000

Conic Equidistant Projection

© Collins Bartholomew Ltd

ANTARCTICA

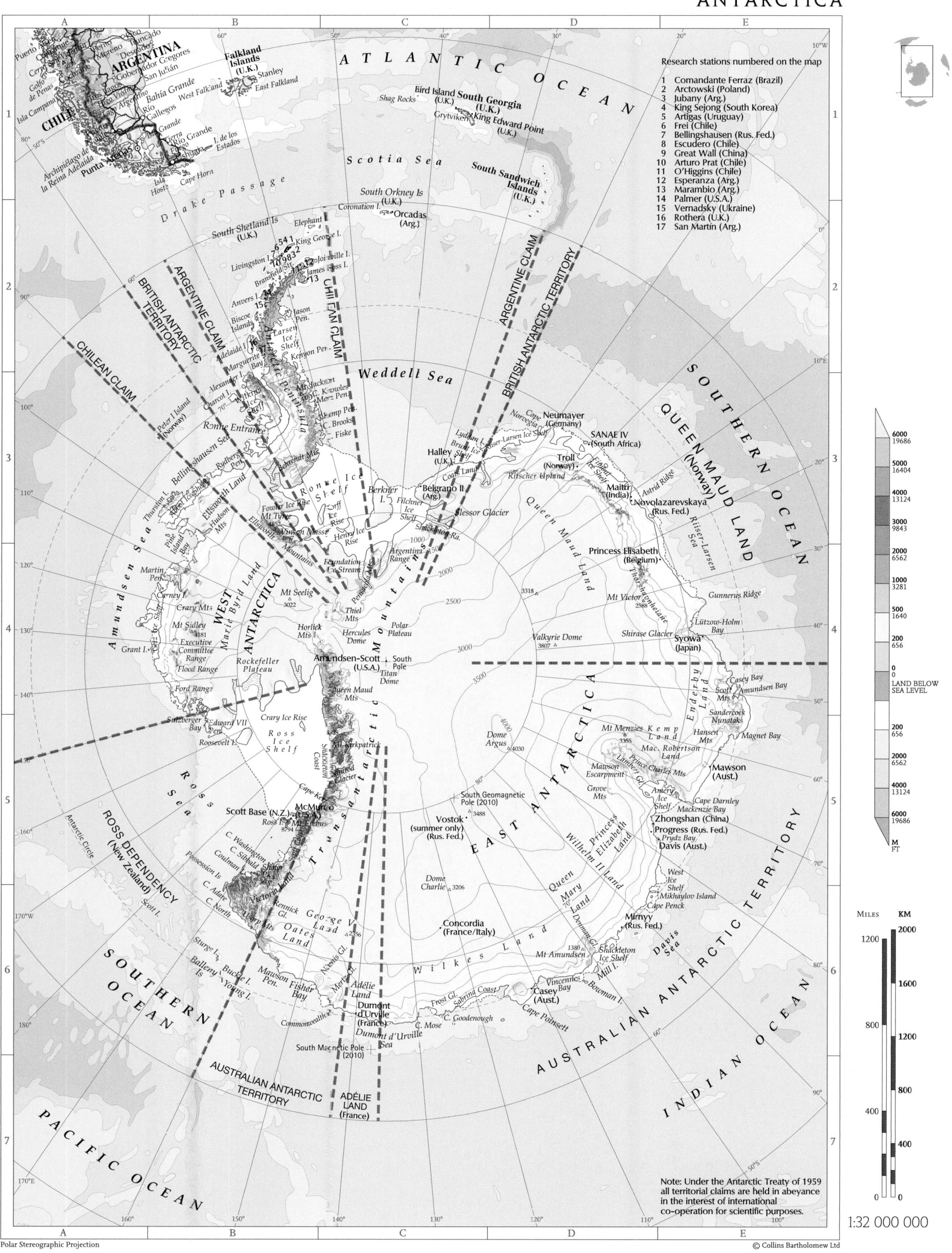

Research stations numbered on the map

1 Comandante Ferraz (Brazil)
2 Arctowski (Poland)
3 Jubany (Arg.)
4 King Sejong (South Korea)
5 Artigas (Uruguay)
6 Frei (Chile)
7 Bellingshausen (Rus. Fed.)
8 Escudero (Chile)
9 Great Wall (China)
10 Arturo Prat (Chile)
11 O'Higgins (Chile)
12 Esperanza (Arg.)
13 Marambio (Arg.)
14 Palmer (U.S.A.)
15 Vernadsky (Ukraine)
16 Rothera (U.K.)
17 San Martin (Arg.)

Note: Under the Antarctic Treaty of 1959
all territorial claims are held in abeyance
in the interest of international
co-operation for scientific purposes.

6000   19686
5000   16404
4000   13124
3000   9843
2000   6562
1000   3281
500   1640
200   656
0
LAND BELOW
SEA LEVEL
200   656
2000   6562
4000   13124
6000   19686
M
FT

MILES   KM

1200   2000
        1600
800   1200
        800
400   400
0   0

1:32 000 000

Polar Stereographic Projection

© Collins Bartholomew Ltd

129

Orthographic Projection

O C E A N

Bering Strait

CENTRAL SIBERIAN
PLATEAU

Nizhnyaya Tunguska

Verkhoyanskiy Khrebet

Khrebet Kolymskiy

BERING
SEA

Arctic Circle

F E D E R A T I O N

Lena

Vilyuy

Aldan

Yakutsk

Mirnyy

Anadyr'

Magadan

Kamchatka Peninsula

Pratilof
Islands

Angara

Bratsk

Kansk

Ust'-Kut

Bodaybo

Lena

Lena

Aldan

Tynda

Stanovoy Khrebet

Amur

Heilong Jiang

Komsomol'sk-
na-Amure

B.goveshchensk

Khabarovsk

Sea
of Okhotsk

Sakhalin

Petropavlovsk-
Kamchatskiy

Aleutian
Islands

Irkutsk

Ulan-Ude

Chita

Da-han

Hulun
Buir

Qiqihar

Jiamusi

Yuzhno-
Sakhalinsk

Korsakov

Kuril Islands

Hoosgol
Nuur

Lake
Baikal

Argun

Hulun
Nur

Daqing

Harbin

Wakkanai

HOKKAIDO

Ullastay

Ulan Bator

Matad

Buir
Nur

Da Hinggan Ling

Changchun

Jilin

Vladivostok

Lake Khanka

Sapporo

Hakodate

MONGOLIA

G O B I   D E S E R T

Dalandzadgad

INNER MONGOLIA

Shenyang

Fushun

Anshan

Ch'ongjin

Sea
of Japan
(East Sea)

Akita

Sendai

Laojunmiao

Wuhai

Baotou
(Huang He)

Hohhot

Datong

Beijing

Tianjin

Dalian

NORTH
KOREA

P'yongyang

Niigata

HONSHŪ

Qilian Shan

Yinchuan

Shijiazhuang

Yellow
(Huang He)

Bo Hai

Korea
Bay

Seoul

Kanazawa

TŌKYŌ

JAPAN

Qinghai
Hu

Xining

Taiyuan

Handan

Jinan

Zibo

Yantai

SOUTH
KOREA

Taejon

Tregu

Kyōto

Yokohama

Guaring
Hu

Lanzhou

Zhengzhou

Jining

Qingdao

Taegu

Kōbe

Ōsaka

C H I N A

Xi'an

Huainan

Nanjing

W xi

Kwangju

Pusan

Kita-kyūshū

Hiroshima

Nagasaki

SHIKOKU

Chengdu

Nanchong

Suizhou

Wuhan

Hefei

Shanghai

Fukuoka

Kumamoto

Neijiang

Chongqing

Yueyang

Hangzhou

East China
Sea

Kagoshima

KYŪSHŪ

Yibin

Guiyang

Changsha

Nanchang

Quzhou

Ningbo

Panzhihua

Qujing

Hengyang

Wenzhou

Bonin Islands
(Japan)

P A C I F I C

Myitkyina

Kunming

Xun Jiang

Liuzhou

Meizhou

Xiamen

zhou

Okinawa

Ryukyu Islands

Volcano Islands
(Japan)

Tropic of Cancer

Nanning

Guangzhou

Shantou

T'aipei

O C E A N

AR
MA)

Ha Nôi

Louangphabang

Zhanjiang

Macao

Hong Kong

Taiwan Strait

T'aitung

TAIWAN

Batan Islands

Northern
Mariana
Islands

Pagan

Chiang
Mai

Vientiane

Huê

Hai Phong

Gulf
of
Tongking

Haikou

Hainan

Luzon Strait

Aparri

Saipan

Tinian

Rota

wlamyaing

Đa Nẵng

Paracel Islands

Luzon

PHILIPPINES

Guam

THAILAND

LAOS

Nakhon
Ratchasima

Mekong

SOUTH

CHINA

SEA

Quezon
City

Manila

Naga

Masbate

Samar

Yap

Bangkok

Tonle
Sap

Nha Trang

Mindoro

Iloilo

Cebu

Caroline Islands

Myeik

CAMBODIA

Phnom
Penh

Ho Chi Minh City

Panay

Negros

Surigao

PALAU
Melekeok

Chuuk

Gulf of
Thailand

Sihanoukville

Spratly Islands

Palawan

Dapitan

Mortlock
Islands

Nakhon Si
Thammarat

Sulu
Sea

Mindanao

Davao

George
Town

Ipoh

Kota Bharu

Kota Kinabalu

Sandakan

Zamboanga

Kuala
Lumpur

BRUNEI
Bandar Seri
Begawan

SABAH

Sulu
Archipelago

Kepulauan
Talaud

Equator

Singapore

Putrajaya

MALAYSIA

SARAWAK

Kuching

Sibu

Sri-Aman

Celebes
Sea

Kepulauan
Sangir

Manado

Halmahera

Bismarck Archipelago

Medan

Strait of Malacca

Sumatra

Pontianak

Balikpapan

Borneo

Palu

Kepulauan
Sula

Laut Maluku

Maluku

Jazirah
Doberai

Manokwari

Jayapura

Bismarck
Sea

New Britain

Siberut

Padang

Ketapang

Barito

Selat Makassar

Parepare

Celebes

Laut Seram

Laut Seram

IRIAN

Puncak Jaya
5030

Kepulauan
Lingga

Bangka

Banjarmasin

Makassar

Buru

Laut Banda

JAYA

New Guinea

Solomon
Sea

Kepulauan Mentawai

Palembang

Laut
Jawa

I N D O N E S I A

Buton

Kepulauan
Aru

Dili

Gulf
of Papua

Bengkulu

Bandar
Lampung

Jakarta

Semarang

Surabaya

Madura

Laut Flores

Wetar

Kepulauan
Tanimbar

Enggano

Bandung

Java

Yogyakarta

Bali

Lombok

Sumbawa

Laut Bali

Flores

Sumba

Laut Sawu

EAST
TIMOR

Timor

Dili

EAST TIMOR

Arafura Sea

Torres Strait

Cape
York
Peninsula

CORAL
SEA

© Collins Bartholomew Ltd

MILES   KM

1000

750

500

250

0

1500

1250

1000

750

500

250

0

1:28 000 000

Conic Equidistant Projection

OCEAN

O. Komsomolets
Severnaya
Zemlya

O. Oktyabr'skoy
Revolyutsii

Ostrov
Bol'shevik

Taymyr Peninsula

Ozero
Taymyr

Gusikha

Novorybnaya

Khatanga

Starry
Kayak

Kheta

New Siberia Islands
(Novosibirskiye Ostrova)

Laptev Sea

(More Laptevykh)

East Siberian Sea
(Vostochno-Sibirskoye More)

Wrangel I.
(Ostrov Vrangelya)

Chukchi
Sea

Bering Strait

U.S.A.

Pribilof
Islands

BERING
SEA

Khrebet Cherskogo

Central Siberian

Plateau

(Sredne-Sibirskoye Ploskogor'ye)

Kamchatka
Peninsula

Sea of Okhotsk
(Okhotskoye More)

Petropavlovsk-
Kamchatskiy

FEDERATION

SIBERIA (SIBIR')

Yakutsk

Stanovoy Khrebet

Stanovoye

Nagor'ye

Lake Baikal

Irkutsk

Ulan-Ude

Sakhalin

Kuril Islands
(Kuril'skiye Ostrova)

Urup
Administered
by Rus. Fed.
Claimed by
Japan

Khabarovsk

Komsomol'sk-
na-Amure

Yuzhno-
Sakhalinsk

Hokkaido

Sapporo

Magadan

MONGOLIA

Ulan Bator
(Ulaanbaatar)

GOBI DESERT

INNER MONGOLIA

Beijing
(Peking)

CHINA

Harbin

Changchun

Shenyang

NORTH
KOREA

P'yongyang

SOUTH
KOREA

Seoul
(Soul)

Vladivostok

Sea of
Japan
(East Sea)

Tokyo

JAPAN

Yellow Sea
(Huang Hai)

Qingdao (Tsingtao)

© Collins Bartholomew Ltd

MILES   KM

800   1400
         1200
600   1000
         800
400   600
200   400
         200
0   0

1:21 000 000

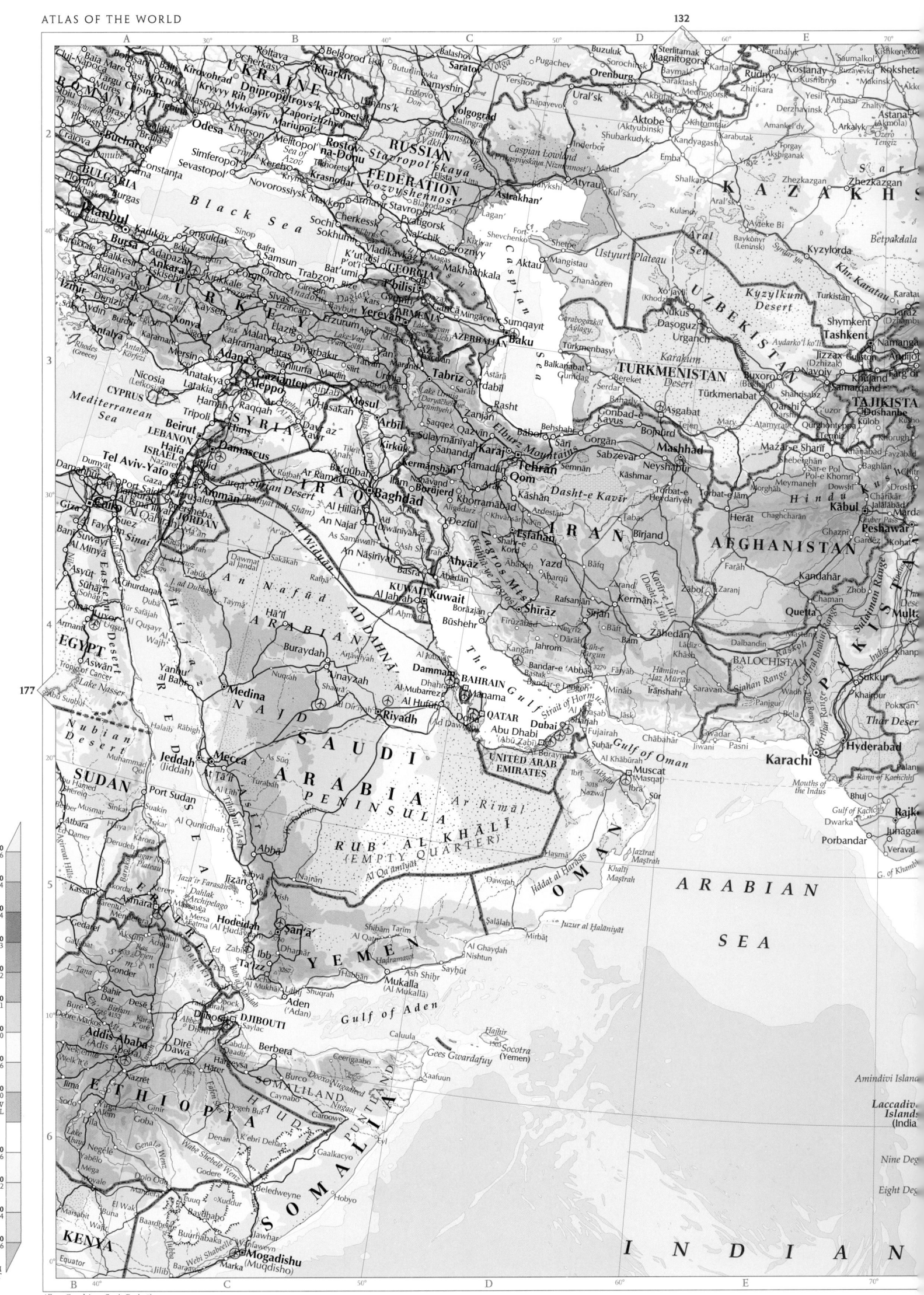

Albers Equal Area Conic Projection

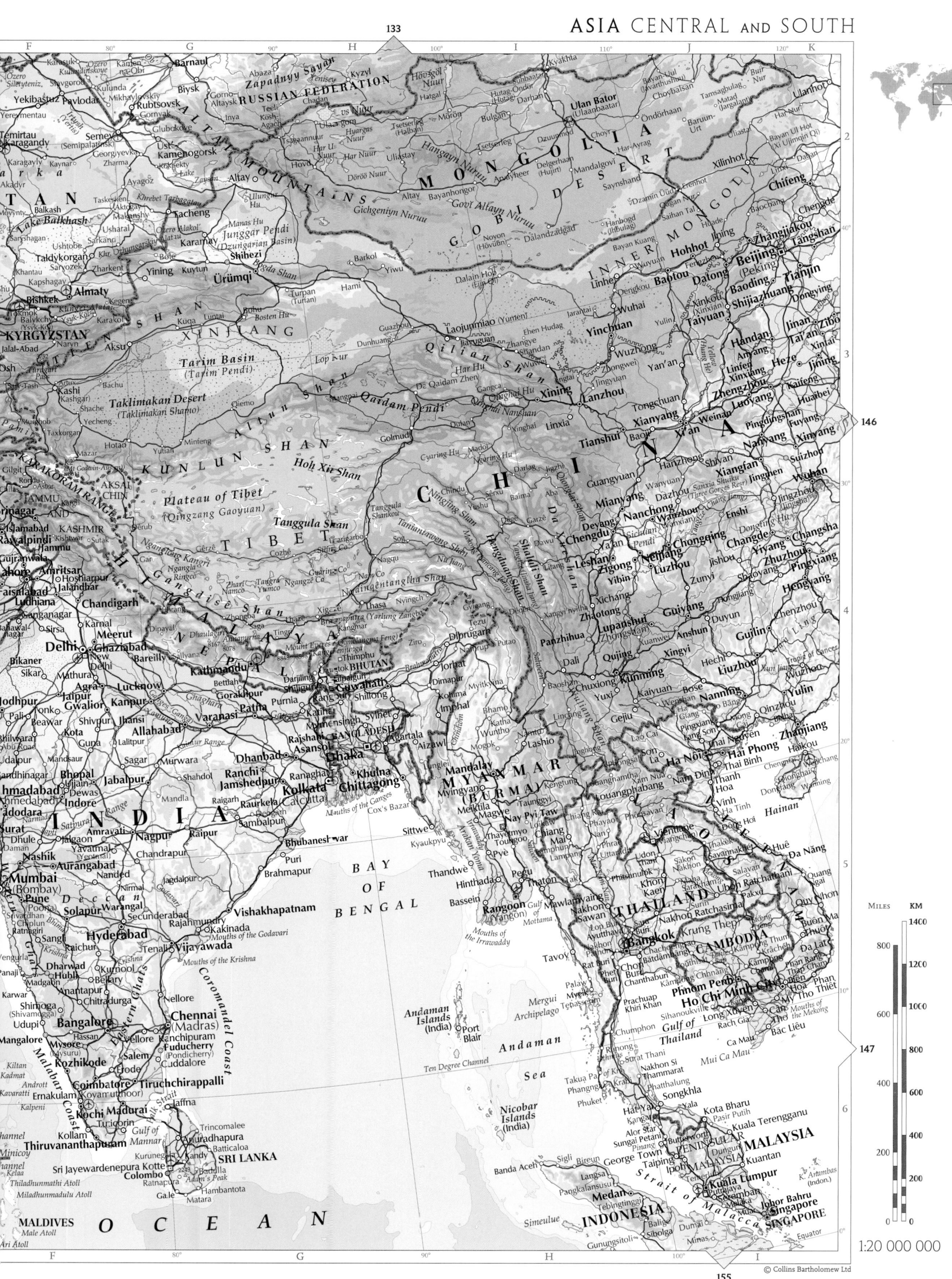

© Collins Bartholomew Ltd

1:20 000 000

THE MIDDLE EAST

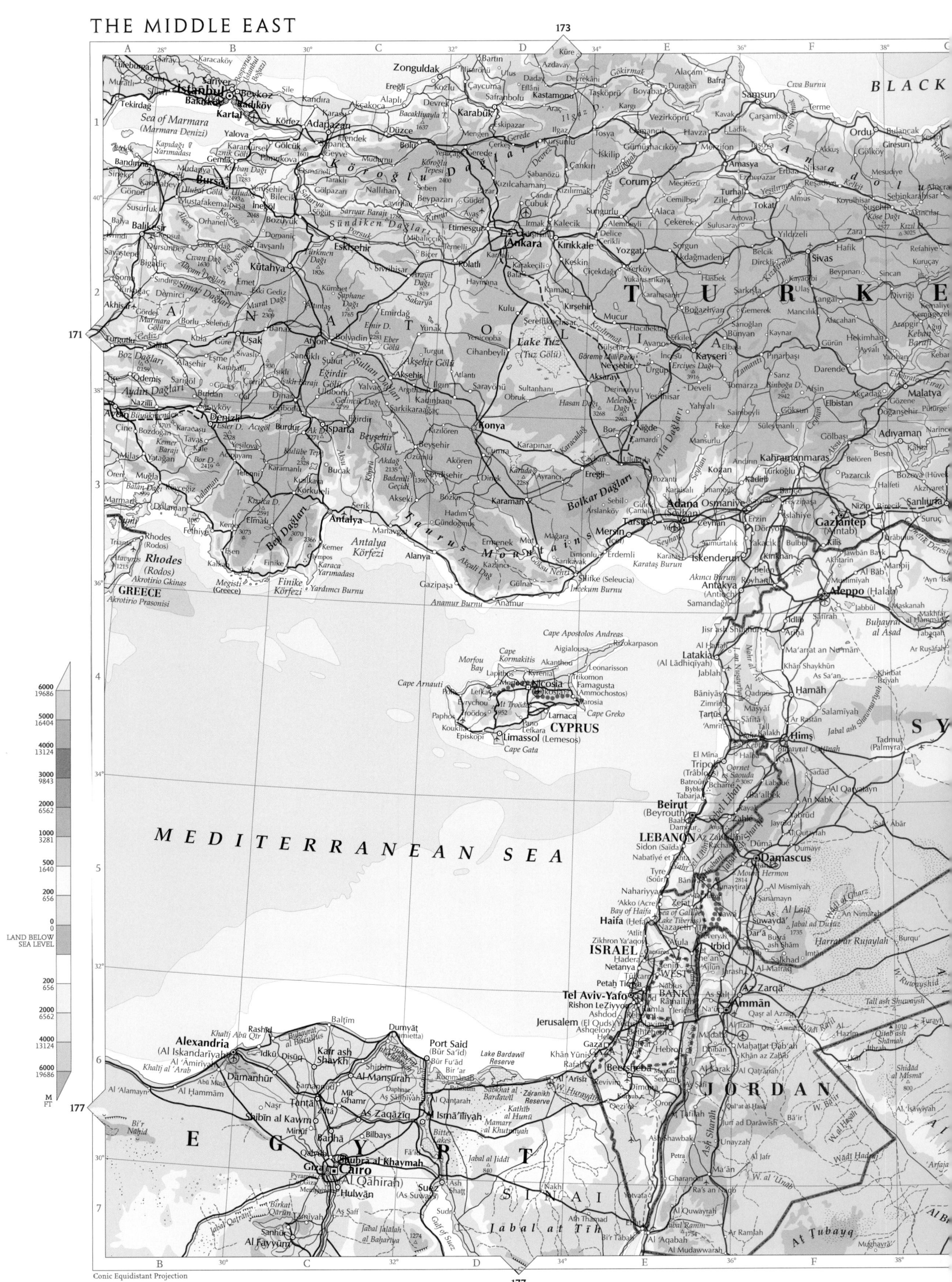

MEDITERRANEAN SEA

BLACK

TURKEY

ANATOLIA

GREECE

CYPRUS

SY

LEBANON

ISRAEL

JORDAN

EGYPT

SINAI

Conic Equidistant Projection

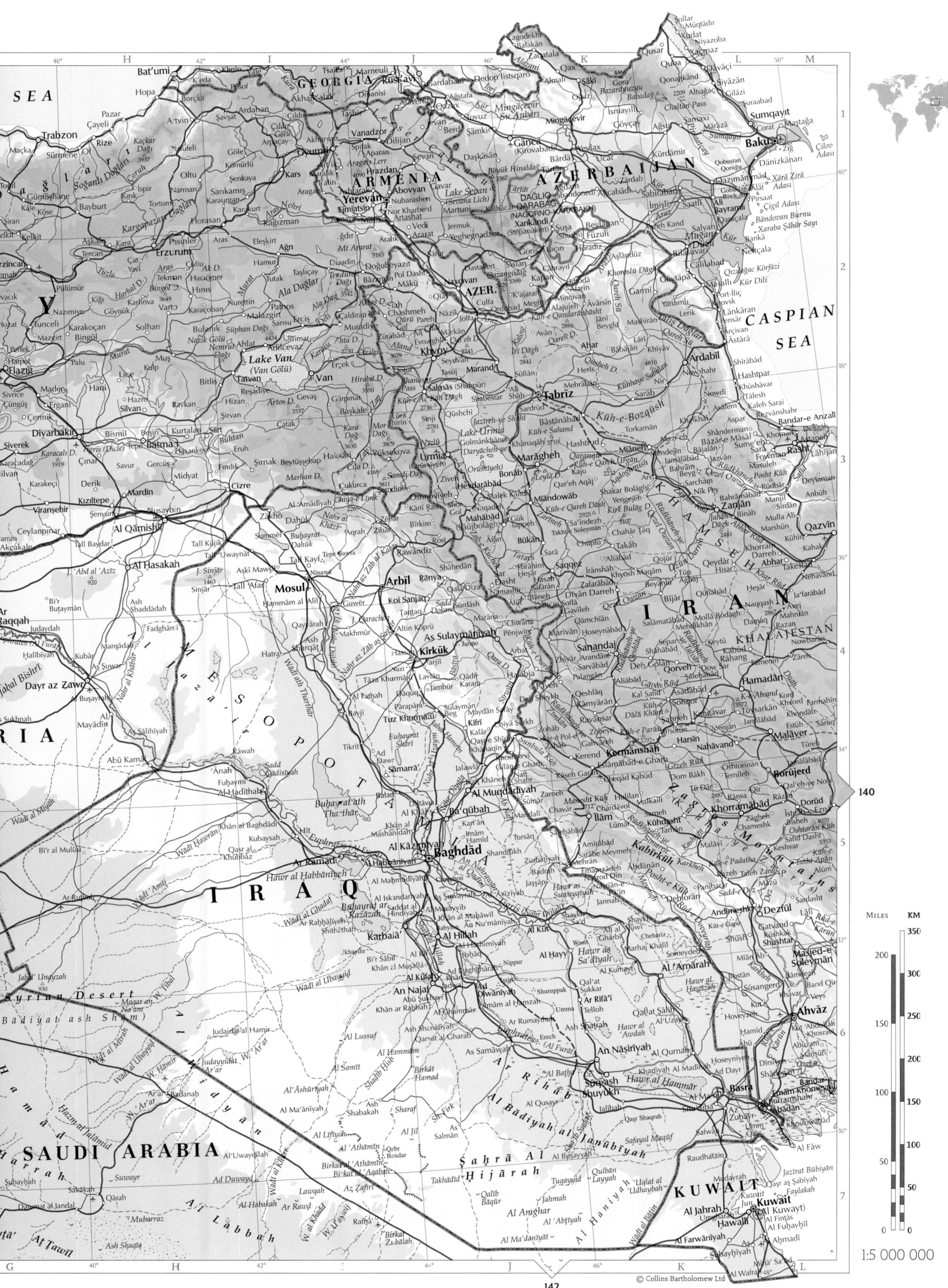

1:5 000 000

© Collins Bartholomew Ltd

140

142

Conic Equidistant Projection

141

132

RUSSIAN FEDERATION

RESP. ALTAY

SEVERNYY KAZAKHSTAN

AKMOLINSKAYA

OBLAST

PAVLODARSKAYA

OBLAST'

ALTAYSKIY KRAY

Astana (Akmola)

Kokshetau

Pavlodar

Semey (Semipalatinsk)

Ust'-Kamenogorsk

VOSTOCHNYY

KAZAKHSTAN

Zyryanovsk

Khrebet Naryn

Altaz Mts

Lake Zaysan (Ozero Zaysan)

KARAGANDINSKAYA OBLAST'
(Kazakhstan)

K A Z A K H S T A N

Temirtau

Karagandy

Shakhtinsk

Zhezkazgan

Khrebet Shyngystau

Khrebet Tarbagatay

Tacheng

Dzungarian Gate

Balkhash

Lake Balkhash
(Ozero Balkash)

Peski Saryyesik-
Atyrau

ZHAMBYLSKAYA

OBLAST'

ALMATINSKAYA

OBLAST'

Betpakdala

Peski Moyynkum

Peski Taukum

Step Zhusandala

Taldykorgan

Khrebet Dzhungarskiy Alatau

Borohoro Shan

Yining

135

Almaty

Kungei Alatau

Ysyk-Köl
(Issyk-Kul')

Terskey Ala-Too

T I E N   S H A N

Tarim Basin
(Tarim pendi)

Bishkek
(Frunze)

YUZHNYY

KAZAKHSTAN

Taraz

Shymkent

Tashkent

K Y R G Y Z S T A N

Talas Ala Range

Kishi Range

Fergana

Osh

Namangan

Andijon

XINJIANG UYGUR ZIZHIQU
(SINKIANG)

Kashi
(Kashgar)

C H I N A

Taklimakan Desert
(Taklimakan Shamo)

Samarqand

Dushanbe

T A J I K I S T A N

Turkestan Range

Gissar Range

Qatorkuhi Zarafshon Range

Pik Lenin

K U N L U N   S H A N

Termiz

Khunjerab Pass

© Collins Bartholomew Ltd

MILES   KM

300   500

400

200   300

200

100   100

0   0

1:7 000 000

139

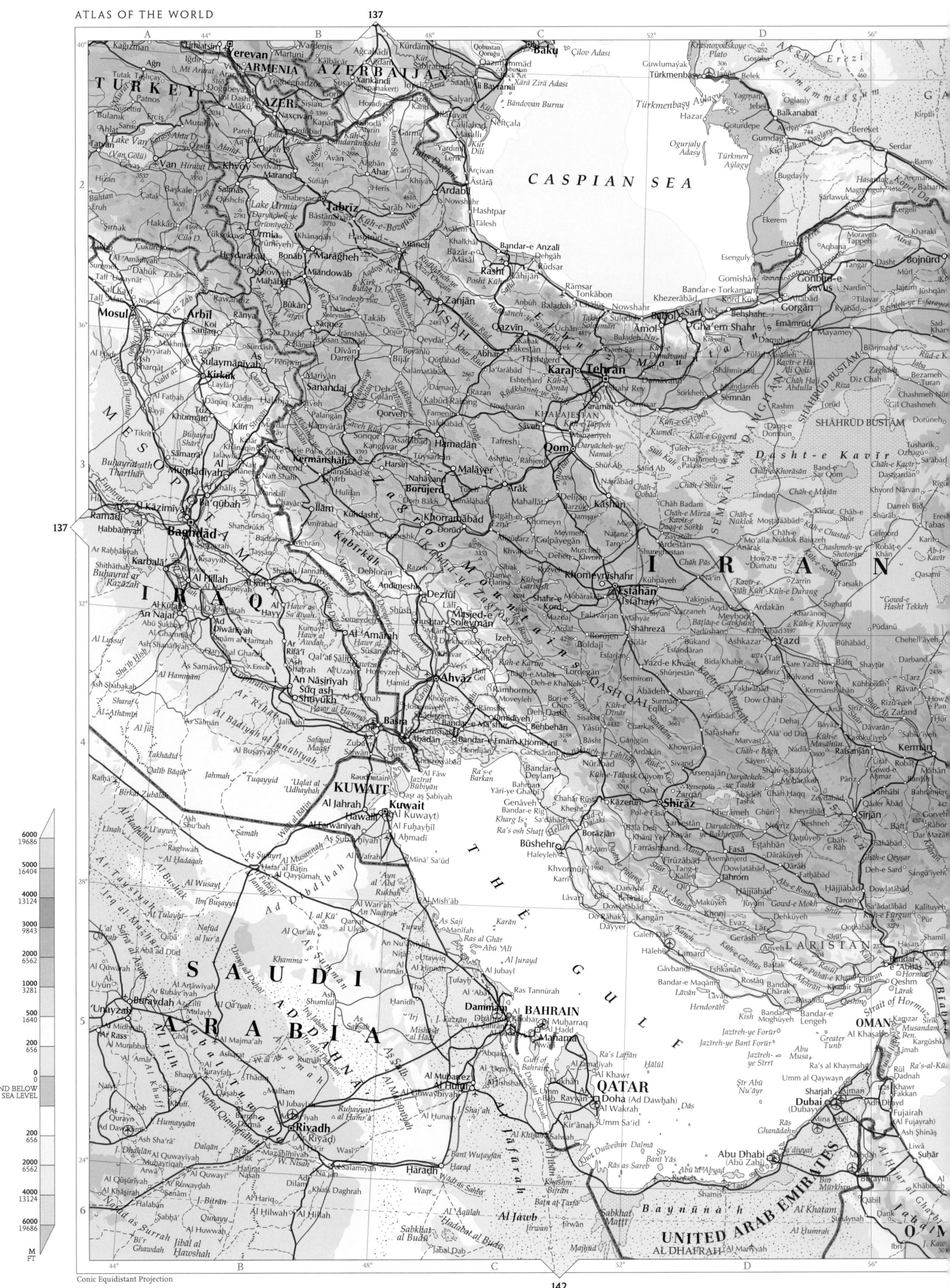

137

CASPIAN SEA

TURKEY

ARMENIA

AZERBAIJAN

AZER.

IRAQ

IRAN

KUWAIT

SAUDI
ARABIA

BAHRAIN

QATAR

OMAN

UNITED ARAB EMIRATES

THE GULF

6000	19686
5000	16404
4000	13124
3000	9843
2000	6562
1000	3281
500	1640
200	656
0	0

LAND BELOW
SEA LEVEL

200	656
2000	6562
4000	13124
6000	19686

M
FT

Conic Equidistant Projection

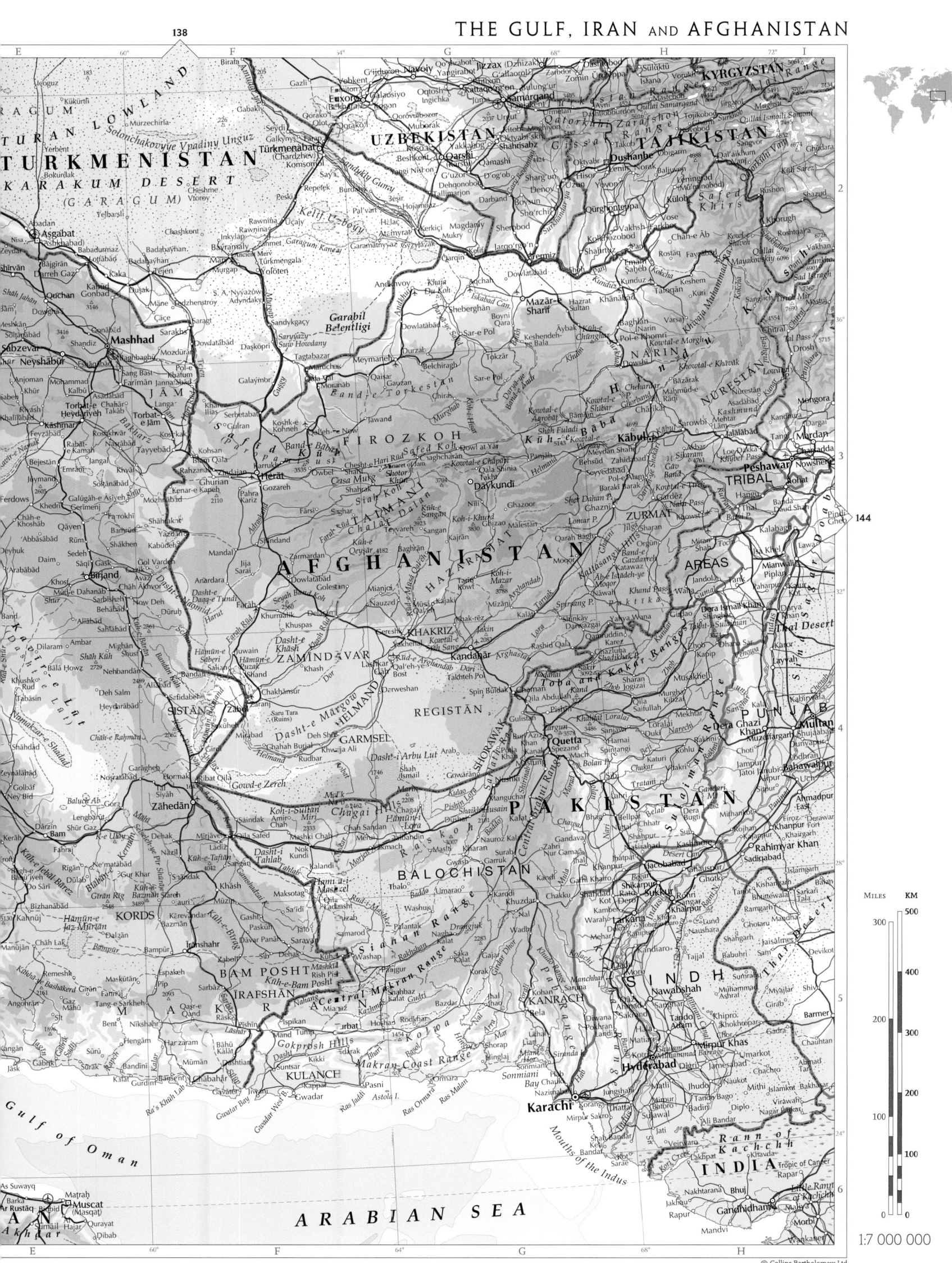

1:7 000 000

© Collins Bartholomew Ltd

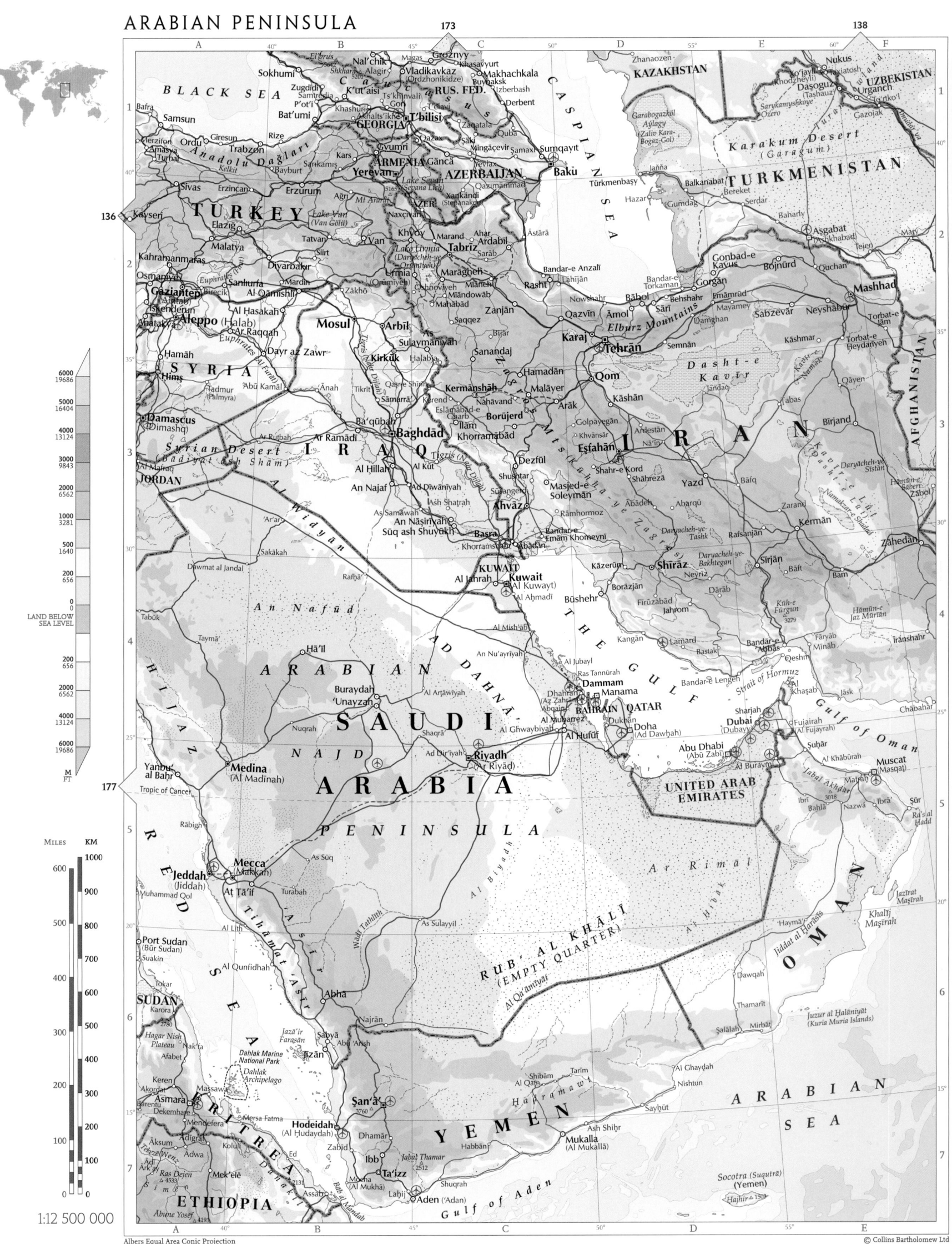

6000
19686

5000
16404

4000
13124

3000
9843

2000
6562

1000
3281

500
1640

200
656

0

LAND BELOW
SEA LEVEL

200
656

2000
6562

4000
13124

6000
19686

M
FT

Miles   KM

600   1000

500   900

400   800

       700

300   600

       500

200   400

       300

100   200

       100

0   0

1:12 500 000

Albers Equal Area Conic Projection

© Collins Bartholomew Ltd

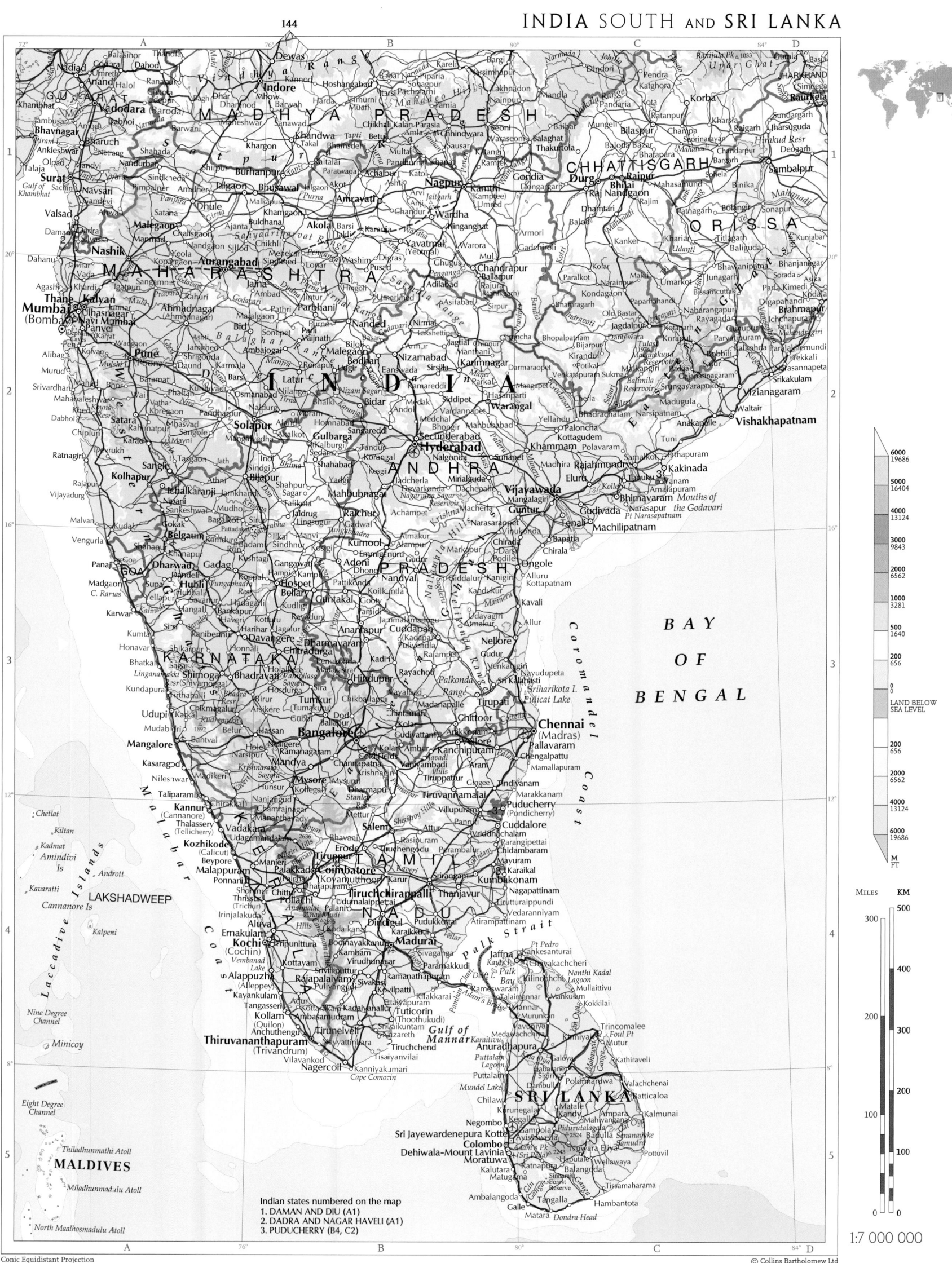

BAY

OF

BENGAL

MALDIVES

Indian states numbered on the map
1. DAMAN AND DIU (A1)
2. DADRA AND NAGAR HAVELI (A1)
3. PUDUCHERRY (B4, C2)

Conic Equidistant Projection

© Collins Bartholomew Ltd

MILES   KM

1:7 000 000

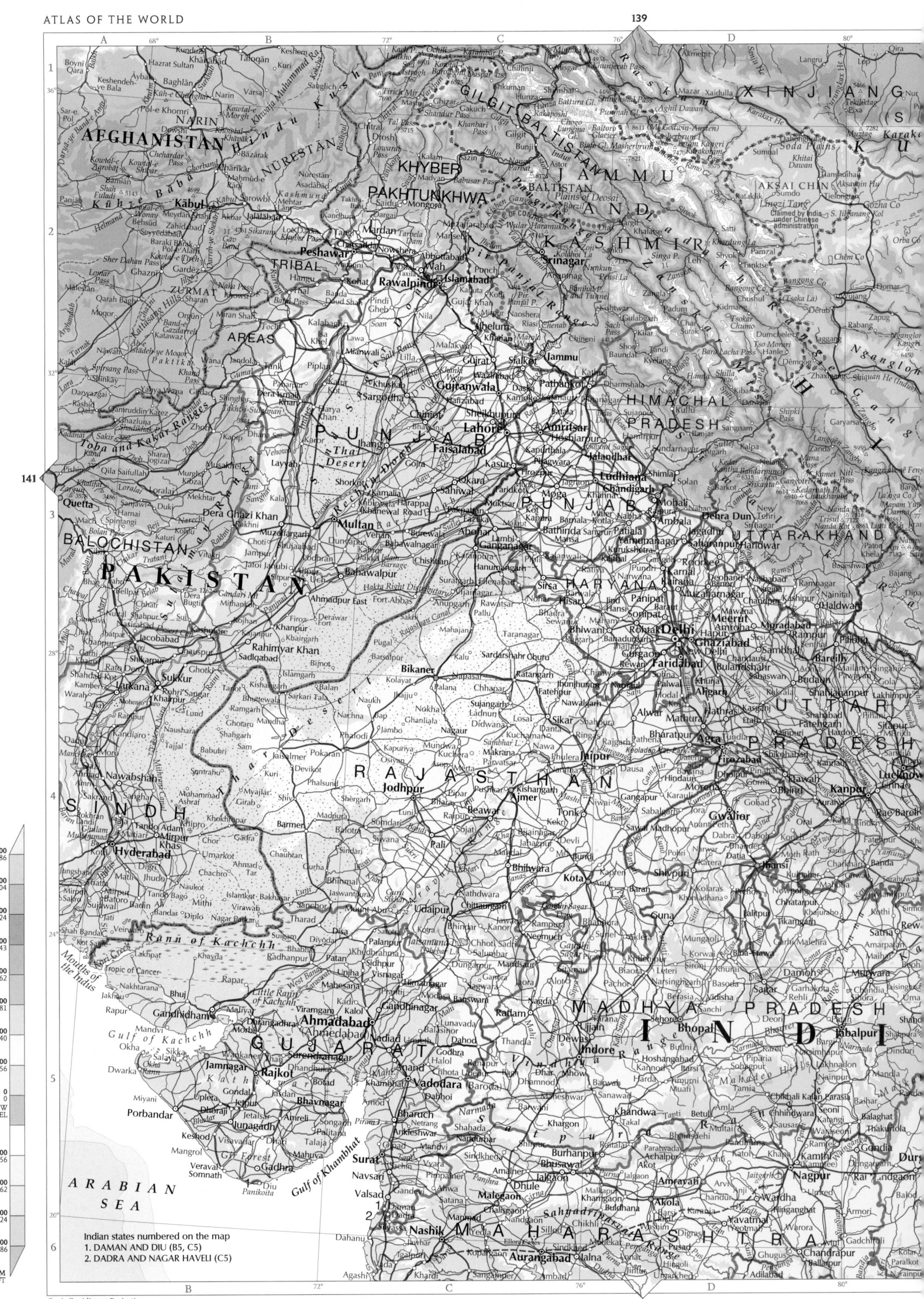

139

141

143

Indian states numbered on the map
1. DAMAN AND DIU (B5, C5)
2. DADRA AND NAGAR HAVELI (C5)

Conic Equidistant Projection

146

147

1:7 000 000

© Collins Bartholomew Ltd

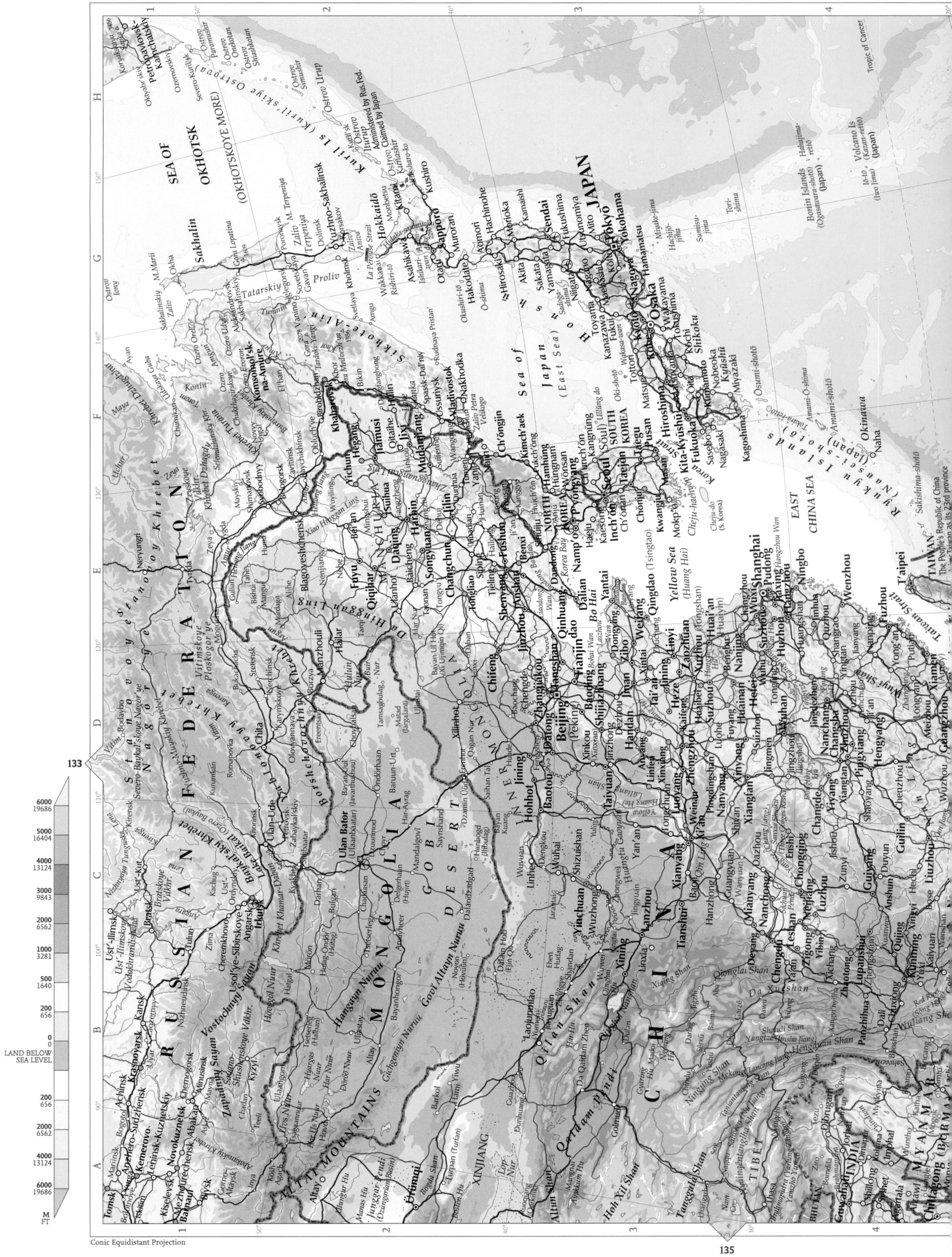

Conic Equidistant Projection

133

135

6000
19686
5000
16404
4000
13124
3000
9843
2000
6562
1000
3281
500
1640
200
656
0
LAND BELOW
SEA LEVEL
200
656
2000
6562
4000
13124
6000
19686
M
FT

PACIFIC

OCEAN

Northern Mariana Islands (U.S.A.)

FEDERATED STATES

OF MICRONESIA

PALAU

PHILIPPINE

SEA

PHILIPPINES

Luzon

Manila

Quezon City

San Pablo

Mindoro

Panay

Negros

Mindanao

Davao

Zamboanga

General Santos

SOUTH

CHINA

SEA

Hong Kong

Macau

Zhanjiang

Haikou

Hainan (China)

Paracel Islands

Spratly Islands

Palawan

Sulu Sea

Celebes Sea

Da Nang

Quy Nhon

Nha Trang

Ho Chi Minh City (Saigon)

Phnom Penh

CAMBODIA

THAILAND

Bangkok

LAOS

VIETNAM

Gulf of Thailand

MALAYSIA

BRUNEI

Bandar Seri Begawan

SARAWAK

SABAH

Kota Kinabalu

Kuching

BORNEO

Banjarmasin

Balikpapan

Samarinda

Makassar

Celebes (Sulawesi)

Manado

Gorontalo

Palu

Kendari

Halmahera

Seram

Ambon

Buru

Maluku

Ternate

Laut Banda (Banda Sea)

INDONESIA

Jakarta

Bandung

Semarang

Surakarta

Surabaya

Java (Jawa)

Laut Jawa (Java Sea)

Denpasar

Bali

Lombok

Sumbawa

Flores

Sumba

Timor

EAST TIMOR

Kupang

Laut Sawu (Savu Sea)

Lesser Sunda Islands

IRIAN JAYA

Pegunungan Maoke

NEW GUINEA

Arafura Sea

AUSTRALIA

Darwin

Arnhem Land

Groote Eylandt

Timor Sea

Singapore

SINGAPORE

Kuala Lumpur

Johor Bahru

George Town

Medan

Padang

Palembang

Bandar Lampung

Sumatra (Sumatera)

Kepulauan Mentawai

Strait of Malacca

Rangoon

Bassein

Mergui Archipelago

Andaman Sea

Phuket

INDIAN

OCEAN

Christmas Island (Aust.)

Cocos Islands (Keeling Is) (Aust.)

MILES   KM

1400

800   1200

1000

600   800

600

400   400

200   200

0   0

1:20 000 000

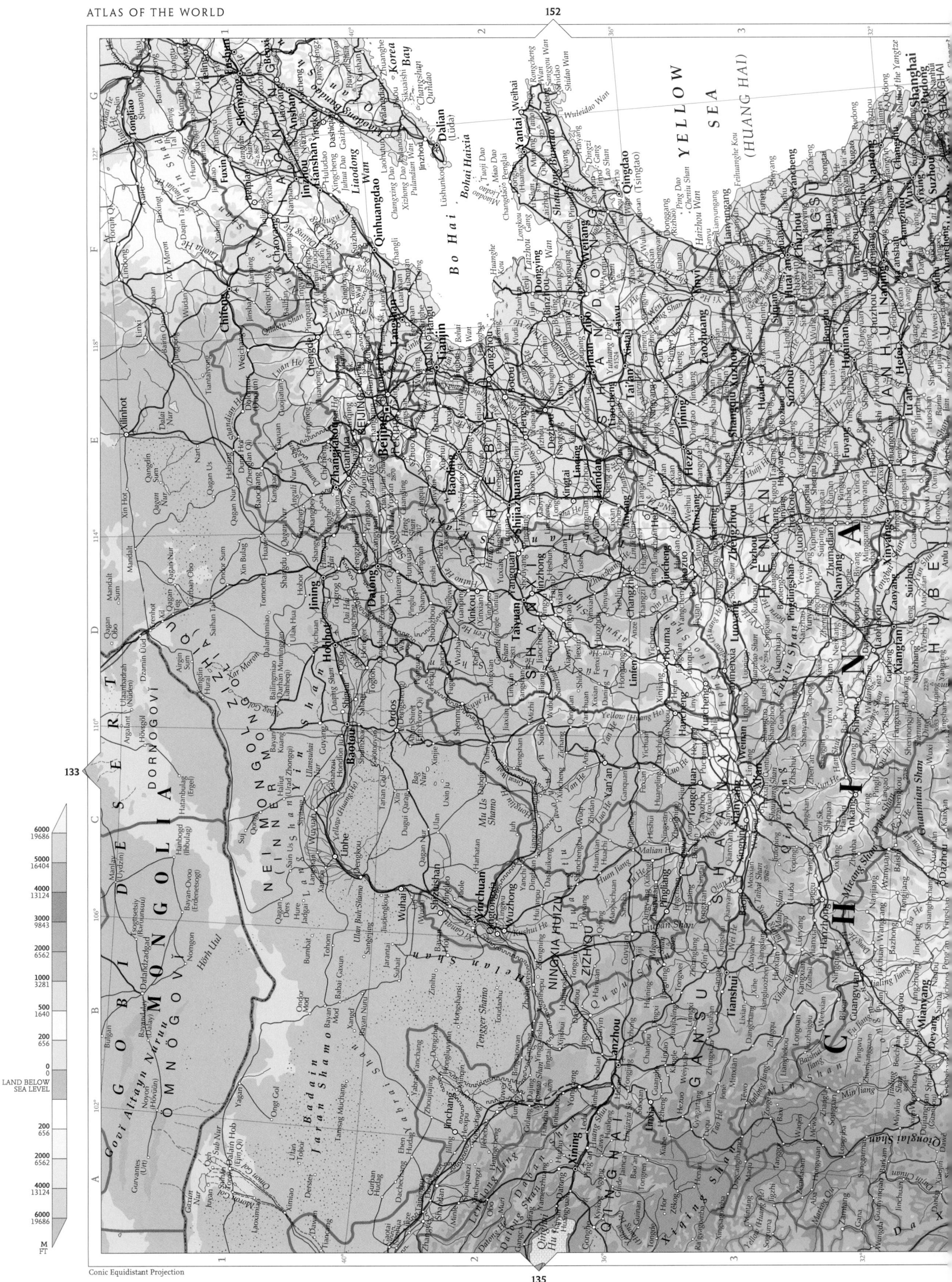

152

133

135

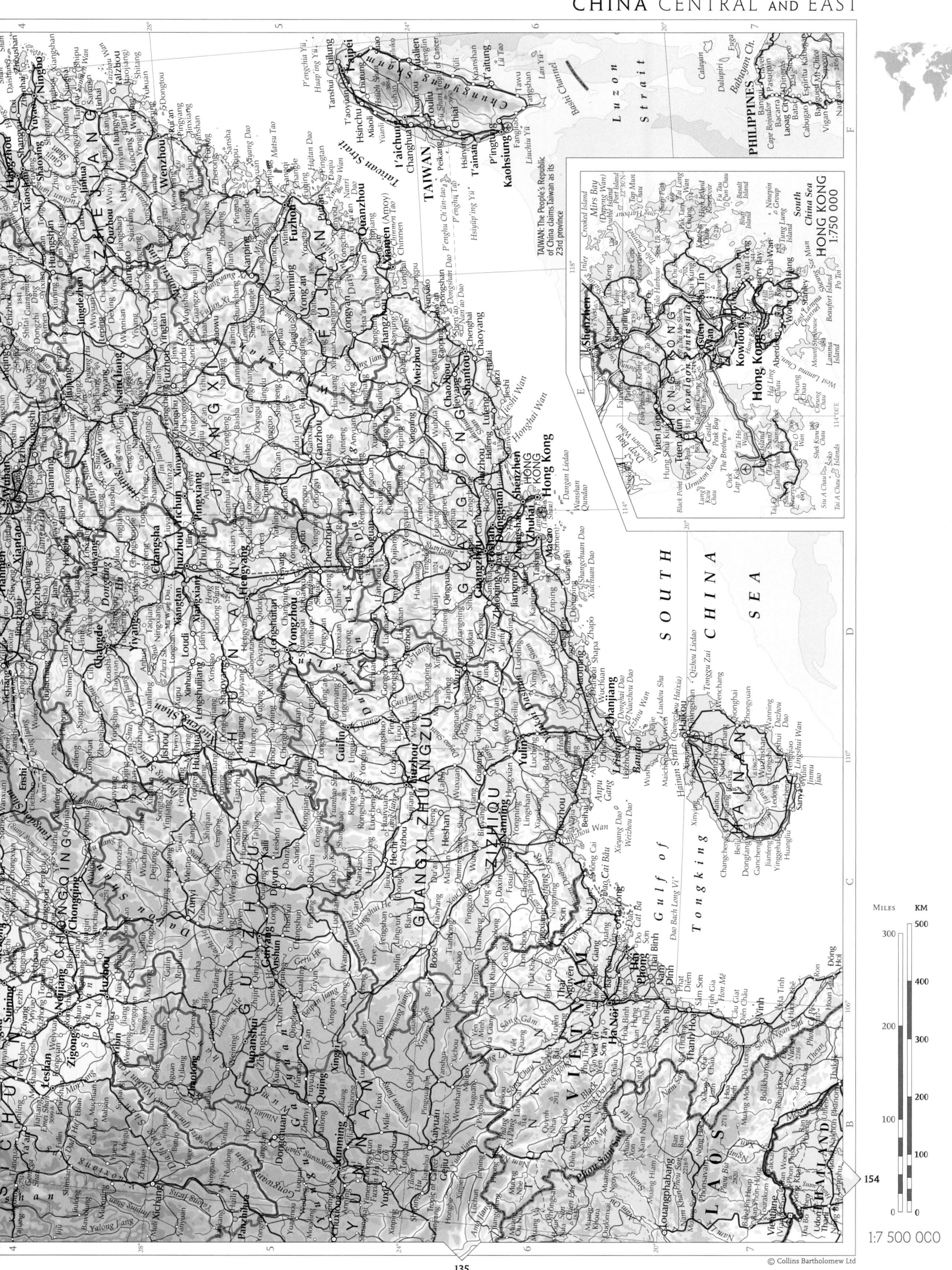

135

MILES   KM

300 — 500

— 400

200 — 300

— 200

100 — 100

0 — 0

1:7 500 000

TAIWAN: The People's Republic of China claims Taiwan as its 23rd province

HONG KONG
1:750 000

154

© Collins Bartholomew Ltd

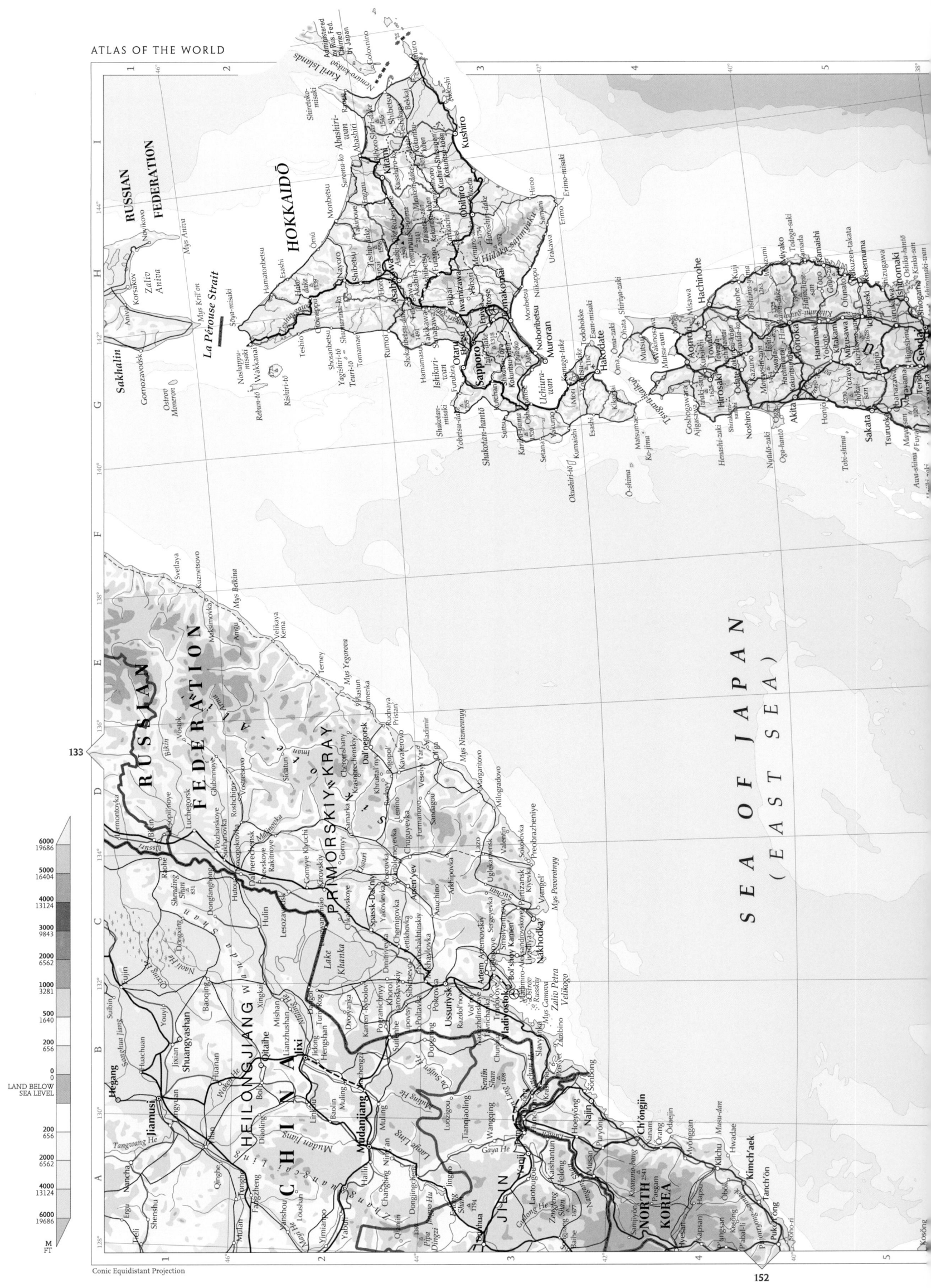

RUSSIAN
FEDERATION

Sakhalin

HOKKAIDŌ

La Pérouse Strait

S E A   O F   J A P A N

(E A S T   S E A)

RUSSIAN
FEDERATION

PRIMORSKIY KRAY

CHINA

HEILONGJIANG

JI LIN

NORTH
KOREA

Vladivostok

Ussuriysk

Nakhodka

Jiamusi

Mudanjiang

Sapporo

Hakodate

Muroran

Hachinohe

Kushiro

Obihiro

Aomori

Akita

Sakata

6000
19686

5000
16404

4000
13124

3000
9843

2000
6562

1000
3281

500
1640

200
656

0

LAND BELOW
SEA LEVEL

200
656

2000
6562

4000
13124

6000
19686

M
FT

133

Conic Equidistant Projection

152

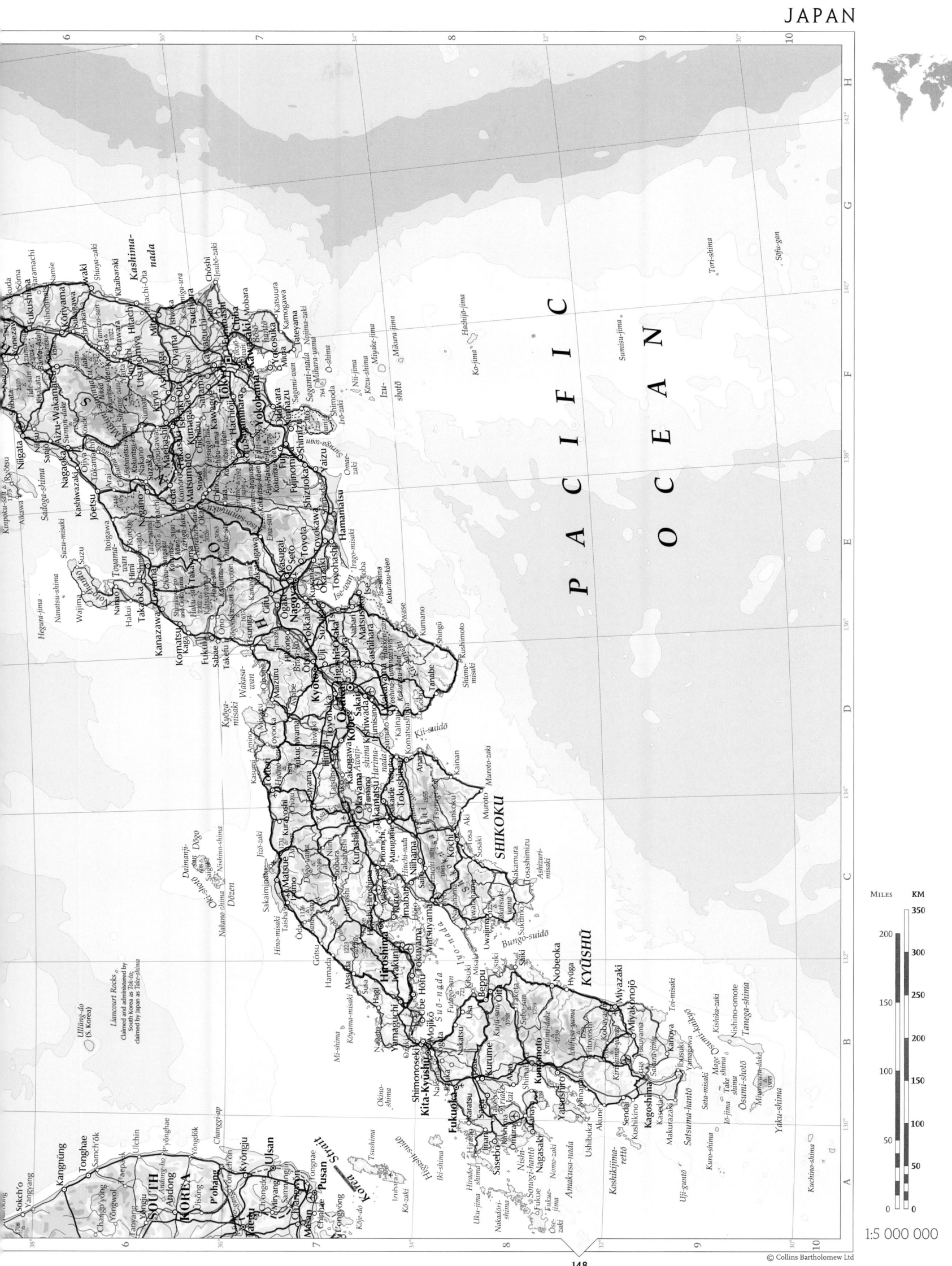

1:5 000 000

© Collins Bartholomew Ltd

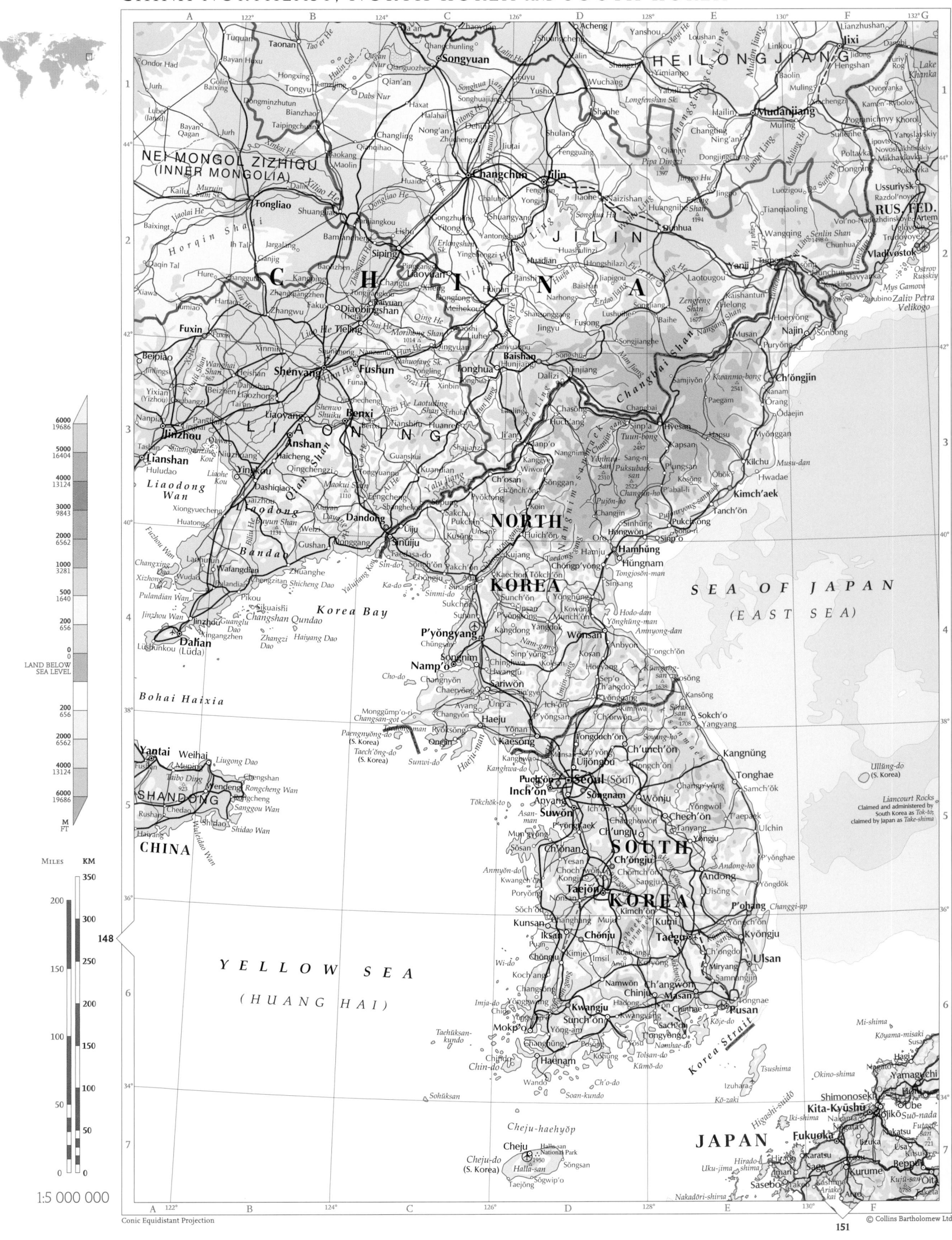

Conic Equidistant Projection

© Collins Bartholomew Ltd

151

1:5 000 000

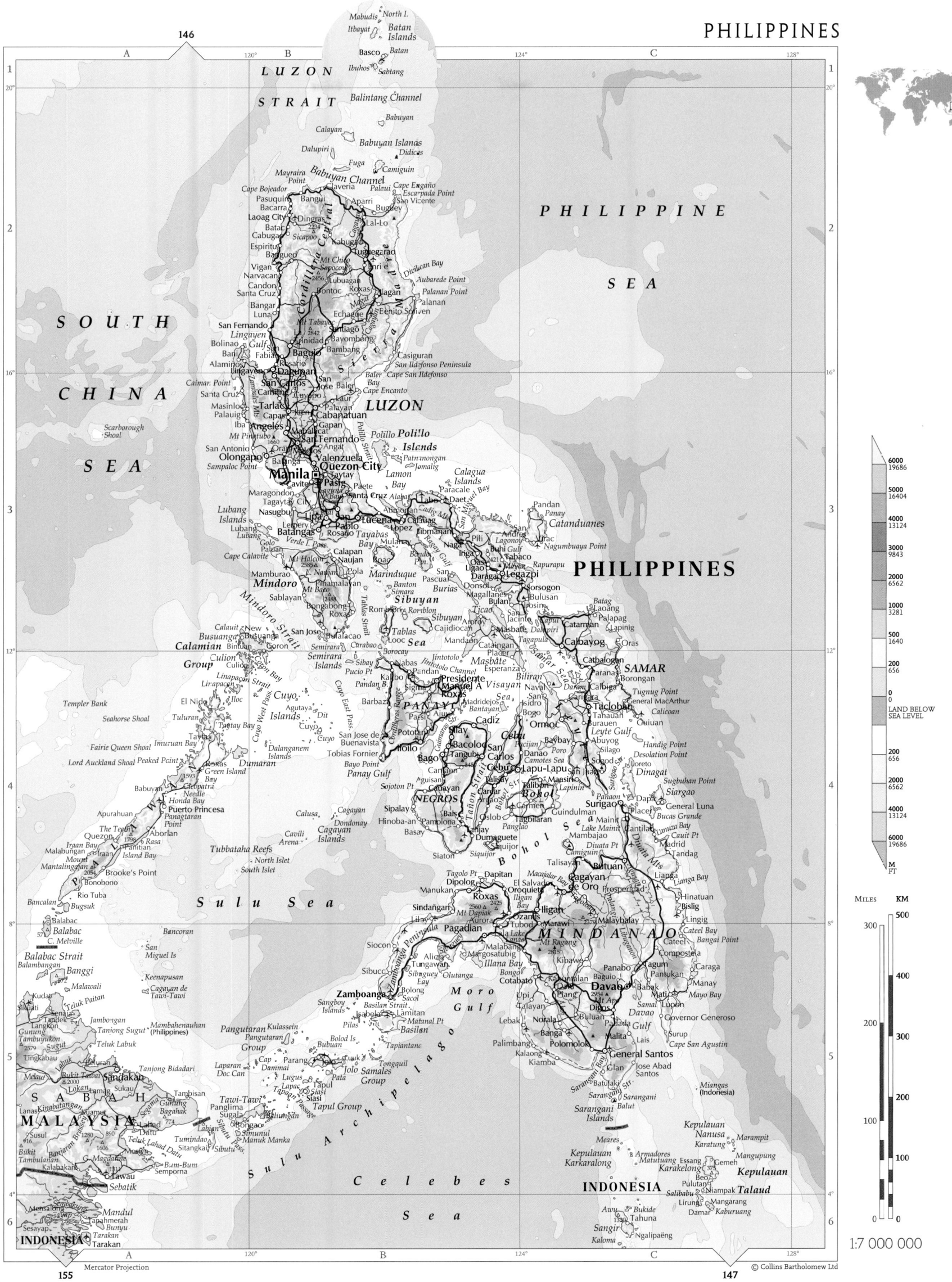

PHILIPPINES

© Collins Bartholomew Ltd

1:7 000 000

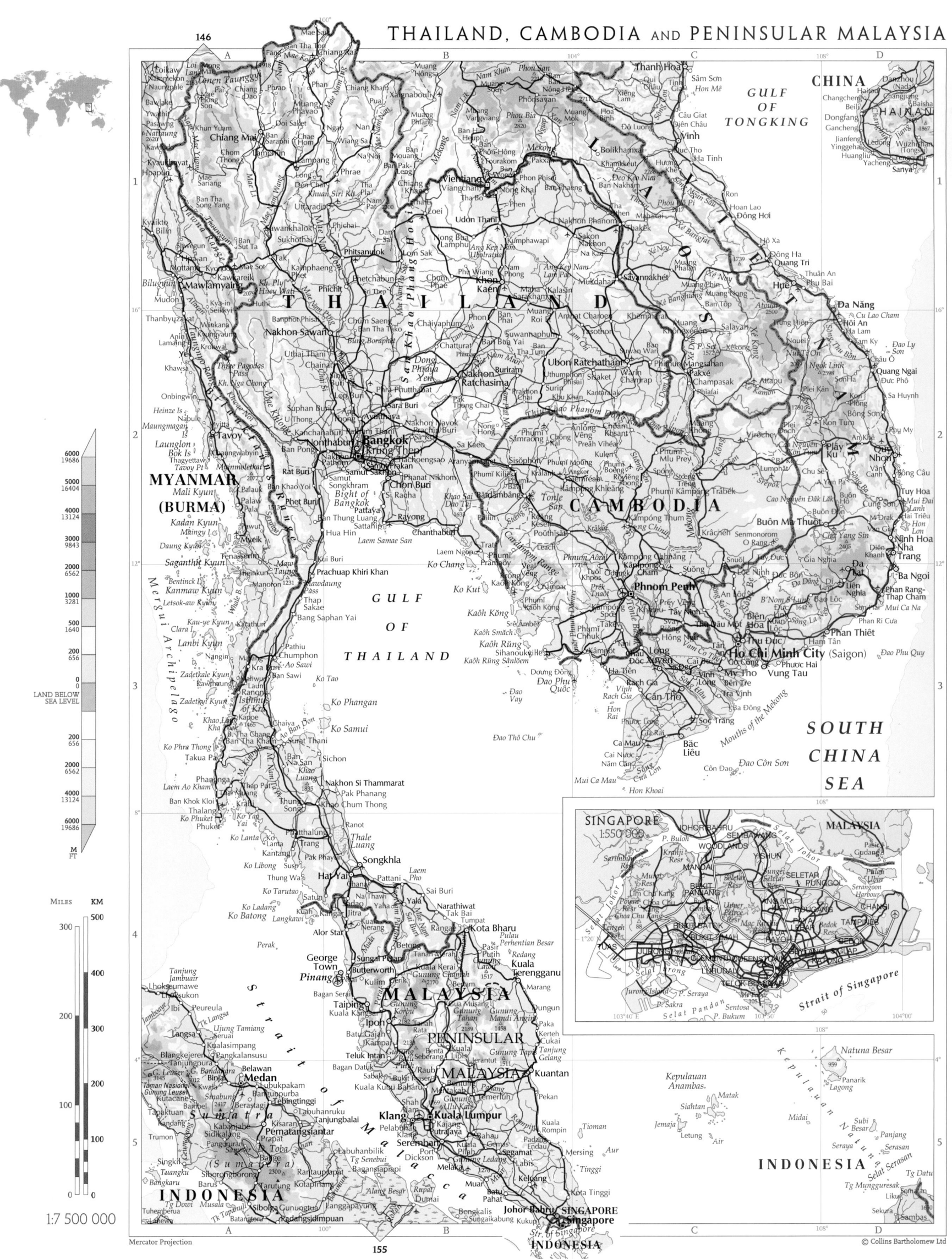

146

CHINA

GULF
OF
TONGKING

LAOS

THAILAND

MYANMAR
(BURMA)

GULF
OF
THAILAND

CAMBODIA

Phnom Penh

Bangkok
Krung Thep

VIETNAM

SOUTH
CHINA
SEA

Ho Chi Minh City (Saigon)

6000
19686
5000
16404
4000
13124
3000
9843
2000
6562
1000
3281
500
1640
0
0
LAND BELOW
SEA LEVEL
200
656
2000
6562
4000
13124
6000
19686
M
FT

MILES   KM
300   500
             400
200   300
100   200
             100
0   0

1:7 500 000

Mercator Projection

SINGAPORE
1:550 000

JOHOR BAHRU

MALAYSIA

Strait of Singapore

MALAYSIA

PENINSULAR
MALAYSIA

Kuala Lumpur

George
Town
Pinang

Medan

Sumatra
(Sumatera)

INDONESIA

Natuna Besar

Kepulauan
Anambas

INDONESIA

Johor Bahru SINGAPORE
Singapore
INDONESIA

155

© Collins Bartholomew Ltd

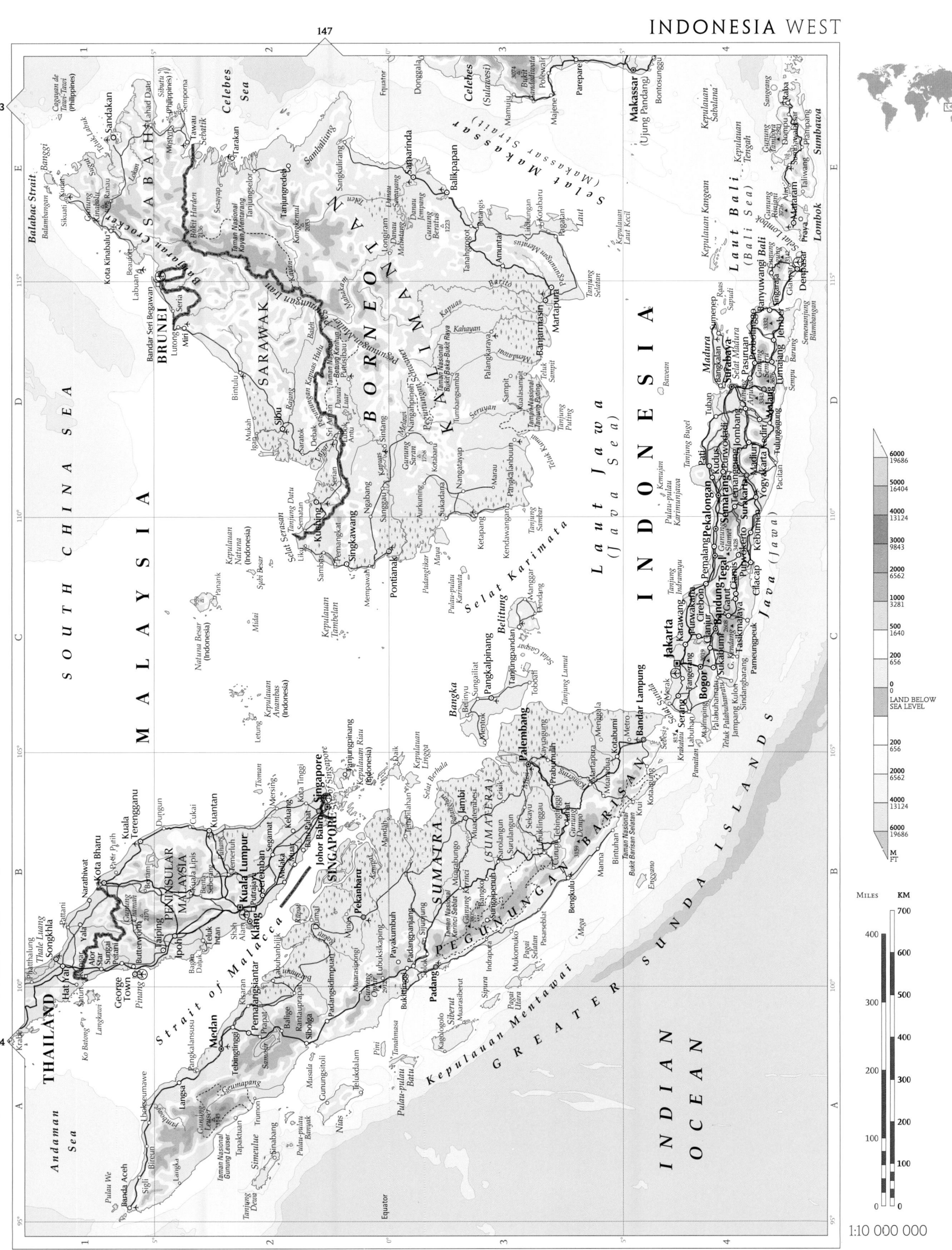

147

Mercator Projection

© Collins Bartholomew Ltd

1:10 000 000

MILES   KM

NORTH AMERICA

Baffin Bay

Greenland

Nordaustlandet

Spitsbergen **Svalbard** (Norway)

Longyearbyen

Greenland Sea

Bjørnøya (Norway)

BARENTS SEA

Zemlya Frantsa-Iosifa

North Cape

Jan Mayen (Norway)

Arctic Circle

Denmark Strait

Reykjavík ICELAND

NORWEGIAN SEA

Trondheim

N O R W A Y

S W E D E N

Gulf of Bothnia

Faroe Islands (Denmark) **Tórshavn**

Bergen

Skagerrak

Kattegat

Oslo □ **Stockholm**

Vänern

Vättern

Gothenburg

Shetland Islands

Orkney Islands

Aalborg

DENMARK

Copenhagen □ Malmö

Odense *Bornholm*

Rockall

Outer Hebrides

SCOTLAND

Glasgow ○ Edinburgh

British Isles

NORTHERN IRELAND

Belfast

UNITED KINGDOM

Manchester ○ Leeds

NORTH SEA

Hamburg

NETHERLANDS

Hannover **Berlin**

Dublin □

IRELAND

Liverpool ○

Birmingham ○

Amsterdam

The Hague ● Essen **Düsseldorf** GERMANY Leipzig

ATLANTIC

WALES

ENGLAND

Cardiff ○

London □

Rotterdam

Brussels Cologne

LILLE **BELGIUM** Frankfurt

English Channel

LUXEMBOURG

Channel Islands

Luxembourg Mannheim

Nuremberg

Stuttgart

OCEAN

Brest ○

Rennes ○

Paris □

Orléans ○

Strasbourg

ZURICH LIECHTEN- Munich

STEIN Innsbruck

Nantes ○

Loire

Dijon ○

Bern

SWITZERLAND

A

Bay of Biscay

Lyon ○

Geneva

Mont Blanc

Milan ○

Turin ○

Genoa

Bordeaux ○

F R A N C E

Toulouse ○

Pyrenees

Marseille

MONACO

Nice

Corsica

○ Flores

Azores (Portugal)

São Jorge

Terceira

A Coruña ○

Bilbao ○

Andorra la Vella ANDORRA

Pico

Ponta Delgada □

São Miguel

○ *Santa Maria*

Oporto ○

P O R T U G A L

Salamanca ○

Zaragoza ○

Barcelona ○

Madrid □

Balearic Islands

Minorca

Majorca

S P A I N

Valencia ○

Ibiza

M E D

Arquipélago dos Açores

Lisbon □

Córdoba ○

Cartagena ○

Seville ○

Málaga ○

Cádiz ○

Gibraltar (U.K.)

Ceuta (Spain)

Melilla (Spain)

Madeira (Portugal)

Ilha de Porto Santo

Funchal □

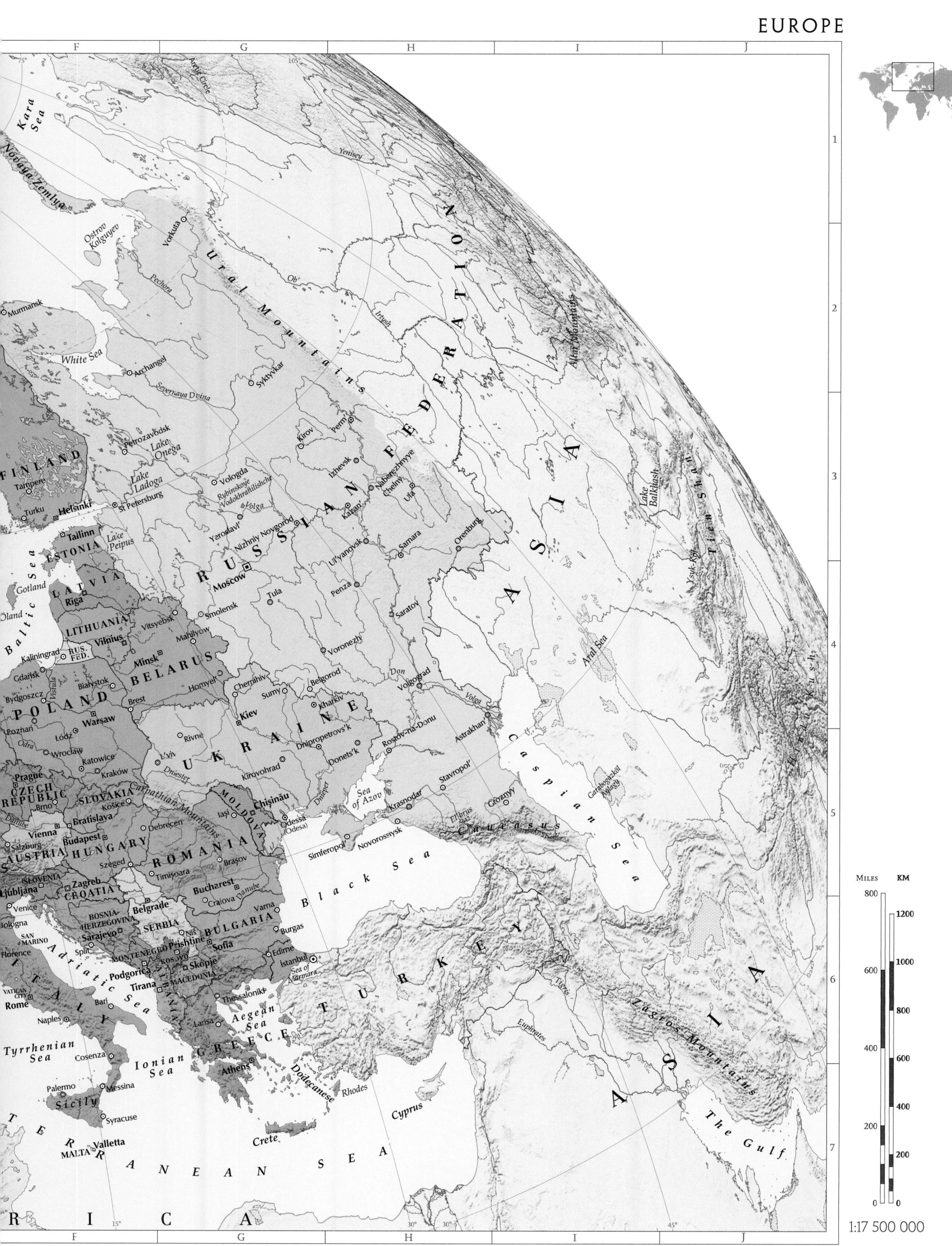

1:17 500 000

© Collins Bartholomew Ltd

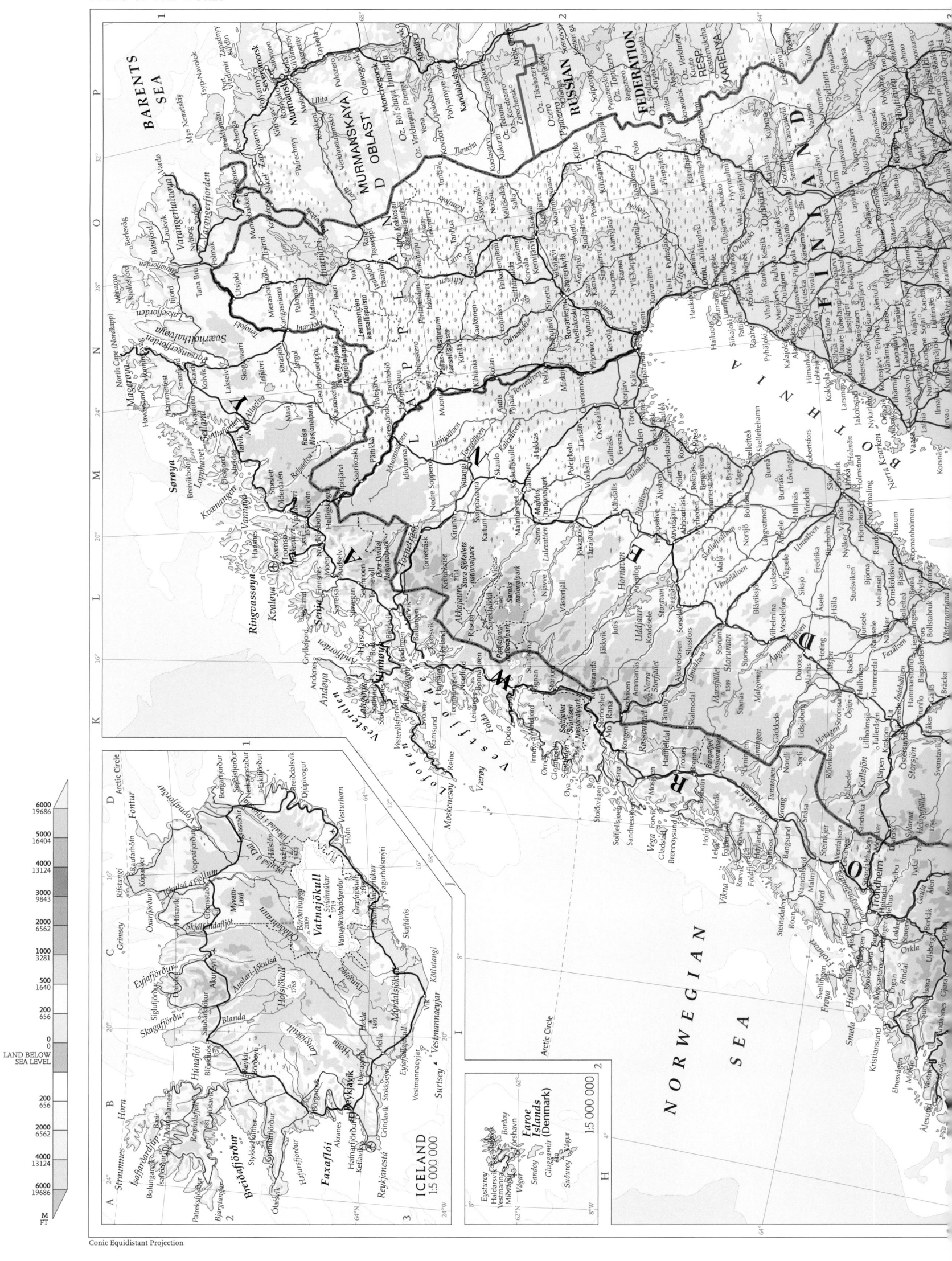

ICELAND
1:5 000 000

Faroe Islands (Denmark)
1:5 000 000

Conic Equidistant Projection

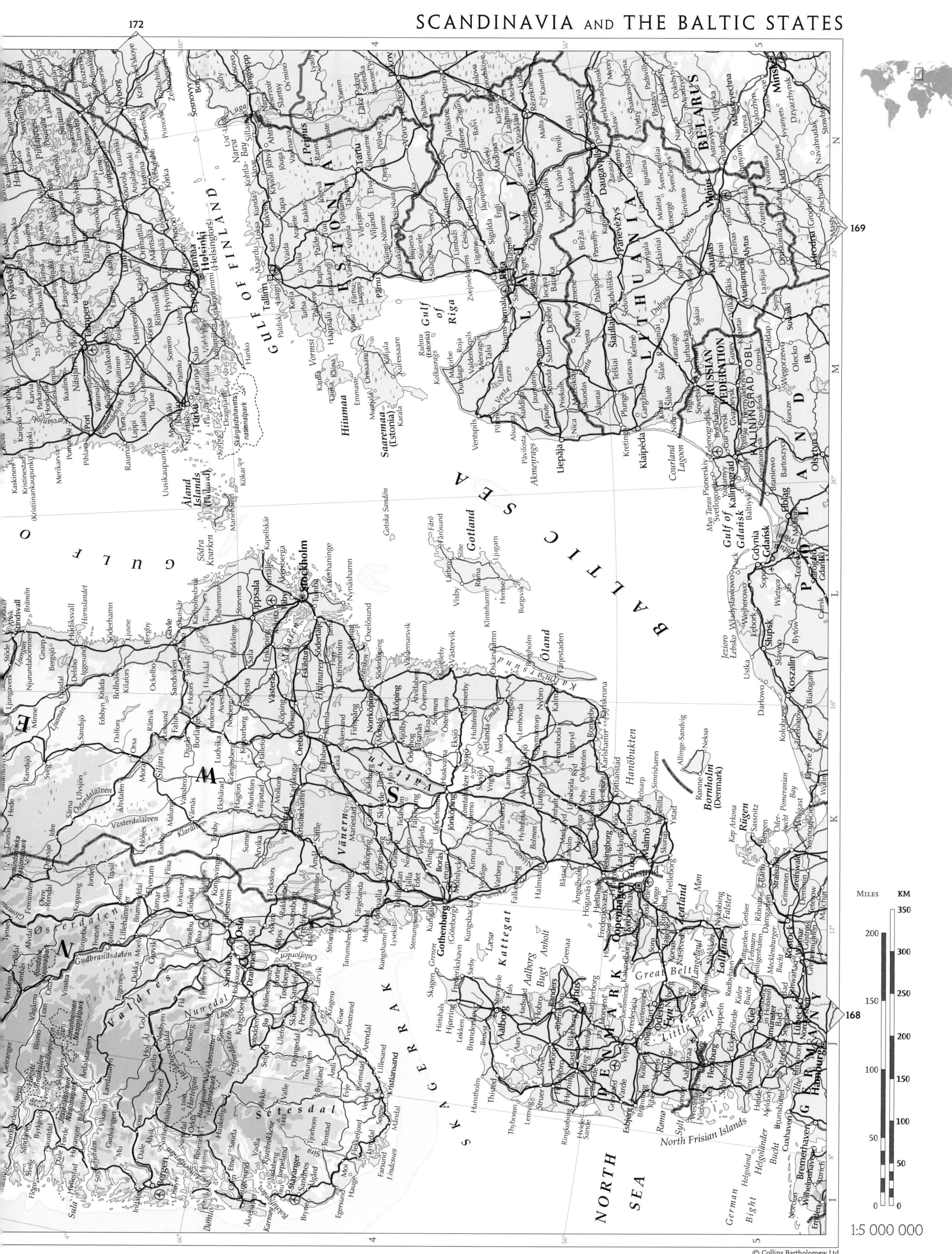

© Collins Bartholomew Ltd

1:5 000 000

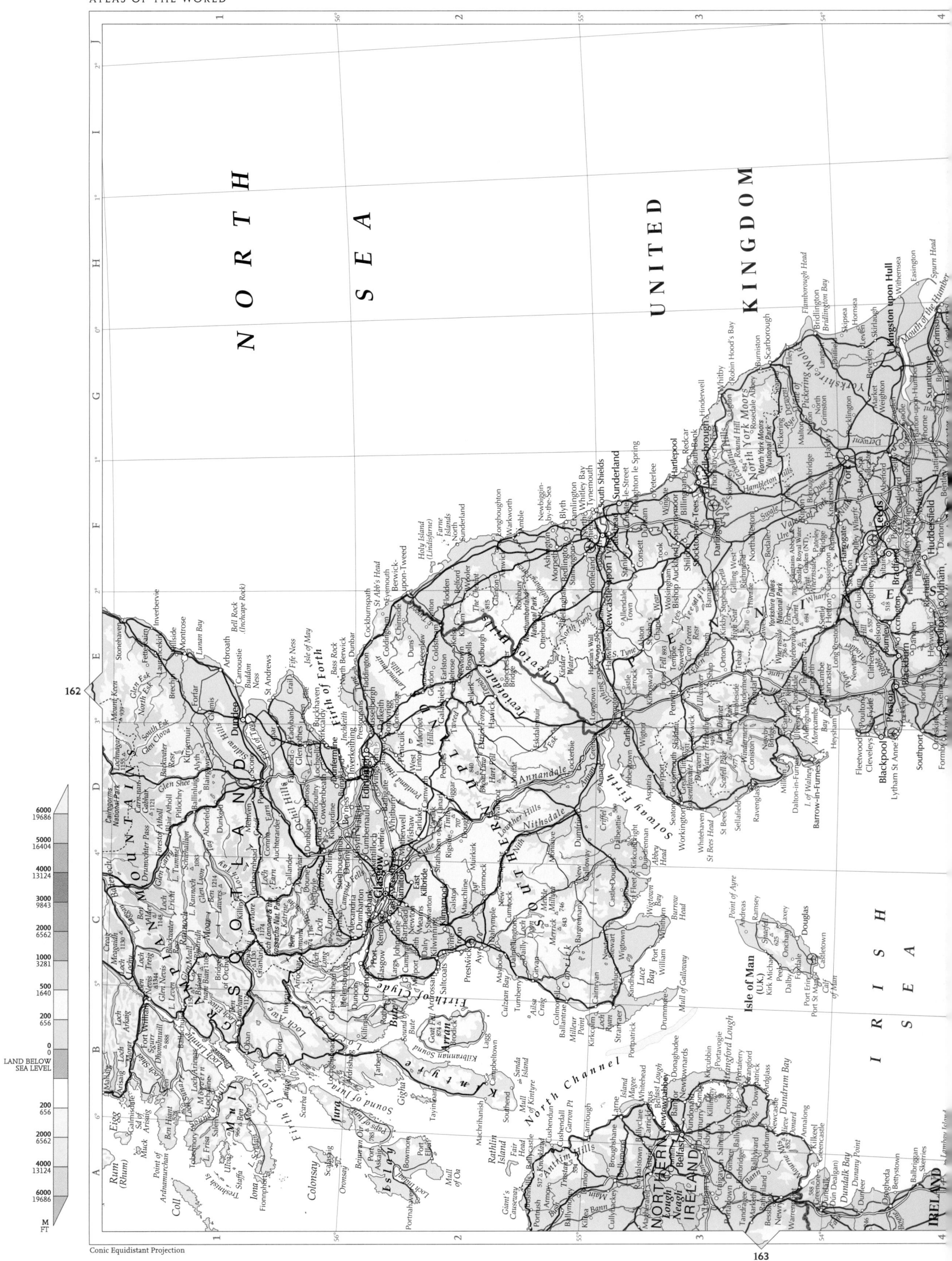

NORTH

SEA

UNITED

KINGDOM

IRISH

SEA

162

163

Conic Equidistant Projection

6000
19686

5000
16404

4000
13124

3000
9843

2000
6562

1000
3281

500
1640

200
656

0
0

LAND BELOW
SEA LEVEL

200
656

2000
6562

4000
13124

6000
19686

M
FT

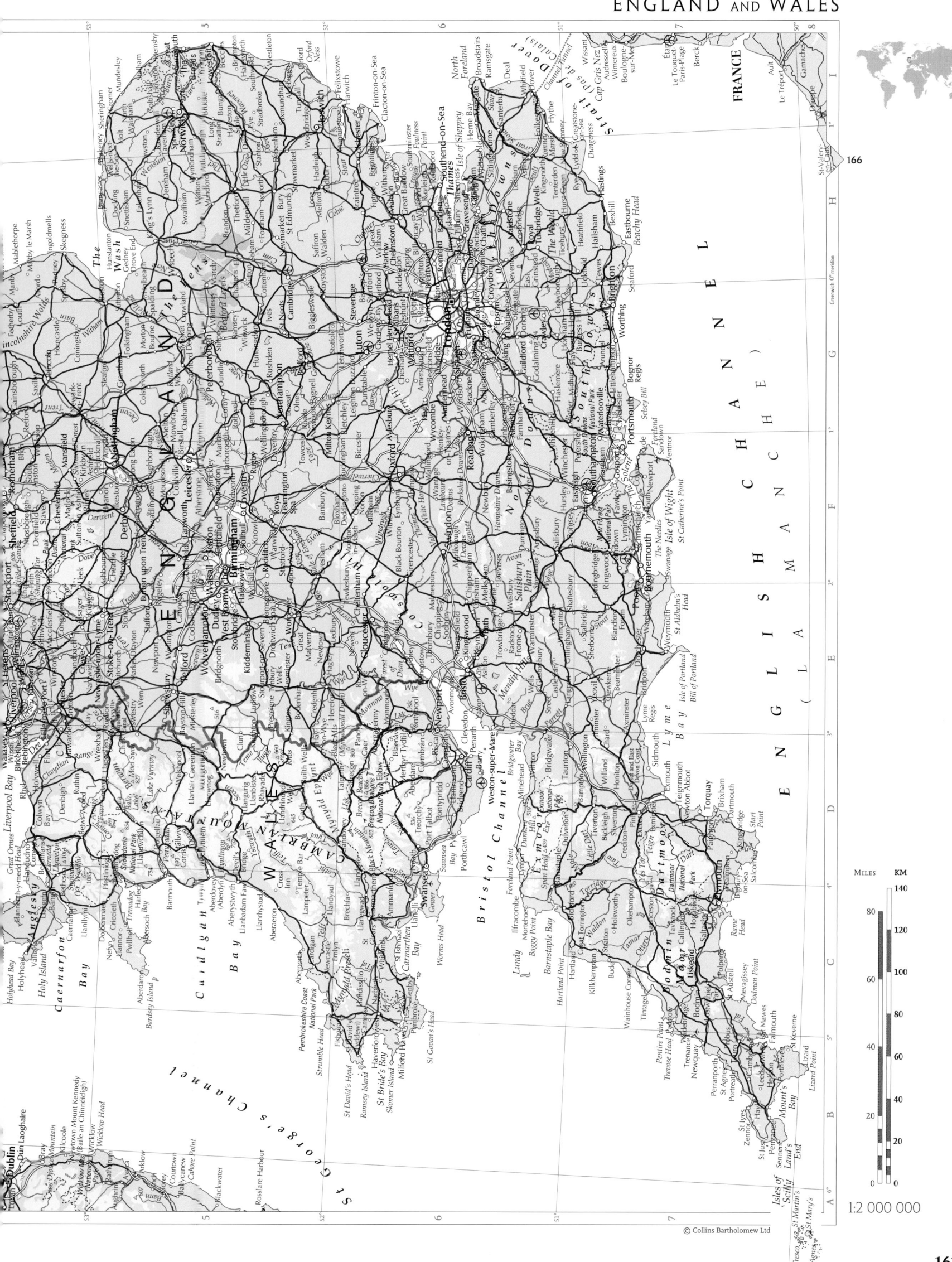

FRANCE

1:2 000 000

© Collins Bartholomew Ltd

SCOTLAND

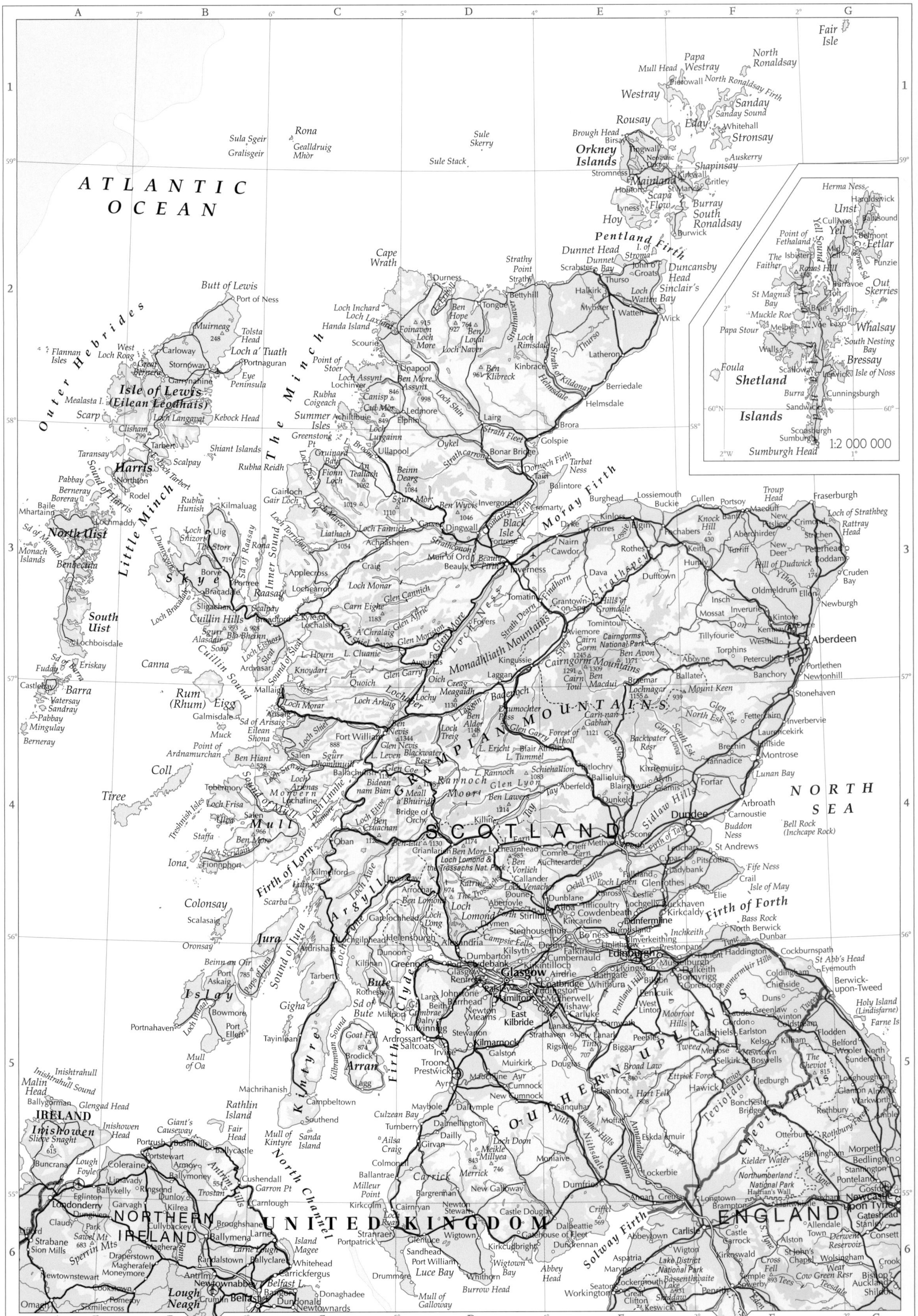

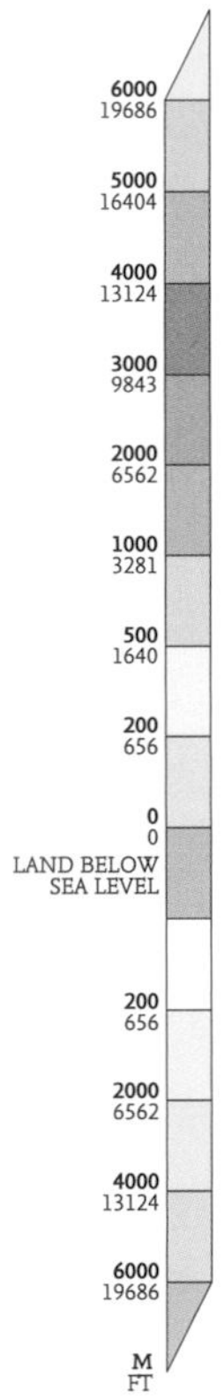

Conic Equidistant Projection

© Collins Bartholomew Ltd

1:2 000 000

M FT	
6000	19686
5000	16404
4000	13124
3000	9843
2000	6562
1000	3281
500	1640
200	656
0	0

LAND BELOW SEA LEVEL

200	656
2000	6562
6000	19686

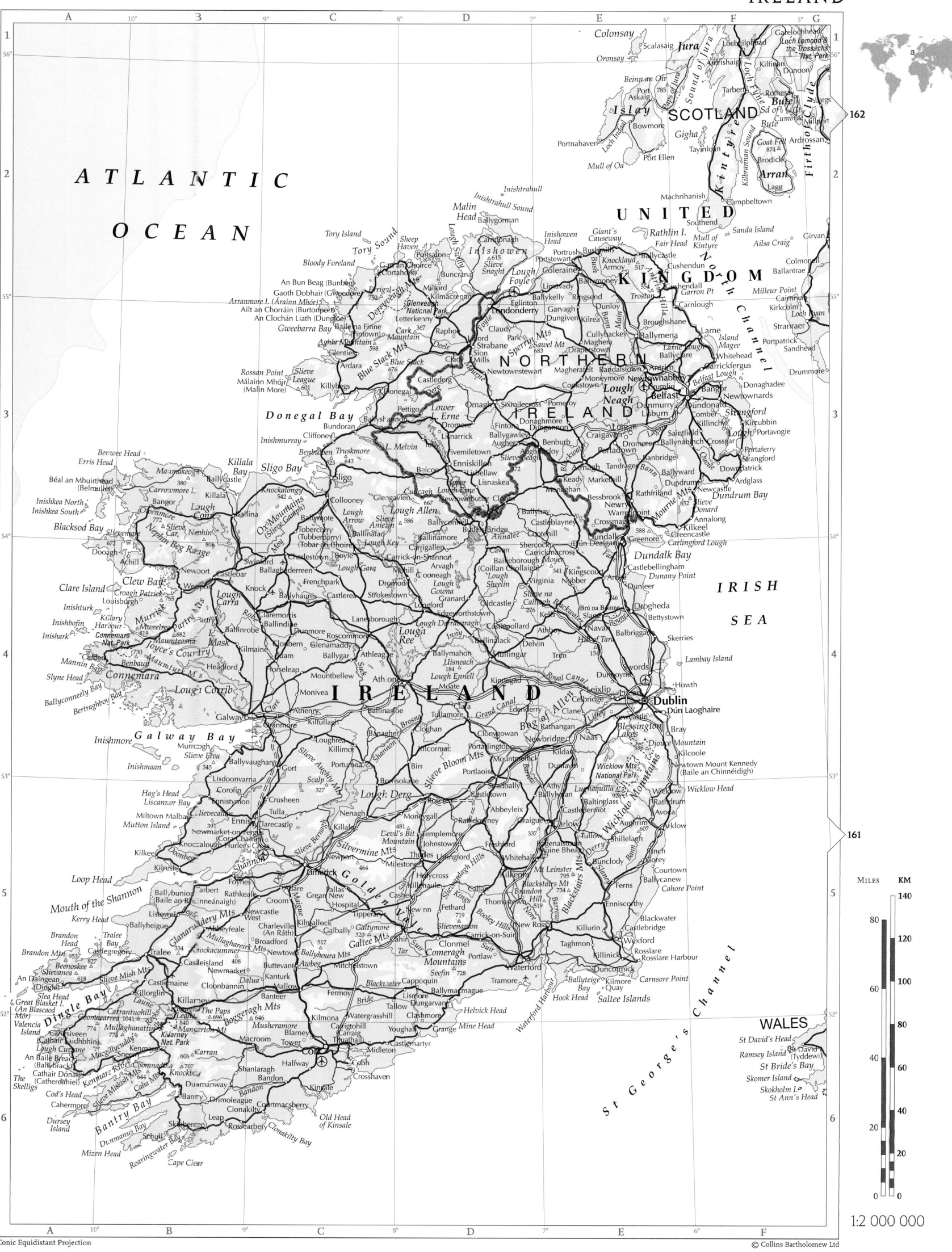

162

161

161

1:2 000 000

Conic Equidistant Projection

© Collins Bartholomew Ltd

NORTH

SEA

NETHERLANDS

BELGIUM

FRANCE

LUXEMBOURG

Conic Equidistant Projection

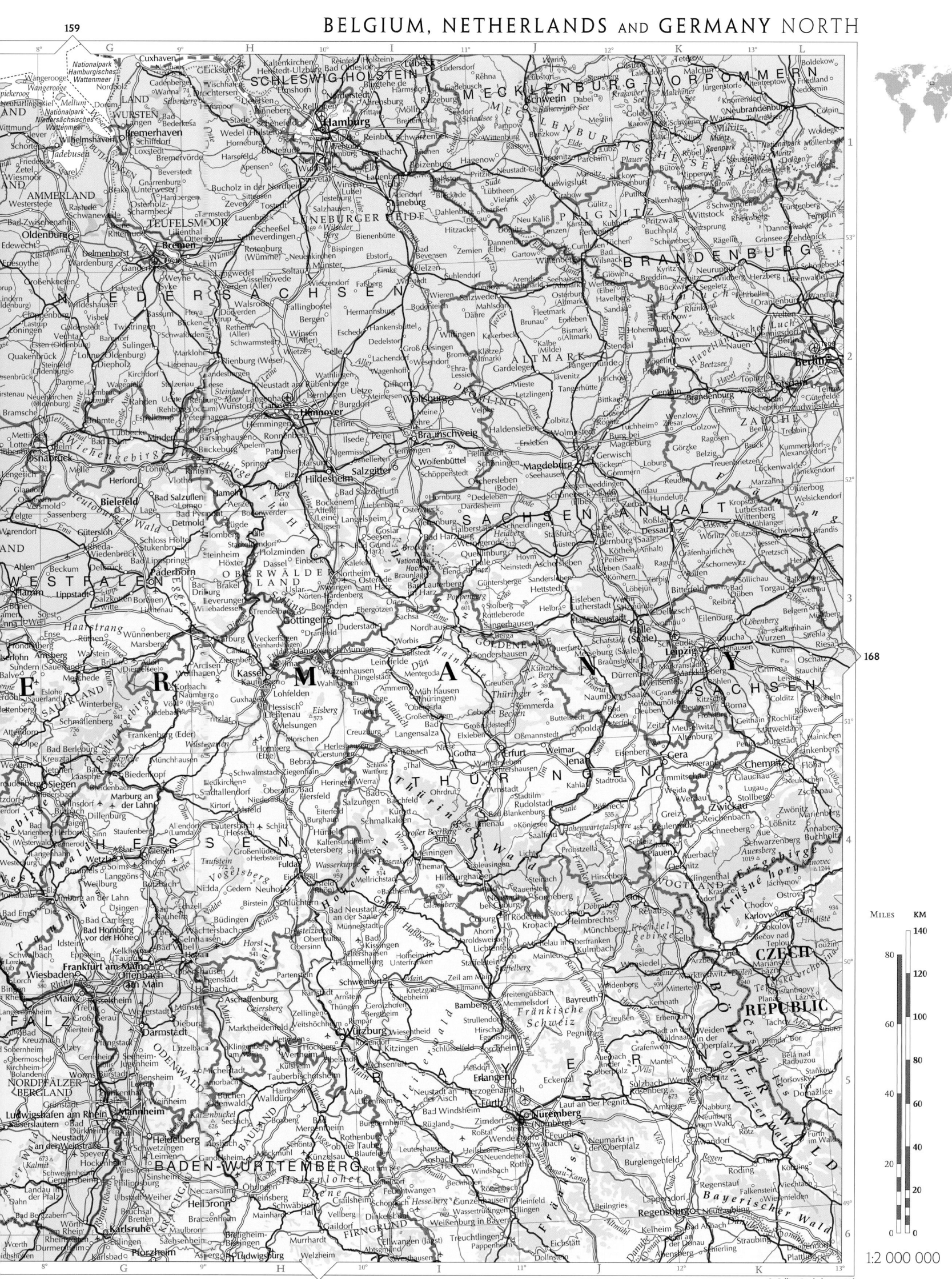

© Collins Bartholomew Ltd

1:2 000 000

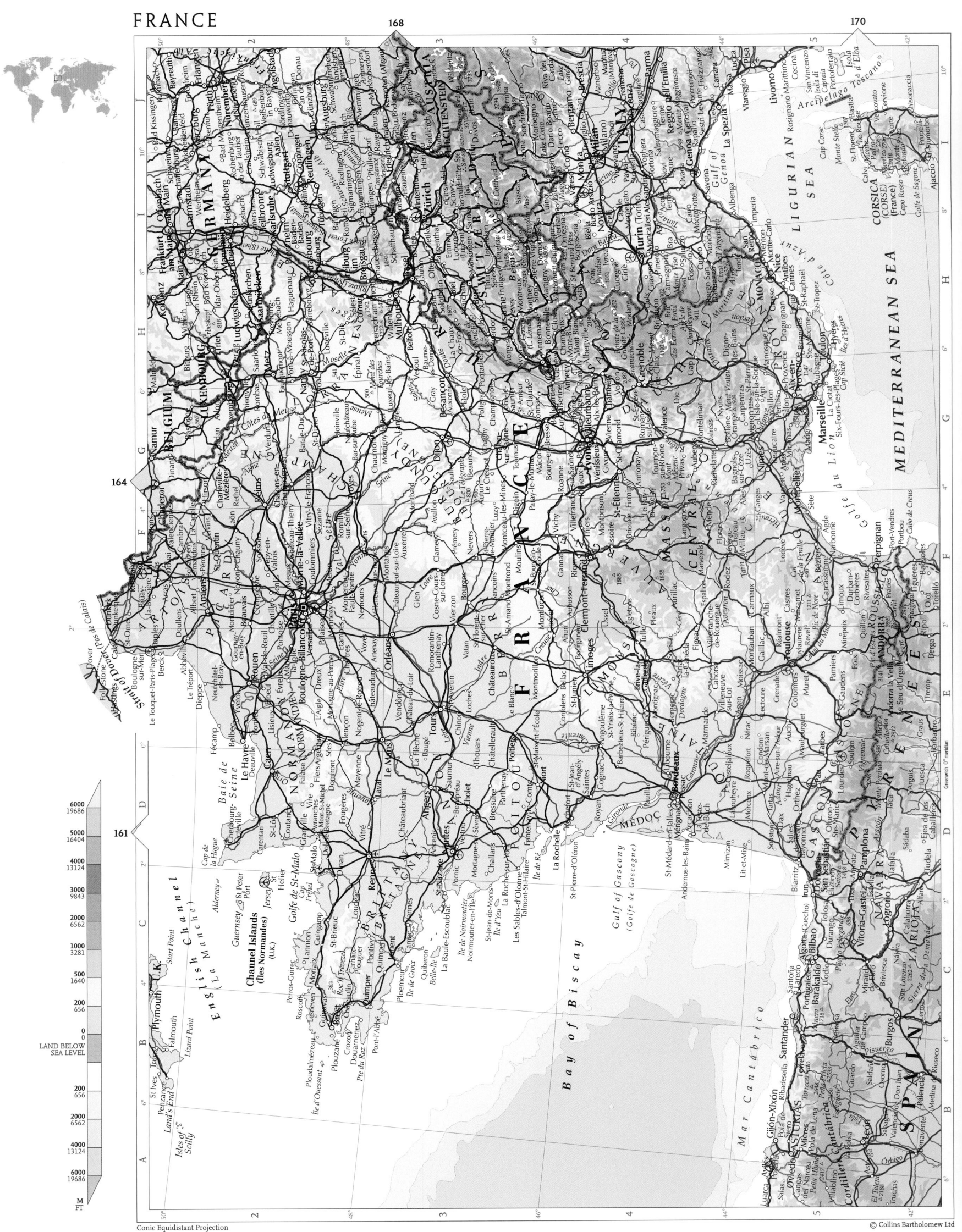

Conic Equidistant Projection

© Collins Bartholomew Ltd

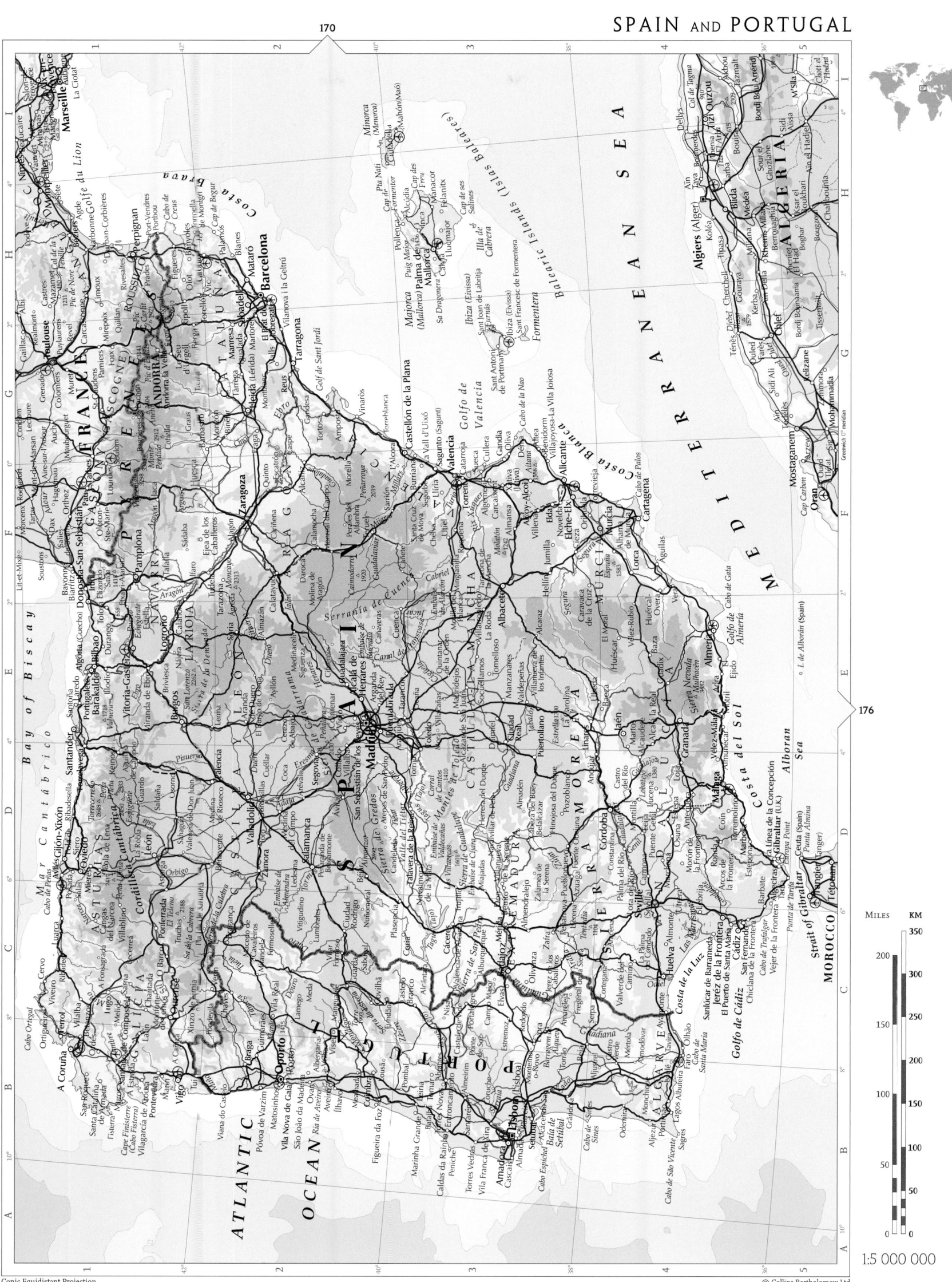

Conic Equidistant Projection

© Collins Bartholomew Ltd

1:5 000 000

Conic Equidistant Projection

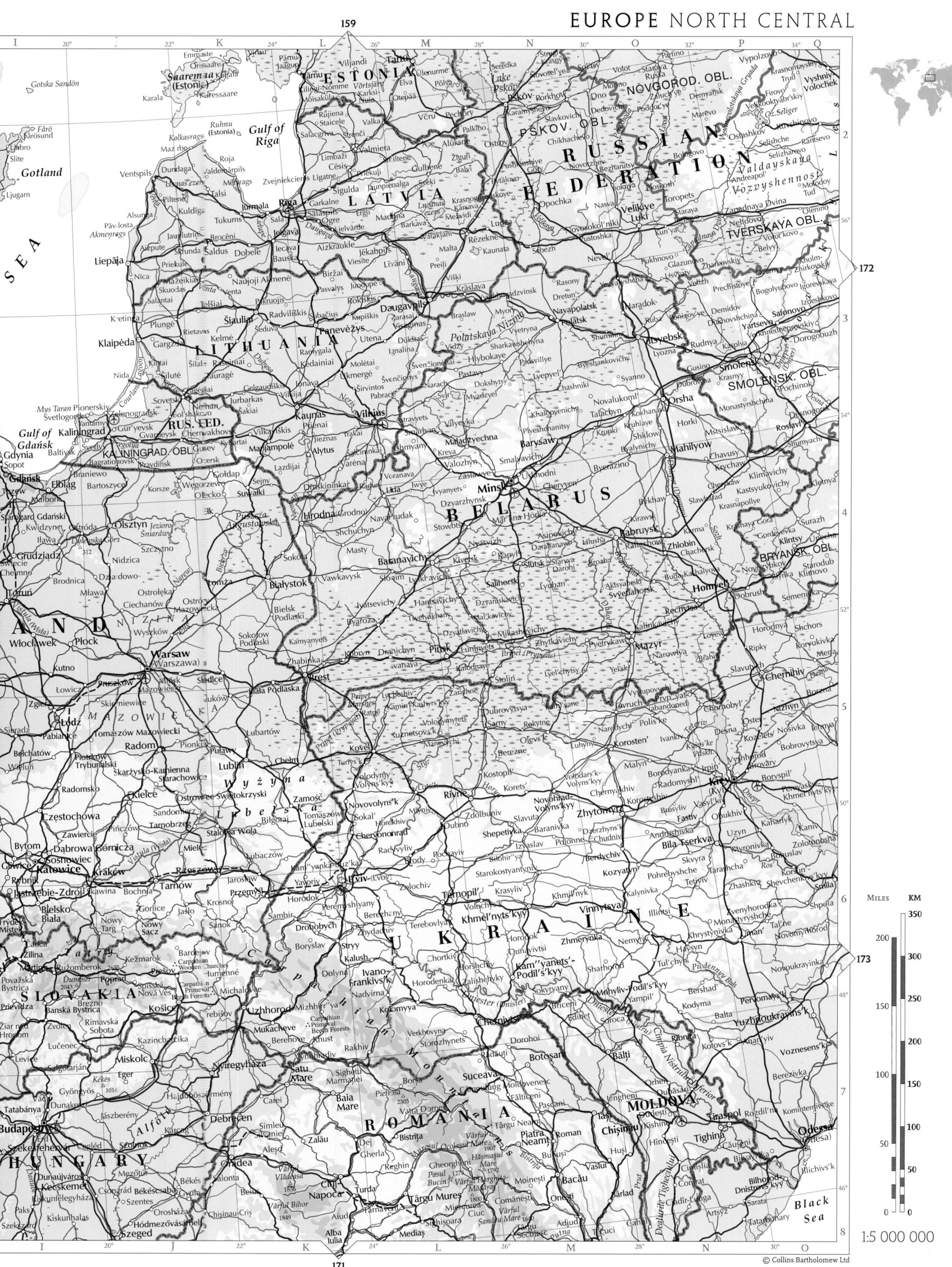

172

173

MILES   KM

350

200   300

250

150   200

100   150

50   100

50

0   0

1:5 000 000

167

176

Conic Equidistant Projection

© Collins Bartholomew Ltd

1:5 000 000

Conic Equidistant Projection

CASPIAN
SEA

KAZAKHSTAN

BLACK SEA

Sea of Azov

Divisions of the Russian Federation numbered on the map

1. RESPUBLIKA ADYGEYA (G6)
2. RESPUBLIKA SEVERNAYA OSETIYA-ALANIYA (NORTH OSSETIA) (H7)
3. RESPUBLIKA INGUSHETIYA (INGUSHETIA) (H7)

POLAND

UKRAINE

MOLDOVA

ROMANIA

BULGARIA

TURKEY

GEORGIA

ARMENIA

AZERBAIJAN

MILES   KM

300   500

          400

200   300

100   200

          100

0   0

1:7 000 000

© Collins Bartholomew Ltd

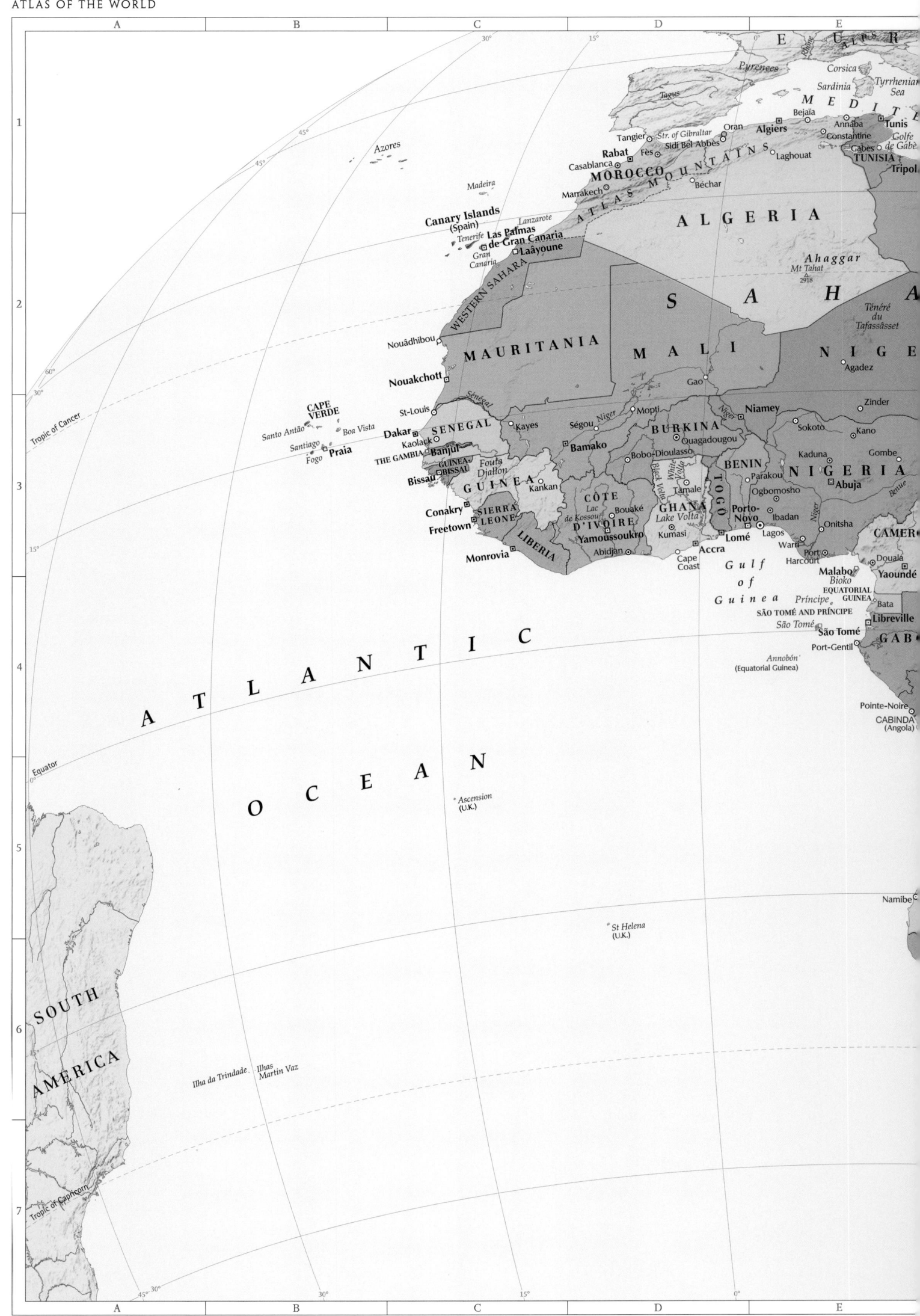

A B C D E

1

2

3

4

5

6

7

Pyrenees
Corsica
Sardinia
Tyrrhenian Sea
M E D I T E
Tagus
Str. of Gibraltar Oran **Algiers** Annaba **Tunis**
Tangier Sidi Bel Abbès Constantine
Rabat Fès Laghouat **TUNISIA** *Golfe de Gabè*
Casablanca Tripol
Béchar
Marrakech
MOROCCO A T L A S   M O U N T A I N S
Madeira
A L G E R I A
Canary Islands *Lanzarote*
(Spain)
Tenerife **Las Palmas** *Ahaggar*
de Gran Canaria *Mt Tahat*
Gran **Laâyoune** S A H A *2918*
Canaria
WESTERN SAHARA S A H *Ténéré du Tafassâsset*
Nouâdhibou
MAURITANIA M A L I N I G E
Agadez
Nouakchott Gao
Senegal Zinder
St-Louis Kayes *Niger* Mopti **Niamey**
CAPE Ségou **BURKINA** Sokoto Kano
VERDE *Boa Vista* **Dakar** *Niger* Ouagadougou
Santo Antão Kaolack **SENEGAL** **Bamako** Bobo-Dioulasso *Niger* Kaduna Gombe
Santiago THE GAMBIA **Banjul** **BENIN** **N I G E R I A**
Fogo **Praia** **GUINEA** *Fouta* Parakou
Bissau **BISSAU** *Djallon* Tamale **Abuja**
GUINEA Kankan **GHANA** **TOGO** Ogbomosho
Conakry **SIERRA** CÔTE *Lac* Bouaké *Lake Volta* **Porto-** Ibadan Onitsha **CAMER**
Freetown **LEONE** D'IVOIRE *de Kossou* *Black Volta* Kumasi **Novo** Lagos
Yamoussoukro **Lomé** Warri **Douala**
Monrovia LIBERIA Abidjan **Accra** Port **Yaoundé**
Cape Harcourt
Coast *Gulf* **Malabo** **Bioko**
of EQUATORIAL
Guinea *Príncipe* **GUINEA** Bata
SÃO TOMÉ AND PRÍNCIPE **Libreville**
São Tomé **São Tomé** **GAB**

A T L A N T I C

Annobón
(Equatorial Guinea)

Pointe-Noire
CABINDA
(Angola)

O C E A N

Ascension
(U.K.)

5 Namibe

St Helena
(U.K.)

SOUTH

AMERICA

Ilha da Trindade *Ilhas Martin Vaz*

Tropic of Cancer
Equator
Tropic of Capricorn

Azores

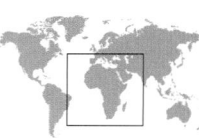

F   G   H   I   J

1

2

3

4

5

6

7

EUROPE

Adriatic Sea
Ionian Sea
Sicily
Crete
Cyprus
MEDITERRANEAN SEA
Black Sea
Volga
Caspian Sea
Aral Sea

A S I A

Zagros Mountains
The Gulf
Gulf of Oman
Dasht-e Kavīr
HIMALAYA
Tropic of Cancer

Mişrātah
Al Baydā'
Gulf of Sirte
Benghazi
Alexandria
Port Said
Shubrā al Khaymah
Giza □ **Cairo**
Suez
Gulf of Suez
Gulf of Aqaba
Red Sea

LIBYA
Libyan Desert
Qattara Depression
Al Minyā
Asyūt
EGYPT
Qina
Luxor
Aswān
Lake Nasser
Nile

SA
R A
Tibesti
Emi Koussi 3415

CHAD
Lake Chad
Abéché
Ndjamena
Maiduguri
Marqua
Sarh
Moundou
Ngaoundéré

Marra Plateau
Baiyuda Desert
Nubian Desert
Omdurman
Wad Medani
Khartoum
El Obeid
Gedaref
Atbara
Port Sudan

SUDAN
White Nile
Blue Nile
Nile
Sobat
Wau
Juba

ERITREA
Asmara
Ras Dejen 4533
Mek'elē
Lake Tana
Bahir Dar
DJIBOUTI
Djibouti
Dirē Dawa
Addis Ababa
Gulf of Aden
Hargeysa

ARABIAN SEA
Rub' al Khālī
Socotra

ETHIOPIA

CENTRAL AFRICAN REPUBLIC
Bossangoa
Bouar
Bangui
Ubangi

DEMOCRATIC REPUBLIC OF THE CONGO
Congo
Mbandaka
Kisangani
Lac Mai-Ndombe
Lomami
Congo
Kananga
Kikwit
Kinshasa
Matadi
Mbuji-Mayi
Kamina

Lake Albert
Lake Edward
Lake Kivu
Bukavu
Kampala
UGANDA
Lake Victoria
Kisumu
Nakuru
Mount Kenya 5199
Nairobi
KENYA
Lake Turkana
Webi Shabeelle
Mogadishu
Kismaayo
SOMALIA

RWANDA
Kigali
BURUNDI
Bujumbura
Kigoma
Mwanza
Arusha
Kilimanjaro
Dodoma
Tabora
Tanga
Mombasa
TANZANIA
Lake Tanganyika
Lake Rukwa
Mbeya
Iringa
Rufiji
Pemba Island
Zanzibar
Zanzibar Island
Dar es Salaam
Mafia Island

INDIAN OCEAN

Victoria Mahé
SEYCHELLES
Aldabra Islands (Seychelles)
Farquhar Group (Seychelles)
Coëtivy
Agalega Islands (Mauritius)
Maldives
Chagos Archipelago

ANGOLA
Luanda
Lobito
Benguela
Lubango
Cuango
Cunene
Cubango

Likasi
Solwezi
Lubumbashi
Chingola
Ndola
Kabwe
Lusaka
Chipata
ZAMBIA
Mongu
Kafue
Zambezi
Kariba
Lake Kariba
Livingstone
Victoria Falls
Lake Nyasa
MALAWI
Lilongwe
Blantyre
Harare
Chitungwiza
Gweru
Mutare
ZIMBABWE
Bulawayo
Francistown
Limpopo
Tete
Quelimane
Nacala
Nampula
Pemba
COMOROS
Moroni
Njazidja
Îles Glorieuses (France)
Tanjona Bobaomby
Antsiranana
Mayotte (France)

NAMIBIA
Windhoek
Namib Desert
Etosha Pan
Okavango Delta
Okavango
Makgadikgadi
Kalahari Desert
BOTSWANA
Gaborone

Orange
Pretoria
Johannesburg
Soweto
Vereeniging
Mbabane
SWAZILAND
Maputo
Xai-Xai
Inhambane
MOZAMBIQUE
Beira
Mozambique Channel
Bassas da India (France)
Île Europa (France)
Île Tromelin (France)
Cargados Carajos Islands (Mauritius)
MADAGASCAR
Mahajanga
Toamasina
Antananarivo
Flanarantsoa
Toliara
Tanjona Vohimena

MAURITIUS
Port Louis
St Denis
Réunion (France)
Rodrigues Island (Mauritius)

REPUBLIC OF SOUTH AFRICA
Kimberley
Bloemfontein
Maseru
LESOTHO
Durban
Great Karoo
Little Karoo
Drakensberg
Cape Town
Khayelitsha
Cape of Good Hope
Cape Agulhas
Port Elizabeth
East London

Equator
0°
15°
Tropic of Capricorn

MILES   KM
1000   1500
750   1250
1000
500   750
250   500
250
0   0

1:28 000 000

© Collins Bartholomew Ltd

ATLANTIC OCEAN

SPAIN

Algiers (Alger)

TUNISIA

MOROCCO

ALGERIA

Grand Erg Occidental

Grand Erg Oriental

Plateau du Tademaït

Madeira (Portugal)
Funchal

Canary Islands (Islas Canarias) (Spain)

La Palma
Santa Cruz de Tenerife
Tenerife
La Gomera
El Hierro
Gran Canaria
Las Palmas de Gran Canaria
Fuerteventura
Lanzarote

Tropic of Cancer

WESTERN SAHARA

Hamada du Drâa

Anti-Atlas

Haut Atlas

MOUNTAINS

El Eglab

S A H A R A

ERG CHECH

Erg Iguidi

Ouarâne

El Mreyyé

El Khnâchîch

Adrar des Ifôghas

MAURITANIA

Nouakchott
Nouâdhibou
Parc National du Banc d'Arguin

M A L I

SENEGAL
Dakar

THE GAMBIA

GUINEA BISSAU
BISSAU

GUINEA
Conakry

SIERRA LEONE
Freetown

LIBERIA
Monrovia

CÔTE D'IVOIRE

GHANA
Accra

TOGO
Lomé

BENIN

BURKINA
Ouagadougou

Bamako

Niamey

N I G E R

NIGERIA
Abuja
Lagos
Benin City
Port Harcourt

Timbuktu (Tombouctou)

Gao

CAMEROON
Yaoundé
Douala

EQUATORIAL GUINEA
Bata
Malabo

SÃO TOMÉ AND PRÍNCIPE
Príncipe
São Tomé

GABON
Libreville
Port-Gentil

GULF OF GUINEA

Gold Coast

Slave Coast

Bight of Benin

Mouths of the Niger

Equator   Greenwich 0° meridian

Elevation scale (M / FT)

| 6000 / 19686 |
| 5000 / 16404 |
| 4000 / 13124 |
| 3000 / 9843 |
| 2000 / 6562 |
| 1000 / 3281 |
| 500 / 1640 |
| 200 / 656 |
| 0 / 0 |
| LAND BELOW SEA LEVEL |
| 200 / 656 |
| 2000 / 6562 |
| 6000 / 19686 |

M / FT

Inset: CAPE VERDE 1:16 000 000

Santo Antão
São Vicente
Santa Luzia
São Nicolau
Sal
Boa Vista
Maio
Santiago
Fogo
Brava
Praia
Mindelo
Ponta do Sol
Ribeira Brava
Vila da Ribeira Brava
Santa Maria
Sal Rei
Curral Velho
Tarrafal
Vila do Maio
Vila Nova Sintra
Ilhéus Secos

Lambert Azimuthal Equal Area Projection

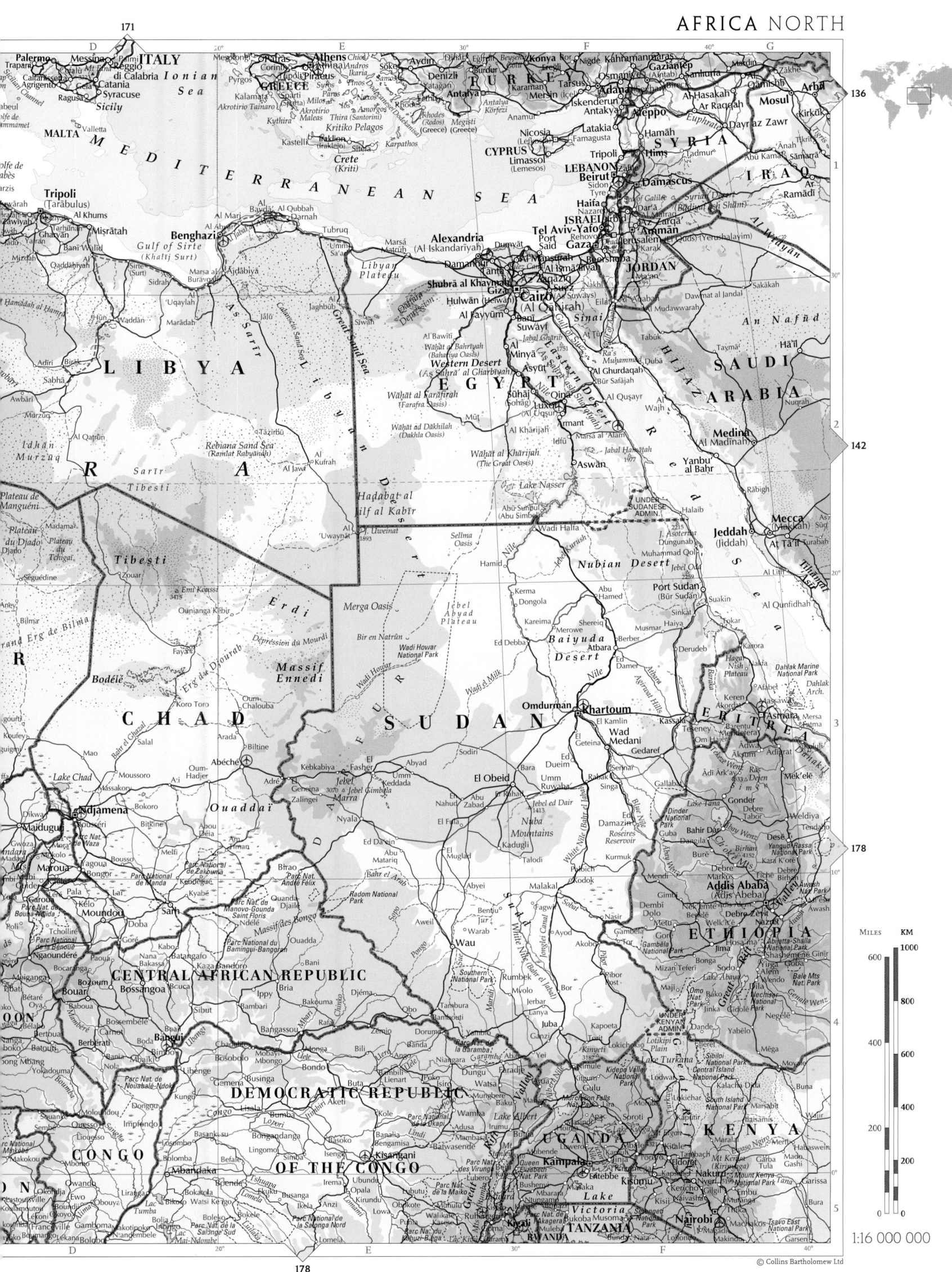

1:16 000 000

© Collins Bartholomew Ltd

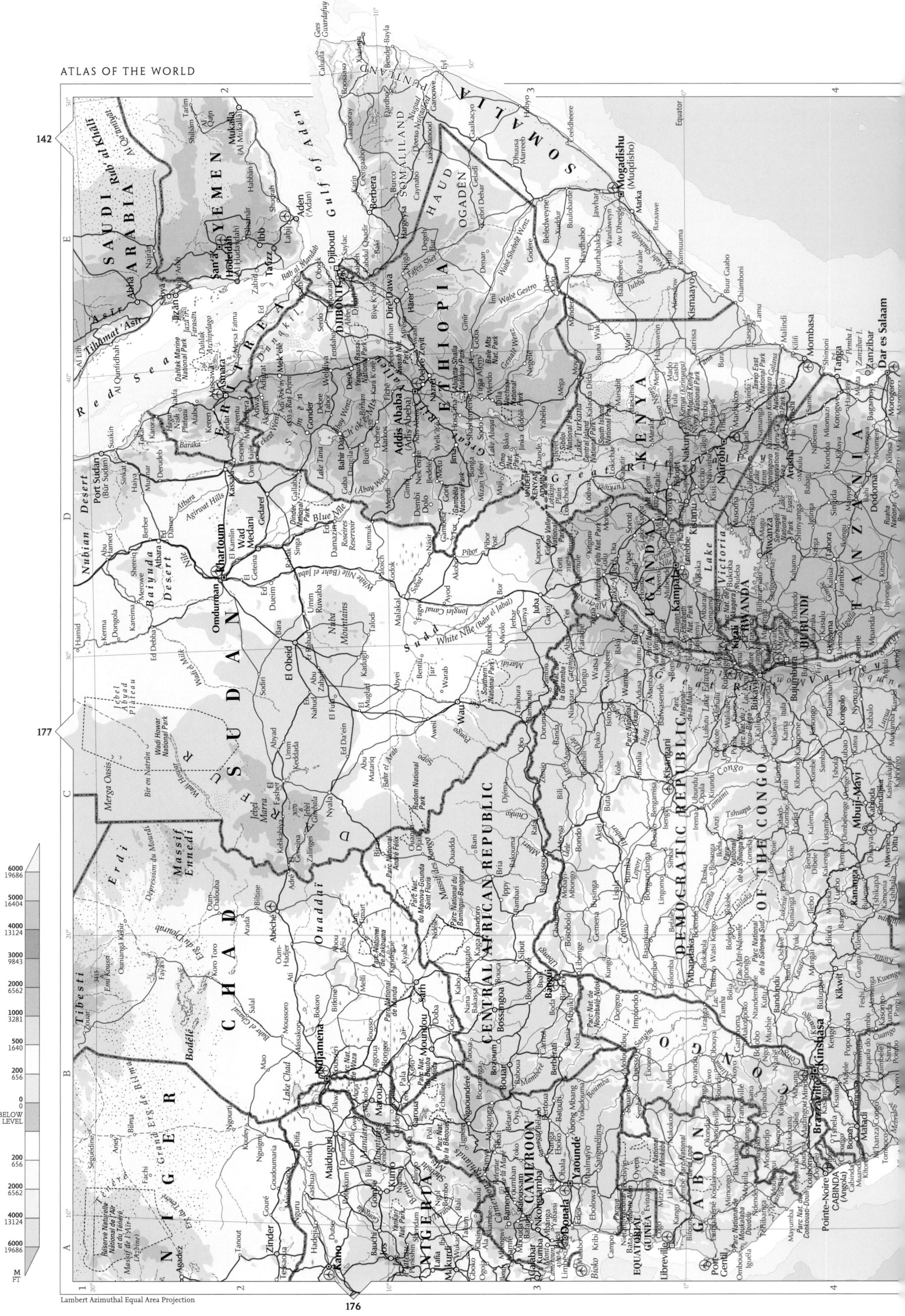

Lambert Azimuthal Equal Area Projection

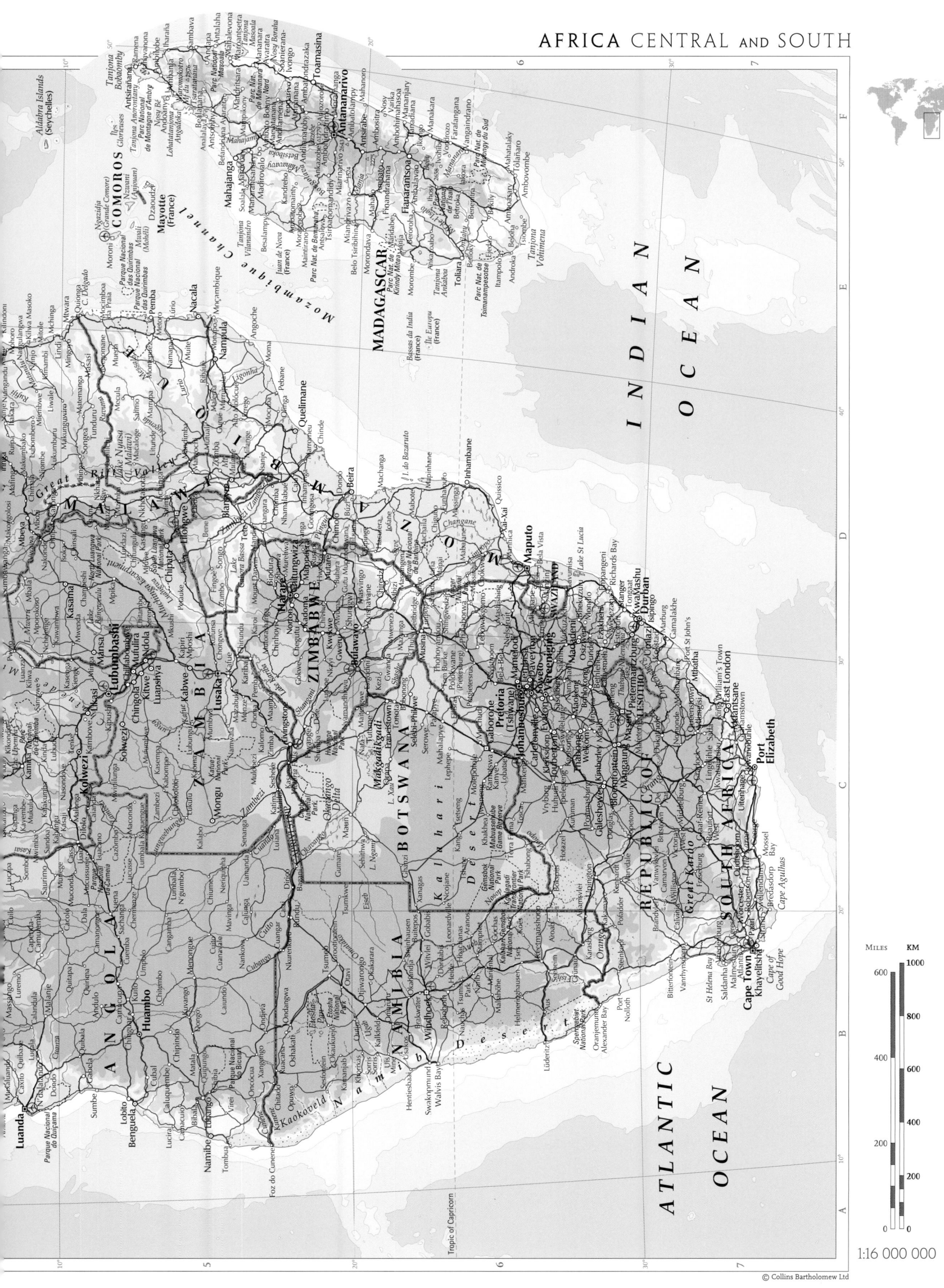

MILES   KM
1000
600
800
400   600
400
200
200
0   0

1:16 000 000

© Collins Bartholomew Ltd

Lambert Azimuthal Equal Area Projection

REPUBLIC OF SOUTH AFRICA

MILES
KM

1:5 000 000

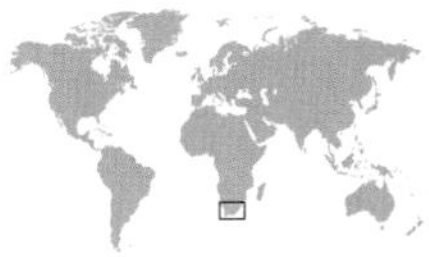

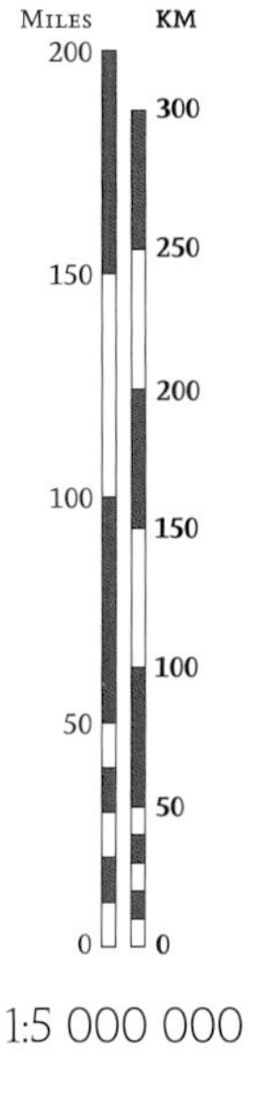

© Collins Bartholomew Ltd

PACIFIC

OCEAN

ARCTIC OCEAN

A S I A

BERING SEA

Aleutian Islands

ALASKA

YUKON

NORTHWEST TERRITORIES

NUN

C A N A D A

BRITISH COLUMBIA

ALBERTA

SASKATCHEWAN

ROCKY MOUNTAINS

WASHINGTON

OREGON

MONTANA

IDAHO

WYOMING

NEVADA

UTAH

CALIFORNIA

ARIZONA

NEW MEXICO

COLORADO

UNITED STATES

MEXICO

Hawaiian Islands (U.S.A.)

HAWAII

Line Islands

Administrative divisions abbreviated on the map:

U.S.A.		CANADA	
CONN.	CONNECTICUT	P.E.I.	PRINCE EDWARD ISLAND
DEL.	DELAWARE		
MD.	MARYLAND		
MASS.	MASSACHUSETTS		
N.H.	NEW HAMPSHIRE		
N.J.	NEW JERSEY		
R.I.	RHODE ISLAND		
VER.	VERMONT		

Orthographic Projection

EUROPE

AFRICA

Arctic Circle

Station Nord

Daneborg

Greenland Sea

Kong Wilhelm Land

Greenland (Kalaallit Nunaat) (Denmark)

Kong Christian IX Land

Kong Frederik VI Kyst

Denmark Strait

Iceland

Nuuk (Godthåb)

Ammassalik

Baffin Bay

Clyde River

Davis Strait

Cape Mercy

Ellesmere Island

Knud Rasmussen Land

Dundas

Nuussuaq

Ilulissat

Somerset Island

Devon Island

Lancaster Sd

Prince Charles I.

Cumberland Sd

Nanortalik

Bothia Pen.

Gulf of Boothia

Melville Peninsula

Foxe Basin

Baffin Island

Iqaluit

Resolution I.

Labrador Sea

Madeira

Repulse Bay

Southampton Island

Cape Dorset

Coral Harbour

Coats I.

Mansel I.

Péninsule d'Ungava

Ungava Bay

Nain

NEWFOUNDLAND AND LABRADOR

Arviat

HUDSON BAY

James Bay

Chisasibi

Belcher Islands

Smallwood Reservoir

Labrador

Azores

Canary Islands

Thompson

MANITOBA

Lake Winnipeg

ONTARIO

Moosonee

QUEBEC

Sept-Îles

Schefferville

Gander

Newfoundland

St John's

Cape Race

Winnipeg

Lake Nipigon

Timmins

Rouyn-Noranda

Chicoutimi

Île d'Anticosti

Gulf of St Lawrence

St Pierre and Miquelon (France)

Thunder Bay

Sault Ste Marie

North Bay

Val-d'Or

St Lawrence

Cabot Str.

Cape Breton I.

Grand Forks

Lake Superior

MICHIGAN

Sudbury

NEW BRUNSWICK

P.E.I.

NOVA SCOTIA

Sable Island

MINNESOTA

Duluth

Lake Huron

Toronto

Ottawa

Montréal

Charlottetown

Fredericton

MAINE

Halifax

Cape Sable

Minneapolis

St Paul

WISCONSIN

L. Michigan

Lansing

Detroit

Ontario

Buffalo

Lake Erie

NEW YORK

VT.

N.H.

Augusta

Concord

MASS.

Boston

Cape Cod

TROPIC OF CANCER

Rochester

Milwaukee

Chicago

Cleveland

Erie

PENNSYLVANIA

Albany

Hartford

CONN.

R.I.

Providence

Sioux Falls

IOWA

Des Moines

ILLINOIS

Indianapolis

Columbus

OHIO

Pittsburgh

New York

Trenton

Philadelphia

Lincoln

Omaha

St Louis

MISSOURI

Cincinnati

WEST VIRGINIA

Baltimore

Washington D.C.

Kansas City

KANSAS

Springfield

KENTUCKY

Charleston

VIRGINIA

Richmond

Bermuda (U.K.)

ATLANTIC

OCEAN

Oklahoma City

ARKANSAS

Nashville

TENNESSEE

Knoxville

Charlotte

N. CAROLINA

Raleigh

Cape Hatteras

OKLAHOMA

Little Rock

Memphis

Columbia

S. CAROLINA

UNITED STATES OF AMERICA

Fort Worth

Dallas

MISS.

Jackson

ALABAMA

Montgomery

GEORGIA

Atlanta

Savannah

Shreveport

Austin

Baton Rouge

LOUISIANA

Mobile

Tallahassee

Jacksonville

San Antonio

Houston

New Orleans

Orlando

Cape Canaveral

Corpus Christi

GULF OF MEXICO

Tampa

FLORIDA

Matamoros

Miami

THE BAHAMAS

Nassau

Cape Verde

Ciudad Victoria

Mérida

Havana

Straits of Florida

Santa Clara

Turks and Caicos Is (U.K.)

Virgin Is (U.K.)

Anguilla (U.K.)

Tampico

Yucatán Channel

CUBA

Holguín

Hispaniola

DOMINICAN REPUBLIC

San Juan

Virgin Is (U.S.A.)

ANTIGUA AND BARBUDA

Montserrat (U.K.)

ST KITTS AND NEVIS

Guadeloupe (France)

DOMINICA

Mexico City

Bahía de Campeche

Yucatán

Cayman Is (U.K.)

Montego Bay

Greater

Santiago

HAITI

Puerto Rico (U.S.A.)

Martinique (France)

ST LUCIA

BARBADOS

Veracruz

Villahermosa

JAMAICA

Kingston

Port-au-Prince

Santo Domingo

Antilles

Lesser Antilles

ST VINCENT AND THE GRENADINES

Puebla

BELIZE

Belmopan

San Pedro Sula

GRENADA

TRINIDAD AND TOBAGO

Oaxaca

Gulf of Tehuantepec

GUATEMALA

HONDURAS

Tegucigalpa

CARIBBEAN SEA

Netherlands Antilles

Port of Spain

Acapulco

Guatemala City

San Salvador

EL SALVADOR

NICARAGUA

Lake Nicaragua

Aruba (Neth.)

Managua

Colón

Panama City

COSTA RICA

PANAMA

Gulf of Panama

San José

SOUTH AMERICA

Amazon (Amazonas)

Equator

© Collins Bartholomew Ltd

MILES   KM

1200   2000

1600

800   1200

800

400

400

1:32 000 000

183

Chamberlin Trimetric Projection

© Collins Bartholomew Ltd

1:17 000 000

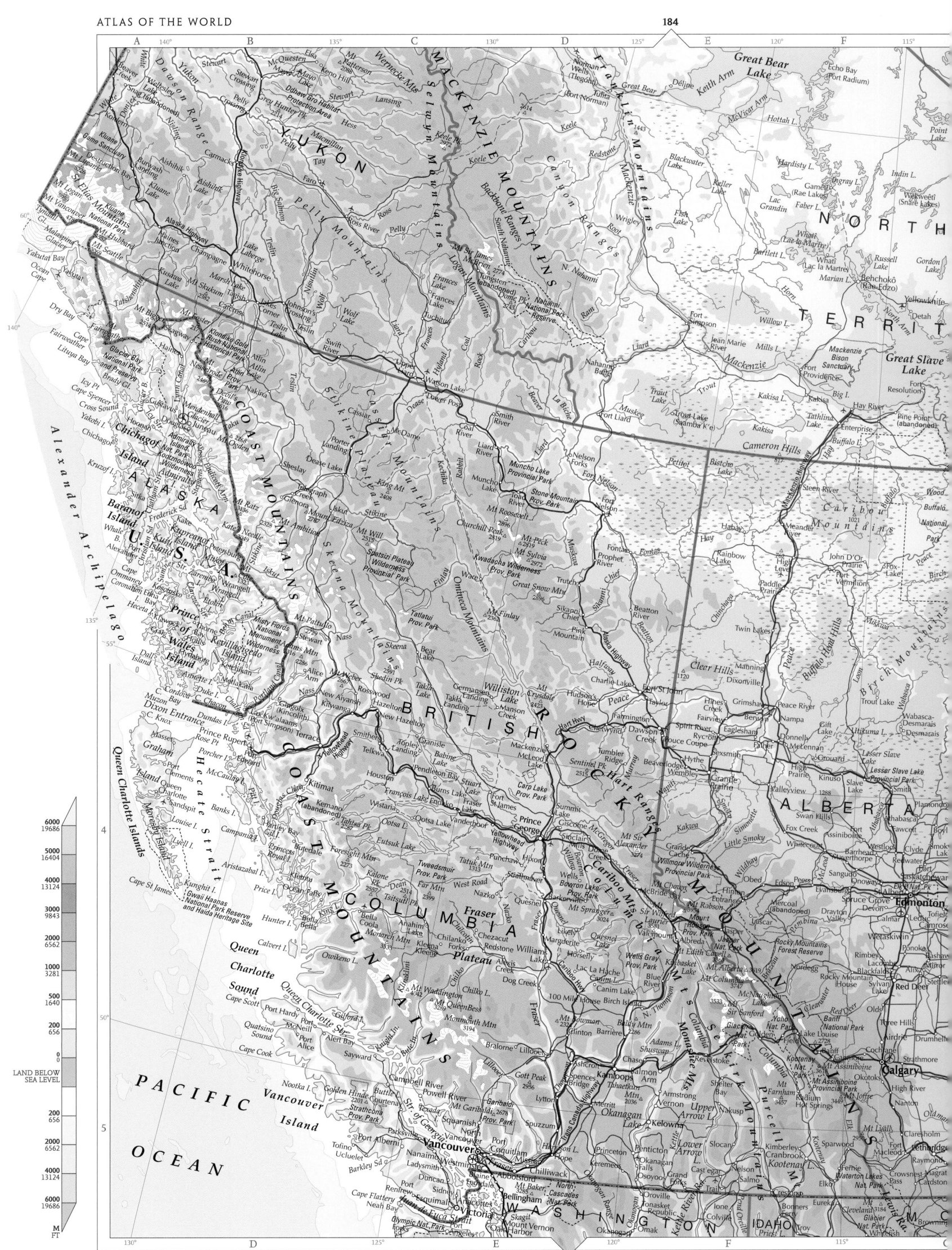

Conic Equidistant Projection

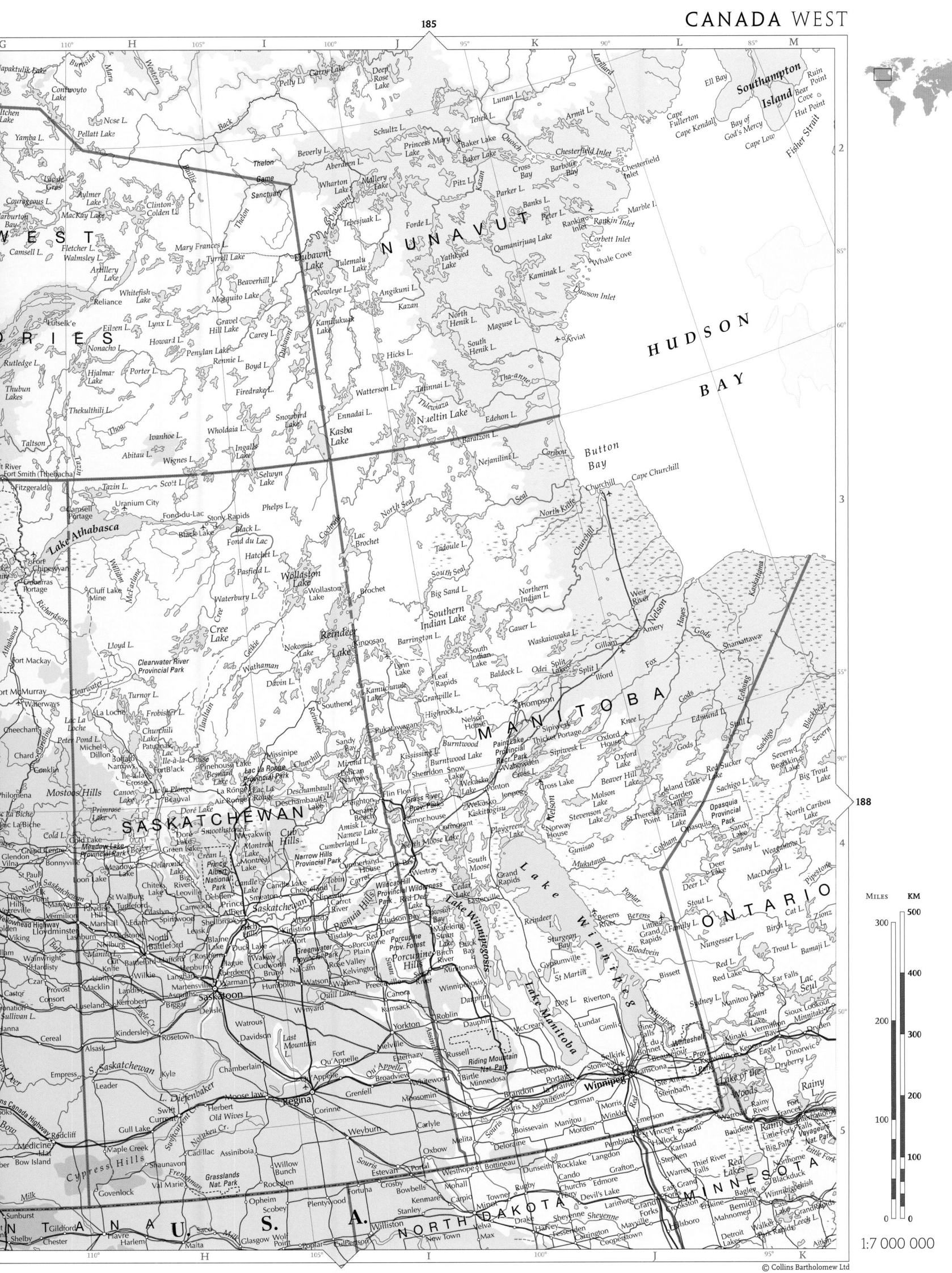

1:7 000 000

© Collins Bartholomew Ltd

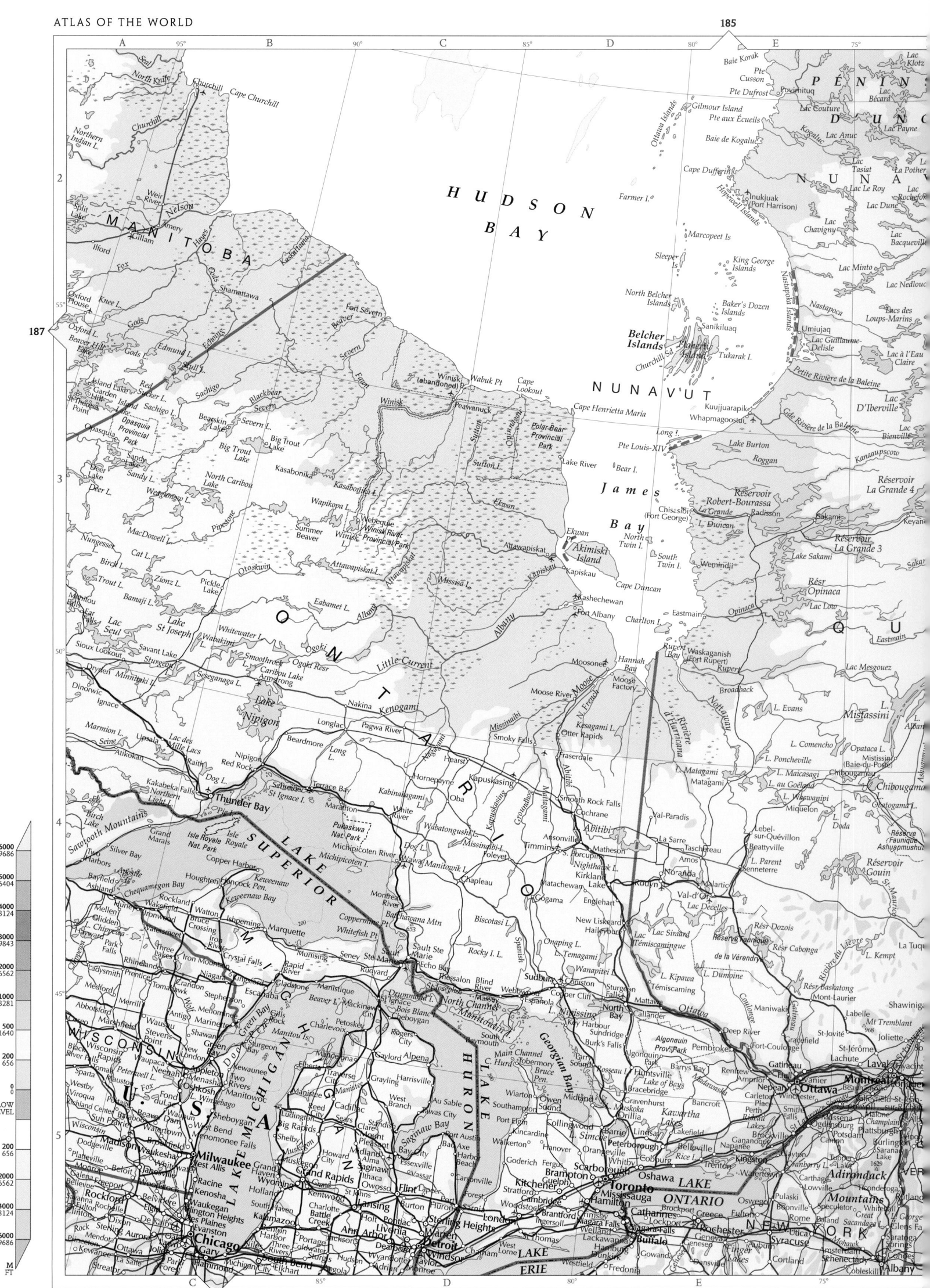

Conic Equidistant Projection

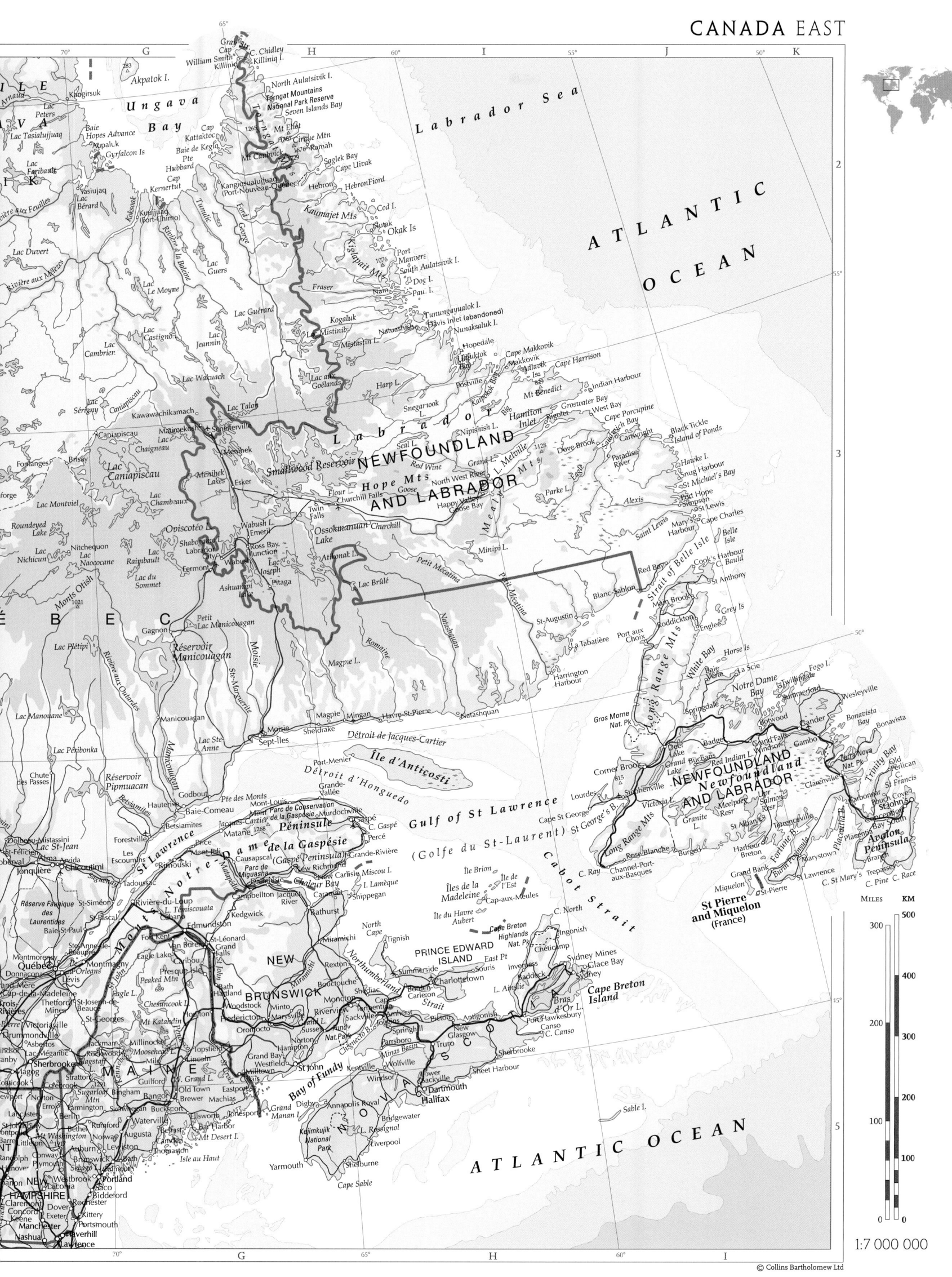

ATLANTIC OCEAN

Labrador Sea

Ungava Bay

NEWFOUNDLAND AND LABRADOR

QUEBEC

Gulf of St Lawrence
(Golfe du St-Laurent)

Cabot Strait

Île d'Anticosti

Détroit de Jacques-Cartier

Newfoundland Labrador

St Pierre and Miquelon (France)

Péninsule de la Gaspésie
(Gaspé Peninsula)

Îles de la Madeleine

PRINCE EDWARD ISLAND

NEW BRUNSWICK

MAINE

NOVA SCOTIA

Cape Breton Island

NEW HAMPSHIRE

Bay of Fundy

ATLANTIC OCEAN

MILES   KM

300   500

400

200   300

200

100

100

0   0

1:7 000 000

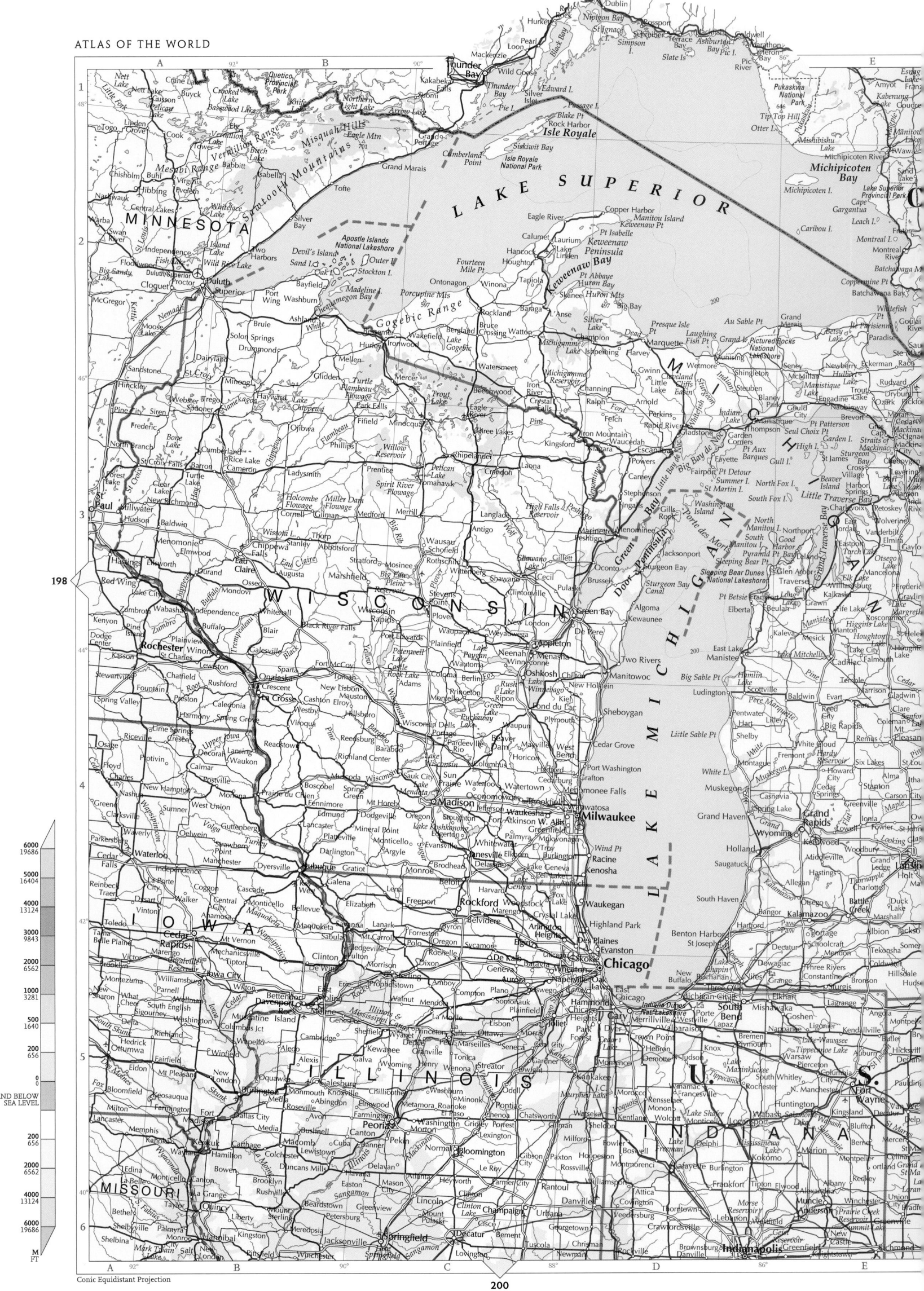

198

200

189

147

© Collins Bartholomew Ltd

1:3 500 000

191

PACIFIC OCEAN

BRITISH COLUMBIA

ALBERTA

SASKATCHEWAN

MANI
CA

WASHINGTON

OREGON

IDAHO

MONTANA

NORTH DAKOTA

SOUTH DAKOTA

WYOMING

NEBRASKA

NEVADA

UTAH

COLORADO

KANS

CALIFORNIA

UNITED STATES

ARIZONA

NEW MEXICO

OKL

TEXAS

BAJA CALIFORNIA

SONORA

CHIHUAHUA

COAHUILA

BAJA CALIFORNIA SUR

SINALOA

DURANGO

NUEVO LEON

MEXICO

Seattle
Tacoma
Portland
Salem
Eugene
Vancouver
Victoria
Calgary
Edmonton
Regina
Saskatoon
Great Falls
Billings
Boise
Salt Lake City
Provo
Denver
Colorado Springs
Pueblo
Sacramento
San Francisco
Oakland
San Jose
Reno
Las Vegas
Los Angeles
San Diego
Tijuana
Phoenix
Tucson
El Paso
Ciudad Juárez
Albuquerque
Santa Fe
Lubbock
San Antonio
Austin
Hermosillo
Chihuahua
Culiacán
Monterrey
Torreón
Gómez Palacio

Tropic of Cancer

6000 19686
5000 16404
4000 13124
3000 9843
2000 6562
1000 3281
500 1640
200 656
0
LAND BELOW SEA LEVEL
200 656
2000 6562
4000 13124
6000 19686
M
FT

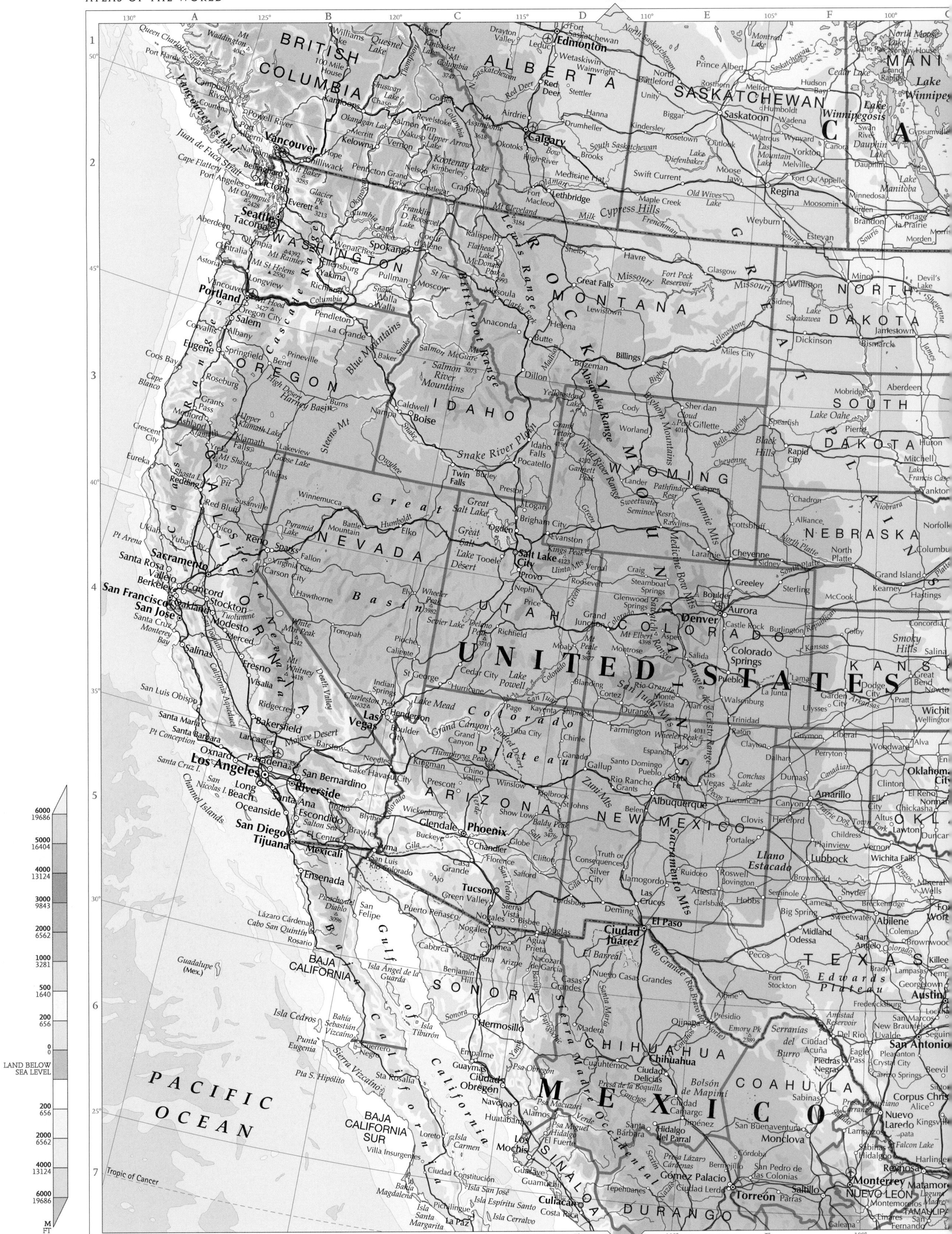

Conic Equidistant Projection

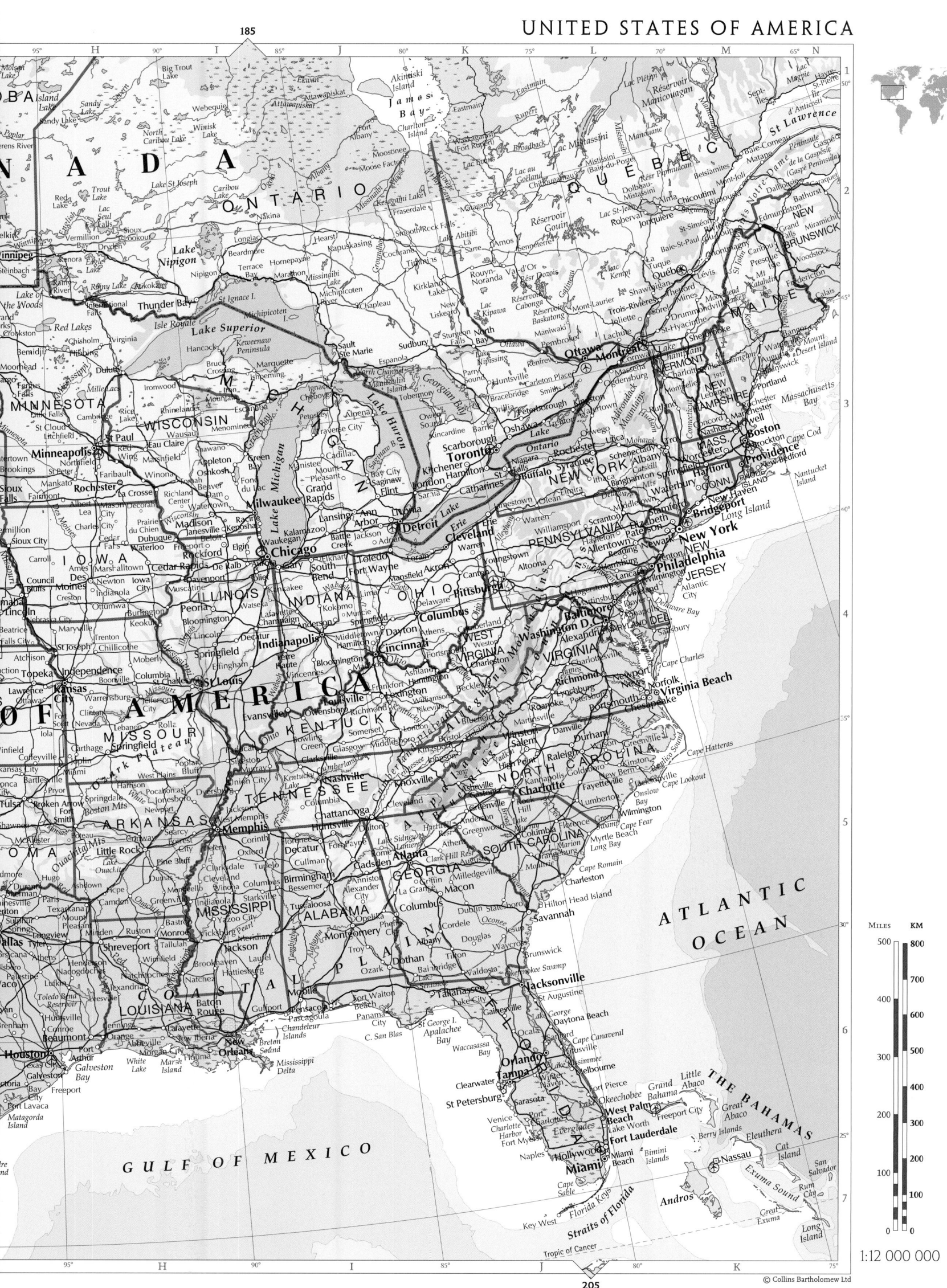

1:12 000 000

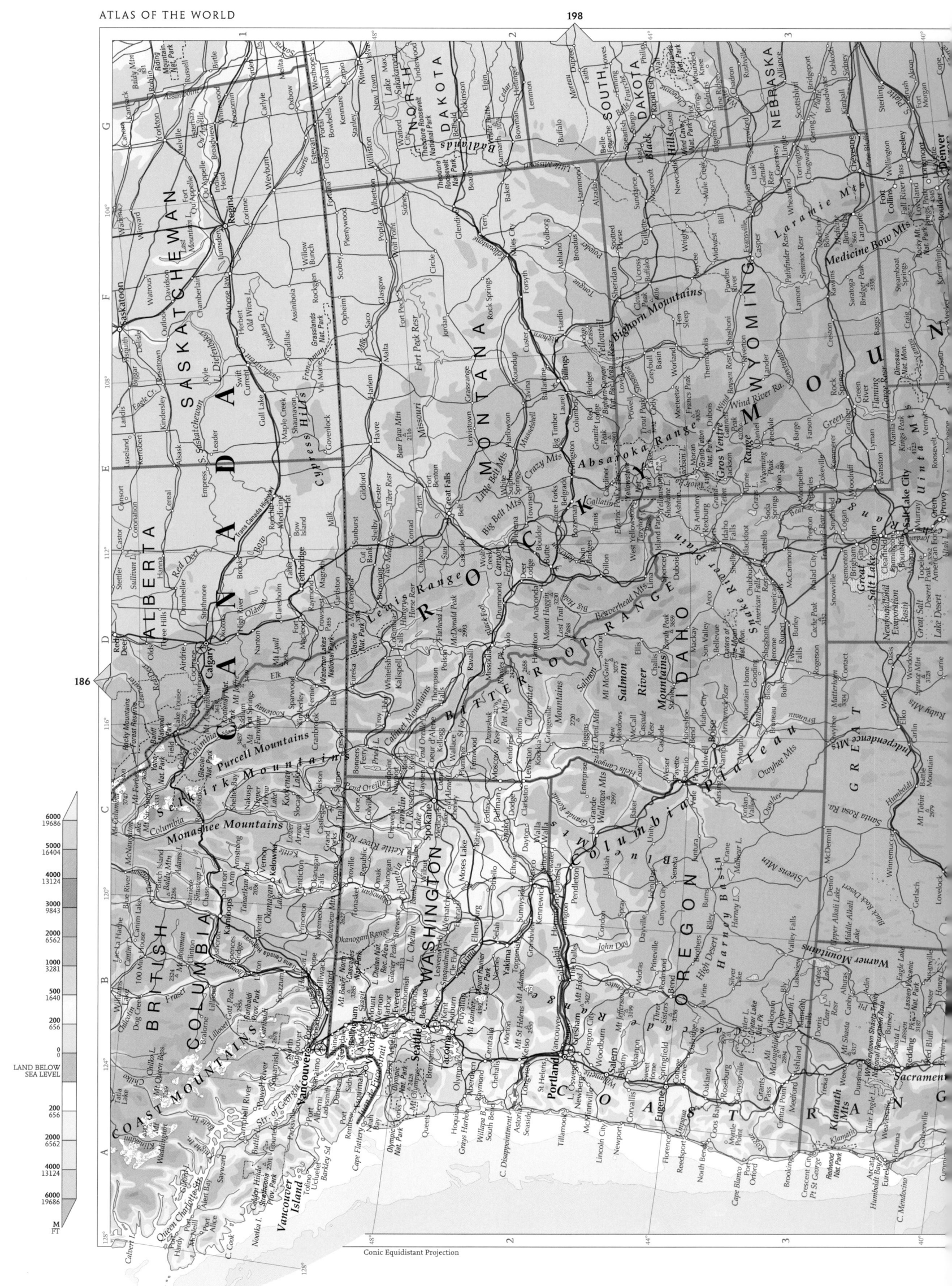

186

6000	19686
5000	16404
4000	13124
3000	9843
2000	6562
1000	3281
500	1640
200	656
0	0

LAND BELOW
SEA LEVEL

200	656
2000	6562
4000	13124
6000	19686

M
FT

Conic Equidistant Projection

199

206

Scale

MILES   KM

300   500

           400

200   300

           200

100

           100

0   0

1:7 000 000

© Collins Bartholomew Ltd

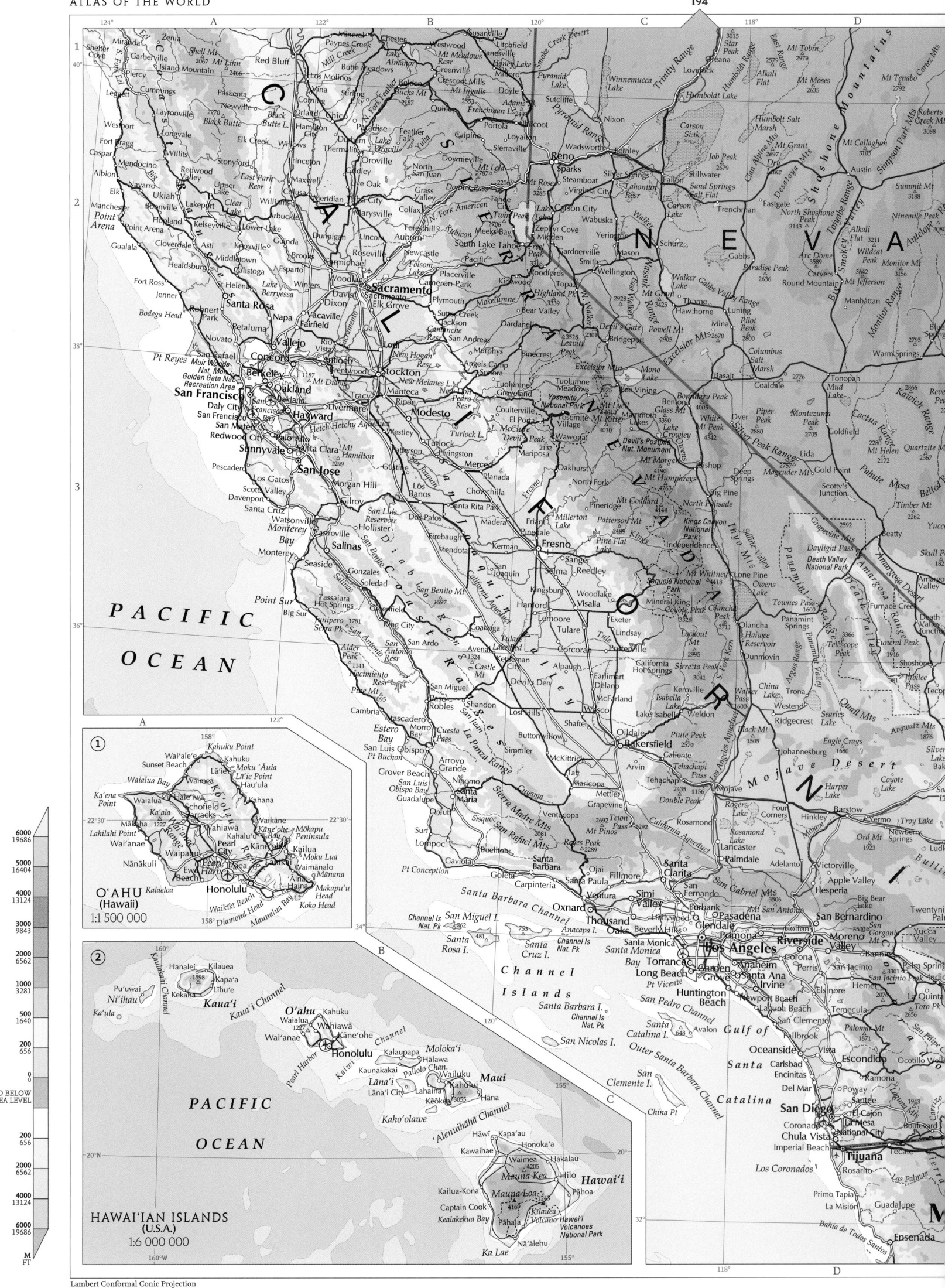

PACIFIC OCEAN

O'AHU
(Hawaii)
1:1 500 000

HAWAI'IAN ISLANDS
(U.S.A.)
1:6 000 000

PACIFIC OCEAN

Lambert Conformal Conic Projection

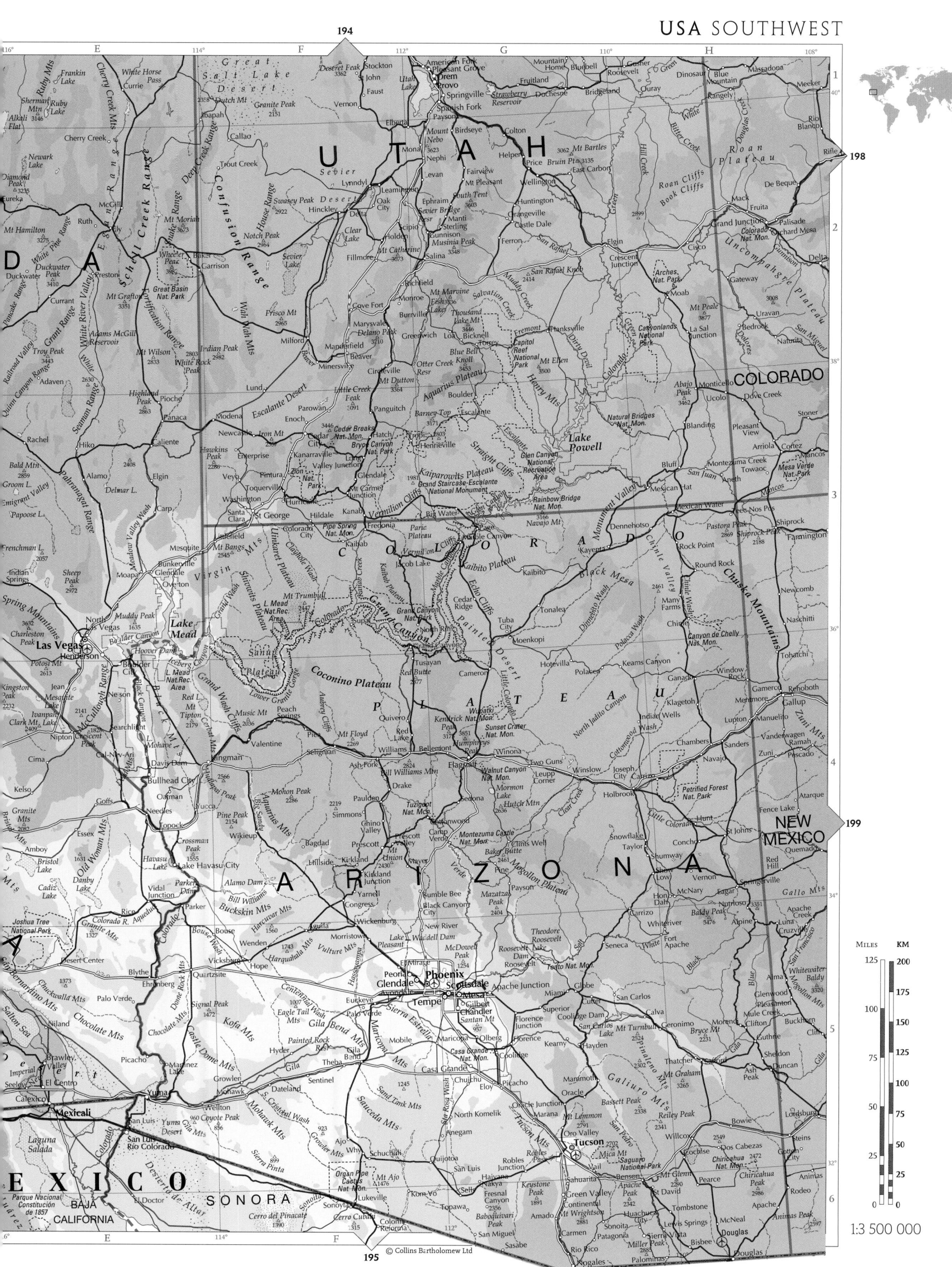

1:3 500 000

200

188

187

194

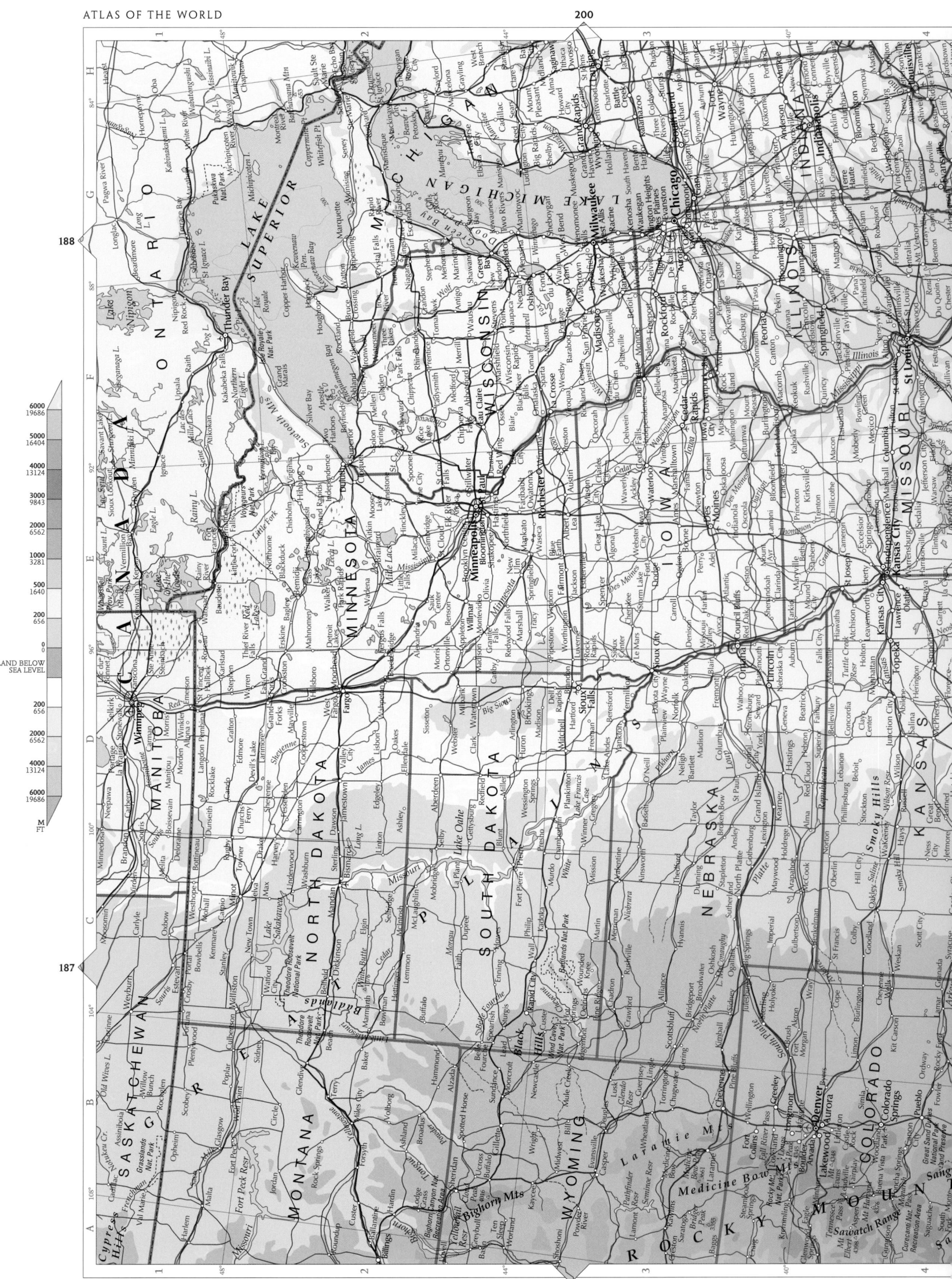

6000
19686

5000
16404

4000
13124

3000
9843

2000
6562

1000
3281

500
1640

200
656

0
0

LAND BELOW
SEA LEVEL

200
656

2000
6562

4000
13124

6000
19686

M
FT

Lambert Conformal Conic Projection

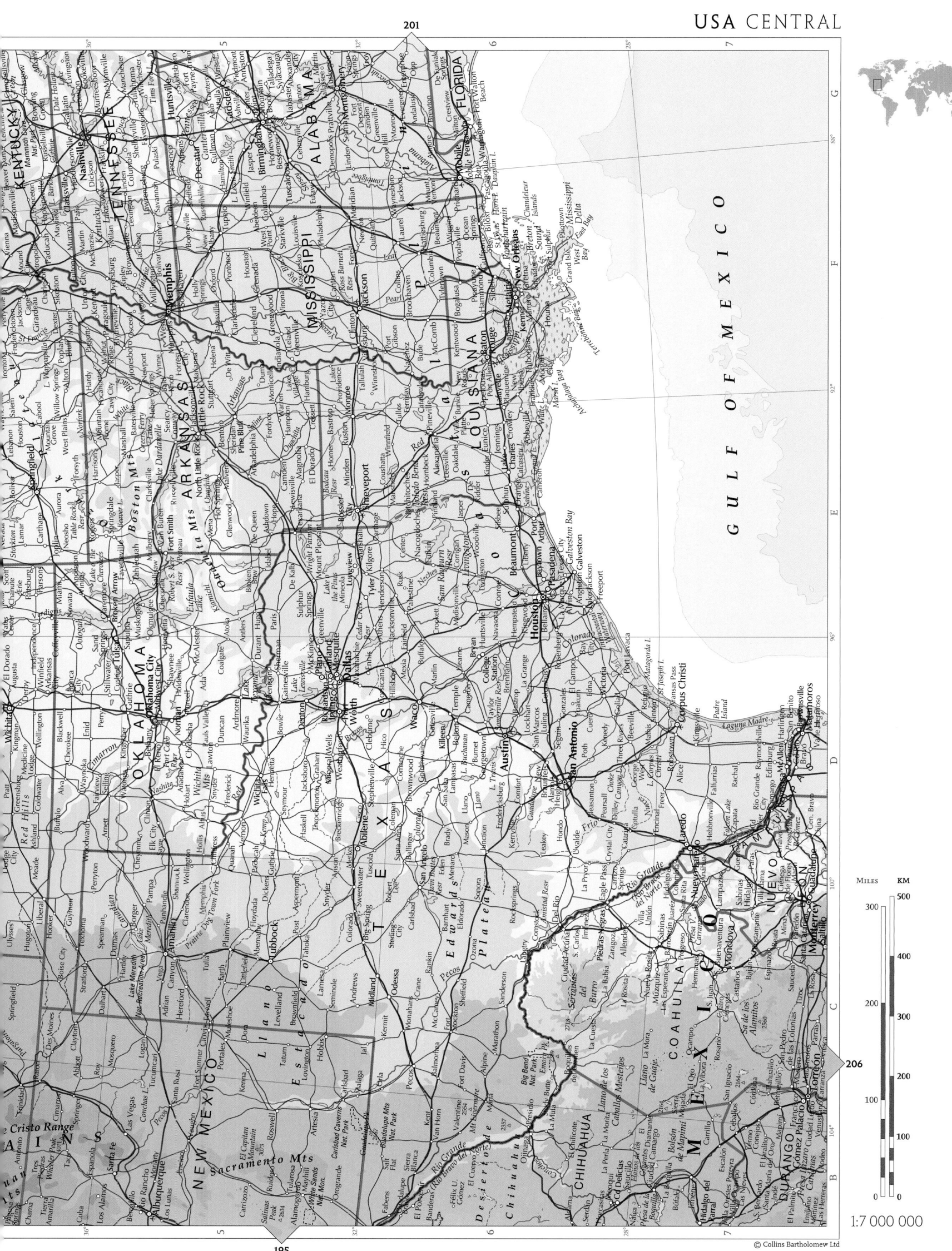

201

195

206

© Collins Bartholomew Ltd

MILES	KM
300	500
	400
200	300
	200
100	100
0	0

1:7 000 000

189

188

6000
19686

5000
16404

4000
13124

3000
9843

2000
6562

1000
3281

500
1640

200
656

0

LAND BELOW
SEA LEVEL

200
656

2000
6562

4000
13124

6000
19686

M
FT

ATLANTIC

OCEAN

THE BAHAMAS

Eleuthera

Grand Bahama

Great Abaco

Andros

Nassau

New Providence

Cat Island

San Salvador

Long Island

Great Exuma

Little Exuma

Tropic of Cancer

NORTH CAROLINA

SOUTH CAROLINA

GEORGIA

ALABAMA

MISSISSIPPI

TENNESSEE

FLORIDA

Chesapeake Beach

Cape Hatteras

Cape Lookout

Cape Fear

Wilmington

Myrtle Beach

Georgetown

Charleston

Hilton Head Island

Savannah

Brunswick

Jacksonville

St Augustine

Daytona Beach

Cape Canaveral

Melbourne

Palm Bay

Vero Beach

Fort Pierce

West Palm Beach

Pompano Beach

Fort Lauderdale

Hollywood

Hialeah

Miami

Homestead

Key Largo

Key West

Dry Tortugas

Marquesas Keys

Naples

Cape Sable

Fort Myers

Port Charlotte

Sarasota

Bradenton

St Petersburg

Clearwater

Tampa

Lakeland

Orlando

Gainesville

Tallahassee

Panama City

Dothan

Montgomery

Mobile

New Orleans

Biloxi

Gulfport

Pascagoula

Jackson

Memphis

Nashville

Chattanooga

Knoxville

Asheville

Greenville

Columbia

Charlotte

Winston-Salem

Greensboro

Durham

Atlanta

Columbus

Macon

Albany

Birmingham

Huntsville

Decatur

Tuscaloosa

Lake Okeechobee

Everglades National Park

Straits of Florida

GULF

OF

MEXICO

Mississippi Delta

MILES   KM

300   500

400

200   300

200

100

100

0   0

1:7 000 000

© Collins Bartholomew Ltd

201

205

199

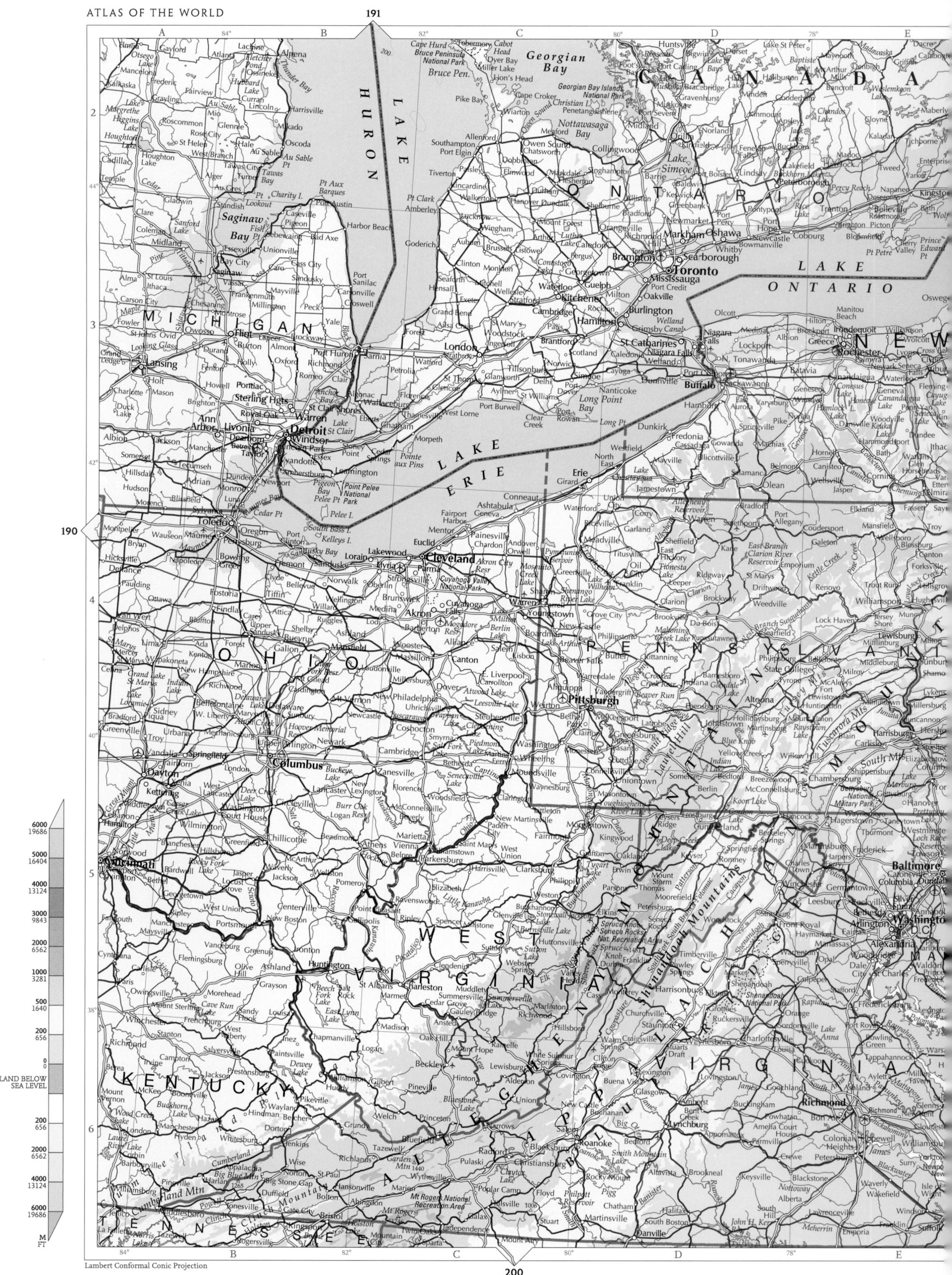

Lambert Conformal Conic Projection

ATLANTIC

OCEAN

ATLANTIC
OCEAN

1:3 500 000

© Collins Bartholomew Ltd

MILES   KM

125   200

        175

100   150

        125

75   100

        75

50

        50

25

        25

0   0

1:3 500 000

192

PACIFIC

OCEAN

GULF OF MEXICO

MEXICO

Mexico City

Guadalajara

Monterrey

U. S.

TEXAS

ARIZONA

NEW MEXICO

OKLAHOMA

ARKANSAS

TENNESSEE

MISSISSIPPI

LOUISIANA

GUATEMALA

BELIZE

EL SALVADOR

HONDURAS

Tropic of Cancer

M	FT
6000	19686
5000	16404
4000	13124
3000	9843
2000	6562
1000	3281
500	1640
200	656
0	0

LAND BELOW
SEA LEVEL

200	656
2000	6562
4000	13124
6000	19686
M	FT

Lambert Azimuthal Equal Area Projection

204

ATLANTIC OCEAN

Bermuda
(U.K.) Hamilton

THE BAHAMAS

Tropic of Cancer

HISPANIOLA

LEEWARD ISLANDS

Turks and Caicos Islands
(U.K.)

CUBA

Cayman Islands
(U.K.)

JAMAICA

HAITI

DOMINICAN
REPUBLIC

Santo
Domingo

Puerto
Rico
(U.S.A.)

San Virgin Is Virgin Is Anguilla
Juan (U.S.A.) (U.K.) The (U.K.)

ANTIGUA
AND
BARBUDA

ST KITTS AND NEVIS

Montserrat
(U.K.)

Guadeloupe
(Fr.)

DOMINICA

Martinique
(Fr.)

ST LUCIA

ST VINCENT &
THE GRENADINES

BARBADOS

WINDWARD ISLANDS

C A R I B B E A N   S E A

Lesser Antilles

GRENADA

Netherlands
Antilles

Aruba
(Neth.)

TRINIDAD
AND
TOBAGO

COSTA RICA

PANAMA

Panama
City

COLOMBIA

VENEZUELA

Caracas

Medellín

Bogotá

Cali

BRAZIL

Pakaraima Mts

1:14 000 000

MILES   KM
500   800

400   600

300   400

200   200

0   0

© Collins Bartholomew Ltd

6000
19686

5000
16404

4000
13124

3000
9843

2000
6562

1000
3281

500
1640

200
656

0
0

LAND BELOW
SEA LEVEL

200
656

2000
6562

4000
13124

6000
19686

M
FT

1:7 000 000

Lambert Conformal Conic Projection

210

199

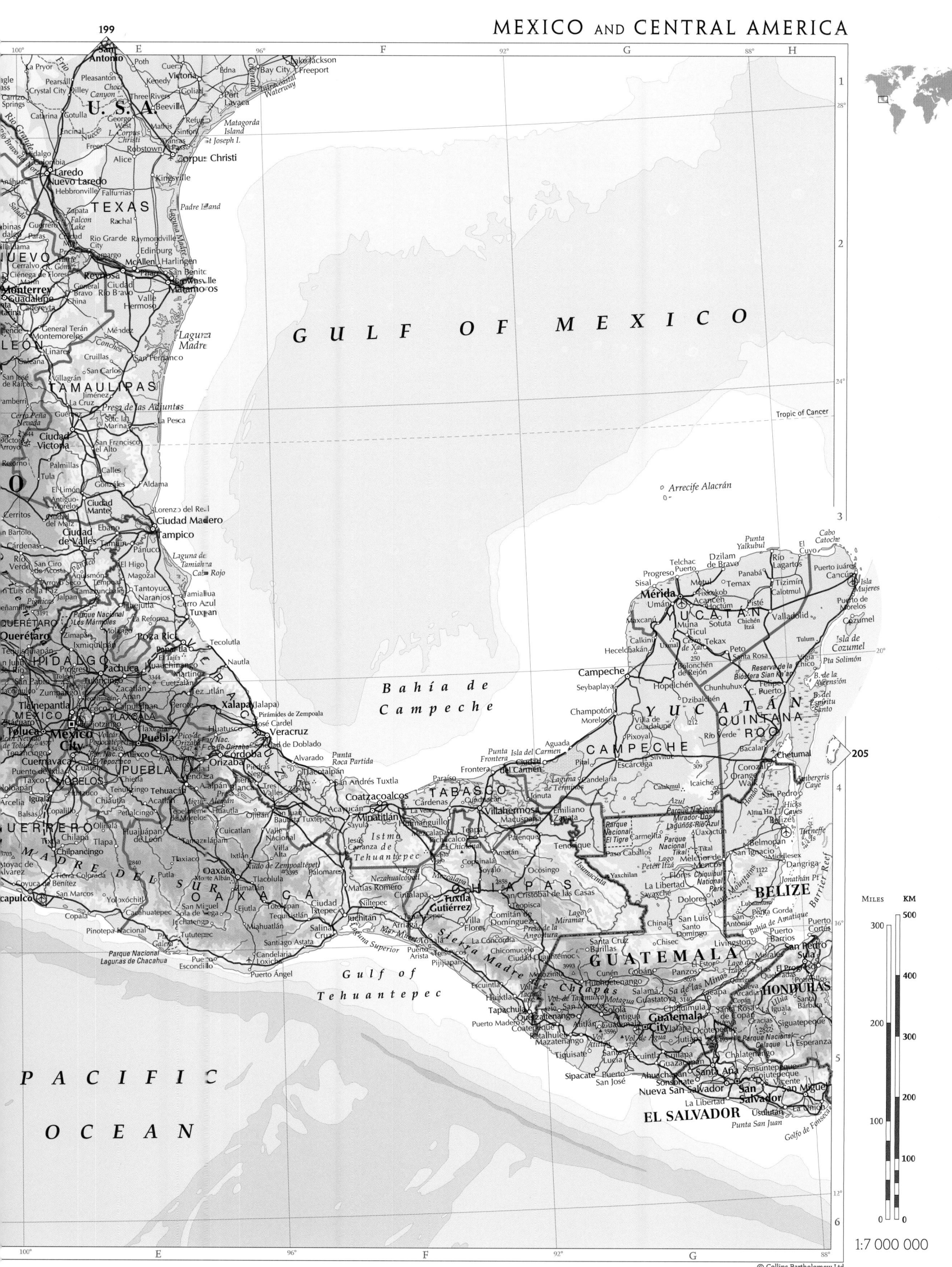

GULF OF MEXICO

U.S.A

San Antonio

TEXAS

Laredo
Nuevo Laredo

Monterrey

LEÓN

TAMAULIPAS

Ciudad Victoria

Ciudad Madero
Tampico

QUERÉTARO

HIDALGO
Pachuca

MÉXICO
Toluca Mexico City
Cuernavaca
PUEBLA Puebla
MORELOS

GUERRERO
Chilpancingo

Acapulco

OAXACA

MADRE DEL SUR

Oaxaca

Istmo de Tehuantepec

Gulf of Tehuantepec

Bahía de Campeche

Tropic of Cancer

Arrecife Alacrán

Mérida

YUCATÁN

Campeche

YUCATÁN

QUINTANA ROO

CAMPECHE

Chetumal

TABASCO
Villahermosa

CHIAPAS
Tuxtla Gutiérrez

San Cristóbal de las Casas

Sierra Madre

Tapachula

GUATEMALA

Guatemala City

EL SALVADOR

San Salvador

HONDURAS

San Pedro Sula

BELIZE

Belmopan

PACIFIC

OCEAN

205

MILES   KM

300   500

400

200   300

200

100   100

0   0

1:7 000 000

© Collins Bartholomew Ltd

207

NORTH
AMERICA

Gulf of Mexico

Cuba

Greater *Ant*

Hispaniola

Gulf of California

Bahía
de
Campeche

Yucatán

Jamaica

C A R I B B E A N

Islas
Revillagigedo

Tropic of Cancer

Barranquilla
Cartagena

Maracaibo

Monteria

San Cristóbal

Isla de Coco

Medellín

Ibagué

Cali

Bogotá

COLOMBIA

Île Clipperton

Isla de Malpelo
(Colombia)

Tunja

Neiva

Pasto

Esmeraldas

Quito

Manta

ECUADOR

Guayaquil

Cuenca

Galapagos
Islands
(Ecuador)

Machala

Iquitos

Ucayali

Piura

Tarapoto

Cruzei
do Sul

Chiclayo

Pucallpa

Trujillo

P E R U

P A C I F I C

Callao

Huancayo

Cusco

Lima

Ica

Juliac

Arequipa

Aric

O C E A N

Iquique

Antofagast

Islas
Desventurados
(Chile)

Copiap

Marquesas Islands

Hiva Oa

La Serena

Îles
du Désappointement

Isla Sala y Gómez

Cerro Aconcagu

Juan Fernández
Islands
(Chile)

Valparaíso

Santiago

Tuamotu Islands

O C E A N I A

Easter Island
(Isla de Pascua)
(Rapa Nui)
(Chile)

Talca

du Roi Georges

Îles

Hao

Henderson Island

Chillán

Rangiroa

Concepción

Îles Gambier

Pitcairn Island

Society
Islands

Tahiti

Mururoa

Valdivia

Puerto Montt

Isla de Chiloé

Archipiélago
de los Chonos

Tubuai Islands

Golfo de Penas

Tropic of Capricorn

Puerto Natales

Punta Arena

Puerto Rico
Anguilla
Antigua
Dominica
Guadeloupe
les
Martinique
St Lucia
SEA
Barbados
Aruba
St Vincent
Lesser Antilles
and the Grenadines
Grenada
Tobago
Maracay
Caracas
Trinidad
Cumaná
Valencia
Ciudad Bolívar
VENEZUELA
Orinoco
Georgetown
Puerto Ayacucho
Paramaribo
SURINAME
Cayenne
Boa Vista
GUYANA
French
Guiana
Negro
Macapá
Japurá
Amazon
Tonantins
Manaus
Santarém
Belém
Carauari
Madeira
Tapajós
São Luís
Parnaíba
Rio Branco
Fortaleza
LVAS
Porto
Velho
BRAZIL
Maraba
Teresina
Xingu
Araguaína
Puerto Maldonado
Trinidad
Natal
Barragem
de Sobradinho
Palmas
João Pessoa
Lake
Titicaca
Represa
Serra da Mesa
Floresta
Recife
La Paz
Juàzeiro
Maceió
BOLIVIA
Cuiabá
Aracaju
Cochabamba
Santa Cruz
Brasília
Salvador
Goiânia
Ilhéus
Sucre
Potosí
Campo
Grande
Patos
de Minas
Tarija
Araçatuba
Uberaba
Teófilo
Otôni
PARAGUAY
Pedro Juan
Caballero
Ribeirão
Preto
Belo
Horizonte
Vitória
Paraná
Maringá
Campinas
San Salvador
de Jujuy
São Paulo
San Miguel
de Tucumán
Asunción
Formosa
Coronel
Oviedo
Santos
Rio
de Janeiro
Nevado
Ojos del Salado
Resistencia
Encarnación
Curitiba
Catamarca
Corrientes
Joinville
Posadas
La Rioja
Florianópolis
Santa Maria
Córdoba
Concordia
Paraná
Lagoa
dos Patos
Porto Alegre
Santa Fé
Mendoza
Rosario
URUGUAY
Rio Grande
San Luis
Buenos
Aires
Montevideo
ARGENTINA
La
Plata
Rio de la Plata
San Rafael
Santa Rosa
euquén
Bahía Blanca
Mar del Plata
Viedma
Golfo San Matías
Trelew
Comodoro Rivadavia
Golfo de San Jorge
Bahía
Grande
o Gallegos
Falkland
Islands
(U.K.)
Stanley
Isla Grande
de Tierra del Fuego
Ushuaia
Isla de los Estados
Cape Horn
Drake Passage
Shag
Rocks
South Georgia
South Georgia
and the
South Sandwich
Islands
(U.K.)
South Orkney
Islands
South Shetland
Islands
Antarctic Peninsula
South
Sandwich
Islands
Traversay Islands
Candlemas Island
Saunders Island
Montagu Island
Southern Thule
Bristol Island

Madeira
Canary
Islands
Gran
Canaria
Tropic of Cancer
Santo Antão
Boa Vista
Cape Verde
São Tiago
Senegal
Niger
Gulf
of
Guinea
A
F
R
I
C
A
Equator
Fernando
de Noronha
(Brazil)
Ascension
Ilha da Trindade
(Brazil)
Ilhas
Martin Vaz
(Brazil)
ATLANTIC
St Helena
OCEAN
Tristan
da Cunha
Cape of Good Hope
Orange
Tropic of Capricorn

ATLANTIC OCEAN

MILES	KM
1200	2000
	1600
800	1200
	800
400	400
0	0

1:32 000 000

© Collins Bartholomew Ltd

A · 85° · B · 80° · C · 75° · D · 70° · E · 65° · F · 60°

CARIBBEAN SEA

NICARAGUA

Bonanza
Prinzapolca
Matagalpa Grande
Siquia
Bluefields
Granada
Juigalpa
Rivas
Liberia
San Juan
La Concepción
Lake Nicaragua

Isla de Providencia (Colombia)
Isla de San Andrés (Colombia)
Islas del Maíz (Corn Is) (Nicaragua)

Pta de Perlas

COSTA RICA
San José
Cartago
Puerto Limón
Bocas del Toro
David
Volcán Barú
Santiago
Chitré
Pen. de Osa
Golfo de Chiriquí
Isla de Coiba
Peninsula de Azuero
Punta Mariato
Punta Mala

PANAMA
Panama Canal
Colón
Panama City
Panama
La Chorrera
Gulf of Panama
Golfo de los Mosquitos

Golfo del Darién
Golfo de Urabá
Turbo
Golfo de Morrosquillo
Sincelejo
Montería
Magangué
El Banco
Golfo de Cupica
Cabo Corrientes

Península de la Guajira
Punta Gallinas
Riohacha
Santa Marta
Parque Nacional Sierra Nevada de Santa Marta
Valledupar
Barranquilla
Cartagena

Aruba (Neth.)
Punta Fijo
Coro
Maracaibo
Cabimas
Lake Maracaibo
Machiques
Maicao
San Carlos del Zulia
El Fuerte

Les   **s**   **er**   **Antilles**

Netherlands Antilles
Bonaire
Willemstad
Isla Orchila (Ven.)
Islas Los Roques (Ven.)
Isla La Tortuga (Ven.)
Isla de Margarita (Ven.)
Los Testigos (Ven.)

St George's
GRENADA
Tobago
Port of Spain
Cumaná
Maturín
Orinoco Delta

Caracas
Maiquetía
Los Teques
Maracay
Valencia
Barquisimeto
Acarigua
Barcelona
Zaraza

VENEZUELA
Barinas
Guanare
San Fernando de Apure
Calabozo
Ciudad Guayana
Ciudad Bolívar
El Callao
El Dorado
San Cristóbal
Cúcuta
Bucaramanga
Socorro
Arauca
Puerto Carreño
Puerto Ayacucho

Pakaraima Mountains
La Gran Sabana
Mt Roraima

COLOMBIA
Medellín
Quibdó
Manizales
Pereira
Armenia
Ibagué
Bogotá
Villavicencio
Tunja
Yopal
Buenaventura
Cali
Palmira
Buga
Tuluá
Popayán
Neiva
San José del Guaviare
Puerto Inírida
Mitú
Florencia
Pasto
Ipiales
Tumaco
Esmeraldas
San Lorenzo
Otavalo
Parque Nacional La Macarena
Parque Nacional Cord. de los Picachos

Boa Vista
Serra Parima
Parque Indígena do Yanomami
Uaupés
Barcelos
Manaus

ECUADOR
Quito
Latacunga
Ambato
Riobamba
Portoviejo
Manta
Jipijapa
Guayaquil
Babahoyo
Milagro
Machala
Cuenca
Azogues
Loja
Golfo de Guayaquil
Isla Puná
Salinas
Playas

Vol. Cotopaxi
Chimborazo
Parque Nacional Sangay
Macas
Zamora
Parque Nacional Podocarpus

Iquitos
Leticia
Tabatinga
Benjamim Constant
Caballococha
Nauta
Requena

P **E** **R** **U**
Tumbes
Sullana
Piura
Paita
Catacaos
Chiclayo
Lambayeque
Cajamarca
Chachapoyas
Tarapoto
Trujillo
Chimbote
Casma
Huaraz
Huánuco
Cerro de Pasco
La Merced
Huancayo
Lima
Callao
Cañete
Chincha Alta
Pisco
Ica
Nazca
Ayacucho
Abancay
Cusco (Cuzco)
Juliaca
Arequipa
Moquegua
Tacna
Ilo
Camana
Mollendo

Pucallpa
Cruzeiro do Sul
Porto Velho
Rio Branco
Guajará-Mirim
Riberalta
Cobija
Puerto Maldonado

S E L V A S

B O L I V I A
La Paz
Cochabamba
Oruro
Sucre
Potosí
Santa Cruz
Lake Titicaca

Iquique
Tocopilla
Antofagasta
Calama
Chuquicamata

San Salvador de Jujuy
San Pedro
ARGENTINA

PACIFIC OCEAN

Tropic of Capricorn

GALAPAGOS IS
(Ecuador)
Isla Darwin
Isla Wolf
I. Pinta
I. Marchena
Pta Albemarle
Parque Nacional Galápagos
Vol. Wolf
I. Santiago
I. Santa Cruz
I. Fernandina
I. Isabela
I. San Cristóbal
Puerto Baquerizo Moreno
Cabo Rosa
I. Española
Isla Floreana

Equator
1:15 000 000

Lambert Azimuthal Equal Area Projection

6000 / 19686
5000 / 16404
4000 / 13124
3000 / 9843
2000 / 6562
1000 / 3281
500 / 1640
200 / 656
0 / 0

LAND BELOW SEA LEVEL

200 / 656
2000 / 6562
4000 / 13124
6000 / 19686
M / FT

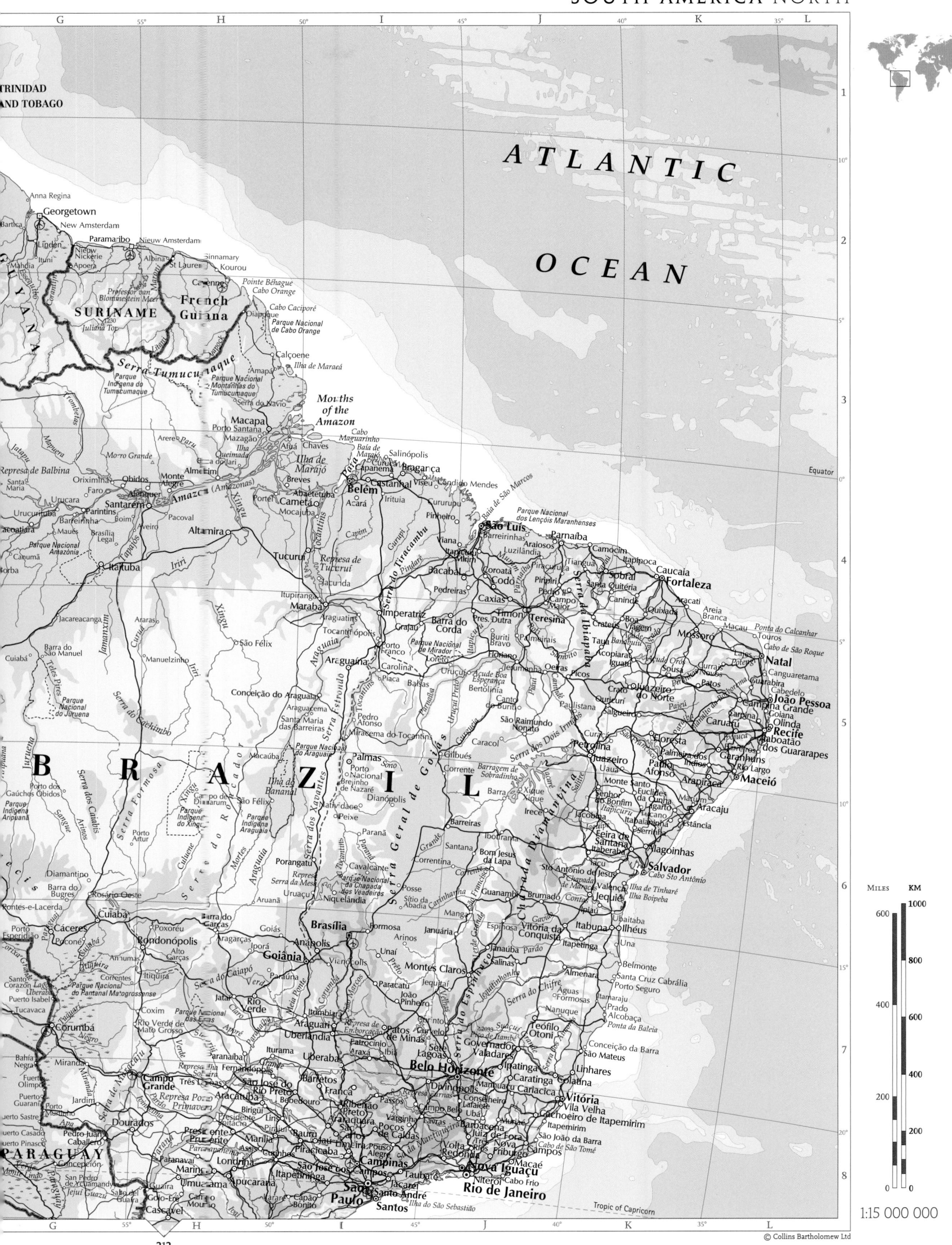

ATLANTIC

OCEAN

TRINIDAD
AND TOBAGO

Georgetown
New Amsterdam
Paramaribo   Nieuw Amsterdam
Nieuw
Nickerie   Albina   St Laurent   Kourou
Apoera   Sinnamary
SURINAME   French
Guiana   Pointe Béhague
Cabo Orange
Parque Nacional
de Cabo Orange

Serra Tumucumaque

Mouths
of the
Amazon

GUYANA

BRAZIL

PARAGUAY

Belém
Macapá

São Luís
Fortaleza
Teresina
Natal
João Pessoa
Recife
Maceió
Aracaju
Salvador

Brasília
Goiânia

Belo Horizonte
Vitória

Rio de Janeiro
São Paulo
Santos

Tropic of Capricorn

MILES   KM
600   1000

800

400   600

400

200
200

0   0

1:15 000 000

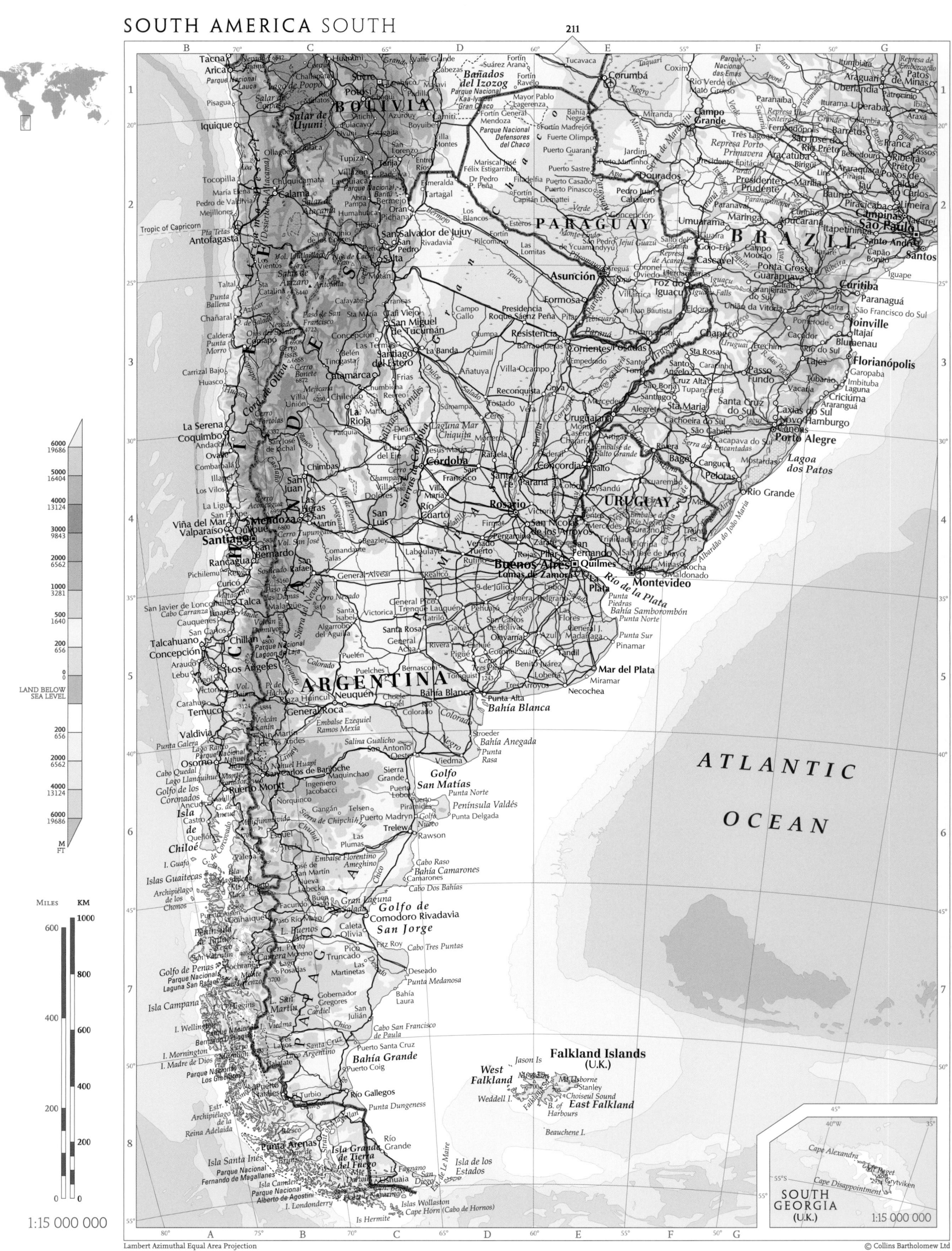

Lambert Azimuthal Equal Area Projection

© Collins Bartholomew Ltd

1:15 000 000

VENEZUELA AND COLOMBIA

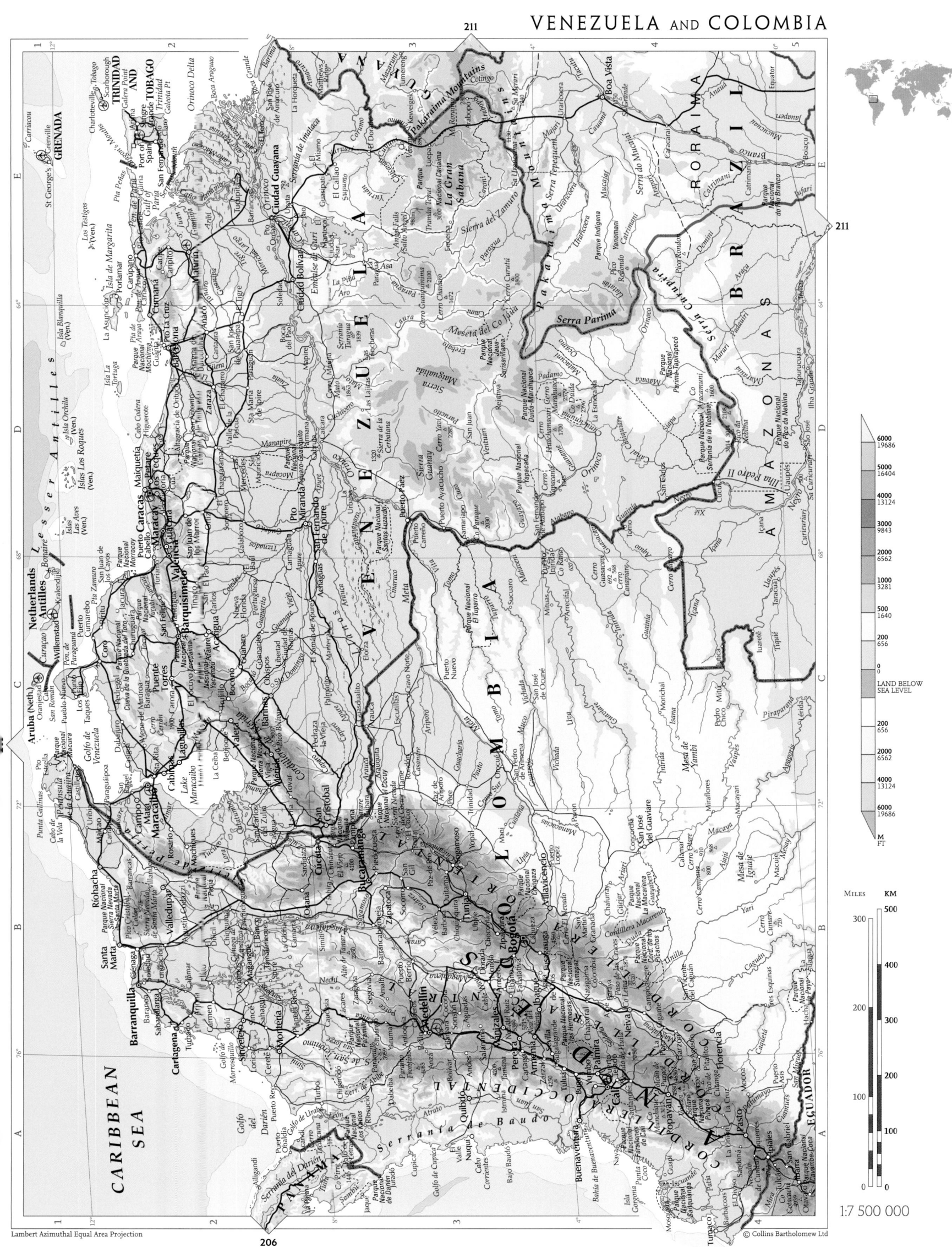

CARIBBEAN
SEA

GRENADA

TRINIDAD
AND
TOBAGO

Netherlands
Antilles

Aruba (Neth.)

VENEZUELA

COLOMBIA

BRAZIL

RORAIMA

AMAZONAS

PANAMA

ECUADOR

	M FT
	6000 19686
	5000 16404
	4000 13124
	3000 9843
	2000 6562
	1000 3281
	500 1640
	200 656
	0
	LAND BELOW SEA LEVEL
	200 656
	2000 6562
	4000 13124
	6000 19686

MILES   KM

300

200

100

0

500

400

300

200

100

0

1:7 500 000

Lambert Azimuthal Equal Area Projection

© Collins Bartholomew Ltd

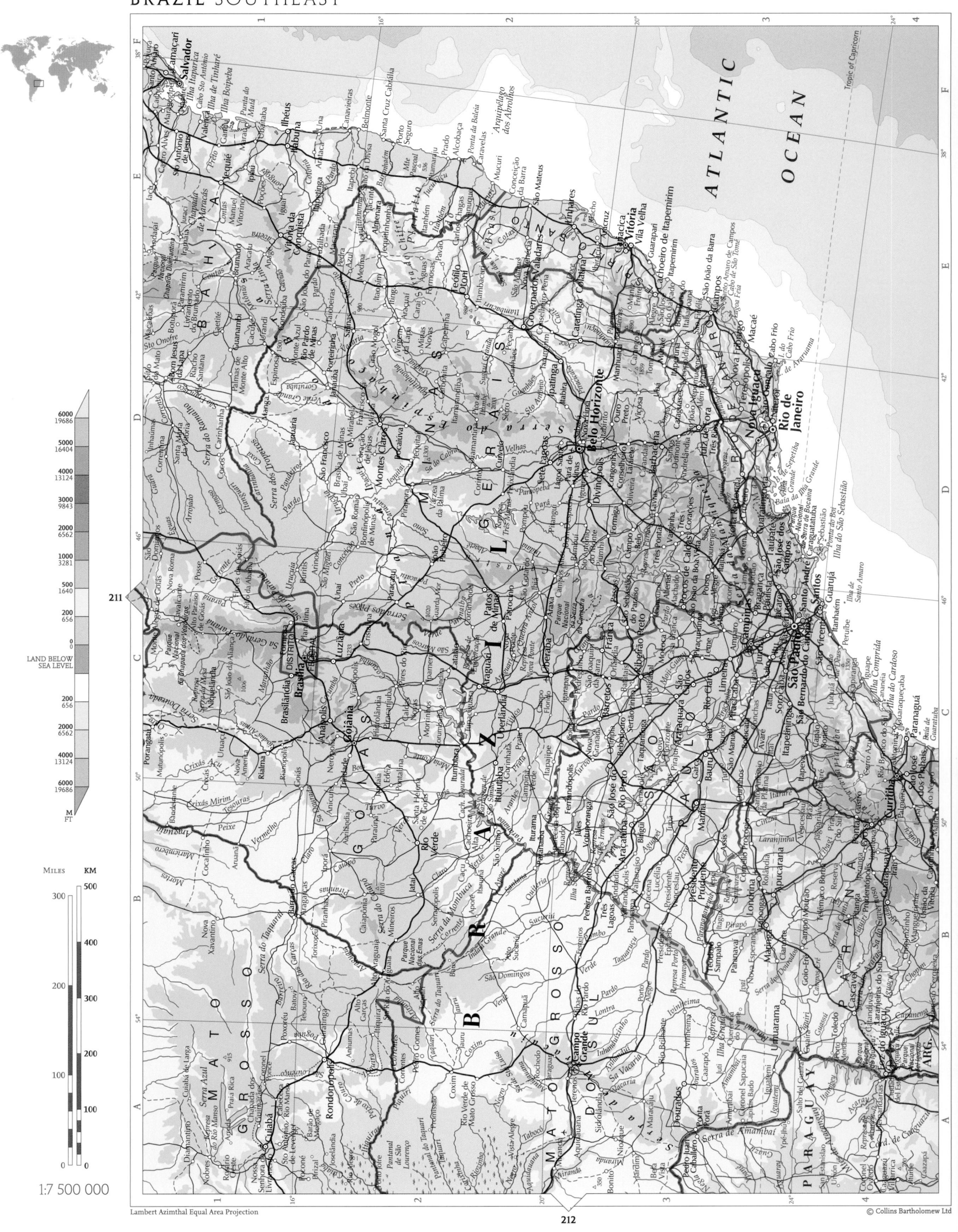

ATLANTIC OCEAN

Tropic of Capricorn

LAND BELOW
SEA LEVEL

MILES   KM

1:7 500 000

Lambert Azimuthal Equal Area Projection

© Collins Bartholomew Ltd

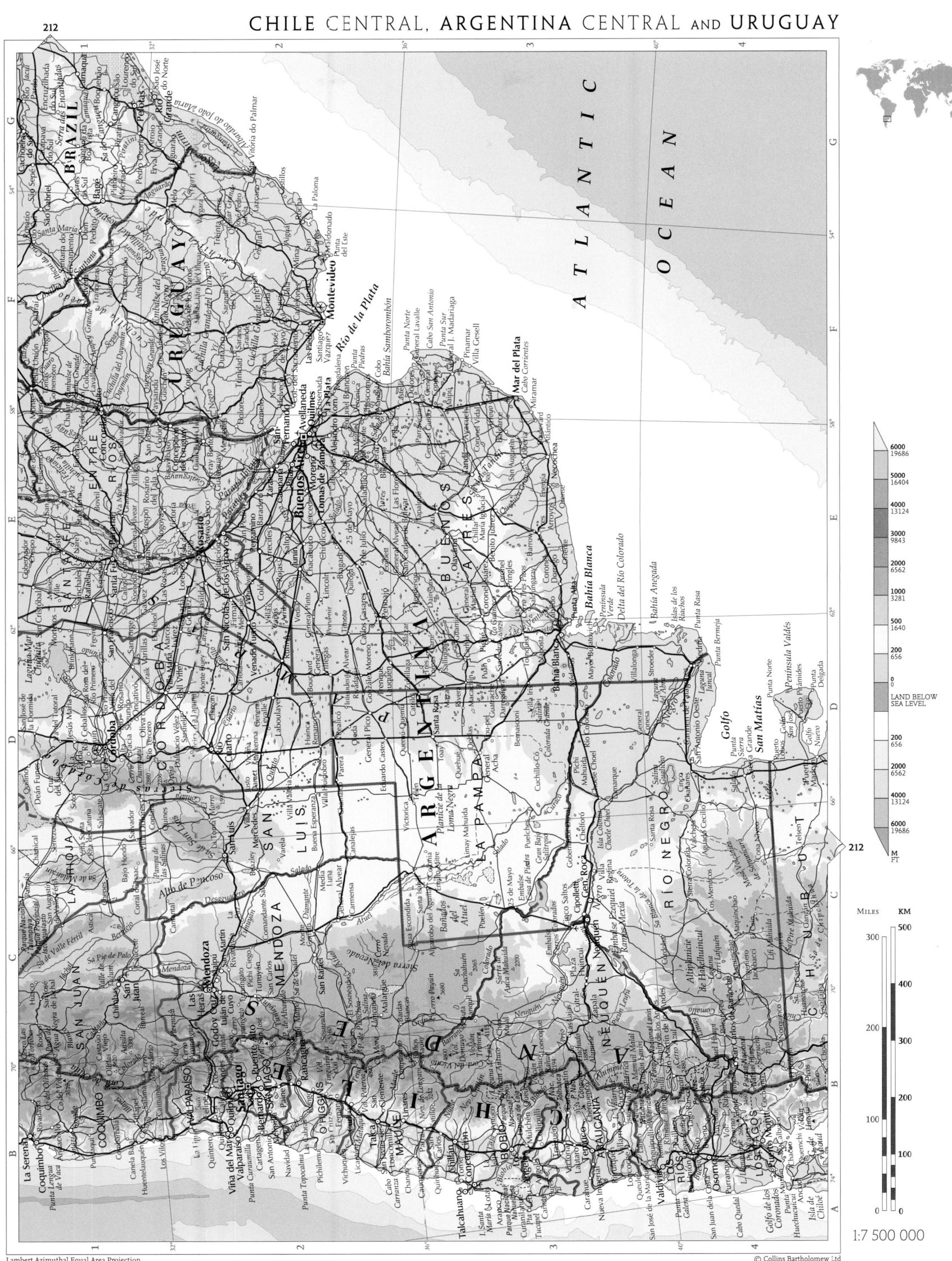

© Collins Bartholomew Ltd

Lambert Azimuthal Equal Area Projection

1:7 500 000

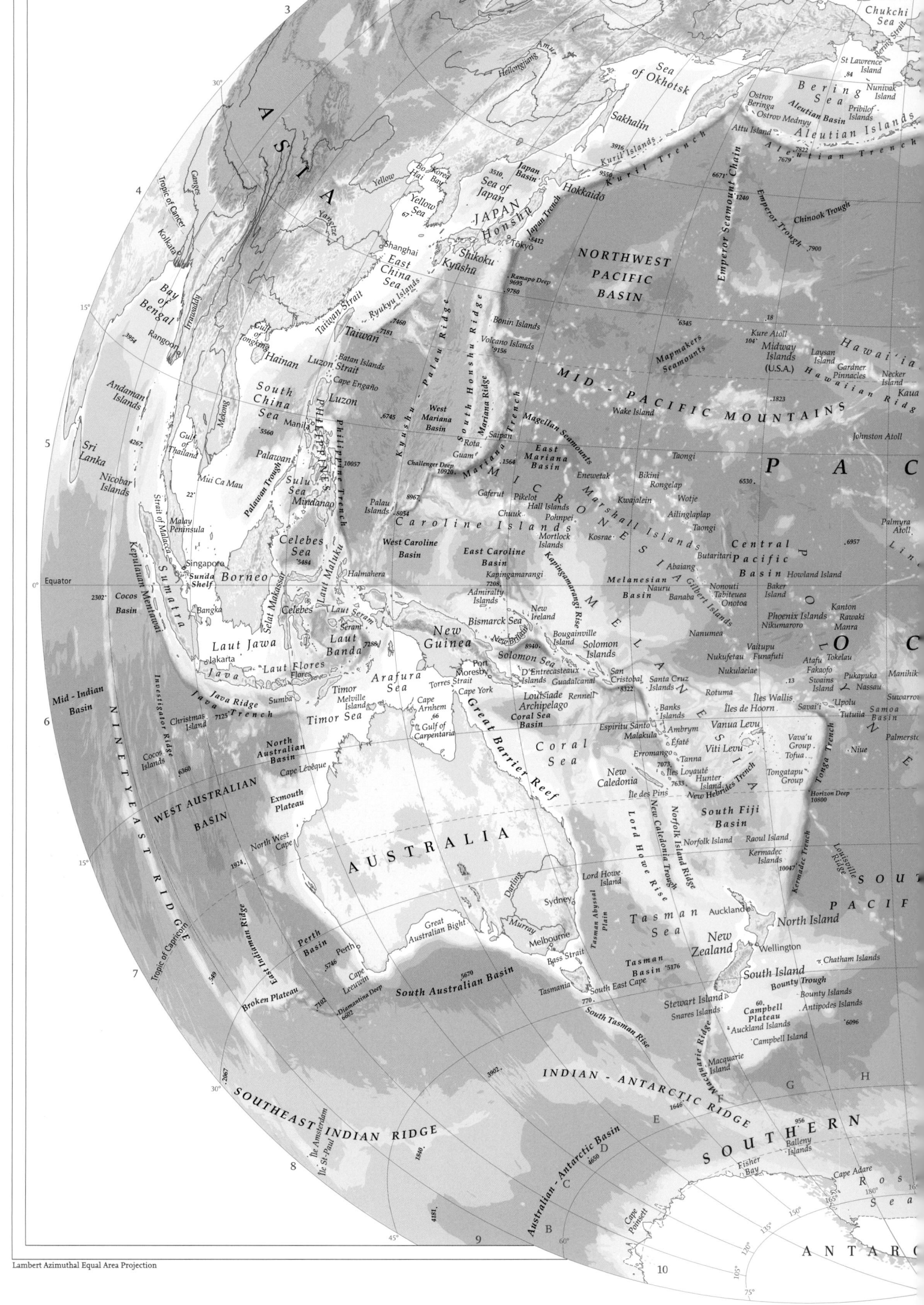

Lambert Azimuthal Equal Area Projection

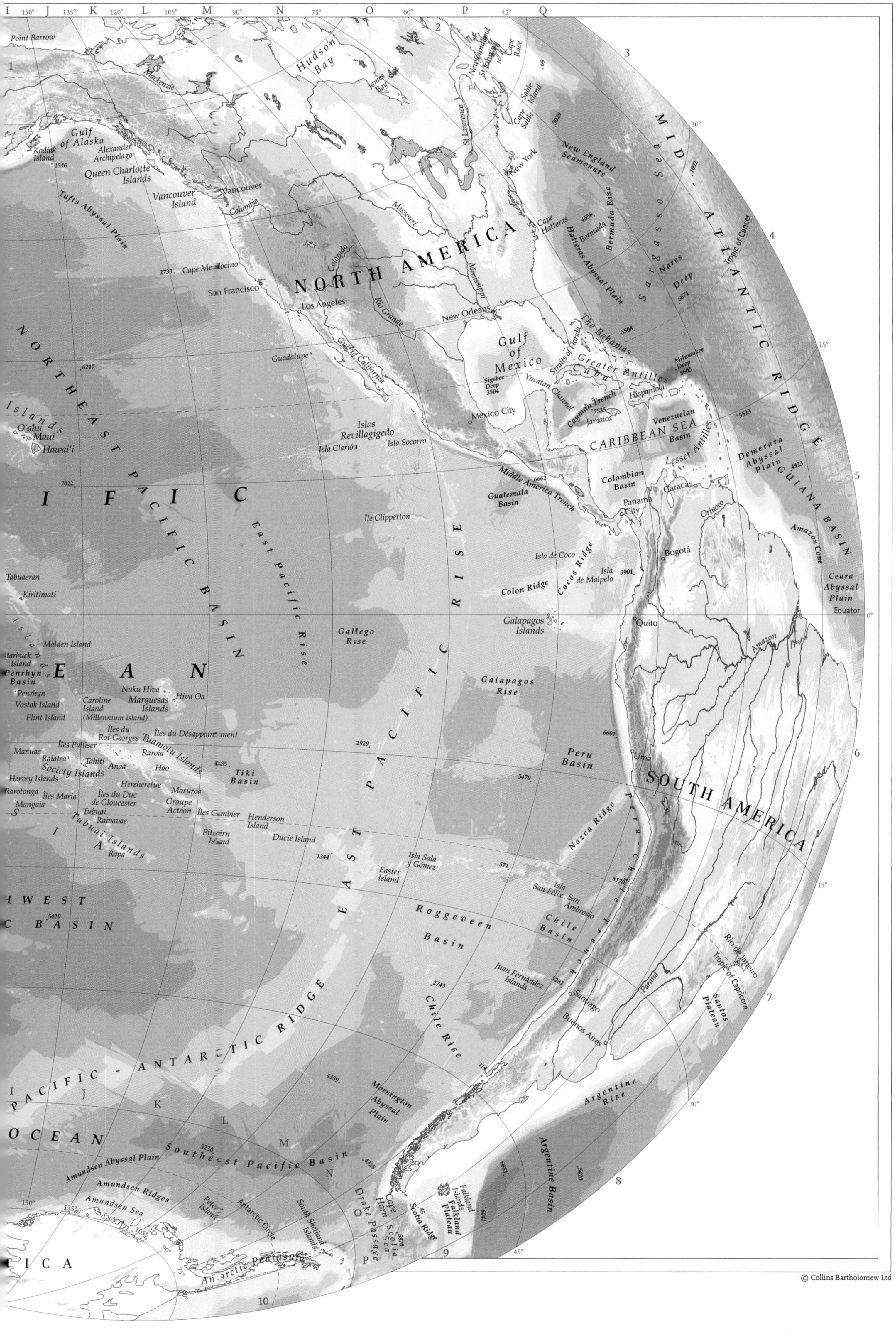

I 150° J 135° K 120° L 105° M 90° N 75° O 60° P 45° Q

Point Barrow

Hudson Bay

Mackenzie

Kodiak Island
Gulf of Alaska
Alexander Archipelago
Queen Charlotte Islands
·1546

Vancouver Island
Vancouver
Columbia

Tufts Abyssal Plain

·2733 Cape Mendocino
San Francisco

NORTH AMERICA
Colorado

St Lawrence
Newfoundland
Cape Race
Sable Island
Cape Sable
New York
Cape Hatteras

New England Seamounts

MID - ATLANTIC RIDGE

Sargasso Sea

Bermuda Rise
·4556 Bermuda
Hatteras Abyssal Plain
Nares Deep ·6677

Tropic of Cancer

Los Angeles
Rio Grande

New Orleans
Gulf of Mexico
Mississippi
·6217

Guadalupe
Gulf of California

Sigsbee Deep 3504
Yucatan Channel
Mexico City

Isles Revillagigedo
Isla Clarión
Isla Socorro

Île Clipperton

The Bahamas
5508
Straits of Florida
Greater Antilles
Cuba
Cayman Trench ·7535
Jamaica
Hispaniola
Milwaukee Deep ·6695

·5523
Venezuelan Basin
CARIBBEAN SEA
Lesser Antilles
Caracas

Demerara Abyssal Plain
·4923
GUIANA BASIN
Amazon Cone

NORTHEAST PACIFIC BASIN

Islands
O'ahu
Maui
Hawai'i
·7022

Tabuaeran
Kiritimati

PACIFIC

East Pacific Rise

Gallego Rise

Middle America Trench ·6662
Guatemala Basin

Isla de Coco
Colon Ridge
Cocos Ridge
Isla de Malpelo
Galapagos Islands

Colombian Basin
Panama City
Bogotá
·3901

Ceara Abyssal Plain

Equator 0°
Quito

OCEAN

Starbuck Island
Malden Island
Penrhyn Basin
Penrhyn
Vostok Island
Flint Island
Caroline Island (Millennium island)
Nuku Hiva
Marquesas Islands
Hiva Oa

Galapagos Rise

Peru Basin
·6601

EAST PACIFIC RISE

Manuae
Raiatea
Tahiti
Society Islands
Hervey Islands
Rarotonga
Îles Maria
Mangaia
Îles Palliser
Anaa
Hao
Hereheretue
Moruroa
Groupe Actéon
Îles Gambier
Tubuai
Raivavae
Rapa

Îles du Roi-Georges
Tuamotu Islands
Raroia
Îles du Désappointment
·85

·1929

Lima
SOUTH AMERICA

Tiki Basin
5470

Îles du Duc de Gloucester
Henderson Island
Pitcairn Island
Ducie Island
·1344

Isla Sala y Gómez
·571
Easter Island
San Félix
Isla San Ambrosio
Chile Basin
·8170

Nazca Ridge

Tubuai Islands

SOUTHWEST PACIFIC BASIN
·5420

Roggeveen Basin

Juan Fernández Islands
·5252
Santiago
Valparaíso
Buenos Aires
Río de Janeiro
Tropic of Capricorn
Santos Plateau

·2743
Chile Rise

Rio Grande

PACIFIC - ANTARCTIC RIDGE

·4359
Mornington Abyssal Plain
·114

Argentine Rise

OCEAN
Amundsen Abyssal Plain
Amundsen Ridges
Amundsen Sea
Peter I Island
Antarctic Circle
South Shetland Islands
South Orkney Sea
Drake Passage
Cape Horn
Scotia Ridge
·45
Falkland Islands / Islas Malvinas
Falkland Plateau
Argentine Basin
·5420
·6041

Southeast Pacific Basin
·5230
·425
·6601

Antarctic Peninsula

© Collins Bartholomew Ltd

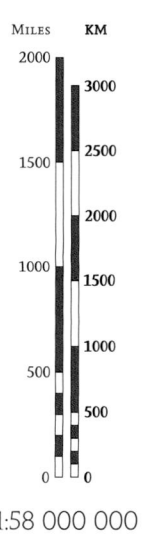

MILES   KM
2000 — 3000
1500 — 2500
        2000
1000 — 1500
        1000
500 — 500
0 —   0

1:58 000 000

217

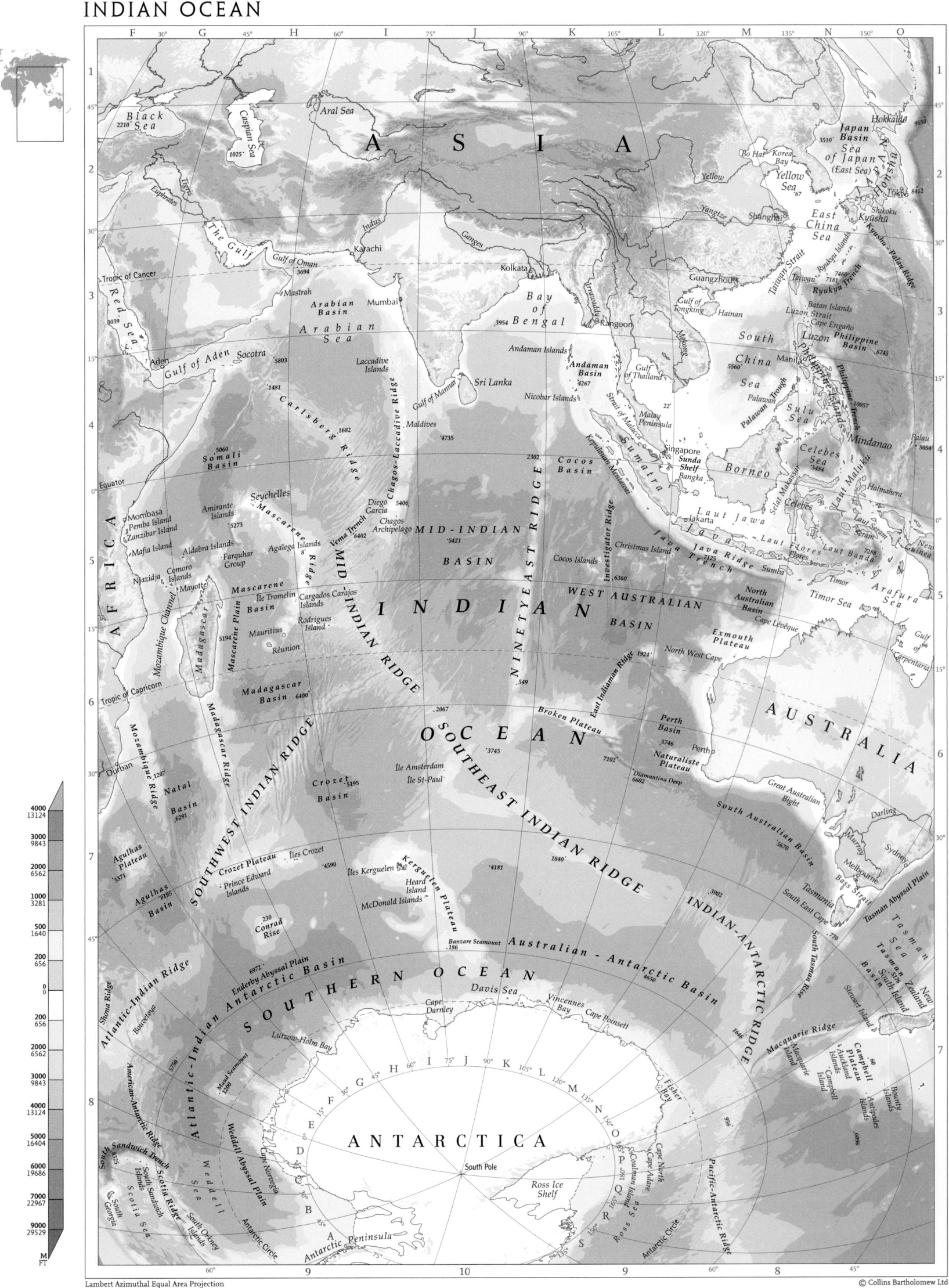

Lambert Azimuthal Equal Area Projection

© Collins Bartholomew Ltd

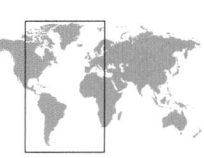

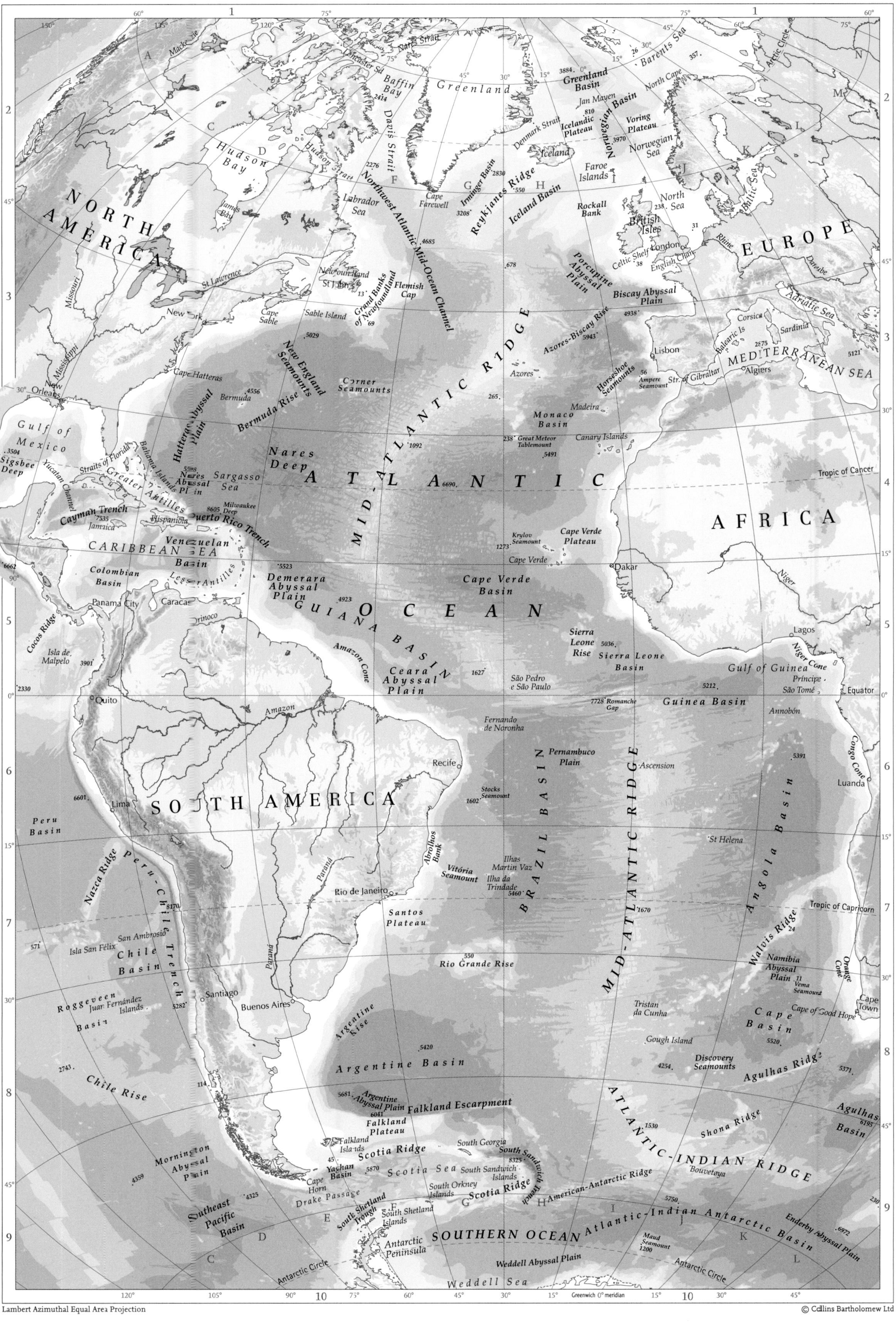

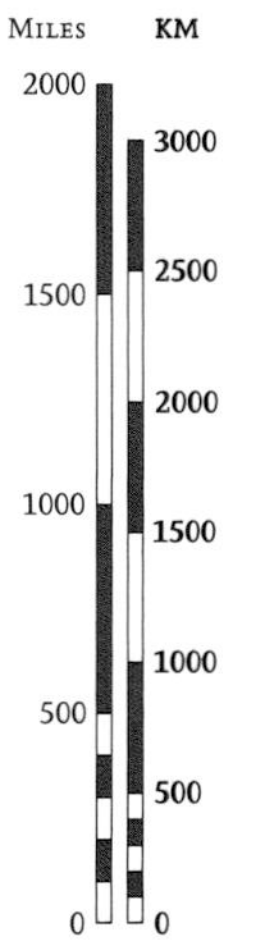

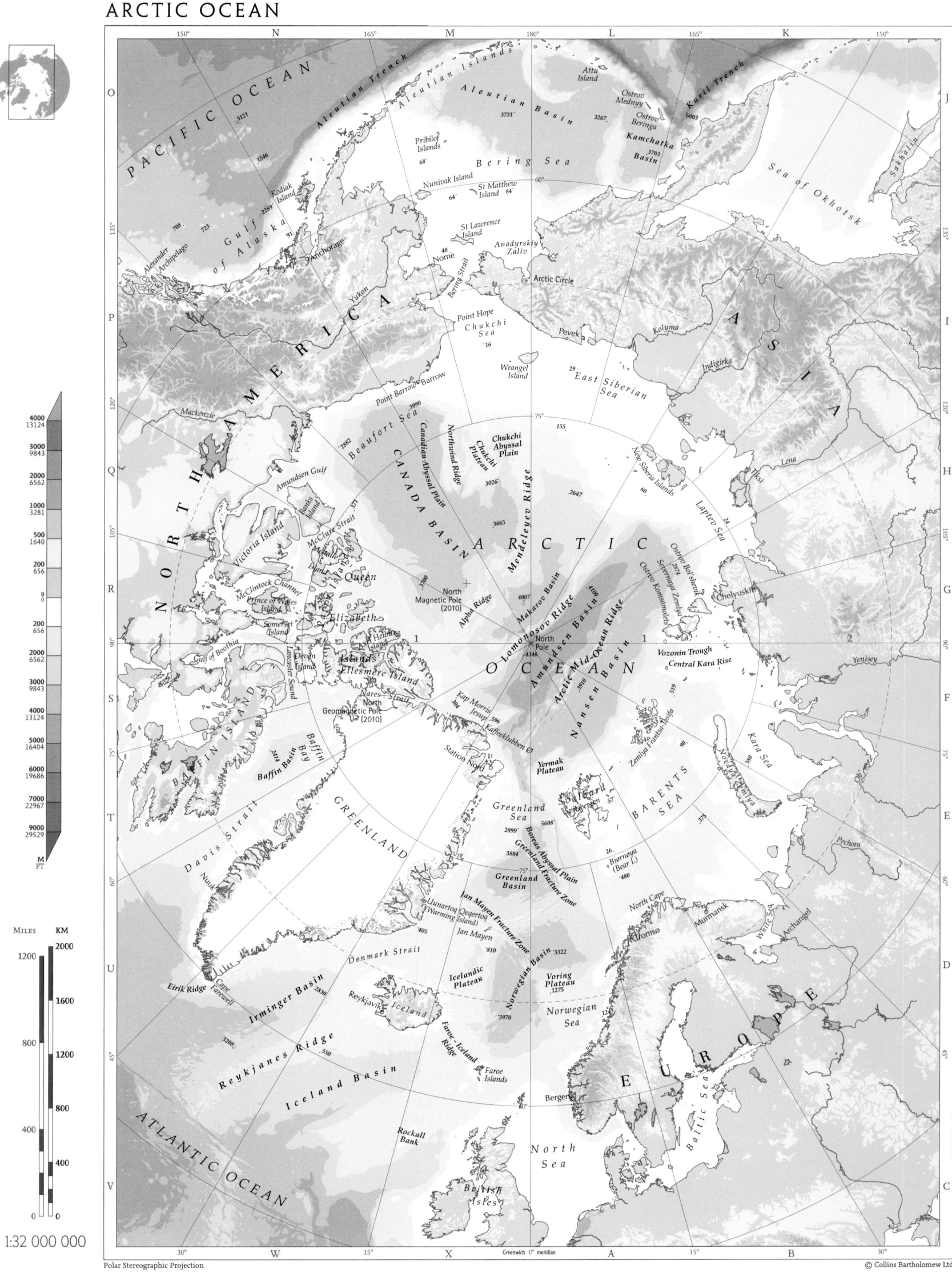

PACIFIC OCEAN

Aleutian Trench

Aleutian Islands

Attu Island

Ostrov Mednyy

Kuril Trench

6003

Aleutian Basin

3731'

3267'

Ostrov Beringa

3703

Kamchatka Basin

Sea of Okhotsk

Sakhalin

5121'

1526'

Pribilof Islands

68'

Bering Sea

706'

723'

Kodiak Island

2289'

Gulf of Alaska

31'

Anchorage

Nunivak Island

64'

St Matthew Island

84'

60°

St Lawrence Island

Nome

40'

Anadyrskiy Zaliv

Bering Strait

Arctic Circle

Yukon

ASIA

NORTH AMERICA

Point Hope

Chukchi Sea

16'

Pevek

Kolyma

Indigirka

Mackenzie

Point Barrow Barrow

3990

Wrangel Island

29'

East Siberian Sea

155'

Lena

120°

135°

105°

Beaufort Sea

Canadian Abyssal Plain

Northwind Ridge

Chukchi Plateau

Chukchi Abyssal Plain

Nov Siberia Islands

60'

Laptev Sea

2882'

3026'

2647'

Amundsen Gulf

Banks Island

CANADA BASIN

371'

3665'

ARCTIC

Mendeleyev Ridge

2574

Ostrov Bol'shevik

Severnaya Zemlya

Ostrov Komsomolets

Chelyuskin

90°

McClure Strait

Melville Island

3590

North Magnetic Pole (2010)

4007

Makarov Basin

4100

Arctic Mid-Ocean Ridge

Amundsen Basin

Vozonin Trough

Central Kara Rise

Yenisey

Victoria Island

McClintock Channel

Prince of Wales Island

Queen

Alpha Ridge

OCEAN

519

Nansen Basin

Somerset Island

Elizabeth

1

North Pole

4346

1

Kara Sea

Gulf of Boothia

Devon Island

Axel Heiberg Island

Lomonosov Ridge

3910

90

2

Prince Patrick Island

Ellesmere Island

Nares Strait

Kap Morris Jesup

596

North Geomagnetic Pole (2010)

304

Kaffeeklubben Ø

Zemlya Frantsa-Iosifa

Novaya Zemlya

350

BAFFIN ISLAND

Lancaster Sound

Station Nord

Yermak Plateau

57

BARENTS SEA

Baffin Bay

2414

Baffin Basin

Greenland Sea

Svalbard

Hinlopen

Ob'

Davis Strait

GREENLAND

5608

2899'

Boreas Abyssal Plain

Greenland Fracture Zone

26'

Bjørnøya (Bear I.)

375

Pechora

3884'

Greenland Basin

480'

North Cape

White Sea

Archangel

Nuuk

Ulunartoq Qeqertoq ('Warming Island')

Jan Mayen Fracture Zone

Tromsø

Murmansk

EUROPE

405'

Jan Mayen

810'

Norwegian Basin

3322

Vøring Plateau

1275

Denmark Strait

Eirik Ridge

Cape Farewell

2830'

Icelandic Plateau

Norwegian Sea

126'

Baltic Sea

Reykjavik

Iceland

3970'

Bergen

Irminger Basin

Reykjanes Ridge

Faroe-Iceland Ridge

550'

Faroe Islands

North Sea

3208'

Iceland Basin

Rockall Bank

ATLANTIC OCEAN

British Isles

Polar Stereographic Projection

© Collins Bartholomew Ltd

4000 13124
3000 9843
2000 6562
1000 3281
500 1640
200 656
0
200 656
2000 6562
3000 9843
4000 13124
5000 16404
6000 19686
7000 22967
9000 29529

M FT

MILES   KM
1200   2000
        1600
800   1200
        800
400   400
0   0

1:32 000 000

The index includes all names shown on the reference maps in the atlas. Each entry includes the country or geographical area in which the feature is located, a page number and an alphanumeric reference. Additional entry details and aspects of the index are explained below.

REFERENCING

Names are referenced by page number and by grid reference. The grid reference relates to the alphanumeric values which appear in the margin of each map. These reflect the graticule on the map – the letter relates to longitude divisions, the number to latitude divisions.

Names are generally referenced to the largest scale map page on which they appear. For large geographical features, including countries, the reference is to the largest scale map on which the feature appears in its entirety, or on which the majority of it appears.

Rivers are referenced to their lowest downstream point – either their mouth or their confluence with another river. The river name will generally be positioned as close to this point as possible.

Entries relating to names appearing on insets are indicated by a small box symbol: □ followed by a grid reference if the inset has its own alphanumeric values.

ALTERNATIVE NAMES

Alternative names appear as cross-references and refer the user to the index entry for the form of the name used on the map.

For rivers with multiple names – for example those which flow through several countries – all alternative name forms are included within the main index entries, with details of the countries in which each form applies.

ADMINISTRATIVE QUALIFIERS

Administrative divisions are included in an entry to differentiate duplicate names – entries of exactly the same name and feature type within the one country – where these division names are shown on the maps. In such cases, duplicate names are alphabetized in the order of the administrative division names. Additional qualifiers are included for names within selected geographical areas, to indicate more clearly their location.

DESCRIPTORS

Entries, other than those for towns and cities, include a descriptor indicating the type of geographical feature. Descriptors are not included where the type of feature is implicit in the name itself, unless there is a town or city of exactly the same name.

NAME FORMS AND ALPHABETICAL ORDER

Name forms are as they appear on the maps, with additional alternative forms included as cross-references. Names appear in full in the index, although they may appear in abbreviated form on the maps.

The German character ß is alphabetized as 'ss'. Names beginning with Mac or Mc are alphabetized exactly as they appear. The terms Saint, Sainte, etc, are abbreviated to St, Ste, etc, but alphabetized as if in the full form.

NUMERICAL ENTRIES

Entries beginning with numerals appear at the beginning of the index, in numerical order. Elsewhere, numerals are alphabetized before 'a'.

PERMUTED TERMS

Names beginning with generic, geographical terms are permuted – the descriptive term is placed after, and the index alphabetized by, the main part of the name. For example, Lake Superior is indexed as Superior, Lake; Mount Everest as Everest, Mount. This policy is applied to all languages. Permuting has not been applied to names of towns, cities or administrative divisions beginning with such geographical terms. These remain in their full form, for example, Lake Isabella, USA.

INDEX ABBREVIATIONS

admin. dist.	administrative district	IN	Indiana	Phil.	Philippines
admin. div.	administrative division	Indon.	Indonesia	plat.	plateau
admin. reg.	administrative region	is	islands	P.N.G.	Papua New Guinea
Afgh.	Afghanistan	Kazakh.	Kazakhstan	Port.	Portugal
AK	Alaska	KS	Kansas	prov.	province
AL	Alabama	KY	Kentucky	pt	point
Alg.	Algeria	Kyrg.	Kyrgyzstan	Qld	Queensland
Alta	Alberta	l.	lake	Que.	Québec
AR	Arkansas	LA	Louisiana	r.	river
Arg.	Argentina	lag.	lagoon	reg.	region
aut. comm.	autonomous community	Lith.	Lithuania	res.	reserve
aut. reg.	autonomous region	Lux.	Luxembourg	resr	reservoir
aut. rep.	autonomous republic	MA	Massachusetts	RI	Rhode Island
AZ	Arizona	Madag.	Madagascar	r. mouth	river mouth
Azer.	Azerbaijan	Man.	Manitoba	Rus. Fed.	Russian Federation
b.	bay	MD	Maryland	S.	South
Bangl.	Bangladesh	ME	Maine	S.A.	South Australia
B.C.	British Columbia	Mex.	Mexico	S. Africa	Republic of South Africa
Bol.	Bolivia	MI	Michigan	salt l.	salt lake
Bos.-Herz.	Bosnia-Herzegovina	MN	Minnesota	Sask.	Saskatchewan
Bulg.	Bulgaria	MO	Missouri	SC	South Carolina
c.	cape	Mont.	Montana	SD	South Dakota
CA	California	Moz.	Mozambique	sea chan.	sea channel
Cent. Afr. Rep.	Central African Republic	MS	Mississippi	Sing.	Singapore
CO	Colorado	MT	Montana	Switz.	Switzerland
Col.	Colombia	mt.	mountain	Tajik.	Tajikistan
CT	Connecticut	mts	mountains	Tanz.	Tanzania
Czech Rep.	Czech Republic	N.	North, Northern	Tas.	Tasmania
DC	District of Columbia	nat. park	national park	terr.	territory
DE	Delaware	N.B.	New Brunswick	Thai.	Thailand
Dem. Rep. Congo	Democratic Republic of the Congo	NC	North Carolina	TN	Tennessee
depr.	depression	ND	North Dakota	Trin. and Tob.	Trinidad and Tobago
des.	desert	NE	Nebraska	Turkm.	Turkmenistan
Dom. Rep.	Dominican Republic	Neth.	Netherlands	TX	Texas
Equat. Guinea	Equatorial Guinea	Neth. Antilles	Netherlands Antilles	U.A.E.	United Arab Emirates
esc.	escarpment	Nfld.	Newfoundland	U.K.	United Kingdom
est.	estuary	NH	New Hampshire	Ukr.	Ukraine
Eth.	Ethiopia	NJ	New Jersey	U.S.A.	United States of America
Fin.	Finland	NM	New Mexico	UT	Utah
FL	Florida	N.S.	Nova Scotia	Uzbek.	Uzbekistan
for.	forest	N.S.W.	New South Wales	VA	Virginia
Fr. Guiana	French Guiana	N.W.T.	Northwest Territories	Venez.	Venezuela
Fr. Polynesia	French Polynesia	N.Z.	New Zealand	Vic.	Victoria
g.	gulf	NV	Nevada	vol.	volcano
GA	Georgia	NY	New York	vol. crater	volcanic crater
Guat.	Guatemala	OH	Ohio	VT	Vermont
h.	hill	OK	Oklahoma	W.	Western
hd	headland	Ont.	Ontario	WA	Washington
HI	Hawaii	OR	Oregon	W.A.	Western Australia
Hond.	Honduras	PA	Pennsylvania	WI	Wisconsin
i.	island	Pak.	Pakistan	WV	West Virginia
IA	Iowa	Para.	Paraguay	WY	Wyoming
ID	Idaho	P.E.I.	Prince Edward Island	Y.T.	Yukon
IL	Illinois	pen.	peninsula		

9 de Julio

1

215 E2 9 de Julio Arg.
215 E2 25 de Mayo Arg.
215 C3 25 de Mayo Arg.
186 E4 100 Mile House Canada

A

159 J5 Aabenraa Denmark
164 E4 Aachen Germany
159 J4 Aalborg Denmark
159 J4 Aalborg Bugt b. Denmark
168 E6 Aalen Germany
164 C4 Aalst Belgium
159 J4 Aars Denmark
164 C4 Aarschot Belgium
148 A3 Aba China
178 D3 Aba Dem. Rep. Congo
176 C4 Aba Nigeria
140 B5 Abā ad Dūd Saudi Arabia
140 C4 Ābādān Iran
138 D5 Abadan Turkm.
140 D4 Ābādeh Iran
140 D4 Ābādeh Ţashk Iran
176 B1 Abadla Alg.
214 D2 Abaeté r. Brazil
211 I4 Abaetetuba Brazil
216 G6 Abaiang atoll Kiribati
195 E4 Abajo Peak U.S.A.
176 C4 Abakaliki Nigeria
146 B1 Abakan Rus. Fed.
146 A1 Abakanskiy Khrebet mts Rus. Fed.
173 E7 Abana Turkey
210 D6 Abancay Peru
140 D4 Abarkūh, Kavīr-e des. Iran
140 D4 Abarqū Iran
150 I2 Abashiri Japan
150 I2 Abashiri-wan b. Japan
124 E3 Abau P.N.G.
139 H2 Abay Kazakh.
178 D3 Ābaya Hāyk' Eth.
Ābaya Hāyk' l. Eth. see Ābaya, Lake
178 D3 Ābaya, Lake l. Eth.
177 F3 Ābay Wenz r. Eth. alt. Azraq, Bahr el (Sudan), conv. Blue Nile
132 K4 Abaza Rus. Fed.
141 E3 'Abbāsābād Iran
170 C4 Abbasanta Sardinia Italy
190 C2 Abbaye, Point U.S.A.
178 E2 Abbe, Lake Eth.
166 E1 Abbeville France
199 E6 Abbeville LA U.S.A.
201 D5 Abbeville SC U.S.A.
163 B5 Abbeyfeale Ireland
162 E6 Abbey Head U.K.
163 D5 Abbeyleix Ireland
160 D3 Abbeytown U.K.
158 L2 Abborrträsk Sweden
129 B3 Abbot Ice Shelf ice feature Antarctica
186 E5 Abbotsford Canada
190 B3 Abbotsford U.S.A.
195 F4 Abbott U.S.A.
144 C2 Abbottabad Pak.
137 H3 'Abd al 'Azīz, Jabal h. Syria
137 K5 Ābdānān Iran
138 C1 Abdulino Rus. Fed.
177 E3 Abéché Chad
128 D4 Abel Tasman National Park N.Z.
176 B4 Abengourou Côte d'Ivoire
Åbenrå Denmark see Aabenraa
165 J6 Abensberg Germany
176 C4 Abeokuta Nigeria
161 C5 Aberaeron U.K.
162 F3 Aberchirder U.K.
127 H5 Abercrombie r. Australia
161 D6 Aberdare U.K.
161 C5 Aberdaron U.K.
127 I4 Aberdeen Australia
187 H4 Aberdeen Canada
149 □ Aberdeen Hong Kong China
180 F6 Aberdeen S. Africa
162 F3 Aberdeen U.K.
203 E5 Aberdeen MD U.S.A.
199 F5 Aberdeen MS U.S.A.
198 D2 Aberdeen SD U.S.A.
194 B2 Aberdeen WA U.S.A.
187 I2 Aberdeen Lake Canada
161 C5 Aberdovey U.K.
Aberdyfi U.K. see Aberdovey
162 E4 Aberfeldy U.K.
160 F4 Aberford U.K.
162 D4 Aberfoyle U.K.
161 D6 Abergavenny U.K.
199 C5 Abernathy U.S.A.
161 C5 Aberporth U.K.
161 C5 Abersoch U.K.
161 C5 Aberystwyth U.K.
142 B6 Abhā Saudi Arabia
140 C2 Abhar Iran
140 C2 Abhar Rūd r. Iran
Ab-i Bazuft r. Iran see Bāzoft, Āb-e
213 A2 Abibe, Serranía de mts Col.
176 B4 Abidjan Côte d'Ivoire
Ab-i-Istada l. Afgh. see Īstādeh-ye Moqor, Āb-e
178 D3 Abijatta-Shalla National Park Eth.
140 E3 Ab-i-Kavīr salt flat Iran
198 D4 Abilene KS U.S.A.
199 D5 Abilene TX U.S.A.
161 F6 Abingdon U.K.
190 B5 Abingdon IL U.S.A.
202 C6 Abingdon VA U.S.A.
173 F6 Abinsk Rus. Fed.
Ab-i-Safed r. Afgh. see Safīd, Daryā-ye
Abiseo, Parque Nacional nat. park Peru see Río Abiseo, Parque Nacional
187 H2 Abitau Lake Canada
188 D4 Abitibi r. Canada
188 E4 Abitibi, Lake Canada
144 C3 Abohar India
176 B4 Aboisso Côte d'Ivoire
176 C4 Abomey Benin
177 D4 Abong Mbang Cameroon
153 A4 Aborlan Phil.
177 D3 Abou Déia Chad
137 J1 Abovyan Armenia
162 F3 Aboyne U.K.
142 C4 Abqaiq Saudi Arabia
Abra, Lago del l. Arg. see Abra, Laguna del
215 D4 Abra, Laguna del l. Arg.
167 B3 Abrantes Port.
212 C2 Abra Pampa Arg.
206 A2 Abreojos, Punta pt Mex.
214 E2 Abrolhos, Arquipélago dos is Brazil
219 G7 Abrolhos Bank sea feature S. Atlantic Ocean
194 E2 Absaroka Range mts U.S.A.
165 H6 Abtsgmünd Germany
140 D5 Abū al Abyaḑ i. U.A.E.
140 C5 Abū 'Alī i. Saudi Arabia
Abu al Jirab i. U.A.E. see Abū al Abyaḑ
142 B6 Abū 'Arīsh Saudi Arabia
142 D5 Abu Dhabi U.A.E.
177 F3 Abu Hamed Sudan
176 C4 Abuja Nigeria
137 H4 Abū Kamāl Syria
177 E3 Abu Matariq Sudan
136 B6 Abū Mīnā tourist site Egypt
140 D5 Abu Musa i. U.A.E.
210 E6 Abunã r. Bol.
210 E5 Abunã Brazil
142 A7 Ābune Yosēf mt. Eth.
136 C6 Abū Qīr, Khalīj b. Egypt
135 F4 Abu Road India
Abu Simbel Egypt see Abū Sunbul
137 J6 Abū Şukhayr Iraq
177 F2 Abū Sunbul Egypt
128 C5 Abut Head N.Z.
153 C4 Abuyog Phil.
177 E3 Abu Zabad Sudan
Abū Ȥabī U.A.E. see Abu Dhabi
137 L6 Abūzam Iran
177 E3 Abyad Sudan
177 E4 Abyei Sudan
140 C2 Ābyek Iran
138 D1 Abzakovo Rus. Fed.
138 D2 Abzanovo Rus. Fed.
203 I2 Acadia National Park U.S.A.
206 D3 Acambaro Mex.
207 G3 Acancéh Mex.
213 A2 Acandí Col.
167 B1 A Cañiza Spain
206 C3 Acaponeta Mex.
207 E4 Acapulco Mex.
211 I4 Acará Brazil
211 J4 Acaraú r. Brazil
214 A4 Acaray r. Para.
212 E3 Acaray, Represa de resr Para.
213 C2 Acarigua Venez.
207 E4 Acatlán Mex.
207 E4 Acatzingo Mex.
207 F4 Acayucán Mex.
176 B4 Accra Ghana
160 E4 Accrington U.K.
213 C3 Achaguas Venez.
144 D5 Achalpur India
143 B2 Achampet India
133 S3 Achayvayam Rus. Fed.
152 D1 Acheng China
164 A4 Achicourt France
163 B4 Achill Ireland
163 A4 Achill Island Ireland
162 C2 Achiltibuie U.K.
165 H1 Achim Germany
146 B1 Achinsk Rus. Fed.
162 C3 Achnasheen U.K.
162 C3 A'Chralaig mt. U.K.
173 F6 Achuyevo Rus. Fed.
136 B3 Acıgöl l. Turkey
136 B3 Acıpayam Turkey
170 F6 Acireale Sicily Italy
198 E3 Ackley U.S.A.
205 J4 Acklins Island Bahamas
161 I5 Acle U.K.
215 B2 Aconcagua r. Chile
215 B2 Aconcagua, Cerro mt. Arg.
211 K5 Acopiara Brazil
156 A6 Açores, Arquipélago dos is N. Atlantic Ocean
167 B1 A Coruña Spain
206 H6 Acoyapa Nicaragua
170 C2 Acqui Terme Italy
126 A4 Acraman, Lake salt flat Australia
Acre Israel see 'Akko
170 G5 Acri Italy
168 I7 Ács Hungary
123 J6 Actéon, Groupe is Fr. Polynesia
202 B4 Ada OH U.S.A.
199 D5 Ada OK U.S.A.
167 D2 Adaja r. Spain
212 E8 Adam, Mount h. Falkland Is
127 H6 Adaminaby Australia
138 D2 Adamovka Rus. Fed.
203 G3 Adams MA U.S.A.
190 C4 Adams WI U.S.A.
194 B2 Adams, Mount U.S.A.
143 B4 Adam's Bridge sea feature India/Sri Lanka
186 F4 Adams Lake Canada
197 E2 Adams McGill Reservoir U.S.A.
186 C3 Adams Mountain U.S.A.
196 B2 Adams Peak U.S.A.
'Adan Yemen see Aden
136 E3 Adana Turkey
136 C1 Adapazarı Turkey
163 C5 Adare Ireland
129 B6 Adare, Cape c. Antarctica
197 E2 Adaven U.S.A.
137 J5 Ad Daghghārah Iraq
142 C5 Ad Dahnā' des. Saudi Arabia
176 A2 Ad Dakhla W. Sahara
Ad Dammām Saudi Arabia see Dammam
140 B5 Ad Dawādimī Saudi Arabia
Ad Dawḩah Qatar see Ad Dawḩah
140 C5 Ad Dawḩah Qatar
137 I4 Ad Dawr Iraq
137 K6 Ad Dayr Iraq
140 B5 Ad Dibdibah plain Saudi Arabia
140 B6 Ad Dilam Saudi Arabia
142 C5 Ad Dir'īyah Saudi Arabia
178 D3 Addis Ababa Eth.
203 J2 Addison U.S.A.
137 J6 Ad Dīwānīyah Iraq
161 G6 Addlestone U.K.
181 F6 Addo Elephant National Park S. Africa
137 I6 Ad Duwayd well Saudi Arabia
201 D6 Adel GA U.S.A.
198 E3 Adel IA U.S.A.
126 C5 Adelaide Australia
201 E7 Adelaide Bahamas
181 G6 Adelaide S. Africa
129 B2 Adelaide Island i. Antarctica
124 D3 Adelaide River Australia
196 D4 Adelanto U.S.A.
129 C6 Adélie Land reg. Antarctica
129 C7 Adélie Land reg. Antarctica
127 H5 Adelong Australia
139 G4 Adelunga Uzbek.
142 C7 Aden Yemen
142 C7 Aden, Gulf of Somalia/Yemen
164 E4 Adenau Germany
165 I1 Adendorf Germany
140 D5 Adh Dhayd U.A.E.
147 F7 Adi i. Indon.
178 D2 Ādī Ārk'ay Eth.
178 D2 Ādīgrat Eth.
144 D6 Adilabad India
137 I2 Adilcevaz Turkey
194 B3 Adin U.S.A.
177 D2 Adīrī Libya
203 F2 Adirondack Mountains U.S.A.
Ādīs Ābeba Eth. see Addis Ababa
178 D3 Ādīs Alem Eth.
136 G3 Adıyaman Turkey
169 M7 Adjud Romania
207 E3 Adjuntas, Presa de las resr Mex.
189 I3 Adlavik Islands Canada
124 C3 Admiralty Gulf Australia
185 J2 Admiralty Inlet Canada
186 C3 Admiralty Island U.S.A.
186 C3 Admiralty Island National Monument-Kootznoowoo Wilderness nat. park U.S.A.
124 E2 Admiralty Islands P.N.G.
143 B3 Adoni India
165 K4 Adorf Germany
165 G3 Adorf (Diemelsee) Germany
166 D5 Adour r. France
167 E4 Adra Spain
170 F6 Adrano Sicily Italy
176 B2 Adrar Alg.
176 A2 Adrar mts Alg.
176 A2 Adrar, Ḑhar hills Mauritania
141 F3 Adraskand r. Afgh.
139 G4 Adrasmon Tajik.
177 E3 Adré Chad
191 E5 Adrian MI U.S.A.
199 C5 Adrian TX U.S.A.
170 E2 Adriatic Sea Europe
143 B4 Adur India
178 C3 Adusa Dem. Rep. Congo
178 D2 Ādwa Eth.
133 O3 Adycha r. Rus. Fed.
173 F6 Adygeya, Respublika aut. rep. Rus. Fed.
173 F6 Adygeysk Rus. Fed.
173 H6 Adyk Rus. Fed.
176 B4 Adzopé Côte d'Ivoire
171 K5 Aegean Sea Greece/Turkey
165 H2 Aerzen Germany
167 B1 A Estrada Spain
178 D2 Afabet Eritrea
137 J3 Āfān Iran
141 F3 Afghanistan country Asia
158 J3 Åfjord Norway
178 E3 Afmadow Somalia
184 C4 Afognak Island U.S.A.
167 C1 A Fonsagrada Spain
174 Africa
136 F3 'Afrīn, Nahr r. Syria/Turkey
136 F2 Afşin Turkey
164 D2 Afsluitdijk barrage Neth.
194 E3 Afton U.S.A.
211 H4 Afuá Brazil
136 E5 'Afula Israel
136 C2 Afyon Turkey
165 K4 Aga Germany
176 C3 Agadez Niger
176 B1 Agadir Morocco
175 I5 Agalega Islands Mauritius
138 D1 Agapovka Rus. Fed.
145 G5 Agartala India
144 C6 Agashi India
191 F2 Agate Canada
171 L6 Agathonisi i. Greece
176 B4 Agboville Côte d'Ivoire
137 K1 Ağcabädi Azer.
137 K2 Ağdam (abandoned) Azer.
166 F5 Agde France
166 E4 Agen France
180 C4 Aggeneys S. Africa
164 F4 Agger r. Germany
144 D1 Aghil Dawan China
163 C3 Aghla Mountain h. Ireland
171 K7 Agia Varvara Greece
136 G2 Ağın Turkey
171 J6 Agios Dimitrios Greece
171 K5 Agios Efstratios i. Greece
171 L5 Agios Fokas, Akrotirio pt Greece
171 J5 Agios Konstantinos Greece
171 K7 Agios Nikolaos Greece
171 J4 Agiou Orous, Kolpos b. Greece
177 F3 Agirwat Hills Sudan
181 F3 Agisanang S. Africa
176 B4 Agnibilékrou Côte d'Ivoire
171 K2 Agnita Romania
148 A2 Agong China
144 D4 Agra India
173 H7 Agrakhanskiy Poluostrov pen. Rus. Fed.
167 F2 Ágreda Spain
137 I2 Ağrı Turkey
171 J7 Agria Gramvousa i. Greece
170 E6 Agrigento Sicily Italy
171 I5 Agrinio Greece
215 B3 Agrio r. Arg.
170 F4 Agropoli Italy
137 J1 Ağstafa Azer.
137 L1 Ağsu Azer.
207 G5 Agua, Volcán de vol. Guat.
206 C3 Agua Brava, Laguna lag. Mex.
207 G4 Aguada Mex.
213 B3 Aguadas Col.
213 B3 Agua de Dios Col.
205 K5 Aguadilla Puerto Rico
215 D4 Aguado Cecilio Arg.
206 I6 Aguadulce Panama
215 C3 Agua Escondida Arg.
206 C3 Aguamilpa, Presa l. Mex.
204 D4 Aguanaval r. Mex.
215 C1 Agua Negra, Paso del pass Arg./Chile
214 B3 Aguapeí r. Brazil
204 C2 Agua Prieta Mex.
214 A3 Aguaray Guazú r. Para.
213 D2 Aguaro-Guariquito, Parque Nacional nat. park Venez.
206 C2 Aguaruto Mex.
206 D3 Aguascalientes Mex.
206 D3 Aguascalientes state Mex.
214 E2 Águas Formosas Brazil
214 C3 Agudos Brazil
197 F5 Aguila U.S.A.
167 D1 Aguilar de Campoo Spain
167 F4 Águilas Spain
153 B4 Aguisan Phil.
180 D7 Agulhas, Cape S. Africa
219 K9 Agulhas Basin sea feature Southern Ocean
214 D3 Agulhas Negras mt. Brazil
218 F7 Agulhas Plateau sea feature Southern Ocean
219 J8 Agulhas Ridge sea feature S. Atlantic Ocean
155 E4 Agung, Gunung vol. Indon.
153 C4 Agusan r. Phil.
213 B2 Agustín Codazzi Col.
153 B4 Agutaya Phil.
176 C2 Ahaggar plat. Alg.
176 C2 Ahaggar, Parc National de l' nat. park Alg.
176 C2 Ahaggar, Tassili oua-n- plat. Alg.
140 B2 Ahar Iran
128 C5 Ahaura N.Z.
164 F2 Ahaus Germany
128 F3 Ahimanawa Range mts N.Z.
128 D1 Ahipara N.Z.
128 D1 Ahipara Bay N.Z.
184 B4 Ahklun Mountains U.S.A.
137 I2 Ahlat Turkey
165 F3 Ahlen Germany
144 C5 Ahmadabad India
143 A2 Ahmadnagar India
144 B3 Ahmadpur East Pak.
144 B4 Ahmad Tar Pak.
Ahmedabad India see Ahmadabad
Ahmednagar India see Ahmadnagar
165 I4 Ahorn Germany
140 C4 Ahram Iran
165 I1 Ahrensburg Germany
137 I2 Ahta Dağı mt. Turkey
158 N3 Ähtäri Fin.
159 N4 Ahtme Estonia
137 L6 Āhū Iran
207 G5 Ahuachapán El Salvador
206 D3 Ahualulco Mex.
166 F3 Ahun France
128 B6 Ahuriri r. N.Z.
140 C4 Ahvāz Iran
144 C5 Ahwa India
180 B3 Ai-Ais Namibia
180 B3 Ai-Ais Hot Springs Game Park Namibia
180 B4 Ai-Ais/Richtersveld Transfrontier Park Namibia
148 D1 Aibag Gol r. China
Aidin Turkm. see Aýdyň
Aiea U.S.A. see 'Aiea
196 □1 'Aiea U.S.A.
136 E4 Aigialousa Cyprus
171 J6 Aigina i. Greece
171 J5 Aigio Greece
166 H4 Aigle de Chambeyron mt. France
215 F2 Aiguá Uruguay
152 C3 Ai He r. China
151 F5 Aikawa Japan
201 D5 Aiken U.S.A.
206 J6 Ailigandi Panama
Ailinglapalap atoll Marshall Is see Ailinglaplap
216 G6 Ailinglaplap atoll Marshall Is
164 A5 Ailly-sur-Noye France
191 G4 Ailsa Craig Canada
162 C5 Ailsa Craig i. U.K.
163 C3 Ailt an Chorráin Ireland
214 E2 Aimorés, Serra dos hills Brazil
196 □1 'Āina Haina U.S.A.
176 C1 Aïn Beïda Alg.
176 B2 'Aïn Ben Tili Mauritania
167 H4 Aïn Defla Alg.
167 H5 Aïn el Hadjel Alg.
176 B1 Aïn Sefra Alg.
189 H4 Ainslie, Lake Canada
198 D3 Ainsworth U.S.A.
Aintab Turkey see Gaziantep
167 H4 Aïn Taya Alg.
167 G5 Aïn Tédélès Alg.
213 B4 Aipe Col.
154 C5 Air i. Indon.
176 C3 Aïr, Massif de l' mts Niger
186 G4 Airdrie Canada
162 E5 Airdrie U.K.
164 D6 Aire r. France
166 D5 Aire-sur-l'Adour France
176 C3 Aïr et du Ténéré, Réserve Naturelle Nationale de l' nature res. Niger
185 K3 Air Force Island Canada
148 D1 Airgin Sum China
187 H3 Air Ronge Canada
148 F2 Ai Shan h. China
186 B2 Aishihik Canada
186 B2 Aishihik Lake Canada
166 G2 Aisne r. France
167 F3 Aitana mt. Spain
124 E2 Aitape P.N.G.
198 E2 Aitkin U.S.A.
138 C1 Aitovo Rus. Fed.
123 I5 Aitutaki i. Pacific Ocean
169 K7 Aiud Romania
166 G5 Aix-en-Provence France
166 G4 Aix-les-Bains France
145 F5 Aiyar Reservoir India
145 H5 Aizawl India
159 N4 Aizkraukle Latvia
159 M4 Aizpute Latvia
151 F6 Aizu-Wakamatsu Japan
170 C4 Ajaccio Corsica France
213 B4 Ajajú r. Col.
207 E4 Ajalpán Mex.
143 A1 Ajanta India
158 K2 Ajaureforsen Sweden
128 D5 Ajax, Mount N.Z.
177 E1 Ajdābiyā Libya
150 G4 Ajigasawa Japan
176 C2 Ajjer, Tassili n' plat. Alg.
136 E5 'Ajlūn Jordan
140 D5 'Ajman U.A.E.
144 C4 Ajmer India
197 F5 Ajo U.S.A.
197 F5 Ajo, Mount U.S.A.
153 B4 Ajuy Phil.
150 H3 Akabira Japan
139 H2 Akadyr Kazakh.
178 D4 Akagera, Parc National de l' Rwanda
143 B2 Akalkot India
150 I3 Akan Kokuritsu-kōen nat. park Japan
136 D4 Akanthou Cyprus
128 D5 Akaroa N.Z.
128 D5 Akaroa Harbour N.Z.
145 H4 Akas reg. India
139 H3 Akbakay Kazakh.
139 I3 Akbalyk Kazakh.
141 H3 Akbar Afgh.
145 E4 Akbarpur India
138 E3 Akbasty Kazakh.
139 H2 Akbaur Kazakh.
139 G2 Akbeyit Kazakh.
167 I4 Akbou Alg.
139 I2 Akbulak Vostochnyy Kazakhstan Kazakh.
139 K2 Akbulak Vostochnyy Kazakhstan Kazakh.
138 B2 Akbulak Kazakh.
138 C2 Akbulak Rus. Fed.
136 F2 Akçadağ Turkey
137 G3 Akçakale Turkey
136 C1 Akçakoca Turkey
136 D3 Akçali Dağ mt. Turkey
136 C3 Akdağ mt. Turkey
136 C3 Ak Dağı mt. Turkey
137 H2 Ak Dağı mts Turkey
136 E2 Akdağmadeni Turkey
138 D4 Akdepe Turkm.
159 L4 Åkersberga Sweden
164 C2 Akersloot Neth.
138 E3 Akespe Kazakh.
178 C3 Aketi Dem. Rep. Congo
138 C4 Akgyr Erezi hills Turkm.
137 I1 Akhalk'alak'i Georgia
173 G7 Akhalts'ikhe Georgia
145 G5 Akhaura Bangl.
177 E1 Akhḑar, Al Jabal al mts Libya
142 E5 Akhḑar, Jabal mts Oman
136 A2 Akhisar Turkey
136 F3 Akhtarīn Syria
173 H5 Akhtubinsk Rus. Fed.
138 D1 Akhunovo Rus. Fed.
137 I1 Akhuryan Armenia
151 C8 Aki Japan
188 D3 Akimiski Island Canada
136 E3 Akıncı Burun pt Turkey
136 G1 Akıncılar Turkey
150 G5 Akita Japan
176 A3 Akjoujt Mauritania
138 E2 Akkabak Kazakh.
158 L2 Akkajaure l. Sweden
Akkala Uzbek. see Oqqal'a
138 E2 Akkarga Kazakh.
139 G3 Akkense Kazakh.
138 D2 Akkermanovka Rus. Fed.
150 I3 Akkeshi Japan
138 B3 Akkistau Kazakh.
136 E5 'Akko Israel
139 G2 Akkol' Akmolinskaya Oblast' Kazakh.
139 H3 Akkol' Almatinskaya Oblast' Kazakh.
138 B3 Akkol' Atyrauskaya Oblast' Kazakh.
139 G4 Akkol' Zhambylskaya Oblast' Kazakh.
139 G4 Akkol', Ozero salt l. Kazakh.
139 I2 Akku Kazakh.
138 F3 Akkum Kazakh.
136 F1 Akkuş Turkey
Akkyr, Gory hills Turkm. see Akgyr Erezi
144 D4 Aklera India
159 M4 Akmeņrags pt Latvia
144 D1 Akmeqit China
Akmola Kazakh. see Astana
139 G2 Akmolinskaya Oblast' admin. div. Kazakh.
139 I4 Ak-Moyun Kyrg.
151 D7 Akō Japan
177 F4 Akobo Sudan
144 D5 Akola India
139 I4 Akongkür China
178 D2 Akordat Eritrea
136 D3 Akören Turkey
144 D5 Akot India
139 H2 Akoy Kazakh.
189 G1 Akpatok Island Canada
139 I4 Akqi China
158 B2 Akranes Iceland
159 I4 Åkrehamn Norway
194 G3 Akron CO U.S.A.
202 C4 Akron OH U.S.A.
202 C4 Akron City Reservoir U.S.A.
144 D2 Aksai Chin terr. Asia
136 E2 Aksaray Turkey
139 H4 Ak-Say r. Kyrg.
173 F6 Aksay Rus. Fed.
144 D2 Aksayqin Hu l. China
136 C2 Akşehir Turkey
136 C2 Akşehir Gölü l. Turkey
136 C3 Akseki Turkey
138 C1 Aksenovo Rus. Fed.
140 D4 Aks-e Rostam r. Iran
138 C2 Akshatau Kazakh.
139 H3 Akshatau Kazakh.
139 I4 Akshi Kazakh.
138 E2 Akshiganak Kazakh.
138 B4 Akshukyr Kazakh.
138 C3 Akshymyrau Kazakh.
139 J4 Aksu Xinjiang China
139 J4 Aksu Xinjiang China
139 I3 Aksu Almatinskaya Oblast' Kazakh.
139 I1 Aksu Pavlodarskaya Oblast' Kazakh.
139 G1 Aksu Severnyy Kazakhstan Kazakh.
139 G4 Aksu Yuzhnyy Kazakhstan Kazakh.
138 C2 Aksu Zapadnyy Kazakhstan Kazakh.
139 I3 Aksu r. Kazakh.
136 C3 Aksu r. Turkey
138 F2 Aksuat Kustanayskaya Oblast' Kazakh.
139 J3 Aksuat Vostochnyy Kazakhstan Kazakh.
139 H2 Aksu-Ayuly Kazakh.
139 J4 Aksu He r. China
178 D2 Āksum Eth.
139 F3 Aksumbe Kazakh.
139 J3 Aksüme China
139 H3 Aksuyek Kazakh.
145 F1 Aktag mt. China
138 F2 Aktas Kazakh.
137 J2 Aktas Dağı mt. Turkey
137 I1 Aktaş Gölü l. Georgia
Aktash Uzbek. see Oqtosh
139 G3 Aktau Karagandinskaya Oblast' Kazakh.
139 H2 Aktau Karagandinskaya Oblast' Kazakh.
138 B4 Aktau Mangistauskaya Oblast' Kazakh.
139 I5 Akto China
138 D2 Aktobe Kazakh.
139 H2 Aktogay Karagandinskaya Oblast' Kazakh.
139 H1 Aktogay Pavlodarskaya Oblast' Kazakh.
139 I3 Aktogay Vostochnyy Kazakhstan Kazakh.
169 N4 Aktsyabrski Belarus
139 F2 Aktuma (abandoned) Kazakh.
138 D3 Aktumsyk Kazakh.
138 D3 Aktumsyk, Mys pt Uzbek.
139 H4 Ak-Tüz Kyrg.
138 D2 Aktyubinskaya Oblast' admin. div. Kazakh.
151 B8 Akune Japan
176 C4 Akure Nigeria
158 C2 Akureyri Iceland
145 G1 Akxokesay China
Akyab Myanmar see Sittwe
139 H3 Akzhal Karagandinskaya Oblast' Kazakh.
139 J2 Akzhal Vostochnyy Kazakhstan Kazakh.
138 D2 Akzhar Aktyubinskaya Oblast' Kazakh.
138 F3 Akzhar Kzyl-Ordinskaya Oblast' Kazakh.
139 J3 Akzhar Vostochnyy Kazakhstan Kazakh.
139 G4 Akzhar Zhambylskaya Oblast' Kazakh.
139 F3 Akzhaykyn, Ozero salt l. Kazakh.
136 G3 Akziyaret Turkey
159 J3 Ål Norway
140 C5 Al 'Abā Saudi Arabia
201 C6 Alabama r. U.S.A.
201 C5 Alabama state U.S.A.
201 C5 Alabaster U.S.A.
153 B3 Alabat i. Phil.
137 J7 Al 'Abţīyah well Iraq
139 G4 Ala-Buka Kyrg.
177 E1 Al Abyār Libya
136 E1 Alaca Turkey
136 F2 Alacahan Turkey
136 E1 Alaçam Turkey
136 B2 Alaçam Dağları mts Turkey
207 G3 Alacrán, Arrecife rf Mex.
137 I2 Ala Dag mt. Turkey
137 I2 Ala Dağlar mts Turkey
136 E3 Ala Dağları mts Turkey
139 J4 Ala'er China
173 H7 Alagir Rus. Fed.
211 K6 Alagoinhas Brazil
167 F2 Alagón Spain
153 C5 Alah r. Phil.
158 M3 Alahärmä Fin.
137 K7 Al Aḩmadī Kuwait
139 G5 Alai Range mts Asia
158 M3 Alajärvi Fin.
206 H6 Alajuela Costa Rica
137 K2 Alajujeh Iran
144 D3 Alaknanda r. India
139 J3 Alakol', Ozero salt l. Kazakh.
158 O2 Alakurtti Rus. Fed.
136 B6 Al 'Alamayn Egypt
210 F4 Alalaú r. Brazil
137 I3 Al 'Amādīyah Iraq
140 B5 Al 'Amār Saudi Arabia
137 K6 Al 'Amārah Iraq
Alamdo China see Alando
137 J7 Al Amghar waterhole Iraq
153 A2 Alaminos Phil.
136 B6 Al 'Āmirīyah Egypt
206 D2 Alamítos, Sierra de los mt. Mex.
197 E3 Alamo U.S.A.
197 F4 Alamo Dam U.S.A.
195 F5 Alamogordo U.S.A.
199 D6 Alamo Heights U.S.A.
195 E6 Alamos Sonora Mex.
206 B2 Alamos Sonora Mex.
195 F4 Alamosa U.S.A.
143 B3 Alampur India
158 K2 Alanäs Sweden
Åland is Fin. see Åland Islands
165 J1 Aland r. Germany
143 B2 Aland India
140 B2 Aland r. Iran
159 L3 Åland Islands Fin.
145 H3 Alando China
154 B5 Alang Besar i. Indon.
Alaniya aut. rep. Rus. Fed. see Severnaya Osetiya-Alaniya, Respublika
190 E3 Alanson U.S.A.
136 D3 Alanya Turkey
140 D4 'Alā' od Dīn Iran
136 C1 Alaplı Turkey
143 B4 Alappuzha India
Alappuzha India see Alappuzha
136 E7 Al 'Aqabah Jordan
140 C6 Al 'Āqūlah well Saudi Arabia
167 E3 Alarcón, Embalse de resr Spain
136 D6 Al 'Arīsh Egypt
142 C4 Al Arţāwīyah Saudi Arabia
155 E4 Alas Indon.
136 B2 Alaşehir Turkey
137 I6 Al 'Āshūrīyah well Iraq
184 D3 Alaska state U.S.A.
184 D4 Alaska, Gulf of U.S.A.
186 E3 Alaska Highway Canada/U.S.A.
184 B4 Alaska Peninsula U.S.A.
184 D3 Alaska Range mts U.S.A.
137 L2 Älät Azer.
Alat Uzbek. see Olot
137 I6 Al 'Athāmīn hills Iraq
172 H4 Alatyr' Rus. Fed.
172 H4 Alatyr' r. Rus. Fed.
210 C4 Alausí Ecuador
137 J1 Alaverdi Armenia
158 N2 Alavieska Fin.
158 M3 Alavus Fin.
126 D5 Alawoona Australia
137 K1 Alazani r. Azer./Georgia
137 J5 Al 'Azīzīyah Iraq
177 D1 Al 'Azīzīyah Libya
170 C2 Alba Italy
136 F3 Al Bāb Syria
167 F3 Albacete Spain
126 D5 Albacutya, Lake dry lake Australia
137 J6 Al Bādiyah al Janūbīyah des. Iraq
171 J1 Alba Iulia Romania
188 F3 Albanel, Lac Canada
171 I4 Albania country Europe
124 B5 Albany Australia
188 C3 Albany r. Canada
201 C6 Albany GA U.S.A.
190 E5 Albany IN U.S.A.
200 C4 Albany KY U.S.A.
203 G3 Albany NY U.S.A.
194 B2 Albany OR U.S.A.
127 H1 Albany Downs Australia
215 G2 Albardão do João Maria coastal area Brazil
140 B5 Al Barrah Saudi Arabia
Al Başrah Iraq see Al Başrah
137 K6 Al Başrah Iraq
137 J6 Al Baţḩa' marsh Iraq
124 E3 Albatross Bay Australia
127 F8 Albatross Island Australia
Al Bawiti Egypt see Al Bawīţī
177 E2 Al Bawīţī Egypt
177 E1 Al Bayḑā' Libya
210 □ Albemarle, Punta pt Galapagos Is Ecuador
201 E5 Albemarle Sound sea chan. U.S.A.
170 C2 Albenga Italy
167 D3 Alberche r. Spain
124 D4 Alberga watercourse Australia
167 B2 Albergaria-a-Velha Port.
127 G4 Albert Australia
166 F2 Albert France
126 C5 Albert, Lake Australia
178 D3 Albert, Lake Dem. Rep. Congo/Uganda
186 F4 Alberta prov. Canada
202 E6 Alberta U.S.A.
186 F4 Alberta, Mount Canada
180 D7 Albertinia S. Africa
164 D4 Albert Kanaal canal Belgium
198 E3 Albert Lea U.S.A.
178 D3 Albert Nile r. Sudan/Uganda
212 B8 Alberto de Agostini, Parque Nacional nat. park Chile
181 H3 Alberton S. Africa
166 H4 Albertville France
164 E6 Albestroff France
166 F5 Albi France
211 H2 Albina Suriname
196 A2 Albion CA U.S.A.
203 I2 Albion ME U.S.A.
190 E4 Albion MI U.S.A.
202 D3 Albion NY U.S.A.
167 E5 Alborán, Isla de i. Spain
167 D4 Alboran Sea Europe
Ålborg Denmark see Aalborg
Ålborg Bugt b. Denmark see Aalborg Bugt
Alborz, Reshteh-ye mts Iran see Elburz Mountains
186 F4 Albreda Canada
140 C5 Al Budayyi' Bahrain
Al Budayyi Bahrain see Al Budayyi'
167 B4 Albufeira Port.
195 F5 Albuquerque U.S.A.
206 I5 Albuquerque, Cayos de is Col.
142 E5 Al Buraymī Oman
167 C3 Alburquerque Spain
127 G6 Albury Australia
137 H4 Al Buşayrah Syria
136 G7 Al Busayţā' plain Saudi Arabia
137 K6 Al Buşayyah Iraq
140 B4 Al Bushūk well Saudi Arabia
167 B3 Alcácer do Sal Port.
167 E2 Alcalá de Henares Spain
167 E4 Alcalá la Real Spain
170 E6 Alcamo Sicily Italy
167 F2 Alcañiz Spain
167 C3 Alcántara Spain
167 E3 Alcaraz Spain
167 D4 Alcaudete Spain
167 E3 Alcázar de San Juan Spain
173 F5 Alchevs'k Ukr.
215 D2 Alcira Arg.
214 E2 Alcobaça Brazil
215 E2 Alcorta Arg.
Alcoy Spain see Alcoy-Alcoi
167 F3 Alcoy-Alcoi Spain
167 H3 Alcúdia Spain
179 E4 Aldabra Islands Seychelles
206 C1 Aldama Chihuahua Mex.
207 E3 Aldama Tamaulipas Mex.
133 N4 Aldan Rus. Fed.
133 O3 Aldan r. Rus. Fed.
164 D1 Aldeboarn Neth.
161 I5 Aldeburgh U.K.
166 D2 Alderney i. U.K.
196 B4 Alder Peak U.S.A.
161 G6 Aldershot U.K.
202 C6 Alderson U.S.A.
140 D6 Al Dhafrah reg. U.A.E.
160 D3 Aldingham U.K.
161 F5 Aldridge U.K.
190 B5 Aledo U.S.A.
176 A3 Aleg Mauritania
214 E3 Alegre Brazil
212 E3 Alegrete Brazil
215 E2 Alejandro Korn Arg.
172 E2 Alekhovshchina Rus. Fed.
138 B4 Aleksandra Bekovicha-Cherkasskogo, Zaliv b. Kazakh.
172 F3 Aleksandrov Rus. Fed.
173 I5 Aleksandrov Gay Rus. Fed.
138 C1 Aleksandrovka Orenburgskaya Oblast' Rus. Fed.
138 D1 Aleksandrovka Respublika Bashkortostan Rus. Fed.
173 H6 Aleksandrovskoye Rus. Fed.
133 P4 Aleksandrovsk-Sakhalinskiy Rus. Fed.
132 F1 Aleksandry, Zemlya i. Rus. Fed.
139 G1 Alekseyevka Kokshetauskaya Oblast' Kazakh.
173 F5 Alekseyevka Belgorodskaya Oblast' Rus. Fed.
173 F5 Alekseyevka Belgorodskaya Oblast' Rus. Fed.
173 G5 Alekseyevskaya Rus. Fed.
172 F4 Aleksin Rus. Fed.
171 I3 Aleksinac Serbia
178 B4 Alèmbé Gabon
136 E1 Alembeyli Turkey
214 D3 Além Paraíba Brazil
158 J3 Ålen Norway
166 E2 Alençon France
211 H4 Alenquer Brazil
196 □2 'Alenuihāhā Channel U.S.A.
136 F3 Aleppo Syria
185 M1 Alert Canada
210 D6 Alerta Peru
186 D4 Alert Bay Canada
166 G4 Alès France
169 K7 Aleşd Romania
170 C2 Alessandria Italy
158 I3 Ålesund Norway
216 G2 Aleutian Basin sea feature Bering Sea
182 C2 Aleutian Islands AK U.S.A.
184 C4 Aleutian Range mts U.S.A.
216 H2 Aleutian Trench sea feature N. Pacific Ocean
133 Q4 Alevina, Mys c. Rus. Fed.
203 J2 Alexander U.S.A.
186 B3 Alexander Archipelago is U.S.A.
180 B4 Alexander Bay b. Namibia/S. Africa
180 B4 Alexander Bay S. Africa
201 C5 Alexander City U.S.A.
129 B3 Alexander Island i. Antarctica
127 F6 Alexandra Australia
128 B6 Alexandra N.Z.
212 □ Alexandra, Cape Atlantic Ocean
171 J4 Alexandreia Greece
203 F2 Alexandria Canada
177 E1 Alexandria Egypt
171 K3 Alexandria Romania
181 G6 Alexandria S. Africa
162 D5 Alexandria U.K.
190 E5 Alexandria IN U.S.A.
199 E6 Alexandria LA U.S.A.
198 E2 Alexandria MN U.S.A.
202 E5 Alexandria VA U.S.A.
203 F2 Alexandria Bay U.S.A.
126 C5 Alexandrina, Lake Australia
171 K4 Alexandroupoli Greece
189 I3 Alexis r. Canada
190 B5 Alexis U.S.A.
186 E4 Alexis Creek Canada
139 J1 Aley r. Rus. Fed.
139 J1 Aleysk Rus. Fed.
164 F4 Alf Germany
167 F1 Alfaro Spain
137 K7 Al Farwānīyah Kuwait
137 I4 Al Fatḩah Iraq
137 L7 Al Fāw Iraq
177 F2 Al Fayyūm Egypt
165 H3 Alfeld (Leine) Germany
214 D3 Alfenas Brazil

137 L7 Al Finṭās Kuwait
169 J7 Alföld plain Hungary
161 H4 Alford U.K.
203 F2 Alfred Canada
203 H3 Alfred U.S.A.
137 L7 Al Fuḩayḩil Kuwait
142 E4 Al Fujayrah U.A.E.
Al Fujayrah U.A.E. see Al Fujayrah
137 J6 Al Furāt r. Iraq/Syria
alt. Firat (Turkey),
conv. Euphrates
138 D2 Alga Kazakh.
138 C2 Algabas Kazakh.
159 I4 Ålgård Norway
215 C3 Algarrobo del Aguila Arg.
167 B4 Algarve reg. Port.
172 G4 Algasovo Rus. Fed.
167 D4 Algeciras Spain
167 F3 Algemesí Spain
Alger Alg. see Algiers
191 E3 Alger U.S.A.
176 B2 Algeria country Africa
165 H2 Algermissen Germany
137 J6 Al Ghammis Iraq
140 B5 Al Ghāṭ Saudi Arabia
142 D6 Al Ghaydah Yemen
170 C4 Alghero Sardinia Italy
177 F2 Al Ghurdaqah Egypt
142 C4 Al Ghwaybiyah Saudi Arabia
176 C1 Algiers Alg.
181 F6 Algoa Bay S. Africa
190 D3 Algoma U.S.A.
198 E3 Algona U.S.A.
191 F4 Algonac U.S.A.
191 H3 Algonquin Park Canada
191 H3 Algonquin Provincial Park Canada
167 E1 Algorta Spain
137 I7 Al Habakah well Saudi Arabia
137 I5 Al Ḩabbānīyah Iraq
165 H2 Al Ḩadaqah well Saudi Arabia
140 C5 Al Ḩadd Bahrain
140 A4 Al Ḩadhālīl plat. Saudi Arabia
137 I4 Al Ḩadīthah Iraq
140 A3 Al Ḩaḍr Iraq
136 F4 Al Ḩaffah Syria
140 B5 Al Ḩā'ir Saudi Arabia
141 E6 Al Hajar Oman
140 E5 Al Hajar al Gharbī mts Oman
137 G6 Al Ḩamād plain Saudi Arabia
167 F4 Alhama de Murcia Spain
136 B6 Al Ḩammām Egypt
137 I6 Al Ḩammām well Iraq
137 J7 Al Ḩaniyah esc. Iraq
140 B6 Al Ḩariq Saudi Arabia
137 G6 Al Ḩarrah reg. Saudi Arabia
137 H3 Al Ḩasakah Syria
137 J5 Al Hāshimīyah Iraq
137 K5 Al Ḩayy Iraq
136 F6 Al Hazim Jordan
137 J5 Al Ḩillah Iraq
140 B6 Al Ḩillah Saudi Arabia
140 B6 Al Ḩinnāh Saudi Arabia
140 C5 Al Ḩinnāh Saudi Arabia
176 B1 Al Hoceima Morocco
Al Ḩudaydah Yemen see Hodeidah
142 C4 Al Ḩufūf Saudi Arabia
140 D6 Al Ḩumrah reg. U.A.E.
140 C5 Al Ḩunayy Saudi Arabia
140 B6 Al Huwwah Saudi Arabia
140 D2 'Alīābād Iran
141 E3 'Alīābād Iran
141 F4 'Alīābād Iran
137 K3 'Alīābād Iran
137 K4 'Alīābād Iran
171 L5 Aliağa Turkey
171 J4 Aliakmonas r. Greece
137 K5 'Alī al Gharbī Iraq
143 A2 Alibag India
144 B4 Ali Bandar Pak.
137 L2 Äli Bayramlı Azer.
167 F3 Alicante Spain
181 G6 Alice S. Africa
199 D7 Alice U.S.A.
186 D3 Alice Arm Canada
124 D4 Alice Springs Australia
201 E7 Alice Town Bahamas
139 H5 Alichur Tajik.
139 H5 Alichur r. Tajik.
153 B5 Alicia Phil.
144 D3 Aligarh India
140 C3 Alīgūdarz Iran
146 E1 Alihe China
178 B4 Alima r. Congo
159 K4 Alingsås Sweden
136 B2 Aliova r. Turkey
144 B3 Alipur Pak.
145 G4 Alipur Duar India
202 C4 Aliquippa U.S.A.
178 E2 Ali Sabieh Djibouti
136 F6 Al 'Isāwīyah Saudi Arabia
137 J2 'Alī Shāh Iran
Al Iskandarīyah Egypt see Alexandria
137 J5 Al Iskandarīyah Iraq
177 F1 Al Ismā'īlīyah Egypt
171 K5 Aliveri Greece
181 G5 Aliwal North S. Africa
186 G4 Alix Canada
136 F6 Al Jafr Jordan
140 C5 Al Jāfūrah des. Saudi Arabia
177 E2 Al Jaghbūb Libya
137 K7 Al Jahrah Kuwait
140 C5 Al Jamalīyah Qatar
140 C6 Al Jawb reg. Saudi Arabia
177 E2 Al Jawf Libya
177 D1 Al Jawsh Libya
137 G3 Al Jazā'ir reg. Iraq/Syria
167 B4 Aljezur Port.
140 C5 Al Jibān reg. Saudi Arabia
137 I6 Al Jil well Iraq
140 B5 Al Jilh esc. Saudi Arabia
140 C5 Al Jishshah Saudi Arabia
Al Jīzah Egypt see Giza
136 F6 Al Jīzah Jordan
142 C4 Al Jubayl Saudi Arabia
140 B5 Al Jubaylah Saudi Arabia
140 C5 Al Jurayd i. Saudi Arabia
140 C5 Al Jurayfah Saudi Arabia
167 B4 Aljustrel Port.
136 E4 Al Karak Jordan
139 G1 Alkaterek Kazakh.
137 J5 Al Kāẓimīyah Iraq
142 E5 Al Khābūrah Oman
137 J5 Al Khāliṣ Iraq
177 F2 Al Khārijah Egypt
142 E4 Al Khaṣab Oman
140 A6 Al Khāṣirah Saudi Arabia
140 D6 Al Khatam reg. U.A.E.
140 C5 Al Khawr Qatar
140 C5 Al Khiṣah well Saudi Arabia
140 B5 Al Khobar Saudi Arabia
140 B5 Al Khuff reg. Saudi Arabia
137 D1 Al Khums Libya
137 J5 Al Kifl Iraq
140 C5 Al Kir'ānah Qatar
164 C2 Alkmaar Neth.
137 J5 Al Kūfah Iraq
177 E2 Al Kufrah Libya
137 K5 Al Kumayt Iraq
137 J5 Al Kūt Iraq
Al Kuwayt Kuwait see Kuwait
137 H7 Al Labbah plain Saudi Arabia
Al Lādhiqīyah Syria see
Al Lādhiqīyah
136 E4 Al Lādhiqīyah Syria
203 I1 Allagash U.S.A.
203 I1 Allagash r. U.S.A.
203 I1 Allagash Lake U.S.A.
145 E4 Allahabad India
136 F5 Al Lajā lava field Syria
133 G3 Allakh-Yun' Rus. Fed.
181 H1 Allanridge S. Africa
181 H1 Alldays S. Africa
136 E4 Allegan U.S.A.
202 D4 Allegheny r. U.S.A.
202 C6 Allegheny Mountains U.S.A.
202 D4 Allegheny Reservoir U.S.A.

163 C3 Allen, Lough l. Ireland
201 D5 Allendale U.S.A.
160 E3 Allendale Town U.K.
206 D3 Allende Coahuila Mex.
207 D2 Allende Nuevo León Mex.
165 G4 Allendorf (Lumda) Germany
191 G2 Allenford Canada
203 F4 Allentown U.S.A.
165 I2 Aller r. Germany
198 C3 Alliance NE U.S.A.
202 C4 Alliance OH U.S.A.
137 L6 Al Lifīyah well Iraq
159 K5 Allinge-Sandvig Denmark
191 H2 Alliston Canada
142 B5 Al Lith Saudi Arabia
162 E4 Alloa U.K.
127 J2 Alloa Australia
143 C3 Allur India
143 C3 Alluru Kottapatnam India
137 I6 Al Lussuf well Iraq
189 F4 Alma Canada
190 E4 Alma MI U.S.A.
198 D2 Alma NE U.S.A.
197 H4 Alma NM U.S.A.
137 I6 Al Ma'āniyah Iraq
Alma-Ata Kazakh. see Almaty
167 B3 Almada Port.
137 J7 Al Ma'dānīyāt well Iraq
167 D2 Almadén Spain
137 K6 Al Madīnah Iraq
136 F5 Al Mafraq Jordan
137 J5 Al Maḩmūdīyah Iraq
140 B5 Al Majma'ah Saudi Arabia
137 K3 Almalı Azer.
140 C5 Al Malsūnīyah reg. Saudi Arabia
Almalyk Uzbek. see Olmaliq
Al Manāmah Bahrain see Manama
196 B3 Almanor, Lake U.S.A.
167 F3 Almansa Spain
177 F1 Al Manṣūrah Egypt
167 D2 Almanzor mt. Spain
137 K6 Al Ma'qil Iraq
140 De Al Mariyyah U.A.E.
177 E1 Al Marj Libya
214 C1 Almas, Rio das r. Brazil
139 H₃ Almatinskaya Oblast' admin. div.
    Kazakh.
139 I4 Almaty Kazakh.
Al Mawṣil Iraq see Al Mawṣil
137 I3 Al Mawṣil Iraq
137 H4 Al Mayādīn Syria
140 B5 Al Mazāḩimīyah Saudi Arabia
167 E2 Almazán Spain
133 M3 Almaznyy Rus. Fed.
211 H4 Almeirim Brazil
167 B3 Almeirim Port.
164 E2 Almelo Neth.
214 E2 Almenara Brazil
167 C2 Almendra, Embalse de resr Spain
167 C3 Almendralejo Spain
164 D1 Almere Neth.
167 E4 Almería Spain
167 F4 Almería, Golfo de b. Spain
132 G₄ Al'met'yevsk Rus. Fed.
159 K4 Älmhult Sweden
140 B5 Al Midhnab Saudi Arabia
167 D3 Almina, Punta pt Morocco
177 F2 Al Minyā Egypt
206 I6 Almirante Panama
125 J3 Alofi Niue
145 H4 Alon Myanmar
145 H3 Along India
171 J5 Alonnisos i. Greece
147 E2 Alor i. Indon.
147 E2 Alor, Kepulauan is Indon.
155 B3 Alor Star Malaysia
144 C3 Alot India
158 O3 Alozero Rus. Fed.
196 C3 Alpaugh U.S.A.
164 E2 Alpen Germany
191 F2 Alpena U.S.A.
220 P7 Alpha Ridge sea feature Arctic Ocean
197 H3 Alpine AZ U.S.A.
199 C6 Alpine TX U.S.A.
194 E3 Alpine WY U.S.A.
174 F4 Alpine National Park Australia
166 D3 Alps mts Europe
142 C5 Al 'Aqīlah reg. Saudi Arabia
177 D1 Al Qaddāḩīyah Libya
136 F4 Al Qadmūs Syria
Al Qāhirah Egypt see Cairo
140 B5 Al Qā'īyah well Saudi Arabia
137 H3 Al Qāmishlī Syria
136 D4 Al Qanṭarah Egypt
136 F4 Al Qar'ah well Saudi Arabia
136 F4 Al Qaryatayn Syria
140 B5 Al Qaṣab Saudi Arabia
142 C5 Al Qaṭn Yemen
136 F6 Al Qaṭrānah Jordan
177 D2 Al Qaṭrūn Libya
140 B5 Al Qayṣūmah Saudi Arabia
177 E2 Al Qubbah Libya
Alqueva, Barragem de resr Port.
136 F5 Al Qunayṭirah Syria
142 B6 Al Qunfidhah Saudi Arabia
140 A5 Al Qurayn Saudi Arabia
137 K6 Al Qurnah Iraq
177 F2 Al Quṣayr Egypt
137 J6 Al Quṣayr Iraq
140 B5 Al Qūṣūrīyah Saudi Arabia
136 F5 Al Quṭayfah Syria
140 A5 Al Quwarah Saudi Arabia
140 B5 Al Quwayy' Saudi Arabia
136 E1 Al Quwayyah Saudi Arabia
136 E2 Al Quwayrah Jordan
166 H3 Alsace reg. France
161 E4 Alsager U.K.
137 I6 Al Samīt well Iraq
187 H4 Alsask Canada
165 H3 Alsfeld Germany
165 J3 Alsleben (Saale) Germany
160 E3 Alston U.K.
127 J2 Alstonville Australia
159 M4 Alsunga Latvia
158 M4 Alta Norway
128 B4 Alta, Mount N.Z.
159 M4 Alta r. Norway
158 M3 Altafjorden sea chan. Norway
215 D3 Alta Gracia Arg.
213 D3 Altagracia de Orituco Venez.
135 G2 Altai Mountains China/Mongolia
201 D5 Altamaha r. U.S.A.
211 H4 Altamira Brazil
206 H5 Altamira Costa Rica
170 G4 Altamura Italy
148 C1 Altan Shiret China
206 C2 Altata Mex.
202 D3 Altavista U.S.A.
135 G3 Altay China
190 O5 Altay Mongolia
146 B2 Altay Mongolia
139 K2 Altay, Respublika aut. rep. Rus. Fed.
139 K2 Altayskiy Rus. Fed.

139 J. Altayskiy Kray admin. div. Rus. Fed.
167 F3 Altea Spain
158 M1 Alteidet Norway
164 E4 Altenahr Germany
164 F2 Altenberge Germany
165 K4 Altenburg Germany
164 F4 Altenkirchen (Westerwald) Germany
345 H1 Altenqoke China
165 L1 Altentreptow Germany
137 L1 Altıağaç Azer.
137 J4 Altın Köprü Iraq
171 L5 Altınoluk Turkey
136 C2 Altıntaş Turkey
210 E7 Altiplano plain Bol.
165 J2 Altmark reg. Germany
165 I5 Altmühl r. Germany
214 E2 Alto Araguaia Brazil
215 C2 Alto de Pencoso hills Arg.
213 E3 Alto de Tamar mt. Col.
214 E2 Alto Garças Brazil
210 E6 Alto Molócuè Moz.
179 D5 Alton Canada
200 E4 Alton IL U.S.A.
199 F4 Altona Germany
203 H3 Altona U.S.A.
202 D4 Altoona U.S.A.
214 C1 Alto Paraíso de Goiás Brazil
214 C1 Alto Sucuriú Brazil
213 I2 Alto Taquari Brazil
165 I6 Altötting Germany
161 E4 Altrincham U.K.
165 K1 Alt Schwerin Germany
207 G4 Altun Mex.
146 B3 Altun Shan mts China
194 E3 Alturas U.S.A.
199 D5 Altus U.S.A.
130 B5 Altynasar tourist site Kazakh.
136 G1 Alucra Turkey
159 N4 Alūksne Latvia
137 L5 Alūm Iran
202 E4 Alum Creek Lake U.S.A.
215 E3 Aluminé Arg.
215 E3 Aluminé, Lago l. Arg.
73 E6 Alupka Ukr.
77 D1 Al 'Uqaylah Libya
140 C5 Al 'Uqayr Saudi Arabia
140 C5 Al Uqṣur Egypt see Luxor
73 E6 Alushta Ukr.
137 J1 'Alūt Iran
143 B4 Aluva India
77 E2 Al 'Uwaynāt Libya
77 C2 Al 'Uwayqīlah Saudi Arabia
37 H6 Al 'Uzayr Iraq
99 D4 Alva U.S.A.
137 L4 Alvand, Kūh-e mt. Iran
207 F4 Alvarado, Paso de pass Chile
215 C2 Alvarado, Paso de pass Chile
210 F4 Alvares Brazil
59 J3 Alvdal Norway
215 B3 Älvdalen Sweden
59 I4 Älvdalen Sweden
59 J3 Älvik Sweden
59 J3 Älvik Sweden
158 M2 Älvsbyn Sweden
37 K7 Al Wafrah Kuwait
34 B4 Al Wajh Saudi Arabia
40 C5 Al Wakrah Qatar
219 J3 Al Wannān Saudi Arabia
44 D4 Alwar India
40 B5 Al Wari'ah Saudi Arabia
37 H5 Al Widyān plat. Iraq/Saudi Arabia
40 B4 Al Wusayl well Saudi Arabia
24 D3 Alyangula Australia
62 E4 Alyth U.K.
59 N5 Alytus Lith.
94 F2 Alzada U.S.A.
64 E5 Alzette r. Lux.
65 G5 Alzey Germany
213 E3 Amacuro r. Guyana/Venez.
24 D4 Amadeus, Lake salt flat Australia
85 K3 Amadjuak Lake Canada
97 G6 Amado U.S.A.
67 B3 Amadora Port.
59 I4 Amål Sweden
43 C2 Amalapuram India
46 D1 Amalat r. Rus. Fed.
213 B3 Amalfi Col.
80 F3 Amalfi Italy
71 I6 Amaliada Greece
44 C5 Amalner India
214 A3 Amambaí Brazil
214 A3 Amambaí r. Brazil
214 A3 Amambaí, Serra de hills Brazil/Para.
46 E4 Amami-Ō-shima i. Japan
46 E4 Amami-shotō is Japan
38 D2 Amangel'dy Kazakh.
38 F2 Amankel'dy Kazakh.
48 C3 Amankol China
46 E2 Amankol China
48 E3 Amanotke Kazakh.
70 G5 Amantea Italy
127 I2 Amanzimtoti S. Africa
211 H4 Amapá Brazil
211 H4 Amapá state Brazil
67 C3 Amareleja Port.
96 D3 Amargosa Desert U.S.A.
96 D3 Amargosa Range mts U.S.A.
96 C3 Amargosa Valley U.S.A.
99 C5 Amarillo U.S.A.
170 F4 Amaro, Monte mt. Italy
136 C1 Amasya Turkey
207 H4 Amatán Mex.
206 G5 Amatique, Bahía de b. Guat.
206 C3 Amatlán de Cañas Mex.
64 D4 Amay Belgium
211 H4 Amazon r. S. America
alt. Amazonas
211 I3 Amazon, Mouths of the Brazil
213 D4 Amazonas state Brazil
211 H4 Amazonas r. S. America
conv. Amazon
211 F5 Amazon Cone sea feature
    S. Atlantic Ocean
211 G4 Amazônia, Parque Nacional nat. park
    Brazil
144 C3 Ambad India
143 B2 Ambajogai India
144 D3 Ambala India
143 C5 Ambalangoda Sri Lanka
143 C5 Ambalavao Madag.
179 E6 Ambanja Madag.
141 E4 Ambar Iran
179 D5 Ambasamudram India
210 B4 Ambato Ecuador
179 E5 Ambato Boeny Madag.
179 E5 Ambato Finandrahana Madag.
179 E5 Ambatolampy Madag.
179 E5 Ambatomainty Madag.
179 E5 Ambatondrazaka Madag.
165 J5 Amberg Germany
207 H4 Ambergris Caye i. Belize
166 G4 Ambérieu-en-Bugey France
191 G3 Amberley Canada
145 E5 Ambikapur India
179 E5 Ambilobe Madag.
160 E2 Amble U.K.
160 F3 Ambleside U.K.
164 F2 Amblève r. Belgium
179 E6 Amboasary Madag.
179 E5 Ambohidratrimo Madag.
179 E6 Ambohimahasoa Madag.
146 D4 Ambon Indon.
147 E7 Ambon i. Indon.
179 E6 Ambositra Madag.
179 E6 Ambovombe Madag.
197 E4 Amboy CA U.S.A.
190 C5 Amboy IL U.S.A.
179 B4 Ambriz Angola
25 G3 Ambrym i. Vanuatu

143 B3 Ambur India
Amdo China see Lharigarbo
206 C3 Ameca Mex.
164 D1 Ameland i. Neth.
202 E6 Amelia Court House U.S.A.
203 G4 Amenia U.S.A.
196 B2 American, North Fork r. U.S.A.
219 H9 American-Antarctic Ridge sea feature
    S. Atlantic Ocean
194 D3 American Falls U.S.A.
194 D3 American Falls Reservoir U.S.A.
197 G1 American Fork U.S.A.
123 H4 American Samoa Pacific Ocean
201 C5 Americus U.S.A.
164 D2 Amersfoort Neth.
181 H3 Amersfoort S. Africa
161 G6 Amersham U.K.
187 K3 Amery Canada
129 D5 Amery Ice Shelf ice feature Antarctica
198 E3 Ames U.S.A.
161 F6 Amesbury U.K.
203 H3 Amesbury U.S.A.
136 E3 Amethi India
171 J5 Amfissa Greece
133 O3 Amga Rus. Fed.
146 F2 Amgu Rus. Fed.
176 C2 Amguid Alg.
146 F1 Amgun' r. Rus. Fed.
189 H4 Amherst Canada
203 G3 Amherst MA U.S.A.
203 I2 Amherst ME U.S.A.
202 D6 Amherst VA U.S.A.
191 F4 Amherstburg Canada
170 D3 Amiata, Monte mt. Italy
166 F2 Amiens France
137 H5 'Amij, Wādī watercourse Iraq
143 A4 Amindivi Islands India
151 D7 Amino Japan
179 B5 Aminuis Namibia
140 B3 Amir'ābād Iran
219 M6 Amirante Islands Seychelles
218 H5 Amirante Trench sea feature
    Indian Ocean
141 F4 Amir Chah Pak.
187 K3 Amisk Lake Canada
199 C6 Amistad Reservoir Mex./U.S.A.
127 J1 Amity Point Australia
144 D5 Amla India
159 J4 Amli Norway
161 C4 Amlwch U.K.
136 E3 'Ammān Jordan
144 B1 Ammanford U.K.
158 O2 Ämmänsaari Fin.
158 L2 Ammarnäs Sweden
185 O3 Ammassalik Greenland
165 I1 Ammerland reg. Germany
165 I3 Ammern Germany
141 D2 Ammersee l. Germany
154 C2 Amnat Charoen Thai.
152 D4 Amnyong-dan hd N. Korea
144 C5 Amod India
149 B6 Amo Jiang r. China
140 D2 Amol Iran
165 H5 Amorbach Germany
171 K6 Amorgos i. Greece
188 E4 Amos Canada
Amoy China see Xiamen
143 C5 Ampara India
214 C2 Amparo Brazil
168 E6 Amper r. Germany
219 I3 Ampere Seamount sea feature
    N. Atlantic Ocean
167 G2 Amposta Spain
144 D5 Amravati India
144 B5 Amreli India
144 B4 Amri Pak.
136 E4 'Amrit Syria
144 D3 Amritsar India
144 D3 Amroha India
158 L2 Åmsele Sweden
164 C2 Amstelveen Neth.
164 C2 Amsterdam Neth.
    (City Plan 107)
181 I3 Amsterdam S. Africa
203 F3 Amsterdam U.S.A.
218 J7 Amsterdam, Île i. Indian Ocean
177 E3 Amstetten Austria
178 B3 Am Timan Chad
138 D3 Amudar'ya r. Turkm./Uzbek.
185 I2 Amund Ringnes Island Canada
217 D6 Amundsen, Mount mt. Antarctica
217 K10 Amundsen Abyssal Plain sea feature
    Southern Ocean
220 B1 Amundsen Basin sea feature
    Arctic Ocean
129 E4 Amundsen Bay b. Antarctica
184 F2 Amundsen Gulf Canada
217 K10 Amundsen Ridges sea feature
    Southern Ocean
129 C4 Amundsen-Scott research stn
    Antarctica
217 L10 Amundsen Sea Antarctica
155 E3 Amuntai Indon.
146 E2 Amur r. China/Rus. Fed.
    alt. Heilong Jiang
133 O4 Amur r. Rus. Fed.
173 F6 Amursk Rus. Fed.
190 E1 Amyot Canada
145 H6 An Myanmar
217 J7 Anaa atoll Fr. Polynesia
133 M2 Anabar r. Rus. Fed.
133 M2 Anabarskiy Zaliv b. Rus. Fed.
126 D4 Ana Branch r. Australia
196 C4 Anacapa Island U.S.A.
213 D2 Anaco Venez.
194 D2 Anaconda U.S.A.
194 B2 Anacortes U.S.A.
199 D5 Anadarko U.S.A.
136 F1 Anadolu Dağları mts Turkey
133 S3 Anadyr' Rus. Fed.
133 T3 Anadyr' r. Rus. Fed.
133 T3 Anadyrskiy Zaliv b. Rus. Fed.
171 K6 Anafi i. Greece
214 A2 Anafé Brazil
137 H4 'Anah Iraq
196 D5 Anaheim U.S.A.
186 D4 Anahim Lake Canada
207 D2 Anáhuac Mex.
143 B4 Anaimalai Hills India
143 B4 Anai Mudi mt. India
143 C2 Anakapalle India
179 E5 Anakao Madag.
210 F4 Anamã Brazil
155 C2 Anambas, Kepulauan is Indon.
190 B4 Anamosa U.S.A.
136 D3 Anamur Turkey
136 D3 Anamur Burnu pt Turkey
151 D7 Anan Japan
144 C5 Anand India
145 F5 Anandapur India
143 B3 Anantapur India
144 D2 Anantnag India
173 E6 Anan'yiv Ukr.
173 F6 Anapa Rus. Fed.
214 D1 Anápolis Brazil
140 D4 Anār Iran
140 C3 Anārak Iran
141 F3 Anār Dareh Afgh.
141 F3 Anardara Afgh.
148 D4 Anatolia reg. Turkey see Anadolu
125 G4 Anatom i. Vanuatu
215 D3 Añatuya Arg.
213 E4 Anauá r. Brazil
163 C5 An Baile Breac Ireland
163 C6 An Blascaod Mór Ireland
163 C2 An Bun Beag Ireland
152 D4 Anbyon N. Korea
166 D3 Ancenis France
148 B4 Anchang China
184 D3 Anchorage U.S.A.

163 F3 Annalong U.K.
162 E6 Annan U.K.
162 E5 Annandale vol. U.K.
162 E5 Annan r. U.K.
202 E5 Annapolis U.S.A.
189 G5 Annapolis Royal Canada
145 E3 Annapurna I mt. Nepal
140 C5 An Naqirah well Saudi Arabia
191 F4 Ann Arbor U.S.A.
211 G2 Anna Regina Guyana
137 K6 An Nāṣirīyah Iraq
164 H4 Annecy France
166 H3 Annemasse France
186 C4 Annette Island U.S.A.
136 F5 An Nimārah Syria
149 B5 Anning China
201 C5 Anniston U.S.A.
174 E4 Annobón i. Equat. Guinea
166 G4 Annonay France
142 C4 An Nu'ayriyah Saudi Arabia
137 J5 An Nu'mānīyah Iraq
179 E5 Anorontany, Tanjona hd Madag.
149 D6 Anpu China
149 E4 Anpu Gang b. China
149 E4 Anqing China
148 F2 Anqiu China
An Ráth Ireland see Charleville
149 D5 Anren China
164 D4 Ans Belgium
149 C2 Ansai China
165 I5 Ansbach Germany
127 G7 Anser Group i.s Australia
152 B3 Anshan China
149 B5 Anshun China
215 C1 Ansilta mt. Arg.
215 F1 Ansina Uruguay
149 C3 Ansley U.S.A.
199 D5 Anson U.S.A.
176 C3 Ansongo Mali
184 C3 Ansonville Canada
202 C5 Ansted U.S.A.
144 D4 Anta India
210 D6 Antabamba Peru
136 F3 Antakya Turkey
179 F5 Antalaha Madag.
136 C3 Antalya Turkey
136 C3 Antalya Körfezi g. Turkey
179 E5 Antananarivo Madag.
129 Antarctica
129 B2 Antarctic Peninsula pen. Antarctica
162 C3 An Teallach mt. U.K.
196 D2 Antelope Range mts U.S.A.
167 D2 Antequera Spain
197 H5 Anthony U.S.A.
176 B2 Anti-Atlas mts Morocco
172 E3 Andreapol' Rus. Fed.
160 C3 Antibes France
189 H4 Anticosti, Île d' i. Canada
190 D3 Antigo U.S.A.
189 H4 Antigonish Canada
205 L5 Antigua i. Antigua and Barbuda
    Antigua Guat. see Antigua Guatemala
    Antigua and Barbuda country
    Caribbean Sea
207 G5 Antigua Guatemala Guat.
207 E3 Antiguo-Morelos Mex.
171 J7 Antikythira i. Greece
171 J7 Antikythiro, Steno sea chan. Greece
    Antioch Turkey see Antakya
196 B3 Antioch CA U.S.A.
190 C4 Antioch IL U.S.A.
213 B3 Antioquia Col.
123 G7 Antipodes Islands N.Z.
171 K5 Antipsara i. Greece
160 C3 Antlers U.S.A.
212 B2 Antofagasta Chile
212 B3 Antofalla, Volcán vol. Arg.
164 B4 Antoing Belgium
206 C3 Antón Panama
214 C4 Antonina Brazil
214 E1 Antônio r. Brazil
195 H4 Antonito U.S.A.
163 E3 Antrim U.K.
163 E3 Antrim Hills U.K.
179 E5 Antsalova Madag.
179 E5 Antsirabe Madag.
179 E5 Antsiranana Madag.
179 E5 Antsohihy Madag.
164 C3 Antsla Estonia
158 N3 Anttola Fin.
215 B3 Antuco Chile
215 B3 Antuco, Volcán vol. Chile
164 B3 Antwerp Belgium
164 B3 Antwerpen Belgium see Antwerp
124 C2 Anuc, Lac l. Canada
150 C3 Anuchino Rus. Fed.
143 C4 Anuppur India
143 C4 Anuradhapura Sri Lanka
145 H5 Anveh Iran
129 A2 Anvers Island i. Antarctica
149 F5 Anxi Fujian China
    Anxian China see Anchang
149 E4 Anxiang China
148 E2 Anxin China
126 A4 Anxious Bay Australia
148 E2 Anyang China
152 D5 Anyang S. Korea
171 K6 Andros i. Greece
148 B3 A'nyêmaqên Shan mts China
149 E4 Anyi China
149 E5 Anyuan China
148 B3 Anyue China
133 R3 Anyuysk Rus. Fed.
213 B3 Anzá Col.
138 F2 Anzhero-Sudzhensk Rus. Fed.
146 A1 Anzhu, Ostrova is Rus. Fed.
178 B4 Anzi Dem. Rep. Congo
125 G3 Anzio Italy
154 A3 Ao Kham, Laem pt Thai.
    Aomen China see Macao
150 G4 Aomori Japan
144 D3 Aonla India
128 C5 Aoraki N.Z.
128 C5 Aoraki/Mount Cook National Park
    N.Z.
154 C2 Aôral, Phnum mt. Cambodia
170 B2 Aosta Italy
176 B2 Aoukâr reg. Mali/Mauritania
211 G8 Apa r. Brazil
197 H5 Apache U.S.A.
197 H5 Apache Creek U.S.A.
197 G6 Apache Junction U.S.A.
197 G6 Apache Peak U.S.A.
201 C6 Apalachee Bay U.S.A.
201 C6 Apalachicola U.S.A.
213 C4 Apaporis r. Col.
214 C2 Aparecida do Tabuado Brazil
153 B2 Aparri Phil.
206 D4 Apatzingán Mex.
164 D2 Apeldoorn Neth.
165 H2 Apensen Germany
144 E3 Api mt. Nepal
214 E1 Apia Samoa
128 B5 Apiti N.Z.
69 A6 Apo, Mount vol. Phil.
214 B2 Apolda Germany
210 E7 Apolo Bol.
201 D6 Apopka, Lake U.S.A.
214 B2 Aporé Brazil
214 B1 Aporé r. Brazil
200 B2 Apostle Islands U.S.A.
190 B2 Apostle Islands National Lakeshore
    nature res. U.S.A.
163 D3 Annalee r. Ireland

Apostolos Andreas, Cape

136 E4 Apostolos Andreas, Cape Cyprus
202 B6 Appalachia U.S.A.
202 C6 Appalachian Mountains U.S.A.
170 E3 Appennino Abruzzese mts Italy
170 D2 Appennino Tosco-Emiliano mts Italy
170 E3 Appennino Umbro-Marchigiano mts Italy
127 I5 Appin Australia
164 E1 Appingedam Neth.
162 C3 Applecross U.K.
198 D2 Appleton MN U.S.A.
190 C3 Appleton WI U.S.A.
196 D4 Apple Valley U.S.A.
202 D6 Appomattox U.S.A.
170 E4 Aprilia Italy
173 F6 Apsheronsk Rus. Fed.
126 D6 Apsley Australia
191 H3 Apsley Canada
166 G5 Apt France
214 B3 Apucarana Brazil
153 A4 Apurahuan Phil.
213 D3 Apure r. Venez.
210 D6 Apurimac r. Peru
177 F2 Aqaba, Gulf of Asia
137 I6 'Aqabah, Birkat al well Iraq
139 I4 Aqal China
140 D2 Aqbana Iran
141 G2 Aqchah Afgh.
140 B2 Aq Chai r. Iran
140 D3 'Aqdā Iran
140 B2 Aqdoghmish r. Iran
137 J3 Āq Kān Dāgh, Kūh-e mt. Iran
138 C5 Āq Qal'eh Iran
137 I3 'Aqrah Iraq
141 G3 Aqrobāţ, Kowtal-e Afgh.
197 F4 Aquarius Mountains U.S.A.
197 G3 Aquarius Plateau U.S.A.
170 G4 Aquaviva delle Fonti Italy
214 A3 Aquidauana Brazil
214 A2 Aquidauana r. Brazil
206 D4 Aquila Mex.
213 D4 Aquio r. Col.
207 E3 Aquismón Mex.
166 D4 Aquitaine reg. France
145 F4 Ara India
141 G4 Arab Afgh.
201 C5 Arab U.S.A.
177 E3 Arab, Bahr el watercourse Sudan
136 B6 'Arab, Khalīj al b. Egypt
141 E3 'Arabābād Iran
218 I4 Arabian Basin sea feature Indian Ocean
Arabian Gulf g. Asia see The Gulf
142 B4 Arabian Peninsula Saudi Arabia
134 E5 Arabian Sea Indian Ocean
213 E3 Arabopó Venez.
213 E3 Arabopó r. Venez.
136 D1 Araç Turkey
213 E4 Araça r. Brazil
211 K6 Aracaju Brazil
213 D4 Aracamuni, Cerro h. Venez.
214 A4 Aracanguy, Montes de hills Para.
211 K4 Aracati Brazil
214 E1 Aracatu Brazil
214 B3 Araçatuba Brazil
167 C4 Aracena Spain
214 E2 Aracruz Brazil
214 D2 Araçuaí Brazil
214 D2 Araçuaí r. Brazil
171 I1 Arad Romania
177 E3 Arada Chad
124 D2 Arafura Sea Australia/Indon.
216 D6 Arafura Shelf sea feature Australia/Indon.
214 B1 Aragarças Brazil
137 I1 Aragats Armenia
137 J1 Aragats Lerr mt. Armenia
167 F2 Aragón aut. comm. Spain
167 F1 Aragón r. Spain
211 I5 Araguacema Brazil
213 D2 Aragua de Barcelona Venez.
211 I5 Araguaia r. Brazil
211 H6 Araguaia, Parque Indígena nat. park Brazil
211 H6 Araguaia, Parque Nacional do nat. park Brazil
211 I5 Araguaína Brazil
213 E2 Araguao, Boca r. mouth Venez.
213 E2 Araguao, Caño r. Venez.
214 C2 Araguari Brazil
214 C2 Araguari r. Brazil
211 I5 Araguatins Brazil
173 H7 Aragvi r. Georgia
151 F6 Arai Japan
Árainn Mhór i. Ireland see Arranmore Island
211 J4 Araiosos Brazil
176 C2 Arak Alg.
140 C3 Arāk Iran
145 H5 Arakan Yoma mts Myanmar
143 B3 Arakkonam India
137 J2 Aralık Turkey
138 D3 Aral Sea salt l. Kazakh./Uzbek.
138 E3 Aral'sk Kazakh.
Aral'skoye More salt l. Kazakh./Uzbek. see Aral Sea
138 B2 Aralsor, Ozero l. Kazakh.
138 C2 Aralsor, Ozero salt l. Kazakh.
140 B5 Aramah plat. Saudi Arabia
207 E2 Aramberri Mex.
144 D6 Aran r. India
167 E2 Aranda de Duero Spain
137 K4 Arandān Iran
171 I2 Aranđelovac Serbia
143 B3 Arani India
163 B4 Aran Islands Ireland
167 E2 Aranjuez Spain
179 B6 Aranos Namibia
199 D7 Aransas Pass U.S.A.
214 B2 Arantes r. Brazil
125 H1 Aranuka atoll Kiribati
154 B2 Aranyaprathet Thai.
151 B8 Arao Japan
176 B3 Araouane Mali
198 D3 Arapahoe U.S.A.
213 E4 Arapari r. Brazil
215 F1 Arapey Grande r. Uruguay
136 G2 Arapgir Turkey
211 K5 Arapiraca Brazil
171 K4 Arapis, Akrotirio pt Greece
214 B3 Arapongas Brazil
145 F4 A Rapti Doon r. Nepal
142 B3 'Ar'ar Saudi Arabia
137 I6 'Ar'ar, Wādī watercourse Iraq/Saudi Arabia
212 G3 Araranguá Brazil
214 C3 Araraquara Brazil
211 H5 Araras Brazil
214 B4 Araras, Serra das mts Brazil
137 J2 Ararat Armenia
126 E6 Ararat Australia
137 J2 Ararat, Mount Turkey
145 F4 Araria India
214 D3 Araruama, Lago de lag. Brazil
137 I2 Aras Turkey
137 I1 Aras r. Turkey alt. Araks (Armenia/Turkey), alt. Araz (Azerbaijan)
214 E1 Arataca Brazil
213 C3 Arauca Col.
213 C3 Arauca r. Venez.
215 B3 Araucanía admin. reg. Chile
215 B3 Arauco Chile
213 C3 Arauquita Col.
213 C2 Araure Venez.
144 C4 Aravalli Range mts India
159 N4 Aravete Estonia
125 F2 Arawa P.N.G.
214 C2 Araxá Brazil
213 D2 Araya, Península de pen. Venez.
213 D2 Araya, Punta de pt Venez.
136 C2 Arayıt Dağı mt. Turkey
137 L2 Araz r. Asia alt. Araks (Armenia/Turkey), alt. Aras (Turkey)
137 J4 Arbat Iraq
172 I3 Arbazh Rus. Fed.
137 J3 Arbīl Iraq
159 K4 Arboga Sweden
187 I4 Arborfield Canada
162 F4 Arbroath U.K.
196 A2 Arbuckle U.S.A.
141 F4 Arbu Lut, Dasht-e des. Afgh.
166 D4 Arcachon France
201 D7 Arcadia U.S.A.
194 A3 Arcata U.S.A.
196 D2 Arc Dome mt. U.S.A.
207 D4 Arcelia Mex.
172 G1 Archangel Rus. Fed.
124 E3 Archer r. Australia
197 H2 Arches National Park U.S.A.
Archman Turkm. see Arçman
137 L2 Ärçivan Azer.
126 A2 Arckaringa watercourse Australia
138 D5 Arçman Turkm.
194 D3 Arco U.S.A.
167 D4 Arcos de la Frontera Spain
185 J2 Arctic Bay Canada
220 B1 Arctic Mid-Ocean Ridge sea feature Arctic Ocean
220 Arctic Ocean
184 E3 Arctic Red r. Canada
129 E2 Arctowski research stn Antarctica
140 C2 Ardabīl Iran
137 I1 Ardahan Turkey
140 C4 Ardakān Iran
140 D3 Ardakān Iran
140 C4 Ardal Iran
159 I3 Årdalstangen Norway
163 C3 Ardara Ireland
171 K4 Ardas r. Bulg.
172 G4 Ardatov Nizhegorodskaya Oblast' Rus. Fed.
172 H4 Ardatov Respublika Mordoviya Rus. Fed.
191 G3 Ardbeg Canada
163 E4 Ardee Ireland
126 B4 Arden, Mount h. Australia
164 D5 Ardennes dept Belgium
164 C5 Ardennes, Canal des France
140 D3 Ardestān Iran
163 F3 Ardglass U.K.
167 C3 Ardila r. Port.
127 G5 Ardlethan Australia
199 D5 Ardmore U.S.A.
162 B4 Ardnamurchan, Point of U.K.
162 C4 Ardrishaig U.K.
126 B5 Ardrossan Australia
162 D5 Ardrossan U.K.
162 C3 Ardvasar U.K.
215 E2 Areco r. Arg.
212 E3 Areguá Para.
211 K4 Areia Branca Brazil
164 E4 Aremberg h. Germany
153 B4 Arena rf Phil.
196 A2 Arena, Point U.S.A.
206 B3 Arena, Punta pt Mex.
206 H5 Arenal Hond.
167 D2 Arenas de San Pedro Spain
159 J4 Arendal Norway
165 J2 Arendsee (Altmark) Germany
161 D5 Arenig Fawr h. U.K.
171 J6 Areopoli Greece
206 C2 Areponapuchi Mex.
210 D7 Arequipa Peru
211 H4 Arere Brazil
167 D2 Arévalo Spain
170 D3 Arezzo Italy
136 G6 'Arfajah well Saudi Arabia
148 D1 Argalant Mongolia
167 E2 Arganda del Rey Spain
153 B4 Argao Phil.
170 D2 Argenta Italy
166 D2 Argentan France
170 D3 Argentario, Monte h. Italy
170 B2 Argentera, Cima dell' mt. Italy
164 F5 Argenthal Germany
212 C5 Argentina country S. America
129 C4 Argentina Range mts Antarctica
219 F9 Argentine Abyssal Plain sea feature S. Atlantic Ocean
219 G8 Argentine Basin sea feature S. Atlantic Ocean
219 F8 Argentine Rise sea feature S. Atlantic Ocean
212 B8 Argentino, Lago l. Arg.
171 K2 Argeș r. Romania
141 G4 Arghandāb, Rūd-e r. Afgh.
141 G4 Arghastan r. Afgh.
136 C2 Argıthanı Turkey
171 J6 Argolikos Kolpos b. Greece
171 J6 Argos Greece
171 I5 Argostoli Greece
167 F1 Arguis Spain
146 E1 Argun' r. China/Rus. Fed.
173 H7 Argun Rus. Fed.
129 D5 Argus, Dome ice feature Antarctica
196 D4 Argus Range mts U.S.A.
190 C4 Argyle U.S.A.
124 C3 Argyle, Lake Australia
162 C4 Argyll reg. U.K.
159 J4 Århus Denmark
128 E3 Aria N.Z.
127 G5 Ariah Park Australia
151 B8 Ariake-kai b. Japan
179 B6 Ariamsvlei Namibia
170 F4 Ariano Irpino Italy
213 B4 Ariari r. Col.
215 D2 Arias Arg.
135 F6 Ari Atoll Maldives
213 E2 Aribí r. Venez.
176 B3 Aribinda Burkina
212 B1 Arica Chile
162 C4 Arienas, Loch l. U.K.
136 F4 Arīḩā Syria
194 G4 Arikaree r. U.S.A.
213 E2 Arima Trin. and Tob.
214 C1 Arinos Brazil
211 G6 Arinos r. Brazil
206 D4 Ario de Rosáles Mex.
213 C3 Ariporo r. Col.
210 F5 Aripuanã Brazil
210 F5 Aripuanã r. Brazil
210 G6 Aripuanã, Parque Indígena nat. park Brazil
210 F5 Ariquemes Brazil
214 B2 Ariranhá r. Brazil
180 B1 Aris Namibia
162 C4 Arisaig U.K.
162 C4 Arisaig, Sound of sea chan. U.K.
186 D4 Aristazabal Island Canada
212 C2 Arizaro, Salar de salt flat Arg.
197 G4 Arizona state U.S.A.
192 D5 Arizpe Mex.
140 B5 'Arjah Saudi Arabia
158 L2 Arjeplog Sweden
213 B2 Arjona Col.
155 D4 Arjuna, Gunung vol. Indon.
173 G5 Arkadak Rus. Fed.
199 E5 Arkadelphia U.S.A.
162 C4 Arkaig, Loch l. U.K.
139 F2 Arkalyk Kazakh.
199 F5 Arkansas r. U.S.A.
199 E5 Arkansas state U.S.A.
199 D4 Arkansas City U.S.A.
145 G1 Arkatag Shan mts China
172 G2 Arkhangel'skaya Oblast' admin. div. Rus. Fed.
172 F4 Arkhangel'skoye Rus. Fed.
150 C3 Arkhipovka Rus. Fed.
163 E5 Arklow Ireland
171 L6 Arkoi i. Greece
168 F3 Arkona, Kap c. Germany
132 J2 Arkticheskogo Instituta, Ostrova is Rus. Fed.
203 F3 Arkville U.S.A.
166 G5 Arles France
181 G4 Arlington S. Africa
194 B2 Arlington OR U.S.A.
198 D2 Arlington SD U.S.A.
202 E5 Arlington VA U.S.A.
190 D4 Arlington Heights U.S.A.
176 C3 Arlit Niger
164 D5 Arlon Belgium
153 C5 Armadores i. Indon.
163 E3 Armagh U.K.
177 F2 Armant Egypt
137 J1 Armavir Armenia
173 G6 Armavir Rus. Fed.
137 J1 Armenia country Asia
213 B3 Armenia Col.
213 B3 Armero Col.
127 I3 Armidale Australia
187 K2 Armit Lake Canada
144 E5 Armori India
163 E2 Armoy U.K.
186 F4 Armstrong B.C. Canada
188 C3 Armstrong Ont. Canada
150 E1 Armu r. Rus. Fed.
143 B2 Armur India
173 E6 Armyans'k Ukr.
Arnaoutis, Cape c. Cyprus see Arnauti, Cape
189 F1 Arnaud r. Canada
136 D4 Arnauti, Cape Cyprus
159 J3 Årnes Norway
199 D4 Arnett U.S.A.
164 D3 Arnhem Neth.
124 D3 Arnhem, Cape Australia
124 D3 Arnhem Bay Australia
124 D3 Arnhem Land reg. Australia
170 D3 Arno r. Italy
126 B4 Arno Bay Australia
161 F4 Arnold U.K.
190 D2 Arnold U.S.A.
191 H1 Arnoux, Lac l. Canada
191 I3 Arnprior Canada
165 G3 Arnsberg Germany
165 I4 Arnstadt Germany
191 H3 Arnstein Canada
165 H5 Arnstein Germany
191 H1 Arntfield Canada
213 E3 Aro r. Venez.
179 B6 Aroab Namibia
165 H3 Arolsen Germany
170 C2 Arona Italy
203 J1 Aroostook Canada
203 I1 Aroostook r. Canada/U.S.A.
125 H2 Arorae i. Kiribati
153 B3 Aroroy Phil.
173 G7 Arpa r. Armenia/Turkey
137 I1 Arpaçay Turkey
141 G5 Arra r. Pak.
137 I5 Ar Raḩḩālīyah Iraq
137 I5 Ar Ramādī Iraq
136 E7 Ar Ramlah Jordan
162 C5 Arran i. U.K.
163 C3 Arranmore Island Ireland
137 G4 Ar Raqqah Syria
166 F1 Arras France
140 A5 Ar Rass Saudi Arabia
136 F4 Ar Rastān Syria
137 I7 Ar Rawḑ well Saudi Arabia
140 C5 Ar Rayyān Qatar
213 C4 Arrecifal Col.
215 E2 Arrecifes Arg.
207 F4 Arriagá Chiapas Mex.
206 D3 Arriaga San Luis Potosí Mex.
215 E2 Arribeños Arg.
137 K6 Ar Rifā'ī Iraq
137 J6 Ar Rihāb salt flat Iraq
142 D5 Ar Rimāl reg. Saudi Arabia
197 H3 Arriola U.S.A.
Ar Riyāḑ Saudi Arabia see Riyadh
162 D4 Arrochar U.K.
215 G2 Arroio Grande Brazil
214 D1 Arrojado r. Brazil
163 C3 Arrow, Lough l. Ireland
190 B1 Arrow Lake Canada
194 D3 Arrowrock Reservoir U.S.A.
128 C5 Arrowsmith, Mount N.Z.
196 B4 Arroyo Grande U.S.A.
215 E2 Arroyo Seco Arg.
207 E3 Arroyo Seco Mex.
140 B5 Ar Rubay'iyah Saudi Arabia
214 A1 Arruda Brazil
137 J6 Ar Rumaythah Iraq
136 G4 Ar Ruşāfah Syria
141 E6 Ar Rustāq Oman
137 H5 Ar Ruţbah Iraq
140 B6 Ar Ruwayḑah Saudi Arabia
Års Denmark see Aars
140 B2 Ars Iran
140 D4 Arsenajān Iran
150 C2 Arsen'yev Rus. Fed.
139 H2 Arshaly Akmolinskaya Oblast' Kazakh.
139 J2 Arshaly Vostochnyy Kazakhstan Kazakh.
143 B3 Arsikere India
172 I3 Arsk Rus. Fed.
136 E3 Arslanköy Turkey
178 E2 Arta Djibouti
171 I5 Arta Greece
137 J2 Artashat Armenia
206 D4 Arteaga Mex.
150 C3 Artem Rus. Fed.
173 F5 Artemivs'k Ukr.
150 C3 Artemovskiy Rus. Fed.
166 E2 Artenay France
195 F5 Artesia U.S.A.
191 G4 Arthur Canada
126 B3 Arthur, Lake salt flat Australia
202 C4 Arthur, Lake U.S.A.
127 G8 Arthur Lake Australia
124 F4 Arthur Point Australia
128 C5 Arthur's Pass N.Z.
128 C5 Arthur's Pass National Park N.Z.
201 F7 Arthur's Town Bahamas
129 E2 Artigas research stn Antarctica
215 F1 Artigas Uruguay
137 I1 Art'ik Armenia
187 H2 Artillery Lake Canada
181 G2 Artisia Botswana
166 F1 Artois reg. France
164 A4 Artois, Collines d' hills France
137 I2 Artos Dağı mt. Turkey
136 F1 Artova Turkey
173 D6 Artsyz Ukr.
129 B2 Arturo Prat research stn Antarctica
139 I5 Artux China
137 H1 Artvin Turkey
147 F7 Aru, Kepulauan is Indon.
178 D3 Arua Uganda
214 B1 Aruanã Brazil
213 C1 Aruba terr. Caribbean Sea
210 F4 Arumã Brazil
145 F4 Arun r. Nepal
145 H3 Arunachal Pradesh state India
161 G7 Arundel U.K.
178 D4 Arusha Tanz.
198 B4 Arvada U.S.A.
163 D4 Arvagh Ireland
146 C2 Arvayheer Mongolia
144 D5 Arvi India
187 K2 Arviat Canada
189 F4 Arvida Canada
158 L2 Arvidsjaur Sweden
159 K4 Arvika Sweden
196 C4 Arvin U.S.A.
140 B6 Arwā' Saudi Arabia
139 G1 Arykbalyk Kazakh.
139 G4 Arys Kazakh.
139 G4 Arys' r. Kazakh.
138 F3 Arys, Ozero salt l. Kazakh.
172 G4 Arzamas Rus. Fed.
165 K4 Arzberg Germany
167 F5 Arzew Alg.
173 H6 Arzgir Rus. Fed.
165 K4 Aš Czech Rep.
213 E3 Asa r. Venez.
176 C4 Asaba Nigeria
136 G3 Asad, Buḩayrat al resr Syria
141 H3 Asadābād Afgh.
140 C3 Asadābād Iran
140 D4 Asadābād Iran
141 E3 Asadābād Iran
154 A5 Asahan r. Indon.
150 H3 Asahi-dake vol. Japan
151 C7 Asahi-gawa r. Japan
150 H3 Asahikawa Japan
139 H4 Asaka Uzbek.
140 C2 Asālem Iran
152 D5 Asan-man b. S. Korea
145 F5 Asansol India
164 F4 Asbach Germany
189 F4 Asbestos Canada
180 E4 Asbestos Mountains S. Africa
203 F4 Asbury Park U.S.A.
170 F4 Ascea Italy
210 F7 Ascensión Bol.
175 C5 Ascension i. S. Atlantic Ocean
207 H4 Ascensión, Bahía de la b. Mex.
165 H5 Aschaffenburg Germany
164 F3 Ascheberg Germany
165 J3 Aschersleben Germany
170 E3 Ascoli Piceno Italy
159 K4 Åseda Sweden
158 L2 Åsele Sweden
171 K3 Asenovgrad Bulg.
138 D5 Aşgabat Turkm.
140 B5 Asharat Saudi Arabia
124 B4 Ashburton watercourse Australia
128 C5 Ashburton N.Z.
190 D1 Ashburton Bay Canada
139 F3 Ashchikol', Ozero salt l. Kazakh.
139 G4 Ashchysay Kazakh.
186 E4 Ashcroft Canada
136 E6 Ashdod Israel
199 E5 Ashdown U.S.A.
201 E5 Asheboro U.S.A.
201 D5 Asheville U.S.A.
127 I2 Ashford Australia
161 H6 Ashford U.K.
197 F4 Ash Fork U.S.A.
Ashgabat Turkm. see Aşgabat
150 H3 Ashibetsu Japan
151 F6 Ashikaga Japan
160 F2 Ashington U.K.
172 I3 Ashit r. Rus. Fed.
151 C8 Ashizuri-misaki pt Japan
140 D4 Ashkazar Iran
199 D4 Ashland KS U.S.A.
202 B5 Ashland KY U.S.A.
203 I1 Ashland ME U.S.A.
194 F2 Ashland MT U.S.A.
203 H3 Ashland NH U.S.A.
202 B4 Ashland OH U.S.A.
194 B3 Ashland OR U.S.A.
202 E6 Ashland VA U.S.A.
190 B2 Ashland WI U.S.A.
127 H2 Ashley Australia
198 D2 Ashley U.S.A.
124 C3 Ashmore and Cartier Islands terr. Australia
172 C4 Ashmyany Belarus
197 H5 Ash Peak U.S.A.
136 E6 Ashqelon Israel
137 I6 Ash Shabakah Iraq
137 H3 Ash Shaddādah Syria
137 J6 Ash Shanāfīyah Iraq
137 H7 Ash Shaqīq well Saudi Arabia
140 B5 Ash Sha'rā' Saudi Arabia
136 E6 Ash Sharāh reg. Jordan
142 E4 Ash Shāriqah U.A.E.
137 I4 Ash Sharqāţ Iraq
137 K6 Ash Shaţrah Iraq
136 D7 Ash Shaţţ Egypt
136 E6 Ash Shawbak Jordan
142 C7 Ash Shiḩr Yemen
140 E5 Ash Shināş Oman
140 B4 Ash Shu'bah Saudi Arabia
140 B5 Ash Shumlūl Saudi Arabia
202 C4 Ashtabula U.S.A.
137 J1 Ashtarak Armenia
143 A2 Ashti Maharashtra India
144 D5 Ashti Maharashtra India
140 C3 Ashtiān Iran
180 D6 Ashton S. Africa
194 E2 Ashton U.S.A.
160 E4 Ashton-under-Lyne U.K.
185 L4 Ashuanipi Lake Canada
188 F4 Ashuapmushuan r. Canada
188 F4 Ashuapmushuan, Réserve Faunique nature res. Canada
201 C5 Ashville U.S.A.
136 F4 'Āşī, Nahr al r. Asia
130 Asia
206 D3 Asientos Mex.
143 B2 Asifabad India
143 D2 Asika India
170 C4 Asinara, Golfo dell' b. Sardinia Italy
132 J4 Asino Rus. Fed.
172 D4 Asipovichy Belarus
142 B5 'Asīr reg. Saudi Arabia
137 H2 Aşkale Turkey
138 D1 Askarovo Rus. Fed.
159 J4 Asker Norway
159 K4 Askersund Sweden
159 J4 Askim Norway
137 I3 Askī Mawşil Iraq
146 B1 Askiz Rus. Fed.
137 K2 Aşlāndūz Iran
178 D2 Asmara Eritrea
159 K4 Åsnen l. Sweden
144 C4 Asop India
137 L3 Aspar Iran
139 H4 Aspara Kazakh.
160 D3 Aspatria U.K.
195 F4 Aspen U.S.A.
165 H6 Asperg Germany
199 C5 Aspermont U.S.A.
128 B6 Aspiring, Mount N.Z.
187 H4 Asquith Canada
139 G4 Assa Kazakh.
136 F4 As Sa'an Syria
178 E2 Assab Eritrea
140 C5 As Sabsab well Saudi Arabia
136 C7 Aş Şaff Egypt
136 E6 Aş Şāfī Jordan
136 F3 As Safīrah Syria
Aş Şaḩrā' al Gharbīyah des. Egypt see Western Desert
Aş Şaḩrā' ash Sharqīyah des. Egypt see Eastern Desert
140 C5 As Saji well Saudi Arabia
138 D4 Assake-Audan, Vpadina depr. Uzbek.
140 B5 As Salamiyah Saudi Arabia
136 D6 Aş Şāliḩīyah Egypt
137 H4 Aş Şāliḩīyah Syria
137 J6 As Salmān Iraq
136 E5 As Salţ Jordan
145 H4 Assam state India
137 J6 As Samāwah Iraq
136 F5 Aş Şanamayn Syria
177 E2 As Sarīr reg. Libya
203 F5 Assateague Island U.S.A.
203 F6 Assateague Island National Seashore nature res. U.S.A.
164 E1 Assen Neth.
164 B3 Assenede Belgium
164 D4 Assesse Belgium
145 F5 Assia Hills India
177 D1 As Sidrah Libya
187 H5 Assiniboia Canada
187 J5 Assiniboine r. Canada
186 F4 Assiniboine, Mount Canada
214 B3 Assis Brazil
170 E3 Assisi Italy
165 G4 Aßlar Germany
137 K7 Aş Şubayḩīyah Kuwait
140 B4 Aş Şufayrī well Saudi Arabia
137 G4 As Sukhnah Syria
137 J4 As Sulaymānīyah Iraq
142 C5 As Sulayyil Saudi Arabia
140 C5 Aş Şulb reg. Saudi Arabia
140 B5 Aş Şummān plat. Saudi Arabia
142 B5 As Sūq Saudi Arabia
137 H4 Aş Şuwar Syria
136 F5 As Suwaydā' Syria
141 E6 As Suwayq Oman
137 J5 Aş Şuwayrah Iraq
As Suways Egypt see Suez
162 C2 Assynt, Loch l. U.K.
171 L7 Astakida i. Greece
134 F1 Astana Kazakh.
137 L3 Astaneh Iran
137 L2 Āstārā Azer.
170 C2 Asti Italy
196 A2 Asti U.S.A.
215 C1 Astica Arg.
141 F5 Astola Island Pak.
144 C2 Astor Pak.
144 C2 Astor r. Pak.
167 C1 Astorga Spain
194 B2 Astoria U.S.A.
159 K4 Åstorp Sweden
134 C2 Astrakhan' Rus. Fed.
173 H6 Astrakhanskaya Oblast' admin. div. Rus. Fed.
172 C4 Astravyets Belarus
129 D3 Astrid Ridge sea feature Southern Ocean
167 C1 Asturias aut. comm. Spain
171 L6 Astypalaia i. Greece
139 J2 Asubulak Kazakh.
212 E3 Asunción Para.
177 F2 Aswān Egypt
177 F2 Asyūţ Egypt
125 I4 Ata i. Tonga
213 D4 Atabapo r. Col./Venez.
139 G4 Atabay Kazakh.
Atacama, Desierto de des. Chile see Atacama Desert
212 C2 Atacama, Salar de salt flat Chile
212 C2 Atacama Desert Chile
125 I2 Atafu atoll Tokelau
139 G4 Atakent Kazakh.
176 C4 Atakpamé Togo
171 J5 Atalanti Greece
206 I7 Atalaya Panama
210 D6 Atalaya Peru
138 F5 Atamyrat Turkm.
139 G1 Atansor, Ozero salt l. Kazakh.
176 A2 Atâr Mauritania
154 A1 Ataran r. Myanmar
197 H4 Atarque U.S.A.
196 B4 Atascadero U.S.A.
139 G2 Atasu Kazakh.
147 E7 Ataúro, Ilha de i. East Timor
171 L6 Atavyros mt. Greece
138 D4 Ataýap Turkm.
177 F3 Atbara Sudan
177 F3 Atbara r. Sudan
139 G2 Atbasar Kazakh.
139 H4 At-Bashy Kyrg.
199 F6 Atchafalaya Bay U.S.A.
198 E4 Atchison U.S.A.
206 C3 Atenguillo Mex.
170 E3 Aterno r. Italy
170 F3 Atessa Italy
164 B4 Ath Belgium
186 G4 Athabasca Canada
184 G4 Athabasca r. U.S.A.
187 G3 Athabasca, Lake Canada
137 I6 'Athāmīn, Birkat al well Iraq
163 E4 Athboy Ireland
163 C4 Athenry Ireland
191 J3 Athens Canada
171 J6 Athens Greece (City Plan 107)
201 C5 Athens AL U.S.A.
201 D5 Athens GA U.S.A.
202 B5 Athens OH U.S.A.
201 C5 Athens TN U.S.A.
199 E5 Athens TX U.S.A.
161 F5 Atherstone U.K.
164 A5 Athies France
Athina Greece see Athens
163 C4 Athleague Ireland
163 D4 Athlone Ireland
143 A2 Athni India
128 B6 Athol N.Z.
203 G3 Athol U.S.A.
162 E4 Atholl, Forest of reg. U.K.
171 K4 Athos mt. Greece
136 E7 Ath Thamad Egypt
163 E5 Athy Ireland
177 D3 Ati Chad
210 D7 Atico Peru
188 B4 Atikokan Canada
189 H3 Atikonak Lake Canada
153 B3 Atimonan Phil.
143 B4 Atirampattinam India
207 G5 Atitlán Guat.
207 G5 Átitlán, Volcán vol. Guat.
140 B5 'Atk, Wādī al watercourse Saudi Arabia
133 Q3 Atka Rus. Fed.
173 H5 Atkarsk Rus. Fed.
207 E4 Atlacomulco Mex.
201 C5 Atlanta GA U.S.A.
190 C5 Atlanta IL U.S.A.
191 E3 Atlanta MI U.S.A.
136 D2 Atlantı Turkey
198 E3 Atlantic U.S.A.
203 F5 Atlantic City U.S.A.
218 D9 Atlantic-Indian-Antarctic Basin sea feature S. Atlantic Ocean
218 D8 Atlantic-Indian Ridge sea feature Southern Ocean
219 Atlantic Ocean
180 C6 Atlantis S. Africa
173 B1 Atlas Mountains Africa
176 C1 Atlas Saharien mts Alg.
186 C3 Atlin Canada
186 C3 Atlin Lake Canada
186 C3 Atlin Provincial Park Canada
136 E5 'Atlit Israel
207 E4 Atlixco Mex.
143 B3 Atmakur Andhra Pradesh India
143 B3 Atmakur India
199 G6 Atmore U.S.A.
199 D5 Atoka U.S.A.
206 D3 Atotonilco el Alto Mex.
154 C1 Atouat mt. Laos
207 D4 Atoyac de Álvarez Mex.
145 G4 Atrai r. India
140 E2 Atrak, Rūd-e r. Iran/Turkm. alt. Atrek, alt. Etrek
213 A3 Atrato r. Col.
140 D2 Atrek r. Iran/Turkm. alt. Atrak, Rūd-e, alt. Etrek
203 F5 Atsion U.S.A.
136 E6 Aţ Ţafīlah Jordan
142 B5 Aţ Ţā'if Saudi Arabia
201 C5 Attalla U.S.A.
154 C2 Attapu Laos
188 D3 Attawapiskat Canada
188 C3 Attawapiskat r. Canada
188 C3 Attawapiskat Lake Canada
137 G7 Aţ Ţawīl mts Saudi Arabia
140 A4 At Taysīyah plat. Saudi Arabia
165 F3 Attendorn Germany
168 F7 Attersee l. Austria
190 D5 Attica IN U.S.A.
202 B4 Attica OH U.S.A.
164 C5 Attigny France
203 H3 Attleboro U.S.A.
161 I5 Attleborough U.K.
136 F7 Aţ Ţubayq reg. Saudi Arabia
216 G2 Attu Island U.S.A.
140 B5 Aţ Ţulayḩī well Saudi Arabia
177 F2 Aţ Ţūr Egypt
143 B4 Attur India
215 C2 Atuel r. Arg.
215 C3 Atuel, Bañados del marsh Arg.
159 K4 Åtvidaberg Sweden
202 C4 Atwood Lake U.S.A.
138 B3 Atyrau Kazakh.
138 B3 Atyrauskaya Oblast' admin. div. Kazakh.
165 I5 Aub Germany
166 G5 Aubagne France
164 D5 Aubange Belgium
153 B2 Aubarede Point Phil.
166 G4 Aubenas France
164 D5 Auboué France
197 F4 Aubrey Cliffs mts U.S.A.
184 F3 Aubry Lake Canada
126 C5 Auburn Australia
191 G4 Auburn Canada
201 C5 Auburn AL U.S.A.
196 B2 Auburn CA U.S.A.
190 E5 Auburn IN U.S.A.
203 H2 Auburn ME U.S.A.
198 E3 Auburn NE U.S.A.
202 E3 Auburn NY U.S.A.
194 B2 Auburn WA U.S.A.
166 F4 Aubusson France
215 C3 Auca Mahuida, Sierra de mt. Arg.
166 E5 Auch France
162 E4 Auchterarder U.K.
128 E2 Auckland N.Z. (City Plan 102)
125 G7 Auckland Islands N.Z.
203 H2 Audet Canada
161 I7 Audresselles France
164 A4 Audruicq France
165 K4 Aue Germany
165 K4 Auerbach Germany
165 J5 Auerbach in der Oberpfalz Germany
165 K4 Auersberg mt. Germany
139 J2 Auezov Kazakh.
163 D3 Augher U.K.
163 E3 Aughnacloy U.K.
163 E5 Aughrim Ireland
180 D4 Augrabies S. Africa
180 D4 Augrabies Falls S. Africa
180 D4 Augrabies Falls National Park S. Africa
191 F3 Au Gres U.S.A.
168 E6 Augsburg Germany
170 F6 Augusta Sicily Italy
201 D5 Augusta GA U.S.A.
199 D4 Augusta KS U.S.A.
203 I2 Augusta ME U.S.A.
190 B3 Augusta WI U.S.A.
169 K4 Augustowska, Puszcza for. Poland
124 B4 Augustus, Mount Australia
138 F1 Auliyekol' Kazakh.
164 B4 Aulnoye-Aymeries France
161 I7 Ault France
Auminzatau, Gory hill Uzbek. see Ovminzatov tog'lari
179 B6 Auob watercourse Namibia
189 G2 Aupaluk Canada
154 C5 Aur i. Malaysia
159 M3 Aura Fin.
144 D4 Auraiya India
144 C6 Aurangabad India
164 F1 Aurich Germany
214 B2 Aurilândia Brazil
166 F4 Aurillac France
155 D3 Aurkuning Indon.
153 B5 Aurora Phil.
194 F4 Aurora CO U.S.A.
190 C5 Aurora IL U.S.A.
203 I2 Aurora ME U.S.A.
199 E4 Aurora MO U.S.A.
179 B6 Aus Namibia
191 F3 Au Sable U.S.A.
191 E3 Au Sable r. U.S.A.
203 G2 Ausable r. U.S.A.
Ausable Forks U.S.A. see Au Sable Forks
203 G2 Au Sable Forks U.S.A.
190 D2 Au Sable Point MI U.S.A.
191 F3 Au Sable Point MI U.S.A.
162 F1 Auskerry i. U.K.
158 C2 Austari-Jökulsá r. Iceland
198 E3 Austin MN U.S.A.
196 D2 Austin NV U.S.A.
199 D6 Austin TX U.S.A.
Australes, Îles is Fr. Polynesia see Tubuai Islands
124 C4 Australia country Oceania
219 P9 Australian-Antarctic Basin sea feature Southern Ocean
129 B7 Australian Antarctic Territory reg. Antarctica
127 H5 Australian Capital Territory admin. div. Australia
168 F7 Austria country Europe
158 K1 Austvågøy i. Norway
206 C4 Autlán Mex.
158 N2 Autti Fin.
166 G3 Autun France
166 F4 Auvergne reg. France
166 F3 Auxerre France
164 A4 Auxi-le-Château France
166 G3 Auxonne France
203 F3 Ava U.S.A.
166 F3 Avallon France
196 C5 Avalon U.S.A.
189 J4 Avalon Peninsula Canada
206 D2 Ávalos Mex.
140 B2 Avān Iran
136 E2 Avanos Turkey
214 C3 Avaré Brazil
137 K2 Āvārsin Iran
196 D4 Avawatz Mountains U.S.A.
141 F3 Avaz Iran
211 G4 Aveiro Brazil
167 B2 Aveiro Port.
167 B2 Aveiro, Ria de est. Port.
137 L4 Āvej Iran
215 E2 Avellaneda Arg.
170 F4 Avellino Italy
196 B3 Avenal U.S.A.
126 F6 Avenel Australia
164 C2 Avenhorn Neth.
170 F4 Aversa Italy
164 B4 Avesnes-sur-Helpe France
159 L3 Avesta Sweden
166 F4 Aveyron r. France
170 E3 Avezzano Italy
162 E3 Aviemore U.K.
170 F4 Avigliano Italy
166 G5 Avignon France
167 D2 Ávila Spain
167 D1 Avilés Spain
164 A4 Avion France
143 C5 Avissawella Sri Lanka
172 H2 Avnyugskiy Rus. Fed.
127 G8 Avoca Tas. Australia
126 E6 Avoca Vic. Australia
126 E6 Avoca r. Australia
163 E5 Avoca Ireland
198 E3 Avoca U.S.A.
170 F6 Avola Sicily Italy
161 E6 Avon r. England U.K.
161 F5 Avon r. England U.K.
161 F7 Avon r. England U.K.
190 B5 Avon U.S.A.
197 F5 Avondale U.S.A.
161 E6 Avonmouth U.K.
201 D7 Avon Park U.S.A.
166 D2 Avranches France
164 A5 Avre r. France
125 G2 Avuavu Solomon Is
151 D7 Awaji-shima i. Japan
128 E3 Awakino N.Z.
Awālī Bahrain see 'Awālī
140 C5 'Awālī Bahrain
128 D1 Awanui N.Z.
128 B6 Awarua Point N.Z.
178 E3 Āwash Eth.
178 D3 Āwash r. Eth.
150 F5 Awa-shima i. Japan
178 D3 Āwash National Park Eth.
180 A2 Awasib Mountains Namibia
139 J4 Awat China
128 D4 Awatere r. N.Z.
177 D2 Awbārī Libya
176 D2 Awbārī, Idhān des. Libya
163 C5 Awbeg r. Ireland
137 K6 'Awdah, Hawr al imp. l. Iraq
178 E3 Aw Dheegle Somalia
162 C4 Awe, Loch l. U.K.
177 E4 Aweil Sudan
176 C4 Awka Nigeria
153 C6 Awu vol. Indon.
126 F6 Axedale Australia
185 I2 Axel Heiberg Island Canada

Borok

172 F3 Borok Rus. Fed.
176 B3 Boromo Burkina
153 C4 Borongan Phil.
160 F3 Boroughbridge U.K.
172 E3 Borovichi Rus. Fed.
139 K1 Borovlyanka Rus. Fed.
172 I3 Borovoy Kirovskaya Oblast' Rus. Fed.
172 E1 Borovoy Respublika Kareliya Rus. Fed.
172 J2 Borovoy Respublika Komi Rus. Fed.
138 F1 Borovskoy Kazakh.
163 C5 Borrisokane Ireland
124 D3 Borroloola Australia
158 J3 Børsa Norway
169 L7 Borşa Romania
138 D4 Borsakelmas sho'rxogi salt marsh Uzbek.
173 C5 Borshchiv Ukr.
146 C2 Borshchovochnyy Khrebet mts Rus. Fed.
139 J3 Bortala He r. China
140 C4 Borūjen Iran
140 C3 Borūjerd Iran
162 B3 Borve U.K.
173 B5 Boryslav Ukr.
173 D5 Boryspil' Ukr.
173 E5 Borzna Ukr.
146 D1 Borzya Rus. Fed.
139 H3 Bosaga Kazakh.
170 G2 Bosanska Dubica Bos.-Herz.
170 G2 Bosanska Gradiška Bos.-Herz.
170 G2 Bosanska Krupa Bos.-Herz.
170 G2 Bosanski Novi Bos.-Herz.
170 G2 Bosansko Grahovo Bos.-Herz.
190 B4 Boscobel U.S.A.
149 C6 Bose China
181 F4 Boshof S. Africa
138 E1 Boskol' Kazakh.
170 G2 Bosnia-Herzegovina country Europe
178 B3 Bosobolo Dem. Rep. Congo
151 G7 Bōsō-hantō pen. Japan
136 B1 Bosporus str. Turkey
178 B3 Bossangoa Centr. Afr. Rep.
178 B3 Bossembélé Centr. Afr. Rep.
199 E5 Bossier City U.S.A.
180 B2 Bossiesvlei Namibia
145 F1 Bostan China
137 K6 Bostān Iran
138 B2 Bostandyk Kazakh.
146 A2 Bosten Hu l. China
161 G5 Boston U.K.
203 H3 Boston U.S.A.
126 A5 Boston Bay Australia
191 H1 Boston Creek Canada
203 H3 Boston-Logan airport U.S.A.
199 E5 Boston Mountains U.S.A.
160 F4 Boston Spa U.K.
190 D5 Boswell U.S.A.
144 B5 Botad India
139 H2 Botakara Kazakh.
127 I4 Botany Bay Australia
158 L3 Boteå Sweden
171 K3 Botev mt. Bulg.
171 J3 Botevgrad Bulg.
181 G3 Bothaville S. Africa
158 L3 Bothnia, Gulf of Fin./Sweden
158 M2 Bothnian Bay g. Fin./Sweden
127 G9 Bothwell Australia
173 H5 Botkul', Ozero l. Kazakh./Rus. Fed.
169 M7 Botoşani Romania
148 E2 Botou China
Bô Trach Vietnam see Hoan Lao
181 G4 Botshabelo S. Africa
179 C6 Botswana country Africa
170 G5 Botte Donato, Monte mt. Italy
160 G4 Bottesford U.K.
198 C1 Bottineau U.S.A.
164 E3 Bottrop Germany
214 C3 Botucatu Brazil
214 D1 Botuporã Brazil
189 J4 Botwood Canada
176 B4 Bouaflé Côte d'Ivoire
176 B4 Bouaké Côte d'Ivoire
178 B3 Bouar Centr. Afr. Rep.
176 B1 Bouârfa Morocco
177 D4 Bouba Ndjida, Parc National de nat. park Cameroon
178 B3 Bouca Centr. Afr. Rep.
164 B4 Bouchain France
203 G2 Boucherville Canada
191 J2 Bouchette Canada
176 B3 Boucle du Baoulé, Parc National de la nat. park Mali
189 H4 Bouctouche Canada
125 F2 Bougainville Island P.N.G.
176 B3 Bougouni Mali
164 D5 Bouillon Belgium
167 H4 Bouira Alg.
176 A2 Boujdour W. Sahara
194 F3 Boulder CO U.S.A.
194 D2 Boulder MT U.S.A.
197 G3 Boulder UT U.S.A.
197 E3 Boulder Canyon gorge U.S.A.
197 E4 Boulder City U.S.A.
196 D5 Boulevard U.S.A.
215 E3 Boulevard Atlántico Arg.
124 D4 Boulia Australia
166 F2 Boulogne-Billancourt France
166 E1 Boulogne-sur-Mer France
176 B3 Boulsa Burkina
178 B4 Boumango Gabon
177 D4 Boumba r. Cameroon
167 H4 Boumerdes Alg.
176 B4 Bouna Côte d'Ivoire
203 H2 Boundary Mountains U.S.A.
196 C3 Boundary Peak U.S.A.
176 B4 Boundiali Côte d'Ivoire
178 B4 Boundji Congo
Boung r. Vietnam see Thu Bôn, Sông
149 A6 Boun Nua Laos
194 E3 Bountiful U.S.A.
125 H6 Bounty Islands N.Z.
216 G9 Bounty Trough sea feature S. Pacific Ocean
176 B3 Bourem Mali
166 E4 Bourganeuf France
166 G3 Bourg-en-Bresse France
166 F3 Bourges France
203 F2 Bourget Canada
191 J1 Bourgmont Canada
Bourgogne reg. France see Burgundy
127 F3 Bourke Australia
191 G1 Bourkes Canada
161 G5 Bourne U.K.
161 F7 Bournemouth U.K.
164 F2 Bourtanger Moor reg. Germany
176 C1 Bou Saâda Alg.
170 C6 Bou Salem Tunisia
197 E5 Bouse U.S.A.
197 E5 Bouse Wash watercourse U.S.A.
177 D3 Bousso Chad
164 B4 Boussu Belgium
176 A3 Boutilimit Mauritania
219 J9 Bouvetøya terr. Atlantic Ocean
164 C5 Bouy France
165 H3 Bovenden Germany
Boven Kapuas, Pegunungan mts Malaysia see Kapuas Hulu, Pegunungan
215 E1 Bovril Arg.
187 G4 Bow r. Canada
198 C1 Bowbells U.S.A.
124 E4 Bowen Australia
190 B5 Bowen U.S.A.
127 H6 Bowen, Mount Australia
127 I1 Bowenville Australia
216 G2 Bowers Ridge sea feature Bering Sea
197 H5 Bowie AZ U.S.A.
199 D5 Bowie TX U.S.A.
187 G5 Bow Island Canada
200 C4 Bowling Green KY U.S.A.
198 F4 Bowling Green MO U.S.A.
202 B4 Bowling Green OH U.S.A.
202 E5 Bowling Green VA U.S.A.
198 C2 Bowman U.S.A.
186 E4 Bowman, Mount Canada
129 D6 Bowman Island i. Antarctica
191 H4 Bowmanville Canada
162 B5 Bowmore U.K.
127 I5 Bowral Australia
127 J3 Bowraville Australia
186 E4 Bowron r. Canada
186 E4 Bowron Lake Provincial Park Canada
165 H5 Boxberg Germany
148 F2 Boxing China
164 D3 Boxtel Neth.
136 E1 Boyabat Turkey
171 J3 Boyana tourist site Bulg.
127 J3 Boyd r. Australia
187 I2 Boyd Lake Canada
186 G4 Boyle Canada
163 C4 Boyle Ireland
163 E4 Boyne r. Ireland
141 G2 Boyni Qara Afgh.
201 D7 Boynton Beach U.S.A.
194 E3 Boysen Reservoir U.S.A.
139 F5 Boysun Uzbek.
210 F8 Boyuibe Bol.
137 J1 Böyük Hinaldağ mt. Azer.
138 B3 Bozashchy, Poluostrov pen. Kazakh.
171 L5 Bozcaada i. Turkey
171 L5 Bozdağ mt. Turkey
136 A2 Boz Dağları mts Turkey
136 B3 Bozdoğan Turkey
161 G5 Bozeat U.K.
194 E2 Bozeman U.S.A.
148 E3 Bozhou China
136 D3 Bozkır Turkey
178 B3 Bozoum Centr. Afr. Rep.
136 G3 Bozova Turkey
140 B2 Bozqūsh, Kūh-e mts Iran
139 H2 Bozshakol' (abandoned) Kazakh.
139 F2 Boztumsyk Kazakh.
136 C2 Bozüyük Turkey
139 J2 Bozymbay Kazakh.
170 B2 Bra Italy
170 G3 Brač i. Croatia
162 B3 Bracadale U.K.
162 B3 Bracadale, Loch b. U.K.
170 E3 Bracciano, Lago di l. Italy
191 H3 Bracebridge Canada
158 K3 Bräcke Sweden
165 H5 Brackenheim Germany
161 G6 Bracknell U.K.
170 G4 Bradano r. Italy
201 D7 Bradenton U.S.A.
205 L5 Brades Montserrat
191 H3 Bradford Canada
160 F4 Bradford U.K.
202 A4 Bradford OH U.S.A.
202 D4 Bradford PA U.S.A.
203 G3 Bradford VT U.S.A.
199 D6 Brady U.S.A.
186 B3 Brady Glacier U.S.A.
162 □ Brae U.K.
126 C4 Braemar Australia
162 E3 Braemar U.K.
167 B2 Braga Port.
215 E2 Bragado Arg.
211 I4 Bragança Brazil
167 C2 Bragança Port.
214 C3 Bragança Paulista Brazil
173 D5 Brahin Belarus
165 I1 Brahlstorf Germany
145 G5 Brahmanbaria Bangl.
143 D1 Brahmani r. India
143 D2 Brahmapur India
145 G4 Brahmaputra r. Asia alt. Dihang (India), alt. Yarlung Zangbo (China)
171 L2 Brăila Romania
164 B5 Braine France
164 C4 Braine-le-Comte Belgium
198 E2 Brainerd U.S.A.
161 H6 Braintree U.K.
181 H1 Brak r. S. Africa
165 G1 Brake (Unterweser) Germany
164 B4 Brakel Belgium
165 H3 Brakel Germany
179 B6 Brakwater Namibia
186 E4 Bralorne Canada
159 J5 Bramming Denmark
159 L3 Brämön i. Sweden
191 H4 Brampton Canada
160 E3 Brampton England U.K.
161 I5 Brampton England U.K.
165 G2 Bramsche Germany
161 H5 Brancaster U.K.
189 J4 Branch Canada
213 E4 Branco r. Brazil
159 J3 Brandbu Norway
159 J5 Brande Denmark
165 K2 Brandenburg Germany
165 K2 Brandenburg land Germany
181 G4 Brandfort S. Africa
165 L3 Brandis Germany
187 J5 Brandon Canada
161 H5 Brandon U.K.
198 D3 Brandon SD U.S.A.
203 G3 Brandon VT U.S.A.
163 A5 Brandon Head Ireland
163 E5 Brandon Hill Ireland
163 A5 Brandon Mountain h. Ireland
180 D5 Brandvlei S. Africa
201 D6 Branford U.S.A.
169 I3 Braniewo Poland
129 B2 Bransfield Strait str. Antarctica
191 G4 Brantford Canada
126 D6 Branxholme Australia
189 H4 Bras d'Or Lake Canada
214 E2 Brasil, Planalto do plat. Brazil
214 C1 Brasilândia Brazil
210 E6 Brasileia Brazil
214 C1 Brasília Brazil
214 D2 Brasília de Minas Brazil
211 G4 Brasília Legal Brazil
169 M3 Braslaw Belarus
171 K2 Braşov Romania
153 A5 Brassey, Banjaran mts Sabah Malaysia
203 I2 Brassua Lake U.S.A.
168 H6 Bratislava Slovakia
146 C1 Bratsk Rus. Fed.
146 C1 Bratskoye Vodokhranilishche resr Rus. Fed.
203 G3 Brattleboro U.S.A.
168 F6 Braunau am Inn Austria
165 G4 Braunfels Germany
165 I3 Braunlage Germany
165 J3 Braunsbedra Germany
165 I2 Braunschweig Germany
176 □ Brava i. Cape Verde
159 L4 Bråviken inlet Sweden
207 E2 Bravo del Norte, Río r. Mex./U.S.A. alt. Rio Grande
197 E5 Brawley U.S.A.
163 E4 Bray Ireland
186 F4 Brazeau r. Canada
211 H6 Brazil country S. America
219 H7 Brazil Basin sea feature S. Atlantic Ocean
199 D5 Brazos r. U.S.A.
178 B4 Brazzaville Congo
171 H2 Brčko Bos.-Herz.
128 A6 Breaksea Sound inlet N.Z.
128 E1 Bream Bay N.Z.
128 E1 Bream Head N.Z.
161 C6 Brechfa U.K.
162 F4 Brechin U.K.
164 C3 Brecht Belgium
198 D2 Breckenridge MN U.S.A.
199 D5 Breckenridge TX U.S.A.
168 H6 Břeclav Czech Rep.
161 D6 Brecon U.K.
161 D6 Brecon Beacons reg. U.K.
161 D6 Brecon Beacons National Park U.K.
164 C3 Breda Neth.
180 D7 Bredasdorp S. Africa
127 H5 Bredbo Australia
165 K2 Breddin Germany
164 E3 Bredevoort Neth.
158 K2 Bredviken Norway
138 E1 Bredy Rus. Fed.
164 D3 Bree Belgium
202 D5 Breezewood U.S.A.
168 D7 Bregenz Austria
158 A2 Breiðafjörður b. Iceland
158 D2 Breiðdalsvík Iceland
165 G4 Breidenbach Germany
168 C6 Breisach am Rhein Germany
165 I1 Breitenfelde Germany
165 I5 Breitengüßbach Germany
158 M1 Breivikbotn Norway
211 I6 Brejinho de Nazaré Brazil
158 J3 Brekstad Norway
165 G1 Bremen Germany
201 C5 Bremen GA U.S.A.
190 D5 Bremen IN U.S.A.
165 G1 Bremerhaven Germany
194 B2 Bremerton U.S.A.
165 H1 Bremervörde Germany
164 F4 Bremm Germany
199 D6 Brenham U.S.A.
158 K2 Brenna Norway
168 E7 Brenner Pass Austria/Italy
191 H2 Brent Canada
170 D2 Brenta r. Italy
161 H6 Brentwood U.K.
196 B3 Brentwood CA U.S.A.
203 G4 Brentwood NY U.S.A.
170 D2 Brescia Italy
170 D1 Bressanone Italy
162 □ Bressay i. U.K.
166 D3 Bressuire France
173 B4 Brest Belarus
166 B2 Brest France
Bretagne reg. France see Brittany
164 A5 Breteuil France
199 F6 Breton Sound b. U.S.A.
128 E1 Brett, Cape N.Z.
165 G5 Bretten Germany
161 E4 Bretton U.K.
201 D5 Brevard U.S.A.
211 H4 Breves Brazil
190 E2 Brevort U.S.A.
127 G2 Brewarrina Australia
203 I2 Brewer U.S.A.
194 C1 Brewster U.S.A.
199 G6 Brewton U.S.A.
181 H3 Breyten S. Africa
169 I6 Brezno Slovakia
170 G2 Brezovo Polje plain Croatia
178 C3 Bria Centr. Afr. Rep.
166 H4 Briançon France
127 G5 Bribbaree Australia
127 J1 Bribie Island Australia
173 C5 Briceni Moldova
166 H4 Bric Froid mt. France/Italy
163 C5 Bride r. Ireland
197 G1 Bridgeland U.S.A.
161 D6 Bridgend U.K.
162 D4 Bridge of Orchy U.K.
196 C2 Bridgeport CA U.S.A.
203 G4 Bridgeport CT U.S.A.
198 C3 Bridgeport NE U.S.A.
194 E2 Bridger U.S.A.
194 F3 Bridger Peak U.S.A.
203 F5 Bridgeton U.S.A.
205 M6 Bridgetown Barbados
127 G9 Bridgewater Australia
189 H5 Bridgewater Canada
203 J1 Bridgewater U.S.A.
126 D7 Bridgewater, Cape Australia
161 E5 Bridgnorth U.K.
203 H2 Bridgton U.S.A.
161 D6 Bridgwater U.K.
161 D6 Bridgwater Bay U.K.
160 G3 Bridlington U.K.
160 G3 Bridlington Bay U.K.
127 G8 Bridport Australia
161 E7 Bridport U.K.
168 C7 Brig Switz.
160 G4 Brigg U.K.
194 D3 Brigham City U.S.A.
127 G6 Bright Australia
161 I6 Brightlingsea U.K.
191 I3 Brighton Canada
128 C6 Brighton N.Z.
161 G7 Brighton U.K.
191 F4 Brighton U.S.A.
166 H5 Brignoles France
176 A3 Brikama Gambia
165 G3 Brilon Germany
170 G4 Brindisi Italy
215 D1 Brinkmann Arg.
126 C4 Brinkworth Australia
189 H4 Brion, Île i. Canada
166 F4 Brioude France
189 F3 Brisay Canada
127 J1 Brisbane Australia
203 J1 Bristol Canada
161 E6 Bristol U.K.
203 G4 Bristol CT U.S.A.
203 F4 Bristol PA U.S.A.
202 B6 Bristol TN U.S.A.
184 B4 Bristol Bay U.S.A.
161 C6 Bristol Channel est. U.K.
209 G7 Bristol Island S. Sandwich Is
197 E4 Bristol Lake U.S.A.
197 E4 Bristol Mountains U.S.A.
129 D3 British Antarctic Territory reg. Antarctica
186 D3 British Columbia prov. Canada
185 J1 British Empire Range mts Canada
130 C7 British Indian Ocean Territory terr. Indian Ocean
212 D4 British Isles is Europe
181 G2 Brits S. Africa
180 E5 Britstown S. Africa
166 C2 Brittany reg. France
166 E4 Brive-la-Gaillarde France
167 E1 Briviesca Spain
161 D7 Brixham U.K.
Brlik Kazakh. see Birlik
168 H6 Brno Czech Rep.
201 D5 Broad r. U.S.A.
203 F3 Broadalbin U.S.A.
188 E3 Broadback r. Canada
126 F6 Broadford Australia
163 C5 Broadford Ireland
162 C3 Broadford U.K.
162 E5 Broad Law h. U.K.
161 I6 Broadstairs U.K.
194 F2 Broadus U.S.A.
187 I4 Broadview Canada
127 J2 Broadwater Australia
198 C3 Broadwater U.S.A.
128 D1 Broadwood N.Z.
159 M4 Brocēni Latvia
187 I3 Brochet Canada
187 I3 Brochet, Lac l. Canada
165 I3 Brocken mt. Germany
184 G2 Brock Island Canada
202 E3 Brockport U.S.A.
203 H3 Brockton U.S.A.
191 J3 Brockville Canada
191 F4 Brockway MI U.S.A.
202 D4 Brockway PA U.S.A.
185 J2 Brodeur Peninsula Canada
190 C4 Brodhead U.S.A.
162 C5 Brodick U.K.
169 I4 Brodnica Poland
173 C5 Brody Ukr.
199 E4 Broken Arrow U.S.A.
127 I4 Broken Bay Australia
198 D3 Broken Bow NE U.S.A.
199 E5 Broken Bow OK U.S.A.
126 D3 Broken Hill Australia
218 K7 Broken Plateau sea feature Indian Ocean
165 I2 Brome Germany
161 G6 Bromley U.K.
161 E5 Bromsgrove U.K.
159 J4 Brønderslev Denmark
181 H2 Bronkhorstspruit S. Africa
158 K2 Brønnøysund Norway
190 E5 Bronson U.S.A.
161 I5 Brooke U.K.
153 A4 Brooke's Point Phil.
190 C4 Brookfield U.S.A.
199 F6 Brookhaven U.S.A.
194 A3 Brookings OR U.S.A.
198 D2 Brookings SD U.S.A.
203 H3 Brookline U.S.A.
190 A5 Brooklyn IA U.S.A.
190 B5 Brooklyn IL U.S.A.
198 E2 Brooklyn Center U.S.A.
202 D6 Brookneal U.S.A.
187 G4 Brooks Canada
196 A2 Brooks CA U.S.A.
203 I2 Brooks ME U.S.A.
129 C3 Brooks, Cape c. Antarctica
184 D3 Brooks Range mts U.S.A.
201 D6 Brooksville U.S.A.
202 D4 Brookville U.S.A.
162 C3 Broom, Loch inlet U.K.
124 C3 Broome Australia
162 E2 Brora U.K.
159 K5 Brösarp Sweden
163 D4 Brosna r. Ireland
194 B3 Brothers U.S.A.
149 □ Brothers, The is Hong Kong China
160 E3 Brough U.K.
162 E1 Brough Head U.K.
163 E3 Broughshane U.K.
126 C4 Broughton r. Australia
Broughton Island Canada see Qikiqtarjuaq
127 J4 Broughton Islands Australia
169 O5 Brovary Ukr.
159 J4 Brovst Denmark
126 C4 Brown, Mount h. Australia
199 C5 Brownfield U.S.A.
194 D1 Browning U.S.A.
190 D6 Brownsburg U.S.A.
203 F5 Browns Mills U.S.A.
201 B5 Brownsville TN U.S.A.
199 D7 Brownsville TX U.S.A.
203 I2 Brownville U.S.A.
203 I2 Brownville Junction U.S.A.
199 D6 Brownwood U.S.A.
169 N4 Brozha Belarus
166 F1 Bruay-la-Bussière France
190 C2 Bruce Crossing U.S.A.
188 D4 Bruce Peninsula Canada
191 G3 Bruce Peninsula National Park Canada
165 G5 Bruchsal Germany
165 K2 Brück Germany
168 G7 Bruck an der Mur Austria
161 E6 Brue r. U.K.
Bruges Belgium see Brugge
164 B3 Brugge Belgium
165 G5 Brühl Germany
164 E4 Brühl Germany
197 G2 Bruin Point mt. U.S.A.
145 I3 Bruint India
180 C2 Brukkaros Namibia
190 B2 Brule U.S.A.
189 H3 Brûlé, Lac l. Canada
164 C5 Brûly Belgium
214 E1 Brumado Brazil
159 J3 Brumunddal Norway
163 E4 Brú Na Bóinne tourist site Ireland
165 J2 Brunau Germany
194 D3 Bruneau U.S.A.
194 D3 Bruneau r. U.S.A.
155 D2 Brunei country Asia
158 K3 Brunflo Sweden
170 D1 Brunico Italy
128 C5 Brunner, Lake N.Z.
187 H4 Bruno Canada
168 D4 Brunsbüttel Germany
201 D6 Brunswick GA U.S.A.
203 I3 Brunswick ME U.S.A.
202 C4 Brunswick OH U.S.A.
212 B8 Brunswick, Península de pen. Chile
Brunswick Head Australia see Brunswick Heads
127 J2 Brunswick Heads Australia
168 H6 Bruntál Czech Rep.
129 C3 Brunt Ice Shelf ice feature Antarctica
181 I4 Bruntville S. Africa
127 G9 Bruny Island Australia
194 G3 Brush U.S.A.
164 C4 Brussels Belgium (City Plan 107)
191 G4 Brussels Canada
190 D3 Brussels U.S.A.
169 N5 Brusyliv Ukr.
127 G6 Bruthen Australia
Bruxelles Belgium see Brussels
202 A4 Bryan OH U.S.A.
199 D6 Bryan TX U.S.A.
126 C4 Bryan, Mount h. Australia
172 E4 Bryansk Rus. Fed.
172 E4 Bryanskaya Oblast' admin. div. Rus. Fed.
173 H6 Bryanskoye Rus. Fed.
197 F3 Bryce Canyon National Park U.S.A.
197 H5 Bryce Mountain U.S.A.
159 I4 Bryne Norway
173 F6 Bryukhovetskaya Rus. Fed.
168 H5 Brzeg Poland
178 E3 Bu'aale Somalia
125 F2 Buala Solomon Is
176 A3 Buba Guinea-Bissau
137 L7 Būbiyān, Jazīrat Kuwait
153 B5 Bubuan i. Phil.
136 C3 Bucak Turkey
213 B3 Bucaramanga Col.
153 C4 Bucas Grande i. Phil.
127 H6 Buchan Australia
176 A4 Buchanan Liberia
190 D5 Buchanan MI U.S.A.
202 D6 Buchanan VA U.S.A.
199 D6 Buchanan, Lake U.S.A.
185 K2 Buchan Gulf Canada
189 I4 Buchans Canada
171 L2 Bucharest Romania
165 I1 Büchen Germany
165 H5 Buchen (Odenwald) Germany
165 K1 Buchholz Germany
165 H1 Bucholz in der Nordheide Germany
196 B4 Buchon, Point U.S.A.
169 L7 Bucin, Pasul pass Romania
127 F3 Buckambool Mountain h. Australia
165 H2 Bückeburg Germany
165 H2 Bücken Germany
197 F5 Buckeye U.S.A.
202 B5 Buckeye Lake U.S.A.
202 C5 Buckhannon U.S.A.
202 C5 Buckhannon r. U.S.A.
162 E4 Buckhaven U.K.
191 H3 Buckhorn Canada
197 H5 Buckhorn U.S.A.
191 H3 Buckhorn Lake Canada
202 B6 Buckhorn Lake U.S.A.
162 F3 Buckie U.K.
191 J3 Buckingham Canada
161 G5 Buckingham U.K.
202 D6 Buckingham U.S.A.
124 D3 Buckingham Bay Australia
124 E4 Buckland Tableland reg. Australia
126 B4 Buckleboo Australia
129 B6 Buckle Island i. Antarctica
197 F4 Buckskin Mountains U.S.A.
196 B2 Bucks Mountain U.S.A.
203 I2 Bucksport U.S.A.
165 K2 Bückwitz Germany
București Romania see Bucharest
202 B4 Bucyrus U.S.A.
169 O4 Buda-Kashalyova Belarus
169 I7 Budapest Hungary
144 D3 Budaun India
126 F3 Budda Australia
162 F4 Buddon Ness pt U.K.
170 C4 Buddusò Sardinia Italy
161 C7 Bude U.K.
199 F6 Bude U.S.A.
173 H6 Budennovsk Rus. Fed.
127 J1 Buderim Australia
165 H4 Büdingen Germany
144 D5 Budni India
172 E3 Budogoshch' Rus. Fed.
145 H2 Budongquan China
170 C4 Budoni Sardinia Italy
140 C6 Budū, Ḥadabat al plain Saudi Arabia
140 C6 Budū', Sabkhat al salt pan Saudi Arabia
176 C4 Buea Cameroon
196 B4 Buellton U.S.A.
215 D2 Buena Esperanza Arg.
213 A4 Buenaventura Col.
204 C3 Buenaventura Mex.
213 A4 Buenaventura, Bahía de b. Col.
195 F4 Buena Vista CO U.S.A.
202 D6 Buena Vista VA U.S.A.
167 E2 Buendía, Embalse de resr Spain
215 B4 Bueno r. Chile
215 E2 Buenos Aires Arg. (City Plan 116)
215 E3 Buenos Aires prov. Arg.
212 B7 Buenos Aires, Lago l. Arg./Chile
212 C7 Buen Pasto Arg.
206 C2 Búfalo Mex.
186 G3 Buffalo r. Canada
202 D3 Buffalo NY U.S.A.
199 D4 Buffalo OK U.S.A.
198 C2 Buffalo SD U.S.A.
199 D6 Buffalo TX U.S.A.
190 B3 Buffalo WI U.S.A.
194 F2 Buffalo WY U.S.A.
190 B3 Buffalo r. U.S.A.
127 G6 Buffalo, Mount h. Australia
186 F3 Buffalo Head Hills Canada
186 F2 Buffalo Lake Canada
187 H3 Buffalo Narrows Canada
180 B4 Buffels watercourse S. Africa
181 G1 Buffels Drift S. Africa
201 D5 Buford U.S.A.
171 K2 Buftea Romania
169 J4 Bug r. Poland
213 A4 Buga Col.
213 A3 Bugalagrande Col.
127 H3 Bugaldie Australia
Bugdayli Turkm. see Bugdaýly
138 C5 Bugdaýly Turkm.
155 D4 Bugel, Tanjung pt Indon.
164 C3 Buggenhout Belgium
170 G2 Bugojno Bos.-Herz.
153 A4 Bugsuk i. Phil.
153 B2 Buguey Phil.
138 C1 Buguruslan Rus. Fed.
140 D4 Būhābād Iran
179 D5 Buhera Zimbabwe
153 B3 Buhi Phil.
194 D3 Buhl ID U.S.A.
190 A2 Buhl MN U.S.A.
137 I3 Bühtan r. Turkey
169 M7 Buhuşi Romania
161 D5 Builth Wells U.K.
176 B4 Bui National Park Ghana
172 I4 Buinsk Rus. Fed.
137 K4 Bu'in Soflā Iran
146 D2 Buir Nur l. Mongolia
179 B6 Buitepos Namibia
171 I3 Bujanovac Serbia
178 C4 Bujumbura Burundi
146 D1 Bukachacha Rus. Fed.
145 G1 Buka Daban mt. China
125 F2 Buka Island P.N.G.
140 B2 Būkān Iran
140 D4 Būkand Iran
139 J1 Bukanskoye Rus. Fed.
Bukantau, Gory hill Uzbek. see Bo'kantov tog'lari
178 C4 Bukavu Dem. Rep. Congo
Bukhara Uzbek. see Buxoro
153 C6 Bukide i. Indon.
155 D3 Bukit Baka-Bukit Raya, Taman Nasional nat. park Indon.
155 B4 Bukit Barisan Selatan, Taman Nasional Indon.
154 B5 Bukit Fraser Malaysia
155 B3 Bukittinggi Indon.
178 D4 Bukoba Tanz.
139 K2 Buktyrma, Vodokhranilishche resr Kazakh.
154 □ Bukum, Pulau i. Sing.
147 F7 Bula Indon.
172 I4 Bula r. Rus. Fed.
168 D7 Bülach Switz.
127 J4 Bulahdelah Australia
Bulahdelal Australia see Bulahdelah
153 B3 Bulalacao Phil.
153 B3 Bulan Phil.
136 G1 Bulancak Turkey
144 D3 Bulandshahr India
137 I2 Bulanık Turkey
179 C6 Bulawayo Zimbabwe
139 G1 Bulayevo Kazakh.
136 F3 Bulbul Syria
136 B2 Buldan Turkey
144 D5 Buldhana India
181 I2 Bulembu Swaziland
146 C2 Bulgan Mongolia
148 B1 Bulgan Mongolia
171 K3 Bulgaria country Europe
126 E1 Bullawarra, Lake salt flat Australia
126 D3 Bullea, Lake salt flat Australia
128 D4 Buller r. N.Z.
127 G6 Buller, Mount Australia
197 E4 Bullhead City U.S.A.
196 D4 Bullion Mountains U.S.A.
126 E2 Bulloo watercourse Australia
126 E2 Bulloo Downs Australia
126 E2 Bulloo Lake salt flat Australia
180 B2 Büllsport Namibia
154 □ Buloh, Pulau i. Sing.
126 E6 Buloke, Lake dry lake Australia
181 G4 Bultfontein S. Africa
153 C5 Buluan Phil.
124 C2 Bulukumba Indon.
133 N2 Bulun Rus. Fed.
178 B4 Bulungu Dem. Rep. Congo
178 C4 Bulungu Dem. Rep. Congo
139 F5 Bulung'ur Uzbek.
Bulungur Uzbek. see Bulung'ur
153 C3 Bulusan Phil.
178 C3 Bumba Dem. Rep. Congo
148 B1 Bumbat China
197 F4 Bumble Bee U.S.A.
153 A5 Bum-Bum i. Malaysia
178 B4 Buna Dem. Rep. Congo
178 D3 Buna Kenya
178 D4 Bunazi Tanz.
Bunbeg Ireland see An Bun Beag
124 B5 Bunbury Australia
163 E5 Bunclody Ireland
163 D2 Buncrana Ireland
178 D4 Bunda Tanz.
124 F4 Bundaberg Australia
127 G2 Bundaleer Australia
127 I3 Bundarra Australia
144 C4 Bundi India
127 J2 Bundjalung National Park Australia
163 C3 Bundoran Ireland
145 F5 Bundu India
161 I5 Bungay U.K.
127 H5 Bungendore Australia
151 C8 Bungo-suidō sea chan. Japan
178 D3 Bunia Dem. Rep. Congo
178 C4 Bunianga Dem. Rep. Congo
126 E6 Buninyong Australia
176 D3 Buni-Yadi Nigeria
144 C2 Bunji Pak.
197 E3 Bunkerville U.S.A.
199 E6 Bunkie U.S.A.
201 D6 Bunnell U.S.A.
136 E2 Bünyan Turkey
153 A6 Bunyu i. Indon.
154 C2 Buôn Đôn Vietnam
154 D2 Buôn Hô Vietnam
154 D2 Buôn Ma Thuôt Vietnam
Buôn Mê Thuôt Vietnam see Buôn Ma Thuôt
133 O2 Buor-Khaya, Guba b. Rus. Fed.
178 D4 Bura Kenya
Burang China see Jirang
214 E2 Buranhaém r. Brazil
138 C2 Burannoye Rus. Fed.
Burao Somalia see Burco
153 C4 Burauen Phil.
142 B4 Buraydah Saudi Arabia
165 G4 Burbach Germany
196 C4 Burbank U.S.A.
127 G4 Burcher Australia
178 E3 Burco Somalia
164 D1 Burdaard Neth.
138 F5 Burdalyk Turkm.
136 C3 Burdur Turkey
178 D2 Burē Eth.
161 I5 Bure r. U.K.
158 M2 Bureå Sweden
146 F1 Bureinskiy Khrebet mts Rus. Fed.
144 C3 Burewala Pak.
Bûr Fu'ad Egypt see Port Fuad
171 L3 Burgas Bulg.
201 E5 Burgaw U.S.A.
165 J2 Burg bei Magdeburg Germany
165 I5 Burgbernheim Germany
165 I2 Burgdorf Germany
189 I4 Burgeo Canada
181 G5 Burgersdorp S. Africa
181 I2 Burgersfort S. Africa
161 G7 Burgess Hill U.K.
165 H4 Burghaun Germany
168 F6 Burghausen Germany
162 E3 Burghead U.K.
164 B3 Burgh-Haamstede Neth.
170 F6 Burgio, Serra di h. Sicily Italy
165 K5 Burglengenfeld Germany
167 E1 Burgos Spain
165 K4 Burgstädt Germany
159 L4 Burgsvik Sweden
164 E1 Burgum Neth.
166 G3 Burgundy reg. France
146 B3 Burhan Budai Shan mts China
171 L5 Burhaniye Turkey
144 D5 Burhanpur India
145 E5 Burhar-Dhanpuri India
145 F4 Burhi Gandak r. India
153 B3 Burias i. Phil.
138 D2 Buribay Rus. Fed.
206 I6 Burica, Punta pt Costa Rica
145 H4 Buri Dihing r. India
145 E4 Buri Gandak r. Nepal
138 E1 Burili Kazakh.
189 I4 Burin Peninsula Canada
154 B2 Buriram Thai.
211 J5 Buriti Bravo Brazil
214 C1 Buritis Brazil
141 G4 Burj Aziz Khan Pak.
124 D3 Burketown Australia
176 B3 Burkina country Africa
139 I2 Burkitty Kazakh.
191 H3 Burk's Falls Canada
139 I1 Burla Rus. Fed.
139 I1 Burla r. Rus. Fed.
194 D3 Burley U.S.A.
191 H4 Burlington Canada
198 C4 Burlington CO U.S.A.
190 B5 Burlington IA U.S.A.
190 D5 Burlington IN U.S.A.
203 I2 Burlington ME U.S.A.
203 G2 Burlington VT U.S.A.
190 C4 Burlington WI U.S.A.
138 D1 Burly Rus. Fed.
Burma country Asia see Myanmar
199 D6 Burnet U.S.A.
194 B3 Burney U.S.A.
203 I2 Burnham U.S.A.
127 F8 Burnie Australia
160 G3 Burniston U.K.
160 E4 Burnley U.K.
194 C3 Burns U.S.A.
187 H1 Burnside r. Canada
184 F4 Burns Lake Canada
202 C5 Burnsville Lake U.S.A.
162 E4 Burntisland U.K.
187 J3 Burntwood r. Canada
187 I3 Burntwood Lake Canada
126 E5 Buronga Australia
138 E4 Burovoy Uzbek.
132 J5 Burqin China
136 G5 Burqu' Jordan
126 C4 Burra Australia
162 □ Burra i. U.K.
162 □ Burravoe U.K.
162 F2 Burray i. U.K.
171 I4 Burrel Albania
127 H4 Burrendong, Lake Australia
127 H3 Burren Junction Australia
127 I5 Burrewarra Point Australia
167 F3 Burriana Spain
127 H5 Burrinjuck Australia
127 H5 Burrinjuck Reservoir Australia
206 D1 Burro, Serranías del mts Mex.
202 B5 Burr Oak Reservoir U.S.A.
162 D6 Burrow Head U.K.
197 G2 Burrville U.S.A.
136 B1 Bursa Turkey
177 F2 Būr Safājah Egypt
Būr Sa'īd Egypt see Port Said
165 G5 Bürstadt Germany
Būr Sudan Sudan see Port Sudan
126 D4 Burta Australia
190 E3 Burt Lake U.S.A.
191 F4 Burton U.S.A.
188 E3 Burton, Lac l. Canada
Burtonport Ireland see Ailt an Chorráin
161 F5 Burton upon Trent U.K.
158 M2 Burträsk Sweden
203 J1 Burtts Corner Canada
126 E4 Burtundy Australia
147 E7 Buru i. Indon.
136 C6 Burullus, Buḥayrat al lag. Egypt
178 D4 Burundi country Africa
178 C4 Bururi Burundi
186 B2 Burwash Landing Canada
162 F2 Burwick U.K.
139 H3 Burylbaytal Kazakh.
173 E5 Buryn' Ukr.
138 B3 Burynshyk Kazakh.
161 H5 Bury St Edmunds U.K.
144 C2 Burzil Pass Pak.
178 C4 Busanga Dem. Rep. Congo
163 E2 Bush r. U.K.
140 C4 Būshehr Iran
145 E2 Bushêngcaka China
178 D4 Bushenyi Uganda
163 E2 Bushmills U.K.
190 B5 Bushnell U.S.A.
154 □ Busing, Pulau i. Sing.
178 C3 Businga Dem. Rep. Congo
136 F5 Buşrá ash Shām Syria
124 B5 Busselton Australia
164 D2 Bussum Neth.
206 D2 Bustamante Mex.
206 C1 Bustillos, Lago Mex.
170 C2 Busto Arsizio Italy
153 A3 Busuanga i. Phil.
178 C3 Buta Dem. Rep. Congo
215 C3 Buta Ranquil Arg.
178 C4 Butare Rwanda
216 G6 Butaritari atoll Kiribati
126 B4 Bute Australia
162 C5 Bute i. U.K.
162 C5 Bute, Sound of sea chan. U.K.
186 D4 Butedale Canada
186 D4 Bute Inlet Canada
181 H4 Butha-Buthe Lesotho
165 G1 Butjadingen reg. Germany
190 E5 Butler IN U.S.A.
202 D4 Butler PA U.S.A.
163 D3 Butlers Bridge Ireland
147 E7 Buton i. Indon.
165 K1 Bütow Germany
194 D2 Butte U.S.A.
165 J3 Buttelstedt Germany
196 B1 Butte Meadows U.S.A.
155 B1 Butterworth Malaysia
181 H6 Butterworth S. Africa
163 C5 Buttevant Ireland

Córdoba

172 I4 Chistopol' Rus. Fed.
146 D1 Chita Rus. Fed.
179 B5 Chitado Angola
179 D5 Chitambo Zambia
178 C4 Chitato Angola
187 H4 Chitek Lake Canada
179 B5 Chitembo Angola
179 D4 Chitipa Malawi
179 C5 Chitokoloki Zambia
150 G3 Chitose Japan
143 B3 Chitradurga India
144 B2 Chitral Pak.
144 B2 Chitral r. Pak.
206 I7 Chitré Panama
145 G5 Chittagong Bangl.
145 F5 Chittaranjan India
144 C4 Chittaurgarh India
143 B3 Chittoor India
143 B4 Chittur India
179 D5 Chitungulu Zambia
179 D5 Chitungwiza Zimbabwe
179 C5 Chiume Angola
206 A2 Chivato, Punta pt Mex.
179 D5 Chivhu Zimbabwe
215 E2 Chivilcoy Arg.
149 D6 Chixi China
137 J4 Chiyā Surkh Iraq
149 E4 Chizhou China
151 D7 Chizu Japan
139 G1 Chkalovo Kazakh.
172 G3 Chkalovsk Rus. Fed.
150 C2 Chkalovskoye Rus. Fed.
176 C1 Chlef Alg.
167 G4 Chlef, Oued r. Alg.
154 □ Choa Chu Kang Sing.
154 □ Choa Chu Kang h. Sing.
154 C2 Chŏâm Khsant Cambodia
215 B1 Choapa r. Chile
179 C5 Chobe National Park Botswana
152 D5 Choch'iwŏn S. Korea
197 E5 Chocolate Mountains U.S.A.
213 B3 Chocontá Col.
152 C4 Cho-do i. N. Korea
152 D6 Ch'o-do i. S. Korea
165 K4 Chodov Czech Rep.
215 D3 Choele Choel Arg.
144 C2 Chogo Lungma Glacier Pak.
173 H6 Chograyskoye Vodckhranilishche resr Rus. Fed.
187 I4 Choiceland Canada
125 F2 Choiseul i. Solomon Is
212 E8 Choiseul Sound sea chan. Falkland Is
206 B2 Choix Mex.
168 H4 Chojnice Poland
150 G5 Chōkai-san vol. Japan
178 D2 Ch'ok'ē Eth.
199 D6 Choke Canyon Lake U.S.A.
145 F3 Choksum China
133 P2 Chokurdakh Rus. Fed.
179 D6 Chókwé Moz.
166 D3 Cholet France
215 B4 Cholila Arg.
139 H4 Cholpon Kyrg.
139 I4 Cholpon-Ata Kyrg.
206 H5 Choluteca Hond.
179 C5 Choma Zambia
152 E5 Chŏmch'ŏn S. Korea
145 G4 Chomo Lhari mt. Bhutan
154 A1 Chom Thong Thai.
168 F5 Chomutov Czech Rep.
133 L3 Chona r. Rus. Fed.
152 D5 Ch'ŏnan S. Korea
154 B2 Chon Buri Thai.
152 D3 Ch'ŏnch'ŏn N. Korea
210 B4 Chone Ecuador
152 C4 Ch'ŏngch'ŏn-gang r. N. Korea
152 E6 Ch'ongdo S. Korea
152 E3 Ch'ŏngjin N. Korea
152 C4 Chŏngju N. Korea
152 D5 Ch'ŏngju S. Korea
154 B2 Chŏng Kal Cambodia
152 D4 Chŏngp'yŏng N. Korea
149 C4 Chongqing China
149 C4 Chongqing mun. China
149 E5 Chongren China
181 J2 Chonguene Moz.
179 C5 Chongwe Zambia
149 D4 Chongyang China
149 F5 Chongyang Xi r. China
149 E5 Chongyi China
149 C6 Chongzuo China
152 D6 Chŏnju S. Korea
212 B7 Chonos, Archipiélago de los is Chile
145 F3 Cho Oyu mt. China
Chơ Phươc Hai Vietnam see Phươc Hai
214 B4 Chopim r. Brazil
214 B4 Chopimzinho Brazil
203 F5 Choptank r. U.S.A.
144 B4 Chor Pak.
171 K7 Chora Sfakion Greece
160 E4 Chorley U.K.
173 D5 Chornobyl' Ukr.
173 E6 Chornomors'ke Ukr.
173 C5 Chortkiv Ukr.
139 G4 Chorvoq suv ombori resr Kazakh./Uzbek.
152 D4 Ch'ŏrwŏn S. Korea
152 C3 Ch'osan N. Korea
151 G7 Chōshi Japan
215 B3 Chos Malal Arg.
168 G4 Choszczno Poland
210 C5 Chota Peru
194 D2 Choteau U.S.A.
144 B3 Choti Pak.
176 A2 Choûm Mauritania
196 B3 Chowchilla U.S.A.
186 F4 Chown, Mount Canada
139 K2 Choya Rus. Fed.
146 D2 Choybalsan Mongolia
146 C2 Choyr Mongolia
168 H6 Chřiby hills Czech Rep.
190 D6 Chrisman U.S.A.
181 I3 Chrissiesmeer S. Africa
127 D5 Christchurch N.Z.
161 F7 Christchurch U.K.
185 L2 Christian, Cape Canada
181 F3 Christiana S. Africa
191 G3 Christian Island Canada
202 C6 Christiansburg U.S.A.
Christianshåb Greenland see Qasigiannguit
186 C3 Christian Sound sea chan. U.S.A.
187 G3 Christina r. Canada
147 C8 Christmas Island terr. Indian Ocean
168 G6 Chrudim Czech Rep
145 G5 Chuadanga Bangl.
181 J2 Chuali, Lago l. Moz.
148 F4 Chuansha China
194 D3 Chubbuck U.S.A.
151 E6 Chūbu-Sangaku Kokuritsu-kōen nat. park Japan
215 C4 Chubut prov. Arg.
212 C6 Chubut r. Arg.
197 E5 Chuckwalla Mountains U.S.A.
173 D5 Chudniv Ukr.
172 D3 Chudovo Rus. Fed.
184 C4 Chugach Mountains U.S.A.
151 C7 Chūgoku-sanchi mts Japan
150 C2 Chuguyevka Rus. Fed.
194 F3 Chugwater U.S.A.
173 F5 Chuhuyiv Ukr.
197 G5 Chuichu U.S.A.
146 F1 Chukchagirskoye, Ozero l. Rus. Fed.
220 M1 Chukchi Plateau sea feature Arctic Ocean
133 U3 Chukchi Sea Rus. Fed./U.S.A.
172 G3 Chukhloma Rus. Fed.
133 T3 Chukotskiy Poluostrov pen. Rus. Fed.
172 H1 Chulasa Rus. Fed.
196 D5 Chula Vista U.S.A.
132 J4 Chulym Rus. Fed.
145 G4 Chumbi China
212 C3 Chumbicha Arg.
139 K2 Chumek Kazakh.
146 F1 Chumikan Rus. Fed.
154 B1 Chum Phae Thai.
154 A3 Chumphon Thai.
154 B2 Chum Saeng Thai.
133 K3 Chuna r. Rus. Fed.
149 F4 Chun'an China
152 D5 Ch'unch'ŏn S. Korea
141 H3 Chūnghar, Kūh-e h. Afgh.
152 D5 Ch'ungju S. Korea
Chungking China see Chongqing
152 C3 Chŭngsan N. Korea
Chungur, Koh-i- hill Afgh. see Chūnghar, Kūh-e
149 F4 Chungyang Shanmo mts Taiwan
152 F2 Chunhua China
207 G4 Chunhuhux Mex.
133 L3 Chunya r. Rus. Fed.
137 K3 Chūplū Iran
210 D7 Chuquibamba Peru
212 C2 Chuquicamata Chile
168 D7 Chur Switz.
133 O3 Churapcha Rus. Fed.
187 K3 Churchill Canada
187 J3 Churchill r. Man./Sask. Canada
189 H3 Churchill r. Nfld Canada
187 K3 Churchill, Cape Canada
189 H3 Churchill Falls Canada
187 H3 Churchill Lake Canada
186 D3 Churchill Peak Canada
188 E2 Churchill Sound sea chan. Canada
198 D1 Churchs Ferry U.S.A.
202 D5 Churchville U.S.A.
145 F4 Churia Ghati Hills Nepal
172 H3 Churov Rus. Fed.
144 C3 Churu India
Churubay Nura Kazakh. see Abay
213 C2 Churuguara Venez.
154 D2 Chu Sê Vietnam
144 D2 Chushul India
197 H3 Chuska Mountains U.S.A.
139 G4 Chust Uzbek.
189 F4 Chute-des-Passes Canada
191 I2 Chute-Rouge Canada
191 J2 Chute-St-Philippe Canada
149 F5 Chutung Taiwan
123 E2 Chuuk is Micronesia
172 H4 Chuvashskaya Respublika aut. rep. Rus. Fed.
149 A5 Chuxiong China
148 F3 Chuzhou China
148 F3 Chuzhou China
137 J4 Chwārtā Iraq
173 D6 Ciadîr-Lunga Moldova
155 C4 Ciamis Indon.
155 C4 Cianjur Indon.
214 B3 Cianorte Brazil
195 E6 Cibuta Mex.
170 F2 Čićarija mts Croatia
136 E1 Çiçekdağı Turkey
173 E7 Cide Turkey
169 J4 Ciechanów Poland
205 I4 Ciego de Ávila Cuba
213 B2 Ciénaga Col.
207 D2 Ciénega de Flores Mex.
205 H4 Cienfuegos Cuba
167 F3 Cieza Spain
167 E3 Cifuentes Spain
137 L2 Çigil Adası i. Azer.
167 E3 Cigüela r. Spain
136 D2 Cihanbeyli Turkey
206 C4 Cihuatlán Mex.
167 D3 Cíjara, Embalse de resr Spain
155 C4 Cilacap Indon.
137 I1 Çıldır Turkey
137 I1 Çıldır Gölü l. Turkey
149 D4 Cili China
138 C4 Çilmämmetgum des. Turkm.
137 J3 Cilo Dağı mt. Turkey
137 M1 Çilov Adası i. Azer.
197 E4 Cima U.S.A.
195 F4 Cimarron U.S.A.
199 D4 Cimarron r. U.S.A.
164 D5 Cimetière d'Ossuaire tourist site France
173 D6 Cimișlia Moldova
170 D2 Cimone, Monte mt. Italy
137 H3 Çınar Turkey
213 C3 Cinaruco r. Venez.
167 G2 Cinca r. Spain
202 A5 Cincinnati U.S.A.
203 F3 Cincinnatus U.S.A.
215 D4 Cinco Chañares Arg.
215 C3 Cinco Saltos Arg.
161 E6 Cinderford U.K.
136 B3 Çine Turkey
164 D4 Ciney Belgium
207 F4 Cintalapa Mex.
166 I5 Cinto, Monte mt. France
214 B3 Cinzas r. Brazil
149 E5 Ciping China
215 C3 Cipolletti Arg.
184 D3 Circle AK U.S.A.
194 F2 Circle MT U.S.A.
202 B5 Circleville OH U.S.A.
197 F2 Circleville UT U.S.A.
155 C4 Cirebon Indon.
161 F6 Cirencester U.K.
170 B2 Ciriè Italy
170 G5 Cirò Marina Italy
189 H2 Cirque Mountain Canada
190 C6 Cisco IL U.S.A.
199 D5 Cisco TX U.S.A.
197 H2 Cisco UT U.S.A.
205 H5 Cisne, Islas del Hond.
213 B3 Cisneros Col.
201 E7 Cistern Point Bahamas
170 G3 Čitluk Bos.-Herz.
180 C6 Citrusdal S. Africa
170 E3 Città di Castello Italy
171 K2 Ciucaș, Vârful mt. Romania
206 D1 Ciudad Acuña Mex.
206 D4 Ciudad Altamirano Mex.
213 E3 Ciudad Bolívar Venez.
206 C2 Ciudad Camargo Mex.
206 B2 Ciudad Constitución Mex.
207 G5 Ciudad Cuauhtémoc Mex.
207 G4 Ciudad del Carmen Mex.
214 A4 Ciudad del Este Para.
206 C1 Ciudad Delicias Mex.
207 E3 Ciudad del Maíz Mex.
213 C2 Ciudad de Nutrias Venez.
207 E3 Ciudad de Valles Mex.
213 E2 Ciudad Guayana Venez.
206 C1 Ciudad Guerrero Mex.
206 D4 Ciudad Guzmán Mex.
206 D4 Ciudad Hidalgo Mex.
207 F4 Ciudad Ixtepec Mex.
204 C2 Ciudad Juárez Mex.
206 D2 Ciudad Lerdo Mex.
207 E3 Ciudad Madero Mex.
207 E3 Ciudad Mante Mex.
207 E4 Ciudad Mendoza Mex.
207 E2 Ciudad Mier Mex.
206 B2 Ciudad Obregón Mex.
213 E3 Ciudad Piar Venez.
167 E3 Ciudad Real Spain
207 E2 Ciudad Río Bravo Mex.
167 C2 Ciudad Rodrigo Spain
207 E3 Ciudad Victoria Mex.
167 H2 Ciutadella Spain
136 F1 Civa Burnu pt Turkey
136 B2 Civan Dağ mt. Turkey
170 E1 Cividale del Friuli Italy
170 E3 Civita Castellana Italy
170 E3 Civitanova Marche Italy
170 D3 Civitavecchia Italy
136 B2 Çivril Turkey
149 F4 Cixi China
148 E2 Cixian China
137 I3 Cizre Turkey
161 I6 Clacton-on-Sea U.K.
163 D3 Clady U.K.
187 G3 Claire, Lake Canada
194 B3 Clair Engle Lake resr U.S.A.
202 D4 Clairton U.S.A.
166 F3 Clamecy France
196 D2 Clan Alpine Mountains U.S.A.
163 E4 Clane Ireland
201 C5 Clanton U.S.A.
180 C6 Clanwilliam S. Africa
163 D4 Clara Ireland
154 A3 Clara Island Myanmar
126 E4 Clare N.S.W. Australia
126 C4 Clare S.A. Australia
163 C4 Clare r. Ireland
190 E4 Clare U.S.A.
163 C5 Clarecastle Ireland
163 A4 Clare Island Ireland
203 G3 Claremont U.S.A.
199 E4 Claremore U.S.A.
163 C4 Claremorris Ireland
127 J2 Clarence r. Australia
128 D5 Clarence N.Z.
201 F7 Clarence Town Bahamas
199 C5 Clarendon U.S.A.
189 J4 Clarenville Canada
186 G5 Claresholm Canada
198 E3 Clarinda U.S.A.
202 C5 Clarington U.S.A.
202 D4 Clarion U.S.A.
202 D4 Clarion r. U.S.A.
217 L4 Clarión, Isla i. Mex.
198 D2 Clark U.S.A.
181 H5 Clarkebury S. Africa
127 H8 Clarke Island Australia
201 D5 Clark Hill Reservoir U.S.A.
197 E4 Clark Mountain U.S.A.
191 G3 Clark Point Canada
202 C5 Clarksburg U.S.A.
199 F5 Clarksdale U.S.A.
192 D2 Clarks Fork r. U.S.A.
203 F4 Clarks Summit U.S.A.
194 C2 Clarkston U.S.A.
199 E5 Clarksville AR U.S.A.
190 A4 Clarksville IA U.S.A.
201 C4 Clarksville TN U.S.A.
214 B1 Claro r. Goiás Brazil
214 B2 Claro r. Goiás Brazil
163 D5 Clashmore Ireland
163 D3 Claudy U.K.
153 B2 Claveria Phil.
164 D4 Clavier Belgium
203 G2 Clayburg U.S.A.
198 D4 Clay Center U.S.A.
197 F3 Clayhole Wash watercourse U.S.A.
201 D5 Clayton GA U.S.A.
195 G4 Clayton NM U.S.A.
203 E2 Clayton NY U.S.A.
203 I1 Clayton Lake U.S.A.
202 C6 Claytor Lake U.S.A.
163 B6 Clear, Cape Ireland
191 G4 Clear Creek Canada
197 G4 Clear Creek r. U.S.A.
184 D4 Cleare, Cape U.S.A.
202 D4 Clearfield PA U.S.A.
194 E3 Clearfield UT U.S.A.
202 B4 Clear Fork Reservoir U.S.A.
186 F3 Clear Hills Canada
198 E3 Clear Lake IA U.S.A.
190 A3 Clear Lake WI U.S.A.
196 A2 Clear Lake l. CA U.S.A.
197 F2 Clear Lake l. UT U.S.A.
194 B3 Clear Lake Reservoir U.S.A.
186 F4 Clearwater r. Alta Canada
187 H3 Clearwater r. Sask. Canada
201 D7 Clearwater U.S.A.
149 □ Clear Water Bay Hong Kong China
194 D2 Clearwater Mountains U.S.A.
187 H3 Clearwater River Provincial Park Canada
199 D5 Cleburne U.S.A.
194 B2 Cle Elum U.S.A.
160 G4 Cleethorpes U.K.
154 □ Clementi Sing.
202 C5 Clendenin U.S.A.
202 C4 Clendening Lake U.S.A.
153 A4 Cleopatra Needle mt. Phil.
191 H1 Cléricy Canada
124 E4 Clermont Australia
164 A5 Clermont France
201 D6 Clermont U.S.A.
164 D5 Clermont-en-Argonne France
166 F4 Clermont-Ferrand France
164 E4 Clervaux Lux.
170 D1 Cles Italy
126 E4 Cleve Australia
161 E6 Clevedon U.K.
199 F5 Cleveland MS U.S.A.
202 C4 Cleveland OH U.S.A.
201 C5 Cleveland TN U.S.A.
194 D1 Cleveland, Mount U.S.A.
190 D2 Cleveland Cliffs Basin l. U.S.A.
160 F3 Cleveland Hills U.K.
160 D4 Cleveleys U.K.
163 B4 Clew Bay Ireland
201 D7 Clewiston U.S.A.
163 A4 Clifden Ireland
197 H5 Cliff U.S.A.
163 C3 Cliffoney Ireland
128 E4 Clifford Bay N.Z.
127 I1 Clifton Australia
197 H5 Clifton U.S.A.
202 D6 Clifton Forge U.S.A.
202 B6 Clinch r. U.S.A.
202 B6 Clinch Mountain mts U.S.A.
186 E4 Clinton B.C. Canada
191 G4 Clinton Ont. Canada
203 G4 Clinton CT U.S.A.
190 B5 Clinton IA U.S.A.
190 C5 Clinton IL U.S.A.
203 H3 Clinton MA U.S.A.
203 I2 Clinton ME U.S.A.
198 E4 Clinton MO U.S.A.
199 F5 Clinton MS U.S.A.
201 E5 Clinton NC U.S.A.
199 D5 Clinton OK U.S.A.
187 H2 Clinton-Colden Lake Canada
190 C5 Clinton Lake U.S.A.
190 C3 Clintonville U.S.A.
197 G4 Clints Well U.S.A.
204 C6 Clipperton, Île terr. Pacific Ocean
162 B3 Clisham h. U.K.
160 E4 Clitheroe U.K.
181 G4 Clocolan S. Africa
163 D4 Cloghan Ireland
163 C5 Clonakilty Ireland
163 C5 Clonakilty Bay Ireland
163 C4 Clonbern Ireland
124 E4 Cloncurry Australia
163 D3 Clones Ireland
163 D5 Clonmel Ireland
163 D4 Clonygowan Ireland
163 B5 Cloonbannin Ireland
163 D4 Clooneagh Ireland
165 G2 Cloppenburg Germany
190 A2 Cloquet U.S.A.
194 F2 Cloud Peak U.S.A.
128 E4 Cloudy Bay N.Z.
149 □ Cloudy Hill Hong Kong China
191 J1 Clova Canada
196 A2 Cloverdale U.S.A.
199 C5 Clovis U.S.A.
191 I3 Cloyne Canada
152 C3 Cluanie, Loch l. U.K.
137 H3 Cluff Lake Mine Canada
159 K7 Cluj-Napoca Romania
151 D5 Clun U.K.
126 E6 Clunes Australia
156 H3 Cluses France
151 D4 Clwydian Range hills U.K.
136 G4 Clyde Canada
152 D5 Clyde r. U.K.
202 E3 Clyde NY U.S.A.
202 B4 Clyde OH U.S.A.
152 D5 Clyde, Firth of est. U.K.
152 D5 Clydebank U.K.
135 L2 Clyde River Canada
195 C5 Coachella U.S.A.
206 D4 Coahuayutla de Guerrero Mex.
206 D2 Coahuila state Mex.
136 D2 Coal r. Canada
190 C5 Coal City U.S.A.
206 D4 Coalcomán Mex.
196 D3 Coaldale U.S.A.
199 D5 Coalgate U.S.A.
196 B3 Coalinga U.S.A.
186 D3 Coal River Canada
161 F5 Coalville U.K.
210 F4 Coari Brazil
210 F5 Coari r. Brazil
193 I5 Coastal Plain U.S.A.
186 D4 Coast Mountains Canada
194 B2 Coast Ranges mts U.S.A.
162 E5 Coatbridge U.K.
207 G5 Coatepeque Guat.
203 F5 Coatesville U.S.A.
189 F4 Coaticook Canada
185 J3 Coats Island Canada
129 C3 Coats Land reg. Antarctica
207 F4 Coatzacoalcos Mex.
191 H2 Cobalt Canada
207 G5 Cobán Guat.
127 F3 Cobar Australia
127 H6 Cobargo Australia
127 H6 Cobberas, Mount Australia
126 E7 Cobden Australia
191 I3 Cobden Canada
163 C6 Cobh Ireland
Cóbh Ireland see Cobh
187 J4 Cobham r. Canada
210 E6 Cobija Bol.
203 F3 Cobleskill U.S.A.
191 H4 Cobourg Canada
124 D3 Cobourg Peninsula Australia
127 F5 Cobram Australia
165 I4 Coburg Germany
210 C4 Coca Ecuador
167 D2 Coca Spain
214 B1 Cocalinho Brazil
210 E7 Cochabamba Bol.
215 B4 Cochamó Chile
164 F4 Cochem Germany
Cochin India see Kochi
197 H5 Cochise U.S.A.
186 G4 Cochrane Alta Canada
188 D4 Cochrane Ont. Canada
187 I3 Cochrane r. Canada
212 B7 Cochrane Chile
126 D4 Cockburn Australia
191 F3 Cockburn Island Canada
162 F5 Cockburnspath U.K.
201 F7 Cockburn Town Bahamas
Cockburn Town Turks and Caicos Is see Grand Turk
160 D3 Cockermouth U.K.
180 F6 Cockscomb mt. S. Africa
206 I6 Coclé del Norte Panama
206 H5 Coco r. Hond./Nicaragua
204 G7 Coco, Isla de i. Col.
213 A4 Coco, Punta pt Col.
197 F4 Coconino Plateau U.S.A.
127 G5 Cocopara National Park Australia
127 G4 Cocoparra Range hills Australia
213 B3 Cocorná Col.
214 D1 Cocos Brazil
218 K4 Cocos Basin sea feature Indian Ocean
147 B8 Cocos Islands terr. Indian Ocean
217 N5 Cocos Ridge sea feature N. Pacific Ocean
206 D3 Cocula Mex.
213 B3 Cocuy, Sierra Nevada del mt. Col.
203 H4 Cod, Cape U.S.A.
210 F4 Codajás Brazil
213 D2 Codera, Cabo c. Venez.
128 A7 Codfish Island N.Z.
170 E2 Codigoro Italy
189 H2 Cod Island Canada
171 K2 Codlea Romania
211 J4 Codó Brazil
161 E5 Codsall U.K.
163 A6 Cod's Head Ireland
194 E2 Cody U.S.A.
124 E3 Coen Australia
164 F3 Coesfeld Germany
175 I5 Coëtivy i. Seychelles
194 C2 Coeur d'Alene U.S.A.
194 C2 Coeur d'Alene Lake U.S.A.
164 E2 Coevorden Neth.
181 H5 Coffee Bay S. Africa
199 E4 Coffeyville U.S.A.
126 A5 Coffin Bay Australia
126 A5 Coffin Bay Australia
127 J3 Coffs Harbour Australia
181 G6 Cofimvaba S. Africa
190 B4 Coggon U.S.A.
166 D4 Cognac France
176 C4 Cogo Equat. Guinea
202 E3 Cohocton r. U.S.A.
203 G3 Cohoes U.S.A.
126 F5 Cohuna Australia
206 I7 Coiba, Isla de i. Panama
206 I7 Coiba, Parque Nacional nat. park Panama
212 C8 Coig r. Arg.
162 C2 Coigeach, Rubha pt U.K.
212 B7 Coihaique Chile
143 B4 Coimbatore India
167 B2 Coimbra Port.
167 D4 Coín Spain
210 E7 Coipasa, Salar de salt flat Bol.
213 C2 Cojedes r. Venez.
206 G5 Cojutepeque El Salvador
194 E3 Cokeville U.S.A.
126 E7 Colac Australia
214 E2 Colatina Brazil
165 J2 Colbitz Germany
198 C4 Colby U.S.A.
210 D7 Colca r. Peru
161 H6 Colchester U.K.
190 B5 Colchester U.S.A.
162 F5 Coldingham U.K.
165 K3 Colditz Germany
187 G4 Cold Lake Canada
187 G4 Cold Lake l. Canada
162 F5 Coldstream U.K.
199 D4 Coldwater KS U.S.A.
190 E5 Coldwater MI U.S.A.
190 D1 Coldwell Canada
203 H2 Colebrook U.S.A.
190 E4 Coleman MI U.S.A.
199 D6 Coleman TX U.S.A.
181 H4 Colenso S. Africa
126 D6 Coleraine Australia
163 E2 Coleraine U.K.
128 C5 Coleridge, Lake N.Z.
127 H9 Coles Bay Australia
180 F5 Colesberg S. Africa
196 B2 Colfax CA U.S.A.
194 C2 Colfax WA U.S.A.
162 □ Colgrave Sound str. U.K.
181 G3 Coligny S. Africa
206 D4 Colima Mex.
206 D4 Colima state Mex.
206 D4 Colima, Nevado de vol. Mex.
162 B4 Coll i. U.K.
167 E2 Collado Villalba Spain
127 H2 Collarenebri Australia
201 C5 College Park U.S.A.
199 D6 College Station U.S.A.
127 G2 Collerina Australia
127 H3 Collie Australia
124 C3 Collier Bay Australia
191 G3 Collingwood Canada
128 D4 Collingwood N.Z.
199 F6 Collins U.S.A.
200 B4 Collinsville U.S.A.
215 B3 Collipulli Chile
163 C3 Collooney Ireland
166 H2 Colmar France
167 E2 Colmenar Viejo Spain
162 D5 Colmonell U.K.
161 H6 Colne r. U.K.
127 I4 Colo r. Australia
164 F4 Cologne Germany
190 C3 Coloma U.S.A.
214 C3 Colômbia Brazil
207 E2 Colombia Mex.
213 B3 Colombia country S. America
219 D5 Colombian Basin sea feature S. Atlantic Ocean
143 B5 Colombo Sri Lanka
166 E5 Colomiers France
215 E2 Colón Arg.
215 E2 Colón Arg.
205 H4 Colón Cuba
206 J6 Colón Panama
195 C6 Colonet, Cabo c. Mex.
214 E1 Colônia r. Brazil
215 D3 Colonia Choele Choel, Isla i. Arg.
215 F2 Colonia del Sacramento Uruguay
215 C3 Colonia Emilio Mitre Arg.
215 F1 Colonia Lavalleja Uruguay
202 E6 Colonial Heights U.S.A.
197 F6 Colonia Reforma Mex.
170 G5 Colonna, Capo c. Italy
217 M5 Colon Ridge sea feature Pacific Ocean
162 B4 Colonsay i. U.K.
215 D3 Colorada Grande, Salina salt pan Arg.
215 D3 Colorado r. La Pampa/Río Negro Arg.
215 C1 Colorado r. San Juan Arg.
197 E5 Colorado r. Mex./U.S.A.
199 D6 Colorado r. U.S.A.
195 F4 Colorado state U.S.A.
215 D3 Colorado, Delta del Río Arg.
197 F3 Colorado City AZ U.S.A.
199 C5 Colorado City TX U.S.A.
196 D5 Colorado Desert U.S.A.
197 H2 Colorado National Monument nat. park U.S.A.
197 G3 Colorado Plateau U.S.A.
197 E4 Colorado River Aqueduct canal U.S.A.
195 F4 Colorado Springs U.S.A.
206 D3 Colotlán Mex.
165 L1 Cölpin Germany
161 G5 Colsterworth U.K.
161 I5 Coltishall U.K.
196 D4 Colton CA U.S.A.
203 F2 Colton NY U.S.A.
197 G2 Colton UT U.S.A.
194 B2 Columbia r. Canada/U.S.A.
202 E5 Columbia MD U.S.A.
198 E4 Columbia MO U.S.A.
199 F6 Columbia MS U.S.A.
202 E4 Columbia PA U.S.A.
201 D5 Columbia SC U.S.A.
201 C5 Columbia TN U.S.A.
185 K1 Columbia, Cape Canada
202 E5 Columbia, District of admin. dist. U.S.A.
186 F4 Columbia, Mount Canada
190 E5 Columbia City U.S.A.
203 J2 Columbia Falls ME U.S.A.
194 D1 Columbia Falls MT U.S.A.
186 F4 Columbia Mountains Canada
194 C2 Columbia Plateau U.S.A.
180 B6 Columbine, Cape S. Africa
201 C5 Columbus GA U.S.A.
200 C4 Columbus IN U.S.A.
199 F5 Columbus MS U.S.A.
194 E2 Columbus MT U.S.A.
198 D3 Columbus NE U.S.A.
195 F6 Columbus NM U.S.A.
202 B5 Columbus OH U.S.A.
199 D6 Columbus TX U.S.A.
190 C4 Columbus WI U.S.A.
190 B5 Columbus Junction U.S.A.
201 F7 Columbus Point Bahamas
196 D2 Columbus Salt Marsh U.S.A.
196 A2 Colusa U.S.A.
128 E2 Colville N.Z.
194 C1 Colville U.S.A.
184 C3 Colville r. U.S.A.
128 E2 Colville Channel N.Z.
184 F3 Colville Lake Canada
161 D4 Colwyn Bay U.K.
170 E2 Comacchio Italy
170 E2 Comacchio, Valli di lag. Italy
145 G3 Comai China
215 B4 Comallo r. Arg.
199 D6 Comanche U.S.A.
129 E2 Comandante Ferraz research stn Antarctica
215 C2 Comandante Salas Arg.
169 M7 Comănești Romania
206 H5 Comayagua Hond.
215 B1 Combarbalá Chile
163 F3 Comber U.K.
191 I3 Combermere Canada
145 H6 Combermere Bay Myanmar
164 A4 Combles France
181 J1 Combomune Moz.
127 J3 Comboyne Australia
188 E3 Comencho, Lac l. Canada
163 D5 Comeragh Mountains hills Ireland
199 D6 Comfort U.S.A.
145 G5 Comilla Bangl.
164 A4 Comines Belgium
170 C4 Comino, Capo c. Sardinia Italy
207 F4 Comitán de Domínguez Mex.
203 G4 Commack U.S.A.
191 H3 Commanda Canada
185 J3 Committee Bay Canada
129 B6 Commonwealth Bay b. Antarctica
170 C2 Como Italy
170 C2 Como, Lake Italy
145 G3 Como Chamling l. China
212 C7 Comodoro Rivadavia Arg.
176 B4 Comoé, Parc National de la nat. park Côte d'Ivoire
143 B4 Comorin, Cape India
179 E5 Comoros country Africa
166 F2 Compiègne France
206 C3 Compostela Mex.
153 C5 Compostela Phil.
214 C4 Comprida, Ilha i. Brazil
190 C5 Compton U.S.A.
173 D6 Comrat Moldova
162 E4 Comrie U.K.
199 C6 Comstock U.S.A.
154 C1 Con, Sông r. Vietnam
145 H4 Cona China
176 A4 Conakry Guinea
215 C4 Cona Niyeo Arg.
127 G8 Conara Junction Australia
214 C2 Conceição r. Brazil
214 E2 Conceição da Barra Brazil
211 I5 Conceição do Araguaia Brazil
212 C3 Concepción Arg.
210 F7 Concepción Bol.
215 B3 Concepción Chile
Concepción Panama see La Concepción
212 E2 Concepción Para.
215 E2 Concepción del Uruguay Arg.
196 B4 Conception, Point U.S.A.
189 J4 Conception Bay South Canada
201 F7 Conception Island Bahamas
214 C3 Conchas Brazil
195 F5 Conchas Lake U.S.A.
197 H4 Concho U.S.A.
206 C2 Conchos r. Chihuahua Mex.
207 E2 Conchos r. Tamaulipas Mex.
196 A3 Concord CA U.S.A.
201 D5 Concord NC U.S.A.
203 H3 Concord NH U.S.A.
129 C6 Concordia research stn Antarctica
215 F1 Concordia Arg.
213 B4 Concordia Col.
180 B4 Concordia S. Africa
198 D4 Concordia U.S.A.
139 H5 Concord Peak Afgh.
127 I1 Condamine Australia
154 C3 Côn Đao Vietnam
206 H5 Condega Nicaragua
214 E1 Condeúba Brazil
127 G4 Condobolin Australia
166 E5 Condom France
194 B2 Condon U.S.A.
164 D4 Condroz reg. Belgium
201 C6 Conecuh r. U.S.A.
170 E2 Conegliano Italy
206 D2 Conejos Mex.
202 D4 Conemaugh r. U.S.A.
191 G4 Conestogo Lake Canada
202 E3 Conesus Lake U.S.A.
Coney Island i. Sing. see Serangoon, Pulau
203 G4 Coney Island U.S.A.
124 F3 Conflict Group is P.N.G.
166 E3 Confolens France
197 F2 Confusion Range mts U.S.A.
145 F3 Congdü China
149 D6 Conghua China
149 C5 Congjiang China
161 E4 Congleton U.K.
178 B4 Congo country Africa
178 B3 Congo r. Africa
178 C4 Congo, Democratic Republic of the country Africa
219 J6 Congo Cone sea feature S. Atlantic Ocean
214 D3 Congonhas Brazil
197 F4 Congress U.S.A.
215 B3 Conguillo, Parque Nacional nat. park Chile
161 G4 Coningsby U.K.
188 D4 Coniston Canada
160 D3 Coniston U.K.
187 G3 Conklin Canada
122 B4 Conkouati-Douli, Parc National de nat. park Congo
215 D2 Conlara Arg.
215 D2 Conlara r. Arg.
163 B3 Conn, Lough l. Ireland
202 C4 Conneaut U.S.A.
200 F3 Connecticut r. U.S.A.
203 G4 Connecticut state U.S.A.
202 D4 Connellsville U.S.A.
163 B4 Connemara reg. Ireland
203 I1 Conners Canada
200 C4 Connersville U.S.A.
126 F4 Conoble Australia
Co Nôi Vietnam see Yên Châu
203 E5 Conowingo U.S.A.
194 E1 Conrad U.S.A.
219 L9 Conrad Rise sea feature Southern Ocean
199 E6 Conroe U.S.A.
214 D3 Conselheiro Lafaiete Brazil
214 E2 Conselheiro Pena Brazil
160 F3 Consett U.K.
154 C3 Côn Sơn, Đao i. Vietnam
187 G4 Consort Canada
168 D7 Constance, Lake Germany/Switz.
210 F5 Constância dos Baetas Brazil
171 M2 Constanța Romania
167 D4 Constantina Spain
176 C1 Constantine Alg.
190 E5 Constantine U.S.A.
197 E6 Constitución de 1857, Parque Nacional nat. park Mex.
194 D3 Contact U.S.A.
210 C5 Contamana Peru
214 E1 Contas r. Brazil
197 G6 Continental U.S.A.
203 H3 Contoocook r. U.S.A.
187 G1 Contwoyto Lake Canada
199 E5 Conway AR U.S.A.
203 H3 Conway NH U.S.A.
201 E5 Conway SC U.S.A.
126 A2 Conway, Lake salt flat Australia
161 D4 Conwy U.K.
161 D4 Conwy r. U.K.
124 D4 Coober Pedy Australia
190 A2 Cook U.S.A.
186 D4 Cook, Cape Canada
125 G3 Cook, Grand Récif de rf New Caledonia
Cook, Mount mt. N.Z. see Aoraki
201 C4 Cookeville U.S.A.
181 F6 Cookhouse S. Africa
184 C3 Cook Inlet sea chan. U.S.A.
123 I5 Cook Islands Pacific Ocean
203 F3 Cooksburg U.S.A.
189 I3 Cook's Harbour Canada
163 E3 Cookstown U.K.
128 E4 Cook Strait N.Z.
124 E3 Cooktown Australia
127 G3 Coolabah Australia
127 H3 Coolah Australia
127 G5 Coolamon Australia
127 J2 Coolangatta Australia
124 C5 Coolgardie Australia
197 G5 Coolidge U.S.A.
197 G5 Coolidge Dam U.S.A.
127 H6 Cooma Australia
163 A6 Coomacarrea h. Ireland
126 D4 Coombah Australia
163 B6 Coomnadiha h. Ireland
126 C5 Coonalpyn Australia
127 H3 Coonamble Australia
126 D6 Coonawarra Australia
126 A3 Coondambo Australia
127 F1 Coongoola Australia
126 C2 Cooper Creek watercourse Australia
127 J3 Coopernook Australia
203 I2 Coopers Mills U.S.A.
201 E7 Cooper's Town Bahamas
198 D2 Cooperstown ND U.S.A.
203 F3 Cooperstown NY U.S.A.
194 A3 Coos Bay U.S.A.
127 H5 Cootamundra Australia
163 D3 Cootehill Ireland
215 B3 Copahue, Volcán vol. Chile
207 F4 Copainalá Mex.
207 E4 Copala Mex.
207 E4 Copalillo Mex.
207 G5 Copán tourist site Hond.
194 G4 Cope U.S.A.
159 K5 Copenhagen Denmark
127 I2 Copeton Reservoir Australia
154 C1 Cô Pi, Phou mt. Laos/Vietnam
212 B3 Copiapó Chile
212 B3 Copiapo r. Chile
126 C3 Copley Australia
170 D2 Copparo Italy
191 G2 Copper Cliff Canada
190 D2 Copper Harbor U.S.A.
Coppermine Canada see Kugluktuk
184 G3 Coppermine r. Canada
190 E2 Coppermine Point Canada
180 E4 Copperton S. Africa
145 F3 Coqên China
215 B1 Coquimbo Chile
215 B1 Coquimbo admin. reg. Chile
171 K3 Corabia Romania
214 D2 Coração de Jesus Brazil
210 D7 Coracora Peru
127 J2 Coraki Australia
124 B4 Coral Bay Australia
201 D7 Coral Gables U.S.A.
185 J3 Coral Harbour Canada
125 F3 Coral Sea Coral Sea Is Terr.
216 F6 Coral Sea Basin S. Pacific Ocean
124 F3 Coral Sea Islands Territory terr. Pacific Ocean
190 B5 Coralville Reservoir U.S.A.
126 E7 Corangamite, Lake Australia
211 G3 Corantijn r. Suriname
137 L1 Corat Azer.
164 B5 Corbeny France
215 E2 Corbett Arg.
187 K2 Corbett Inlet Canada
164 A5 Corbie France
202 A6 Corbin U.S.A.
161 G5 Corby U.K.
196 C3 Corcoran U.S.A.
212 B6 Corcovado, Golfo de sea chan. Chile
206 I6 Corcovado, Parque Nacional nat. park Costa Rica
201 D6 Cordele U.S.A.
213 B4 Cordillera de los Picachos, Parque Nacional nat. park Col.
153 B4 Cordilleras Range mts Phil.
215 C4 Córdoba Arg.
215 D1 Córdoba Arg.
215 D2 Córdoba prov. Arg.

Doorn

164 D2 **Doorn** Neth.
190 D3 **Door Peninsula** U.S.A.
164 D3 **Doorwerth** Neth.
178 E3 **Dooxo Nugaaleed** *val.* Somalia
141 F4 **Dor** *watercourse* Afgh.
199 C5 **Dora** U.S.A.
170 C2 **Dora Baltea** *r.* Italy
140 C5 **Do Rāhak** Iran
161 E7 **Dorchester** U.K.
179 B6 **Dordabis** Namibia
166 E4 **Dordogne** *r.* France
164 C3 **Dordrecht** Neth.
181 G5 **Dordrecht** S. Africa
180 C1 **Doreenville** Namibia
187 H4 **Doré Lake** Canada
187 H4 **Doré Lake** *l.* Canada
170 C4 **Dorgali** *Sardinia* Italy
145 H2 **Dorgê Co** *l.* China
141 G4 **Dori** *r.* Afgh.
176 B3 **Dori** Burkina
180 C5 **Doring** *r.* S. Africa
161 G6 **Dorking** U.K.
164 E3 **Dormagen** Germany
164 B5 **Dormans** France
162 D3 **Dornoch Firth** *est.* U.K.
148 C1 **Dornogovĭ** *prov.* Mongolia
164 F1 **Dornum** Germany
172 E4 **Dorogobuzh** Rus. Fed.
169 M7 **Dorohoi** Romania
146 B2 **Döröö Nuur** *salt l.* Mongolia
158 L2 **Dorotea** Sweden
124 B4 **Dorre Island** Australia
127 J3 **Dorrigo** Australia
194 B3 **Dorris** U.S.A.
191 H3 **Dorset** Canada
161 D7 **Dorset and East Devon Coast** *tourist site* U.K.
164 F3 **Dortmund** Germany
202 B6 **Dorton** U.S.A.
136 F3 **Dörtyol** Turkey
140 C3 **Dorūd** Iran
165 G1 **Dorum** Germany
178 C3 **Doruma** Dem. Rep. Congo
140 E3 **Dorūneh** Iran
165 H2 **Dörverden** Germany
141 E4 **Do Sārī** Iran
212 C6 **Dos Bahías, Cabo** *c.* Arg.
197 H5 **Dos Cabezas** U.S.A.
210 C5 **Dos de Mayo** Peru
149 C6 **Đo Sơn** Vietnam
Đo Son Vietnam *see* **Đo Sơn**
196 B3 **Dos Palos** U.S.A.
165 K2 **Dosse** *r.* Germany
176 C3 **Dosso** Niger
138 C3 **Dossor** Kazakh.
139 G4 **Do'stlik** Uzbek.
139 J3 **Dostyk** Kazakh.
201 C6 **Dothan** U.S.A.
166 F1 **Douai** France
176 C4 **Douala** Cameroon
166 B2 **Douarnenez** France
149 □ **Double Island** *Hong Kong* China
196 C4 **Double Peak** U.S.A.
166 H3 **Doubs** *r.* France
128 A6 **Doubtful Sound** *inlet* N.Z.
128 D1 **Doubtless Bay** N.Z.
176 B3 **Douentza** Mali
160 C3 **Douglas** Isle of Man
180 E4 **Douglas** S. Africa
162 E5 **Douglas** U.K.
186 C3 **Douglas** *AK* U.S.A.
197 H6 **Douglas** *AZ* U.S.A.
201 D6 **Douglas** *GA* U.S.A.
194 F3 **Douglas** *WY* U.S.A.
127 H8 **Douglas Apsley National Park** Australia
186 D4 **Douglas Channel** Canada
197 H2 **Douglas Creek** *r.* U.S.A.
166 F1 **Doullens** France
162 D4 **Doune** U.K.
214 C2 **Dourada, Cachoeira** *waterfall* Brazil
214 B2 **Dourada, Serra** *hills* Brazil
214 C1 **Dourada, Serra** *mts* Brazil
214 A3 **Dourados** Brazil
214 A3 **Dourados** *r.* Brazil
214 B3 **Dourados, Serra dos** *hills* Brazil
167 C2 **Douro** *r.* Port. *alt.* **Duero (Spain)**
164 D5 **Douzy** France
161 F4 **Dove** *r. England* U.K.
161 I5 **Dove** *r. England* U.K.
189 I3 **Dove Brook** Canada
197 H3 **Dove Creek** U.S.A.
127 G9 **Dover** Australia
161 I6 **Dover** U.K.
203 F5 **Dover** *DE* U.S.A.
203 H3 **Dover** *NH* U.S.A.
203 F4 **Dover** *NJ* U.S.A.
202 C4 **Dover** *OH* U.S.A.
161 I7 **Dover, Strait of** France/U.K.
203 I2 **Dover-Foxcroft** U.S.A.
161 D5 **Dovey** *r.* U.K.
140 B3 **Doveyrich** *r.* Iran/Iraq
190 D5 **Dowagiac** U.S.A.
140 D4 **Dow Chāhī** Iran
141 E2 **Dowgha'ī** Iran
154 A5 **Dowi, Tanjung** *pt* Indon.
141 F3 **Dowlatābād** Afgh.
141 G2 **Dowlatābād** Afgh.
141 G2 **Dowlatābād** Afgh.
140 D4 **Dowlatābād** Iran
140 D4 **Dowlatābād** Iran
140 E4 **Dowlatābād** Iran
141 F2 **Dowlatābād** Iran
141 G3 **Dowl at Yār** Afgh.
196 B2 **Downieville** U.S.A.
163 F3 **Downpatrick** U.K.
203 F3 **Downsville** U.S.A.
137 K4 **Dow Sar** Iran
141 H3 **Dowshī** Afgh.
196 B1 **Doyle** U.S.A.
203 F4 **Doylestown** U.S.A.
151 C6 **Dōzen** *is* Japan
191 I2 **Dozois, Réservoir** *resr* Canada
176 B2 **Drâa, Hamada du** *plat.* Alg.
214 B3 **Dracena** Brazil
164 E1 **Drachten** Neth.
171 K2 **Drăgănești-Olt** Romania
171 K2 **Drăgășani** Romania
213 E2 **Dragon's Mouths** *str.* Trin. and Tob./Venez.
159 M3 **Dragsfjärd** Fin.
166 H5 **Draguignan** France
173 C4 **Drahichyn** Belarus
127 J2 **Drake** Australia
197 F4 **Drake** *AZ* U.S.A.
187 I5 **Drake** *ND* U.S.A.
181 H5 **Drakensberg** *mts* Lesotho/ S. Africa
181 I2 **Drakensberg** *mts* S. Africa
219 E9 **Drake Passage** *sea chan.* S. Atlantic Ocean
171 K4 **Drama** Greece
159 J4 **Drammen** Norway
159 J4 **Drangedal** Norway
141 G5 **Drangjuk** *h.* Pak.
165 H3 **Dransfeld** Germany
163 E3 **Draperstown** U.K.
144 C2 **Dras** India
168 F7 **Drau** *r.* Austria
186 G4 **Drayton Valley** Canada
170 B6 **Dréan** Alg.
165 H4 **Dreistelzberge** *h.* Germany
168 F5 **Dresden** Germany
172 D4 **Dretun'** Belarus
166 E2 **Dreux** France
159 K3 **Drevsjø** Norway
160 G3 **Driffield** U.K.
202 D4 **Driftwood** U.S.A.
163 B6 **Drimoleague** Ireland
170 G3 **Drniš** Croatia
165 H1 **Drochtersen** Germany
163 E4 **Drogheda** Ireland
173 B5 **Drohobych** Ukr.
161 E5 **Droitwich Spa** U.K.
145 G4 **Drokung** India
165 I2 **Drömling** *reg.* Germany
163 D4 **Dromod** Ireland
163 D3 **Dromore** *Northern Ireland* U.K.
163 E3 **Dromore** *Northern Ireland* U.K.
161 F4 **Dronfield** U.K.
185 P2 **Dronning Louise Land** *reg.* Greenland
164 D2 **Dronten** Neth.
144 B2 **Drosh** Pak.
173 F4 **Droskovo** Rus. Fed.
127 F7 **Drouin** Australia
186 G4 **Drumheller** Canada
194 D2 **Drummond** *MT* U.S.A.
190 B2 **Drummond** *WI* U.S.A.
191 F3 **Drummond Island** U.S.A.
189 F4 **Drummondville** Canada
162 D6 **Drummore** U.K.
162 D4 **Drumochter, Pass of** U.K.
159 N5 **Druskininkai** Lith.
133 P3 **Druzhina** Rus. Fed.
171 K3 **Dryanovo** Bulg.
186 B3 **Dry Bay** U.S.A.
187 K5 **Dryberry Lake** Canada
190 E2 **Dryburg** U.S.A.
188 B4 **Dryden** Canada
196 D2 **Dry Lake** U.S.A.
162 D4 **Drymen** U.K.
124 C3 **Drysdale** *r.* Australia
140 C3 **Dūāb** *r.* Iran
149 C6 **Du'an** China
203 F2 **Duane** U.S.A.
145 G4 **Duars** *reg.* India
205 J5 **Duarte, Pico** *mt.* Dom. Rep.
134 B4 **Ḍubā** Saudi Arabia
142 E4 **Dubai** U.A.E.
169 N7 **Dubăsari** Moldova
187 I2 **Dubawnt** *r.* Canada
187 I2 **Dubawnt Lake** Canada
Dubayy U.A.E. *see* **Dubai**
134 B4 **Dubbagh, Jabal ad** *mt.* Saudi Arabia
127 H4 **Dubbo** Australia
190 D1 **Dublin** Canada
163 E4 **Dublin** Ireland
201 D5 **Dublin** U.S.A.
172 F3 **Dubna** Rus. Fed.
173 C5 **Dubno** Ukr.
194 D2 **Dubois** *ID* U.S.A.
194 E3 **Dubois** *WY* U.S.A.
202 D4 **Du Bois** U.S.A.
173 H5 **Dubovka** Rus. Fed.
173 G6 **Dubovskoye** Rus. Fed.
176 A4 **Dubréka** Guinea
171 H3 **Dubrovnik** Croatia
173 C5 **Dubrovytsya** Ukr.
172 D4 **Dubrowna** Belarus
190 B4 **Dubuque** U.S.A.
159 M5 **Dubysa** *r.* Lith.
154 C3 **Đức Bôn** Vietnam
123 J6 **Duc de Gloucester, Îles du** *is* Fr. Polynesia
149 E4 **Duchang** China
197 G1 **Duchesne** U.S.A.
123 J7 **Ducie Island** *atoll* Pitcairn Is
201 C5 **Duck** *r.* U.S.A.
187 I4 **Duck Bay** Canada
187 H4 **Duck Lake** Canada
190 E4 **Duck Lake** U.S.A.
197 E2 **Duckwater** U.S.A.
197 E2 **Duckwater Peak** U.S.A.
154 C3 **Đức Linh** Vietnam
Đức Pho Vietnam *see* **Đức Phô**
154 D2 **Đức Phô** Vietnam
154 C1 **Duc Tho** Vietnam
Đức Trong Vietnam *see* **Liên Nghia**
213 B4 **Duda** *r.* Col.
164 E5 **Dudelange** Lux.
165 I3 **Duderstadt** Germany
145 E4 **Dudhi** India
145 G4 **Dudhnai** India
132 J3 **Dudinka** Rus. Fed.
161 E5 **Dudley** U.K.
144 D6 **Dudna** *r.* India
162 F3 **Dudwick, Hill of** U.K.
176 B4 **Duékoué** Côte d'Ivoire
167 C2 **Duero** *r.* Spain *alt.* **Douro (Portugal)**
191 H1 **Dufault, Lac** *l.* Canada
164 C3 **Duffel** Belgium
188 E2 **Dufferin, Cape** Canada
202 B6 **Duffield** U.S.A.
125 G2 **Duff Islands** Solomon Is
162 E3 **Dufftown** U.K.
170 B2 **Dufourspitze** *mt.* Italy/Switz.
188 E1 **Dufrost, Pointe** *pt* Canada
170 F3 **Dugi Otok** *i.* Croatia
148 C2 **Dugui Qarag** China
148 D3 **Du He** *r.* China
213 D4 **Duida, Cerro** *mt.* Venez.
210 E3 **Duida-Marahuaca, Parque Nacional** Venez.
164 E3 **Duisburg** Germany
213 B3 **Duitama** Col.
181 I1 **Duiwelskloof** S. Africa
148 B4 **Dujiangyan** China
181 G5 **Dukathole** S. Africa
186 C4 **Duke Island** U.S.A.
140 C5 **Dukhān** Qatar
169 P3 **Dukhovshchina** Rus. Fed.
144 B3 **Duki** Pak.
159 N5 **Dūkštas** Lith.
141 E4 **Dūlab** Iran
146 B3 **Dulan** China
212 D3 **Dulce** *r.* Arg.
206 I6 **Dulce, Golfo** *b.* Costa Rica
206 H5 **Dulce Nombre de Culmí** Hond.
145 E2 **Dulishi Hu** *salt l.* China
181 I2 **Dullstroom** S. Africa
164 F3 **Dülmen** Germany
171 L3 **Dulovo** Bulg.
190 A2 **Duluth** U.S.A.
190 A2 **Duluth/Superior** *airport* U.S.A.
161 D6 **Dulverton** U.K.
136 F5 **Dūmā** Syria
153 B4 **Dumaguete** Phil.
155 B2 **Dumai** Indon.
153 B4 **Dumaran** *i.* Phil.
199 F5 **Dumas** *AR* U.S.A.
199 C5 **Dumas** *TX* U.S.A.
136 F5 **Ḍumayr** Syria
162 D5 **Dumbarton** U.K.
181 I3 **Dumbe** S. Africa
169 I6 **Ďumbier** *mt.* Slovakia
144 D2 **Dumchele** India
145 G5 **Dum-Dum** India
145 H4 **Dum Duma** India
162 E5 **Dumfries** U.K.
145 F4 **Dumka** India
165 G2 **Dümmer** *l.* Germany
188 E4 **Dumoine, Lac** Canada
129 C6 **Dumont d'Urville** *research stn* Antarctica
129 C6 **Dumont d'Urville Sea** *sea* Antarctica
164 E4 **Dümpelfeld** Germany
177 F1 **Dumyāṭ** Egypt
165 I3 **Dün** *ridge* Germany
171 H1 **Duna** *r.* Hungary *alt.* **Donau (Austria/Germany),** *alt.* **Dunaj (Slovakia),** *alt.* **Dunărea (Romania),** *alt.* **Dunav (Serbia),** *conv.* **Danube**
171 L3 **Dunaj** *r.* Slovakia *alt.* **Donau (Austria/Germany),** *alt.* **Duna (Hungary),** *alt.* **Dunărea (Romania),** *alt.* **Dunav (Serbia),** *conv.* **Danube**
168 H7 **Dunajská Streda** Slovakia
169 I7 **Dunakeszi** Hungary
127 G9 **Dunalley** Australia
163 E4 **Dunany Point** Ireland
171 L3 **Dunărea** *r.* Romania *alt.* **Donau (Austria/Germany),** *alt.* **Duna (Hungary),** *alt.* **Dunaj (Slovakia),** *alt.* **Dunav (Serbia),** *conv.* **Danube**
169 I7 **Dunaújváros** Hungary
171 L3 **Dunav** *r.* *alt.* **Donau (Austria/Germany),** *alt.* **Duna (Hungary),** *alt.* **Dunaj (Slovakia),** *alt.* **Dunărea (Romania),** *conv.* **Danube**
173 C5 **Dunayivtsi** Ukr.
128 C6 **Dunback** N.Z.
162 F4 **Dunbar** U.K.
162 E4 **Dunblane** U.K.
163 E4 **Dunboyne** Ireland
186 E5 **Duncan** Canada
197 H5 **Duncan** *AZ* U.S.A.
199 D5 **Duncan** *OK* U.S.A.
188 D3 **Duncan, Cape** Canada
188 E3 **Duncan, Lac** *l.* Canada
202 E4 **Duncannon** U.S.A.
162 E2 **Duncansby Head** U.K.
190 B5 **Duncans Mills** U.S.A.
163 E5 **Duncormick** Ireland
159 M4 **Dundaga** Latvia
191 G3 **Dundalk** Canada
163 E3 **Dundalk** Ireland
202 E5 **Dundalk** U.S.A.
163 E4 **Dundalk Bay** Ireland
185 L2 **Dundas** Greenland
186 C4 **Dundas Island** Canada
Dún Dealgan Ireland *see* **Dundalk**
181 I4 **Dundee** S. Africa
162 F4 **Dundee** U.K.
191 F5 **Dundee** *MI* U.S.A.
202 E3 **Dundee** *NY* U.S.A.
148 E1 **Dund Hot** China
163 F3 **Dundonald** U.K.
127 F1 **Dundoo** Australia
162 E6 **Dundrennan** U.K.
163 F3 **Dundrum** U.K.
163 F3 **Dundrum Bay** U.K.
145 E4 **Dundwa Range** *mts* India/Nepal
188 F2 **Dune, Lac** *l.* Canada
128 C6 **Dunedin** N.Z.
201 D6 **Dunedin** U.S.A.
127 H4 **Dunedoo** Australia
162 E4 **Dunfermline** U.K.
163 E3 **Dungannon** U.K.
144 C5 **Dungarpur** India
163 D5 **Dungarvan** Ireland
161 H7 **Dungeness** *hd* U.K.
212 C8 **Dungeness, Punta** *pt* Arg.
164 F4 **Düngenheim** Germany
163 E3 **Dungiven** U.K.
Dungloe Ireland *see* **An Clochán Liath**
127 I4 **Dungog** Australia
178 C3 **Dungu** Dem. Rep. Congo
155 B2 **Dungun** Malaysia
177 F2 **Dungunab** Sudan
152 E2 **Dunhua** China
146 B2 **Dunhuang** China
126 E6 **Dunkeld** Australia
162 E4 **Dunkeld** U.K.
Dunkerque France *see* **Dunkirk**
161 D6 **Dunkery Hill** U.K.
166 F1 **Dunkirk** France
202 D3 **Dunkirk** U.S.A.
176 B4 **Dunkwa** Ghana
163 E4 **Dún Laoghaire** Ireland
163 E4 **Dunlavin** Ireland
163 E4 **Dunleer** Ireland
163 E2 **Dunloy** U.K.
163 B6 **Dunmanus Bay** Ireland
163 B6 **Dunmanway** Ireland
163 C4 **Dunmore** Ireland
201 E7 **Dunmore Town** Bahamas
196 D3 **Dunmovin** U.S.A.
163 F3 **Dunmurry** U.K.
201 E5 **Dunn** U.S.A.
162 E2 **Dunnet Bay** U.K.
162 E2 **Dunnet Head** U.K.
196 B2 **Dunnigan** U.S.A.
198 C3 **Dunning** U.S.A.
191 H4 **Dunnville** Canada
126 E6 **Dunolly** Australia
162 D5 **Dunoon** U.K.
162 F5 **Duns** U.K.
198 C1 **Dunseith** U.S.A.
194 B3 **Dunsmuir** U.S.A.
161 G6 **Dunstable** U.K.
128 B6 **Dunstan Mountains** N.Z.
164 D5 **Dun-sur-Meuse** France
128 C6 **Duntroon** N.Z.
162 B3 **Dunvegan, Loch** *b.* U.K.
144 B3 **Dunyapur** Pak.
Duolun China *see* **Dolonnur**
154 C3 **Dương Đông** Vietnam
149 D5 **Dupang Ling** *mts* China
191 H1 **Duparquet, Lac** *l.* Canada
171 J3 **Dupnitsa** Bulg.
198 C2 **Dupree** U.S.A.
200 B4 **Du Quoin** U.S.A.
124 C3 **Durack** *r.* Australia
136 E1 **Durağan** Turkey
166 G5 **Durance** *r.* France
191 F4 **Durand** *MI* U.S.A.
190 B3 **Durand** *WI* U.S.A.
206 C2 **Durango** Mex.
206 C2 **Durango** *state* Mex.
167 E1 **Durango** Spain
195 F4 **Durango** U.S.A.
199 D5 **Durant** U.S.A.
215 F2 **Durazno** Uruguay
215 F1 **Durazno, Cuchilla Grande del** *hills* Uruguay
181 I4 **Durban** S. Africa
166 F5 **Durban-Corbières** France
180 C6 **Durbanville** S. Africa
202 D5 **Durbin** U.S.A.
164 D4 **Durbuy** Belgium
164 E4 **Düren** Germany
144 E5 **Durg** India
145 F5 **Durgapur** India
191 G3 **Durham** Canada
160 F3 **Durham** U.K.
196 B2 **Durham** *CA* U.S.A.
201 E4 **Durham** *NC* U.S.A.
203 H3 **Durham** *NH* U.S.A.
173 D6 **Durlești** Moldova
165 G6 **Durmersheim** Germany
171 H3 **Durmitor** *mt.* Montenegro
162 D2 **Durness** U.K.
171 H4 **Durrës** Albania
161 F6 **Durrington** U.K.
163 A6 **Dursey Island** Ireland
136 B2 **Dursunbey** Turkey
141 F3 **Dūruḥ** Iran
136 F5 **Durūz, Jabal ad** *mt.* Syria
124 D2 **D'Urville, Tanjung** *pt* Indon.
128 D4 **D'Urville Island** N.Z.
141 G3 **Durzab** Afgh.
138 D5 **Duşak** Turkm.
141 G4 **Dushai** Pak.
Dushak Turkm. *see* **Duşak**
149 C5 **Dushan** China
139 G5 **Dushanbe** Tajik.
173 H7 **Dushet'i** Georgia
128 A6 **Dusky Sound** *inlet* N.Z.
164 E3 **Düsseldorf** Germany
Dustlik Uzbek. *see* **Do'stlik**
197 F1 **Dutch Mountain** U.S.A.
180 E1 **Dutlwe** Botswana
176 C3 **Dutse** Nigeria
126 B3 **Dutton, Lake** *salt flat* Australia
197 F2 **Dutton, Mount** U.S.A.
172 H3 **Duvannoye** Rus. Fed.
189 F2 **Duvert, Lac** *l.* Canada
140 C5 **Duweihin, Khor** *b.* Saudi Arabia/U.A.E.
139 J5 **Düxanbibazar** China
149 C5 **Duyun** China
141 F5 **Duzab** Pak.
136 C1 **Düzce** Turkey
173 F5 **Dvorichna** Ukr.
150 B2 **Dvoryanka** Rus. Fed.
144 B5 **Dwarka** India
181 G2 **Dwarsberg** S. Africa
190 C5 **Dwight** U.S.A.
164 E2 **Dwingelderveld, Nationaal Park** *nat. park* Neth.
194 C2 **Dworshak Reservoir** U.S.A.
180 D6 **Dwyka** S. Africa
Dyanev Turkm. *see* **Galkynyş**
172 E4 **Dyat'kovo** Rus. Fed.
162 F3 **Dyce** U.K.
190 D5 **Dyer** *IN* U.S.A.
196 C3 **Dyer** *NV* U.S.A.
185 L3 **Dyer, Cape** Canada
191 G3 **Dyer Bay** Canada
201 B4 **Dyersburg** U.S.A.
190 B4 **Dyersville** U.S.A.
162 E3 **Dyke** U.K.
Dykh Tau *mt.* Georgia/Rus. Fed. *see* **Gistola, Gora**
165 K5 **Dyleň** *h.* Czech Rep.
169 I4 **Dylewska Góra** *h.* Poland
126 F2 **Dynevor Downs** Australia
181 H5 **Dyoki** S. Africa
190 A4 **Dysart** U.S.A.
180 E6 **Dysselsdorp** S. Africa
146 D2 **Dzamïn Üüd** Mongolia
179 E5 **Dzaoudzi** Africa
172 G3 **Dzerzhinsk** Rus. Fed.
169 M5 **Dzerzhyns'k** Ukr.
146 E1 **Dzhagdy, Khrebet** *mts* Rus. Fed.
Dzhalal-Abad Kyrg. *see* **Jalal-Abad**
Dzhanga Turkm. *see* **Jaňňa**
Dzhangel'dy Uzbek. *see* **Jongeldi**
173 E6 **Dzhankoy** Ukr.
Dzharkurgan Uzbek. *see* **Jarqo'rg'on**
Dzhebel Turkm. *see* **Jebel**
Dzhigirbent Turkm. *see* **Jigerbent**
Dzhizak Uzbek. *see* **Jizzax**
146 F1 **Dzhugdzhur, Khrebet** *mts* Rus. Fed.
Dzhuma Uzbek. *see* **Juma**
139 I2 **Dzhusaly** Kazakh.
169 J4 **Działdowo** Poland
207 G4 **Dzibalchén** Mex.
207 G3 **Dzilam de Bravo** Mex.
Dzungaria Basin *basin* China *see* **Junggar Pendi**
139 J3 **Dzungarian Gate** *pass* China/Kazakh.
146 C2 **Dzuunmod** Mongolia
172 C4 **Dzyaniskavichy** Belarus
172 C4 **Dzyarzhynsk** Belarus
169 M4 **Dzyatlavichy** Belarus

E

188 C3 **Eabamet Lake** Canada
197 H4 **Eagar** U.S.A.
189 I3 **Eagle** *r.* Canada
195 F4 **Eagle** U.S.A.
203 F3 **Eagle Bay** U.S.A.
196 D4 **Eagle Crags** *mt.* U.S.A.
187 H4 **Eagle Creek** *r.* Canada
187 K5 **Eagle Lake** *l.* Canada
203 I1 **Eagle Lake** U.S.A.
194 B3 **Eagle Lake** *l. CA* U.S.A.
203 I1 **Eagle Lake** *l.* U.S.A.
190 B2 **Eagle Mountain** *h.* U.S.A.
199 C6 **Eagle Pass** U.S.A.
184 D3 **Eagle Plain** Canada
190 C2 **Eagle River** *MI* U.S.A.
190 C3 **Eagle River** *WI* U.S.A.
186 F3 **Eaglesham** Canada
197 F5 **Eagle Tail Mountains** U.S.A.
188 B3 **Ear Falls** Canada
196 C4 **Earlimart** U.S.A.
162 F5 **Earlston** U.K.
191 H2 **Earlton** Canada
162 E4 **Earn** *r.* U.K.
162 D4 **Earn, Loch** *l.* U.K.
199 C5 **Earth** U.S.A.
160 H4 **Easington** U.K.
201 D5 **Easley** U.S.A.
129 D5 **East Antarctica** *reg.* Antarctica
203 F4 **East Ararat** U.S.A.
202 D3 **East Aurora** U.S.A.
199 F6 **East Bay** U.S.A.
203 G2 **East Berkshire** U.S.A.
161 H7 **Eastbourne** U.K.
202 D4 **East Branch Clarion River Reservoir** U.S.A.
203 H4 **East Brooklyn** U.S.A.
128 G2 **East Cape** N.Z.
197 G2 **East Carbon City** U.S.A.
216 E5 **East Caroline Basin** *sea feature* N. Pacific Ocean
190 D5 **East Chicago** U.S.A.
146 E3 **East China Sea** Asia
128 E2 **East Coast Bays** N.Z.
203 G2 **East Corinth** U.S.A.
East Dereham U.K. *see* **Dereham**
208 C5 **Easter Island** S. Pacific Ocean
181 G5 **Eastern Cape** *prov.* S. Africa
177 F2 **Eastern Desert** Egypt
143 C2 **Eastern Ghats** *mts* India
144 B4 **Eastern Nara** *canal* Pak.
Eastern Transvaal *prov.* S. Africa *see* **Mpumalanga**
187 J4 **Easterville** Canada
212 E8 **East Falkland** *i.* Falkland Is
203 H4 **East Falmouth** U.S.A.
196 D2 **Eastgate** U.S.A.
198 D2 **East Grand Forks** U.S.A.
161 G6 **East Grinstead** U.K.
203 G3 **Easthampton** U.S.A.
203 G4 **East Hampton** U.S.A.
202 D4 **East Hickory** U.S.A.
218 K6 **East Indiaman Ridge** *sea feature* Indian Ocean
203 G3 **East Jamaica** U.S.A.
190 E3 **East Jordan** U.S.A.
162 D5 **East Kilbride** U.K.
190 D3 **East Lake** U.S.A.
149 □ **East Lamma Channel** *Hong Kong* China
161 F7 **Eastleigh** U.K.
202 C4 **East Liverpool** U.S.A.
162 B3 **East Loch Tarbert** *inlet* U.K.
181 G6 **East London** S. Africa
202 B5 **East Lynn Lake** U.S.A.
188 E3 **Eastmain** Canada
188 F3 **Eastmain** *r.* Canada
203 G2 **Eastman** Canada
201 D5 **Eastman** U.S.A.
216 F5 **East Mariana Basin** *sea feature* Pacific Ocean
203 I2 **East Millinocket** U.S.A.
190 B5 **East Moline** U.S.A.
190 C5 **Easton** *IL* U.S.A.
203 E5 **Easton** *MD* U.S.A.
203 F4 **Easton** *PA* U.S.A.
217 L8 **East Pacific Ridge** *sea feature* S. Pacific Ocean
217 L4 **East Pacific Rise** *sea feature* N. Pacific Ocean
196 A2 **East Park Reservoir** U.S.A.
189 H4 **East Point** *pt* Canada
201 C5 **East Point** U.S.A.
203 J2 **Eastport** *ME* U.S.A.
190 E3 **Eastport** *MI* U.S.A.
196 D1 **East Range** *mts* U.S.A.
200 B4 **East St Louis** U.S.A.
East Sea *sea* Pacific Ocean *see* **Japan, Sea of**
133 Q2 **East Siberian Sea** Rus. Fed.
127 H7 **East Sister Island** Australia
147 E7 **East Timor** *country* Asia
145 F4 **East Tons** *r.* India
127 F3 **East Toorale** Australia
190 C4 **East Troy** U.S.A.
203 F6 **Eastville** U.S.A.
196 C2 **East Walker** *r.* U.S.A.
203 G3 **East Wallingford** U.S.A.
201 D5 **Eatonton** U.S.A.
190 B3 **Eau Claire** U.S.A.
190 B3 **Eau Claire** *r.* U.S.A.
188 F2 **Eau Claire, Lac à l'** *l.* Canada
147 G6 **Eauripik** *atoll* Micronesia
216 E5 **Eauripik Rise-New Guinea Rise** *sea feature* N. Pacific Ocean
207 E3 **Ebano** Mex.
161 D6 **Ebbw Vale** U.K.
176 D4 **Ebebiyin** Equat. Guinea
180 B2 **Ebenerde** Namibia
202 D4 **Ebensburg** U.S.A.
136 C2 **Eber Gölü** *l.* Turkey
165 I3 **Ebergötzen** Germany
168 F4 **Eberswalde-Finow** Germany
191 F4 **Eberts** Canada
150 G3 **Ebetsu** Japan
149 B4 **Ebian** China
139 J3 **Ebinur Hu** *salt l.* China
170 F4 **Eboli** Italy
176 D4 **Ebolowa** Cameroon
137 J3 **Ebrāhim Ḩeṣār** Iran
167 G2 **Ebro** *r.* Spain
165 I1 **Ebstorf** Germany
171 L4 **Eceabat** Turkey
153 B2 **Echague** Phil.
167 E1 **Echegárate, Puerto** *pass* Spain
206 A1 **Echeverria, Pico** *mt.* Mex.
127 G9 **Echo, Lake** Australia
186 F1 **Echo Bay** *N.W.T.* Canada
191 E2 **Echo Bay** *Ont.* Canada
197 G3 **Echo Cliffs** U.S.A.
188 B3 **Echoing** *r.* Canada
191 J2 **Échouani, Lac** *l.* Canada
164 D3 **Echt** Neth.
164 E5 **Echternach** Lux.
126 F6 **Echuca** Australia
165 G4 **Echzell** Germany
167 D4 **Écija** Spain
165 J5 **Eckental** Germany
190 E2 **Eckerman** U.S.A.
168 D3 **Eckernförde** Germany
185 K2 **Eclipse Sound** *sea chan.* Canada
166 H4 **Écrins, Massif des** *mts* France
210 C4 **Ecuador** *country* S. America
188 E2 **Écueils, Pointe aux** *pt* Canada
178 E2 **Ed** Eritrea
159 J4 **Ed** Sweden
187 H4 **Edam** Canada
164 D2 **Edam** Neth.
162 F1 **Eday** *i.* U.K.
177 E3 **Ed Da'ein** Sudan
177 F3 **Ed Damazin** Sudan
177 F3 **Ed Damer** Sudan
177 F3 **Ed Debba** Sudan
177 F3 **Ed Dueim** Sudan
127 H8 **Eddystone Point** Australia
164 D2 **Ede** Neth.
176 D4 **Edéa** Cameroon
187 J2 **Edehon Lake** Canada
214 C2 **Edéia** Brazil
127 H6 **Eden** Australia
160 E3 **Eden** *r.* U.K.
199 D6 **Eden** U.S.A.
181 F4 **Edenburg** S. Africa
128 B7 **Edendale** N.Z.
163 D4 **Edenderry** Ireland
126 D6 **Edenhope** Australia
201 E4 **Edenton** U.S.A.
181 G3 **Edenville** S. Africa
171 J4 **Edessa** Greece
165 F1 **Edewecht** Germany
203 H4 **Edgartown** U.S.A.
198 D2 **Edgeley** U.S.A.
198 C3 **Edgemont** U.S.A.
190 C4 **Edgerton** U.S.A.
163 D4 **Edgeworthstown** Ireland
139 K2 **Edigan** Rus. Fed.
190 A5 **Edina** U.S.A.
199 D7 **Edinburg** U.S.A.
162 E5 **Edinburgh** U.K.
169 M6 **Edineț** Moldova
173 C7 **Edirne** Turkey
186 F4 **Edith Cavell, Mount** Canada
194 B2 **Edmonds** U.S.A.
186 G4 **Edmonton** Canada
187 J5 **Edmore** U.S.A.
190 B4 **Edmund** U.S.A.
187 K4 **Edmund Lake** Canada
189 G4 **Edmundston** Canada
199 D6 **Edna** U.S.A.
186 C3 **Edna Bay** U.S.A.
171 L5 **Edremit** Turkey
159 K3 **Edsbyn** Sweden
186 F4 **Edson** Canada
215 D2 **Eduardo Castex** Arg.
126 F5 **Edward** *r.* Australia
178 C4 **Edward, Lake** Dem. Rep. Congo/Uganda
190 C1 **Edward Island** Canada
203 F2 **Edwards** U.S.A.
126 A2 **Edward's Creek** Australia
199 C6 **Edwards Plateau** U.S.A.
200 B4 **Edwardsville** U.S.A.
129 B5 **Edward VII Peninsula** *pen.* Antarctica
186 C3 **Edziza, Mount** Canada
164 B3 **Eeklo** Belgium
196 A1 **Eel** *r.* U.S.A.
196 A2 **Eel, South Fork** *r.* U.S.A.
164 E1 **Eemshaven** *pt* Neth.
164 E1 **Eenrum** Neth.
180 D3 **Eenzamheid Pan** *salt pan* S. Africa
125 G3 **Éfaté** *i.* Vanuatu
Efes *tourist site* Turkey *see* **Ephesus**
200 B4 **Effingham** U.S.A.
136 D1 **Eflâni** Turkey
197 E2 **Egan Range** *mts* U.S.A.
191 I3 **Eganville** Canada
169 J7 **Eger** Hungary
159 I4 **Egersund** Norway
165 G3 **Eggegebirge** *hills* Germany
165 J5 **Eggolsheim** Germany
164 C4 **Eghezée** Belgium
158 D2 **Egilsstaðir** Iceland
136 C3 **Eğirdir** Turkey
136 C3 **Eğirdir Gölü** *l.* Turkey
166 F4 **Égletons** France
163 D2 **Eglinton** U.K.
184 F2 **Eglinton Island** Canada
164 C2 **Egmond aan Zee** Neth.
128 D3 **Egmont, Cape** N.Z.
Egmont, Mount *vol.* N.Z. *see* **Taranaki, Mount**
128 E3 **Egmont National Park** N.Z.
136 B2 **Eğrigöz Dağı** *mts* Turkey
160 G3 **Egton** U.K.
214 D1 **Éguas** *r.* Brazil
133 U3 **Egvekinot** Rus. Fed.
177 E2 **Egypt** *country* Africa
148 A2 **Ehen Hudag** China
168 D6 **Ehingen (Donau)** Germany
165 I2 **Ehra-Lessien** Germany
197 E5 **Ehrenberg** U.S.A.
165 I5 **Eibelstadt** Germany
164 E2 **Eibergen** Neth.
165 H4 **Eichenzell** Germany
165 J6 **Eichstätt** Germany
159 I3 **Eidfjord** Norway
159 J3 **Eidsvoll** Norway
164 E4 **Eifel** *hills* Germany
162 B4 **Eigg** *i.* U.K.
143 A4 **Eight Degree Channel** India/Maldives
124 C3 **Eighty Mile Beach** Australia
136 E7 **Eilat** Israel
127 F6 **Eildon** Australia
127 F6 **Eildon, Lake** Australia
187 H2 **Eileen Lake** Canada
165 K3 **Eilenburg** Germany
165 I2 **Eimke** Germany
165 H3 **Einbeck** Germany
164 D3 **Eindhoven** Neth.
168 D7 **Einsiedeln** Switz.
219 G2 **Eirik Ridge** *sea feature* N. Atlantic Ocean
210 E5 **Eirunepé** Brazil
165 H3 **Eisberg** *h.* Germany
179 C5 **Eiseb** *watercourse* Namibia
165 I4 **Eisenach** Germany
165 J4 **Eisenberg** Germany
168 G4 **Eisenhüttenstadt** Germany
168 H7 **Eisenstadt** Austria
165 I4 **Eisfeld** Germany
162 C3 **Eishort, Loch** *inlet* U.K.
165 J3 **Eisleben Lutherstadt** Germany
165 H4 **Eiterfeld** Germany
Eivissa Spain *see* **Ibiza**
Eivissa *i.* Spain *see* **Ibiza**
167 F1 **Ejea de los Caballeros** Spain
179 E6 **Ejeda** Madag.
137 J1 **Ejmiatsin** Armenia
207 E4 **Ejutla** Mex.
159 M4 **Ekenäs** Fin.
138 C5 **Ekerem** Turkm.
164 C3 **Ekeren** Belgium
128 E4 **Eketahuna** N.Z.
133 L3 **Ekonda** Rus. Fed.
159 K3 **Ekshärad** Sweden
159 K4 **Eksjö** Sweden
180 B4 **Eksteenfontein** S. Africa
178 C4 **Ekuku** Dem. Rep. Congo
188 D3 **Ekwan** *r.* Canada
188 D3 **Ekwan Point** Canada
171 J6 **Elafonisou, Steno** *sea chan.* Greece
181 H2 **Elands** *r.* S. Africa
181 H2 **Elandsdoorn** S. Africa
170 B7 **El Aouinet** Alg.
206 A1 **El Arco** Mex.
171 J5 **Elassona** Greece
137 G2 **Elazığ** Turkey
170 D3 **Elba, Isola d'** *i.* Italy
146 F1 **El'ban** Rus. Fed.
213 B2 **El Banco** Col.
192 E5 **El Barreal** *salt l.* Mex.
171 I4 **Elbasan** Albania
136 E2 **Elbaşı** Turkey
213 C2 **El Baúl** Venez.
176 C1 **El Bayadh** Alg.
165 I1 **Elbe** *r.* Germany *alt.* **Labe (Czech Rep.)**
195 F4 **Elbert, Mount** U.S.A.
190 D3 **Elberta** *MI* U.S.A.
197 G2 **Elberta** *UT* U.S.A.
201 D5 **Elberton** U.S.A.
166 E2 **Elbeuf** France
136 F2 **Elbistan** Turkey
169 I3 **Elbląg** Poland
215 B4 **El Bolsón** Arg.
201 E7 **Elbow Cay** *i.* Bahamas
173 G7 **El'brus** *mt.* Rus. Fed.
164 D2 **Elburg** Neth.
167 E2 **El Burgo de Osma** Spain
140 C2 **Elburz Mountains** Iran
215 C4 **El Cain** Arg.
196 D5 **El Cajon** U.S.A.
213 E3 **El Callao** Venez.
199 D6 **El Campo** U.S.A.
195 F5 **El Capitan Mountain** U.S.A.
197 E5 **El Centro** U.S.A.
210 F7 **El Cerro** Bol.
213 D2 **El Chaparro** Venez.
Elche Spain *see* **Elche-Elx**
167 F3 **Elche-Elx** Spain
207 F4 **El Chichónal** *vol.* Mex.
206 C1 **El Chilicote** Mex.
124 D3 **Elcho Island** Australia
213 B3 **El Cocuy** Col.
213 B3 **El Cocuy, Parque Nacional** *nat. park* Col.
207 H3 **El Cuyo** Mex.
167 F3 **Elda** Spain
165 J1 **Elde** *r.* Germany
191 H2 **Eldee** Canada
126 D3 **Elder, Lake** Australia
206 D2 **El Diamante** Mex.
213 B2 **El Difícil** Col.
133 O3 **El'dikan** Rus. Fed.
213 A4 **El Diviso** Col.
197 E6 **El Doctor** Mex.
190 A5 **Eldon** *IA* U.S.A.
198 E4 **Eldon** *MO* U.S.A.
212 F3 **Eldorado** Arg.
206 C2 **El Dorado** Mex.
199 E5 **El Dorado** *AR* U.S.A.
199 D4 **El Dorado** *KS* U.S.A.
199 C6 **Eldorado** U.S.A.
213 E3 **El Dorado** Venez.
178 D3 **Eldoret** Kenya
194 E2 **Electric Peak** U.S.A.
176 B2 **El Eglab** *plat.* Alg.
167 E4 **El Ejido** Spain
172 F4 **Elektrostal'** Rus. Fed.
210 D4 **El Encanto** Col.
165 I3 **Elend** Germany
143 A2 **Elephanta Caves** *tourist site* India
195 F5 **Elephant Butte Reservoir** U.S.A.
129 B2 **Elephant Island** *i.* Antarctica
145 H5 **Elephant Point** Bangl.
137 I2 **Eleşkirt** Turkey
207 G5 **El Estor** Guat.
176 C1 **El Eulma** Alg.
201 E7 **Eleuthera** *i.* Bahamas
170 C6 **El Fahs** Tunisia
177 E3 **El Fasher** Sudan
165 H4 **Elfershausen** Germany
206 B2 **El Fuerte** Mex.
178 C2 **El Fula** Sudan
177 E3 **El Geneina** Sudan
177 F3 **El Geteina** Sudan
162 E3 **Elgin** U.K.
190 C4 **Elgin** *IL* U.S.A.
198 C2 **Elgin** *ND* U.S.A.
197 E3 **Elgin** *NV* U.S.A.
197 G2 **Elgin** *UT* U.S.A.
133 P3 **El'ginskiy** Rus. Fed.
206 D3 **El Gogorrón, Parque Nacional** *nat. park* Mex.
176 C1 **El Goléa** Alg.
177 F4 **Elgon, Mount** Uganda
170 B6 **El Hadjar** Alg.
206 I6 **El Hato del Volcán** Panama
176 A2 **El Hierro** *i.* Canary Is
207 E3 **El Higo** Mex.
176 C2 **El Homr** Alg.
162 F4 **Elie** U.K.
128 C5 **Elie de Beaumont** *mt.* N.Z.
184 B3 **Elim** U.S.A.
189 H2 **Eliot, Mount** Canada
173 H6 **Elista** Rus. Fed.
190 B4 **Elizabeth** *IL* U.S.A.
203 F4 **Elizabeth** *NJ* U.S.A.
202 C5 **Elizabeth** *WV* U.S.A.
201 E4 **Elizabeth City** U.S.A.
203 H4 **Elizabeth Islands** U.S.A.
201 D4 **Elizabethton** U.S.A.
200 C4 **Elizabethtown** *KY* U.S.A.
201 E5 **Elizabethtown** *NC* U.S.A.
203 G2 **Elizabethtown** *NY* U.S.A.
202 E4 **Elizabethtown** *PA* U.S.A.
167 F1 **Elizondo** Spain
176 B1 **El Jadida** Morocco
206 C2 **El Jaralito** Mex.
170 D7 **El Jem** Tunisia
206 H5 **El Jicaral** Nicaragua
186 G4 **Elk** *r.* Canada
169 K4 **Ełk** Poland
196 A2 **Elk** U.S.A.
202 C5 **Elk** *r.* U.S.A.
170 C6 **El Kala** Alg.
177 F3 **El Kamlin** Sudan
199 D5 **Elk City** U.S.A.
196 A2 **Elk Creek** U.S.A.
196 B2 **Elk Grove** U.S.A.
190 E5 **Elkhart** U.S.A.
176 B2 **El Khnâchîch** *esc.* Mali
190 C4 **Elkhorn** U.S.A.
198 D3 **Elkhorn** *r.* U.S.A.

Ferdows

171 L3 Elkhovo Bulg.
202 D5 Elkins U.S.A.
186 G4 Elk Island National Park Canada
191 G2 Elk Lake Canada
190 E3 Elk Lake *l.* U.S.A.
202 E4 Elkland U.S.A.
186 F5 Elko Canada
194 D3 Elko U.S.A.
187 G4 Elk Point Canada
198 E2 Elk River U.S.A.
203 F5 Elkton *MD* U.S.A.
202 D5 Elkton *VA* U.S.A.
187 L2 Ell Bay Canada
185 H2 Ellef Ringnes Island Canada
197 G2 Ellen, Mount U.S.A.
144 C3 Ellenabad India
198 D2 Ellendale U.S.A.
194 B2 Ellensburg U.S.A.
203 F4 Ellenville U.S.A.
127 H6 Ellery, Mount Australia
128 D5 Ellesmere, Lake N.Z.
185 J2 Ellesmere Island Canada
161 E4 Ellesmere Port U.K.
184 H3 Ellice *r.* Canada
202 D3 Ellicottville U.S.A.
207 E3 El Limón Mex.
165 I5 Ellingen Germany
181 G5 Elliot S. Africa
181 H5 Elliotdale S. Africa
191 F2 Elliot Lake Canada
194 D2 Ellis U.S.A.
Ellisras S. Africa *see* Lephalale
126 A4 Elliston Australia
162 F3 Ellon U.K.
144 C5 Ellora Caves *tourist site* India
203 I2 Ellsworth *ME* U.S.A.
190 A3 Ellsworth *WI* U.S.A.
129 B3 Ellsworth Land *reg.* Antarctica
129 B3 Ellsworth Mountains *mts* Antarctica
165 I6 Ellwangen (Jagst) Germany
136 B3 Elmalı Turkey
196 D6 El Maneadero Mex.
213 E3 El Manteco Venez.
176 C1 El Meghaïer Alg.
213 E3 El Miamo Venez.
136 E4 El Mīna Lebanon
El Mina Lebanon *see* El Mīna
190 E3 Elmira *MI* U.S.A.
202 E3 Elmira *NY* U.S.A.
197 F5 El Mirage U.S.A.
167 E4 El Moral Spain
126 F6 Elmore Australia
215 D2 El Morro *mt.* Arg.
176 B2 El Mreyyé *reg.* Mauritania
165 H1 Elmshorn Germany
177 E3 El Muglad Sudan
191 G3 Elmwood Canada
190 C5 Elmwood *IL* U.S.A.
190 A3 Elmwood *WI* U.S.A.
158 I3 Elnesvågen Norway
213 B3 El Nevado, Cerro *mt.* Col.
153 A4 El Nido Phil.
177 F3 El Obeid Sudan
206 D2 El Oro Mex.
213 C3 Elorza Venez.
176 C1 El Oued Alg.
197 G5 Eloy U.S.A.
206 C2 El Palmito Mex.
213 E2 El Pao Venez.
213 C2 El Pao Venez.
190 C5 El Paso *IL* U.S.A.
195 F6 El Paso *TX* U.S.A.
162 C2 Elphin U.K.
196 C3 El Portal U.S.A.
El Porvenír Panama *see* El Porvenir
206 J6 El Porvenir Panama
167 H2 El Prat de Llobregat Spain
El Progreso Guat. *see* Guastatoya
207 H5 El Progreso Hond.
206 B2 El Puerto, Cerro *mt.* Mex.
167 C4 El Puerto de Santa María Spain
El Quds Israel/West Bank *see* Jerusalem
206 J6 El Real Panama
199 D5 El Reno U.S.A.
207 D3 El Retorno Mex.
190 B4 Elroy U.S.A.
206 D3 El Rucio Mex.
186 B2 Elsa Canada
206 D2 El Salado Mex.
206 C3 El Salto Mex.
207 G5 El Salvador *country* Central America
206 D2 El Salvador Mex.
153 C4 El Salvador Phil.
213 C3 El Samán de Apure Venez.
191 F1 Elsas Canada
206 C1 El Sauz Mex.
165 G2 Else *r.* Germany
Elsen Nur *l.* China *see* Dorgê Co
196 D5 Elsinore U.S.A.
213 D2 El Sombrero Venez.
215 C2 El Sosneado Arg.
207 E3 El Tajín *tourist site* Mex.
213 B3 El Tama, Parque Nacional *nat. park* Venez.
170 C6 El Tarf Alg.
167 C1 El Teleno *mt.* Spain
207 E4 El Tepozteco, Parque Nacional *nat. park* Mex.
213 D2 El Tigre Venez.
207 G4 El Tigre, Parque Nacional *nat. park* Guat.
165 I5 Eltmann Germany
213 C2 El Tocuyo Venez.
173 H5 El'ton Rus. Fed.
173 H5 El'ton, Ozero *l.* Rus. Fed.
194 C2 Eltopia U.S.A.
213 E2 El Toro Venez.
215 E2 El Trébol Arg.
206 B3 El Triunfo Mex.
213 C3 El Tuparro, Parque Nacional *nat. park* Col.
212 B8 El Turbio Chile
143 C2 Eluru India
159 N4 Elva Estonia
213 A3 El Valle Col.
162 E5 Elvanfoot U.K.
167 C3 Elvas Port.
159 J3 Elverum Norway
213 B3 El Viejo *mt.* Col.
213 C2 El Vigía Venez.
210 D5 Elvira Brazil
178 E3 El Wak Kenya
190 E5 Elwood U.S.A.
165 I3 Elxleben Germany
161 H5 Ely U.K.
190 B2 Ely *MN* U.S.A.
197 E2 Ely *NV* U.S.A.
202 B4 Elyria U.S.A.
165 G4 Elz Germany
165 H2 Elze Germany
125 G3 Emaé *i.* Vanuatu
140 D2 Emāmrūd Iran
141 H2 Emām Şaḩēb Afgh.
137 K5 Emāmzādeh Naşrod Dīn Iran
159 L4 Emån *r.* Sweden
214 B2 Emas, Parque Nacional das *nat. park* Brazil
139 J3 Emazar Kazakh.
138 D2 Emba Kazakh.
181 H3 Embalenhle S. Africa
187 G3 Embarras Portage Canada
214 C2 Emborcação, Represa de *resr* Brazil
203 F2 Embrun Canada
178 D4 Embu Kenya
164 F1 Emden Germany
149 B4 Emeishan China
149 B4 Emei Shan *mt.* China
124 E4 Emerald *Qld* Australia
126 F6 Emerald *Vic.* Australia
189 G3 Emeril Canada
187 J5 Emerson Canada
136 B2 Emet Turkey
181 I2 eMgwenya S. Africa
197 E1 Emigrant Valley U.S.A.
eMijindini S. Africa *see* eMjindini
177 D3 Emi Koussi *mt.* Chad
206 C2 Emiliano Martínez Mex.
207 G4 Emiliano Zapata Mex.
139 J3 Emin China
171 L3 Emine, Nos *pt* Bulg.
139 J3 Emin He *r.* China
171 L3 Eminska Planina *hills* Bulg.
136 C2 Emirdağ Turkey
136 C2 Emir Dağı *mt.* Turkey
127 G3 Emita Australia
181 I2 eMjindini S. Africa
159 K4 Emmaboda Sweden
159 M4 Emmaste Estonia
127 I2 Emmaville Australia
164 D2 Emmeloord Neth.
164 F4 Emmelshausen Germany
164 E2 Emmen Neth.
168 D7 Emmen Switz.
164 E3 Emmerich Germany
143 B3 Emmiganuru India
199 C4 Emory Peak U.S.A.
206 B2 Empalme Mex.
181 I4 Empangeni S. Africa
212 E3 Empedrado Arg.
216 G3 Emperor Seamount Chain *sea feature* N. Pacific Ocean
216 G3 Emperor Trough *sea feature* N. Pacific Ocean
170 D3 Empoli Italy
198 D4 Emporia *KS* U.S.A.
202 E6 Emporia *VA* U.S.A.
202 D4 Emporium U.S.A.
187 G4 Empress Canada
141 E3 'Emrānī Iran
164 F2 Ems *r.* Germany
191 H3 Emsdale Canada
164 F2 Emsdetten Germany
164 F1 Ems-Jade-Kanal *canal* Germany
164 F2 Emsland *reg.* Germany
181 H4 eMzinoni S. Africa
158 K3 Enafors Sweden
147 F7 Enarotali Indon.
151 E7 Ena-san *mt.* Japan
215 G Encantadas, Serra das *hills* Brazil
206 A2 Encantado, Cerro *mt.* Mex.
153 B3 Encanto, Cape Phil.
206 D1 Encarnación Mex.
212 E3 Encarnación Para.
199 D6 Encinal U.S.A.
196 D5 Encinitas U.S.A.
195 F5 Encino U.S.A.
126 C5 Encounter Bay Australia
214 E1 Encruzilhada Brazil
215 G1 Encruzilhada do Sul Brazil
186 D4 Endako Canada
147 E7 Ende Indon.
124 E3 Endeavour Strait Australia
Endeh Indon. *see* Ende
219 L9 Enderby Abyssal Plain *sea feature* Southern Ocean
129 D4 Enderby Land *reg.* Antarctica
203 E3 Endicott U.S.A.
186 C3 Endicott Arm *est.* U.S.A.
184 C3 Endicott Mountains U.S.A.
138 D2 Energetik Rus. Fed.
215 E3 Energía Arg.
173 E6 Enerhodar Ukr.
216 G Enewetak *atoll* Marshall Is
170 D6 Enfidaville Tunisia
203 G2 Enfield U.S.A.
190 E2 Engadine U.S.A.
158 J3 Engan Norway
153 B2 Engaño, Cape Phil.
150 H2 Engaru Japan
181 G5 Engcobo S. Africa
201 F5 Engelhard U.S.A.
173 H5 Engel's Rus. Fed.
164 C1 Engelschmangat *sea chan.* Neth.
126 A2 Engenina *watercourse* Australia
155 B4 Enggano *i.* Indon.
164 C4 Enghien Belgium
161 E5 England *admin. div.* U.K.
189 I3 Englee Canada
191 H2 Englehart Canada
193 H3 English *r.* Canada
161 D7 English Channel France/U.K.
173 G7 Enguri *r.* Georgia
181 I4 Enhlalakahle S. Africa
199 D4 Enid U.S.A.
150 G3 Eniwa Japan
164 D2 Enkhuizen Neth.
159 L4 Enköping Sweden
170 F6 Enna *Sicily* Italy
187 I2 Ennadai Lake Canada
177 E3 En Nahud Sudan
177 E3 Ennedi, Massif *mts* Chad
163 D4 Ennell, Lough *l.* Ireland
127 F2 Enngonia Australia
198 C2 Enning U.S.A.
163 C5 Ennis Ireland
194 E2 Ennis *MT* U.S.A.
199 D5 Ennis *TX* U.S.A.
163 E5 Enniscorthy Ireland
163 D3 Enniskillen U.K.
163 B5 Ennistymon Ireland
168 G7 Enns *r.* Austria
158 O3 Eno Fin.
158 M1 Enontekiö Fin.
149 D6 Enping China
153 B2 Enrile Phil.
164 D2 Ens Neth.
127 G6 Ensay Australia
164 E2 Enschede Neth.
165 G3 Ense Germany
213 B4 Ensenada Arg.
204 A2 Ensenada Mex.
149 C4 Enshi China
186 F2 Enterprise *N.W.T.* Canada
191 I3 Enterprise *Ont.* Canada
201 C6 Enterprise *AL* U.S.A.
194 C2 Enterprise *OR* U.S.A.
197 F3 Enterprise *UT* U.S.A.
186 F4 Entrance Canada
215 C4 Entre Ríos *prov.* Arg.
210 F8 Entre Ríos Bol.
167 B3 Entroncamento Port.
176 C4 Enugu Nigeria
133 U3 Enurmino Rus. Fed.
210 D5 Envira Brazil
210 D5 Envira *r.* Brazil
128 C5 Enys, Mount N.Z.
151 F7 Enzan Japan
164 D2 Epe Neth.
164 B5 Epernay France
171 L6 Ephesus *tourist site* Turkey
197 G2 Ephraim U.S.A.
203 E4 Ephrata *PA* U.S.A.
194 C2 Ephrata *WA* U.S.A.
125 G3 Epi *i.* Vanuatu
166 H2 Épinal France
136 D4 Episkopi Cyprus
170 E4 Epomeo, Monte *vol.* Italy
161 H6 Epping U.K.
165 G4 Eppstein Germany
161 D5 Eppynt, Mynydd *hills* U.K.
161 G6 Epsom U.K.
215 D3 Epu-pel Arg.
140 D4 Eqlīd Iran
176 C4 Equatorial Guinea *country* Africa
213 E3 Equeipa Venez.
153 A4 Eran Bay Phil.
136 F1 Erbaa Turkey
165 K5 Erbendorf Germany
164 F5 Erbeskopf *h.* Germany
137 I2 Erçek Turkey
137 I2 Erciş Turkey
136 E2 Erciyes Dağı *mt.* Turkey
169 I7 Érd Hungary
145 H2 Erdaogou China
152 D2 Erdao Jiang *r.* China
136 A1 Erdek Turkey
136 E3 Erdemli Turkey
Erdenetsogt Mongolia *see* Bayan-Ovoo
177 E3 Erdi *reg.* Chad
173 H6 Erdniyevskiy Rus. Fed.
214 B4 Eré, Campos *hills* Brazil
213 D3 Erebato *r.* Venez.
129 B5 Erebus, Mount *vol.* Antarctica
137 J6 Erech *tourist site* Iraq
212 F3 Erechim Brazil
146 D2 Ereentsav Mongolia
136 E3 Ereğli Turkey
136 C1 Ereğli Turkey
170 F6 Erei, Monti *mts Sicily* Italy
Eréndira Mex. *see* Carácuaro
148 D1 Erenhot China
140 E3 Eresk Iran
167 D2 Eresma *r.* Spain
171 J5 Eretria Greece
165 J4 Erfurt Germany
137 G2 Ergani Turkey
176 B2 'Erg Chech *des.* Alg./Mali
Ergel Mongolia *see* Hatanbulag
171 L4 Ergene *r.* Turkey
159 N4 Ērgļi Latvia
150 A1 Ergu China
152 C3 Erhulai China
162 D2 Eriboll, Loch *inlet* U.K.
162 D4 Ericht, Loch *l.* U.K.
190 B5 Erie *IL* U.S.A.
199 E4 Erie *KS* U.S.A.
202 C3 Erie *PA* U.S.A.
202 C3 Erie, Lake Canada/U.S.A.
150 H3 Erimo Japan
150 H4 Erimo-misaki *c.* Japan
162 A3 Eriskay *i.* U.K.
178 D2 Eritrea *country* Africa
139 H5 Erkech-Tam Kyrg.
136 E2 Erkilet Turkey
165 J5 Erlangen Germany
124 D4 Erldunda Australia
152 E2 Erlong Shan *mt.* China
152 C2 Erlongshan Shuiku *resr* China
164 D2 Ermelo Neth.
181 H3 Ermelo S. Africa
136 D3 Ermenek Turkey
171 K6 Ermoupoli Greece
143 B4 Ernakulam India
143 B4 Erode India
180 A1 Erongo *admin. reg.* Namibia
164 D3 Erp Neth.
176 B1 Er Rachidia Morocco
177 F3 Er Rahad Sudan
179 D5 Errego Moz.
170 D7 Er Remla Tunisia
163 C2 Errigal *h.* Ireland
163 A4 Erris Head Ireland
203 H2 Errol U.S.A.
125 G3 Erromango *i.* Vanuatu
171 I4 Ersekë Albania
198 D2 Erskine U.S.A.
158 M3 Ersmark Sweden
173 G5 Ertil' Rus. Fed.
126 C3 Erudina Australia
137 I3 Eruh Turkey
215 G2 Erval Brazil
202 D5 Erwin U.S.A.
165 G3 Erwitte Germany
165 J2 Erxleben *Sachsen-Anhalt* Germany
165 J2 Erxleben *Sachsen-Anhalt* Germany
165 K4 Erzgebirge *mts* Czech Rep./Germany
136 F3 Erzin Turkey
137 G2 Erzincan Turkey
137 H2 Erzurum Turkey
150 C4 Esan-misaki *pt* Japan
150 C4 Esashi Japan
150 H2 Esashi Japan
159 J5 Esbjerg Denmark
197 G3 Escalante U.S.A.
197 G3 Escalante *r.* U.S.A.
197 F3 Escalante Desert U.S.A.
206 C2 Escalón Mex.
190 E3 Escanaba U.S.A.
207 G4 Escárcega Mex.
153 B2 Escarpada Point Phil.
167 F2 Escatrón Spain
164 B4 Escaut *r.* Belgium
164 C3 Esch Neth.
164 E2 Esche Germany
165 I2 Eschede Germany
164 C5 Esch-sur-Alzette Lux.
165 I3 Eschwege Germany
164 E4 Eschweiler Germany
196 C5 Escondido U.S.A.
206 C3 Escuinapa Mex.
207 G5 Escuintla Guat.
207 F5 Escuintla Mex.
213 C3 Escutillas Col.
136 B3 Eşen Turkey
138 C5 Esenguly Turkm.
164 F1 Esens Germany
140 C3 Eşfahān Iran
140 C3 Esfandāran Iran
140 E2 Esfarayen, Reshteh-ye *mts* Iran
140 C4 Esfarjan Iran
141 E3 Eshāqābād Iran
140 D5 Eshkanān Iran
181 I4 Eshowe S. Africa
140 C3 Eshtehārd Iran
179 C5 Esigodini Zimbabwe
181 J4 Esikhawini S. Africa
127 J3 Esk Australia
127 G8 Esk *r.* Australia
160 D2 Esk *r.* U.K.
162 E5 Eskdalemuir U.K.
189 G3 Esker Canada
158 D2 Eskifjörður Iceland
136 B2 Eski Gediz Turkey
159 L4 Eskilstuna Sweden
184 E3 Eskimo Lakes Canada
139 H4 Eski-Nookat Kyrg.
136 D1 Eskipazar Turkey
136 C2 Eskişehir Turkey
157 D1 Esla *r.* Spain
140 B3 Eslāmābād-e Gharb Iran
136 B3 Esler Dağı *mt.* Turkey
155 G3 Eslohe (Sauerland) Germany
159 K5 Eslöv Sweden
136 B2 Eşme Turkey
210 C3 Esmeraldas Ecuador
130 E1 Esnagi Lake Canada
154 B4 Esnes France
141 F5 Espakeh Iran
156 F4 Espalion France
191 G2 Espanola Canada
195 F4 Espanola U.S.A.
210 Española, Isla *i. Galapagos Is* Ecuador
136 A2 Esparto U.S.A.
155 G2 Espelkamp Germany
124 C5 Esperance Australia
129 E2 Esperanza *research stn* Antarctica
215 E2 Esperanza Arg.
206 B2 Esperanza Mex.
153 C4 Esperanza Phil.
236 H5 Esperanza, Sierra de la *mts* Hond.
157 B3 Espichel, Cabo *c.* Port.
157 D1 Espigüete *mt.* Spain
236 D2 Espinazo Mex.
214 D2 Espinhaço, Serra do *mts* Brazil
214 D1 Espinosa Brazil
214 E2 Espírito Santo *state* Brazil
153 B2 Espiritu Phil.
125 G3 Espíritu Santo *i.* Vanuatu
207 H4 Espíritu Santo, Bahía del *b.* Mex.
236 B2 Espíritu Santo, Isla *i.* Mex.
159 N3 Espoo Fin.
157 F4 Espuña *mt.* Spain
212 B6 Esquel Arg.
186 E5 Esquimalt Canada
153 B2 Essang Indon.
176 B1 Essaouira Morocco
176 A2 Es Semara W. Sahara
164 C3 Essen Belgium
164 E3 Essen Germany
165 F2 Essen (Oldenburg) Germany
211 G3 Essequibo *r.* Guyana
191 F4 Essex Canada
197 E4 Essex U.S.A.
203 G2 Essex Junction U.S.A.
191 F4 Essexville U.S.A.
133 Q4 Esso Rus. Fed.
189 H4 Est, Île de l' *i.* Canada
203 I1 Est, Lac de l' *l.* Canada
212 D8 Estados, Isla de los *i.* Arg.
140 D4 Eşţahbān Iran
191 G2 Estaire Canada
211 K6 Estância Brazil
167 G1 Estats, Pic d' *mt.* France/Spain
181 H4 Estcourt S. Africa
165 H1 Este *r.* Germany
206 H5 Estelí Nicaragua
167 E1 Estella Spain
167 D4 Estepa Spain
167 D4 Estepona Spain
187 I4 Esterhazy Canada
196 B4 Estero Bay U.S.A.
212 D2 Esteros Para.
187 I5 Estevan Canada
198 E3 Estherville U.S.A.
201 D5 Estill U.S.A.
159 N4 Estonia *country* Europe
164 A5 Estrées-St-Denis France
167 C2 Estrela, Serra da *mts* Port.
167 E3 Estrella *mt.* Spain
197 F5 Estrella, Sierra *mts* U.S.A.
167 C3 Estremoz Port.
211 I5 Estrondo, Serra *hills* Brazil
137 L4 Estūh Iran
126 C2 Etadunna Australia
144 D4 Etah India
164 D5 Étain France
166 F2 Étampes France
166 E1 Étaples France
144 D4 Etawah India
181 I3 Ethandakukhanya S. Africa
180 E4 E'Thembini S. Africa
178 D3 Ethiopia *country* Africa
136 D2 Etimesğut Turkey
162 C4 Etive, Loch *inlet* U.K.
170 F6 Etna, Mount *vol. Sicily* Italy
159 I4 Etne Norway
186 C3 Etolin Island U.S.A.
179 B5 Etosha National Park Namibia
179 B5 Etosha Pan *salt pan* Namibia
138 C5 Etrek Turkm.
171 K3 Etropole Bulg.
143 B4 Ettaiyapuram India
164 E5 Ettelbruck Lux.
164 C3 Etten-Leur Neth.
165 G6 Ettlingen Germany
162 E5 Ettrick Forest *reg.* U.K.
Etxarri Spain *see* Etxarri-Aranatz
167 E1 Etxarri-Aranatz Spain
206 C3 Etzatlán Mex.
127 G4 Euabalong Australia
124 C5 Eucla Australia
202 C4 Euclid U.S.A.
211 K6 Euclides da Cunha Brazil
127 H6 Eucumbene, Lake Australia
126 C5 Eudunda Australia
201 C6 Eufaula U.S.A.
199 E5 Eufaula Lake *resr* U.S.A.
194 B2 Eugene U.S.A.
206 A2 Eugenia, Punta *pt* Mex.
127 H4 Eugowra Australia
127 F2 Eulo Australia
127 H3 Eumungerie Australia
199 E6 Eunice U.S.A.
164 E4 Eupen Belgium
137 J6 Euphrates *r.* Asia
alt. Al Furāt (Iraq/Syria),
alt. Fırat (Turkey)
159 M3 Eura Fin.
166 E2 Eure *r.* France
194 A3 Eureka *CA* U.S.A.
194 E1 Eureka *MT* U.S.A.
197 E2 Eureka *NV* U.S.A.
126 D3 Eurinilla *watercourse* Australia
126 D3 Euriowie Australia
127 F6 Euroa Australia
179 E6 Europa, Île *i.* Indian Ocean
167 D4 Europa Point Gibraltar
156 Europe
164 D4 Euskirchen Germany
126 E5 Euston Australia
201 C5 Eutaw U.S.A.
186 D4 Eutsuk Lake Canada
165 K3 Eutzsch Germany
181 H3 Evander S. Africa
188 E3 Evans, Lac *l.* Canada
195 F4 Evans, Mount U.S.A.
186 F4 Evansburg Canada
127 J2 Evans Head Australia
185 J3 Evans Strait Canada
190 D4 Evanston *IL* U.S.A.
194 E3 Evanston *WY* U.S.A.
191 F3 Evansville Canada
200 C4 Evansville *IN* U.S.A.
190 C4 Evansville *WI* U.S.A.
194 F3 Evansville *WY* U.S.A.
190 E4 Evart U.S.A.
181 G3 Evaton S. Africa
140 D5 Evaz Iran
190 A2 Eveleth U.S.A.
133 Q3 Evensk Rus. Fed.
126 A3 Everard, Lake *salt flat* Australia
124 D4 Everard Range *hills* Australia
164 D3 Everdingen Neth.
145 F4 Everest, Mount China
203 J1 Everett Canada
194 B1 Everett U.S.A.
164 B3 Evergem Belgium
201 D7 Everglades *swamp* U.S.A.
201 D7 Everglades National Park U.S.A.
199 G6 Evergreen U.S.A.
161 F5 Evesham U.K.
161 F5 Evesham, Vale of *val.* U.K.
158 M3 Evijärvi Fin.
176 D4 Evinayong Equat. Guinea
159 I4 Evje Norway
167 C3 Évora Port.
146 F1 Evoron, Ozero *l.* Rus. Fed.
140 B3 Evowghlī Iran
166 E2 Évreux France
171 J6 Evrotas *r.* Greece
136 D4 Evrychou Cyprus
171 K5 Evvoia *i.* Greece
196 'Ewa Beach U.S.A.
178 D3 Ewaso Ngiro *r.* Kenya
162 C3 Ewe, Loch *b.* U.K.
178 B4 Ewo Congo
210 E6 Exaltación Bol.
181 G4 Excelsior S. Africa
196 C2 Excelsior Mountain U.S.A.
196 C2 Excelsior Mountains U.S.A.
198 E4 Excelsior Springs U.S.A.
161 D6 Exe *r.* U.K.
129 B4 Executive Committee Range *mts* Antarctica
127 I5 Exeter Australia
191 G4 Exeter Canada
161 D7 Exeter U.K.
196 C3 Exeter *CA* U.S.A.
203 H3 Exeter *NH* U.S.A.
164 E2 Exloo Neth.
161 D7 Exminster U.K.
161 D6 Exmoor *hills* U.K.
161 D6 Exmoor National Park U.K.
203 F6 Exmore U.S.A.
161 D7 Exmouth U.K.
127 H3 Exmouth, Mount Australia
124 B4 Exmouth Gulf Australia
218 L6 Exmouth Plateau *sea feature* Indian Ocean
167 D3 Extremadura *aut. comm.* Spain
201 E7 Exuma Sound *sea chan.* Bahamas
178 D4 Eyasi, Lake *salt l.* Tanz.
161 I5 Eye U.K.
162 F5 Eyemouth U.K.
162 B2 Eye Peninsula U.K.
158 C3 Eyjafjallajökull *ice cap* Iceland
158 C1 Eyjafjörður *inlet* Iceland
178 E3 Eyl Somalia
161 F6 Eynsham U.K.
126 B2 Eyre (North), Lake *salt flat* Australia
126 B2 Eyre (South), Lake *salt flat* Australia
128 B6 Eyre Mountains N.Z.
126 A4 Eyre Peninsula Australia
165 H2 Eystrup Germany
158 Eysturoy *i.* Faroe Is
181 I4 Ezakheni S. Africa
181 H3 Ezenzeleni S. Africa
215 C3 Ezequiel Ramos Mexía, Embalse *resr* Arg.
149 E4 Ezhou China
172 I2 Ezhva Rus. Fed.
171 L5 Ezine Turkey
136 F1 Ezinepazar Turkey

F

159 J5 Faaborg Denmark
199 B6 Fabens U.S.A.
154 Faber, Mount *h.* Sing.
186 F2 Faber Lake Canada
Fåborg Denmark *see* Faaborg
170 E3 Fabriano Italy
213 B3 Facatativá Col.
164 B4 Faches-Thumesnil France
176 D3 Fachi Niger
203 F4 Factoryville U.S.A.
212 B7 Facundo Arg.
176 C3 Fada-N'Gourma Burkina
137 H4 Fadghāmī Syria
170 D2 Faenza Italy
Faeroes *terr.* Atlantic Ocean *see* Faroe Islands
147 F7 Fafanlap Indon.
178 E3 Fafen Shet' *watercourse* Eth.
171 K2 Făgăraş Romania
159 J3 Fagernes Norway
159 K4 Fagersta Sweden
212 C8 Fagnano, Lago *l.* Arg./Chile
164 C4 Fagne *reg.* Belgium
176 B3 Faguibine, Lac *l.* Mali
158 C3 Fagurhólsmýri Iceland
177 F4 Fagwir Sudan
140 C4 Fahlīān, Rūdkhāneh-ye *watercourse* Iran
141 E4 Fahraj Iran
136 D6 Fā'id Egypt
184 D3 Fairbanks U.S.A.
202 B5 Fairborn U.S.A.
198 D3 Fairbury U.S.A.
202 E5 Fairfax U.S.A.
196 A2 Fairfield *CA* U.S.A.
190 B5 Fairfield *IA* U.S.A.
200 C4 Fairfield *OH* U.S.A.
199 D6 Fairfield *TX* U.S.A.
203 G3 Fair Haven U.S.A.
163 E2 Fair Head U.K.
153 A4 Fairie Queen Shoal *sea feature* Phil.
162 G1 Fair Isle *i.* U.K.
128 C7 Fairlie N.Z.
198 E3 Fairmont *MN* U.S.A.
202 C5 Fairmont *WV* U.S.A.
195 F4 Fairplay U.S.A.
190 D3 Fairport U.S.A.
202 C4 Fairport Harbor U.S.A.
186 F3 Fairview Canada
190 E3 Fairview *MI* U.S.A.
199 D4 Fairview *OK* U.S.A.
197 G2 Fairview *UT* U.S.A.
149 Fairview Park *Hong Kong* China
186 B3 Fairweather, Cape U.S.A.
186 B3 Fairweather, Mount Canada/U.S.A.
147 G6 Fais *i.* Micronesia
144 C3 Faisalabad Pak.
198 C2 Faith U.S.A.
145 E4 Faizabad India
125 I2 Fakaofo *atoll* Tokelau
161 H5 Fakenham U.K.
158 K3 Fåker Sweden
147 F7 Fakfak Indon.
140 D4 Fakhrābād Iran
152 B2 Faku China
161 C7 Fal *r.* U.K.
176 A4 Falaba Sierra Leone
166 D2 Falaise France
145 G4 Falakata India
145 H5 Falam Myanmar
140 C3 Falavarjan Iran
199 D7 Falcon Lake Mex./U.S.A.
199 D7 Falfurrias U.S.A.
186 F3 Falher Canada
165 L3 Falkenberg Germany
159 K4 Falkenberg Sweden
165 K1 Falkenhagen Germany
165 K3 Falkenhain Germany
165 L2 Falkensee Germany
165 K5 Falkenstein Germany
162 E5 Falkirk U.K.
162 E4 Falkland U.K.
219 F9 Falkland Escarpment *sea feature* S. Atlantic Ocean
212 E8 Falkland Islands *terr.* Atlantic Ocean
219 F9 Falkland Plateau *sea feature* S. Atlantic Ocean
212 D8 Falkland Sound *sea chan.* Falkland Is
159 K4 Falköping Sweden
196 D5 Fallbrook U.S.A.
165 H2 Fallingbostel Germany
196 C2 Fallon U.S.A.
203 H4 Fall River U.S.A.
194 F3 Fall River Pass U.S.A.
198 E3 Falls City U.S.A.
161 B7 Falmouth U.K.
202 A5 Falmouth *KY* U.S.A.
203 H3 Falmouth *ME* U.S.A.
190 E3 Falmouth *MI* U.S.A.
180 C7 False Bay S. Africa
206 B3 Falso, Cabo *c.* Mex.
159 J5 Falster *i.* Denmark
169 M7 Fălticeni Romania
159 L3 Falun Sweden
136 D4 Famagusta Cyprus
164 E5 Fameck France
140 C3 Famenin Iran
164 D4 Famenne *val.* Belgium
187 J4 Family Lake Canada
148 F4 Fanchang China
163 E4 Fane *r.* Ireland
154 A1 Fang Thai.
148 D3 Fangcheng China
149 C4 Fangdou Shan *mts* China
149 F6 Fangliao Taiwan
148 E2 Fangshan *Beijing* China
148 D2 Fangshan *Shanxi* China
149 F6 Fangshan Taiwan
148 D3 Fangxian China
152 C1 Fangzheng China
149 Fanling *Hong Kong* China
162 C3 Fannich, Loch *l.* U.K.
141 E5 Fannūj Iran
170 E3 Fano Italy
149 F5 Fanshan China
148 D2 Fanshi China
Fan Si Pan *mt.* Vietnam *see* Phăng Xi Păng
178 C3 Faradje Dem. Rep. Congo
179 E6 Farafangana Madag.
177 E2 Farāfirah, Wāḩāt al *oasis* Egypt
Farafra Oasis *oasis* Egypt *see* Farāfirah, Wāḩāt al
141 F3 Farāh Afgh.
141 F4 Farah Rūd *watercourse* Afgh.
213 A4 Farallones de Cali, Parque Nacional *nat. park* Col.
176 A3 Faranah Guinea
138 E5 Farap Turkm.
142 B6 Farasān, Jazā'ir *is* Saudi Arabia
147 G6 Faraulep *atoll* Micronesia
161 F7 Fareham U.K.
185 N4 Farewell, Cape Greenland
128 D4 Farewell, Cape N.Z.
128 D4 Farewell Spit N.Z.
159 K4 Färgelanda Sweden
198 D2 Fargo U.S.A.
139 G4 Farg'ona Uzbek.
198 E2 Faribault U.S.A.
189 F2 Faribault, Lac *l.* Canada
144 D3 Faridabad India
144 C3 Faridkot India
145 G5 Faridpur Bangl.
176 A3 Farim Guinea-Bissau
141 E3 Farīmān Iran
159 L4 Färjestaden Sweden
139 G5 Farkhor Tajik.
137 L4 Farmahīn Iran
190 C5 Farmer City U.S.A.
188 D2 Farmer Island Canada
186 E3 Farmington Canada
190 B5 Farmington *IA* U.S.A.
190 B5 Farmington *IL* U.S.A.
203 H2 Farmington *ME* U.S.A.
203 H3 Farmington *NH* U.S.A.
197 H3 Farmington *NM* U.S.A.
194 E3 Farmington *UT* U.S.A.
186 D4 Far Mountain Canada
202 D6 Farmville U.S.A.
161 G6 Farnborough U.K.
160 F2 Farne Islands U.K.
161 G6 Farnham U.K.
186 F4 Farnham, Mount Canada
211 G4 Faro Brazil
186 C2 Faro Canada
167 C4 Faro Port.
159 L4 Fårö *i.* Sweden
158 Faroe Islands *terr.* N. Atlantic Ocean
159 L4 Fårösund Sweden
175 I5 Farquhar Group *is* Seychelles
Farquhar Islands Seychelles *see* Farquhar Group
140 D4 Farrāshband Iran
202 C4 Farrell U.S.A.
191 J3 Farrellton Canada
141 E3 Farrokhī Iran
140 D3 Farsakh Iran
171 J5 Farsala Greece
141 F3 Fārsī Afgh.
194 E3 Farson U.S.A.
159 I4 Farsund Norway
138 D5 Fārūj Iran
Farvel, Kap *c.* Greenland *see* Farewell, Cape
199 C5 Farwell U.S.A.
140 E5 Fāryāb Iran
140 D4 Fasā Iran
170 G4 Fasano Italy
165 I2 Faßberg Germany
202 E4 Fassett U.S.A.
173 D5 Fastiv Ukr.
144 D4 Fatehgarh India
144 C4 Fatehpur *Rajasthan* India
144 E4 Fatehpur *Uttar Pradesh* India
144 D4 Fatehpur Sikri India
140 D4 Fatḩābād Iran
191 G3 Fathom Five National Marine Park Canada
176 A3 Fatick Senegal
166 H2 Faulquemont France
181 F4 Fauresmith S. Africa
158 K2 Fauske Norway
197 F1 Faust U.S.A.
170 E6 Favignana, Isola *i. Sicily* Italy
186 G4 Fawcett Canada
161 F7 Fawley U.K.
188 C3 Fawn *r.* Canada
158 B2 Faxaflói *b.* Iceland
158 L3 Faxälven *r.* Sweden
177 D3 Faya Chad
190 D3 Fayette U.S.A.
199 E4 Fayetteville *AR* U.S.A.
201 E5 Fayetteville *NC* U.S.A.
201 C5 Fayetteville *TN* U.S.A.
137 L7 Faylakah *i.* Kuwait
141 H2 Fayzābād Afgh.
176 C4 Fazao Malfakassa, Parc National de *nat. park* Togo
144 C3 Fazilka India
140 C5 Fazrān, Jabal *h.* Saudi Arabia
176 A2 Fdérik Mauritania
163 B5 Feale *r.* Ireland
201 E5 Fear, Cape U.S.A.
196 B2 Feather, North Fork *r.* U.S.A.
196 B2 Feather Falls U.S.A.
128 E4 Featherston N.Z.
127 G6 Feathertop, Mount Australia
166 E2 Fécamp France
215 F1 Federación Arg.
212 E4 Federal Arg.
138 E1 Fedorovka *Kustanayskaya Oblast'* Kazakh.
139 I1 Fedorovka *Pavlodarskaya Oblast'* Kazakh.
138 B2 Fedorovka *Zapadnyy Kazakhstan* Kazakh.
138 C1 Fedorovka Rus. Fed.
168 E3 Fehmarn *i.* Germany
165 K2 Fehrbellin Germany
214 E3 Feia, Lagoa *lag.* Brazil
148 E4 Feidong China
148 F3 Feihuanghe Kou *r. mouth* China
210 D5 Feijó Brazil
128 E4 Feilding N.Z.
211 K6 Feira de Santana Brazil
148 E4 Feixi China
136 F3 Feke Turkey
167 H3 Felanitx Spain
190 D3 Felch U.S.A.
165 L1 Feldberg Germany
168 D7 Feldberg *mt.* Germany
168 D7 Feldkirch Austria
168 G7 Feldkirchen in Kärnten Austria
215 E1 Feliciano *r.* Arg.
207 G4 Felipe C. Puerto Mex.
214 D2 Felixlândia Brazil
161 I6 Felixstowe U.K.
170 D1 Feltre Italy
159 J3 Femunden *l.* Norway
159 K3 Femundsmarka Nasjonalpark *nat. park* Norway
170 D3 Fenaio, Punta del *pt* Italy
197 H4 Fence Lake U.S.A.
191 H3 Fenelon Falls Canada
171 K4 Fengari *mt.* Greece
149 E4 Fengcheng *Jiangxi* China
152 C3 Fengcheng *Liaoning* China
149 C4 Fengdu China
149 C5 Fenggang China
152 D1 Fengguang China
149 F4 Fenghua China
149 C5 Fenghuang China
148 C4 Fengjie China
149 D6 Fengkai China
149 F6 Fenglin Taiwan
152 D2 Fengman China
148 F2 Fengnan China
148 E1 Fengning China
148 E3 Fengqiu China
149 C5 Fengshan China
149 E6 Fengshun China
148 E3 Fengtai China
149 E4 Fengxin China
148 E3 Fengyang China
148 D1 Fengzhen China
148 D2 Fen He *r.* China
145 G5 Feni Bangl.
125 F2 Feni Islands P.N.G.
166 F5 Fenille, Col de la *pass* France
190 B4 Fennimore U.S.A.
179 E5 Fenoarivo Atsinanana Madag.
191 F4 Fenton U.S.A.
125 I3 Fenua Ura *i.* Fr. Polynesia
148 D2 Fenxi China
148 D2 Fenyang China
149 E5 Fenyi China
173 E6 Feodosiya Ukr.
170 B6 Fer, Cap de *c.* Alg.
141 E3 Ferdows Iran

Fergana

Fergana Uzbek. *see* **Farg'ona**
139 H4 **Fergana Too Tizmegi** *mts* Kyrg.
191 G4 **Fergus** Canada
198 D2 **Fergus Falls** U.S.A.
124 F2 **Fergusson Island** P.N.G.
170 C7 **Fériana** Tunisia
171 I3 **Ferizaj** Kosovo
176 B4 **Ferkessédougou** Côte d'Ivoire
170 E3 **Fermo** Italy
189 G3 **Fermont** Canada
167 C2 **Fermoselle** Spain
163 C5 **Fermoy** Ireland
210 □ **Fernandina, Isla** *i. Galapagos Is* Ecuador
201 D6 **Fernandina Beach** U.S.A.
212 B8 **Fernando de Magallanes, Parque Nacional** *nat. park* Chile
219 G6 **Fernando de Noronha** *i.* Brazil
214 B3 **Fernandópolis** Brazil
194 B1 **Ferndale** U.S.A.
161 F7 **Ferndown** U.K.
186 F5 **Fernie** Canada
127 G2 **Fernlee** Australia
196 C2 **Fernley** U.S.A.
203 F4 **Fernridge** U.S.A.
163 E5 **Ferns** Ireland
194 C2 **Fernwood** U.S.A.
170 D2 **Ferrara** Italy
214 B3 **Ferreiros** Brazil
199 F6 **Ferriday** U.S.A.
170 C4 **Ferro, Capo** *c. Sardinia* Italy
167 B1 **Ferrol** Spain
197 G2 **Ferron** U.S.A.
138 D1 **Fershampenuaz** Rus. Fed.
Ferwerd Neth. *see* **Ferwert**
164 D1 **Ferwert** Neth.
176 B1 **Fès** Morocco
178 B4 **Feshi** Dem. Rep. Congo
187 J5 **Fessenden** U.S.A.
198 F4 **Festus** U.S.A.
162 □ **Fethaland, Point of** U.K.
163 D5 **Fethard** Ireland
136 B3 **Fethiye** Turkey
138 C4 **Fetisovo** Kazakh.
162 □ **Fetlar** *i.* U.K.
162 F4 **Fettercairn** U.K.
165 J5 **Feucht** Germany
165 I5 **Feuchtwangen** Germany
189 F2 **Feuilles, Rivière aux** *r.* Canada
136 F3 **Fevzipaşa** Turkey
141 E3 **Feyẕābād** Iran
161 D5 **Ffestiniog** U.K.
179 E6 **Fianarantsoa** Madag.
178 D3 **Fichē** Eth.
165 K4 **Fichtelgebirge** *hills* Germany
181 G4 **Ficksburg** S. Africa
186 F4 **Field** *B.C.* Canada
191 G2 **Field** *Ont.* Canada
171 H4 **Fier** Albania
190 E3 **Fife Lake** U.S.A.
162 F4 **Fife Ness** *pt* U.K.
127 G4 **Fifield** Australia
190 B3 **Fifield** U.S.A.
166 F4 **Figeac** France
167 B2 **Figueira da Foz** Port.
167 H1 **Figueres** Spain
176 B1 **Figuig** Morocco
125 H3 **Fiji** *country* Pacific Ocean
206 H6 **Filadelfia** Costa Rica
212 D2 **Filadelfia** Para.
129 C3 **Filchner Ice Shelf** *ice feature* Antarctica
160 G3 **Filey** U.K.
171 I5 **Filippiada** Greece
159 K4 **Filipstad** Sweden
158 J3 **Fillan** Norway
196 C4 **Fillmore** *CA* U.S.A.
197 F2 **Fillmore** *UT* U.S.A.
129 D3 **Fimbul Ice Shelf** *ice feature* Antarctica
203 F2 **Finch** Canada
162 E3 **Findhorn** *r.* U.K.
137 H3 **Fındık** Turkey
202 B4 **Findlay** U.S.A.
127 H8 **Fingal** Australia
188 E5 **Finger Lakes** U.S.A.
179 D5 **Fingoè** Moz.
136 C3 **Finike** Turkey
136 C3 **Finike Körfezi** *b.* Turkey
167 B1 **Finisterre, Cape** Spain
172 S5 **Finland** *country* Europe
159 M4 **Finland, Gulf of** Europe
186 D3 **Finlay** *r.* Canada
186 D3 **Finlay, Mount** Canada
127 F5 **Finley** Australia
165 J3 **Finne** *ridge* Germany
126 A4 **Finniss, Cape** Australia
158 L1 **Finnsnes** Norway
159 K4 **Finspång** Sweden
163 D3 **Fintona** U.K.
Fintown Ireland *see* **Baile na Finne**
162 C3 **Fionn Loch** *l.* U.K.
162 B4 **Fionnphort** U.K.
128 A6 **Fiordland National Park** N.Z.
196 B3 **Firebaugh** U.S.A.
187 I2 **Firedrake Lake** Canada
203 G4 **Fire Island National Seashore** *nature res.* U.S.A.
Firenze Italy *see* **Florence**
137 J6 **Firk, Sha'īb** *watercourse* Iraq
215 E2 **Firmat** Arg.
166 G4 **Firminy** France
165 I6 **Firngrund** *reg.* Germany
169 P2 **Firovo** Rus. Fed.
144 B3 **Firoz** Pak.
144 D4 **Firozabad** India
141 G3 **Firozkoh** *reg.* Afgh.
144 C3 **Firozpur** India
203 H2 **First Connecticut Lake** U.S.A.
140 D4 **Fīrūzābād** Iran
164 F5 **Fischbach** Germany
179 B6 **Fish** *watercourse* Namibia
180 D5 **Fish** *r.* S. Africa
129 B6 **Fisher Bay** *b.* Antarctica
203 F6 **Fisherman Island** U.S.A.
203 H4 **Fishers Island** U.S.A.
187 M2 **Fisher Strait** Canada
161 C6 **Fishguard** U.K.
186 E2 **Fish Lake** Canada
190 A2 **Fish Lake** *MN* U.S.A.
197 G2 **Fish Lake** *UT* U.S.A.
191 F4 **Fish Point** U.S.A.
129 C3 **Fiske, Cape** *c.* Antarctica
164 B5 **Fismes** France
167 B1 **Fisterra** Spain
Fisterra, Cabo *c.* Spain *see* **Finisterre, Cape**
203 H3 **Fitchburg** U.S.A.
187 G3 **Fitzgerald** Canada
201 D6 **Fitzgerald** U.S.A.
126 B4 **Fitzgerald Bay** Australia
212 C7 **Fitz Roy** Arg.
124 C3 **Fitzroy Crossing** Australia
191 G3 **Fitzwilliam Island** Canada
163 D3 **Fivemiletown** U.K.
170 D2 **Fivizzano** Italy
178 C4 **Fizi** Dem. Rep. Congo
159 J3 **Flå** Norway
181 H5 **Flagstaff** S. Africa
197 G4 **Flagstaff** U.S.A.
203 H2 **Flagstaff Lake** U.S.A.
188 E2 **Flaherty Island** Canada
190 B3 **Flambeau** *r.* U.S.A.
160 G3 **Flamborough Head** U.K.
165 K2 **Fläming** *hills* Germany
194 E3 **Flaming Gorge Reservoir** U.S.A.
180 D5 **Flaminksvlei** *salt pan* S. Africa
164 A3 **Flanders** *reg.* Europe
164 A4 **Flandre** *reg.* France
162 A2 **Flannan Isles** U.K.
158 K2 **Flåsjön** *l.* Sweden
190 E4 **Flat** *r.* U.S.A.
194 D2 **Flathead Lake** U.S.A.
128 E4 **Flat Point** N.Z.
124 E3 **Flattery, Cape** Australia
194 A1 **Flattery, Cape** U.S.A.
165 J2 **Fleetmark** Germany
160 D4 **Fleetwood** U.K.
203 F4 **Fleetwood** U.S.A.
159 I4 **Flekkefjord** Norway
202 E3 **Fleming** U.S.A.
202 B5 **Flemingsburg** U.S.A.
219 G2 **Flemish Cap** *sea feature* N. Atlantic Ocean
159 L4 **Flen** Sweden
168 D3 **Flensburg** Germany
166 D2 **Flers** France
191 G3 **Flesherton** Canada
187 H2 **Fletcher Lake** Canada
191 F3 **Fletcher Pond** *l.* U.S.A.
124 E3 **Flinders** *r.* Australia
124 B5 **Flinders Bay** Australia
126 B5 **Flinders Chase National Park** Australia
126 A4 **Flinders Island** *S.A.* Australia
127 H7 **Flinders Island** *Tas.* Australia
126 C3 **Flinders Ranges** *mts* Australia
126 C3 **Flinders Ranges National Park** Australia
187 I4 **Flin Flon** Canada
161 D4 **Flint** U.K.
191 F4 **Flint** U.S.A.
201 C6 **Flint** *r. GA* U.S.A.
191 F4 **Flint** *r. MI* U.S.A.
217 I5 **Flint Island** Kiribati
127 H1 **Flinton** Australia
159 K3 **Flisa** Norway
160 E2 **Flodden** U.K.
165 L4 **Flöha** Germany
165 L4 **Flöha** *r.* Germany
129 B4 **Flood Range** *mts* Antarctica
190 A2 **Floodwood** U.S.A.
200 B4 **Flora** U.S.A.
166 F4 **Florac** France
164 E5 **Florange** France
210 □ **Floreana, Isla** *i. Galapagos Is* Ecuador
191 F4 **Florence** Canada
170 D3 **Florence** Italy
201 C5 **Florence** *AL* U.S.A.
197 G5 **Florence** *AZ* U.S.A.
198 D4 **Florence** *KS* U.S.A.
202 C5 **Florence** *OH* U.S.A.
194 A3 **Florence** *OR* U.S.A.
201 E5 **Florence** *SC* U.S.A.
197 G5 **Florence Junction** U.S.A.
203 J1 **Florenceville** Canada
213 B4 **Florencia** Col.
164 C4 **Florennes** Belgium
212 C6 **Florentino Ameghino, Embalse** *resr* Arg.
215 E2 **Flores** *r.* Arg.
156 A6 **Flores** *i.* Azores
207 G4 **Flores** Guat.
147 E7 **Flores** *i.* Indon.
147 D7 **Flores, Laut** *sea* Indon.
214 C1 **Flores de Goiás** Brazil
Flores Sea *sea* Indon. *see* **Flores, Laut**
211 K5 **Floresta** Brazil
211 J5 **Floriano** Brazil
212 G3 **Florianópolis** Brazil
215 F2 **Florida** Uruguay
201 D6 **Florida** *state* U.S.A.
205 H4 **Florida, Straits of** Bahamas/U.S.A.
201 D7 **Florida Bay** U.S.A.
201 D7 **Florida City** U.S.A.
125 G2 **Florida Islands** Solomon Is
193 J7 **Florida Keys** *is* U.S.A.
171 I4 **Florina** Greece
159 I3 **Florø** Norway
189 H3 **Flour Lake** Canada
190 A4 **Floyd** *IA* U.S.A.
202 C6 **Floyd** *VA* U.S.A.
197 F4 **Floyd, Mount** U.S.A.
199 C5 **Floydada** U.S.A.
164 D2 **Fluessen** *l.* Neth.
124 E2 **Fly** *r.* P.N.G.
202 C5 **Fly** U.S.A.
171 H3 **Foča** Bos.-Herz.
162 E3 **Fochabers** U.K.
181 G3 **Fochville** S. Africa
171 L2 **Focşani** Romania
149 D6 **Fogang** China
170 F4 **Foggia** Italy
176 □ **Fogo** *i.* Cape Verde
189 J4 **Fogo Island** Canada
162 D2 **Foinaven** *h.* U.K.
166 E5 **Foix** France
158 K2 **Folda** *sea chan.* Norway
158 K2 **Foldereid** Norway
158 J2 **Foldfjorden** *sea chan.* Norway
171 K6 **Folegandros** *i.* Greece
191 F1 **Foleyet** Canada
170 E3 **Foligno** Italy
161 I6 **Folkestone** U.K.
161 G5 **Folkingham** U.K.
201 D6 **Folkston** U.S.A.
159 J3 **Folldal** Norway
170 D3 **Follonica** Italy
196 B2 **Folsom Lake** U.S.A.
173 G6 **Fomin** Rus. Fed.
172 I2 **Fominskiy** Rus. Fed.
187 H3 **Fond-du-Lac** Canada
187 I3 **Fond du Lac** *r.* Canada
190 C4 **Fond du Lac** U.S.A.
167 B2 **Fondevila** Spain
170 E4 **Fondi** Italy
125 H2 **Fongafale** *i.* Tuvalu
170 C4 **Fonni** *Sardinia* Italy
206 H5 **Fonseca, Golfo do** *b.* Central America
189 F3 **Fontanges** Canada
186 E3 **Fontas** Canada
186 E3 **Fontas** *r.* Canada
210 E4 **Fonte Boa** Brazil
166 D3 **Fontenay-le-Comte** France
158 D1 **Fontur** *pt* Iceland
191 H3 **Foot's Bay** Canada
148 C3 **Foping** China
127 H4 **Forbes** Australia
194 C1 **Forbes, Mount** Canada
165 J5 **Forchheim** Germany
189 G2 **Ford** *r.* Canada
190 D2 **Ford** *r.* U.S.A.
159 I3 **Førde** Norway
187 J2 **Forde Lake** Canada
161 H5 **Fordham** U.K.
161 F7 **Fordingbridge** U.K.
129 B4 **Ford Range** *mts* Antarctica
127 F2 **Fords Bridge** Australia
199 E5 **Fordyce** U.S.A.
176 A4 **Forécariah** Guinea
161 F7 **Foreland** *hd* U.K.
161 D6 **Foreland Point** U.K.
186 D4 **Foresight Mountain** Canada
191 G4 **Forest** Canada
199 F5 **Forest** *MS* U.S.A.
202 B4 **Forest** *OH* U.S.A.
203 G3 **Forest Dale** U.S.A.
127 G5 **Forest Hill** Australia
196 B2 **Foresthill** U.S.A.
127 H9 **Forestier, Cape** Australia
127 H9 **Forestier Peninsula** Australia
190 A3 **Forest Lake** U.S.A.
201 C5 **Forest Park** U.S.A.
189 G4 **Forestville** Canada
162 F4 **Forfar** U.K.
194 A2 **Forks** U.S.A.
202 E4 **Forksville** U.S.A.
170 E2 **Forlì** Italy
160 D4 **Formby** U.K.
167 G3 **Formentera** *i.* Spain
167 H3 **Formentor, Cap de** *c.* Spain
214 D3 **Formiga** Brazil
212 E3 **Formosa** Arg.
214 C1 **Formosa** Brazil
211 G6 **Formosa, Serra** *hills* Brazil
214 D1 **Formoso** *r.* Brazil
162 E3 **Forres** U.K.
126 E7 **Forrest** Australia
190 C5 **Forrest** U.S.A.
199 F5 **Forrest City** U.S.A.
190 C4 **Forreston** U.S.A.
158 L3 **Fors** Sweden
124 E3 **Forsayth** Australia
158 M2 **Forsnäs** Sweden
159 M3 **Forssa** Fin.
127 J4 **Forster** Australia
199 E4 **Forsyth** *MO* U.S.A.
194 F2 **Forsyth** *MT* U.S.A.
191 I1 **Forsythe** Canada
144 C3 **Fort Abbas** Pak.
188 D3 **Fort Albany** Canada
211 K4 **Fortaleza** Brazil
197 H5 **Fort Apache** U.S.A.
186 G4 **Fort Assiniboine** Canada
190 C4 **Fort Atkinson** U.S.A.
162 D3 **Fort Augustus** U.K.
181 G6 **Fort Beaufort** S. Africa
194 E2 **Fort Benton** U.S.A.
187 H3 **Fort Black** Canada
196 A2 **Fort Bragg** U.S.A.
Fort-Chimo Canada *see* **Kuujjuaq**
187 G3 **Fort Chipewyan** Canada
199 D5 **Fort Cobb Reservoir** U.S.A.
194 F3 **Fort Collins** U.S.A.
191 I3 **Fort-Coulonge** Canada
203 F2 **Fort Covington** U.S.A.
199 C6 **Fort Davis** U.S.A.
205 L6 **Fort-de-France** Martinique
201 C5 **Fort Deposit** U.S.A.
198 E3 **Fort Dodge** U.S.A.
198 E1 **Fort Frances** U.S.A.
Fort George Canada *see* **Chisasibi**
184 F3 **Fort Good Hope** Canada
162 D4 **Forth** *r.* U.K.
162 F4 **Forth, Firth of** *est.* U.K.
197 E2 **Fortification Range** *mts* U.S.A.
212 D2 **Fortín Capitán Demattei** Para.
212 D2 **Fortín General Mendoza** Para.
212 E2 **Fortín Madrejón** Para.
212 D2 **Fortín Pilcomayo** Arg.
210 F7 **Fortín Ravelo** Bol.
210 F7 **Fortín Suárez Arana** Bol.
203 I1 **Fort Kent** U.S.A.
201 D7 **Fort Lauderdale** U.S.A.
186 E2 **Fort Liard** Canada
187 G3 **Fort Mackay** Canada
186 G5 **Fort Macleod** Canada
190 B5 **Fort Madison** U.S.A.
190 B3 **Fort McCoy** U.S.A.
187 G3 **Fort McMurray** Canada
184 E3 **Fort McPherson** Canada
194 G3 **Fort Morgan** U.S.A.
201 D7 **Fort Myers** U.S.A.
186 E3 **Fort Nelson** Canada
186 E3 **Fort Nelson** *r.* Canada
Fort Norman Canada *see* **Tulita**
201 C5 **Fort Payne** U.S.A.
194 F1 **Fort Peck** U.S.A.
194 F2 **Fort Peck Reservoir** U.S.A.
201 D7 **Fort Pierce** U.S.A.
198 C2 **Fort Pierre** U.S.A.
186 F2 **Fort Providence** Canada
187 I4 **Fort Qu'Appelle** Canada
186 G2 **Fort Resolution** Canada
128 B7 **Fortrose** N.Z.
162 D3 **Fortrose** U.K.
196 A2 **Fort Ross** U.S.A.
Fort Rupert Canada *see* **Waskaganish**
186 E4 **Fort St James** Canada
186 E3 **Fort St John** Canada
186 G4 **Fort Saskatchewan** Canada
199 E4 **Fort Scott** U.S.A.
188 C2 **Fort Severn** Canada
138 B3 **Fort-Shevchenko** Kazakh.
186 E2 **Fort Simpson** Canada
187 G2 **Fort Smith** Canada
199 E5 **Fort Smith** U.S.A.
199 C6 **Fort Stockton** U.S.A.
195 F5 **Fort Sumner** U.S.A.
194 A3 **Fortuna** *CA* U.S.A.
198 C1 **Fortuna** *ND* U.S.A.
189 I4 **Fortune Bay** Canada
186 F3 **Fort Vermilion** Canada
201 C6 **Fort Walton Beach** U.S.A.
190 E5 **Fort Wayne** U.S.A.
162 C4 **Fort William** U.K.
199 D5 **Fort Worth** U.S.A.
184 D3 **Fort Yukon** U.S.A.
140 D5 **Forūr, Jazīreh-ye** *i.* Iran
158 K2 **Forvik** Norway
149 D6 **Foshan** China
170 B2 **Fossano** Italy
127 G7 **Foster** Australia
186 B3 **Foster, Mount** Canada/U.S.A.
185 P2 **Foster Bugt** *b.* Greenland
202 B4 **Fostoria** U.S.A.
161 G4 **Fotherby** U.K.
166 D2 **Fougères** France
162 □ **Foula** *i.* U.K.
161 H6 **Foulness Point** U.K.
143 C4 **Foul Point** Sri Lanka
128 C4 **Foulwind, Cape** N.Z.
176 D4 **Foumban** Cameroon
129 C4 **Foundation Ice Stream** *glacier* Antarctica
176 A3 **Foundiougne** Senegal
190 A4 **Fountain** U.S.A.
160 F3 **Fountains Abbey & Studley Royal Water Garden (NT)** *tourist site* U.K.
166 G2 **Fourches, Mont des** *h.* France
196 D4 **Four Corners** U.S.A.
181 H4 **Fouriesburg** S. Africa
164 C4 **Fourmies** France
171 L6 **Fournoi** *i.* Greece
190 C2 **Fourteen Mile Point** U.S.A.
176 A3 **Fouta Djallon** *reg.* Guinea
128 A7 **Foveaux Strait** N.Z.
201 E7 **Fowl Cay** *i.* Bahamas
195 F4 **Fowler** *CO* U.S.A.
190 D5 **Fowler** *IN* U.S.A.
190 E4 **Fowler** *MI* U.S.A.
129 B3 **Fowler Ice Rise** *ice feature* Antarctica
124 D5 **Fowlers Bay** Australia
137 L3 **Fowman** Iran
187 K3 **Fox** *r.* Canada
190 C4 **Fox** *r.* U.S.A.
186 F4 **Fox Creek** Canada
160 C3 **Foxdale** U.K.
185 J3 **Foxe Basin** *g.* Canada
185 J3 **Foxe Channel** Canada
185 K3 **Foxe Peninsula** Canada
128 C5 **Fox Glacier** N.Z.
186 G3 **Fox Lake** Canada
190 C4 **Fox Lake** U.S.A.
128 E4 **Foxton** N.Z.
162 D3 **Foyers** U.K.
163 D3 **Foyle** *r.* Ireland/U.K.
163 D2 **Foyle, Lough** *b.* Ireland/U.K.
163 B5 **Foynes** Ireland
179 B5 **Foz do Cunene** Angola
214 A4 **Foz do Iguaçu** Brazil
167 G2 **Fraga** Spain
214 C3 **Franca** Brazil
125 G3 **Français, Récif des** *rf* New Caledonia
166 F3 **France** *country* Europe
126 D6 **Frances** Australia
186 D2 **Frances** *r.* Canada
186 D2 **Frances Lake** Canada
186 D2 **Frances Lake** *l.* Canada
190 D5 **Francesville** U.S.A.
178 B4 **Franceville** Gabon
203 H2 **Francis, Lake** U.S.A.
198 D3 **Francis Case, Lake** U.S.A.
206 D2 **Francisco I. Madero** Mex.
206 C2 **Francisco I. Madero** Mex.
214 D2 **Francisco Sá** Brazil
179 C6 **Francistown** Botswana
186 D4 **François Lake** Canada
194 E3 **Francs Peak** U.S.A.
164 D1 **Franeker** Neth.
165 L4 **Frankenberg** Germany
165 G3 **Frankenberg (Eder)** Germany
191 F4 **Frankenmuth** U.S.A.
165 G5 **Frankenthal (Pfalz)** Germany
165 J4 **Frankenwald** *mts* Germany
181 H3 **Frankfort** S. Africa
190 D5 **Frankfort** *IN* U.S.A.
200 C4 **Frankfort** *KY* U.S.A.
190 D3 **Frankfort** *MI* U.S.A.
165 G4 **Frankfurt am Main** Germany
168 G4 **Frankfurt an der Oder** Germany
197 E1 **Frankin Lake** U.S.A.
165 J5 **Fränkische Alb** *hills* Germany
165 J5 **Fränkische Schweiz** *reg.* Germany
194 E3 **Franklin** *ID* U.S.A.
200 C4 **Franklin** *IN* U.S.A.
199 F6 **Franklin** *LA* U.S.A.
203 H3 **Franklin** *MA* U.S.A.
201 D5 **Franklin** *NC* U.S.A.
203 H3 **Franklin** *NH* U.S.A.
203 F4 **Franklin** *NJ* U.S.A.
202 D4 **Franklin** *PA* U.S.A.
201 C5 **Franklin** *TN* U.S.A.
202 E6 **Franklin** *VA* U.S.A.
202 D5 **Franklin** *WV* U.S.A.
184 F3 **Franklin Bay** Canada
194 C1 **Franklin D. Roosevelt Lake** *resr* U.S.A.
127 F9 **Franklin-Gordon National Park** Australia
126 B4 **Franklin Harbor** *b.* Australia
Franklin Harbour *b.* Australia *see* **Franklin Harbor**
186 E2 **Franklin Mountains** Canada
128 A6 **Franklin Mountains** N.Z.
127 G8 **Franklin Sound** *sea chan.* Australia
185 I2 **Franklin Strait** Canada
126 F7 **Frankston** Australia
159 L3 **Fränsta** Sweden
132 G2 **Frantsa-Iosifa, Zemlya** *is* Rus. Fed.
190 E1 **Franz** Canada
128 C5 **Franz Josef Glacier** N.Z.
170 C5 **Frasca, Capo della** *c. Sardinia* Italy
170 E4 **Frascati** Italy
186 E4 **Fraser** *r. B.C.* Canada
189 H2 **Fraser** *r. Nfld* Canada
180 D5 **Fraserburg** S. Africa
162 F3 **Fraserburgh** U.K.
188 D4 **Fraserdale** Canada
125 F4 **Fraser Island** Australia
186 E4 **Fraser Lake** Canada
186 E4 **Fraser Plateau** Canada
128 F3 **Frasertown** N.Z.
190 E2 **Frater** Canada
168 D7 **Frauenfeld** Switz.
215 E2 **Fray Bentos** Uruguay
164 E4 **Frechen** Germany
160 E4 **Freckleton** U.K.
190 E3 **Frederic** *MI* U.S.A.
190 A3 **Frederic** *WI* U.S.A.
159 J5 **Fredericia** Denmark
202 E5 **Frederick** *MD* U.S.A.
199 D5 **Frederick** *OK* U.S.A.
199 D6 **Fredericksburg** *TX* U.S.A.
202 E5 **Fredericksburg** *VA* U.S.A.
186 C3 **Frederick Sound** *sea chan.* U.S.A.
199 F4 **Fredericktown** U.S.A.
189 G4 **Fredericton** Canada
Frederikshåb Greenland *see* **Paamiut**
159 J4 **Frederikshavn** Denmark
159 K5 **Frederiksværk** Denmark
197 F3 **Fredonia** *AZ* U.S.A.
202 D3 **Fredonia** *NY* U.S.A.
158 L2 **Fredrika** Sweden
159 J4 **Fredrikstad** Norway
203 F4 **Freehold** U.S.A.
203 F4 **Freeland** U.S.A.
126 C3 **Freeling Heights** *h.* Australia
196 C2 **Freel Peak** U.S.A.
198 D3 **Freeman** U.S.A.
190 D5 **Freeman, Lake** U.S.A.
190 C4 **Freeport** *IL* U.S.A.
203 H3 **Freeport** *ME* U.S.A.
203 G4 **Freeport** *NY* U.S.A.
199 E6 **Freeport** *TX* U.S.A.
201 E7 **Freeport City** Bahamas
199 D7 **Freer** U.S.A.
181 G4 **Free State** *prov.* S. Africa
176 A4 **Freetown** Sierra Leone
167 C3 **Fregenal de la Sierra** Spain
166 C2 **Fréhel, Cap** *c.* France
129 E2 **Frei (Chile)** *research stn* Antarctica
168 C6 **Freiburg im Breisgau** Germany
164 F5 **Freisen** Germany
168 E6 **Freising** Germany
168 G6 **Freistadt** Austria
166 H5 **Fréjus** France
124 B5 **Fremantle** Australia
190 E4 **Fremont** *MI* U.S.A.
198 D3 **Fremont** *NE* U.S.A.
202 B4 **Fremont** *OH* U.S.A.
197 G2 **Fremont** *r.* U.S.A.
202 B6 **Frenchburg** U.S.A.
202 C4 **French Creek** *r.* U.S.A.
211 H3 **French Guiana** *terr.* S. America
126 F7 **French Island** Australia
187 H5 **Frenchman** *r.* Canada/U.S.A.
196 C2 **Frenchman** U.S.A.
196 B2 **Frenchman Lake** U.S.A.
197 E3 **Frenchman Lake** U.S.A.
127 F9 **Frenchman's Cap** *mt.* Australia
163 C4 **Frenchpark** Ireland
128 D4 **French Pass** N.Z.
123 I5 **French Polynesia** *terr.* Pacific Ocean
121 **French Southern and Antarctic Lands** *terr.* Indian Ocean
203 I1 **Frenchville** U.S.A.
164 F2 **Freren** Germany
163 D5 **Freshford** Ireland
197 G6 **Fresnal Canyon** U.S.A.
206 D3 **Fresnillo** Mex.
196 C3 **Fresno** U.S.A.
196 C3 **Fresno** *r.* U.S.A.
167 H3 **Freu, Cap des** *c.* Spain
165 F4 **Freudenberg** Germany
168 D6 **Freudenstadt** Germany
164 A4 **Frévent** France
127 H9 **Freycinet National Park** Australia
127 H9 **Freycinet Peninsula** Australia
165 K1 **Freyenstein** Germany
166 H2 **Freyming-Merlebach** France
215 D1 **Freyre** Arg.
176 A3 **Fria** Guinea
196 C3 **Friant** U.S.A.
212 C3 **Frias** Arg.
168 C7 **Fribourg** Switz.
165 F1 **Friedeburg** Germany
165 L1 **Friedland** Germany
168 D7 **Friedrichshafen** Germany
203 I3 **Friendship** U.S.A.
165 K2 **Friesack** Germany
164 D1 **Friese Wad** *tidal flat* Neth.
165 F1 **Friesoythe** Germany
161 I6 **Frinton-on-Sea** U.K.
199 D6 **Frio** *r.* U.S.A.
162 B4 **Frisa, Loch** *l.* U.K.
197 F2 **Frisco Mountain** U.S.A.
165 H3 **Fritzlar** Germany
185 L3 **Frobisher Bay** Canada
187 H3 **Frobisher Lake** Canada
158 J3 **Frohavet** *b.* Norway
165 K3 **Frohburg** Germany
164 A5 **Froissy** France
173 G5 **Frolovo** Rus. Fed.
172 J2 **Frolovsk** Rus. Fed.
126 C2 **Frome** *watercourse* Australia
161 E6 **Frome** U.K.
126 C3 **Frome, Lake** *salt flat* Australia
126 C3 **Frome Downs** Australia
165 F3 **Fröndenberg** Germany
207 F4 **Frontera** Mex.
207 F4 **Frontera, Punta** *pt* Mex.
202 D5 **Front Royal** U.S.A.
170 E4 **Frosinone** Italy
202 D5 **Frostburg** U.S.A.
129 C6 **Frost Glacier** *glacier* Antarctica
158 J3 **Frøya** *i.* Norway
164 A4 **Fruges** France
197 H2 **Fruita** U.S.A.
197 G1 **Fruitland** U.S.A.
139 G4 **Frunze** Kyrg.
Frunze Kyrg. *see* **Bishkek**
168 C7 **Frutigen** Switz.
169 I6 **Frýdek-Místek** Czech Rep.
203 H2 **Fryeburg** U.S.A.
149 F5 **Fu'an** China
149 D5 **Fuchuan** China
149 F4 **Fuchun Jiang** *r.* China
162 A3 **Fuday** *i.* U.K.
149 E5 **Fude** China
149 F5 **Fuding** China
167 E2 **Fuenlabrada** Spain
167 D3 **Fuente Obejuna** Spain
152 D2 **Fu'er He** *r.* China
212 E2 **Fuerte Olimpo** Para.
176 A2 **Fuerteventura** *i.* Canary Is
153 B2 **Fuga** *i.* Phil.
148 E3 **Fugou** China
148 D2 **Fugu** China
137 I4 **Fuḩaymī** Iraq
151 F7 **Fuji** Japan
149 E5 **Fujian** *prov.* China
148 C4 **Fu Jiang** *r.* China
151 F7 **Fuji-Hakone-Izu Kokuritsu-kōen** *nat. park* Japan
150 B1 **Fujin** China
151 F7 **Fujinomiya** Japan
151 F7 **Fuji-san** *vol.* Japan
150 H3 **Fukagawa** Japan
151 D7 **Fukuchiyama** Japan
151 A8 **Fukue** Japan
151 A8 **Fukue-jima** *i.* Japan
151 E6 **Fukui** Japan
151 B8 **Fukuoka** Japan
151 G6 **Fukushima** Japan
151 B9 **Fukuyama** Japan
140 D2 **Fūlād Maīalleh** Iran
165 H4 **Fulda** Germany
165 H3 **Fulda** *r.* Germany
161 G6 **Fulham** U.K.
148 E3 **Fuli** China
149 B4 **Fulin** China
149 C4 **Fuling** China
187 L2 **Fullerton, Cape** Canada
190 B5 **Fulton** *IL* U.S.A.
200 B4 **Fulton** *KY* U.S.A.
198 F4 **Fulton** *MO* U.S.A.
203 E3 **Fulton** *NY* U.S.A.
181 J2 **Fumane** Moz.
164 C5 **Fumay** France
151 F7 **Funabashi** Japan
125 H2 **Funafuti** *atoll* Tuvalu
152 B3 **Funan** China
176 A1 **Funchal** Madeira
213 B2 **Fundación** Col.
167 C2 **Fundão** Port.
206 B2 **Fundición** Mex.
189 G5 **Fundy, Bay of** *g.* Canada
189 G4 **Fundy National Park** Canada
196 D3 **Funeral Peak** U.S.A.
Fung Wong Shan *h. Hong Kong* China *see* **Lantau Peak**
179 D6 **Funhalouro** Moz.
148 F3 **Funing** *Jiangsu* China
149 B6 **Funing** *Yunnan* China
148 D3 **Funiu Shan** *mts* China
176 C3 **Funtua** Nigeria
162 □ **Funzie** U.K.
149 F5 **Fuqing** China
150 H3 **Furano** Japan
140 E5 **Fürgun, Kūh-e** *mt.* Iran
172 G3 **Furmanov** Rus. Fed.
150 D3 **Furmanovo** Rus. Fed.
196 D3 **Furnace Creek** U.S.A.
167 G3 **Furnás** *h.* Spain
214 C3 **Furnas, Represa** *resr* Brazil
124 E6 **Furneaux Group** *is* Australia
165 F2 **Fürstenau** Germany
165 L1 **Fürstenberg** Germany
168 G4 **Fürstenwalde** Germany
165 I5 **Fürth** Germany
165 K5 **Furth im Wald** Germany
150 G3 **Furubira** Japan
150 G5 **Furukawa** Japan
185 J3 **Fury and Hecla Strait** Canada
213 B3 **Fusagasugá** Col.
149 C7 **Fushan** *Hainan* China
152 A5 **Fushan** *Shandong* China
152 B3 **Fushun** *Liaoning* China
149 B4 **Fushun** *Sichuan* China
152 D2 **Fusong** China
149 C6 **Fusui** China
151 B8 **Futago-san** *vol.* Japan
125 H3 **Futuna** *i.* Vanuatu
149 E5 **Futun Xi** *r.* China
148 C3 **Fuxian** China
149 B5 **Fuxian Hu** *l.* China
152 A2 **Fuxin** *Liaoning* China
152 A2 **Fuxin** *Liaoning* China
150 F5 **Fuya** Japan
148 E3 **Fuyang** *Anhui* China
149 F4 **Fuyang** *Zhejiang* China
148 E2 **Fuyang He** *r.* China
146 E2 **Fuyu** *Heilong.* China
152 D1 **Fuyu** *Jilin* China
149 B5 **Fuyuan** China
149 F5 **Fuzhou** China
149 E5 **Fuzhou** China
152 A4 **Fuzhou Wan** *b.* China
137 K2 **Füzuli** Azer.
159 J5 **Fyn** *i.* Denmark
162 C5 **Fyne, Loch** *inlet* U.K.
F.Y.R.O.M. (Former Yugoslav Republic of Macedonia) *country* Europe *see* **Macedonia**

G

170 C6 **Gaâfour** Tunisia
178 E3 **Gaalkacyo** Somalia
138 E5 **Gabakly** Turkm.
181 F2 **Gabane** Botswana
196 D2 **Gabbs** U.S.A.
196 C2 **Gabbs Valley Range** *mts* U.S.A.
179 B5 **Gabela** Angola
176 D1 **Gabès** Tunisia
177 D1 **Gabès, Golfe de** *g.* Tunisia
139 H2 **Gabiden Mustafin** Kazakh.
127 H6 **Gabo Island** Australia
178 B4 **Gabon** *country* Africa
179 C6 **Gaborone** Botswana
141 E5 **Gābrīk** Iran
141 E5 **Gābrīk** *watercourse* Iran
171 K3 **Gabrovo** Bulg.
176 A3 **Gabú** Guinea-Bissau
140 C2 **Gach Sār** Iran
140 C4 **Gachsārān** Iran
143 A3 **Gadag** India
144 E5 **Gadchiroli** India
158 K2 **Gäddede** Sweden
165 J1 **Gadebusch** Germany
144 B5 **Gadhra** India
144 B4 **Gadra** Pak.
201 C5 **Gadsden** U.S.A.
143 B2 **Gadwal** India
138 E5 **Gadyn** Turkm.
Gæi'dnuvuop'pi Norway *see* **Geaidnovuohppi**
161 D6 **Gaer** U.K.
171 K2 **Găeşti** Romania
170 E4 **Gaeta** Italy
170 E4 **Gaeta, Golfo di** *g.* Italy
216 E5 **Gaferut** *i.* Micronesia
201 D5 **Gaffney** U.S.A.
176 C1 **Gafsa** Tunisia
172 E4 **Gagarin** Rus. Fed.
139 G4 **Gagarin** Uzbek.
172 H4 **Gagino** Rus. Fed.
176 B4 **Gagnoa** Côte d'Ivoire
189 G3 **Gagnon** Canada
173 G7 **Gagra** Georgia
137 K4 **Gahvāreh** Iran
180 C3 **Gaiab** *watercourse* Namibia
145 G4 **Gaibandha** Bangl.
171 K7 **Gaïdouronisi** *i.* Greece
165 H6 **Gaildorf** Germany
166 E5 **Gaillac** France
145 G3 **Gaindainqoinkor** China
201 D6 **Gainesville** *FL* U.S.A.
201 D5 **Gainesville** *GA* U.S.A.
199 D5 **Gainesville** *TX* U.S.A.
161 G4 **Gainsborough** U.K.
126 A3 **Gairdner, Lake** *salt flat* Australia
162 C3 **Gairloch** U.K.
162 C3 **Gair Loch** *b.* U.K.
152 B3 **Gaizhou** China
143 C2 **Gajapatinagaram** India
141 G5 **Gajar** Pak.
180 E3 **Gakarosa** *mt.* S. Africa
144 C1 **Gakuch** Pak.
145 G3 **Gala** China
Galaasiya Uzbek. *see* **Galaosiyo**
178 D4 **Galana** *r.* Kenya
138 F5 **Galaosiyo** Uzbek.
210 □ **Galápagos, Parque Nacional** *nat. park* Ecuador
217 N6 **Galapagos Islands** Pacific Ocean
217 M6 **Galapagos Rise** *sea feature* Pacific Ocean
162 F5 **Galashiels** U.K.
171 L3 **Galata, Nos** *pt* Bulg.
171 M2 **Galaţi** Romania
171 H4 **Galatina** Italy
202 C6 **Galax** U.S.A.
141 F3 **Galaýmor** Turkm.
163 C5 **Galbally** Ireland
159 J3 **Galdhøpiggen** *mt.* Norway
207 D2 **Galeana** Mex.
140 D5 **Galeh Dār** Iran
190 B4 **Galena** U.S.A.
213 E2 **Galeota Point** Trin. and Tob.
215 B4 **Galera, Punta** *pt* Chile
207 E5 **Galera, Punta** *pt* Mex.
213 E2 **Galera Point** Trin. and Tob.
190 B5 **Galesburg** U.S.A.
180 F4 **Galeshewe** S. Africa
190 B3 **Galesville** U.S.A.
202 E4 **Galeton** U.S.A.
173 G7 **Gali** Georgia
172 G3 **Galich** Rus. Fed.
172 G2 **Galichskaya Vozvyshennost'** *hills* Rus. Fed.
167 C1 **Galicia** *aut. comm.* Spain
136 E5 **Galilee, Sea of** Israel
202 B4 **Galion** U.S.A.
170 C6 **Galite, Canal de la** *sea chan.* Tunisia
197 G5 **Galiuro Mountains** U.S.A.
138 E5 **Galkynyş** Turkm.
177 F3 **Gallabat** Sudan
139 F4 **G'allaorol** Uzbek.
201 C4 **Gallatin** U.S.A.
194 E2 **Gallatin** *r.* U.S.A.
143 C5 **Galle** Sri Lanka
217 L6 **Gallego Rise** *sea feature* Pacific Ocean
212 B8 **Gallegos** *r.* Arg.
213 C1 **Gallinas, Punta** *pt* Col.
171 L4 **Gallipoli** Turkey
202 B5 **Gallipolis** U.S.A.
158 M2 **Gällivare** Sweden
158 K3 **Gällö** Sweden
203 E3 **Gallo Island** U.S.A.
197 H4 **Gallo Mountains** U.S.A.
197 H4 **Gallup** U.S.A.
Gallyaaral Uzbek. *see* **G'allaorol**
162 B4 **Galmisdale** U.K.
127 H5 **Galong** Australia
143 C4 **Galoya** Sri Lanka
143 C5 **Gal Oya** *r.* Sri Lanka
162 D5 **Galston** U.K.
196 B2 **Galt** U.S.A.
176 A2 **Galtat Zemmour** W. Sahara
163 C5 **Galtee Mountains** *hills* Ireland
163 C5 **Galtymore** *h.* Ireland
141 E3 **Galūgāh-e Āsīyeh** Iran
190 B5 **Galva** U.S.A.
199 E6 **Galveston** U.S.A.
199 E6 **Galveston Bay** U.S.A.
215 E2 **Galvez** Arg.
145 E3 **Galwa** Nepal
163 B4 **Galway** Ireland
163 B4 **Galway Bay** Ireland
Gâm *r.* Vietnam *see* **Gâm, Sông**
149 B6 **Gâm, Sông** *r.* Vietnam
161 I8 **Gamaches** France
181 I5 **Gamalakhe** S. Africa
213 B2 **Gamarra** Col.
145 G3 **Gamba** China
Gamba China *see* **Gamba**
178 D3 **Gambēla** Eth.
178 D3 **Gambēla National Park** Eth.
184 A3 **Gambell** U.S.A.
144 D4 **Gambhir** *r.* India
176 A3 **Gambia** *r.* Gambia
176 A3 **Gambia, The** *country* Africa
123 J6 **Gambier, Îles** *is* Fr. Polynesia
126 B5 **Gambier Islands** Australia
189 J4 **Gambo** Canada
178 B4 **Gamboma** Congo
197 H4 **Gamerco** U.S.A.
186 F2 **Gamêtì** Canada
159 L4 **Gamleby** Sweden
158 M2 **Gammelstaden** Sweden
126 C3 **Gammon Ranges National Park** Australia
180 C4 **Gamoep** S. Africa
150 B3 **Gamova, Mys** *pt* Rus. Fed.
143 C5 **Gampola** Sri Lanka
141 F4 **Gamshadzai Kūh** *mts* Iran
171 J3 **Gamzigrad-Romuliana** *tourist site*
148 A3 **Gana** China
197 H4 **Ganado** U.S.A.
191 I3 **Gananoque** Canada
137 K1 **Gäncä** Azer.
149 C7 **Gancheng** China
155 E3 **Gandadiwata, Bukit** *mt.* Indon.
145 G3 **Gandaingoin** China
178 C4 **Gandajika** Dem. Rep. Congo
145 E4 **Gandak Barrage** *dam* Nepal
Gandak Dam Nepal *see* **Gandak Barrage**
144 B3 **Gandari Mountain** Pak.
144 A3 **Gandava** Pak.
189 J4 **Gander** Canada
165 G1 **Ganderkesee** Germany
167 G2 **Gandesa** Spain
144 C5 **Gandevi** India
144 B5 **Gandhidham** India
144 C5 **Gandhinagar** India
144 C4 **Gandhi Sagar** *resr* India
144 C4 **Gandhi Sagar Dam** India
167 F3 **Gandia** Spain
Gand-i-Zureh *plain* Afgh. *see* **Zereh, Gowd-e**
214 E1 **Gandu** Brazil
144 E4 **Ganga** *r.* Bangl./India *alt.* **Padma,** *conv.* **Ganges**
143 C5 **Ganga** *r.* Sri Lanka
215 C4 **Gangán** Arg.
144 C3 **Ganganagar** India
144 D4 **Gangapur** India
145 H5 **Gangaw** Myanmar
143 B3 **Gangawati** India
148 A2 **Gangca** China
144 E3 **Gangdisê Shan** *mts* China
144 E4 **Ganges** *r.* Bangl./India *alt.* **Ganga,** *alt.* **Padma**
166 F5 **Ganges** France
145 G5 **Ganges, Mouths of the** Bangl./India
218 J3 **Ganges Cone** *sea feature* Indian Ocean
144 D3 **Gangoh** India
144 D3 **Gangotri Group** *mts* India
145 G4 **Gangtok** India
148 B3 **Gangu** China
143 D2 **Ganjam** India
140 C4 **Ganjgūn** Iran
152 B2 **Ganjig** China
149 B4 **Ganluo** China
127 G5 **Ganmain** Australia

Grafton

166 F3 Gannat France
194 E3 Gannett Peak U.S.A.
144 C5 Ganora India
148 C2 Ganquan China
180 C7 Gansbaai S. Africa
148 B3 Gansu prov. China
148 B2 Gantang China
126 B6 Gantheaume, Cape Australia
173 G7 Gant'iadi Georgia
149 E5 Ganxian China
180 F3 Ganyesa S. Africa
148 F3 Ganyu China
138 B3 Ganyushkino Kazakh.
149 E5 Ganzhou China
177 F4 Ganzi Sudan
176 B3 Gao Mali
149 E4 Gao'an China
148 E2 Gaocheng China
148 F4 Gaochun China
148 B2 Gaolan China
148 F2 Gaomi China
149 D5 Gaomutang China
148 D3 Gaoping China
148 A2 Gaotai China
148 E2 Gaotang China
163 C2 Gaoth Dobhair Ireland
148 C2 Gaotouyao China
176 B3 Gaoua Burkina
176 A3 Gaoual Guinea
Gaoxian China see Wenjiang
148 E2 Gaoyang China
148 E2 Gaoyi China
148 F3 Gaoyou China
148 F3 Gaoyou Hu l. China
149 D6 Gaozhou China
166 H4 Gap France
153 B3 Gapan Phil.
144 E2 Gar China
163 C4 Gara, Lough l. Ireland
Garabekevyul Turkm. see Garabekewül
138 F5 Garabekewül Turkm.
141 F2 Garabil Belentligi hills Turkm.
138 C4 Garabogaz Turkm.
138 C4 Garabogazköl Turkm.
138 C4 Garabogazköl Aýlagy b. Turkm.
206 J6 Garachiné Panama
141 F4 Garāgheh Iran
Garagum des. Turkm. see Karakum Desert
138 E5 Garagum Kanaly canal Turkm.
127 H2 Garah Australia
138 F5 Garamätnyýaz Turkm.
178 C3 Garamba r. Dem. Rep. Congo
178 C3 Garamba, Parc National de la nat. park Dem. Rep. Congo
211 K5 Garanhuns Brazil
181 G2 Ga-Rankuwa S. Africa
178 D3 Garba Tula Kenya
196 A1 Garberville U.S.A.
140 C3 Garbosh, Kūh-e mt. Iran
165 H2 Garbsen Germany
214 C3 Garça Brazil
214 B1 Garças, Rio das r. Brazil
145 G2 Garco China
170 D2 Garda, Lake Italy
137 J1 Gardabani Georgia
170 B6 Garde, Cap de c. Alg.
165 J2 Gardelegen Germany
198 C4 Garden City U.S.A.
190 D3 Garden Corners U.S.A.
196 C5 Garden Grove U.S.A.
187 K4 Garden Hill Canada
190 E3 Garden Island U.S.A.
180 E6 Garden Route National Park nat. park S. Africa
141 H3 Gardēz Afgh.
Gardez Afgh. see Gardēz
203 I2 Gardiner ME U.S.A.
194 E2 Gardiner MT U.S.A.
203 G4 Gardiners Island U.S.A.
190 C5 Gardner U.S.A.
203 J2 Gardner Lake U.S.A.
123 H2 Gardner Pinnacles is U.S.A.
196 C2 Gardnerville U.S.A.
162 D4 Garelochhead U.K.
190 E2 Gargantua, Cape Canada
137 L6 Gargar Iran
159 M5 Gargždai Lith.
144 D5 Garhakota India
144 A3 Garhi Khairo Pak.
144 D4 Garhi Malehra India
186 E5 Garibaldi, Mount Canada
186 E5 Garibaldi Provincial Park Canada
181 F5 Gariep Dam resr S. Africa
180 B5 Garies S. Africa
170 E4 Garigliano r. Italy
178 D4 Garissa Kenya
159 N4 Garkalne Latvia
202 D4 Garland PA U.S.A.
199 D5 Garland TX U.S.A.
140 C2 Garmī Iran
168 E7 Garmisch-Partenkirchen Germany
Garmo, Qullai mt. Tajik. see Ismoili Somoní, Qullai
140 D3 Garmsar Iran
141 F4 Garmsel reg. Afgh.
198 E4 Garnett U.S.A.
126 E4 Garnpung Lake imp. l. Australia
145 G4 Garo Hills India
166 D4 Garonne r. France
178 E3 Garoowe Somalia
212 G3 Garopaba Brazil
177 D4 Garoua Cameroon
215 D3 Garré Arg.
197 E2 Garrison U.S.A.
163 F2 Garron Point U.K.
141 G4 Garruk Pak.
162 D4 Garry, Loch l. U.K.
187 I1 Garry Lake Canada
162 B2 Garrynahine U.K.
178 E4 Garsen Kenya
138 C4 Garşy Turkm.
161 D5 Garth U.K.
165 J1 Gartow Germany
180 B3 Garub Namibia
155 C4 Garut Indon.
163 E3 Garvagh U.K.
162 D3 Garve U.K.
190 D5 Gary U.S.A.
144 E3 Garyarsa China
151 C7 Garyū-zan mt. Japan
144 D2 Gar Zangbo r. China
146 B3 Garzê China
213 B4 Garzón Col.
Gascogne, Golfe de g. France/Spain see Gascony, Gulf of
198 E4 Gasconade r. U.S.A.
166 D5 Gascony reg. France
166 C4 Gascony, Gulf of France/Spain
124 B4 Gascoyne r. Australia
144 D2 Gasherbrum I mt. China/Pakistan
141 F5 Gasht Iran
176 D3 Gashua Nigeria
141 E3 Gask Iran
155 C3 Gaspar, Selat sea chan. Indon.
189 H4 Gaspé Canada
189 H4 Gaspé, Cap c. Canada
189 G4 Gaspésie, Parc de Conservation de la nature res. Canada
189 G4 Gaspésie, Péninsule de la pen. Canada
164 E2 Gasselte Neth.
201 D5 Gastonia U.S.A.
215 C4 Gastre Arg.
167 E4 Gata, Cabo de c. Spain
136 D4 Gata, Cape Cyprus
172 D3 Gatchina Rus. Fed.
202 B6 Gate City U.S.A.
162 D6 Gatehouse of Fleet U.K.
160 F3 Gateshead U.K.
199 D6 Gatesville U.S.A.
197 H2 Gateway U.S.A.
203 F4 Gateway National Recreational Area park U.S.A.
191 J3 Gatineau Canada
191 J2 Gatineau r. Canada
127 J1 Gatton Australia
206 I7 Gatún, Lago l. Panama
137 L5 Gatvand Iran
125 H3 Gau i. Fiji
187 J3 Gauer Lake Canada
158 J3 Gaula r. Norway
202 C5 Gauley Bridge U.S.A.
164 D5 Gaume reg. Belgium
145 F4 Gauri Sankar mt. China
181 G3 Gauteng prov. S. Africa
141 G3 Gauzan Afgh.
137 J1 Gavar Armenia
141 F5 Gavāter Iran
141 H3 Gāv Band Afgh.
140 D5 Gāvbandī Iran
140 D5 Gāvbūs, Kūh-e mts Iran
171 K7 Gavdos i. Greece
140 B3 Gāveh Rūd r. Iran
214 E1 Gavião r. Brazil
137 K4 Gavīleh Iran
196 B4 Gaviota U.S.A.
159 L3 Gävle Sweden
172 F3 Gavrilov-Yam Rus. Fed.
180 B3 Gawachab Namibia
126 C5 Gawler Australia
126 A4 Gawler Ranges hills Australia
148 A1 Gaxun Nur salt l. China
138 D2 Gay Rus. Fed.
145 F4 Gaya India
176 C3 Gaya Niger
152 E1 Gaya He r. China
190 E3 Gaylord U.S.A.
136 E6 Gaza terr. Asia
136 E6 Gaza Gaza
181 J1 Gaza prov. Moz.
Gaz-Achak Turkm. see Gazojak
139 G4 G'azalkent Uzbek.
Gazandzhyk Turkm. see Bereket
141 H3 Gazdarreh, Band-e Afgh.
Gazgan Uzbek. see G'ozg'on
136 F3 Gaziantep Turkey
141 F3 Gazīk Iran
136 D3 Gazipaşa Turkey
138 E4 Gazli Uzbek.
141 E5 Gaz Māhū Iran
138 E4 Gazojak Turkm.
178 C3 Gbadolite Dem. Rep. Congo
176 A4 Gbangbatok Sierra Leone
176 B4 Gbarnga Liberia
176 C4 Gboko Nigeria
169 I3 Gdańsk Poland
169 I3 Gdańsk, Gulf of Poland/Rus. Fed.
172 C3 Gdov Rus. Fed.
169 I3 Gdynia Poland
158 M1 Geaidnovuohppi Norway
162 C3 Gealldruig Mhòr i. U.K.
165 I3 Gebesee Germany
177 F3 Gedaref Sudan
165 H4 Gedern Germany
164 C5 Gedinne Belgium
136 A2 Gediz r. Turkey
161 H5 Gedney Drove End U.K.
159 J5 Gedser Denmark
164 D3 Geel Belgium
126 F7 Geelong Australia
180 D4 Geel Vloer salt pan S. Africa
164 F2 Geeste Germany
165 I1 Geesthacht Germany
127 G9 Geeveston Australia
176 D3 Geidam Nigeria
165 H4 Geiersberg h. Germany
187 I3 Geikie r. Canada
164 E4 Geilenkirchen Germany
159 J3 Geilo Norway
159 I3 Geiranger Norway
190 E6 Geist Reservoir U.S.A.
165 K3 Geithain Germany
149 B6 Gejiu China
170 F6 Gela Sicily Italy
178 E3 Geladī Eth.
154 B4 Gelang, Tanjung pt Malaysia
164 E3 Geldern Germany
173 F6 Gelendzhik Rus. Fed.
169 K3 Gelgaudiškis Lith.
Gelibolu Turkey see Gallipoli
136 C2 Gelincik Dağı mt. Turkey
140 E3 Gelmord Iran
165 H4 Gelnhausen Germany
164 F3 Gelsenkirchen Germany
154 B5 Gemas Malaysia
153 C5 Gemeh Indon.
178 B3 Gemena Dem. Rep. Congo
136 F2 Gemerek Turkey
136 B1 Gemlik Turkey
170 E1 Gemona del Friuli Italy
179 C6 Gemsbok National Park Botswana
180 D3 Gemsbokplein well S. Africa
178 E3 Genalē Wenz r. Eth.
164 C4 Genappe Belgium
140 C4 Genāveh Iran
215 D3 General Acha Arg.
215 E3 General Alvear Buenos Aires Arg.
215 E1 General Alvear Entre Ríos Arg.
215 C2 General Alvear Mendoza Arg.
215 E2 General Belgrano Arg.
207 E2 General Bravo Mex.
212 B7 General Carrera, Lago l. Chile
206 D2 General Cepeda Mex.
215 F3 General Conesa Buenos Aires Arg.
215 D4 General Conesa Río Negro Arg.
215 F3 General Guido Arg.
215 F3 General Juan Madariaga Arg.
215 E3 General La Madrid Arg.
215 F3 General Lavalle Arg.
215 D2 General Levalle Arg.
153 C4 General Luna Phil.
153 C4 General MacArthur Phil.
215 D2 General Pico Arg.
215 E2 General Pinto Arg.
215 C3 General Roca Arg.
153 C5 General Santos Phil.
207 E2 General Terán Mex.
215 D2 General Villegas Arg.
202 D3 Genesee r. U.S.A.
190 B5 Geneseo IL U.S.A.
202 E3 Geneseo NY U.S.A.
181 G3 Geneva S. Africa
168 C7 Geneva Switz.
190 C5 Geneva IL U.S.A.
198 D3 Geneva NE U.S.A.
202 E3 Geneva NY U.S.A.
202 C4 Geneva OH U.S.A.
166 H3 Geneva, Lake France/Switz.
190 C4 Geneva, Lake U.S.A.
Genève Switz. see Geneva
149 F4 Genglou China
167 D4 Genil r. Spain
164 D4 Genk Belgium
164 D3 Gennep Neth.
127 H6 Genoa Australia
170 C2 Genoa Italy
170 C2 Genoa, Gulf of Italy
Genova Italy see Genoa
Gent Belgium see Ghent
165 K2 Genthin Germany
124 B5 Geographe Bay Australia
132 F2 Georga, Zemlya i. Rus. Fed.
189 G2 George r. Canada
180 E6 George S. Africa
127 H5 George, Lake N.S.W. Australia
126 C6 George, Lake S.A. Australia
201 D6 George, Lake FL U.S.A.
203 G3 George, Lake NY U.S.A.
128 A6 George Sound inlet N.Z.
126 C4 Georgetown Australia
127 G8 George Town Australia
201 F7 George Town Bahamas
191 H4 Georgetown Canada
176 A3 Georgetown Gambia
211 G2 Georgetown Guyana
155 B1 George Town Malaysia
203 F5 Georgetown DE U.S.A.
190 D6 Georgetown IL U.S.A.
200 C4 Georgetown KY U.S.A.
202 B5 Georgetown OH U.S.A.
201 E5 Georgetown SC U.S.A.
199 D6 Georgetown TX U.S.A.
129 C6 George V Land reg. Antarctica
199 D6 George West U.S.A.
173 G7 Georgia country Asia
201 D5 Georgia state U.S.A.
186 E5 Georgia, Strait of Canada
191 G3 Georgian Bay Canada
191 H3 Georgian Bay Islands National Park Canada
124 D4 Georgina watercourse Australia
139 J2 Georgiyevka Vostochnyy Kazakhstan Kazakh.
139 G4 Georgiyevka Yuzhnyy Kazakhstan Kazakh.
173 G6 Georgiyevsk Rus. Fed.
172 H3 Georgiyevskoye Rus. Fed.
165 K4 Gera Germany
164 B4 Geraardsbergen Belgium
211 I6 Geral de Goiás, Serra hills Brazil
128 C6 Geraldine N.Z.
214 C1 Geral do Paraná, Serra hills Brazil
124 B4 Geraldton Australia
140 D5 Gerāsh Iran
137 H3 Gercüş Turkey
136 D1 Gerede Turkey
136 D1 Gerede r. Turkey
141 G4 Gereshk Afgh.
154 B4 Gerik Malaysia
141 E3 Gerīmenj Iran
198 C3 Gering U.S.A.
194 C3 Gerlach U.S.A.
168 C3 German Bight g. Denmark/Ger.
186 E3 Germansen Landing Canada
202 E5 Germantown U.S.A.
168 E5 Germany country Europe
165 G5 Germersheim Germany
181 H3 Germiston S. Africa
165 G5 Gernsheim Germany
164 E4 Gerolstein Germany
165 I5 Gerolzhofen Germany
197 G5 Geronimo U.S.A.
127 I5 Gerringong Australia
165 H4 Gersfeld (Rhön) Germany
165 I4 Gerstungen Germany
165 J2 Gerwisch Germany
145 F2 Gêrzê China
173 E7 Gerze Turkey
164 F3 Gescher Germany
178 E3 Gestro, Wabē r. Eth.
140 D3 Getcheh, Kūh-e hills Iran
164 D4 Gete r. Belgium
202 E5 Gettysburg PA U.S.A.
198 D2 Gettysburg SD U.S.A.
202 E5 Gettysburg National Military Park nat. park U.S.A.
149 C5 Getu He r. China
129 B4 Getz Ice Shelf ice feature Antarctica
155 A2 Geumapang r. Indon.
127 H4 Geurie Australia
137 I2 Gevaş Turkey
171 J4 Gevgelija Macedonia
154 □ Geylang Sing.
181 F3 Geysdorp S. Africa
136 C1 Geyve Turkey
180 F2 Ghaap Plateau S. Africa
137 I5 Ghadaf, Wādī al watercourse Iraq
176 C1 Ghadāmis Libya
140 D2 Gha'em Shahr Iran
139 G4 Ghafurov Tajik.
144 E4 Ghaghara r. India
145 F5 Ghaghra India
176 B4 Ghana country Africa
140 D5 Ghanādah, Rās pt U.A.E.
144 C4 Ghanliala India
179 C6 Ghanzi Botswana
180 E1 Ghanzi admin. dist. Botswana
140 C5 Ghār, Ras al pt Saudi Arabia
136 E6 Gharandal Jordan
176 C1 Ghardaïa Alg.
177 F2 Ghārib, Jabal mt. Egypt
Ghārib, Jabal mt. Egypt see Ḩamāţah, Jabal
139 G5 Gharm Tajik.
177 D1 Gharyān Libya
136 F5 Gharz, Wādī al watercourse Syria
176 D2 Ghāt Libya
144 E3 Ghauspur Pak.
177 D3 Ghazal, Bahr el watercourse Chad
176 E1 Ghazaouet Alg.
144 D3 Ghaziabad India
145 E4 Ghazipur India
144 A3 Ghazluna Pak.
141 H3 Ghaznī Afgh.
141 H3 Ghazni r. Afgh.
141 G3 Ghazoor Afgh.
164 E3 Ghent Belgium
169 L7 Gheorgheni Romania
169 K7 Gherla Romania
170 C3 Ghisonaccia Corsica France
141 C3 Ghizao Afgh.
144 C1 Ghizar Pak.
143 A2 Ghod r. India
145 C4 Ghoraghat Bangl.
141 H3 Ghorband r. Afgh.
144 E4 Ghotaru India
144 E4 Ghotki Pak.
Ghuari r. India see Ghugri
139 H5 Ghŭdara Tajik.
145 F4 Ghugri r. India
144 C6 Ghugus India
144 B4 Ghulam Muhammad Barrage Pak.
140 C4 Ghūrī Iran
141 F3 Ghurian Afgh.
164 A3 Ghyvelde France
Gia Đinh Vietnam see Thu Đuc
173 G6 Giaginskaya Rus. Fed.
154 C2 Gia Nghia Vietnam
171 J4 Giannitsa Greece
181 H4 Giant's Castle mt. S. Africa
163 E2 Giant's Causeway lava field U.K.
155 E4 Gianyar Indon.
154 C3 Gia Rai Vietnam
170 F5 Giarre Sicily Italy
170 B2 Giaveno Italy
180 B2 Gibeon Namibia
167 D4 Gibraltar Europe
167 C5 Gibraltar, Strait of Morocco/Spain
190 C5 Gibson City U.S.A.
124 C4 Gibson Desert Australia
146 B2 Gichgeniyn Nuruu mts Mongolia
143 B3 Giddalur India
178 D3 Gīdolē Eth.
166 F3 Gien France
165 G4 Gießen Germany
165 I2 Gifhorn Germany
186 F3 Gift Lake Canada
151 E7 Gifu Japan
213 B4 Gigante Col.
162 C5 Gigha i. U.K.
138 F4 G'ijduvon Uzbek.
Gijón Spain see Gijón-Xixón
157 D1 Gijón-Xixón Spain
197 F5 Gila r. U.S.A.
197 F5 Gila Bend U.S.A.
197 F5 Gila Bend Mountains U.S.A.
197 E5 Gila Mountains U.S.A.
137 J4 Gīlān-e Gharb Iran
137 L1 Gilāzi Azer.
124 E3 Gilbert r. Australia
197 G5 Gilbert AZ U.S.A.
202 C6 Gilbert WV U.S.A.
125 H2 Gilbert Islands Kiribati
216 G5 Gilbert Ridge sea feature Pacific Ocean
211 I5 Gilbués Brazil
140 E3 Gil Chashmeh Iran
194 E2 Gildford U.S.A.
186 D4 Gilford Island Canada
127 I2 Gilgai Australia
127 H3 Gilgandra Australia
178 D4 Gilgil Kenya
127 H2 Gil Gil Creek r. Australia
144 C2 Gilgit Pak.
144 C2 Gilgit r. Pak.
144 C1 Gilgit and Baltistan admin. div. Pak.
127 G4 Gilgunnia Australia
186 D4 Gil Island Canada
187 K3 Gillam Canada
126 B4 Gilles, Lake salt flat Australia
190 C3 Gillett U.S.A.
194 F2 Gillette U.S.A.
161 E6 Gillingham England U.K.
161 H6 Gillingham England U.K.
160 F3 Gilling West U.K.
190 D3 Gills Rock U.S.A.
190 D5 Gilman IL U.S.A.
190 B3 Gilman WI U.S.A.
188 E2 Gilmour Island Canada
196 B3 Gilroy U.S.A.
203 G3 Gilsum U.S.A.
177 E3 Gimbala, Jebel mt. Sudan
178 D3 Gīmbī Eth.
187 J4 Gimli Canada
210 E6 Ginebra, Laguna l. Bol.
143 C5 Gin Ganga r. Sri Lanka
143 B3 Gingee India
186 D3 Gingolx Canada
178 E3 Gīnīr Eth.
170 G4 Ginosa Italy
170 G4 Gioia del Colle Italy
127 G6 Gippsland reg. Australia
144 B4 Girab India
141 E5 Gīrān Iran
141 E4 Gīrān Rīg mt. Iran
202 C3 Girard U.S.A.
144 B3 Girdao Pak.
141 G5 Girdar Dhor r. Pak.
141 F4 Girdi Iran
136 G1 Giresun Turkey
144 B5 Gir Forest India
145 F4 Giridih India
127 G3 Girilambone Australia
144 C5 Girna r. India
167 H2 Girona Spain
166 D4 Gironde est. France
127 G4 Girral Australia
162 D5 Girvan U.K.
172 E2 Girvas Rus. Fed.
144 E4 Girwan India
128 G3 Gisborne N.Z.
186 E4 Giscome Canada
159 K4 Gislaved Sweden
139 F5 Gissar Range mts Tajik./Uzbek.
173 G7 Gistola, Gora mt. Georgia/Rus. Fed.
178 C4 Gitarama Rwanda
178 C4 Gitega Burundi
170 E3 Giulianova Italy
171 K3 Giurgiu Romania
171 K2 Giuvala, Pasul pass Romania
164 C4 Givet France
166 G4 Givors France
164 C6 Givry-en-Argonne France
181 I1 Giyani S. Africa
177 F2 Giza Egypt
137 K4 Gīzeh Rūd r. Iran
Gizhduvan Uzbek. see G'ijduvon
133 R3 Gizhiga Rus. Fed.
171 I3 Gjakovë Kosovo
171 I3 Gjilan Kosovo
171 I4 Gjirokastër Albania
185 I3 Gjoa Haven Canada
158 J3 Gjøra Norway
159 J3 Gjøvik Norway
171 M6 Gkinas, Akrotirio pt Greece
189 I4 Glace Bay Canada
186 B3 Glacier Bay U.S.A.
186 B3 Glacier Bay National Park and Preserve U.S.A.
186 F4 Glacier National Park Canada
194 D1 Glacier National Park U.S.A.
194 B1 Glacier Peak vol. U.S.A.
158 J2 Gladstad Norway
124 F4 Gladstone Qld Australia
126 C4 Gladstone S.A. Australia
127 H8 Gladstone Tas. Australia
190 D3 Gladstone U.S.A.
190 E4 Gladwin U.S.A.
162 E4 Glamis U.K.
164 F5 Glan r. Germany
153 C5 Glan Phil.
163 B5 Glanaruddery Mountains hills Ireland
165 G2 Glandorf Germany
160 F2 Glanton U.K.
191 G4 Glanworth Canada
162 D5 Glasgow U.K.
200 C4 Glasgow KY U.S.A.
194 F1 Glasgow MT U.S.A.
202 D6 Glasgow VA U.S.A.
187 H4 Glaslyn Canada
196 C3 Glass Mountain U.S.A.
161 E6 Glastonbury U.K.
165 K4 Glauchau Germany
132 G4 Glazov Rus. Fed.
173 F4 Glazunovka Rus. Fed.
169 O3 Glazunovo Rus. Fed.
203 H2 Glen U.S.A.
162 C3 Glen Affric val. U.K.
191 G2 Glen Afton Canada
128 E2 Glen Afton N.Z.
181 H1 Glen Alpine Dam S. Africa
163 C4 Glenamaddy Ireland
190 E3 Glen Arbor U.S.A.
128 C6 Glenavy N.Z.
162 C3 Glen Cannich val. U.K.
195 E4 Glen Canyon gorge U.S.A.
197 G3 Glen Canyon National Recreation Area park U.S.A.
162 E4 Glen Clova val. U.K.
126 D6 Glencoe Australia
191 G4 Glencoe Canada
181 I4 Glencoe S. Africa
162 C4 Glen Coe val. U.K.
191 E2 Glendale Canada
197 F5 Glendale AZ U.S.A.
196 C4 Glendale CA U.S.A.
197 E3 Glendale NV U.S.A.
197 F3 Glendale UT U.S.A.
202 D4 Glendale Lake U.S.A.
127 I4 Glen Davis Australia
194 F2 Glendive U.S.A.
187 G4 Glendon Canada
194 F3 Glendo Reservoir U.S.A.
126 D6 Glenelg r. Australia
162 F4 Glen Esk val. U.K.
162 A5 Glengad Head Ireland
162 C3 Glen Garry val. Scotland U.K.
162 D4 Glen Garry val. Scotland U.K.
163 D3 Glengavlen Ireland
127 I2 Glen Innes Australia
162 D6 Glenluce U.K.
162 D4 Glen Lyon val. U.K.
162 D3 Glen More val. U.K.
127 H1 Glenmorgan Australia
162 D3 Glen Moriston val. U.K.
197 G6 Glenn, Mount U.S.A.
184 D3 Glennallen U.S.A.
162 C4 Glen Nevis val. U.K.
191 F3 Glennie U.S.A.
202 E6 Glenns U.S.A.
186 C3 Glenora Canada
203 F2 Glen Robertson Canada
162 E4 Glenrothes U.K.
203 G3 Glens Falls U.S.A.
162 E4 Glen Shee val. U.K.
162 C3 Glen Shiel val. U.K.
163 C3 Glenties Ireland
163 D2 Glenveagh National Park Ireland
202 C5 Glenville U.S.A.
199 E5 Glenwood AR U.S.A.
197 H5 Glenwood NM U.S.A.
195 F4 Glenwood Springs U.S.A.
190 B2 Glidden U.S.A.
165 I1 Glinde Germany
169 I5 Gliwice Poland
197 G5 Globe U.S.A.
168 H5 Głogów Poland
158 K2 Glomfjord Norway
159 J4 Glomma r. Norway
179 E5 Glorieuses, Îles is Indian Ocean
127 I3 Gloucester Australia
161 E6 Gloucester U.K.
203 H3 Gloucester MA U.S.A.
202 E6 Gloucester VA U.S.A.
203 F3 Gloversville U.S.A.
165 K2 Glöwen Germany
150 D1 Glubinnoye Rus. Fed.
173 G6 Glubokiy Rus. Fed.
139 J2 Glubokoye Kazakh.
165 H1 Glückstadt Germany
158 □ Gluggarnir h. Faroe Is
160 F4 Glusburn U.K.
173 H5 Gmelinka Rus. Fed.
168 G6 Gmünd Austria
168 F7 Gmunden Austria
159 L3 Gnarp Sweden
165 H1 Gnarrenburg Germany
168 H4 Gniezno Poland
143 A3 Goa India
143 A3 Goa state India
180 B3 Goageb Namibia
127 I6 Goalen Head Australia
145 G4 Goalpara India
162 C5 Goat Fell h. U.K.
178 E3 Goba Eth.
179 B6 Gobabis Namibia
180 C3 Gobas Namibia
215 E1 Gobernador Crespo Arg.
215 C3 Gobernador Duval Arg.
212 B7 Gobernador Gregores Arg.
146 C2 Gobi Desert Mongolia
151 D8 Gobō Japan
137 L1 Gobustan Rock Art tourist site Azer.
164 E3 Goch Germany
179 B6 Gochas Namibia
154 C3 Go Công Vietnam
161 G6 Godalming U.K.
143 C2 Godavari r. India
143 C2 Godavari, Mouths of the India
189 G4 Godbout Canada
196 C3 Goddard, Mount U.S.A.
178 E3 Godere Eth.
191 G4 Goderich Canada
144 C5 Godhra India
215 C2 Godoy Cruz Arg.
187 K3 Gods r. Canada
187 K4 Gods Lake Canada
187 L2 God's Mercy, Bay of Canada
Godthåb Greenland see Nuuk
Godwin-Austen, Mount mt. China/Pakistan see K2
164 B3 Goedereede Neth.
188 E4 Goéland, Lac au l. Canada
189 H2 Goélands, Lac aux l. Canada
164 B3 Goes Neth.
191 E2 Goetzville U.S.A.
197 E4 Goffs U.S.A.
191 G2 Gogama Canada
190 C2 Gogebic, Lake U.S.A.
190 C2 Gogebic Range hills U.S.A.
144 D4 Gohad India
211 L5 Goiana Brazil
214 C2 Goiandira Brazil
214 C2 Goiânia Brazil
214 B1 Goiás Brazil
214 B2 Goiás state Brazil
214 B4 Goio-Erê Brazil
144 C3 Gojra Pak.
143 A2 Gokak India
173 C7 Gökçeada i. Turkey
136 B2 Gökçedağ Turkey
145 G3 Gokhar La pass China
136 E1 Gökirmak r. Turkey
138 D4 Goklenkuy, Solonchak salt l. Turkm.
141 F5 Gokprosh Hills Pak.
136 F2 Göksun Turkey
136 E3 Göksu Nehri r. Turkey
179 C5 Gokwe Zimbabwe
159 J3 Gol Norway
144 E3 Gola India
145 H4 Golaghat India
141 E4 Golbāf Iran
136 F3 Gölbaşı Turkey
136 B1 Gölcük Turkey
169 K3 Gołdap Poland
165 K1 Goldberg Germany
127 J2 Gold Coast Australia
176 B4 Gold Coast coastal area Ghana
186 F4 Golden Canada
128 D4 Golden Bay N.Z.
165 I3 Goldene Aue reg. Germany
196 A3 Golden Gate National Recreation Area park U.S.A.
186 D5 Golden Hinde mt. Canada
165 G2 Goldenstedt Germany
163 C5 Golden Vale lowland Ireland
196 D3 Goldfield U.S.A.
196 D3 Gold Point U.S.A.
201 E5 Goldsboro U.S.A.
199 D6 Goldthwaite U.S.A.
137 I1 Göle Turkey
141 F3 Golestān Afgh.
196 C4 Goleta U.S.A.
206 I6 Golfito Costa Rica
199 D6 Goliad U.S.A.
152 A1 Golin Baixing China
136 F1 Gölköy Turkey
165 K2 Golm Germany
137 J3 Golmānkhāneh Iran
146 B3 Golmud China
145 H1 Golmud He r. China
153 B3 Golo i. Phil.
150 I3 Golovnino Rus. Fed.
140 C3 Golpāyegān Iran
136 C1 Gölpazarı Turkey
162 E3 Golspie U.K.
139 H1 Golubovka Kazakh.
141 F3 Gol Vardeh Iran
171 K4 Golyama Syutkya mt. Bulg.
171 K4 Golyam Persenk mt. Bulg.
165 K2 Golzow Germany
178 C4 Goma Dem. Rep. Congo
145 G3 Gomang Co salt l. China
144 E4 Gomati r. India
154 □ Gombak, Bukit h. Sing.
176 D3 Gombe Nigeria
178 D4 Gombe r. Tanz.
177 D3 Gombi Nigeria
206 D2 Gómez Palacio Mex.
140 D2 Gomīshān Iran
165 J2 Gommern Germany
145 F2 Gomo Co salt l. China
141 E2 Gonābād Iran
205 J5 Gonaïves Haiti
181 I1 Gonarezhou National Park Zimbabwe
205 J5 Gonâve, Île de la i. Haiti
140 D2 Gonbad-e Kavus Iran
145 E4 Gonda India
144 B5 Gondal India
178 D2 Gonder Eth.
144 E5 Gondia India
136 A1 Gönen Turkey
149 D4 Gong'an China
149 D5 Gongcheng China
149 A4 Gongga Shan mt. China
148 A2 Gonghe China
148 E1 Gonghui China
139 J4 Gongliu China
214 E1 Gongogi r. Brazil
176 D3 Gongola r. Nigeria
127 G3 Gongolgon Australia
149 B4 Gongquan China
149 B5 Gongwang Shan mts China
Gongxian China see Gongquan
148 D3 Gongyi China
152 C2 Gongzhuling China
181 H6 Gonubie S. Africa
207 E3 Gonzáles Mex.
196 B3 Gonzales CA U.S.A.
199 D6 Gonzales TX U.S.A.
215 D2 González Moreno Arg.
202 E6 Goochland U.S.A.
129 C6 Goodenough, Cape c. Antarctica
124 F2 Goodenough Island P.N.G.
191 H3 Gooderham Canada
190 E3 Good Harbor Bay U.S.A.
180 C7 Good Hope, Cape of S. Africa
194 D3 Gooding U.S.A.
198 C4 Goodland U.S.A.
127 G2 Goodooga Australia
160 G4 Goole U.K.
127 F5 Goolgowi Australia
127 H4 Goolma Australia
127 H4 Gooloogong Australia
126 C5 Goolwa Australia
127 F2 Goombalie Australia
127 I2 Goondiwindi Australia
189 H3 Goose r. Canada
194 B3 Goose Lake U.S.A.
143 B3 Gooty India
168 D6 Göppingen Germany
145 E4 Gorakhpur India
171 H3 Goražde Bos.-Herz.
172 G3 Gorchukha Rus. Fed.
206 I5 Gorda, Punta pt Nicaragua
201 E7 Gorda Cay i. Bahamas
136 B2 Gördes Turkey
169 O4 Gordeyevka Rus. Fed.
127 F9 Gordon r. Australia
162 F5 Gordon U.K.
127 G9 Gordon, Lake Australia
186 G2 Gordon Lake Canada
202 D5 Gordon Lake U.S.A.
202 D5 Gordonsville U.S.A.
177 D4 Goré Chad
178 D3 Gorē Eth.
128 B7 Gore N.Z.
191 F3 Gore Bay Canada
162 E5 Gorebridge U.K.
136 E2 Göreme Milli Parkı nat. park Turkey
163 E5 Gorey Ireland
141 E4 Gorg Iran
140 D2 Gorgān Iran
213 A4 Gorgona, Isla i. Col.
203 H2 Gorham U.S.A.
173 H7 Gori Georgia
164 C3 Gorinchem Neth.
137 K2 Goris Armenia
170 E2 Gorizia Italy
Gor'kiy Rus. Fed. see Nizhniy Novgorod
173 H5 Gor'ko-Solenoye, Ozero l. Rus. Fed.
172 G3 Gor'kovskoye Vodokhranilishche resr Rus. Fed.
139 J1 Gor'koye, Ozero salt l. Rus. Fed.
169 J6 Gorlice Poland
168 G5 Görlitz Germany
144 D4 Gormi India
171 K3 Gorna Oryakhovitsa Bulg.
171 I2 Gornji Milanovac Serbia
170 G3 Gornji Vakuf Bos.-Herz.
139 K2 Gorno-Altaysk Rus. Fed.
150 G1 Gornozavodsk Rus. Fed.
139 J2 Gornyak Rus. Fed.
150 C2 Gornyy Primorskiy Kray Rus. Fed.
173 I5 Gornyy Saratovskaya Oblast' Rus. Fed.
173 H5 Gornyy Balykley Rus. Fed.
150 C2 Gornyye Klyuchi Rus. Fed.
172 G3 Gorodets Rus. Fed.
173 H5 Gorodishche Rus. Fed.
173 G6 Gorodovikovsk Rus. Fed.
124 E2 Goroka P.N.G.
126 D6 Goroke Australia
172 G3 Gorokhovets Rus. Fed.
176 B3 Gorom Gorom Burkina
179 D5 Gorongosa Moz.
147 E6 Gorontalo Indon.
173 F5 Gorshechnoye Rus. Fed.
163 C4 Gort Ireland
Gortahork Ireland see Gort an Choirce
163 C2 Gort an Choirce Ireland
214 D1 Gorutuba r. Brazil
141 E4 Gorveh Iran
173 F6 Goryachiy Klyuch Rus. Fed.
165 K2 Görzke Germany
168 G4 Gorzów Wielkopolski Poland
127 I4 Gosford Australia
160 F2 Gosforth U.K.
190 E5 Goshen IN U.S.A.
203 F4 Goshen NY U.S.A.
150 G4 Goshogawara Japan
165 I3 Goslar Germany
138 C4 Goşoba Turkm.
170 F2 Gospić Croatia
161 F7 Gosport U.K.
171 I4 Gostivar Macedonia
Göteborg Sweden see Gothenburg
159 K4 Götene Sweden
165 I4 Gotha Germany
159 J4 Gothenburg Sweden
198 C3 Gothenburg U.S.A.
159 L4 Gotland i. Sweden
171 J4 Gotse Delchev Bulg.
159 L4 Gotska Sandön i. Sweden
151 C7 Gōtsu Japan
165 H3 Göttingen Germany
186 E4 Gott Peak Canada
138 C5 Goturdepe Turkm.
152 A3 Goubangzi China
164 C2 Gouda Neth.
176 A3 Goudiri Senegal
176 D3 Goudoumaria Niger
190 E1 Goudreau Canada
219 I8 Gough Island S. Atlantic Ocean
188 F4 Gouin, Réservoir resr Canada
190 E2 Goulais River Canada
127 H5 Goulburn Australia
127 I4 Goulburn r. N.S.W. Australia
126 F6 Goulburn r. Vic. Australia
124 D3 Goulburn Islands Australia
190 E2 Gould City U.S.A.
176 B3 Goundam Mali
167 G4 Gouraya Alg.
176 D3 Gouré Niger
180 D7 Gourits r. S. Africa
176 B3 Gourma-Rharous Mali
166 E2 Gournay-en-Bray France
127 H6 Gourock Range mts Australia
164 A5 Goussainville France
203 F2 Gouverneur U.S.A.
187 H5 Govenlock Canada
214 E2 Governador Valadares Brazil
153 C5 Governor Generoso Phil.
201 E7 Governor's Harbour Bahamas
146 B2 Govĭ Altayn Nuruu mts Mongolia
145 E4 Govind Ballash Pant Sagar resr India
144 D3 Govind Sagar resr India
Govurdak Turkm. see Magdanly
202 D3 Gowanda U.S.A.
141 G4 Gowārān Afgh.
140 D4 Gowd-e Aḩmar Iran
161 C6 Gower pen. U.K.
191 G2 Gowganda Canada
163 D4 Gowna, Lough l. Ireland
212 E3 Goya Arg.
137 K1 Göyçay Azer.
138 C4 Goýmatdag hills Turkm.
137 H2 Göynük Turkey
150 G5 Goyō-zan mt. Japan
137 L2 Göytäpä Azer.
Gōzareh Afgh. see Gozareh
141 F3 Gozareh Afgh.
136 G2 Gözene Turkey
138 F4 G'ozg'on Uzbek.
144 E2 Gozha Co salt l. China
170 F6 Gozo i. Malta
180 F6 Graaff-Reinet S. Africa
Graaf-Reinet S. Africa see Graaff-Reinet
180 C6 Graafwater S. Africa
165 I4 Grabfeld plain Germany
176 B4 Grabo Côte d'Ivoire
180 C7 Grabouw S. Africa
165 J1 Grabow Germany
170 F2 Gračac Croatia
191 I2 Gracefield Canada
138 C1 Grachevka Rus. Fed.
139 I2 Grachi (abandoned) Kazakh.
206 G5 Gracias Hond.
165 K3 Gräfenhainichen Germany
165 J5 Grafenwöhr Germany
127 J2 Grafton Australia

Hjälmaren

177 F3 Haiya Sudan
148 A2 Haiyan Qinghai China
149 F4 Haiyan Zhejiang China
152 A5 Haiyang China
152 B4 Haiyang Dao i. China
148 B2 Haiyuan China
148 F3 Haizhou Wan b. China
140 C3 Hāj Ali Qoli, Kavīr-e salt l. Iran
169 J7 Hajdúböszörmény Hungary
170 C7 Hajeb El Ayoun Tunisia
142 D7 Ḥajhir mt. Yemen
140 D3 Haji Abdulla, Chāh well Iran
150 F5 Hajiki-zaki pt Japan
145 F4 Hajipur India
140 D4 Ḥājjīābād Iran
140 D4 Ḥājjīābād Iran
145 H5 Haka Myanmar
196 □2 Hakalau U.S.A.
215 C4 Hakelhuincul, Altiplanicie de plat. Arg.
137 I3 Hakkâri Turkey
158 M2 Hakkas Sweden
151 D7 Hakken-zan mt. Japan
150 H2 Hako-dake mt. Japan
150 G4 Hakodate Japan
180 B1 Hakos Mountains Namibia
144 C3 Hakra Right Distributary watercourse Pak.
180 D3 Hakseen Pan salt pan S. Africa
151 E6 Hakui Japan
151 E6 Haku-san vol. Japan
151 E6 Haku-san Kokuritsu-kōen nat. park Japan
144 B4 Hala Pak.
Ḥalab Syria see Aleppo
140 B6 Halabān Saudi Arabia
137 J4 Ḥalabja Iraq
138 F5 Halaç Turkm.
152 C1 Halahai China
177 F2 Halaib Sudan
154 D2 Ha Lam Vietnam
142 E6 Ḥalānīyāt, Juzur al is Oman
Halawa U.S.A. see Hālawa
196 □2 Hālawa U.S.A.
136 F4 Halba Lebanon
Halban Mongolia see Tsetserleg
165 J3 Halberstadt Germany
153 B3 Halcon, Mount Phil.
158 □ Haldarsvik Faroe Is
159 J4 Halden Norway
165 J2 Haldensleben Germany
145 G5 Haldi r. India
145 G5 Haldia India
145 G4 Haldibari India
144 D3 Haldwani India
191 F3 Hale U.S.A.
140 D5 Hāleh Iran
196 □1 Hale'iwa U.S.A.
Haleiwa U.S.A. see Hale'iwa
161 E5 Halesowen U.K.
161 I5 Halesworth U.K.
140 C4 Haleyleh Iran
136 F3 Halfeti Turkey
128 B7 Halfmoon Bay N.Z.
186 E3 Halfway r. Canada
163 C6 Halfway Ireland
164 C2 Halfweg Neth.
145 E4 Halia India
137 G4 Ḥalībīyah Syria
191 H3 Haliburton Canada
189 H5 Halifax Canada
160 F4 Halifax U.K.
202 D6 Halifax U.S.A.
148 C1 Haliut China
162 E2 Halkirk U.K.
158 L3 Hälla Sweden
152 D7 Halla-san mt. S. Korea
152 D7 Halla-san National Park nat. park S. Korea
126 A5 Hall Bay Australia
185 J3 Hall Beach Canada
164 C4 Halle Belgium
164 E3 Halle Neth.
165 J3 Halle (Saale) Germany
159 K4 Hällefors Sweden
168 F7 Hallein Austria
165 J3 Halle-Neustadt Germany
129 C3 Halley research stn Antarctica
122 E2 Hall Islands Micronesia
158 L2 Hällnäs Sweden
198 D1 Hallock U.S.A.
185 L3 Hall Peninsula Canada
159 K4 Hallsberg Sweden
124 C3 Halls Creek Australia
191 H3 Halls Lake Canada
164 B4 Halluin France
158 K3 Hallviken Sweden
147 E6 Halmahera i. Indon.
159 K4 Halmstad Sweden
144 C5 Halol India
149 C6 Ha Long Vietnam
159 J4 Hals Denmark
158 D2 Hälslón resr Iceland
158 N3 Halsua Fin.
164 F3 Haltern Germany
160 E3 Haltwhistle U.K.
140 D5 Ḥālūl i. Qatar
164 F3 Halver Germany
164 B5 Ham France
151 C7 Hamada Japan
140 C3 Hamadān Iran
136 F4 Ḥamāh Syria
150 G3 Hamamasu Japan
151 E7 Hamamatsu Japan
159 J3 Hamar Norway
158 K1 Hamarøy Norway
Ḥamāṭah, Jabal mt. Egypt see Gharib, Jabal
177 F2 Ḥamāṭah, Jabal mt. Egypt
150 H2 Hamatonbetsu Japan
143 C5 Hambantota Sri Lanka
165 G1 Hambergen Germany
160 F3 Hambleton Hills U.K.
165 H1 Hamburg Germany
181 G6 Hamburg S. Africa
199 F5 Hamburg AR U.S.A.
202 D3 Hamburg NY U.S.A.
203 F4 Hamburg PA U.S.A.
165 G1 Hamburgisches Wattenmeer, Nationalpark nat. park Germany
203 G4 Hamden U.S.A.
159 N3 Hämeenlinna Fin.
165 H2 Hameln Germany
124 B4 Hamersley Range mts Australia
152 D4 Hamhŭng N. Korea
146 B2 Hami China
140 C4 Ḥamīd Iran
177 F2 Hamid Sudan
126 E6 Hamilton Australia
205 L2 Hamilton Bermuda
191 H4 Hamilton Canada
128 E2 Hamilton N.Z.
162 D5 Hamilton U.K.
201 C5 Hamilton AL U.S.A.
190 B5 Hamilton IL U.S.A.
194 D2 Hamilton MT U.S.A.
203 F3 Hamilton NY U.S.A.
202 A5 Hamilton OH U.S.A.
196 B3 Hamilton, Mount CA U.S.A.
197 E2 Hamilton, Mount NV U.S.A.
196 A2 Hamilton City U.S.A.
159 N3 Hamina Fin.
137 H6 Ḥāmir, Wādī al watercourse Saudi Arabia
144 D3 Hamirpur India
152 D4 Hamju N. Korea
126 C5 Hamley Bridge Australia
190 D3 Hamlin Lake U.S.A.
165 F3 Hamm Germany
137 I3 Ḥammām al 'Alīl Iraq
170 D6 Hammamet Tunisia
177 D1 Hammamet, Golfe de g. Tunisia
137 K6 Ḥammār, Hawr al imp. l. Iraq
158 L3 Hammarstrand Sweden
165 H4 Hammelburg Germany
158 K3 Hammerdal Sweden
158 M1 Hammerfest Norway
164 E3 Hamminkeln Germany
126 C4 Hammond Australia
190 D5 Hammond IN U.S.A.
199 F6 Hammond LA U.S.A.
194 F2 Hammond MT U.S.A.
191 E3 Hammond Bay U.S.A.
202 E3 Hammondsport U.S.A.
203 F5 Hammonton U.S.A.
164 D4 Hamoir Belgium
128 C6 Hampden N.Z.
143 E3 Hampi India
161 F6 Hampshire Downs hills U.K.
189 G4 Hampton Canada
199 E5 Hampton AR U.S.A.
203 H3 Hampton NH U.S.A.
203 E6 Hampton VA U.S.A.
177 E2 Ḥamrā', Al Ḥamādah al plat. Libya
137 J4 Ḥamrīn, Jabal hills Iraq
154 C3 Ham Tân Vietnam
144 E2 Hamta Pass India
137 I2 Hamur Turkey
139 C4 Hamza Uzbek.
164 C4 Han, Grotte de tourist site Belgium
Hana U.S.A. see Hāna
196 □2 Hāna U.S.A.
180 E1 Hanahai watercourse Botswana/Namibia
196 □2 Hanalei U.S.A.
150 G5 Hanamaki Japan
165 G4 Hanau Germany
146 C2 Hanbogd Mongolia
148 E3 Hancheng China
202 E5 Hancock MD U.S.A.
190 C2 Hancock MI U.S.A.
203 F4 Hancock NY U.S.A.
162 C2 Handa Island U.K.
148 E2 Handan China
178 C4 Handeni Tanz.
153 C4 Handig Point Phil.
196 C3 Hanford U.S.A.
143 A3 Hangal India
152 C5 Han-gang r. S. Korea
146 B2 Hangayn Nuruu mts Mongolia
148 E2 Hangu China
144 B2 Hangu Pak.
152 B2 Hanguang China
149 F4 Hangzhou China
149 F4 Hangzhou Wan b. China
137 I2 Hani Turkey
140 C5 Ḥanīdh Saudi Arabia
Hanjiang China see Yangzhou
148 B2 Hanjiaoshui China
165 I1 Hankensbüttel Germany
180 F6 Hankey S. Africa
159 N4 Hanko Fin.
197 G2 Hanksville U.S.A.
144 D2 Hanle India
128 D5 Hanmer Springs N.Z.
187 G4 Hanna Canada
188 D3 Hannah Bay Canada
190 B6 Hannibal U.S.A.
165 H2 Hannover Germany
165 H3 Hannoversch Münden Germany
164 D4 Hannut Belgium
159 K5 Hanöbukten b. Sweden
149 B6 Ha Nôi Vietnam
191 G3 Hanover Canada
180 F5 Hanover S. Africa
203 G3 Hanover NH U.S.A.
202 E5 Hanover PA U.S.A.
129 E4 Hansen Mountains mts Antarctica
149 D4 Hanshou China
148 E4 Han Shui r. China
144 D3 Hansi India
158 L1 Hansnes Norway
126 B4 Hanson, Lake salt flat Australia
202 B6 Hansonville U.S.A.
159 J4 Hanstholm Denmark
164 E4 Han-sur-Nied France
172 C4 Hantsavichy Belarus
144 C3 Hanumangarh India
127 G5 Hanwood Australia
148 C3 Hanyin China
140 B5 Hanzaram Iran
148 C3 Hanzhong China
123 J6 Hao atoll Fr. Polynesia
145 G5 Haora India
158 N2 Haparanda Sweden
145 H4 Hapoli India
189 H3 Happy Valley-Goose Bay Canada
152 E3 Hapsu N. Korea
144 D3 Hapur India
143 C5 Haputale Sri Lanka
140 C5 Ḥaraḍ well Saudi Arabia
140 C5 Ḥaraḍh Saudi Arabia
172 D4 Haradok Belarus
151 G6 Haramachi Japan
144 C2 Haramukh mt. India
144 C3 Harappa Road Pak.
179 D5 Harare Zimbabwe
142 E6 Ḥarāsīs, Jiddat al des. Oman
146 C2 Har-Ayrag Mongolia
176 A4 Harbel Liberia
146 E2 Harbin China
191 F4 Harbor Beach U.S.A.
190 E3 Harbor Springs U.S.A.
189 I4 Harbour Breton Canada
212 E8 Harbours, Bay of Falkland Is
197 F5 Harcuvar Mountains U.S.A.
144 D4 Harda India
159 I4 Hardangerfjorden sea chan. Norway
159 I3 Hardangervidda plat. Norway
159 I3 Hardangervidda Nasjonalpark nat. park Norway
180 B2 Hardap admin. reg. Namibia
180 B2 Hardap Dam Namibia
155 E2 Harden, Bukit mt. Indon.
164 E2 Hardenberg Neth.
164 D2 Harderwijk Neth.
180 C5 Hardeveld mts S. Africa
165 H5 Hardheim Germany
194 F2 Hardin U.S.A.
181 H5 Harding S. Africa
187 G4 Hardisty Canada
186 F2 Hardisty Lake Canada
144 E4 Hardoi India
203 G2 Hardwick U.S.A.
126 B5 Hardwicke Bay Australia
199 F4 Hardy U.S.A.
Hardy, Mt mt. N.Z. see Rangipoua
190 E4 Hardy Reservoir U.S.A.
164 B4 Harelbeke Belgium
164 F1 Haren Neth.
164 F2 Haren (Ems) Germany
178 E3 Hārer Eth.
203 F4 Harford U.S.A.
178 E3 Hargeysa Somalia
169 L7 Harghita-Mădăraş, Vârful mt. Romania
137 H2 Harhal Dağları mts Turkey
148 C2 Harhatan China
146 B2 Har Hu l. China
176 B2 Ḥaricha, Ḥamâda El des. Mali
144 D3 Haridwar India
143 A3 Harihar India
128 C5 Harihari N.Z.
151 D7 Harima-nada b. Japan
145 G5 Haringhat r. Bangl.
164 C3 Haringvliet est. Neth.
141 G3 Hari Rūd r. Afgh./Iran
159 M3 Harjavalta Fin.
198 E3 Harlan IA U.S.A.
202 B6 Harlan KY U.S.A.
161 C5 Harlech U.K.
194 E1 Harlem U.S.A.
161 I5 Harleston U.K.
164 D1 Harlingen Neth.
199 D7 Harlingen U.S.A.
161 H6 Harlow U.K.
194 E2 Harlowton U.S.A.
164 B5 Harly France
203 I2 Harmony ME U.S.A.
190 A4 Harmony MN U.S.A.
165 I1 Harmsdorf Germany
144 A3 Harnai Pak.
164 A4 Harnes France
194 B3 Harney Basin U.S.A.
194 C3 Harney Lake U.S.A.
159 L3 Härnösand Sweden
146 E2 Har Nur China
146 B2 Har Nuur l. Mongolia
162 □ Haroldswick U.K.
176 B4 Harper Liberia
196 D4 Harper Lake U.S.A.
202 E5 Harpers Ferry U.S.A.
189 H2 Harp Lake Canada
165 G2 Harpstedt Germany
137 G2 Harput Turkey
197 F5 Harquahala Mountains U.S.A.
137 G3 Harran Turkey
188 E3 Harricana, Rivière d' r. Canada
201 C5 Harriman U.S.A.
203 G3 Harriman Reservoir U.S.A.
127 I3 Harrington Australia
203 F5 Harrington U.S.A.
189 I3 Harrington Harbour Canada
162 B3 Harris reg. U.K.
126 A3 Harris, Lake salt flat Australia
162 A3 Harris, Sound of sea chan. U.K.
200 B4 Harrisburg IL U.S.A.
202 E4 Harrisburg PA U.S.A.
181 H4 Harrismith S. Africa
199 E4 Harrison AR U.S.A.
190 E3 Harrison MI U.S.A.
189 I3 Harrison, Cape Canada
184 C2 Harrison Bay U.S.A.
202 D5 Harrisonburg U.S.A.
186 E5 Harrison Lake Canada
198 E4 Harrisonville U.S.A.
191 F3 Harrisville MI U.S.A.
203 F2 Harrisville NY U.S.A.
202 C5 Harrisville WV U.S.A.
160 F4 Harrogate U.K.
165 K2 Harsefeld Germany
140 B3 Harsīn Iran
136 G1 Harşit r. Turkey
171 L2 Hârşova Romania
158 L1 Harstad Norway
165 H2 Harsum Germany
190 D4 Hart U.S.A.
126 B3 Hart, Lake salt flat Australia
152 B2 Hartao China
180 D4 Hartbees watercourse S. Africa
168 G7 Hartberg Austria
159 I3 Harteigan mt. Norway
162 E5 Hart Fell h. U.K.
203 G4 Hartford CT U.S.A.
190 D4 Hartford MI U.S.A.
198 D3 Hartford SD U.S.A.
190 C4 Hartford WI U.S.A.
186 E3 Hart Highway Canada
189 G4 Hartland Canada
161 C7 Hartland U.K.
203 I2 Hartland U.S.A.
161 C6 Hartland Point U.K.
160 F3 Hartlepool U.K.
199 C5 Hartley U.S.A.
186 D4 Hartley Bay Canada
159 N3 Hartola Fin.
186 E4 Hart Ranges mts Canada
168 E6 Härtsfeld hills Germany
180 F3 Hartswater S. Africa
201 D5 Hartwell Reservoir U.S.A.
146 B2 Har Us Nuur l. Mongolia
141 F3 Harut watercourse Afgh.
190 C4 Harvard U.S.A.
195 F4 Harvard, Mount U.S.A.
203 J2 Harvey Canada
190 D2 Harvey MI U.S.A.
198 C2 Harvey ND U.S.A.
161 I6 Harwich U.K.
127 J2 Harwood Australia
144 C3 Haryana state India
165 I3 Harz hills Germany
136 E6 Ḥasā, Qal'at al tourist site Jordan
136 F6 Ḥaşāh, Wādī al watercourse Jordan
139 I5 Hasalbag China
136 E2 Hasan Dağı mts Turkey
137 H3 Hasankeyf Turkey
140 E5 Ḥasan Langī Iran
143 B2 Hasanparti India
140 B2 Ḥasan Sālārān Iran
138 D5 Hasardag mt. Turkm.
136 E5 Hasbani r. Lebanon
136 E2 Hasbek Turkey
143 C1 Hasdo r. India
165 F2 Hase r. Germany
164 F2 Haselünne Germany
165 I4 Hasenkopf h. Germany
140 C3 Hashtgerd Iran
140 C2 Hashtpar Iran
140 B2 Hashtrud Iran
140 E3 Hasht Tekkeh, Gowd-e waterhole Iran
199 D5 Haskell U.S.A.
161 G6 Haslemere U.K.
169 L7 Hăşmaşul Mare mt. Romania
143 B3 Hassan India
137 J4 Hassan Iraq
144 C2 Hassan Abdal Pak.
197 F5 Hassayampa watercourse U.S.A.
165 I4 Haßberge hills Germany
164 D4 Hasselt Belgium
164 E2 Hasselt Neth.
176 C1 Hassi Messaoud Alg.
159 K4 Hässleholm Sweden
126 F7 Hastings Australia
127 J3 Hastings r. Australia
128 F3 Hastings N.Z.
161 H7 Hastings U.K.
190 E4 Hastings MI U.S.A.
190 A3 Hastings MN U.S.A.
198 D3 Hastings NE U.S.A.
148 C1 Hatanbulag Mongolia
197 F3 Hatch U.S.A.
187 I3 Hatchet Lake Canada
201 B5 Hatchie r. U.S.A.
126 E4 Hatfield Australia
160 C4 Hatfield U.K.
146 C1 Hatgal Mongolia
144 D4 Hathras India
145 F4 Hatia Nepal
154 C3 Ha Tiên Vietnam
154 C1 Ha Tinh Vietnam
137 I4 Hatra Iraq
126 E5 Hattah Australia
126 E5 Hattah-Kulkyne National Park Australia
201 F5 Hatteras, Cape U.S.A.
219 E4 Hatteras Abyssal Plain sea feature S. Atlantic Ocean
158 K2 Hattfjelldal Norway
143 C2 Hatti r. India
201 B6 Hattiesburg U.S.A.
164 F3 Hattingen Germany
154 B4 Hat Yai Thai.
154 C3 Hâu, Sông r. Vietnam
178 E3 Haud reg. Eth.
159 I4 Hauge Norway
159 I4 Haugesund Norway
Hâu Giang, Sông r. Vietnam see Hâu, Sông
128 E3 Hauhungaroa mt. N.Z.
159 I4 Haukeligrend Norway
158 N2 Haukipudas Fin.
159 O3 Haukivesi l. Fin.
187 H3 Haultain r. Canada
128 E2 Hauraki Gulf N.Z.
128 A7 Hauroko, Lake N.Z.
203 I3 Haut, Isle au i. U.S.A.
176 B1 Haut Atlas mts Morocco
189 G4 Hauterive Canada
176 B1 Hauts Plateaux Alg.
196 □ Hau'ula U.S.A.
Hauula U.S.A. see Hau'ula
205 H4 Havana Cuba
190 B5 Havana U.S.A.
161 G7 Havant U.K.
197 E4 Havasu, Lake U.S.A.
165 K2 Havel r. Germany
164 D4 Havelange Belgium
165 K2 Havelberg Germany
165 K2 Havelländisches Luch marsh Germany
191 I3 Havelock Canada
201 E5 Havelock U.S.A.
128 F3 Havelock North N.Z.
161 C6 Haverfordwest U.K.
203 H3 Haverhill U.S.A.
143 A3 Haveri India
164 D4 Haversin Belgium
164 F3 Havixbeck Germany
168 G6 Havlíčkův Brod Czech Rep.
158 N1 Havøysund Norway
171 L5 Havran Turkey
194 E1 Havre U.S.A.
189 H4 Havre Aubert, Île du i. Canada
203 E5 Havre de Grace U.S.A.
189 H3 Havre-St-Pierre Canada
171 L4 Havsa Turkey
136 E1 Havza Turkey
196 □2 Hawai'i i. U.S.A.
196 □1 Hawai'ian Islands N. Pacific Ocean
216 H4 Hawaiian Ridge sea feature N. Pacific Ocean
Hawaii Volcanoes National Park U.S.A. see Hawai'i Volcanoes National Park
196 □2 Hawai'i Volcanoes National Park U.S.A.
137 K7 Ḥawallī Kuwait
161 D4 Hawarden U.K.
128 B6 Hawea, Lake N.Z.
128 E3 Hawera N.Z.
160 E3 Hawes U.K.
Hawi U.S.A. see Hāwī
196 □2 Hāwī U.S.A.
162 F5 Hawick U.K.
137 K6 Ḥawizah, Hawr al imp. l. Iraq
128 B6 Hawkdun Range mts N.Z.
128 F3 Hawke Bay N.Z.
189 I3 Hawke Island Canada
126 C3 Hawker Australia
126 D2 Hawkers Gate Australia
203 F2 Hawkesbury Canada
197 F3 Hawkins Peak U.S.A.
191 F3 Hawks U.S.A.
203 J2 Hawkshaw Canada
203 F4 Hawley U.S.A.
137 I5 Ḥawrān, Wādī watercourse Iraq
140 B6 Ḥawshah, Jibāl al mts Saudi Arabia
180 C7 Hawston S. Africa
196 C2 Hawthorne U.S.A.
152 C1 Haxat China
160 F3 Haxby U.K.
126 F5 Hay Australia
186 F2 Hay r. Canada
190 B3 Hay r. U.S.A.
Haya China see Yagan
150 G5 Hayachine-san mt. Japan
197 G5 Hayden AZ U.S.A.
194 C2 Hayden ID U.S.A.
187 K3 Hayes r. Man. Canada
185 I3 Hayes r. Nunavut Canada
161 B7 Hayle U.K.
142 E6 Haymā' Oman
136 D2 Haymana Turkey
202 E5 Haymarket U.S.A.
203 I2 Haynesville U.S.A.
161 D5 Hay-on-Wye U.K.
173 C7 Hayrabolu Turkey
186 F2 Hay River Canada
198 D4 Hays U.S.A.
173 D5 Haysyn Ukr.
196 A3 Hayward CA U.S.A.
190 B2 Hayward WI U.S.A.
161 G7 Haywards Heath U.K.
138 C5 Hazar Turkm.
141 G3 Hazarajat reg. Afgh.
202 B6 Hazard U.S.A.
145 F5 Hazaribagh India
145 E5 Hazaribagh Range mts India
164 A4 Hazebrouck France
186 D3 Hazelton Canada
184 G2 Hazen Strait Canada
164 C2 Hazerswoude-Rijndijk Neth.
203 F4 Hazleton U.S.A.
138 E4 Hazorasp Uzbek.
141 G2 Hazrat Sultan Afgh.
137 H2 Hazro Turkey
215 D2 H. Bouchard Arg.
163 B4 Headford Ireland
196 A2 Healdsburg U.S.A.
126 F6 Healesville Australia
161 F4 Heanor U.K.
218 I8 Heard Island Indian Ocean
199 D6 Hearne U.S.A.
188 D4 Hearst Canada
161 H7 Heathfield U.K.
203 E6 Heathsville U.S.A.
199 D7 Hebbronville U.S.A.
148 E2 Hebei prov. China
127 G2 Hebel Australia
192 D3 Heber City U.S.A.
199 E5 Heber Springs U.S.A.
148 E3 Hebi China
189 H2 Hebron Canada
190 D5 Hebron IN U.S.A.
198 D3 Hebron NE U.S.A.
203 G3 Hebron NY U.S.A.
136 E6 Hebron West Bank
189 H2 Hebron Fiord inlet Canada
186 C4 Hecate Strait Canada
207 G3 Hecelchakán Mex.
186 C3 Heceta Island U.S.A.
149 C5 Hechi China
149 C4 Hechuan China
159 K3 Hede Sweden
159 K3 Hedemora Sweden
194 C2 He Devil Mountain U.S.A.
149 D6 Hedi Shuiku resr China
190 A5 Hedrick U.S.A.
164 D2 Heeg Neth.
164 F2 Heek Germany
164 C4 Heer Belgium
164 E2 Heerde Neth.
164 D2 Heerenveen Neth.
164 C2 Heerhugowaard Neth.
164 D4 Heerlen Neth.
Hefa Israel see Haifa
148 E4 Hefei China
149 D4 Hefeng China
150 B1 Hegang China
151 E6 Hegura-jima i. Japan
165 J3 Heidberg h. Germany
168 D3 Heide Germany
179 B6 Heide Namibia
165 G5 Heidelberg Germany
181 H3 Heidelberg Gauteng S. Africa
180 D7 Heidelberg W. Cape S. Africa
181 G3 Heilbron S. Africa
165 H5 Heilbronn Germany
168 E3 Heiligenhafen Germany
149 □ Hei Ling Chau i. Hong Kong China
152 E1 Heilongjiang prov. China
146 E2 Heilong Jiang r. China/Rus. Fed. alt. Amur
165 I5 Heilsbronn Germany
158 J3 Heimdal Norway
159 N3 Heinola Fin.
154 A2 Heinze Islands Myanmar
152 B3 Heishan China
164 D3 Heist-op-den-Berg Belgium
164 E3 Hejian China
149 B4 Hejiang China
149 D6 He Jiang r. China
148 D3 Hejin China
136 F2 Hekimhan Turkey
158 C3 Hekla vol. Iceland
148 B2 Hekou Gansu China
149 B6 Hekou Yunnan China
158 K3 Helagsfjället mt. Sweden
148 B2 Helan Shan mts China
165 J3 Helbra Germany
145 H4 Helem India
196 D3 Helen, Mount U.S.A.
199 F5 Helena AR U.S.A.
194 E2 Helena MT U.S.A.
162 D4 Helensburgh U.K.
136 E6 Ḥelez Israel
168 C3 Helgoland i. Germany
168 D3 Helgoländer Bucht g. Germany
127 J1 Helidon Australia
158 B3 Hella Iceland
158 L1 Helland Norway
140 C4 Helleh r. Iran
164 C3 Hellevoetsluis Neth.
158 M1 Helligskogen Norway
167 F3 Hellín Spain
194 C2 Hells Canyon gorge U.S.A.
141 F4 Helmand r. Afgh.
141 F4 Helmand reg. Afgh.
141 F4 Helmand, Hāmūn salt flat Afgh./Iran
165 J4 Helmbrechts Germany
165 I3 Helme r. Germany
179 B6 Helmeringhausen Namibia
164 D3 Helmond Neth.
162 E2 Helmsdale U.K.
162 E2 Helmsdale r. U.K.
165 J2 Helmstedt Germany
152 E2 Helong China
197 G2 Helper U.S.A.
159 K4 Helsingborg Sweden
Helsingfors Fin. see Helsinki
159 K4 Helsingør Denmark
159 N3 Helsinki Fin.
161 B7 Helston U.K.
160 D3 Helvellyn h. U.K.
163 D5 Helvick Head Ireland
161 G6 Hemel Hempstead U.K.
196 D5 Hemet U.S.A.
202 E3 Hemlock Lake U.S.A.
165 H2 Hemmingen Germany
203 G2 Hemmingford Canada
165 H1 Hemmoor Germany
199 D6 Hempstead U.S.A.
161 I5 Hemsby U.K.
159 L4 Hemse Sweden
148 A3 Henan China
148 D3 Henan prov. China
167 E2 Henares r. Spain
150 F4 Henashi-zaki pt Japan
136 C1 Hendek Turkey
215 E3 Henderson Arg.
200 C4 Henderson KY U.S.A.
201 E4 Henderson NC U.S.A.
197 E3 Henderson NV U.S.A.
203 E3 Henderson NY U.S.A.
199 E5 Henderson TX U.S.A.
123 J7 Henderson Island Pitcairn Is
201 D5 Hendersonville NC U.S.A.
201 C4 Hendersonville TN U.S.A.
140 C4 Hendījān Iran
160 D5 Hendon U.K.
140 D5 Hendorābī i. Iran
141 E5 Hengām Iran
146 B4 Hengduan Shan mts China
164 E2 Hengelo Neth.
152 F1 Hengshan Heilong. China
149 D5 Hengshan Hunan China
148 C2 Hengshan Shaanxi China
149 D5 Heng Shan mt. Hunan China
148 D2 Heng Shan mts Shanxi China
148 E2 Hengshui China
149 C6 Hengxian China
149 D5 Hengyang Hunan China
149 D5 Hengyang Hunan China
173 E6 Heniches'k Ukr.
128 C6 Henley N.Z.
161 G6 Henley-on-Thames U.K.
203 F5 Henlopen, Cape U.S.A.
164 F4 Hennef (Sieg) Germany
181 G3 Hennenman S. Africa
165 L2 Hennigsdorf Berlin Germany
203 H3 Henniker U.S.A.
199 D5 Henrietta U.S.A.
188 D2 Henrietta Maria, Cape Canada
197 G3 Henrieville U.S.A.
190 C5 Henry U.S.A.
129 C3 Henry Ice Rise ice feature Antarctica
185 L3 Henry Kater, Cape Canada
197 G2 Henry Mountains U.S.A.
191 G4 Hensall Canada
165 H1 Henstedt-Ulzburg Germany
179 B6 Hentiesbaai Namibia
127 G5 Henty Australia
Henzada Myanmar see Hinthada
187 H4 Hepburn Canada
149 E5 Heping China
149 C6 Hepu China
148 D2 Hequ China
141 F3 Herāt Afgh.
166 F5 Hérault r. France
187 H5 Herbert Canada
165 G4 Herborn Germany
165 H4 Herbstein Germany
129 C4 Hercules Dome ice feature Antarctica
164 F3 Herdecke Germany
165 F4 Herdorf Germany
206 H6 Heredia Costa Rica
161 E5 Hereford U.K.
199 C5 Hereford U.S.A.
123 J6 Hereheretue atoll Fr. Polynesia
164 C4 Herent Belgium
165 G2 Herford Germany
165 I4 Heringen (Werra) Germany
198 D4 Herington U.S.A.
140 B2 Herīs Iran
168 D7 Herisau Switz.
203 F3 Herkimer U.S.A.
165 I3 Herleshausen Germany
206 D2 Hermanas Mex.
162 □ Herma Ness hd U.K.
165 I2 Hermannsburg Germany
180 C7 Hermanus S. Africa
181 H5 Hermes, Cape S. Africa
127 G3 Hermidale Australia
194 C2 Hermiston U.S.A.
212 C9 Hermite, Islas is Chile
124 E2 Hermit Islands P.N.G.
136 E5 Hermon, Mount Lebanon/Syria
206 B1 Hermosillo Mex.
212 F3 Hernandarias Para.
164 F3 Herne Germany
161 I6 Herne Bay U.K.
159 J4 Herning Denmark
190 D1 Heron Bay Canada
206 D3 Herradura Mex.
167 D3 Herrera del Duque Spain
127 G8 Herrick Australia
165 I5 Herrieden Germany
202 E4 Hershey U.S.A.
161 G6 Hertford U.K.
181 F4 Hertzogville S. Africa
164 D4 Herve Belgium
125 F4 Hervey Bay Australia
217 I7 Hervey Islands Cook Is
165 K2 Herzberg Brandenburg Germany
165 L3 Herzberg Brandenburg Germany
164 F2 Herzlake Germany
165 J5 Herzogenaurach Germany
165 K1 Herzsprung Germany
137 L4 Ḥeşār Iran
164 C4 Hesbaye reg. Belgium
164 F1 Hesel Germany
149 C6 Heshan China
148 C3 Heshui China
148 D2 Heshun China
197 E4 Hesperia U.S.A.
186 C2 Hess r. Canada
165 I5 Heßdorf Germany
165 I5 Hesselberg h. Germany
165 H4 Hessen land Germany
165 H3 Hessisch Lichtenau Germany
149 B6 Het r. Laos
196 B3 Hetch Hetchy Aqueduct canal U.S.A.
164 D3 Heteren Neth.
198 C2 Hettinger U.S.A.
160 E3 Hetton U.K.
165 J3 Hettstedt Germany
160 E3 Hexham U.K.
148 F4 Hexian China
148 B2 Hexipu China
180 C6 Hex River Pass S. Africa
148 D3 Heyang China
140 B2 Ḥeydarābād Iran
141 F4 Ḥeydarābād Iran
160 E3 Heysham U.K.
149 E6 Heyuan China
126 D7 Heywood Australia
160 E4 Heywood U.K.
190 C5 Heyworth U.S.A.
148 E3 Heze China
149 B5 Hezhang China
148 B3 Hezheng China
149 D5 Hezhou China
148 B3 Hezuo China
201 D7 Hialeah U.S.A.
198 E4 Hiawatha U.S.A.
190 A2 Hibbing U.S.A.
127 F9 Hibbs, Point Australia
201 D5 Hickory U.S.A.
128 G2 Hicks Bay N.Z.
207 G4 Hicks Cayes is Belize
187 J2 Hicks Lake Canada
202 A4 Hicksville U.S.A.
199 D5 Hico U.S.A.
150 H3 Hidaka-sanmyaku mts Japan
207 E2 Hidalgo Mex.
207 E3 Hidalgo state Mex.
206 C2 Hidalgo del Parral Mex.
214 C2 Hidrolândia Brazil
151 C7 Higashi-Hiroshima Japan
150 G5 Higashine Japan
151 D7 Higashi-Ōsaka Japan
151 A8 Higashi-suidō sea chan. Japan
203 F3 Higgins Bay U.S.A.
190 E3 Higgins Lake U.S.A.
194 B3 High Desert U.S.A.
190 C3 High Falls Reservoir U.S.A.
190 E3 High Island U.S.A.
149 □ High Island Reservoir Hong Kong China
190 D4 Highland Park U.S.A.
196 C2 Highland Peak CA U.S.A.
197 E3 Highland Peak NV U.S.A.
186 F3 High Level Canada
145 F5 High Level Canal India
201 E5 High Point U.S.A.
186 F3 High Prairie Canada
186 G4 High River Canada
201 E7 High Rock Bahamas
187 I3 Highrock Lake Canada
127 F9 High Rocky Point Australia
160 E3 High Seat h. U.K.
203 F4 Hightstown U.S.A.
161 G6 High Wycombe U.K.
206 B2 Higuera de Zaragoza Mex.
213 D2 Higuerote Venez.
159 M4 Hiiumaa i. Estonia
142 A4 Hijaz reg. Saudi Arabia
197 E3 Hiko U.S.A.
151 E7 Hikone Japan
128 G2 Hikurangi mt. N.Z.
197 F3 Hildale U.S.A.
165 I4 Hildburghausen Germany
165 I4 Hilders Germany
165 H2 Hildesheim Germany
145 G4 Hili Bangl.
198 D4 Hill City U.S.A.
197 H2 Hill Creek r. U.S.A.
164 C2 Hillegom Neth.
159 K5 Hillerød Denmark
198 D2 Hillsboro ND U.S.A.
203 H3 Hillsboro NH U.S.A.
202 B5 Hillsboro OH U.S.A.
199 D5 Hillsboro TX U.S.A.
190 B4 Hillsboro WI U.S.A.
202 C5 Hillsboro WV U.S.A.
190 E5 Hillsdale MI U.S.A.
203 G3 Hillsdale NY U.S.A.
202 E4 Hillsgrove U.S.A.
162 F4 Hillside U.K.
197 F4 Hillside U.S.A.
127 F4 Hillston Australia
202 C6 Hillsville U.S.A.
127 I5 Hilltop Australia
196 □2 Hilo U.S.A.
181 I4 Hilton S. Africa
202 E3 Hilton U.S.A.
191 F2 Hilton Beach Canada
201 D5 Hilton Head Island U.S.A.
137 G3 Hilvan Turkey
164 D2 Hilversum Neth.
144 D3 Himachal Pradesh state India
144 D2 Himalaya mts Asia
145 F3 Himalchul mt. Nepal
158 M2 Himanka Fin.
171 H4 Himarë Albania
144 C5 Himatnagar India
151 D7 Himeji Japan
150 G5 Himekami-dake mt. Japan
181 H4 Himeville S. Africa
151 E6 Himi Japan
136 F4 Ḥimş Syria
153 C4 Hinatuan Phil.
169 N7 Hînceşti Moldova
124 E3 Hinchinbrook Island Australia
161 F5 Hinckley U.K.
190 A2 Hinckley MN U.S.A.
197 F2 Hinckley UT U.S.A.
203 F3 Hinckley Reservoir U.S.A.
144 D3 Hindan r. India
144 D4 Hindaun India
160 G3 Hinderwell U.K.
160 E4 Hindley U.K.
202 B6 Hindman U.S.A.
126 D6 Hindmarsh, Lake dry lake Australia
143 D1 Hindola India
141 G3 Hindu Kush mts Afgh./Pak.
143 B3 Hindupur India
186 F3 Hines Creek Canada
201 D6 Hinesville U.S.A.
144 D5 Hinganghat India
141 G5 Hinglaj Pak.
141 G5 Hingol r. Pak.
144 D6 Hingoli India
137 H2 Hinis Turkey
196 D4 Hinkley U.S.A.
158 K1 Hinnøya i. Norway
153 B4 Hinobaan Phil.
167 D3 Hinojosa del Duque Spain
151 C7 Hino-misaki pt Japan
203 G3 Hinsdale U.S.A.
164 F1 Hinte Germany
147 B5 Hinthada Myanmar
186 F4 Hinton Canada
202 C6 Hinton U.S.A.
164 C2 Hippolytushoef Neth.
137 J2 Hirabit Dağ mt. Turkey
151 A8 Hirado Japan
151 A8 Hirado-shima i. Japan
143 C1 Hirakud Reservoir India
150 H3 Hiroo Japan
150 G4 Hirosaki Japan
151 C7 Hiroshima Japan
165 J5 Hirschaid Germany
165 J4 Hirschberg Germany
168 E7 Hirschberg mt. Germany
166 G2 Hirson France
159 J4 Hirtshals Denmark
144 C3 Hisar India
137 L3 Hisar Iran
141 G3 Hisar, Koh-i- mts Afgh.
136 D1 Hisarönü Turkey
137 J6 Hisb, Sha'īb watercourse Iraq
139 G5 Hisor Tajik.
205 J4 Hispaniola i. Caribbean Sea
145 F4 Hisua India
137 I5 Hīt Iraq
151 G6 Hitachi Japan
151 G6 Hitachi-Ōta Japan
151 B8 Hitoyoshi Japan
158 J3 Hitra i. Norway
165 J1 Hitzacker Germany
151 C7 Hiuchi-nada b. Japan
123 J5 Hiva Oa i. Fr. Polynesia
186 E4 Hixon Canada
137 I2 Hizan Turkey
159 K4 Hjälmaren l. Sweden

I

243

Khorixas

244

Lake Meredith National Recreation Area

138 F5 Koson Uzbek.
152 E3 Kosŏng N. Korea
152 E4 Kosŏng N. Korea
139 G4 Kosonsoy Uzbek.
171 I3 Kosovo country Europe
123 F2 Kosrae atoll Micronesia
139 I5 Kosrap China
165 J5 Kösseine h. Germany
138 C3 Kosshagyl Kazakh.
176 B4 Kossou, Lac de l. Côte d'Ivoire
138 E1 Kostanay Kazakh.
138 E1 Kostanayskaya Oblast' admin. div. Kazakh.
171 J3 Kostenets Bulg.
181 G2 Koster S. Africa
171 J3 Kostinbrod Bulg.
132 J3 Kostino Rus. Fed.
172 D1 Kostomuksha Rus. Fed.
173 C5 Kostopil' Ukr.
172 G3 Kostroma Rus. Fed.
172 G3 Kostroma r. Rus. Fed.
172 G3 Kostromskaya Oblast' admin. div. Rus. Fed.
168 G4 Kostrzyn Poland
173 F5 Kostyantynivka Ukr.
168 H3 Koszalin Poland
168 H7 Kőszeg Hungary
143 C1 Kota Chhattisgarh India
144 C4 Kota Rajasthan India
155 B4 Kotaagung Indon.
144 C4 Kota Barrage India
155 D3 Kotabaru Indon.
155 E3 Kotabaru Indon.
155 B1 Kota Bharu Malaysia
155 B3 Kotabumi Indon.
144 C4 Kota Dam India
155 E1 Kota Kinabalu Sabah Malaysia
139 I3 Kotanemel', Gora mt. Kazakh.
143 C2 Kotaparh India
154 B5 Kotapinang Indon.
144 C4 Kotari r. India
155 B2 Kota Tinggi Malaysia
172 I3 Kotel'nich Rus. Fed.
173 G6 Kotel'nikovo Rus. Fed.
133 O2 Kotel'nyy, Ostrov i. Rus. Fed.
144 D3 Kotgarh India
165 J3 Köthen (Anhalt) Germany
144 E4 Kothi India
159 N3 Kotka Fin.
144 C3 Kot Kapura India
172 H2 Kotlas Rus. Fed.
144 C2 Kotli Pak.
184 B3 Kotlik U.S.A.
158 C3 Kötlutangi pt Iceland
159 O4 Kotly Rus. Fed.
170 G2 Kotor Varoš Bos.-Herz.
176 B4 Kotouba Côte d'Ivoire
173 H5 Kotovo Rus. Fed.
173 G4 Kotovsk Rus. Fed.
173 D6 Kotovs'k Ukr.
144 C4 Kotra India
144 E6 Kotri r. India
144 B4 Kotri Pak.
144 A5 Kot Sarae Pak.
143 C2 Kottagudem India
143 B4 Kottarakara India
143 B4 Kottayam India
143 B3 Kotturu India
Koturdepe Turkm. see Goturdepe
133 L2 Kotuy r. Rus. Fed.
176 B3 Kotzyrtas Kazakh.
184 B3 Kotzebue U.S.A.
184 B3 Kotzebue Sound sea chan. U.S.A.
165 K5 Kötzting Germany
176 A3 Koubia Guinea
176 B3 Koudougou Burkina
180 E6 Koueveldberge mts S. Africa
177 D3 Koufey Niger
171 L7 Koufonisi i. Greece
180 E6 Kougaberge mts S. Africa
136 D4 Kouklia Cyprus
178 B4 Koulamoutou Gabon
176 B3 Koulikoro Mali
125 G4 Koumac New Caledonia
176 A3 Koundâra Guinea
176 B3 Koupéla Burkina
211 H2 Kourou Fr. Guiana
176 B3 Kouroussa Guinea
177 D3 Kousséri Cameroon
176 B3 Koutiala Mali
159 N3 Kouvola Fin.
158 O2 Kovdor Rus. Fed.
158 P2 Kovdozero, Ozero l. Rus. Fed.
173 C5 Kovel' Ukr.
172 G3 Kovernino Rus. Fed.
143 B4 Kovilpatti India
172 G3 Kovrov Rus. Fed.
172 G4 Kovylkino Rus. Fed.
172 F2 Kovzhskoye, Ozero l. Rus. Fed.
128 C5 Kowhitirangi N.Z.
149 □ Kowloon Peak h. Hong Kong China
149 □ Kowloon Peninsula Hong Kong China
152 D4 Kowŏn N. Korea
139 J5 Koxlax China
139 I5 Koxtag China
151 B7 Kōyama-misaki pt Japan
Koyamutthoor India see Coimbatore
138 F1 Koybagar, Ozero l. Kazakh.
136 B3 Köyceğiz Turkey
132 I2 Koygorodok Rus. Fed.
Koymatdag, Gory hills Turkm. see Goymatdag
143 A2 Koyna Reservoir India
172 H1 Koynas Rus. Fed.
184 C3 Koyukuk r. U.S.A.
136 F1 Koyulhisar Turkey
172 F3 Koza Rus. Fed.
151 A7 Kō-zaki pt Japan
136 E3 Kozan Turkey
171 I4 Kozani Greece
170 G2 Kozara mts Bos.-Herz.
173 D5 Kozelets' Ukr.
172 E4 Kozel'sk Rus. Fed.
143 A4 Kozhikode India
Kozhikode India see Kozhikode
136 C1 Kozlu Turkey
172 H3 Koz'modem'yansk Rus. Fed.
139 G4 Kozmoldak Kazakh.
171 J4 Kožuf mts Greece/Macedonia
151 F7 Kōzu-shima i. Japan
173 D5 Kozyatyn Ukr.
176 C4 Kpalimé Togo
154 A3 Kra, Isthmus of Thai.
154 A3 Krabi Thai.
154 A3 Kra Buri Thai.
154 C2 Krâchéh Cambodia
158 L2 Kraddsele Sweden
159 J4 Kragerø Norway
164 D2 Kraggenburg Neth.
171 I2 Kragujevac Serbia
165 G5 Kraichgau reg. Germany
155 C4 Krakatau i. Indon.
154 C2 Krâkôr Cambodia
169 I5 Kraków Poland
165 K1 Krakower See l. Germany
154 B2 Krâlănh Cambodia
213 C1 Kralendijk Neth. Antilles
173 F5 Kramators'k Ukr.
158 L3 Kramfors Sweden
164 C3 Krammer est. Neth.
171 J6 Kranidi Greece
170 F1 Kranj Slovenia
154 □ Kranji Reservoir Sing.
181 I4 Kranskop S. Africa
172 H2 Krasavino Rus. Fed.
132 G2 Krasino Rus. Fed.
150 B3 Kraskino Rus. Fed.
159 N5 Krāslava Latvia
165 K4 Kraslice Czech Rep.
169 O4 Krasnapollye Belarus
172 D4 Krasnaya Gora Rus. Fed.
139 H2 Krasnaya Polyana Kazakh.
173 H5 Krasnoarmeysk Rus. Fed.
173 F5 Krasnoarmiys'k Ukr.
172 H2 Krasnoborsk Rus. Fed.
173 F6 Krasnodar Rus. Fed.
173 F6 Krasnodarskiy Kray admin. div. Rus. Fed.
173 F5 Krasnodon Ukr.
172 D3 Krasnogorodskoye Rus. Fed.
173 G6 Krasnogvardeyskoye Rus. Fed.
173 F5 Krasnohrad Ukr.
173 E6 Krasnohvardiys'ke Ukr.
138 C2 Krasnokholm Rus. Fed.
169 O2 Krasnomayskiy Rus. Fed.
173 E6 Krasnoperekops'k Ukr.
150 D2 Krasnorechenskiy Rus. Fed.
159 O3 Krasnosel'skoye Rus. Fed.
172 G4 Krasnoslobodsk Rus. Fed.
138 D1 Krasnousol'skiy Rus. Fed.
138 C5 Krasnovodsk, Mys pt Turkm.
Krasnovodskiy Zaliv b. Turkm. see Türkmenbaşy Aýlagy
138 C4 Krasnovodskoye Plato plat. Turkm.
146 E1 Krasnoyarsk Rus. Fed.
169 O3 Krasnyy Rus. Fed.
172 H3 Krasnyye Baki Rus. Fed.
173 H6 Krasnyye Barrikady Rus. Fed.
172 F3 Krasnyy Kholm Rus. Fed.
173 H5 Krasnyy Kut Rus. Fed.
172 G3 Krasnyy Luch Rus. Fed.
173 F5 Krasnyy Lyman Ukr.
139 I3 Krasnyy Oktyabr' Kazakh.
139 C1 Krasnyy Yar Kazakh.
173 I5 Krasnyy Yar Astrakhanskaya Oblast' Rus. Fed.
173 E1 Krasnyy Yar Samarskaya Oblast' Rus. Fed.
173 H5 Krasnyy Yar Volgogradskaya Oblast' Rus. Fed.
173 C5 Krasyliv Ukr.
173 H7 Kraynovka Rus. Fed.
164 E3 Krefeld Germany
173 E5 Kremenchuk Ukr.
173 E5 Kremenchuts'ke Vodoskhovyshche resr Ukr.
173 C5 Kremenskaya Rus. Fed.
168 C6 Křemešník h. Czech Rep.
194 F3 Kremmling U.S.A.
168 C6 Krems an der Donau Austria
133 T3 Kresta, Zaliv g. Rus. Fed.
172 E3 Krestsy Rus. Fed.
159 N5 Kretinga Lith.
164 E4 Kreuzau Germany
165 F4 Kreuztal Germany
169 N3 Kreva Belarus
176 C4 Kribi Cameroon
181 H3 Kriel S. Africa
171 I4 Krikellos Greece
150 H2 Kril'on, Mys c. Rus. Fed.
135 F5 Krishna r. India
135 C5 Krishna, Mouths of the India
143 B3 Krishnagiri India
145 C5 Krishnanagar India
143 B3 Krishnaraja Sagara l. India
159 I4 Kristiansand Norway
159 K4 Kristianstad Sweden
158 I3 Kristiansund Norway
Kristiinankaupunki Fin. see Kristinestad
159 K4 Kristinehamn Sweden
159 M3 Kristinestad Fin.
Kriti i. Greece see Crete
Kritiko Pelagos sea Greece
Krivoy Rog Ukr. see Kryvyy Rih
171 G3 Križevci Croatia
170 F1 Krk i. Croatia
158 K3 Krokom Sweden
158 J3 Krokstadøra Norway
158 K2 Krokstranda Norway
173 E5 Krolevets' Ukr.
165 J4 Kronach Germany
154 B3 Krŏng Kaôh Kŏng Cambodia
158 N3 Kronoby Fin.
185 C3 Kronprins Frederik Bjerge nunataks Greenland
154 A1 Kronwa Myanmar
181 G3 Kroonstad S. Africa
173 G6 Kropotkin Rus. Fed.
165 K3 Kropstädt Germany
169 J6 Krosno Poland
168 H5 Krotoszyn Poland
181 I2 Kruger National Park S. Africa
169 N3 Kruhlaye Belarus
155 B4 Krui Indon.
180 F7 Kruisfontein S. Africa
171 H4 Krujë Albania
171 K4 Krumovgrad Bulg.
Krung Thep Thai. see Bangkok
169 N3 Krupki Belarus
171 I3 Kruševac Serbia
165 K4 Krušné hory mts Czech Rep.
186 B3 Kruzof Island U.S.A.
172 D4 Krychaw Belarus
219 H4 Krylov Seamount sea feature N. Atlantic Ocean
173 F6 Krymsk Rus. Fed.
Kryms'kyy Pivostriv pen. Ukr. see Crimea
139 H3 Krypsalo (abandoned) Kazakh.
173 E6 Kryvyy Rih Ukr.
176 B2 Ksabi Alg.
176 C1 Ksar el Boukhari Alg.
176 B1 Ksar el Kebir Morocco
173 F5 Kshenskiy Rus. Fed.
170 D7 Ksour Essaf Tunisia
172 H3 Kstovo Rus. Fed.
140 B5 Kū', Jabal al h. Saudi Arabia
154 A4 Kuah Malaysia
154 B4 Kuala Kangsar Malaysia
154 B4 Kuala Kerai Malaysia
154 B4 Kuala Kubu Baharu Malaysia
154 B4 Kuala Lipis Malaysia
155 B2 Kuala Lumpur Malaysia
154 B4 Kuala Nerang Malaysia
154 B4 Kuala Pilah Malaysia
154 B4 Kuala Rompin Malaysia
155 D3 Kualasampit Indon.
154 A4 Kualasimpang Indon.
155 B1 Kuala Terengganu Malaysia
155 A5 Kuamut Sabah Malaysia
152 C3 Kuandian China
149 F6 Kuanshan Taiwan
155 B2 Kuantan Malaysia
173 G6 Kuban' r. Rus. Fed.
137 G4 Kubār Syria
137 I5 Kubaysah Iraq
172 F3 Kubenskoye, Ozero l. Rus. Fed.
138 D4 Kubla Ustyurt Uzbek.
172 H4 Kubnya r. Rus. Fed.
171 L3 Kubrat Bulg.
144 C4 Kuchaman India
144 C4 Kuchera India
155 D2 Kuching Sarawak Malaysia
151 A10 Kuchino-shima i. Japan
139 I1 Kuchukskoye, Ozero salt l. Rus. Fed.
171 H4 Kuçovë Albania
143 A3 Kudal India
155 E1 Kudat Sabah Malaysia
143 B3 Kudligi India
143 A3 Kudremukh mt. India
155 D4 Kudus Indon.
168 F7 Kufstein Austria
185 J3 Kugaaruk Canada
172 I3 Kugesi Rus. Fed.
184 G3 Kugluktuk Canada
184 E3 Kugmallit Bay Canada
140 E5 Kūhak Iran
141 F5 Kūh, Ra's-al- pt Iran
145 E3 Kuhanbokano mt. China
165 K1 Kuhbier Germany
140 E4 Kūhbonān Iran
140 B3 Kūhdasht Iran
140 E2 Kūh-e Shāh Jahān mts Iran
137 L3 Kūhīn Iran
158 O2 Kuhmo Fin.
159 N3 Kuhmoinen Fin.
140 D3 Kūhpāyeh Iran
165 K3 Kühren Germany
154 A2 Kui Buri Thai.
180 B2 Kuis Namibia
180 A1 Kuiseb Pass Namibia
179 B5 Kuito Angola
186 C3 Kuiu Island U.S.A.
158 N2 Kuivaniemi Fin.
150 G5 Kujang India
152 C4 Kujang N. Korea
151 G4 Kuji Japan
151 B8 Kujū-san vol. Japan
191 F1 Kukatush Canada
171 I3 Kukës Albania
172 I3 Kukmor Rus. Fed.
154 B5 Kukup Malaysia
138 D5 Kükürtli Turkm.
140 D5 Kūl r. Iran
136 B2 Kula Turkey
145 G3 Kula Kangri mt. Bhutan
138 B3 Kulaly, Ostrov i. Kazakh.
139 H4 Kulan Kazakh.
139 H4 Kulanak Kyrg.
141 F5 Kulanch reg. Pak.
138 D3 Kulandy Kazakh.
138 D3 Kulandy, Poluostrov pen. Kazakh.
139 G2 Kulanotpes watercourse Kazakh.
139 I4 Kulansarak China
141 G4 Kulao r. Pak.
133 O2 Kular Rus. Fed.
153 B5 Kulassein i. Phil.
145 H4 Kulaura Bangl.
159 M4 Kuldīga Latvia
Kul'dzhuktau, Gory hill Uzbek. see Quljuqtov tog'lari
180 D1 Kule Botswana
172 G4 Kulebaki Rus. Fed.
154 C2 Kulen Cambodia
138 E1 Kulevchi Rus. Fed.
172 H2 Kulikovo Rus. Fed.
154 B4 Kulim Malaysia
Kulkuduk Uzbek. see Ko'lquduq
126 F3 Kulkyne watercourse Australia
144 D3 Kullu India
165 J4 Kulmbach Germany
139 G5 Kŭlob Tajik.
137 H2 Kulp Turkey
144 D4 Kulpahar India
203 F4 Kulpsville U.S.A.
138 C3 Kul'sary Kazakh.
165 H5 Külsheim Germany
136 D3 Kulu Turkey
136 C3 Kulübe Tepe mt. Turkey
139 I1 Kulunda Rus. Fed.
139 J1 Kulunda r. Rus. Fed.
139 I1 Kulundinskaya Step' plain Kazakh./Rus. Fed.
139 I1 Kulundinskoye, Ozero salt l. Rus. Fed.
140 D4 Kūlvand Iran
126 E5 Kulwin Australia
139 H1 Kulykol' Kazakh.
173 H6 Kuma r. Rus. Fed.
151 F6 Kumagaya Japan
155 D3 Kumai, Teluk b. Indon.
150 F3 Kumaishi Japan
138 E2 Kumak Rus. Fed.
138 D2 Kumak r. Rus. Fed.
151 B8 Kumamoto Japan
151 E8 Kumano Japan
171 I3 Kumanovo Macedonia
176 B4 Kumasi Ghana
176 C4 Kumba Cameroon
143 B4 Kumbakonam India
136 C2 Kümbet Turkey
180 E1 Kumchuru Botswana
140 D3 Kumel well Iran
138 C1 Kumertau Rus. Fed.
152 D5 Kŭm-gang r. S. Korea
152 E4 Kŭmgang-san mt. N. Korea
152 E6 Kŭmho-gang r. S. Korea
152 E5 Kumi S. Korea
159 K4 Kumla Sweden
165 J2 Kummersdorf-Alexanderdorf Germany
176 D3 Kumo Nigeria
152 D6 Kŭmō-do i. S. Korea
139 I3 Kumola watercourse Kazakh.
154 A1 Kumphawapi Thai.
180 C4 Kums Namibia
143 A3 Kumta India
173 H7 Kumukh Rus. Fed.
141 H3 Kunar r. Afgh.
146 G2 Kunashir, Ostrov i. Rus. Fed.
145 E2 Kunchuk Tso salt l. China
159 N4 Kunda Estonia
145 E4 Kunda India
143 A3 Kundapura India
144 B2 Kundar r. Afgh./Pak.
141 H2 Kunduz Afgh.
141 H2 Kunduz r. Afgh.
139 H2 Kundykol' Pavlodarskaya Oblast' Kazakh.
179 B5 Kunene r. Angola/Namibia alt. Cunene
Künes China see Xinyuan
139 J4 Künes Chang China
139 J4 Künes He r. China
159 J4 Kungälv Sweden
139 I4 Kungei Alatau mts Kazakh./Kyrg.
186 C4 Kunghit Island Canada
Kungrad Uzbek. see Qo'ng'irot
138 E4 Kungradkol' Uzbek.
159 K4 Kungsbacka Sweden
159 J4 Kungshamn Sweden
178 B3 Kungu Dem. Rep. Congo
144 D6 Kuni r. India
151 B8 Kunimi-dake mt. Japan
145 F5 Kunjabar India
145 G4 Kunlui r. India/Nepal
139 I5 Kunlun Shan mts China
145 H2 Kunlun Shankou pass China
149 B5 Kunming China
144 D4 Kuno r. India
152 D6 Kunsan S. Korea
148 F4 Kunshan China
124 C3 Kununurra Australia
144 D4 Kunwari r. India
172 D3 Kun'ya Rus. Fed.
Kunyu Shan hill China see Taibo Ding
165 H5 Künzelsau Germany
165 J3 Künzels-Berg h. Germany
149 F5 Kuocang Shan mts China
159 N3 Kuohijärvi l. Fin.
158 O2 Kuolayarvi Rus. Fed.
158 N3 Kuopio Fin.
158 M3 Kuortane Fin.
170 G2 Kupa r. Croatia/Slovenia
147 E8 Kupang Indon.
159 N5 Kupiškis Lith.
186 C3 Kupreanof Island U.S.A.
173 F5 Kup"yans'k Ukr.
139 J4 Kuqa China
137 L2 Kür r. Azer.
137 J1 Kura r. Azer./Georgia
173 C7 Kura r. Georgia/Rus. Fed.
139 H4 Kuragaty Kazakh.
173 H7 Kurakh Rus. Fed.
138 C2 Kurashasayskiy Kazakh.
151 C7 Kurashiki Japan
145 E5 Kurasia India
151 C7 Kurayoshi Japan
136 E1 Kurban Dağı mt. Turkey
173 E5 Kurchatov Rus. Fed.
137 L1 Kürdämir Azer.
139 H4 Kurday (abandoned) Kazakh.
137 L2 Kür Dili pt Azer.
143 A2 Kurduvadi India
171 K3 Kŭrdzhali Bulg.
151 C7 Kure Japan
136 C1 Küre Turkey
123 G1 Kure Atoll U.S.A.
159 M4 Kuressaare Estonia
132 H4 Kurgan Rus. Fed.
173 G6 Kurganinsk Rus. Fed.
141 F2 Kuri Afgh.
144 B4 Kuri India
Kuria Muria Islands is Oman see Ḥalānīyāt, Juzur al
145 G4 Kuri Chhu r. Bhutan
158 M3 Kurikka Fin.
150 G5 Kurikoma-yama vol. Japan
216 E2 Kuril Basin sea feature Sea of Okhotsk
146 G2 Kuril Islands Rus. Fed.
138 B2 Kurilovka Rus. Fed.
146 G2 Kuril'sk Japan
Kuril'skiye Ostrova is Rus. Fed. see Kuril Islands
216 E3 Kuril Trench sea feature N. Pacific Ocean
138 B1 Kurmanayevka Rus. Fed.
177 F3 Kurmuk Sudan
143 B3 Kurnool India
136 E6 Kurnub tourist site Israel
151 E6 Kurobe Japan
150 G4 Kuroishi Japan
151 G6 Kuroiso Japan
151 A9 Kuro-shima i. Japan
172 I4 Kurovskoye Rus. Fed.
128 C6 Kurow N.Z.
144 B2 Kurram r. Afgh./Pak.
127 I4 Kurri Kurri Australia
139 J2 Kurshim Kazakh.
173 F5 Kursk Rus. Fed.
173 H6 Kurskaya Rus. Fed.
173 F5 Kurskaya Oblast' admin. div. Rus. Fed.
136 D1 Kurşunlu Turkey
137 H3 Kurtalan Turkey
139 I3 Kurtty r. Kazakh.
136 G2 Kuruçay Turkey
144 D3 Kurukshetra India
146 A2 Kuruktag mts China
180 E3 Kuruman S. Africa
180 D3 Kuruman watercourse S. Africa
151 B8 Kurume Japan
146 D1 Kurumkan Rus. Fed.
143 C5 Kurunegala Sri Lanka
177 F2 Kurush, Jebel hills Sudan
139 J2 Kur'ya Rus. Fed.
138 B4 Kuryk Kazakh.
171 L6 Kuşadası Turkey
171 L6 Kuşadası Körfezi b. Turkey
186 B2 Kusawa Lake Canada
164 F5 Kusel Germany
136 A1 Kuş Gölü l. Turkey
173 F6 Kushchevskaya Rus. Fed.
151 B9 Kushikino Japan
151 D8 Kushimoto Japan
150 I3 Kushiro Japan
150 I3 Kushiro-Shitsugen Kokuritsu-kōen nat. park Japan
137 L5 Kūshkak Iran
143 B3 Kushtagi India
145 G5 Kushtia Bangl.
148 C2 Kushui He r. China
184 C3 Kuskokwim r. U.S.A.
184 B4 Kuskokwim Bay U.S.A.
184 C3 Kuskokwim Mountains U.S.A.
138 F1 Kusmuryn Kazakh.
152 C4 Kusŏng N. Korea
150 I3 Kussharo-ko l. Japan
165 F1 Küstenkanal canal Germany
140 C4 Kut Iran
154 B3 Kut, Ko i. Thai.
137 L6 Kūt 'Abdollāh Iran
155 A2 Kutacane Indon.
136 B2 Kütahya Turkey
173 G7 K'ut'aisi Georgia
173 H6 Kutan Rus. Fed.
150 G3 Kutchan Japan
137 L5 Küt-e Gapu tourist site Iran
170 G2 Kutina Croatia
170 G2 Kutjevo Croatia
169 I4 Kutno Poland
178 B4 Kutu Dem. Rep. Congo
145 G5 Kutubdia Island Bangl.
184 G2 Kuujjua r. Canada
189 G2 Kuujjuaq Canada
188 E2 Kuujjuarapik Canada
Kuuli-Mayak Turkm. see Guwlumaýak
158 O2 Kuusamo Fin.
159 N3 Kuusankoski Fin.
138 D2 Kuvandyk Rus. Fed.
179 B5 Kuvango Angola
172 E3 Kuvshinovo Rus. Fed.
137 K7 Kuwait country Asia
137 K7 Kuwait Kuwait
137 K7 Kuwait Jun b. Kuwait
151 E7 Kuwana Japan
172 G1 Kuya Rus. Fed.
132 I4 Kuybyshev Novosibirskaya Oblast' Rus. Fed.
Kuybyshev Samarskaya Oblast' Rus. Fed. see Samara
172 I4 Kuybyshevskoye Vodokhranilishche resr Rus. Fed.
148 D2 Kuye He r. China
139 H3 Kuygan Kazakh.
135 G2 Kuytun China
139 J3 Kuytun He r. China
171 M6 Kuyucak Turkey
138 F2 Kuyukkol', Ozero salt l. Kazakh.
139 K2 Kuyus Rus. Fed.
138 D3 Kuyushe Kazakh.
139 O3 Kuznechnoye Rus. Fed.
172 H4 Kuznetsk Rus. Fed.
150 F1 Kuznetsovo Rus. Fed.
173 C5 Kuznetsovs'k Ukr.
158 M1 Kvænangen sea chan. Norway
158 L1 Kvaløya i. Norway
158 K2 Kvalsund Norway
138 D1 Kvarkeno Rus. Fed.
170 F2 Kvarnerić sea chan. Croatia
184 C4 Kvichak Bay U.S.A.
186 D3 Kwadacha Wilderness Provincial Park Canada
149 □ Kwai Tau Leng h. Hong Kong China
216 G6 Kwajalein atoll Marshall Is
154 A5 Kwala Indon.
181 I4 KwaMashu S. Africa
181 H2 KwaMhlanga S. Africa
152 D5 Kwangch'ŏn S. Korea
152 D6 Kwangju S. Korea
178 B4 Kwango r. Dem. Rep. Congo
178 D4 Kwangwazi Tanz.
152 D6 Kwangyang S. Korea
152 E3 Kwanmo-bong mt. N. Korea
181 F6 Kwanobuhle S. Africa
181 F6 KwaNojoli S. Africa
181 G6 Kwanonqubela S. Africa
180 F5 KwaNonzame S. Africa
181 G6 Kwatinidubu S. Africa
181 H3 KwaZamokuhle S. Africa
180 F6 KwaZamukucinga S. Africa
181 F6 Kwazamuxolo S. Africa
181 H3 KwaZanele S. Africa
181 I4 KwaZulu-Natal prov. S. Africa
179 C5 Kwekwe Zimbabwe
180 F1 Kweneng admin. dist. Botswana
178 B4 Kwenge r. Dem. Rep. Congo
181 G5 Kwezi-Naledi S. Africa
169 I4 Kwidzyn Poland
184 B4 Kwigillingok U.S.A.
124 E2 Kwikila P.N.G.
178 B4 Kwilu r. Angola/Dem. Rep. Congo
147 F7 Kwoka mt. Indon.
149 □ Kwun Tong Hong Kong China
177 D4 Kyabé Chad
126 F6 Kyabram Australia
154 A1 Kya-in Seikkyi Myanmar
146 C1 Kyakhta Rus. Fed.
126 E5 Kyalite Australia
126 A4 Kyancutta Australia
172 F1 Kyanda Rus. Fed.
154 A1 Kyaukhnyat Myanmar
145 H6 Kyaukpyu Myanmar
145 H5 Kyauktaw Myanmar
159 M5 Kybartai Lith.
126 D6 Kybybolite Australia
149 C6 Ky Cung, Sông r. Vietnam
144 D2 Kyelang India
148 A2 Kyikug China
Kyiv Ukr. see Kiev
173 D5 Kyivs'ke Vodoskhovyshche resr Ukr.
Kyklades is Greece see Cyclades
187 H4 Kyle Canada
162 C3 Kyle of Lochalsh U.K.
164 E5 Kyll r. Germany
171 J6 Kyllini mt. Greece
126 F6 Kyneton Australia
178 D3 Kyoga, Lake Uganda
151 D7 Kyōga-misaki pt Japan
127 J2 Kyogle Australia
154 A1 Kyondo Myanmar
152 E6 Kyŏngju S. Korea
151 D7 Kyōto Japan
171 I6 Kyparissia Greece
171 I6 Kyparissiakos Kolpos b. Greece
171 K5 Kyra Panagia i. Greece
136 D4 Kyrenia Cyprus
139 H4 Kyrgyzstan country Asia
165 K2 Kyritz Germany
138 B2 Kyrkopa Kazakh.
158 J3 Kyrksæterøra Norway
132 G3 Kyrta Rus. Fed.
172 H1 Kyssa Rus. Fed.
133 O3 Kytalyktakh Rus. Fed.
171 J6 Kythira i. Greece
171 K6 Kythnos i. Greece
154 A1 Kyungyaung Myanmar
151 B8 Kyūshū i. Japan
216 E4 Kyushu-Palau Ridge sea feature N. Pacific Ocean
171 J3 Kyustendil Bulg.
127 G5 Kywong Australia
158 N3 Kyyjärvi Fin.
138 C3 Kyzan Kazakh.
146 B1 Kyzyl Rus. Fed.
139 J3 Kyzyl-Adyr Kyrg.
139 H5 Kyzylart Pass Kyrg.
139 H3 Kyzylbelen, Gora h. Kazakh.
139 G2 Kyzyldikan Kazakh.
139 J3 Kyzylkesek Kazakh.
138 C2 Kyzylkoga Kazakh.
138 E2 Kyzylkol', Ozero l. Kazakh.
138 E4 Kyzylkum Desert Uzbek.
139 H4 Kyzyl-Kyya Kyrg.
138 F3 Kyzylorda Kazakh.
138 E3 Kyzylordinskaya Oblast' admin. div. Kazakh.
138 C4 Kyzylsay Kazakh.
139 I4 Kyzyl-Suu Kyrg.
139 H5 Kyzyl-Suu r. Kyrg.
138 D3 Kyzyltas Kazakh.
139 H3 Kyzyltau Kazakh.
138 F2 Kyzyluy Kazakh.
138 B3 Kyzylysor Kazakh.
138 C2 Kyzylzhar Aktyubinskaya Oblast' Kazakh.
139 G2 Kyzylzhar Karagandinskaya Oblast' Kazakh.

L

164 F4 Laacher See l. Germany
159 N4 Laagri Estonia
206 I6 La Amistad, Parque Internacional nat. park Costa Rica/Panama
207 F5 La Angostura, Presa de resr Mex.
158 N1 Laanila Fin.
206 D3 La Ardilla, Cerro mt. Mex.
178 E3 Laascaanood Somalia
178 E2 Laasgoray Somalia
213 E2 La Asunción Venez.
176 A2 Laâyoune W. Sahara
173 G6 Laba r. Rus. Fed.
206 D1 La Babia Mex.
212 D3 La Banda Arg.
194 E3 La Barge U.S.A.
125 H3 Labasa Fiji
166 C3 La Baule-Escoublac France
176 A3 Labé Guinea
188 F4 Labelle Canada
190 B5 La Belle U.S.A.
186 B2 Laberge, Lake Canada
153 A5 Labian, Tanjung pt Sabah Malaysia
186 E2 La Biche r. Canada
187 G4 La Biche, Lac l. Canada
173 G6 Labinsk Rus. Fed.
154 B5 Labis Malaysia
153 B3 Labo Phil.
206 C2 La Boquilla Mex.
136 F4 Laboué Lebanon
166 D4 Labouheyre France
215 D2 Laboulaye Arg.
189 H3 Labrador reg. Canada
189 G3 Labrador City Canada
185 M3 Labrador Sea Canada/Greenland
210 F5 Lábrea Brazil
155 E1 Labuan Malaysia
155 C4 Labuhan Indon.
155 B2 Labuhanbilik Indon.
154 A5 Labuhanruku Indon.
153 A5 Labuk r. Sabah Malaysia
155 E1 Labuk, Teluk b. Sabah Malaysia
147 E7 Labuna Indon.
126 A3 Labyrinth, Lake salt flat Australia
132 H3 Labytnangi Rus. Fed.
171 H4 Laç Albania
215 D1 La Calera Arg.
215 B2 La Calera Chile
166 F2 La Capelle France
215 B4 Lácar, Lago l. Arg.
215 D2 La Carlota Arg.
167 D3 La Carolina Spain
171 L2 Lăcăuţi, Vârful mt. Romania
203 I1 Lac-Baker Canada
134 F5 Laccadive Islands India
187 J4 Lac du Bonnet Canada
206 H5 La Ceiba Hond.
213 C2 La Ceiba Venez.
126 C6 Lacepede Bay Australia
203 E4 Laceyville U.S.A.
203 H1 Lac-Frontière Canada
172 F2 Lacha, Ozero l. Rus. Fed.
168 C7 La Chaux-de-Fonds tourist site Switz.
165 I2 Lachendorf Germany
191 F3 Lachine U.S.A.
126 F5 Lachlan r. Australia
206 J6 La Chorrera Panama
188 F4 Lachute Canada
137 K2 Laçın Azer.
166 G5 La Ciotat France
206 C3 La Ciudad Mex.
202 D3 Lackawanna U.S.A.
187 G4 Lac La Biche Canada
186 E4 Lac La Hache Canada
Lac La Martre Canada see Whatì
187 H3 Lac La Ronge Provincial Park Canada
189 F4 Lac-Mégantic Canada
203 G2 Lacolle Canada
206 B1 La Colorada Mex.
186 G4 Lacombe Canada
206 I6 La Concepción Panama
206 C1 La Concordia Mex.
170 C5 Laconi Sardinia Italy
203 H3 Laconia U.S.A.
191 I1 La Corne Canada
190 B4 La Crescent U.S.A.
190 B4 La Crosse U.S.A.
213 A4 La Cruz Col.
206 C3 La Cruz Sinaloa Mex.
207 E2 La Cruz Tamaulipas Mex.
206 H5 La Cruz Nicaragua
206 D1 La Cuesta Mex.
198 E4 La Cygne U.S.A.
144 D2 Ladakh Range mts India
154 A4 Ladang, Ko i. Thai.
136 E1 Ladik Turkey
180 D6 Ladismith S. Africa
141 F4 Lādīz Iran
144 C4 Ladnun India
172 D2 Ladoga, Lake Rus. Fed.
213 B3 La Dorada Col.
Ladozhskoye Ozero l. Rus. Fed. see Ladoga, Lake
145 H4 Ladu mt. India
172 E2 Ladva Rus. Fed.
172 E2 Ladva-Vetka Rus. Fed.
185 J2 Lady Ann Strait Canada
162 E4 Ladybank U.K.
181 G4 Ladybrand S. Africa
191 G2 Lady Evelyn Lake Canada
181 G5 Lady Frere S. Africa
181 G5 Lady Grey S. Africa
186 E5 Ladysmith Canada
181 H4 Ladysmith S. Africa
190 B3 Ladysmith U.S.A.
124 E2 Lae P.N.G.
154 B2 Laem Ngop Thai.
159 I3 Lærdalsøyri Norway
210 F8 La Esmeralda Bol.
213 D4 La Esmeralda Venez.
159 J4 Læsø i. Denmark
206 G5 La Esperanza Hond.
215 D1 La Falda Arg.
194 F4 Lafayette CO U.S.A.
190 D5 Lafayette IN U.S.A.
199 E6 Lafayette LA U.S.A.
201 C5 La Fayette U.S.A.
164 B5 La Fère France
164 B5 La-Ferté-Milon France
164 B6 La Ferté-sous-Jouarre France
140 C5 Laffān, Ra's pt Qatar
176 C4 Lafia Nigeria
166 D3 La Flèche France
202 A6 La Follette U.S.A.
191 H2 Laforce Canada
191 G2 Laforest Canada
189 F3 Laforge Canada
213 B2 La Fría Venez.
140 D5 Lāft Iran
170 C6 La Galite i. Tunisia
173 H6 Lagan' Rus. Fed.
132 C4 Lagan Sweden
163 E3 Lagan r. U.K.
211 K6 Lagarto Brazil
165 G3 Lage Germany
159 J4 Lågen r. Norway
162 C5 Lagg U.K.
162 D3 Laggan U.K.
162 D4 Laggan, Loch l. U.K.
176 C1 Laghouat Alg.
145 F2 Lagkor Co salt l. China
213 B2 La Gloria Col.
Lago Agrio Ecuador see Nueva Loja
214 D2 Lagoa Santa Brazil
137 K1 Lagodekhi Georgia
176 A2 La Gomera i. Canary Is
154 D5 Lagong i. Indon.
153 B3 Lagonoy Gulf Phil.
212 B7 Lago Posadas Arg.
215 B4 Lago Ranco Chile
176 C4 Lagos Nigeria
167 B4 Lagos Port.
206 D3 Lagos de Moreno Mex.
188 E3 La Grande r. Canada
194 C2 La Grande U.S.A.
188 E3 La Grande 3, Réservoir resr Canada
188 F3 La Grande 4, Réservoir resr Canada
124 C3 La Grange Australia
Lagrange Australia see La Grange
201 C5 La Grange GA U.S.A.
203 I2 La Grange ME U.S.A.
190 D5 La Grange MI U.S.A.
190 B5 La Grange MO U.S.A.
199 D6 La Grange TX U.S.A.
190 E5 Lagrange U.S.A.
213 E3 La Gran Sabana plat. Venez.
212 G3 Laguna Brazil
196 D5 Laguna Beach U.S.A.
215 B3 Laguna de Laja, Parque Nacional nat. park Chile
206 I5 Laguna de Perlas Nicaragua
196 D5 Laguna Mountains U.S.A.
210 C5 Lagunas Peru
212 A7 Laguna San Rafael, Parque Nacional nat. park Chile
207 E5 Lagunas de Chacahua, Parque Nacional nat. park Mex.
213 C2 Lagunillas Venez.
La Habana Cuba see Havana
155 E1 Lahad Datu Sabah Malaysia
153 A5 Lahad Datu, Teluk b. Sabah Malaysia
196 □2 Lahaina U.S.A.
155 B3 Lahat Indon.
154 A5 Lahewa Indon.
142 B7 Laḥij Yemen
140 C2 Lāhījān Iran
196 □1 Lahilahi Point U.S.A.
165 F4 Lahn r. Germany
164 F4 Lahnstein Germany
159 K4 Laholm Sweden
196 C2 Lahontan Reservoir U.S.A.
144 C3 Lahore Pak.
213 E3 La Horqueta Venez.
144 B3 Lahri Pak.
159 N3 Lahti Fin.
206 C4 La Huerta Mex.
177 D4 Laï Chad
148 F3 Lai'an China
149 C6 Laibin China
127 J1 Laidley Australia
196 □1 Lā'ie U.S.A.
196 □1 Lā'ie Point U.S.A.
149 C4 Laifeng China
166 E2 L'Aigle France
206 G5 La Iguala Hond.
158 M3 Laihia Fin.
145 H4 Laimakuri India
180 D6 Laingsburg S. Africa
158 M2 Lainioälven r. Sweden
162 D2 Lairg U.K.
153 C5 Lais Phil.
159 M3 Laitila Fin.
170 D1 Laives Italy
148 E2 Laiwu China
148 F2 Laiyang China
148 E2 Laiyuan China
148 F2 Laizhou China
148 F2 Laizhou Wan b. China
215 B3 Laja r. Chile
215 B3 Laja, Laguna de l. Chile
124 D3 Lajamanu Australia
211 K5 Lajes Brazil
212 F3 Lajes Brazil
206 C2 La Joya Mex.
195 G4 La Junta U.S.A.
194 E2 Lake U.S.A.
198 D3 Lake Andes U.S.A.
136 D6 Lake Bardawil Reserve nature res. Egypt
126 E6 Lake Bolac Australia
127 G4 Lake Cargelligo Australia
127 J3 Lake Cathie Australia
199 E6 Lake Charles U.S.A.
194 B1 Lake Chelan National Recreation Area park U.S.A.
201 D6 Lake City FL U.S.A.
190 E3 Lake City MI U.S.A.
190 A3 Lake City MN U.S.A.
201 E5 Lake City SC U.S.A.
160 D3 Lake District National Park U.K.
126 B2 Lake Eyre National Park Australia
191 H3 Lakefield Canada
190 C4 Lake Geneva U.S.A.
Lake Harbour Canada see Kimmirut
197 E4 Lake Havasu City U.S.A.
196 C4 Lake Isabella U.S.A.
199 E6 Lake Jackson U.S.A.
201 D6 Lakeland U.S.A.
190 C2 Lake Linden U.S.A.
186 F4 Lake Louise Canada
197 E4 Lake Mead National Recreation Area park U.S.A.
199 C5 Lake Meredith National Recreation Area park U.S.A.

Lumijoki

158 N2 Lumijoki Fin.
154 C2 Lumphăt Cambodia
128 B6 Lumsden N.Z.
155 C3 Lumut, Tanjung pt Indon.
153 B2 Luna Phil.
197 H5 Luna U.S.A.
162 F4 Lunan Bay U.K.
187 K2 Lunan Lake Canada
191 F5 Luna Pier U.S.A.
144 C5 Lunavada India
144 B4 Lund Pak.
159 K5 Lund Sweden
197 F2 Lund U.S.A.
187 J4 Lundar Canada
179 D5 Lundazi Zambia
161 C6 Lundy i. U.K.
Lundy Island U.K. see Lundy
160 E3 Lune r. U.K.
165 I1 Lüneburg Germany
165 I1 Lüneburger Heide reg. Germany
164 F3 Lünen Germany
166 H2 Lunéville France
179 C5 Lunga r. Zambia
145 E2 Lungdo China
145 E3 Lunggar China
176 A4 Lungi Sierra Leone
149 □ Lung Kwu Chau i. Hong Kong China
145 H5 Lunglei India
179 C5 Lungwebungu r. Zambia
144 C4 Luni India
144 C4 Luni r. India
144 B3 Luni r. Pak.
196 C2 Luning U.S.A.
172 H4 Lunino Rus. Fed.
173 C4 Luninyets Belarus
144 C1 Lunkho mt. Afgh./Pak.
164 F2 Lünne Germany
176 A4 Lunsar Sierra Leone
181 H2 Lunsklip S. Africa
135 G2 Luntai China
149 C5 Luocheng China
148 C3 Luochuan China
149 C5 Luodian China
149 D6 Luoding China
149 D6 Luodou Sha i. China
148 E3 Luohe China
148 D3 Luo He r. Henan China
148 C3 Luo He r. Shaanxi China
148 D3 Luoning China
149 B5 Luoping China
148 E3 Luoshan China
149 E4 Luotian China
Luoto Fin. see Larsmo
148 D3 Luoyang China
149 F5 Luoyuan China
152 F2 Luozigou China
179 C5 Lupane Zimbabwe
149 B5 Lupanshui China
155 D2 Lupar r. Sarawak Malaysia
171 J2 Lupeni Romania
179 D5 Lupilichi Moz.
153 C5 Lupon Phil.
197 H4 Lupton U.S.A.
148 B3 Luqu China
148 E2 Luquan Hebei China
149 B5 Luquan Yunnan China
137 J3 Lūrā Shīrīn Iran
179 B4 Luremo Angola
162 C2 Lurgainn, Loch l. U.K.
163 E3 Lurgan U.K.
179 E5 Lúrio Moz.
179 D5 Lurio r. Moz.
154 □ Lurudal Sing.
179 C5 Lusaka Zambia
178 C4 Lusambo Dem. Rep. Congo
124 F2 Lusancay Islands and Reefs P.N.G.
186 F4 Luscar Canada
187 H4 Luseland Canada
148 D3 Lushi China
171 H4 Lushnjë Albania
152 D2 Lushuihe China
152 A4 Lüshunkou China
181 H5 Lusikisiki S. Africa
194 F3 Lusk U.S.A.
141 E4 Lūt, Kavīr-e des. Iran
149 F6 Lü Tao i. Taiwan
191 G4 Luther Lake Canada
161 G6 Luton U.K.
155 D2 Lutong Sarawak Malaysia
187 G2 Łutselk'e Canada
173 C5 Luts'k Ukr.
164 D2 Luttelgeest Neth.
164 E2 Luttenberg Neth.
165 H5 Lützelbach Germany
129 E4 Lützow-Holm Bay b. Antarctica
180 D4 Lutzputs S. Africa
180 C5 Lutzville S. Africa
153 B5 Luuk Phil.
159 N3 Luumäki Fin.
178 E3 Luuq Somalia
198 D3 Luverne U.S.A.
178 C4 Luvua r. Dem. Rep. Congo
181 I1 Luvuvhu r. S. Africa
178 D3 Luwero Uganda
147 E7 Luwuk Indon.
164 E5 Luxembourg country Europe
164 E5 Luxembourg Lux.
166 H3 Luxeuil-les-Bains France
149 B5 Luxi Yunnan China
Luxi China see Wuxi
181 F5 Luxolweni S. Africa
177 F2 Luxor Egypt
148 E3 Luyi China
164 D3 Luyksgestel Neth.
172 H2 Luza Rus. Fed.
172 I2 Luza r. Rus. Fed.
Luzern Switz. see Lucerne
149 C5 Luzhai China
149 B5 Luzhi China
149 B4 Luzhou China
214 C2 Luziânia Brazil
211 J4 Luzilândia Brazil
153 B3 Luzon i. Phil.
153 B1 Luzon Strait Phil.
166 F3 Luzy France
173 C5 L'viv Ukr.
Lvov Ukr. see L'viv
172 D3 Lyady Rus. Fed.
172 C4 Lyakhavichy Belarus
186 G5 Lyall, Mount Canada
Lyangar Uzbek. see Langar
158 L2 Lycksele Sweden
161 H7 Lydd U.K.
129 C3 Lyddan Island i. Antarctica
Lydenburg S. Africa see Mashishing
161 E6 Lydney U.K.
173 D5 Lyel'chytsy Belarus
196 C3 Lyell, Mount U.S.A.
186 C4 Lyell Island Canada
172 D4 Lyepyel' Belarus
202 E4 Lykens U.S.A.
194 E3 Lyman U.S.A.
161 E7 Lyme Bay U.K.
161 E7 Lyme Regis U.K.
161 F7 Lymington U.K.
202 D6 Lynchburg U.S.A.
203 H2 Lynchville U.S.A.
127 H4 Lyndhurst N.S.W. Australia
126 C3 Lyndhurst S.A. Australia
203 G2 Lyndonville U.S.A.
162 E2 Lyness U.K.
159 I4 Lyngdal Norway
203 H3 Lynn U.S.A.
186 B3 Lynn Canal sea chan. U.S.A.
197 F2 Lynndyl U.S.A.
187 I3 Lynn Lake Canada
161 D6 Lynton U.K.
187 H2 Lynx Lake Canada
166 G4 Lyon France
203 G2 Lyon Mountain U.S.A.
201 D5 Lyons GA U.S.A.
202 E3 Lyons NY U.S.A.
203 F3 Lyons Falls U.S.A.
172 D4 Lyozna Belarus
125 F2 Lyra Reef P.N.G.
159 J4 Lysekil Sweden
172 H3 Lyskovo Rus. Fed.
154 D2 Ly Sơn, Đao i. Vietnam
132 G4 Lys'va Rus. Fed.
173 F5 Lysychans'k Ukr.
160 D4 Lytham St Anne's U.K.
186 E4 Lytton Canada
172 D4 Lyuban' Belarus
173 C5 Lyubeshiv Ukr.
172 E4 Lyudinovo Rus. Fed.
172 H3 Lyunda r. Rus. Fed.

M

149 B6 Ma, Sông r. Laos/Vietnam
179 C5 Maamba Zambia
Ma'an Jordan see Ma'ān
136 E6 Ma'ān Jordan
158 N3 Maaninka Fin.
158 O2 Maaninkavaara Fin.
148 F4 Ma'anshan China
159 N4 Maardu Estonia
136 F4 Ma'arrat an Nu'mān Syria
164 D2 Maarssen Neth.
164 E3 Maas r. Neth. alt. Meuse (Belgium/France)
164 D3 Maaseik Belgium
153 C4 Maasin Phil.
164 D4 Maasmechelen Belgium
164 D3 Maas-Schwalm-Nette, Naturpark nat. park Neth.
181 H1 Maasstroom S. Africa
164 D4 Maastricht Neth.
127 G9 Maatsuyker Group is Australia
153 B3 Mabalacat Phil.
179 D6 Mabalane Moz.
210 G2 Mabaruma Guyana
191 I3 Maberly Canada
149 B4 Mabian China
161 H4 Mablethorpe U.K.
181 H2 Mabopane S. Africa
179 D6 Mabote Moz.
180 E2 Mabuasehube Game Reserve nature res. Botswana
153 B1 Mabudis i. Phil.
180 F2 Mabule Botswana
180 E2 Mabutsane Botswana
212 B7 Macá, Monte mt. Chile
215 D3 Macachín Arg.
214 E3 Macaé Brazil
153 C4 Macajalar Bay Phil.
127 G6 Macalister r. Australia
179 D5 Macaloge Moz.
184 H3 MacAlpine Lake Canada
181 J1 Macandze Moz.
149 D6 Macao China
211 H3 Macapá Brazil
210 C4 Macará Ecuador
214 E1 Macarani Brazil
213 B4 Macarena, Cordillera mts Col.
213 E2 Macareo, Caño r. Venez.
126 E7 Macarthur Australia
210 C4 Macas Ecuador
Macassar Strait str. Indon. see Makassar, Selat
211 K5 Macau Brazil
211 H6 Macaúba Brazil
214 D1 Macaúbas Brazil
213 B4 Macaya r. Col.
213 B4 Macayari Col.
181 J2 Maccaretane Moz.
161 E4 Macclesfield U.K.
124 C4 Macdonald, Lake salt flat Australia
124 D4 Macdonnell Ranges mts Australia
188 B3 MacDowell Lake Canada
162 F3 Macduff U.K.
167 C2 Macedo de Cavaleiros Port.
126 F6 Macedon mt. Australia
171 I4 Macedonia country Europe
171 J4 Macedonia reg. Greece/Macedonia
211 K5 Maceió Brazil
176 B4 Macenta Guinea
170 E3 Macerata Italy
126 B4 Macfarlane, Lake salt flat Australia
163 B6 Macgillycuddy's Reeks mts Ireland
144 A3 Mach Pak.
210 C4 Machachi Ecuador
214 D3 Machado Brazil
179 D6 Machaila Moz.
178 D4 Machakos Kenya
210 C4 Machala Ecuador
179 D6 Machanga Moz.
181 J2 Machatuine Moz.
164 C5 Machault France
148 E4 Macheng China
143 B2 Macherla India
143 C2 Machhakund Reservoir India
203 J2 Machias ME U.S.A.
202 D3 Machias NY U.S.A.
203 I1 Machias r. U.S.A.
143 C2 Machilipatnam India
213 B2 Machiques Venez.
162 C5 Machrihanish U.K.
210 D6 Machu Picchu tourist site Peru
161 D5 Machynlleth U.K.
181 J2 Macia Moz.
171 M2 Măcin Romania
176 B3 Macina Mali
127 I2 Macintyre r. Australia
127 I2 Macintyre Brook r. Australia
197 H2 Mack U.S.A.
137 G1 Maçka Turkey
124 E4 Mackay Australia
194 D2 Mackay U.S.A.
124 C4 Mackay, Lake salt flat Australia
187 G2 MacKay Lake Canada
186 E3 Mackenzie B.C. Canada
190 C1 Mackenzie Ont. Canada
184 E3 Mackenzie r. Canada
129 E5 Mackenzie Bay b. Antarctica
184 E3 Mackenzie Bay Canada
186 F2 Mackenzie Bison Sanctuary nature res. Canada
184 G2 Mackenzie King Island Canada
186 C2 Mackenzie Mountains Canada
190 E3 Mackinac, Straits of lake channel U.S.A.
190 E3 Mackinac Island U.S.A.
190 C5 Mackinaw r. U.S.A.
190 E3 Mackinaw City U.S.A.
187 H4 Macklin Canada
127 J3 Macksville Australia
127 J2 Maclean Australia
181 H5 Maclear S. Africa
127 J3 Macleay r. Australia
124 B4 MacLeod, Lake dry lake Australia
186 C2 Macmillan r. Canada
190 B5 Macomb U.S.A.
170 C4 Macomer Sardinia Italy
166 G3 Mâcon France
201 D5 Macon GA U.S.A.
198 E4 Macon MO U.S.A.
179 C5 Macondo Angola
127 G3 Macquarie r. N.S.W. Australia
127 G8 Macquarie r. Tas. Australia
127 I4 Macquarie, Lake b. Australia
127 F9 Macquarie Harbour Australia
135 F7 Macquarie Island S. Pacific Ocean
127 G3 Macquarie Marshes Australia
127 H4 Macquarie Mountain Australia
216 F9 Macquarie Ridge sea feature S. Pacific Ocean
154 □ MacRitchie Reservoir Sing.
129 E5 Mac. Robertson Land reg. Antarctica
163 C6 Macroom Ireland
213 C1 Macuira, Parque Nacional nat. park Col.
213 B4 Macuje Col.
124 D4 Macumba watercourse Australia
210 D6 Macusani Peru
207 F4 Macuspana Mex.
206 B2 Macuzari, Presa resr Mex.
203 I2 Macwahoc U.S.A.
136 E6 Mādabā Jordan
181 I3 Madadeni S. Africa
177 D3 Madagali Nigeria
179 E6 Madagascar country Africa
218 H6 Madagascar i. Africa
218 H6 Madagascar Basin sea feature Indian Ocean
218 G7 Madagascar Ridge sea feature Indian Ocean
143 B3 Madakasira India
177 D2 Madama Niger
171 K4 Madan Bulg.
143 B3 Madanapalle India
124 E2 Madang P.N.G.
176 C3 Madaoua Niger
145 G5 Madaripur Bangl.
191 I3 Madawaska Canada
191 I3 Madawaska r. Canada
203 I1 Madawaska U.S.A.
176 A1 Madeira terr. Atlantic Ocean
210 F5 Madeira r. Brazil
219 H3 Madeira, Arquipélago da is Port.
189 H4 Madeleine, Îles de la is Canada
190 B2 Madeline Island U.S.A.
137 G2 Maden Turkey
139 I3 Madeniyet Kazakh.
204 C3 Madera Mex.
196 B3 Madera U.S.A.
143 A3 Madgaon India
145 F4 Madhepura India
143 C2 Madhira India
145 F4 Madhubani India
144 D5 Madhya Pradesh state India
181 F3 Madibogo S. Africa
210 E6 Madidi r. Bol.
126 B2 Madigan Gulf salt flat Australia
143 A3 Madikeri India
181 G2 Madikwe Game Reserve nature res. S. Africa
178 B4 Madingou Congo
Madini r. Bol. see Madidi
179 E5 Madirovalo Madag.
200 C4 Madison IN U.S.A.
203 I2 Madison ME U.S.A.
198 D2 Madison MN U.S.A.
198 D3 Madison NE U.S.A.
198 D2 Madison SD U.S.A.
190 C4 Madison WI U.S.A.
202 C5 Madison WV U.S.A.
192 D2 Madison r. U.S.A.
200 C4 Madisonville KY U.S.A.
199 E6 Madisonville TX U.S.A.
155 D4 Madiun Indon.
191 I3 Madoc Canada
178 D3 Mado Gashi Kenya
146 B3 Madoi China
159 N4 Madona Latvia
144 B4 Madpura India
171 L5 Madra Dağı mts Turkey
Madras India see Chennai
194 B2 Madras U.S.A.
207 E2 Madre, Laguna lag. Mex.
199 D7 Madre, Laguna lag. U.S.A.
153 B2 Madre, Sierra mt. Phil.
207 F4 Madre de Chiapas, Sierra mts Mex.
210 D6 Madre de Dios r. Peru
212 A8 Madre de Dios, Isla i. Chile
206 D4 Madre del Sur, Sierra mts Mex.
206 B1 Madre Occidental, Sierra mts Mex.
206 D2 Madre Oriental, Sierra mts Mex.
153 C4 Madrid Phil.
167 E2 Madrid Spain (City Plan 112)
153 B4 Madridejos Phil.
167 E3 Madridejos Spain
143 C2 Madugula India
155 D4 Madura i. Indon.
155 D4 Madura, Selat sea chan. Indon.
143 B4 Madurai India
145 E4 Madwas India
144 C2 Madyan Pak.
173 H7 Madzhalis Rus. Fed.
151 F6 Maebashi Japan
154 A1 Mae Hong Son Thai.
154 A1 Mae Sai Thai.
154 A1 Mae Sariang Thai.
154 A1 Mae Sot Thai.
205 I5 Maestra, Sierra mts Cuba
179 E5 Maevatanana Madag.
125 G3 Maéwo i. Vanuatu
187 I4 Mafeking Canada
181 G4 Mafeteng Lesotho
127 G6 Maffra Australia
179 D4 Mafia Island Tanz.
181 F2 Mafikeng S. Africa
179 D4 Mafinga Tanz.
214 C4 Mafra Brazil
181 I5 Magabeni S. Africa
133 Q4 Magadan Rus. Fed.
178 D4 Magadi Kenya
181 J1 Magaiza Moz.
153 B3 Magallanes Phil.
212 B8 Magallanes, Estrecho de Chile
213 B2 Magangué Col.
136 D3 Mağara Turkey
173 H7 Magas Rus. Fed.
153 B2 Magat r. Phil.
215 F2 Magdalena Arg.
210 F6 Magdalena Bol.
213 B3 Magdalena r. Col.
204 B2 Magdalena Mex.
195 E6 Magdalena r. Mex.
195 F5 Magdalena U.S.A.
206 A2 Magdalena, Bahía b. Mex.
212 B6 Magdalena, Isla i. Chile
206 A2 Magdalena, Isla i. Mex.
153 A5 Magdaline, Gunung mt. Sabah Malaysia
138 F5 Magdanly Turkm.
165 J2 Magdeburg Germany
163 F3 Magee, Island pen. U.K.
216 E4 Magellan Seamounts sea feature N. Pacific Ocean
158 N1 Magerøya i. Norway
151 B9 Mage-shima i. Japan
170 C2 Maggiorasca, Monte mt. Italy
170 C2 Maggiore, Lake Italy
176 A3 Maghama Mauritania
163 E3 Maghera U.K.
163 E3 Magherafelt U.K.
160 E4 Maghull U.K.
194 D3 Magna U.S.A.
170 F6 Magna Grande mt. Sicily Italy
129 E5 Magnet Bay b. Antarctica
124 E3 Magnetic Island Australia
158 P1 Magnetity Rus. Fed.
138 D1 Magnitogorsk Rus. Fed.
199 E5 Magnolia U.S.A.
189 F4 Magog Canada
207 E3 Magozal Mex.
189 H3 Magpie Canada
190 E1 Magpie r. Canada
189 H3 Magpie, Lac l. Canada
186 G5 Magrath Canada
196 D3 Magruder Mountain U.S.A.
176 A3 Magṭa' Laḥjar Mauritania
138 D5 Magtymguly Turkm.
178 D4 Magu Tanz.
149 B6 Maguan China
211 I4 Maguarinho, Cabo c. Brazil
181 J2 Magude Moz.
203 J2 Magundy Canada
187 J2 Maguse Lake Canada
145 H5 Magwe Myanmar
145 H5 Magyichaung Myanmar
140 B2 Mahābād Iran
143 A2 Mahabaleshwar India
145 F4 Mahabharat Range mts Nepal
179 E6 Mahabo Madag.
143 A2 Mahad India
144 D5 Mahadeo Hills India
178 D3 Mahagi Dem. Rep. Congo
179 E5 Mahajamba r. Madag.
144 C3 Mahajan India
179 E5 Mahajanga Madag.
155 D2 Mahakam r. Indon.
179 C6 Mahalapye Botswana
179 E5 Mahalevona Madag.
140 C3 Mahallāt Iran
144 D3 Maham India
140 E4 Māhān Iran
143 D1 Mahanadi r. India
179 E5 Mahanoro Madag.
144 C6 Maharashtra state India
143 C1 Mahasamund India
154 B1 Maha Sarakham Thai.
179 E6 Mahatalaky Madag.
136 F6 Maḥaṭṭat Ḍab'ah Jordan
179 E5 Mahavanona Madag.
179 E5 Mahavavy r. Madag.
143 C5 Mahaweli Ganga r. Sri Lanka
154 C1 Mahaxai Laos
143 C2 Mahbubabad India
143 B2 Mahbubnagar India
140 D5 Maḥḍah Oman
211 G2 Mahdia Guyana
170 D7 Mahdia Tunisia
175 I5 Mahé i. Seychelles
143 D2 Mahendragiri mt. India
144 C5 Mahesana India
144 C5 Maheshwar India
144 C5 Mahi r. India
141 E4 Māhī watercourse Iran
128 F3 Mahia Peninsula N.Z.
172 D4 Mahilyow Belarus
143 C5 Mahiyangana Sri Lanka
181 I4 Mahlabatini S. Africa
165 J2 Mahlsdorf Germany
Maḥmūd-e 'Erāqī Afgh. see Maḥmūd-e Rāqī
141 H3 Maḥmūd-e Rāqī Afgh.
137 L4 Mahniān Iran
198 D2 Mahnomen U.S.A.
144 D4 Mahoba India
167 I3 Mahón Spain
202 D4 Mahoning Creek Lake U.S.A.
145 H5 Mahudaung mts Myanmar
144 B5 Mahuva India
171 L4 Mahya Dağı mt. Turkey
140 C3 Mahyār Iran
145 H4 Maibang India
213 B2 Maicao Col.
188 E4 Maicasagi, Lac l. Canada
149 C6 Maichen China
161 G6 Maidenhead U.K.
187 H4 Maidstone Canada
161 H6 Maidstone U.K.
177 D3 Maiduguri Nigeria
213 D3 Maigualida, Sierra mts Venez.
163 C5 Maigue r. Ireland
144 E4 Maihar India
148 C3 Maiji Shan mt. China
144 E5 Maikala Range hills India
178 C4 Maiko, Parc National de la nat. park Dem. Rep. Congo
144 E3 Mailani India
165 H5 Main r. Germany
163 E3 Main r. U.K.
189 I3 Main Brook Canada
191 F3 Main Channel lake channel Canada
178 B4 Mai-Ndombe, Lac l. Dem. Rep. Congo
165 J5 Main-Donau-Kanal canal Germany
Maindong China see Coqên
191 I4 Main Duck Island Canada
203 I2 Maine state U.S.A.
176 D3 Maïné-Soroa Niger
154 A2 Maingy Island Myanmar
165 H5 Mainhardt Germany
153 C4 Mainit Phil.
153 C4 Mainit, Lake Phil.
162 E1 Mainland i. Orkney, Scotland U.K.
162 □ Mainland i. Shetland, Scotland U.K.
165 J4 Mainleus Germany
145 E5 Mainpat reg. India
144 D4 Mainpuri India
127 J1 Main Range National Park Australia
179 E5 Maintirano Madag.
165 G4 Mainz Germany
176 □ Maio i. Cape Verde
215 C2 Maipó, Volcán vol. Chile
215 F3 Maipú Arg.
215 C2 Maipú Arg.
213 D2 Maiquetía Venez.
145 G5 Maiskhal Island Bangl.
179 C6 Maitengwe Botswana
127 I4 Maitland N.S.W. Australia
126 B5 Maitland S.A. Australia
129 D3 Maitri research stn Antarctica
206 I5 Maíz, Islas del is Nicaragua
145 G3 Maizhokunggar China
151 D7 Maizuru Japan
171 H3 Maja Jezercë mt. Albania
143 B2 Majalgaon India
213 E4 Majari r. Brazil
136 E5 Majdel Aanjar tourist site Lebanon
155 E3 Majene Indon.
140 C6 Majhūd well Saudi Arabia
178 D3 Majī Eth.
148 E2 Majia He r. China
149 D6 Majiang China
167 H3 Major, Puig mt. Spain
167 H3 Majorca i. Spain
145 H4 Majuli Island India
Majuro atoll Marshall Is see Taongi
181 G4 Majwemasweu S. Africa
178 B4 Makabana Congo
Makaha U.S.A. see Mākaha
196 □1 Mākaha U.S.A.
147 D7 Makale Indon.
145 F4 Makalu mt. China
178 C4 Makamba Burundi
139 J3 Makanshy Kazakh.
Makapuu Hd hd U.S.A. see Makapu'u Head
196 □1 Makapu'u Head U.S.A.
220 M1 Makarov Basin sea feature Arctic Ocean
170 G3 Makarska Croatia
172 I2 Makar-Yb Rus. Fed.
155 E4 Makassar Indon.
155 E3 Makassar, Selat str. Indon.
138 C3 Makat Kazakh.
181 J3 Makatini Flats lowland S. Africa
176 A4 Makeni Sierra Leone
179 C6 Makgadikgadi depr. Botswana
173 H7 Makhachkala Rus. Fed.
138 B3 Makhambet Kazakh.
136 G4 Makhfar al Ḥammām Syria
137 I4 Makhmūr Iraq
139 G1 Makhorovka Kazakh.
178 D4 Makindu Kenya
139 G1 Makinsk Kazakh.
173 F5 Makiyivka Ukr.
Makkah Saudi Arabia see Mecca
189 I2 Makkovik Canada
189 I2 Makkovik, Cape Canada
164 D1 Makkum Neth.
171 I1 Makó Hungary
178 B3 Makokou Gabon
179 D4 Makongolosi Tanz.
180 E2 Makopong Botswana
178 B4 Makotipoko Congo
141 F5 Makran reg. Iran/Pak.
144 C4 Makrana India
141 F5 Makran Coast Range mts Pak.
143 C2 Makri India
171 K6 Makronisos i. Greece
172 E3 Maksatikha Rus. Fed.
150 E1 Maksimovka Rus. Fed.
141 F4 Maksotag Iran
140 B2 Mākū Iran
145 H4 Makum India
179 D4 Makumbako Tanz.
179 D5 Makunguwiro Tanz.
151 B9 Makurazaki Japan
176 C4 Makurdi Nigeria
140 D4 Makūyeh Iran
181 F3 Makwassie S. Africa
158 L2 Malå Sweden
206 I7 Mala, Punta pt Panama
153 C5 Malabang Phil.
143 A3 Malabar Coast India
176 C4 Malabo Equat. Guinea
153 A4 Malabuñgan Phil.
155 A2 Malacca, Strait of Indon./Malaysia
194 D3 Malad City U.S.A.
172 C4 Maladzyechna Belarus
167 D4 Málaga Spain
203 F5 Malaga NJ U.S.A.
195 F5 Malaga NM U.S.A.
163 C3 Málainn Mhóir Ireland
125 G2 Malaita i. Solomon Is
177 F4 Malakal Sudan
125 G3 Malakula i. Vanuatu
144 C2 Malakwal Pak.
147 E7 Malamala Indon.
141 G5 Malan, Ras pt Pak.
155 D4 Malang Indon.
179 B4 Malanje Angola
215 C1 Malanzán, Sierra de mts Arg.
143 B4 Malappuram India
159 L4 Mälaren l. Sweden
215 C2 Malargüe Arg.
191 H1 Malartic Canada
191 H1 Malartic, Lac l. Canada
186 A3 Malaspina Glacier U.S.A.
136 G2 Malatya Turkey
144 C3 Malaut India
137 K5 Malāvī Iran
153 A5 Malawali i. Sabah Malaysia
179 D5 Malawi country Africa
Malawi, Lake l. Africa see Nyasa, Lake
172 E3 Malaya Vishera Rus. Fed.
153 C4 Malaybalay Phil.
140 C3 Malāyer Iran
131 F6 Malay Peninsula pen. Asia
139 I3 Malaysary Kazakh.
155 C2 Malaysia country Asia
155 B2 Malaysia, Semenanjung Malaysia
137 I2 Malazgirt Turkey
169 I3 Malbork Poland
164 E5 Malborn Germany
165 K1 Malchin Germany
165 K1 Malchiner See l. Germany
145 G4 Maldah India
164 B3 Maldegem Belgium
199 F4 Malden U.S.A.
123 I4 Malden Island Kiribati
130 C6 Maldives country Indian Ocean
161 H6 Maldon U.K.
215 F2 Maldonado Uruguay
130 C6 Male Maldives
171 J6 Maleas, Akrotirio pt Greece
135 F6 Male Atoll Maldives
181 F4 Malebogo S. Africa
143 B2 Malegaon Maharashtra India
144 C5 Malegaon Maharashtra India
168 H6 Malé Karpaty hills Slovakia
137 K3 Malek Kandī Iran
178 B4 Malele Dem. Rep. Congo
179 D5 Malema Moz.
144 C3 Maler Kotla India
141 G3 Mālestān Afgh.
173 H7 Malgobek Rus. Fed.
158 L2 Malgomaj l. Sweden
140 B5 Malham Saudi Arabia
194 C3 Malheur Lake U.S.A.
176 B3 Mali country Africa
178 C4 Mali Dem. Rep. Congo
176 A3 Mali Guinea
148 C3 Malian He r. China
144 E4 Malihabad India
141 F4 Malik Naro mt. Pak.
154 A2 Mali Kyun i. Myanmar
147 E7 Malili Indon.
155 C4 Malimping Indon.
178 E4 Malindi Kenya
163 D2 Malin Head Ireland
Malin More Ireland see Málainn Mhóir
139 G2 Malinovka Kazakh.
150 D2 Malinovka r. Rus. Fed.
139 I2 Malinovoye Ozero Rus. Fed.
149 B6 Malipo China
170 F2 Mali Raginac mt. Croatia
153 C5 Malita Phil.
154 A3 Maliwun Myanmar
144 B5 Maliya India
137 K5 Malkaili Iran
143 C2 Malkangiri India
144 D5 Malkapur India
173 C7 Malkara Turkey
169 M4 Mal'kavichy Belarus
171 L4 Malko Tŭrnovo Bulg.
127 H6 Mallacoota Australia
127 H6 Mallacoota Inlet b. Australia
162 C4 Mallaig U.K.
126 C5 Mallala Australia
126 E5 Mallee Cliffs National Park Australia
187 J2 Mallery Lake Canada
Mallorca i. Spain see Majorca
163 C5 Mallow Ireland
161 D5 Mallwyd U.K.
158 J2 Malm Norway
158 M2 Malmberget Sweden
164 E4 Malmédy Belgium
180 C6 Malmesbury S. Africa
161 E6 Malmesbury U.K.
159 K5 Malmö Sweden
172 I3 Malmyzh Rus. Fed.
125 G3 Malo i. Vanuatu
153 B3 Malolos Phil.
203 F2 Malone U.S.A.
149 B5 Malong China
179 C5 Malonga Dem. Rep. Congo
172 F2 Maloshuyka Rus. Fed.
159 I3 Måløy Norway
172 F4 Maloyaroslavets Rus. Fed.
210 B3 Malpelo, Isla de i. Col.
143 A3 Malprabha r. India
170 F7 Malta country Europe
159 N4 Malta Latvia
194 F1 Malta U.S.A.
170 F6 Malta Channel Italy/Malta
179 B6 Maltahöhe Namibia
161 F4 Maltby U.K.
161 H4 Maltby le Marsh U.K.
160 G3 Malton U.K.
Maluku is Indon. see Moluccas
147 E7 Maluku, Laut sea Indon.
159 K3 Malung Sweden
181 H4 Maluti Mountains Lesotho
125 G2 Malu'u Solomon Is
143 A2 Malvan India
199 E5 Malvern U.S.A.
173 D5 Malyn Ukr.
133 R3 Malyy Anyuy r. Rus. Fed.
Malyy Balkhan, Khrebet h. Turkm. see Kiçi Balkan Daglary
173 H6 Malyye Derbety Rus. Fed.
138 B1 Malyy Irgiz r. Rus. Fed.
Malyy Kavkaz mts Asia see Lesser Caucasus
133 P2 Malyy Lyakhovskiy, Ostrov i. Rus. Fed.
181 H3 Mamafubedu S. Africa
143 C3 Mamallapuram India
153 A5 Mambahenauhan i. Phil.
153 C4 Mambajao Phil.
178 C3 Mambasa Dem. Rep. Congo
178 B3 Mambéré r. Centr. Afr. Rep.
153 B3 Mamburao Phil.
181 H2 Mamelodi S. Africa
Mamfé Cameroon see Mamfe
176 C4 Mamfe Cameroon
139 F1 Mamlyutka Kazakh.
197 G5 Mammoth U.S.A.
200 C4 Mammoth Cave National Park U.S.A.
196 C3 Mammoth Lakes U.S.A.
210 E6 Mamoré r. Bol./Brazil
176 A3 Mamou Guinea
179 E5 Mampikony Madag.
176 B4 Mampong Ghana
215 B3 Mamuil Malal, Paso pass Arg./Chile
155 E3 Mamuju Indon.
180 D1 Mamuno Botswana
138 D2 Mamyt Kazakh.
176 B4 Man Côte d'Ivoire
160 C3 Man, Isle of terr. Irish Sea
213 B3 Manacacias r. Col.
210 F4 Manacapuru Brazil
167 H3 Manacor Spain
147 E6 Manado Indon.
206 H5 Managua Nicaragua
206 H5 Managua, Lago de l. Nicaragua
179 E6 Manakara Madag.
128 D5 Manakau mt. N.Z.
140 C5 Manama Bahrain
124 E2 Manam Island P.N.G.
213 E2 Manamo, Caño r. Venez.
Manana i. U.S.A. see Mānana
196 □1 Mānana i. U.S.A.
179 E6 Mananara r. Madag.
179 E5 Mananara Avaratra Madag.
179 E5 Mananara Nord, Parc National de nat. park Madag.
126 E5 Manangatang Australia
179 E6 Mananjary Madag.
143 B4 Mananthavady India
144 D3 Mana Pass India
213 D2 Manapire r. Venez.
128 A6 Manapouri, Lake N.Z.
179 E5 Manarantsandry Madag.
145 G4 Manas r. Bhutan
146 A2 Manas Hu l. China
145 F3 Manaslu mt. Nepal
202 E5 Manassas U.S.A.
145 G4 Manas Wildlife Sanctuary nature res. Bhutan
147 E7 Manatuto East Timor
210 F4 Manaus Brazil
136 C3 Manavgat Turkey
128 E4 Manawatu r. N.Z.
153 C5 Manay Phil.
136 F3 Manbij Syria
161 H4 Manby U.K.
190 E3 Mancelona U.S.A.
161 E4 Manchester U.K.
196 A2 Manchester CA U.S.A.
203 G4 Manchester CT U.S.A.
190 B4 Manchester IA U.S.A.
202 B6 Manchester KY U.S.A.
191 E4 Manchester MI U.S.A.
203 H3 Manchester NH U.S.A.
202 B5 Manchester OH U.S.A.
201 C5 Manchester TN U.S.A.
203 G3 Manchester VT U.S.A.
144 A4 Manchhar Lake Pak.
136 F2 Mancılık Turkey
197 H3 Mancos U.S.A.
197 H3 Mancos r. U.S.A.
141 F5 Mand Pak.
140 D4 Mand, Rūd-e r. Iran
177 D4 Manda, Parc National de nat. park Chad
179 E6 Mandabe Madag.
155 B2 Mandah Indon.
154 □ Mandai Sing.
141 F3 Mandal Afgh.
144 C4 Mandal India
159 I4 Mandal Norway
147 G7 Mandala, Puncak mt. Indon.
147 B4 Mandalay Myanmar
146 C2 Mandalgovĭ Mongolia
137 J5 Mandalī Iraq
148 D1 Mandalt China
148 D1 Mandalt Sum China
198 C2 Mandan U.S.A.
153 B3 Mandaon Phil.
177 D3 Mandara Mountains Cameroon/Nigeria
170 C5 Mandas Sardinia Italy
178 E3 Mandera Kenya
197 F2 Manderfield U.S.A.
164 E4 Manderscheid Germany
205 I5 Mandeville Jamaica
128 B6 Mandeville N.Z.
144 B4 Mandha India
176 B3 Mandiana Guinea
154 B4 Mandi Angin, Gunung mt. Malaysia
179 D5 Mandié Moz.
179 D5 Mandimba Moz.
181 I4 Mandini S. Africa
145 F5 Mandira Dam India
144 E5 Mandla India
179 E5 Mandritsara Madag.
144 C4 Mandsaur India
153 A6 Mandul i. Indon.
124 B5 Mandurah Australia
170 G4 Manduria Italy
144 B5 Mandvi Gujarat India
144 C5 Mandvi Gujarat India
143 B3 Mandya India
138 E5 Mäne Turkm.
143 B2 Maner r. India
170 D2 Manerbio Italy
137 K5 Manesht Kūh mt. Iran
169 L5 Manevychi Ukr.
170 F4 Manfredonia Italy
170 G4 Manfredonia, Golfo di g. Italy
214 D1 Manga Brazil
176 B3 Manga Burkina
178 B4 Mangai Dem. Rep. Congo
123 I5 Mangaia i. Cook Is
128 E3 Mangakino N.Z.
143 C2 Mangalagiri India
145 H4 Mangaldai India
171 M3 Mangalia Romania
143 A3 Mangalore India
143 A2 Mangalvedha India
145 G4 Mangan India
143 C2 Mangapet India
153 C6 Mangarang Indon.
181 G4 Mangaung S. Africa
128 E3 Mangaweka N.Z.
145 G4 Mangde Chhu r. Bhutan
163 B6 Mangerton Mountain h. Ireland
155 C3 Manggar Indon.
138 B4 Mangistau Kazakh.
Mangit Uzbek. see Mang'it
138 E4 Mang'it Uzbek.
146 B3 Mangnai China
179 D5 Mangochi Malawi
147 E7 Mangole i. Indon.
161 E6 Mangotsfield U.K.
144 B5 Mangrol India
167 C2 Mangualde Port.
141 G4 Manguchar Pak.
215 G2 Mangueira, Lagoa l. Brazil
214 B4 Mangueirinha Brazil
177 D2 Manguéni, Plateau du Niger
146 E1 Mangui China
153 C5 Mangupung i. Indon.
138 B3 Mangystau, Poluostrov pen. Kazakh.
138 B3 Mangystau, Zaliv b. Kazakh.
138 C3 Mangystauskaya Oblast' admin. div. Kazakh.
198 D4 Manhattan KS U.S.A.
196 D2 Manhattan NV U.S.A.
179 D6 Manhica Moz.
181 J3 Manhoca Moz.
214 D3 Manhuaçu Brazil
214 E2 Manhuaçu r. Brazil
213 B3 Mani Col.
179 E5 Mania r. Madag.
170 E1 Maniago Italy
210 F5 Manicoré Brazil
189 G3 Manicouagan Canada
189 G3 Manicouagan r. Canada
189 G3 Manicouagan, Petit Lac l. Canada
189 G3 Manicouagan, Réservoir resr Canada
140 C5 Manīfah Saudi Arabia
123 I4 Manihiki atoll Cook Is
Manikgarh India see Rajura
144 E4 Manikpur India
153 B3 Manila Phil.
194 E3 Manila U.S.A.
127 H4 Manildra Australia
127 I3 Manilla Australia
145 H4 Manipur state India
171 L5 Manisa Turkey
190 D3 Manistee U.S.A.
190 E3 Manistee r. U.S.A.

Meekatharra

190 D3 Manistique U.S.A.
190 E2 Manistique Lake U.S.A.
188 B2 Manitoba prov. Canada
187 J4 Manitoba, Lake Canada
187 H4 Manito Lake Canada
187 J5 Manitou Canada
191 G3 Manitou, Lake Canada
202 E3 Manitou Beach U.S.A.
188 B3 Manitou Falls Canada
190 D2 Manitou Island U.S.A.
200 C2 Manitou Islands U.S.A.
191 F3 Manitoulin Island Canada
191 G3 Manitowaning Canada
190 E1 Manitowik Lake Canada
190 D3 Manitowoc U.S.A.
191 J2 Maniwaki Canada
213 B3 Manizales Col.
179 E6 Manja Madag.
181 J2 Manjacaze Moz.
143 B4 Manjeri India
152 D3 Man Jiang r. China
137 L3 Manjil Iran
143 B2 Manjra r. India
198 E3 Mankato U.S.A.
181 I3 Mankayane Swaziland
176 B4 Mankono Côte d'Ivoire
143 C4 Mankulam Sri Lanka
148 C1 Manlay Mongolia
127 I4 Manly Australia
144 C5 Manmad India
155 B3 Manna Indon.
126 C4 Mannahill Australia
143 B4 Mannar Sri Lanka
143 B4 Mannar, Gulf of India/Sri Lanka
143 B3 Manneru r. India
165 G5 Mannheim Germany
163 A4 Mannin Bay Ireland
186 F3 Manning Canada
201 D5 Manning U.S.A.
161 I6 Manningtree U.K.
170 C4 Mannu, Capo c. Sardinia Italy
126 C5 Mannum Australia
147 F7 Manokwari Indon.
178 C4 Manono Dem. Rep. Congo
154 A3 Manoron Myanmar
166 G5 Manosque France
185 K4 Manouane, Lac l. Canada
178 C3 Manovo-Gounda Saint Floris, Parc National du nat. park Centr. Afr. Rep.
152 D3 Manp'o N. Korea
125 I2 Manra i. Kiribati
167 G2 Manresa Spain
144 C3 Mansa India
179 C5 Mansa Zambia
176 A3 Mansa Konko Gambia
144 C2 Mansehra Pak.
185 K3 Mansel Island Canada
127 G6 Mansfield Australia
161 F4 Mansfield U.K.
199 E5 Mansfield LA U.S.A.
202 B4 Mansfield OH U.S.A.
202 E4 Mansfield PA U.S.A.
186 E3 Manson Creek Canada
137 L6 Manşūrī Iran
136 E3 Mansurlu Turkey
210 B4 Manta Ecuador
210 B4 Manta, Bahía de b. Ecuador
153 A4 Mantalingajan, Mount Phil.
196 B3 Manteca U.S.A.
213 C3 Mantecal Venez.
165 K5 Mantel Germany
201 F5 Manteo U.S.A.
166 E2 Mantes-la-Jolie France
143 B3 Manthani India
197 G2 Manti U.S.A.
214 D3 Mantiqueira, Serra da mts Brazil
190 E3 Manton U.S.A.
Mantoua Italy see Mantova
170 D2 Mantova Italy
159 N3 Mäntsälä Fin.
159 N3 Mänttä Fin.
172 H3 Manturovo Rus. Fed.
159 N3 Mäntyharju Fin.
158 N2 Mäntyjärvi Fin.
210 D6 Manu, Parque Nacional nat. park Peru
217 I7 Manuae atoll Fr. Polynesia
Manua Islands American Samoa see Manu'a Islands
123 H4 Manu'a Islands American Samoa
197 H4 Manuelito U.S.A.
215 F2 Manuel J. Cobo Arg.
214 E1 Manuel Vitorino Brazil
211 H5 Manuelzinho Brazil
147 E7 Manui i. Indon.
141 E5 Manūjān Iran
153 B4 Manukan Phil.
128 E2 Manukau N.Z.
128 E2 Manukau Harbour N.Z.
153 A5 Manuk Manka i. Phil.
126 C4 Manunda watercourse Australia
124 E2 Manus Island P.N.G.
143 B3 Manvi India
181 F2 Manyana Botswana
173 G6 Manych-Gudilo, Ozero l. Rus. Fed.
197 H3 Many Farms U.S.A.
178 D4 Manyoni Tanz.
167 E3 Manzanares Spain
205 I4 Manzanillo Cuba
206 C4 Manzanillo Mex.
206 J6 Manzanillo, Punta pt Panama
140 C3 Manzariyeh Iran
146 D2 Manzhouli China
136 D6 Manzilah, Buḩayrat al lag. Egypt
181 I3 Manzini Swaziland
177 D3 Mao Chad
Maó Spain see Mahón
148 D4 Maocifan China
148 C2 Maojiachuan China
147 F7 Maoke, Pegunungan mts Indon.
181 G3 Maokeng S. Africa
152 B3 Maokui Shan mt. China
152 B2 Maolin China
148 B2 Maomao Shan mt. China
149 D6 Maoming China
149 □ Ma On Shan h. Hong Kong China
179 D6 Mapai Moz.
144 E3 Mapam Yumco l. China
124 C2 Mapane Indon.
178 B3 Mapé, Retenue de la resr Cameroon
181 F5 Maphodi S. Africa
206 D2 Mapimí Mex.
179 D6 Mapinhane Moz.
213 D3 Mapire Venez.
190 E4 Maple r. U.S.A.
187 H5 Maple Creek Canada
216 G4 Mapmaker Seamounts sea feature N. Pacific Ocean
181 G4 Mapoteng Lesotho
211 G4 Mapuera r. Brazil
181 J2 Mapulanguene Moz.
181 H1 Mapungubwe National Park S. Africa
179 D6 Maputo Moz.
181 J2 Maputo prov. Moz.
181 J3 Maputo r. Moz.
181 G4 Maputsoe Lesotho
137 H6 Maqar an Na'am well Iraq
148 B3 Maqu China
145 E3 Maquan He r. China
178 B4 Maquela do Zombo Angola
215 C4 Maquinchao Arg.
215 C4 Maquinchao r. Arg.
190 B4 Maquoketa U.S.A.
190 B4 Maquoketa r. U.S.A.
141 G5 Mar r. Pak.
214 D3 Mar, Serra do mts Brazil
187 H1 Mara r. Canada
145 E5 Mara India
181 H1 Mara S. Africa
210 E4 Maraã Brazil
211 I5 Marabá Brazil
211 H3 Maracá, Ilha de i. Brazil
213 C2 Maracaibo Venez.
213 C2 Maracaibo, Lake inlet Venez.
214 A3 Maracaju Brazil
214 A3 Maracaju, Serra de hills Brazil
214 E1 Maracás, Chapada de hills Brazil
213 D2 Maracay Venez.
177 D2 Marādah Libya
176 C3 Maradi Niger
140 B2 Marāgheh Iran
214 E1 Maragogipe Brazil
153 B3 Maragondon Phil.
213 D4 Marahuaca, Cerro mt. Venez.
211 I4 Marajó, Baía de est. Brazil
211 I3 Marajó, Ilha de i. Brazil
181 G2 Marakele National Park S. Africa
143 B3 Marakkanam India
178 D3 Maralal Kenya
144 C2 Marala Weir Pak.
137 I1 Maralik Armenia
124 D5 Maralinga Australia
125 G2 Maramasike i. Solomon Is
129 E2 Marambio research stn Antarctica
153 C5 Marampit i. Indon.
137 J4 Marāna Iraq
197 G5 Marana U.S.A.
140 B2 Marand Iran
154 B4 Marang Malaysia
154 A3 Marang Myanmar
214 C1 Maranhão r. Brazil
210 C4 Marañón r. Peru
181 K2 Marão Moz.
167 C2 Marão mt. Port.
213 E4 Marari r. Brazil
128 A6 Mararoa r. N.Z.
190 C1 Marathon Canada
201 D7 Marathon FL U.S.A.
199 C6 Marathon TX U.S.A.
214 E1 Maraú Brazil
155 E3 Marau Indon.
213 E4 Marauiá r. Brazil
153 C4 Marawi Phil.
137 L1 Märäzä Azer.
167 D4 Marbella Spain
124 B4 Marble Bar Australia
197 G3 Marble Canyon U.S.A.
197 G3 Marble Canyon gorge U.S.A.
181 H2 Marble Hall S. Africa
203 H3 Marblehead U.S.A.
187 K2 Marble Island Canada
181 I5 Marburg S. Africa
202 E5 Marburg, Lake U.S.A.
165 G4 Marburg an der Lahn Germany
168 H7 Marcali Hungary
161 H5 March U.K.
126 C4 Marchant Hill Australia
164 D4 Marche-en-Famenne Belgium
167 D4 Marchena Spain
210 □ Marchena, Isla i. Galapagos Is Ecuador
Mar Chiquita, Lago l. Arg. see Mar Chiquita, Laguna
215 D1 Mar Chiquita, Laguna l. Arg.
168 G6 Marchtrenk Austria
201 D7 Marco U.S.A.
164 B4 Marcoing France
188 E2 Marcopeet Islands Canada
215 D2 Marcos Juárez Arg.
203 G2 Marcy, Mount U.S.A.
144 C2 Mardan Pak.
215 F3 Mar del Plata Arg.
137 H3 Mardin Turkey
125 G4 Maré i. New Caledonia
162 C3 Maree, Loch l. U.K.
190 A5 Marengo IA U.S.A.
190 C4 Marengo IL U.S.A.
170 E6 Marettimo, Isola i. Sicily Italy
172 E3 Marevo Rus. Fed.
199 B6 Marfa U.S.A.
126 B2 Margaret watercourse Australia
124 B5 Margaret River Australia
213 E2 Margarita, Isla de i. Venez.
150 D3 Margaritovo Rus. Fed.
127 G9 Margate Australia
181 I5 Margate S. Africa
161 I6 Margate U.K.
178 C3 Margherita Peak mt. Dem. Rep. Congo/Uganda
Margilan Uzbek. see Marg'ilon
139 G4 Marg'ilon Uzbek.
Margo, Dasht-i des. Afgh. see Mārgow, Dasht-e
153 B4 Margosatubig Phil.
141 F4 Mārgow, Dasht-e des. Afgh.
164 D4 Margraten Neth.
190 E3 Margrethe, Lake U.S.A.
186 E4 Marguerite Canada
129 B2 Marguerite Bay b. Antarctica
145 G3 Margyang China
137 K5 Marhaj Khalīl Iraq
139 H4 Marhamat Uzbek.
137 I3 Marhan Dāgh h. Iraq
173 E6 Marhanets' Ukr.
217 I7 Maria atoll Fr. Polynesia
212 C2 María Elena Chile
215 E3 María Ignacia Arg.
127 H9 Maria Island Australia
124 D3 Maria Island Australia
216 E4 Mariana Ridge sea feature N. Pacific Ocean
216 E5 Mariana Trench sea feature N. Pacific Ocean
145 H4 Mariani India
186 F2 Marian Lake Canada
199 F5 Marianna AR U.S.A.
201 C6 Marianna FL U.S.A.
168 F6 Mariánské Lázně Czech Rep.
206 C3 Marías, Islas is Mex.
206 I7 Mariato, Punta pt Panama
128 D1 Maria van Diemen, Cape N.Z.
170 F1 Maribor Slovenia
197 F5 Maricopa AZ U.S.A.
196 C4 Maricopa CA U.S.A.
197 F5 Maricopa Mountains U.S.A.
177 E4 Maridi watercourse Sudan
129 B4 Marie Byrd Land reg. Antarctica
205 L5 Marie-Galante i. Guadeloupe
159 L3 Mariehamn Fin.
214 B1 Mariembero r. Brazil
165 L4 Marienberg Germany
164 F1 Marienhafe Germany
179 B6 Mariental Namibia
159 K4 Mariestad Sweden
201 C5 Marietta GA U.S.A.
202 C5 Marietta OH U.S.A.
166 G5 Marignane France
146 G1 Marii, Mys pt Rus. Fed.
146 A1 Mariinsk Rus. Fed.
138 E1 Mariinskoye Rus. Fed.
159 M5 Marijampolė Lith.
214 C3 Marília Brazil
207 D2 Marín Mex.
141 G4 Marin mt. Pak.
167 B1 Marín Spain
170 G5 Marina di Gioiosa Ionica Italy
172 D4 Mar''ina Horka Belarus
153 B3 Marinduque i. Phil.
190 D3 Marinette U.S.A.
214 B3 Maringá Brazil
167 B3 Marinha Grande Port.
200 B4 Marion IL U.S.A.
190 E5 Marion IN U.S.A.
203 J2 Marion ME U.S.A.
202 B4 Marion OH U.S.A.
201 E5 Marion SC U.S.A.
202 C6 Marion VA U.S.A.
201 D5 Marion, Lake U.S.A.
126 B5 Marion Bay Australia
213 D3 Maripa Venez.
196 C3 Mariposa U.S.A.
Mariscal Estigarribia Para. see Mariscal José Félix Estigarribia
Mariscal José Félix Estigarribia
212 D2 Mariscal José Félix Estigarribia Para.
166 H4 Maritime Alps mts France/Italy
171 K3 Maritsa r. Bulg.
172 I3 Mari-Turek Rus. Fed.
173 F6 Mariupol' Ukr.
213 E2 Mariusa, Caño r. Venez.
140 B3 Marīvān Iran
172 I3 Mariy El, Respublika aut. rep. Rus. Fed.
178 E3 Marka Somalia
139 K2 Markakol', Ozero l. Kazakh.
137 J2 Mārkān Iran
143 B3 Markapur India
159 K4 Markaryd Sweden
191 G3 Markdale Canada
181 H1 Marken S. Africa
164 D2 Markermeer l. Neth.
161 G5 Market Deeping U.K.
161 E5 Market Drayton U.K.
161 G5 Market Harborough U.K.
163 E3 Markethill U.K.
160 G4 Market Weighton U.K.
133 M3 Markha r. Rus. Fed.
191 H4 Markham Canada
Markhamet Uzbek. see Marhamat
139 I5 Markit China
173 F5 Markivka Ukr.
165 K3 Markkleeberg Germany
165 H2 Marklohe Germany
148 A3 Markog Qu r. China
165 K3 Markranstädt Germany
173 H5 Marks Rus. Fed.
165 H5 Marktheidenfeld Germany
168 E7 Marktoberdorf Germany
165 K4 Marktredwitz Germany
190 B6 Mark Twain Lake U.S.A.
164 F3 Marl Germany
203 H3 Marlborough U.S.A.
161 F6 Marlborough Downs hills U.K.
164 B5 Marle France
199 D6 Marlin U.S.A.
202 C5 Marlinton U.S.A.
127 H6 Marlo Australia
166 E4 Marmande France
136 B1 Marmara, Sea of g. Turkey
Marmara Denizi g. Turkey see Marmara, Sea of
136 B2 Marmara Gölü l. Turkey
136 B3 Marmaris Turkey
198 C2 Marmarth U.S.A.
202 C5 Marmet U.S.A.
188 B4 Marmion Lake Canada
170 D1 Marmolada mt. Italy
166 F2 Marne-la-Vallée France
137 J1 Marneuli Georgia
165 J1 Marnitz Germany
126 E6 Marnoo Australia
179 E5 Maroantsetra Madag.
165 I4 Maroldsweisach Germany
179 E5 Maromokotro mt. Madag.
179 D5 Marondera Zimbabwe
211 H2 Maroni r. Fr. Guiana
127 J1 Maroochydore Australia
123 I6 Marotiri is Fr. Polynesia
177 D3 Maroua Cameroon
179 E5 Marovoay Madag.
137 H4 Marqādah Syria
181 G4 Marquard S. Africa
123 I5 Marquesas Islands Fr. Polynesia
201 D7 Marquesas Keys is U.S.A.
190 D2 Marquette U.S.A.
164 B4 Marquion France
Marquises, Îles is Fr. Polynesia see Marquesas Islands
126 E3 Marra Australia
127 G3 Marra r. Australia
177 E3 Marra, Jebel Sudan
181 J2 Marracuene Moz.
176 B1 Marrakech Morocco
181 K2 Marrangua, Lagoa l. Moz.
127 G5 Marrar Australia
127 F8 Marrawah Australia
126 C2 Marree Australia
199 F6 Marrero U.S.A.
179 D5 Marromeu Moz.
179 D5 Marrupa Moz.
177 F2 Marsá al 'Alam Egypt
177 D1 Marsa al Burayqah Libya
178 D3 Marsabit Kenya
170 E6 Marsala Sicily Italy
177 E1 Marsá Maţrūḩ Egypt
165 G3 Marsberg Germany
170 E3 Marsciano Italy
127 G4 Marsden Australia
164 C2 Marsdiep sea chan. Neth.
166 G5 Marseille France
190 C5 Marseilles U.S.A.
158 K2 Marsfjället mt. Sweden
187 H4 Marshall Canada
199 E5 Marshall AR U.S.A.
200 C4 Marshall IL U.S.A.
190 E4 Marshall MI U.S.A.
198 E2 Marshall MN U.S.A.
198 E4 Marshall MO U.S.A.
199 E5 Marshall TX U.S.A.
127 G7 Marshall Bay Australia
123 F2 Marshall Islands country N. Pacific Ocean
198 E3 Marshalltown U.S.A.
190 B3 Marshfield U.S.A.
201 E7 Marsh Harbour Bahamas
203 J1 Mars Hill U.S.A.
199 F6 Marsh Island U.S.A.
186 C2 Marsh Lake Canada
137 L3 Marshūn Iran
194 C3 Marsing U.S.A.
159 L4 Märsta Sweden
145 F4 Marsyangdi r. Nepal
Martaban Myanmar see Mottama
155 D3 Martapura Indon.
155 B3 Martapura Indon.
191 H2 Marten River Canada
187 H4 Martensville Canada
207 E2 Marte R. Gómez, Presa resr Mex.
203 H4 Martha's Vineyard i. U.S.A.
168 C7 Martigny Switz.
169 I6 Martin Slovakia
198 C3 Martin SD U.S.A.
201 B4 Martin TN U.S.A.
201 C5 Martin, Lake U.S.A.
207 E3 Martínez Mex.
197 E5 Martinez Lake U.S.A.
205 L6 Martinique terr. Caribbean Sea
129 B4 Martin Peninsula pen. Antarctica
202 D4 Martinsburg PA U.S.A.
202 E5 Martinsburg WV U.S.A.
202 C4 Martins Ferry U.S.A.
202 D6 Martinsville U.S.A.
219 H7 Martin Vaz, Ilhas is S. Atlantic Ocean
138 D2 Martok Kazakh.
128 E4 Marton N.Z.
167 G2 Martorell Spain
167 E4 Martos Spain
137 J1 Martuni Armenia
141 F3 Maruchak Afgh.
151 C7 Marugame Japan
128 D5 Maruia r. N.Z.
211 K6 Maruim Brazil
173 G7 Marukhis Ugheltekhili pass Georgia/Rus. Fed.
127 H5 Marulan Australia
139 K1 Marushka Rus. Fed.
140 D4 Marvast Iran
166 F4 Marvejols France
197 G2 Marvine, Mount U.S.A.
187 G4 Marwayne Canada
141 F2 Mary Turkm.
125 F4 Maryborough Qld Australia
126 E6 Maryborough Vic. Australia
180 E3 Marydale S. Africa
172 I4 Mar'yevka Rus. Fed.
187 H2 Mary Frances Lake Canada
203 E5 Maryland state U.S.A.
160 D3 Maryport U.K.
189 I3 Mary's Harbour Canada
189 J4 Marystown Canada
197 F2 Marysvale U.S.A.
189 G4 Marysville Canada
196 B2 Marysville CA U.S.A.
198 D4 Marysville KS U.S.A.
202 B4 Marysville OH U.S.A.
198 E3 Maryville MO U.S.A.
201 C5 Maryville TN U.S.A.
165 K2 Marzahna Germany
206 H6 Masachapa Nicaragua
136 E6 Masada tourist site Israel
140 D4 Masāhūn, Kūh-e mt. Iran
178 D4 Masaka Uganda
181 G5 Masakhane S. Africa
137 L2 Masallı Azer.
124 C2 Masamba Indon.
152 E6 Masan S. Korea
203 I1 Masardis U.S.A.
179 D5 Masasi Tanz.
210 F7 Masavi Bol.
206 H6 Masaya Nicaragua
153 B3 Masbate Phil.
153 B4 Masbate i. Phil.
176 C1 Mascara Alg.
218 H6 Mascarene Basin sea feature Indian Ocean
218 H6 Mascarene Plain sea feature Indian Ocean
218 H5 Mascarene Ridge sea feature Indian Ocean
203 G2 Mascouche Canada
181 G4 Maseru Lesotho
181 H4 Mashai Lesotho
149 C6 Mashan China
144 D2 Masherbrum mt. Pak.
141 E2 Mashhad Iran
144 C4 Mashi r. India
137 K2 Mashīrān Iran
181 I2 Mashishing S. Africa
141 F4 Mashkel, Hamun-i- salt flat Pak.
141 F5 Mashkel, Rudi-i r. Pak.
141 F4 Mashki Chah Pak.
141 F5 Māshkīd r. Iran
158 M1 Masi Norway
206 B2 Masiáca Mex.
181 G5 Masibambane S. Africa
181 G4 Masilo S. Africa
178 D3 Masindi Uganda
153 A3 Masinloc Phil.
180 E5 Masinyusane S. Africa
Maşīrah i. Oman see Maşīrah, Jazīrat
142 E5 Maşīrah, Jazīrat i. Oman
142 E6 Maşīrah, Khalīj b. Oman
137 J1 Masis Armenia
140 C4 Masjed-e Soleymān Iran
163 B4 Mask, Lough l. Ireland
136 G3 Maskanah Syria
141 E5 Maskūtān Iran
141 G4 Maslti Pak.
179 E5 Masoala, Parc National Madag.
179 F5 Masoala, Tanjona c. Madag.
190 E4 Mason MI U.S.A.
196 C2 Mason NV U.S.A.
199 D6 Mason TX U.S.A.
128 A7 Mason Bay N.Z.
198 E3 Mason City IA U.S.A.
190 C5 Mason City IL U.S.A.
202 D5 Masontown U.S.A.
Masqaţ Oman see Muscat
170 D2 Massa Italy
203 G3 Massachusetts state U.S.A.
203 H3 Massachusetts Bay U.S.A.
197 H1 Massadona U.S.A.
170 G4 Massafra Italy
177 D3 Massakory Chad
170 D3 Massa Marittimo Italy
179 D6 Massangena Moz.
179 B4 Massango Angola
178 D2 Massawa Eritrea
203 G2 Massawippi, Lac l. Canada
203 F2 Massena U.S.A.
186 C4 Masset Canada
191 F2 Massey Canada
166 F4 Massif Central mts France
202 C4 Massillon U.S.A.
179 D6 Massinga Moz.
179 D6 Massingir Moz.
181 J1 Massingir, Barragem de resr Moz.
181 J2 Massintonto r. Moz./S. Africa
191 J3 Masson-Angers Canada
137 L1 Maştağa Azer.
139 G5 Mastchoh Tajik.
138 B2 Masteksay Kazakh.
128 E4 Masterton N.Z.
201 E7 Mastic Point Bahamas
144 C1 Mastuj Pak.
141 G4 Mastung Pak.
172 C4 Masty Belarus
151 B7 Masuda Japan
137 L3 Masuleh Iran
179 D6 Masvingo Zimbabwe
136 F4 Maşyāf Syria
191 G2 Matachewan Canada
206 C1 Matachic Mex.
213 D4 Matacuni r. Venez.
146 D2 Matad Mongolia
178 B4 Matadi Dem. Rep. Congo
206 H5 Matagalpa Nicaragua
188 E4 Matagami Canada
188 E4 Matagami, Lac l. Canada
199 D6 Matagorda Island U.S.A.
154 C5 Matak i. Indon.
139 H2 Matak Kazakh.
128 F2 Matakana Island N.Z.
179 B5 Matala Angola
143 C5 Matale Sri Lanka
176 A3 Matam Senegal
206 D2 Matamoros Coahuila Mex.
207 E2 Matamoros Tamaulipas Mex.
153 B5 Matanal Point Phil.
179 D4 Matandu r. Tanz.
189 G4 Matane Canada
144 B2 Matanni Pak.
205 H4 Matanzas Cuba
189 G4 Matapédia r. Canada
215 B2 Mataquito r. Chile
143 C5 Matara Sri Lanka
155 E4 Mataram Indon.
210 D7 Matarani Peru
124 D3 Mataranka Australia
167 H2 Mataró Spain
181 H5 Matatiele S. Africa
128 B7 Mataura N.Z.
128 B7 Mataura r. N.Z.
125 J3 Matā'utu Wallis and Futuna Is
213 C3 Mataveni r. Col.
128 F3 Matawai N.Z.
210 F6 Mategua Bol.
206 D3 Matehuala Mex.
179 D5 Matemanga Tanz.
170 G4 Matera Italy
170 C6 Mateur Tunisia
188 D4 Matheson Canada
199 D6 Mathis U.S.A.
126 F5 Mathoura Australia
144 D4 Mathura India
153 C5 Mati Phil.
145 G4 Matiali India
149 D5 Matianxu China
144 B4 Matiari Pak.
207 F4 Matías Romero Mex.
189 G3 Matimekosh Canada
191 F2 Matinenda Lake Canada
203 I3 Matinicus Island U.S.A.
145 G5 Matla r. India
181 G2 Matlabas S. Africa
181 G2 Matlabas r. S. Africa
144 B4 Matli Pak.
161 F4 Matlock U.K.
213 D3 Mato r. Venez.
213 D3 Mato, Cerro mt. Venez.
210 G7 Mato Grosso Brazil
214 A1 Mato Grosso state Brazil
214 A1 Mato Grosso, Planalto do plat. Brazil
214 A3 Mato Grosso do Sul state Brazil
181 J2 Matola Moz.
167 B2 Matosinhos Port.
142 E5 Maţraḩ Oman
180 C6 Matroosberg mt. S. Africa
151 C7 Matsue Japan
150 G4 Matsumae Japan
151 E6 Matsumoto Japan
151 E7 Matsusaka Japan
149 F5 Matsu Tao i. Taiwan
151 C8 Matsuyama Japan
188 D4 Mattagami r. Canada
191 H2 Mattawa Canada
203 I2 Mattawamkeag U.S.A.
168 C7 Matterhorn mt. Italy/Switz.
194 D3 Matterhorn mt. U.S.A.
123 G2 Matthew Island S. Pacific Ocean
213 E3 Matthews Ridge Guyana
205 J4 Matthew Town Bahamas
140 D6 Maţţī, Sabkhat salt pan Saudi Arabia
200 B4 Mattoon U.S.A.
143 C5 Matugama Sri Lanka
125 H3 Matuku i. Fiji
213 E2 Maturín Venez.
153 C5 Matutuang i. Indon.
181 G4 Matwabeng S. Africa
144 E4 Mau India
145 E4 Mau Aimma India
164 B4 Maubeuge France
166 E5 Maubourguet France
162 D5 Mauchline U.K.
219 J10 Maud Seamount sea feature S. Atlantic Ocean
211 G4 Maués Brazil
145 E4 Mauganj India
196 □2 Maui i. U.S.A.
165 G6 Maulbronn Germany
215 B2 Maule admin. reg. Chile
215 B2 Maule r. Chile
215 B4 Maullín Chile
163 B3 Maumakeogh h. Ireland
202 B4 Maumee U.S.A.
202 B4 Maumee r. U.S.A.
191 F5 Maumee Bay U.S.A.
163 B4 Maumtrasna h. Ireland
163 B4 Maumturk Mountains hills Ireland
179 C5 Maun Botswana
196 □2 Mauna Kea vol. U.S.A.
196 □2 Mauna Loa vol. U.S.A.
196 □1 Maunalua Bay U.S.A.
145 E4 Maunath Bhanjan India
181 G1 Maunatlala Botswana
128 E2 Maungaturoto N.Z.
145 H5 Maungdaw Myanmar
154 A2 Maungmagan Islands Myanmar
184 F3 Maunoir, Lac l. Canada
126 B5 Maupertuis Bay Australia
124 D4 Maurice, Lake salt flat Australia
164 D3 Maurik Neth.
176 A3 Mauritania country Africa
175 I6 Mauritius country Indian Ocean
190 B4 Mauston U.S.A.
213 D4 Mavaca r. Venez.
179 C5 Mavinga Angola
181 G5 Mavuya S. Africa
144 D3 Mawana India
178 B4 Mawanga Dem. Rep. Congo
149 D4 Ma Wang Dui tourist site China
154 A3 Mawdaung Pass Myanmar/Thai.
128 G3 Mawhai Point N.Z.
154 A1 Mawlamyaing Myanmar
129 E5 Mawson research stn Antarctica
129 D5 Mawson Escarpment esc. Antarctica
129 B6 Mawson Peninsula pen. Antarctica
154 A3 Maw Taung mt. Myanmar
198 C2 Max U.S.A.
207 G3 Maxcanú Mex.
170 C5 Maxia, Punta mt. Sardinia Italy
190 D5 Maxinkuckee, Lake U.S.A.
158 M3 Maxmo Fin.
191 F2 Maxton U.S.A.
196 A2 Maxwell U.S.A.
162 F4 May, Isle of i. U.K.
155 C3 Maya i. Indon.
146 F1 Maya r. Rus. Fed.
205 J4 Mayaguana i. Bahamas
205 K5 Mayagüez Puerto Rico
176 C3 Mayahi Niger
141 H2 Mayakovskiy, Qullai mt. Tajik.
139 G4 Mayakum Kazakh.
178 B4 Mayama Congo
140 D2 Mayamey Iran
207 G4 Maya Mountains Belize
149 C5 Mayang China
148 B3 Mayanhe China
150 F5 Maya-san mt. Japan
162 D5 Maybole U.K.
137 J4 Maydān Sarāy Iraq
Maydā Shahr Afgh. see Meydān Shahr
127 G9 Maydena Australia
164 F4 Mayen Germany
166 D2 Mayenne France
166 D2 Mayenne r. France
197 F4 Mayer U.S.A.
186 F4 Mayerthorpe Canada
128 C5 Mayfield N.Z.
200 B4 Mayfield U.S.A.
195 F5 Mayhill U.S.A.
152 E1 Mayi He r. China
139 I3 Maykamys Kazakh.
139 H2 Maykayyn Kazakh.
139 G5 Maykhura Tajik.
173 G6 Maykop Rus. Fed.
139 H4 Mayluu-Suu Kyrg.
138 E3 Maylybas Kazakh.
139 K1 Mayma Rus. Fed.
139 G4 Maymak Kazakh.
Maymyo Myanmar see Pyin-U-Lwin
146 B1 Mayna Rus. Fed.
143 A2 Mayni India
191 I3 Maynooth Canada
186 B2 Mayo Canada
153 C5 Mayo Bay Phil.
178 B4 Mayoko Congo
186 B2 Mayo Lake Canada
153 B3 Mayon vol. Phil.
215 D3 Mayor Buratovich Arg.
128 F2 Mayor Island N.Z.
212 D1 Mayor Pablo Lagerenza Para.
179 E5 Mayotte terr. Africa
153 B2 Mayraira Point Phil.
146 E1 Mayskiy Rus. Fed.
202 B5 Maysville U.S.A.
178 B4 Mayumba Gabon
145 E3 Mayum La pass China
143 B4 Mayuram India
191 F4 Mayville MI U.S.A.
198 D2 Mayville ND U.S.A.
202 D3 Mayville NY U.S.A.
190 C4 Mayville WI U.S.A.
198 C3 Maywood U.S.A.
215 D3 Maza Arg.
172 F3 Maza Rus. Fed.
179 C5 Mazabuka Zambia
211 H4 Mazagão Brazil
166 F5 Mazamet France
144 D1 Mazar China
141 G3 Mazar, Koh-i- mt. Afgh.
170 E6 Mazara del Vallo Sicily Italy
141 G2 Mazār-e Sharīf Afgh.
213 E3 Mazaruni r. Guyana
206 B1 Mazatán Mex.
207 G5 Mazatenango Guat.
206 C3 Mazatlán Mex.
197 G4 Mazatzal Peak U.S.A.
140 C3 Mazdaj Iran
159 M4 Mažeikiai Lith.
137 G2 Mazgirt Turkey
140 A5 Maz̧hūr, 'Irq al des. Saudi Arabia
159 M4 Mazirbe Latvia
178 D4 Mazomora Tanz.
169 J4 Mazowiecka, Nizina lowland Poland
137 L3 Mazr'eh Iran
137 L5 Māzū Iran
179 C6 Mazunga Zimbabwe
173 D4 Mazyr Belarus
181 I3 Mbabane Swaziland
176 B4 Mbahiakro Côte d'Ivoire
178 B3 Mbaïki Centr. Afr. Rep.
179 D4 Mbala Zambia
178 D3 Mbale Uganda
176 D4 Mbalmayo Cameroon
178 B4 Mbandaka Dem. Rep. Congo
176 C4 Mbanga Cameroon
178 B4 M'banza Congo Angola
178 D4 Mbarara Uganda
178 C3 Mbari r. Centr. Afr. Rep.
181 J3 Mbaswana S. Africa
176 D4 Mbengwi Cameroon
179 D4 Mbeya Tanz.
179 D5 Mbinga Tanz.
179 D6 Mbizi Zimbabwe
178 B3 Mbomo Congo
176 D4 Mbouda Cameroon
176 A3 Mbour Senegal
176 A3 Mbout Mauritania
179 D4 Mbozi Tanz.
178 C4 Mbuji-Mayi Dem. Rep. Congo
178 D4 Mbulu Tanz.
178 D4 Mbuyuni Tanz.
203 J2 McAdam Canada
199 E5 McAlester U.S.A.
202 E4 McAlevys Fort U.S.A.
127 H5 McAlister mt. Australia
199 D7 McAllen U.S.A.
202 B5 McArthur U.S.A.
191 I3 McArthur Mills Canada
186 E4 McBride Canada
194 C2 McCall U.S.A.
199 C6 McCamey U.S.A.
194 D3 McCammon U.S.A.
186 C4 McCauley Island Canada
184 H2 McClintock Channel Canada
196 B3 McClure, Lake U.S.A.
184 F2 McClure Strait Canada
199 F6 McComb U.S.A.
198 C3 McConaughy, Lake U.S.A.
202 E5 McConnellsburg U.S.A.
202 C5 McConnelsville U.S.A.
198 C3 McCook U.S.A.
187 J4 McCreary Canada
197 E4 McCullough Range mts U.S.A.
186 D3 McDame Canada
194 C3 McDermitt U.S.A.
218 I8 McDonald Islands Indian Ocean
194 D2 McDonald Peak U.S.A.
126 C2 McDonnell Creek watercourse Australia
180 B4 McDougall's Bay S. Africa
197 G5 McDowell Peak U.S.A.
196 C4 McFarland U.S.A.
187 H3 McFarlane r. Canada
197 E2 McGill U.S.A.
184 C3 McGrath U.S.A.
186 E4 McGregor r. Canada
180 C6 McGregor S. Africa
190 A2 McGregor U.S.A.
191 G2 McGregor Bay Canada
194 D2 McGuire, Mount U.S.A.
179 D4 Mchinga Tanz.
203 G2 McIndoe Falls U.S.A.
198 C2 McIntosh U.S.A.
125 I2 McKean i. Kiribati
202 A6 McKee U.S.A.
202 D4 McKeesport U.S.A.
203 F3 McKeever U.S.A.
201 B4 McKenzie U.S.A.
184 C3 McKinley, Mount U.S.A.
199 D5 McKinney U.S.A.
196 C4 McKittrick U.S.A.
198 C2 McLaughlin U.S.A.
186 F3 McLennan Canada
186 F4 McLeod r. Canada
186 E3 McLeod Lake Canada
194 B3 McLoughlin, Mount U.S.A.
190 E2 McMillan U.S.A.
194 B2 McMinnville OR U.S.A.
201 C5 McMinnville TN U.S.A.
129 B5 McMurdo research stn Antarctica
197 H4 McNary U.S.A.
186 F4 McNaughton Lake Canada
197 H6 McNeal U.S.A.
198 D4 McPherson U.S.A.
127 J2 McPherson Range mts Australia
186 B2 McQuesten r. Canada
201 D5 McRae U.S.A.
186 E1 McVicar Arm b. Canada
181 G6 Mdantsane S. Africa
170 B6 M'Daourouch Alg.
154 D2 M'Đrak Vietnam
154 C1 Mê, Hon i. Vietnam
197 E3 Mead, Lake resr U.S.A.
199 C4 Meade U.S.A.
187 H4 Meadow Lake Canada
187 H4 Meadow Lake Provincial Park Canada
197 E3 Meadow Valley Wash r. U.S.A.
202 C4 Meadville U.S.A.
191 G3 Meaford Canada
150 I3 Meaken-dake vol. Japan
162 A2 Mealasta Island U.K.
167 B2 Mealhada Port.
162 D4 Meall a' Bhuiridh mt. U.K.
189 I3 Mealy Mountains Canada
127 H1 Meandarra Australia
186 F3 Meander River Canada
153 C5 Meares i. Indon.
166 F2 Meaux France
178 B4 Mebridege r. Angola
142 A5 Mecca Saudi Arabia
203 H2 Mechanic Falls U.S.A.
202 B4 Mechanicsburg U.S.A.
190 B5 Mechanicsville U.S.A.
164 C3 Mechelen Belgium
164 D4 Mechelen Neth.
176 B1 Mecheria Alg.
164 E4 Mechernich Germany
136 E1 Mecitözü Turkey
164 F4 Meckenheim Germany
168 E3 Mecklenburger Bucht b. Germany
165 J1 Mecklenburgische Seenplatte reg. Germany
165 K1 Mecklenburg-Vorpommern land Germany
179 D5 Mecula Moz.
167 C2 Meda Port.
143 B2 Medak India
155 A2 Medan Indon.
215 D3 Médanos Arg.
212 C7 Medanosa, Punta pt Arg.
143 C4 Medawachchiya Sri Lanka
143 B2 Medchal India
203 J2 Meddybemps Lake U.S.A.
167 H4 Médéa Alg.
165 G3 Medebach Germany
213 B3 Medellín Col.
161 F4 Meden r. U.K.
176 D1 Medenine Tunisia
176 A3 Mederdra Mauritania
194 B3 Medford OR U.S.A.
190 B3 Medford WI U.S.A.
203 F5 Medford Farms U.S.A.
171 M2 Medgidia Romania
190 B5 Media U.S.A.
215 C2 Media Luna Arg.
169 L7 Mediaş Romania
194 C2 Medical Lake U.S.A.
194 F3 Medicine Bow U.S.A.
194 F3 Medicine Bow Mountains U.S.A.
194 F3 Medicine Bow Peak U.S.A.
187 G4 Medicine Hat Canada
199 D4 Medicine Lodge U.S.A.
214 E2 Medina Brazil
142 A5 Medina Saudi Arabia
202 D3 Medina NY U.S.A.
202 C4 Medina OH U.S.A.
167 E2 Medinaceli Spain
167 D2 Medina del Campo Spain
167 D2 Medina de Rioseco Spain
145 F5 Medinipur India
156 F5 Mediterranean Sea Africa/Europe
170 B6 Medjerda, Monts de la mts Alg.
138 D2 Mednogorsk Rus. Fed.
166 D4 Médoc reg. France
172 H3 Medvedevo Rus. Fed.
173 H5 Medveditsa r. Rus. Fed.
170 F2 Medvednica mts Croatia
133 R2 Medvezh'i, Ostrova is Rus. Fed.
146 F2 Medvezh'ya, Gora mt. China/Rus. Fed.
172 E2 Medvezh'yegorsk Rus. Fed.
161 H6 Medway r. U.K.
124 B4 Meekatharra Australia

Murray Bridge

126 C5 Murray Bridge Australia
180 E5 Murraysburg S. Africa
126 D5 Murrayville Australia
165 H6 Murrhardt Germany
127 H5 Murringo Australia
163 B4 Murrisk reg. Ireland
163 B4 Murroogh Ireland
127 H5 Murrumbateman Australia
126 F5 Murrumbidgee r. Australia
179 D5 Murrupula Moz.
127 I3 Murrurundi Australia
170 G1 Murska Sobota Slovenia
126 D2 Murteree, Lake salt flat Australia
126 E6 Murtoa Australia
143 A2 Murud India
152 A2 Muruin Sum Shuiku resr China
143 C4 Murunkan Sri Lanka
128 F3 Murupara N.Z.
123 J6 Mururoa atoll Fr. Polynesia
144 E5 Murwara India
127 J2 Murwillumbah Australia
138 E5 Murzechirla Turkm.
177 D2 Murzūq Libya
177 D2 Murzūq, Idhān des. Libya
168 G7 Mürzzuschlag Austria
137 H2 Muş Turkey
144 B3 Musakhel Pak.
171 J3 Musala mt. Bulg.
155 A2 Musala i. Indon.
152 E2 Musan N. Korea
140 E5 Musandam Peninsula Oman
Musa Qala, Rūd-i r. Afgh. see Mūsá Qal'eh, Rūd-e
141 G3 Mūsá Qal'eh Afgh.
141 G3 Mūsá Qal'eh, Rūd-e r. Afgh.
142 E5 Muscat Oman
190 B5 Muscatine U.S.A.
190 B4 Muscoda U.S.A.
203 I3 Muscongus Bay U.S.A.
124 D4 Musgrave Ranges mts Australia
163 C5 Musheramore h. Ireland
178 B4 Mushie Dem. Rep. Congo
143 B2 Musi r. India
155 B3 Musi r. Indon.
197 F4 Music Mountain U.S.A.
181 I1 Musina S. Africa
197 G2 Musinia Peak U.S.A.
186 E2 Muskeg r. Canada
203 H4 Muskeget Channel U.S.A.
190 D4 Muskegon U.S.A.
190 D4 Muskegon r. U.S.A.
202 C5 Muskingum r. U.S.A.
199 E5 Muskogee U.S.A.
191 H3 Muskoka Canada
191 H3 Muskoka, Lake Canada
186 E3 Muskwa r. Canada
136 F3 Muslimīyah Syria
177 F3 Musmar Sudan
178 D4 Musoma Tanz.
124 E2 Mussau Island P.N.G.
162 E5 Musselburgh U.K.
164 F2 Musselkanaal Neth.
194 E2 Musselshell r. U.S.A.
136 B1 Mustafakemalpaşa Turkey
138 C2 Mustayevo Rus. Fed.
159 M4 Mustjala Estonia
152 E3 Musu-dan pt N. Korea
127 I4 Muswellbrook Australia
177 E2 Mūţ Egypt
136 D3 Mut Turkey
214 E1 Mutá, Ponta do pt Brazil
179 D5 Mutare Zimbabwe
147 E7 Mutis, Gunung mt. Indon.
126 D4 Mutooroo Australia
179 D5 Mutorashanga Zimbabwe
150 G4 Mutsu Japan
150 G4 Mutsu-wan b. Japan
128 B7 Muttonbird Islands N.Z.
163 B5 Mutton Island Ireland
179 D5 Mutuali Moz.
214 C1 Mutunópolis Brazil
143 C4 Mutur Sri Lanka
158 N1 Mutusjärvi r. Fin.
158 N2 Muurola Fin.
148 C2 Mu Us Shamo des. China
179 B4 Muxaluando Angola
172 E2 Muyezerskiy Rus. Fed.
178 D4 Muyinga Burundi
Muynak Uzbek. see Mo'ynoq
178 C4 Muyumba Dem. Rep. Congo
148 D4 Muyuping China
144 C2 Muzaffarabad Pak.
144 B3 Muzaffargarh Pak.
144 D3 Muzaffarnagar India
145 F4 Muzaffarpur India
181 J1 Muzamane Moz.
139 J4 Muzat He r. China
141 F5 Mūzīn Iran
186 C4 Muzon, Cape U.S.A.
206 D2 Múzquiz Mex.
144 E2 Muztag mt. Xinjiang China
145 F1 Muz Tag mt. Xinjiang/Xizang China
Muztag mt. China see Muz Tag
139 H5 Muztagata mt. China
177 E4 Mvolo Sudan
178 D4 Mvomero Tanz.
179 D5 Mvuma Zimbabwe
179 E5 Mwali i. Comoros
179 C4 Mwanza Dem. Rep. Congo
178 D4 Mwanza Tanz.
178 C4 Mwaro Burundi
163 B4 Mweelrea h. Ireland
178 C4 Mweka Dem. Rep. Congo
179 C5 Mwenda Zambia
178 C4 Mwene-Ditu Dem. Rep. Congo
179 D6 Mwenezi Zimbabwe
179 C4 Mweru, Lake Dem. Rep. Congo/Zambia
179 C4 Mwimba Dem. Rep. Congo
179 C5 Mwinilunga Zambia
172 C4 Myadzyel Belarus
145 H5 Myaing Myanmar
144 B4 Myajlar India
127 J4 Myall Lake Australia
146 B2 Myanmar country Asia
162 E2 Mybster U.K.
171 J6 Mycenae tourist site Greece
145 H5 Myebon Myanmar
154 A2 Myeik Myanmar
154 A3 Myeik Kyunzu is Myanmar
147 B4 Myingyan Myanmar
154 A2 Myinmoletkat mt. Myanmar
146 B4 Myitkyina Myanmar
154 A2 Myitta Myanmar
145 H5 Myittha r. Myanmar
173 E6 Mykolayiv Ukr.
171 K6 Mykonos Greece
171 K6 Mykonos i. Greece
132 G3 Myla Rus. Fed.
145 G4 Mymensingh Bangl.
159 M3 Mynämäki Fin.
139 H3 Mynaral Kazakh.
Myohaung Myanmar see Mrauk-U
152 E3 Myŏnggan N. Korea
172 C4 Myory Belarus
158 C3 Mýrdalsjökull ice cap Iceland
158 K1 Myre Norway
158 M2 Myrheden Sweden
173 E5 Myrhorod Ukr.
173 D5 Myronivka Ukr.
201 E5 Myrtle Beach U.S.A.
127 G6 Myrtleford Australia
194 A3 Myrtle Point U.S.A.
171 J6 Myrtoo Pelagos sea Greece
139 G4 Myrzakent Kazakh.
168 G4 Myślibórz Poland
143 B3 Mysore India
133 T3 Mys Shmidta Rus. Fed.
203 F5 Mystic Islands U.S.A.
Mysuru India see Mysore
154 C3 My Tho Vietnam
171 L5 Mytilini Greece
172 F4 Mytishchi Rus. Fed.
158 C2 Mývatn-Laxá nature res. Iceland
181 G5 Mzamomhle S. Africa
165 K5 Mže r. Czech Rep.
179 D5 Mzimba Malawi
179 D5 Mzuzu Malawi

N

149 B6 Na, Nam r. China/Vietnam
165 J5 Naab r. Germany
196 □2 Nā'ālehu U.S.A.
Naalehu U.S.A. see Nā'ālehu
159 M3 Naantali Fin.
163 E4 Naas Ireland
180 B4 Nababeep S. Africa
143 C2 Nabarangapur India
151 E7 Nabari Japan
153 B4 Nabas Phil.
136 E5 Nabatîyé et Tahta Lebanon
Nabatiyet et Tahta Lebanon see Nabatîyé et Tahta
165 K5 Nabburg Germany
178 D4 Naberera Tanz.
132 G4 Naberezhnyye Chelny Rus. Fed.
177 D1 Nabeul Tunisia
144 D3 Nabha India
127 J4 Nabiac Australia
147 F7 Nabire Indon.
136 E5 Nāblus West Bank
181 H2 Naboomspruit S. Africa
154 A2 Nabule Myanmar
179 E5 Nacala Moz.
206 H5 Nacaome Hond.
194 B2 Naches U.S.A.
144 B4 Nachna India
196 B4 Nacimiento Reservoir U.S.A.
199 E6 Nacogdoches U.S.A.
204 C2 Nacozari de García Mex.
144 C5 Nadiad India
140 D4 Nadīk Iran
176 B1 Nador Morocco
140 D3 Nadūshan Iran
173 C5 Nadvirna Ukr.
132 E3 Nadvoitsy Rus. Fed.
132 I3 Nadym Rus. Fed.
159 J5 Næstved Denmark
171 I5 Nafpaktos Greece
171 J6 Nafplio Greece
137 J5 Naft, Āb r. Iraq
140 C4 Naft-e Safīd Iran
137 J5 Naft Khāneh Iraq
140 B3 Naft Shahr Iran
140 A5 Nafy Saudi Arabia
145 G2 Nag, Co l. China
153 B3 Naga Phil.
188 D4 Nagagami r. Canada
151 C8 Nagahama Japan
145 H4 Naga Hills India
151 G5 Nagai Japan
145 H4 Nagaland state India
126 F6 Nagambie Australia
151 F6 Nagano Japan
151 F6 Nagaoka Japan
145 H4 Nagaon India
143 B4 Nagapattinam India
144 D2 Nagar India
143 B2 Nagarjuna Sagar Reservoir India
144 B4 Nagar Parkar Pak.
145 G3 Nagarzê China
151 A8 Nagasaki Japan
151 B7 Nagato Japan
144 C4 Nagaur India
143 C2 Nagavali r. India
144 C5 Nagda India
143 B4 Nagercoil India
141 G5 Nagha Kalat Pak.
144 D3 Nagina India
145 E3 Nagma Nepal
Nagorno-Karabakh aut. reg. Azer. see Dağlıq Qarabağ
172 I3 Nagorsk Rus. Fed.
151 E7 Nagoya Japan
144 D5 Nagpur India
145 H3 Nagqu China
153 C3 Nagumbuaya Point Phil.
132 F1 Nagurskoye Rus. Fed.
170 G1 Nagyatád Hungary
168 H7 Nagykanizsa Hungary
146 E4 Naha Japan
144 D3 Nahan India
141 F5 Nahang r. Iran/Pak.
186 E2 Nahanni Butte Canada
Nahanni National Park Canada see Nahanni National Park Reserve
186 D2 Nahanni National Park Reserve Canada
136 E5 Nahariyya Israel
140 C3 Nahāvand Iran
165 F5 Nahe r. Germany
215 B3 Nahuelbuta, Parque Nacional nat. park Chile
215 B4 Nahuel Huapí, Lago l. Arg.
215 B4 Nahuel Huapí, Parque Nacional nat. park Arg.
201 D6 Nahunta U.S.A.
206 C2 Naica Mex.
145 H2 Naij Tal China
189 H2 Nain Canada
140 D3 Nā'īn Iran
144 D3 Nainital India
144 E5 Nainpur India
162 E3 Nairn U.K.
191 G2 Nairn Centre Canada
178 D4 Nairobi Kenya
178 D4 Naivasha Kenya
152 D2 Naizishan China
140 B5 Na'jān Saudi Arabia
142 B4 Najd reg. Saudi Arabia
167 E1 Nájera Spain
144 D3 Najibabad India
152 F2 Najin N. Korea
142 B6 Najrān Saudi Arabia
151 A8 Nakadōri-shima i. Japan
154 C1 Na Kae Thai.
151 B8 Nakama Japan
151 C8 Nakamura Japan
133 L3 Nakanno Rus. Fed.
151 F6 Nakano Japan
151 C6 Nakano-shima i. Japan
141 H3 Naka Pass Afgh.
151 B8 Nakatsu Japan
151 E7 Nakatsugawa Japan
178 D2 Nakfa Eritrea
177 F1 Nakhl Egypt
146 F2 Nakhodka Rus. Fed.
154 B2 Nakhon Nayok Thai.
154 B2 Nakhon Pathom Thai.
154 C1 Nakhon Phanom Thai.
154 B2 Nakhon Ratchasima Thai.
154 B2 Nakhon Sawan Thai.
154 A3 Nakhon Si Thammarat Thai.
144 B5 Nakhtarana India
186 C3 Nakina B.C. Canada
188 C3 Nakina Ont. Canada
184 C4 Naknek U.S.A.
179 D4 Nakonde Zambia
159 J5 Nakskov Denmark
152 E6 Naktong-gang r. S. Korea
178 D4 Nakuru Kenya
186 F4 Nakusp Canada
141 G5 Nal Pak.
141 G5 Nal r. Pak.
181 J2 Nalázi Moz.
145 G4 Nalbari India
173 G7 Nal'chik Rus. Fed.
143 B2 Naldurg India
143 B2 Nalgonda India
143 B3 Nallamala Hills India
136 C1 Nallıhan Turkey
139 G1 Nalobino Kazakh.
176 D1 Nālūt Libya
181 J2 Namaacha Moz.
181 H3 Namahadi S. Africa
140 C3 Namak, Daryācheh-ye salt flat Iran
141 E3 Namak, Kavīr-e salt flat Iran
141 E4 Namakzar-e Shadad salt flat Iran
178 D4 Namanga Kenya
139 G4 Namangan Uzbek.
179 D5 Namapa Moz.
180 B4 Namaqualand reg. S. Africa
137 J3 Namashīr Iran
124 F2 Namatanai P.N.G.
127 J1 Nambour Australia
127 J3 Nambucca Heads Australia
154 C3 Năm Căn Vietnam
Namch'ŏn N. Korea see P'yŏngsan
146 B3 Nam Co salt l. China
158 K2 Namdalen val. Norway
158 J2 Namdalseid Norway
149 C6 Nam Đinh Vietnam
190 B3 Namekagon r. U.S.A.
152 D4 Nam-gang r. N. Korea
152 E6 Namhae-do i. S. Korea
179 B6 Namib Desert Namibia
179 B5 Namibe Angola
179 B6 Namibia country Africa
219 J8 Namibia Abyssal Plain sea feature S. Atlantic Ocean
151 G6 Namie Japan
145 H3 Namjagbarwa Feng mt. China
145 G3 Namka China
147 E7 Namlea Indon.
154 A1 Nammekon Myanmar
127 H3 Namoi r. Australia
186 F3 Nampa Canada
144 E3 Nampa mt. Nepal
194 C3 Nampa U.S.A.
176 B3 Nampala Mali
154 B1 Nam Pat Thai.
154 B1 Nam Phong Thai.
152 C4 Namp'o N. Korea
179 D5 Nampula Moz.
135 H4 Namrup India
145 H4 Namsai India
145 E3 Namsê La pass Nepal
158 J2 Namsen r. Norway
145 G3 Namsi La pass Bhutan
158 J2 Namsos Norway
133 N3 Namtsy Rus. Fed.
146 B4 Namtu Myanmar
164 C4 Namur Belgium
179 C5 Namwala Zambia
152 D6 Namwŏn S. Korea
154 B1 Nan Thai.
154 B2 Nan, Mae Nam r. Thai.
178 B3 Nana Bakassa Centr. Afr. Rep.
186 E5 Nanaimo Canada
196 □1 Nānākuli U.S.A.
Nanakuli U.S.A. see Nānākuli
152 E3 Nanam N. Korea
149 F5 Nan'an China
180 B2 Nananib Plateau Namibia
151 E6 Nanao Japan
149 E6 Nan'ao Dao i. China
151 E6 Nanatsu-shima i. Japan
148 C4 Nanbu China
150 A1 Nancha China
149 E4 Nanchang Jiangxi China
149 E4 Nanchang Jiangxi China
149 E5 Nancheng China
149 C4 Nanchong China
149 C4 Nanchuan China
166 H2 Nancy France
144 E3 Nanda Devi mt. India
144 E3 Nanda Kot mt. India
149 C5 Nandan China
143 B2 Nanded India
127 I3 Nandewar Range mts Australia
144 C5 Nandgaon India
149 D6 Nandu Jiang r. China
144 C5 Nandurbar India
143 B3 Nandyal India
149 D6 Nanfeng Guangdong China
149 E5 Nanfeng Jiangxi China
145 H3 Nang China
177 D4 Nanga Eboko Cameroon
155 D3 Nangahpinoh Indon.
152 E2 Nangang Shan mts China/N. Korea
144 C2 Nanga Parbat mt. Pak.
155 D3 Nangatayap Indon.
154 A3 Nangin Myanmar
152 D3 Nangnim N. Korea
152 D3 Nangnim-sanmaek mts N. Korea
148 E2 Nangong China
179 D4 Nangulangwa Tanz.
148 A2 Nanhua China
148 F4 Nanhui China
143 B3 Nanjangud India
148 C3 Nanjiang China
149 E5 Nanjing Fujian China
148 F3 Nanjing Jiangsu China
149 E5 Nankang China
Nanking Jiangsu China see Nanjing
151 C8 Nankoku Japan
179 B5 Nankova Angola
148 E2 Nanle China
148 F4 Nanling China
149 D5 Nan Ling mts China
149 C6 Nanliu Jiang r. China
149 C6 Nanning China
154 B1 Na Noi Thai.
185 N3 Nanortalik Greenland
149 C5 Nanpan Jiang r. China
152 A3 Nanpiao China
149 F5 Nanping China
149 F5 Nanri Dao i. China
Nansei-shotō is Japan see Ryukyu Islands
220 B1 Nansen Basin sea feature Arctic Ocean
185 I1 Nansen Sound sea chan. Canada
166 D3 Nantes France
164 A5 Nanteuil-le-Haudouin France
143 C4 Nanthi Kadal Lagoon lag. Sri Lanka
191 G4 Nanticoke Canada
203 F5 Nanticoke r. U.S.A.
186 G4 Nanton Canada
148 F4 Nantong China
149 F6 Nant'ou Taiwan
203 H4 Nantucket U.S.A.
203 H4 Nantucket Island U.S.A.
203 H4 Nantucket Sound g. U.S.A.
161 E4 Nantwich U.K.
125 H2 Nanumanga i. Tuvalu
125 H2 Nanumea atoll Tuvalu
214 E2 Nanuque Brazil
153 C5 Nanusa, Kepulauan is Indon.
149 B4 Nanxi China
149 D4 Nanxian China
149 E5 Nanxiong China
148 D3 Nanyang China
152 C3 Nanzamu China
148 D4 Nanzhang China
148 D3 Nanzhao China
167 G3 Nao, Cabo de la c. Spain
189 F3 Naococane, Lac l. Canada
145 G4 Naogaon Bangl.
150 C1 Naoli He r. China
141 F3 Naomid, Dasht-e des. Afgh./Iran
144 C2 Naoshera India
149 D6 Naozhou Dao i. China
196 A2 Napa U.S.A.
203 J1 Napadogan Canada
187 G1 Napaktulik Lake Canada
191 I3 Napanee Canada
144 C4 Napasar India
185 M3 Napasoq Greenland
190 C5 Naperville U.S.A.
128 F3 Napier N.Z.
203 G2 Napierville Canada
170 F4 Naples Italy
201 D7 Naples FL U.S.A.
203 H3 Naples ME U.S.A.
149 B6 Napo China
210 D4 Napo r. Ecuador/Peru
202 A4 Napoleon U.S.A.
Napoli Italy see Naples
215 D3 Naposta Arg.
215 D3 Naposta r. Arg.
190 E5 Nappanee U.S.A.
137 J3 Naqadeh Iran
137 L4 Naqqash Iran
151 D7 Nara Japan
176 B3 Nara Mali
169 M3 Narach Belarus
126 D6 Naracoorte Australia
127 G4 Naradhan Australia
144 C4 Naraina India
144 E6 Narainpur India
206 B2 Naranjo Mex.
207 E3 Naranjos Mex.
143 D2 Narasannapeta India
143 C2 Narasapatnam, Point India
143 C2 Narasapur India
143 C2 Narasaraopet India
145 F5 Narasinghapur India
154 B4 Narathiwat Thai.
143 A2 Narayangaon India
161 C6 Narberth U.K.
166 F5 Narbonne France
167 C1 Narcea r. Spain
140 D2 Nardin Iran
170 H4 Nardò Italy
215 E1 Nare Arg.
144 B3 Narechi r. Pak.
219 E4 Nares Abyssal Plain sea feature N. Atlantic Ocean
219 E4 Nares Deep sea feature N. Atlantic Ocean
185 L1 Nares Strait Canada/Greenland
169 J4 Narew r. Poland
152 D2 Narhong China
144 A3 Nari r. Pak.
179 B6 Narib Namibia
180 B5 Nariep S. Africa
173 H6 Narimanov Rus. Fed.
141 H2 Narin Afgh.
141 H3 Narin reg. Afgh.
136 G3 Narince Turkey
145 H1 Narin Gol watercourse China
151 G7 Narita Japan
206 B2 Narizon, Punta pt Mex.
144 C5 Narmada r. India
137 H1 Narman Turkey
144 D3 Narnaul India
170 E3 Narni Italy
169 N5 Narodychi Ukr.
172 F4 Naro-Fominsk Rus. Fed.
127 I6 Narooma Australia
172 G4 Narovchat Rus. Fed.
173 D5 Narowlya Belarus
159 M3 Närpes Fin.
127 H3 Narrabri Australia
203 H4 Narragansett Bay U.S.A.
127 G2 Narran r. Australia
127 G5 Narrandera Australia
127 G2 Narran Lake Australia
127 H4 Narromine Australia
187 I4 Narrow Hills Provincial Park Canada
202 C6 Narrows U.S.A.
203 F4 Narrowsburg U.S.A.
Narsimhapur India see Narsinghpur
145 G5 Narsingdi Bangl.
144 D5 Narsinghgarh India
144 D5 Narsinghpur India
143 C2 Narsipatnam India
148 E1 Nart China
151 D7 Naruto Japan
159 O4 Narva Estonia
159 N4 Narva Bay Estonia/Rus. Fed.
153 B2 Narvacan Phil.
159 O4 Narva Reservoir resr Estonia/Rus. Fed.
158 L1 Narvik Norway
144 D3 Narwana India
144 D4 Narwar India
132 G3 Nar'yan-Mar Rus. Fed.
139 H4 Naryn Kyrg.
139 H4 Naryn r. Kyrg.
139 K2 Naryn, Khrebet mts Kazakh.
139 J4 Narynkol Kazakh.
158 L3 Näsåker Sweden
Na Scealagae Skelligs is Ireland see The Skelligs
197 H3 Naschitti U.S.A.
128 C6 Naseby N.Z.
144 C5 Nashik India
190 A4 Nashua IA U.S.A.
203 H3 Nashua NH U.S.A.
201 C4 Nashville U.S.A.
136 F5 Naşīb Syria
159 M3 Näsijärvi l. Fin.
177 F4 Nasir Sudan
177 F2 Nāşir, Buḩayrat resr Egypt
144 B3 Nasirabad Pak.
179 C5 Nasondoye Dem. Rep. Congo
136 C6 Naşr Egypt
140 C3 Naşrābād Iran
141 E3 Naşrābād Iran
137 K5 Naşrīān-e Pā'īn Iran
186 D3 Nass r. Canada
201 E7 Nassau Bahamas
216 H6 Nassau i. Cook Is
159 K4 Nässjö Sweden
185 M3 Nassuttooq inlet Greenland
188 E2 Nastapoca r. Canada
188 E2 Nastapoka Islands Canada
151 F6 Nasu-dake vol. Japan
153 B3 Nasugbu Phil.
169 O2 Nasva Rus. Fed.
179 C6 Nata Botswana
178 D4 Nata Tanz.
213 B4 Natagaima Col.
211 K5 Natal Brazil
Natal prov. S. Africa see KwaZulu-Natal
218 G7 Natal Basin sea feature Indian Ocean
140 C3 Naţanz Iran
189 H3 Natashquan Canada
189 H3 Natashquan r. Canada
199 F6 Natchez U.S.A.
199 E6 Natchitoches U.S.A.
168 G4 Natecka, Puszcza for. Poland
126 F6 Nathalia Australia
154 B4 Na Thawi Thai.
144 C4 Nathdwara India
167 H2 Nati, Punta pt Spain
126 D6 Natimuk Australia
196 D5 National City U.S.A.
176 C3 Natitingou Benin
211 I6 Natividade Brazil
206 B1 Nátora Mex.
150 G5 Natori Japan
178 D4 Natron, Lake salt l. Tanz.
154 A1 Nattaung mt. Myanmar
189 H2 Natuashish Canada
155 C2 Natuna, Kepulauan is Indon.
155 C2 Natuna Besar i. Indon.
203 F2 Natural Bridge U.S.A.
197 G3 Natural Bridges National Monument nat. park U.S.A.
218 L7 Naturaliste Plateau sea feature Indian Ocean
197 H2 Naturita U.S.A.
190 E2 Naubinway U.S.A.
179 B6 Nauchas Namibia
165 K2 Nauen Germany
203 G4 Naugatuck U.S.A.
153 B3 Naujan Phil.
153 B3 Naujan, Lake Phil.
159 M4 Naujoji Akmenė Lith.
144 C4 Naukh India
144 B4 Naukot Pak.
165 H3 Naumburg (Hessen) Germany
165 J3 Naumburg (Saale) Germany
154 A1 Naungpale Myanmar
136 E6 Na'ūr Jordan
141 G4 Nauroz Kalat Pak.
125 G2 Nauru country Pacific Ocean
144 B4 Naushara Pak.
159 I3 Naustdal Norway
210 D4 Nauta Peru
180 C3 Naute Dam Namibia
207 E3 Nautla Mex.
141 G3 Nauzad Afgh.
145 G5 Navadwip India
172 C4 Navahrudak Belarus
197 H4 Navajo U.S.A.
195 F4 Navajo Lake U.S.A.
197 G3 Navajo Mountain U.S.A.
153 C4 Naval Phil.
167 D3 Navalmoral de la Mata Spain
167 D3 Navalvillar de Pela Spain
163 E4 Navan Ireland
172 D4 Navapolatsk Belarus
133 S3 Navarin, Mys c. Rus. Fed.
212 C9 Navarino, Isla i. Chile
167 F1 Navarra aut. comm. Spain
126 E6 Navarre Australia
196 A2 Navarro U.S.A.
172 G4 Navashino Rus. Fed.
199 D6 Navasota U.S.A.
162 D2 Naver, Loch l. U.K.
158 K3 Näverede Sweden
215 B2 Navidad Chile
143 A2 Navi Mumbai India
172 E4 Navlya Rus. Fed.
171 M2 Năvodari Romania
Navoi Uzbek. see Navoiy
138 F4 Navoiy Uzbek.
206 B2 Navojoa Mex.
172 G3 Navoloki Rus. Fed.
144 C5 Navsari India
144 C4 Nawa India
136 F5 Nawá Syria
145 G4 Nawabganj Bangl.
144 B4 Nawabshah Pak.
145 F4 Nawada India
141 G3 Nāwah Afgh.
144 C4 Nawalgarh India
137 J2 Naxçıvan Azer.
149 B4 Naxi China
171 K6 Naxos Greece
171 K6 Naxos i. Greece
Nây, Mui pt Vietnam see Đai Lanh, Mui
213 A4 Naya Col.
143 D1 Nayagarh India
206 C3 Nayar Mex.
206 C3 Nayarit state Mex.
150 H2 Nayoro Japan
135 H5 Nay Pyi Taw Myanmar
143 B3 Nayudupeta India
214 E1 Nazaré Brazil
143 B4 Nazareth India
136 E5 Nazareth Israel
199 B7 Nazas Mex.
206 C2 Nazas r. Mex.
210 D6 Nazca Peru
217 N7 Nazca Ridge sea feature S. Pacific Ocean
Nazerat Israel see Nazareth
137 J2 Nāzīk Iran
137 I2 Nazik Gölü l. Turkey
141 F4 Nāzīl Iran
136 B3 Nazilli Turkey
141 G5 Nazimabad Pak.
137 G2 Nazımiye Turkey
145 H4 Nazira India
186 E4 Nazko Canada
186 E4 Nazko r. Canada
137 J3 Nāzlū r. Iran
173 H7 Nazran' Rus. Fed.
178 D3 Nazrēt Eth.
142 E5 Nazwá Oman
179 C4 Nchelenge Zambia
179 C6 Ncojane Botswana
179 B4 N'dalatando Angola
178 C3 Ndélé Centr. Afr. Rep.
178 B4 Ndendé Gabon
177 D3 Ndjamena Chad
179 C5 Ndola Zambia
181 I4 Ndwedwe S. Africa
127 G1 Neabul Creek r. Australia
163 E3 Neagh, Lough l. U.K.
194 A1 Neah Bay U.S.A.
124 D4 Neale, Lake salt flat Australia
126 B2 Neales watercourse Australia
171 J5 Nea Liosia Greece
171 J6 Neapoli Greece
161 D6 Neath U.K.
161 D6 Neath r. U.K.
127 G1 Nebine Creek r. Australia
Nebitdag Turkm. see Balkanabat
213 D4 Neblina, Pico da mt. Brazil
197 G2 Nebo, Mount U.S.A.
172 E3 Nebolchi Rus. Fed.
198 C3 Nebraska state U.S.A.
198 E3 Nebraska City U.S.A.
170 F6 Nebrodi, Monti mts Sicily Italy
199 E6 Neches r. U.S.A.
213 B3 Nechí r. Col.
178 D3 Nechisar National Park Eth.
165 G5 Neckar r. Germany
165 H5 Neckarsulm Germany
123 H2 Necker Island U.S.A.
215 E3 Necochea Arg.
165 L1 Neddemin Germany
188 F2 Nedlouc, Lac l. Canada
Nêdong China see Zêtang
158 M1 Nedre Soppero Sweden
197 E4 Needles U.S.A.
144 C4 Neemuch India
190 C3 Neenah U.S.A.
187 J4 Neepawa Canada
185 J2 Neergaard Lake Canada
164 D3 Neerijnen Neth.
164 D3 Neerpelt Belgium
137 L2 Neftçala Azer.
138 B1 Neftegorsk Rus. Fed.
132 G4 Neftekamsk Rus. Fed.
173 H6 Neftekumsk Rus. Fed.
132 I3 Nefteyugansk Rus. Fed.
161 C5 Nefyn U.K.
170 C6 Nefza Tunisia
178 B4 Negage Angola
178 D3 Negēlē Eth.
214 A3 Negla r. Para.
179 D5 Negomane Moz.
143 B5 Negombo Sri Lanka
171 J4 Negotino Macedonia
210 C5 Negra, Cordillera mts Peru
210 B5 Negra, Punta pt Peru
170 B7 Négrine Alg.
210 B4 Negritos Peru
215 D4 Negro r. Arg.
214 A2 Negro r. Brazil
210 F4 Negro r. S. America
215 F2 Negro r. Uruguay
153 B4 Negros i. Phil.
171 M3 Negru Vodă Romania
137 L4 Nehavand Iran
141 F4 Nehbandān Iran
146 E2 Nehe China
149 B4 Neijiang China
187 H4 Neilburg Canada
148 B1 Nei Mongol Zizhiqu aut. reg. China
165 J3 Neinstedt Germany
168 G5 Neiße r. Germany/Poland
213 B4 Neiva Col.
148 D3 Neixiang China
187 J3 Nejanilini Lake Canada
140 D2 Neka Iran
178 D3 Nek'emtē Eth.
159 K5 Neksø Denmark
144 D3 Nelang India
172 E3 Nelidovo Rus. Fed.
198 D3 Neligh U.S.A.
133 P3 Nel'kan Rus. Fed.
146 F1 Nel'kan Rus. Fed.
143 B3 Nelligere India
143 B3 Nellore India
186 F5 Nelson Canada
187 K3 Nelson r. Canada
128 D4 Nelson N.Z.
160 E4 Nelson U.K.
197 E4 Nelson U.S.A.
126 D7 Nelson, Cape Australia
212 B8 Nelson, Estrecho str. Chile
127 J4 Nelson Bay Australia
186 E3 Nelson Forks Canada
187 J3 Nelson House Canada
181 I2 Nelspruit S. Africa
176 B3 Néma Mauritania
172 I3 Nema Rus. Fed.
190 A2 Nemadji r. U.S.A.
172 B4 Neman Rus. Fed.
141 E4 Ne'matābād Iran
172 G3 Nemda r. Rus. Fed.
172 J2 Nemed Rus. Fed.
191 F2 Nemegos Canada
158 O1 Nemetskiy, Mys c. Rus. Fed.
166 F2 Nemours France
137 I2 Nemrut Dağı mt. Turkey
150 I3 Nemuro Japan
150 I3 Nemuro-kaikyō sea chan. Japan
173 D5 Nemyriv Ukr.
163 C5 Nenagh Ireland
125 G3 Nendo i. Solomon Is
161 H5 Nene r. U.K.
146 E2 Nenjiang China
164 E5 Nennig tourist site Germany
172 F1 Nenoksa Rus. Fed.
162 E1 Neolithic Orkney tourist site U.K.
199 E4 Neosho U.S.A.
198 E4 Neosho r. U.S.A.
145 E3 Nepal country Asia
191 J3 Nepean Canada
197 G2 Nephi U.S.A.
163 B3 Nephin h. Ireland
163 B3 Nephin Beg Range hills Ireland
178 C3 Nepoko r. Dem. Rep. Congo
Neptune U.S.A. see Neptune City
203 F4 Neptune City U.S.A.
166 E4 Nérac France
127 J1 Nerang Australia
146 D1 Nerchinsk Rus. Fed.
172 G3 Nerekhta Rus. Fed.
170 G3 Neretva r. Bos.-Herz./Croatia
179 C5 Neriquinha Angola
159 M5 Neris r. Lith.
172 F3 Nerl' r. Rus. Fed.
214 C2 Nerópolis Brazil
146 E1 Neryungri Rus. Fed.
164 D1 Nes Neth.
159 J3 Nes Norway
159 J3 Nesbyen Norway
158 D2 Neskaupstaður Iceland
164 A5 Nesle France
158 K2 Nesna Norway
162 D3 Ness, Loch l. U.K.
198 D4 Ness City U.S.A.
165 I4 Nesse r. Germany
186 C3 Nesselrode, Mount Canada/U.S.A.
171 K4 Nestos r. Greece
136 E5 Netanya Israel
187 M2 Netchek, Cape Canada
164 D2 Netherlands country Europe
213 C1 Netherlands Antilles terr. Caribbean Sea
165 G4 Netphen Germany
145 G4 Netrakona Bangl.
144 C5 Netrang India
185 K3 Nettilling Lake Canada
190 A1 Nett Lake U.S.A.
190 A1 Nett Lake l. U.S.A.
165 L1 Neubrandenburg Germany
168 C7 Neuchâtel Switz.
168 C7 Neuchâtel, Lac de l. Switz.
165 I5 Neuendettelsau Germany
164 E2 Neuenhaus Germany
165 H1 Neuenkirchen Germany
165 G2 Neuenkirchen (Oldenburg) Germany
164 D5 Neufchâteau Belgium
166 G2 Neufchâteau France
166 E2 Neufchâtel-en-Bray France
165 F1 Neuharlingersiel Germany
165 H1 Neuhaus (Oste) Germany
165 H4 Neuhof Germany
165 J1 Neu Kaliß Germany
165 H4 Neukirchen Hessen Germany
165 K4 Neukirchen Sachsen Germany
165 J5 Neumarkt in der Oberpfalz Germany
129 D3 Neumayer research stn Antarctica
168 D3 Neumünster Germany
165 K5 Neunburg vorm Wald Germany
168 H7 Neunkirchen Austria
164 F5 Neunkirchen Germany
215 C3 Neuquén Arg.
215 C3 Neuquén prov. Arg.
215 C3 Neuquén r. Arg.
165 K2 Neuruppin Germany
201 E5 Neuse r. U.S.A.
168 H7 Neusiedler See l. Austria/Hungary
164 E3 Neuss Germany
164 F4 Neustadt (Wied) Germany
165 H2 Neustadt am Rübenberge Germany
165 I5 Neustadt an der Aisch Germany
165 K5 Neustadt an der Waldnaab Germany
165 G5 Neustadt an der Weinstraße Germany
165 J4 Neustadt bei Coburg Germany
165 J1 Neustadt-Glewe Germany
165 L1 Neustrelitz Germany
165 K6 Neutraubling Germany
164 F4 Neuwied Germany
165 H1 Neu Wulmstorf Germany
199 E4 Nevada U.S.A.
196 D2 Nevada state U.S.A.
167 E4 Nevada, Sierra mts Spain
196 B1 Nevada, Sierra mts U.S.A.
215 C2 Nevado, Cerro mt. Arg.
215 C3 Nevado, Sierra del mts Arg.
207 E4 Nevado de Toluca, Volcán vol. Mex.
172 D3 Nevel' Rus. Fed.
166 F3 Nevers France
127 G3 Nevertire Australia
171 H3 Nevesinje Bos.-Herz.
173 G6 Nevinnomyssk Rus. Fed.
162 C3 Nevis, Loch inlet U.K.
136 E2 Nevşehir Turkey
150 C2 Nevskoye Rus. Fed.
197 E5 New r. CA U.S.A.
202 C6 New r. WV U.S.A.
186 D3 New Aiyansh Canada
200 C4 New Albany IN U.S.A.
199 F5 New Albany MS U.S.A.
203 E4 New Albany PA U.S.A.
211 G2 New Amsterdam Guyana
127 G2 New Angledool Australia
203 F5 Newark DE U.S.A.
203 F5 Newark MD U.S.A.
203 F4 Newark NJ U.S.A.
202 E3 Newark NY U.S.A.
202 B4 Newark OH U.S.A.
203 F4 Newark airport U.S.A.
197 E2 Newark Lake U.S.A.
161 G4 Newark-on-Trent U.K.
203 E3 Newark Valley U.S.A.
203 H4 New Bedford U.S.A.
194 B2 Newberg U.S.A.
203 F3 New Berlin U.S.A.
201 E5 New Bern U.S.A.
190 E2 Newberry MI U.S.A.
201 D5 Newberry SC U.S.A.
196 D4 Newberry Springs U.S.A.
160 F2 Newbiggin-by-the-Sea U.K.
201 F7 New Bight Bahamas
191 I3 Newboro Canada
203 G3 New Boston MA U.S.A.
202 B5 New Boston OH U.S.A.
199 D6 New Braunfels U.S.A.
163 E4 Newbridge Ireland
124 E2 New Britain i. P.N.G.
203 G4 New Britain U.S.A.
216 F6 New Britain Trench sea feature Pacific Ocean
189 G4 New Brunswick prov. Canada
203 F4 New Brunswick U.S.A.
190 D5 New Buffalo U.S.A.
162 F3 Newburgh U.K.
203 F4 Newburgh U.S.A.
161 F6 Newbury U.K.
203 H3 Newburyport U.S.A.
153 A3 New Busuanga Phil.
160 E3 Newby Bridge U.K.
125 G4 New Caledonia terr. S. Pacific Ocean

Nunivak Island

216 F7 New Caledonia Trough sea feature Tasman Sea
189 G4 New Carlisle Canada
127 I4 Newcastle Australia
191 H4 Newcastle Canada
163 E4 Newcastle Ireland
181 H3 Newcastle S. Africa
163 F3 Newcastle U.K.
196 B2 Newcastle CA U.S.A.
190 E6 New Castle IN U.S.A.
202 B4 Newcastle OH U.S.A.
202 C4 New Castle PA U.S.A.
197 F3 Newcastle UT U.S.A.
202 C6 New Castle VA U.S.A.
194 F3 Newcastle WY U.S.A.
161 C5 Newcastle Emlyn U.K.
161 E4 Newcastle-under-Lyme U.K.
160 F3 Newcastle upon Tyne U.K.
163 B5 Newcastle West Ireland
203 F6 New Church U.S.A.
197 H3 Newcomb U.S.A.
162 D5 New Cumnock U.K.
162 F3 New Deer U.K.
144 D3 New Delhi India
203 J1 New Denmark Canada
196 B3 New Don Pedro Reservoir U.S.A.
127 I3 New England Range mts Australia
219 F3 New England Seamounts sea feature N. Atlantic Ocean
161 E6 Newent U.K.
161 F7 New Forest National Park nat. park U.K.
185 M5 Newfoundland i. Canada
Newfoundland prov. Canada see Newfoundland and Labrador
189 I4 Newfoundland and Labrador prov. Canada
194 D3 Newfoundland Evaporation Basin salt l. U.S.A.
162 D5 New Galloway U.K.
125 F2 New Georgia i. Solomon Is
125 F2 New Georgia Islands Solomon Is
125 F2 New Georgia Sound sea chan. Solomon Is
189 H4 New Glasgow Canada
124 E2 New Guinea i. Asia
202 B4 New Hampshire U.S.A.
203 G3 New Hampshire state U.S.A.
190 A4 New Hampton U.S.A.
124 F2 New Hanover i. P.N.G.
181 I4 New Hanover S. Africa
203 G4 New Haven U.S.A.
186 D3 New Hazelton Canada
216 G7 New Hebrides Trench sea feature Pacific Ocean
196 B2 New Hogan Reservoir U.S.A.
190 C4 New Holstein U.S.A.
199 F6 New Iberia U.S.A.
181 I2 Newington S. Africa
163 D5 Newinn Ireland
124 F2 New Ireland i. P.N.G.
203 F5 New Jersey state U.S.A.
202 E6 New Kent U.S.A.
162 E5 New Lanark U.K.
202 B5 New Lexington U.S.A.
190 B4 New Lisbon U.S.A.
191 H2 New Liskeard Canada
203 G4 New London CT U.S.A.
190 B5 New London IA U.S.A.
190 B6 New London MO U.S.A.
190 C3 New London WI U.S.A.
124 B4 Newman Australia
190 D6 Newman U.S.A.
191 H3 Newmarket Canada
163 B5 Newmarket Ireland
161 H5 Newmarket U.K.
202 D5 New Market U.S.A.
163 C5 Newmarket-on-Fergus Ireland
202 C5 New Martinsville U.S.A.
194 C2 New Meadows U.S.A.
196 B3 New Melones Lake U.S.A.
195 F5 New Mexico state U.S.A.
201 C5 Newnan U.S.A.
127 G9 New Norfolk Australia
199 F6 New Orleans U.S.A.
203 F4 New Paltz U.S.A.
202 C4 New Philadelphia U.S.A.
162 F3 New Pitsligo U.K.
128 E3 New Plymouth N.Z.
163 B4 Newport Ireland
163 C5 Newport Ireland
161 E5 Newport England U.K.
161 F7 Newport England U.K.
161 D6 Newport Wales U.K.
199 F5 Newport AR U.S.A.
202 A5 Newport KY U.S.A.
203 I2 Newport ME U.S.A.
191 F5 Newport MI U.S.A.
203 G3 Newport NH U.S.A.
194 A2 Newport OR U.S.A.
203 H4 Newport RI U.S.A.
203 G2 Newport VT U.S.A.
194 C1 Newport WA U.S.A.
196 D5 Newport Beach U.S.A.
193 K4 Newport News U.S.A.
202 E6 Newport News airport U.S.A.
161 G5 Newport Pagnell U.K.
201 E7 New Providence i. Bahamas
161 B7 Newquay U.K.
189 G4 New Richmond Canada
190 A3 New Richmond U.S.A.
197 F5 New River U.S.A.
199 F6 New Roads U.S.A.
161 H7 New Romney U.K.
163 E5 New Ross Ireland
163 E3 Newry U.K.
190 A5 New Sharon U.S.A.
133 P2 New Siberia Islands Rus. Fed.
201 D6 New Smyrna Beach U.S.A.
127 G4 New South Wales state Australia
144 D3 New Tehri India
149 □ New Territories reg. Hong Kong China
160 E4 Newton U.K.
198 E3 Newton IA U.S.A.
198 D4 Newton KS U.S.A.
203 H3 Newton MA U.S.A.
199 F5 Newton MS U.S.A.
203 F4 Newton NJ U.S.A.
161 D7 Newton Abbot U.K.
162 F3 Newtonhill U.K.
162 D5 Newton Mearns U.K.
162 D6 Newton Stewart U.K.
163 C5 Newtown Ireland
161 E5 Newtown England U.K.
161 D5 Newtown Wales U.K.
198 C1 New Town U.S.A.
163 F3 Newtownabbey U.K.
163 F3 Newtownards U.K.
163 D3 Newtownbutler U.K.
163 E4 Newtown Mount Kennedy Ireland
162 F5 Newtown St Boswells U.K.
163 D3 Newtownstewart U.K.
198 E2 New Ulm U.S.A.
196 A2 Newville U.S.A.
186 E5 New Westminster Canada
203 G4 New York U.S.A. (City Plan 114)
203 E3 New York state U.S.A.
128 New Zealand country Oceania
132 G3 Neya Rus. Fed.
141 E3 Ney Bīd Iran
140 D4 Neyrīz Iran
141 E2 Neyshābūr Iran
143 B4 Neyyattinkara India
207 F4 Nezahualcóyotl, Presa resr Mex.
155 D2 Ngabang Indon.
178 B4 Ngabé Congo
154 A2 Nga Chong, Khao mt. Myanmar/Thai.
153 C6 Ngalipaëng Indon.
179 C6 Ngami, Lake Botswana
145 F3 Ngamring China
145 E3 Ngangla Ringco salt l. China
144 E2 Nganglong Kangri mt. China
144 E2 Nganglong Kangri mts China
145 F3 Ngangzê Co salt l. China
Ngan Sau, Sông r. Vietnam see Ngan Sâu, Sông
154 C1 Ngan Sâu, Sông r. Vietnam
149 B6 Ngân Sơn Vietnam
154 A1 Ngao Thai.
177 D4 Ngaoundéré Cameroon
128 E2 Ngaruawahia N.Z.
128 F3 Ngaruroro r. N.Z.
128 E3 Ngauruhoe vol. N.Z.
179 E5 Ngazidja i. Comoros
154 B1 Ngiap r. Laos
178 B4 Ngo Congo
Ngoc Linh mt. Vietnam see Ngok Linh
145 F3 Ngoin, Co salt l. China
154 C2 Ngok Linh mt. Vietnam
176 D4 Ngol Bembo Nigeria
145 F2 Ngoqumaima China
146 B3 Ngoring Hu l. China
178 D4 Ngorongoro Conservation Area nature res. Tanz.
177 D3 Ngourti Niger
177 D3 Nguigmi Niger
147 F6 Ngulu atoll Micronesia
154 B1 Ngum, Nam r. Laos
176 D3 Nguru Nigeria
Nguyên Binh Vietnam see Ngân Sơn
Ngwaketse admin. dist. Botswana see Southern
181 G3 Ngwathe S. Africa
181 I3 Ngwavuma r. Swaziland
181 I4 Ngwelezana S. Africa
179 D5 Nhamalabué Moz.
154 D2 Nha Trang Vietnam
126 D6 Nhill Australia
181 I3 Nhlangano Swaziland
149 B6 Nho Quan Vietnam
124 D3 Nhulunbuy Australia
187 I4 Niacam Canada
176 B3 Niafounké Mali
191 H4 Niagara r. Canada/U.S.A.
190 D3 Niagara U.S.A.
191 H4 Niagara Falls Canada
202 D3 Niagara Falls U.S.A.
176 C3 Niamey Niger
153 C5 Niampak Indon.
179 D4 Niangandu Tanz.
178 C3 Niangara Dem. Rep. Congo
155 A2 Nias i. Indon.
159 M4 Nīca Latvia
206 H5 Nicaragua country Central America
206 H6 Nicaragua, Lake Nicaragua
170 G5 Nicastro Italy
166 H5 Nice France
189 F3 Nichicun, Lac l. Canada
145 E4 Nichlaul India
201 E7 Nicholl's Town Bahamas
191 F2 Nicholson Canada
135 H6 Nicobar Islands India
136 D4 Nicosia Cyprus
206 H6 Nicoya, Golfo de b. Costa Rica
206 H6 Nicoya, Península de pen. Costa Rica
203 I1 Nictau Canada
159 M5 Nida Lith.
160 F4 Nidd r. U.K.
165 H4 Nidda Germany
165 H4 Nidder r. Germany
169 I4 Nidzica Poland
168 D3 Niebüll Germany
164 E5 Niederanven Lux.
165 H4 Niederaula Germany
168 F7 Niedere Tauern mts Austria
165 G2 Niedersachsen land Germany
164 E1 Niedersächsisches Wattenmeer, Nationalpark Germany
176 D4 Niefang Equat. Guinea
176 B3 Niellé Côte d'Ivoire
165 H2 Nienburg (Weser) Germany
164 E3 Niers r. Germany
165 G5 Nierstein Germany
211 G2 Nieuw Amsterdam Suriname
164 C2 Nieuwe-Niedorp Neth.
164 E1 Nieuwe Pekela Neth.
164 C3 Nieuwerkerk aan de IJssel Neth.
211 G2 Nieuw Nickerie Suriname
164 E1 Nieuwolda Neth.
180 C5 Nieuwoudtville S. Africa
164 A3 Nieuwpoort Belgium
164 C3 Nieuw-Vossemeer Neth.
136 E3 Niğde Turkey
176 C3 Niger country Africa
176 C4 Niger r. Africa
176 C4 Niger, Mouths of the Nigeria
219 J5 Niger Cone sea feature S. Atlantic Ocean
176 C4 Nigeria country Africa
191 G1 Nighthawk Lake Canada
171 J4 Nigrita Greece
151 G6 Nihonmatsu Japan
151 F6 Niigata Japan
151 C8 Niihama Japan
196 □2 Ni'ihau i. U.S.A.
Niihau i. U.S.A. see Ni'ihau
151 F7 Nii-jima i. Japan
150 H3 Niikappu Japan
151 C7 Niimi Japan
151 F6 Niitsu Japan
164 D2 Nijkerk Neth.
164 D3 Nijmegen Neth.
164 E2 Nijverdal Neth.
158 O1 Nikel' Rus. Fed.
176 C4 Nikki Benin
151 F6 Nikkō Kokuritsu-kōen nat. park Japan
139 F1 Nikolayevka Kazakh.
138 E1 Nikolayevka Chelyabinskaya Oblast' Rus. Fed.
172 H4 Nikolayevka Ul'yanovskaya Oblast' Rus. Fed.
173 H5 Nikolayevsk Rus. Fed.
172 H4 Nikol'sk Rus. Fed.
172 H3 Nikol'sk Rus. Fed.
133 F4 Nikol'skoye Rus. Fed.
173 E6 Nikopol' Ukr.
137 L3 Nīk Pey Iran
136 F1 Niksar Turkey
141 F5 Nīkshahr Iran
171 H3 Nikšić Montenegro
125 I2 Nikumaroro atoll Kiribati
125 H2 Nikunau i. Kiribati
144 C2 Nila Pak.
145 F5 Nilagiri India
197 E5 Niland U.S.A.
143 E2 Nilanga India
178 F1 Nile r. Africa
190 D5 Niles U.S.A.
143 A3 Nileshwar India
143 E4 Nilgiri Hills India
141 G3 Nīlī Afgh.
139 J4 Nilka China
158 O3 Nilsiä Fin.
207 F4 Niltepec Mex.
Nimach India see Neemuch
127 J2 Nimbin Australia
166 G5 Nîmes France
127 H6 Nimmitabel Australia
129 C5 Nimrod Glacier glacier Antarctica
177 F4 Nimule Sudan
137 I3 Nīnawá tourist site Iraq
127 H2 Nindigully Australia
143 A4 Nine Degree Channel India
126 E3 Nine Mile Lake salt flat Australia
196 D2 Ninemile Peak U.S.A.
149 □ Ninepin Group is Hong Kong China
218 J7 Ninetyeast Ridge sea feature Indian Ocean
127 C7 Ninety Mile Beach Australia
128 D1 Ninety Mile Beach N.Z.
Nineveh Iraq see Nīnawá
203 F3 Nineveh U.S.A.
152 E1 Ning'an China
149 F4 Ningbo China
148 F1 Ningcheng China
149 F5 Ningde China
149 E5 Ningdu China
Ninggang China see Longshi
149 F4 Ningguo China
149 F4 Ninghai China
148 E2 Ninghe China
149 E5 Ninghua China
146 B3 Ningjing Shan mts China
148 E3 Ningling China
149 C6 Ningming China
149 B5 Ningnan China
148 C3 Ningqiang China
148 C3 Ningshan China
148 D2 Ningwu China
148 B2 Ningxia Huizu Zizhiqu aut. reg. China
148 C3 Ningxian China
149 D4 Ningxiang China
148 E3 Ningyang China
149 D5 Ningyuan China
149 C6 Ninh Binh Vietnam
154 D2 Ninh Hoa Vietnam
129 C6 Ninnis Glacier glacier Antarctica
150 G4 Ninohe Japan
214 A3 Nioaque Brazil
198 C3 Niobrara r. U.S.A.
145 H4 Nioko India
176 A3 Niokolo Koba, Parc National du nat. park Senegal
176 B3 Niono Mali
176 B3 Nioro Mali
166 D3 Niort France
143 A2 Nipani India
187 I4 Nipawin Canada
188 C4 Nipigon Canada
188 C4 Nipigon, Lake Canada
190 C1 Nipigon Bay Canada
189 H3 Nipishish Lake Canada
191 H2 Nipissing Canada
191 G2 Nipissing, Lake Canada
196 B4 Nipomo U.S.A.
137 J5 Nippur tourist site Iraq
197 E4 Nipton U.S.A.
214 C1 Niquelândia Brazil
140 B2 Nir Iran
143 A2 Nira r. India
143 B2 Nirmal India
143 B2 Nirmal Range hills India
171 I3 Niš Serbia
167 C3 Nisa Port.
138 D5 Nisa tourist site Turkm.
140 B5 Nisah, Wādī watercourse Saudi Arabia
170 F6 Niscemi Sicily Italy
151 B9 Nishino-omote Japan
151 C6 Nishino-shima i. Japan
151 A8 Nishi-Sonogi-hantō pen. Japan
151 D7 Nishiwaki Japan
144 D6 Nishtun Yemen
186 B2 Nisling r. Canada
164 C3 Nispen Neth.
159 K4 Nissan r. Sweden
169 M6 Nistru r. Ukr. alt. Dnister (Ukr.), conv. Dniester
169 N7 Nistrului Inferior, Cîmpia lowland Moldova
186 C2 Nisutlin r. Canada
171 L6 Nisyros i. Greece
140 C5 Niţā Saudi Arabia
189 F3 Nitchequon Canada
214 D3 Niterói Brazil
162 E5 Nith r. U.K.
162 E5 Nithsdale val. U.K.
144 D3 Niti Pass China
168 I6 Nitra Slovakia
202 C5 Nitro U.S.A.
Niuafo'ou i. Tonga see Niuafo'ou
123 H4 Niuafo'ou i. Tonga
125 I3 Niuatoputopu i. Tonga
125 J3 Niue terr. Pacific Ocean
125 H3 Niulakita i. Tuvalu
149 B5 Niulan Jiang r. China
125 H2 Niutao i. Tuvalu
152 B3 Niuzhuang China
158 N3 Nivala Fin.
164 C4 Nivelles Belgium
172 J2 Nivshera Rus. Fed.
144 C4 Niwai India
196 C2 Nixon U.S.A.
145 E1 Niya He r. China
137 L1 Niyazoba Azer.
143 B2 Nizamabad India
143 B2 Nizam Sagar l. India
172 H3 Nizhegorodskaya Oblast' admin. div. Rus. Fed.
133 R3 Nizhnekolymsk Rus. Fed.
146 B1 Nizhneudinsk Rus. Fed.
132 I3 Nizhnevartovsk Rus. Fed.
133 O2 Nizhneyansk Rus. Fed.
172 G4 Nizhniy Lomov Rus. Fed.
172 G3 Nizhniy Novgorod Rus. Fed.
172 J2 Nizhniy Odes Rus. Fed.
172 H3 Nizhniy Yenansk Rus. Fed.
139 I1 Nizhnyaya Suyetka Rus. Fed.
146 C1 Nizhnyaya Tunguska r. Rus. Fed.
173 D5 Nizhyn Ukr.
Nizina reg. Poland see Mazowiecka, Nizina
136 F3 Nizip Turkey
150 D3 Nizmennyy, Mys pt Rus. Fed.
158 M1 Njallavarri mt. Norway
158 L2 Njavve Sweden
179 D4 Njinjo Tanz.
179 D4 Njombe Tanz.
159 L3 Njurundabommen Sweden
176 D4 Nkambe Cameroon
181 I4 Nkandla S. Africa
176 B4 Nkawkaw Ghana
179 C5 Nkayi Zimbabwe
179 D5 Nkhata Bay Malawi
179 D5 Nkhotakota Malawi
176 C4 Nkongsamba Cameroon
181 G5 Nkululeko S. Africa
179 B5 Nkurenkuru Namibia
181 G6 Nkwenkwezi S. Africa
145 I4 Noa Dihing r. India
145 G5 Noakhali Bangl.
143 D1 Noamundi India
163 E4 Nobber Ireland
151 B8 Nobeoka Japan
150 G3 Noboribetsu Japan
214 A1 Nobres Brazil
126 E1 Noccundra Australia
206 D3 Nochistlán Mex.
126 E1 Nockatunga Australia
210 F6 Noel Kempff Mercado, Parque Nacional nat. park Bol.
191 G2 Noelville Canada
204 B2 Nogales Mex.
192 D5 Nogales U.S.A.
151 B8 Nōgata Japan
138 C2 Nogayty Kazakh.
166 E2 Nogent-le-Rotrou France
166 F2 Nogent-sur-Oise France
172 F4 Noginsk Rus. Fed.
151 E7 Nōgōhaku-san mt. Japan
215 E2 Nogoyá Arg.
215 E2 Nogoya r. Arg.
144 C3 Nohar India
150 G4 Noheji Japan
164 F5 Nohfelden Germany
166 C3 Noirmoutier, Île de i. France
166 C3 Noirmoutier-en-l'Île France
164 E5 Noisseville France
151 F7 Nojima-zaki c. Japan
144 C4 Nokha India
141 F4 Nok Kundi Pak.
187 I3 Nokomis Lake Canada
178 B3 Nola Centr. Afr. Rep.
172 I3 Nolinsk Rus. Fed.
203 H4 No Mans Land i. U.S.A.
184 B3 Nome U.S.A.
148 B1 Nomgon Mongolia
145 I1 Nomhon China
181 G5 Nomonde S. Africa
151 A8 Nomo-zaki pt Japan
172 G3 Nomzha Rus. Fed.
187 H2 Nonacho Lake Canada
181 I4 Nondweni S. Africa
152 C1 Nong'an China
154 B1 Nong Bua Lamphu Thai.
154 C1 Nông Hèt Laos
154 B2 Nong Hong Thai.
154 B1 Nong Khai Thai.
181 I3 Nongoma S. Africa
126 B4 Nonning Australia
164 E5 Nonnweiler Germany
206 C2 Nonoava Mex.
125 H2 Nonouti atoll Kiribati
152 D5 Nonsan S. Korea
154 B2 Nonthaburi Thai.
180 F5 Nonzwakazi S. Africa
127 G2 Noorama Creek watercourse Australia
164 B3 Noordbeveland i. Neth.
164 E1 Noordbroek-Uiterburen Neth.
164 C2 Noorderhaaks i. Neth.
164 D2 Noordoost Polder Neth.
164 C2 Noordwijk-Binnen Neth.
186 D5 Nootka Island Canada
139 G5 Norak Tajik.
153 C5 Norala Phil.
191 H1 Noranda Canada
159 K3 Norberg Sweden
132 D2 Nordaustlandet i. Svalbard
186 F4 Nordegg Canada
164 F1 Norden Germany
132 K2 Nordenshel'da, Arkhipelag is Rus. Fed.
164 F1 Norderland reg. Germany
164 F1 Norderney Germany
164 F1 Norderney i. Germany
165 I1 Norderstedt Germany
159 I3 Nordfjordeid Norway
158 K2 Nordfold Norway
168 C3 Nordfriesische Inseln Germany
165 I3 Nordhausen Germany
165 G1 Nordholz Germany
164 F2 Nordhorn Germany
Nordkapp c. Norway see North Cape
158 L1 Nordkjosbotn Norway
158 K2 Nordli Norway
168 E6 Nördlingen Germany
158 L3 Nordmaling Sweden
165 F5 Nordpfälzer Bergland reg. Germany
164 F3 Nordrhein-Westfalen land Germany
163 D5 Nore r. Ireland
166 F5 Nore, Pic de mt. France
198 D3 Norfolk NE U.S.A.
203 F2 Norfolk NY U.S.A.
203 E6 Norfolk VA U.S.A.
125 G4 Norfolk Island terr. Pacific Ocean
216 G7 Norfolk Island Ridge sea feature Tasman Sea
199 E4 Norfork Lake U.S.A.
164 E1 Norg Neth.
159 I3 Norheimsund Norway
151 E6 Norikura-dake vol. Japan
132 J3 Noril'sk Rus. Fed.
137 J1 Nor Kharberd Armenia
191 H3 Norland Canada
190 C5 Normal U.S.A.
199 D5 Norman U.S.A.
201 D5 Norman, Lake resr U.S.A.
124 F2 Normanby Island P.N.G.
Normandes, Îles is English Chan. see Channel Islands
Normandie reg. France see Normandy
166 D2 Normandy reg. France
124 E3 Normanton Australia
126 C5 Normanville Australia
186 D1 Norman Wells Canada
215 B4 Norquinco Arg.
158 M3 Norra Kvarken str. Fin./Sweden
158 K2 Norra Storfjället mts Sweden
164 A4 Norrent-Fontes France
202 B6 Norris Lake U.S.A.
203 F4 Norristown U.S.A.
159 L4 Norrköping Sweden
159 L4 Norrtälje Sweden
124 C5 Norseman Australia
158 L2 Norsjö Sweden
125 G3 Norsup Vanuatu
215 F3 Norte, Punta pt Buenos Aires Arg.
212 D6 Norte, Punta pt Chubut Arg.
165 H3 Nörten-Hardenberg Germany
129 B6 North, Cape c. Antarctica
189 H4 North, Cape Canada
203 G3 North Adams U.S.A.
160 F3 Northallerton U.K.
182 North America
124 B4 Northampton Australia
161 G5 Northampton U.K.
203 G3 Northampton U.S.A.
202 E5 North Anna r. U.S.A.
203 I2 North Anson U.S.A.
186 G2 North Arm b. Canada
201 D5 North Augusta U.S.A.
189 H2 North Aulatsivik Island Canada
218 L5 North Australian Basin sea feature Indian Ocean
187 H4 North Battleford Canada
191 H2 North Bay Canada
188 E2 North Belcher Islands Canada
194 A3 North Bend U.S.A.
162 F4 North Berwick U.K.
203 H3 North Berwick U.S.A.
127 F3 North Bourke Australia
190 A3 North Branch U.S.A.
189 H4 North Cape Canada
158 N1 North Cape Norway
128 D1 North Cape N.Z.
188 B3 North Caribou Lake Canada
201 D5 North Carolina state U.S.A.
194 B1 North Cascades National Park U.S.A.
191 F2 North Channel lake channel Canada
163 E2 North Channel U.K.
203 H2 North Conway U.S.A.
198 C2 North Dakota state U.S.A.
161 H6 North Downs hills U.K.
202 D3 North East U.S.A.
217 I4 Northeast Pacific Basin sea feature Pacific Ocean
201 E7 Northeast Providence Channel Bahamas
165 H3 Northeim Germany
201 F7 North End Point Bahamas
Northern prov. S. Africa see Limpopo
180 D4 Northern Cape prov. S. Africa
187 J3 Northern Indian Lake Canada
163 D3 Northern Ireland prov. U.K.
188 B4 Northern Light Lake Canada
147 G5 Northern Mariana Islands terr. Pacific Ocean
124 D3 Northern Territory admin. div. Australia
162 F4 North Esk r. U.K.
203 G3 Northfield MA U.S.A.
198 E2 Northfield MN U.S.A.
203 G2 Northfield VT U.S.A.
161 I6 North Foreland c. U.K.
196 C3 North Fork U.S.A.
190 E3 North Fox Island U.S.A.
188 D3 North French r. Canada
220 T1 North Geomagnetic Pole (2010)
160 G3 North Grimston U.K.
203 J2 North Head Canada
128 E2 North Head hd N.Z.
187 J2 North Henik Lake Canada
203 G3 North Hudson U.S.A.
128 E3 North Island N.Z.
153 B1 North Island Phil.
153 B4 North Islet rf Phil.
197 G4 North Jadito Canyon gorge U.S.A.
190 D5 North Judson U.S.A.
187 J3 North Knife r. Canada
145 E4 North Koel r. India
197 G5 North Komelik U.S.A.
152 D3 North Korea country Asia
145 H4 North Lakhimpur India
197 E3 North Las Vegas U.S.A.
199 E5 North Little Rock U.S.A.
179 D5 North Luangwa National Park Zambia
220 O1 North Magnetic Pole (2010)
190 E5 North Manchester U.S.A.
190 D3 North Manitou Island U.S.A.
187 I4 North Moose Lake Canada
186 D2 North Nahanni r. Canada
187 K5 Northome U.S.A.
North Ossetia aut. rep. Rus. Fed. see Severnaya Osetiya-Alaniya, Respublika
196 C3 North Palisade mt. U.S.A.
198 C3 North Platte U.S.A.
198 C3 North Platte r. U.S.A.
127 G7 North Point Australia
149 □ North Point Hong Kong China
220 A1 North Pole Arctic Ocean
190 E3 Northport U.S.A.
197 F3 North Rim U.S.A.
162 F1 North Ronaldsay i. U.K.
162 F1 North Ronaldsay Firth sea chan. U.K.
196 B2 North San Juan U.S.A.
187 G4 North Saskatchewan r. Canada
161 H1 North Sea Europe
187 I3 North Seal r. Canada
160 F2 North Shields U.K.
196 D2 North Shoshone Peak U.S.A.
184 D3 North Slope plain U.S.A.
161 H4 North Somercotes U.K.
127 J1 North Stradbroke Island Australia
203 H2 North Stratford U.S.A.
160 F2 North Sunderland U.K.
128 E3 North Taranaki Bight b. N.Z.
186 F4 North Thompson r. Canada
162 A3 Northton U.K.
202 D3 North Tonawanda U.S.A.
128 A7 North Trap rf N.Z.
203 G2 North Troy U.S.A.
188 D3 North Twin Island Canada
160 E2 North Tyne r. U.K.
162 A3 North Uist i. U.K.
160 E2 Northumberland National Park U.K.
189 H4 Northumberland Strait Canada
186 E5 North Vancouver Canada
203 F3 Northville U.S.A.
161 I5 North Walsham U.K.
180 F3 North West prov. S. Africa
219 F1 Northwest Atlantic Mid-Ocean Channel sea chan. N. Atlantic Ocean
124 B4 North West Cape Australia
North West Frontier prov. Pak. see Khyber Pakhtunkhwa
216 F3 Northwest Pacific Basin sea feature N. Pacific Ocean
201 E7 Northwest Providence Channel Bahamas
189 I3 North West River Canada
187 G2 Northwest Territories admin. div. Canada
161 E4 Northwich U.K.
203 F5 North Wildwood U.S.A.
220 N1 Northwind Ridge sea feature Arctic Ocean
203 H2 North Woodstock U.S.A.
160 G3 North York Moors moorland U.K.
160 G3 North York Moors National Park U.K.
189 G4 Norton Canada
160 G3 Norton U.K.
198 D4 Norton KS U.S.A.
202 B6 Norton VA U.S.A.
203 G2 Norton VT U.S.A.
179 D5 Norton Zimbabwe
184 B3 Norton Sound sea chan. U.S.A.
129 D3 Norvegia, Cape c. Antarctica
203 G4 Norwalk CT U.S.A.
202 B4 Norwalk OH U.S.A.
159 J3 Norway country Europe
203 H2 Norway U.S.A.
191 I3 Norway Bay Canada
187 J4 Norway House Canada
219 I1 Norwegian Basin sea feature N. Atlantic Ocean
185 I2 Norwegian Bay Canada
157 D3 Norwegian Sea N. Atlantic Ocean
191 G4 Norwich Canada
161 I5 Norwich U.K.
203 G4 Norwich CT U.S.A.
203 F3 Norwich NY U.S.A.
203 H3 Norwood MA U.S.A.
203 F2 Norwood NY U.S.A.
202 A5 Norwood OH U.S.A.
187 H1 Nose Lake Canada
150 G2 Noshappu-misaki hd Japan
150 G4 Noshiro Japan
173 D5 Nosivka Ukr.
172 H3 Noskovo Rus. Fed.
180 D2 Nosop watercourse Africa alt. Nossob
132 G3 Nosovaya Rus. Fed.
141 E4 Nosratābād Iran
162 □ Noss, Isle of i. U.K.
214 A1 Nossa Senhora do Livramento Brazil
159 K4 Nossebro Sweden
180 C2 Nossob watercourse Africa alt. Nosop
179 E6 Nosy Varika Madag.
197 F2 Notch Peak U.S.A.
168 H4 Noteć r. Poland
170 F6 Noto, Golfo di g. Sicily Italy
159 J4 Notodden Norway
151 E6 Noto-hantō pen. Japan
189 G4 Notre-Dame, Monts mts Canada
189 J4 Notre Dame Bay Canada
191 J3 Notre-Dame-de-la-Salette Canada
203 H2 Notre-Dame-des-Bois Canada
191 J2 Notre-Dame-du-Laus Canada
191 H2 Notre-Dame-du-Nord Canada
191 G3 Nottawasaga Bay Canada
188 E3 Nottaway r. Canada
161 F5 Nottingham U.K.
202 E6 Nottoway r. U.S.A.
164 F5 Nottuln Germany
187 H5 Notukeu Creek r. Canada
178 B3 Nouabalé-Ndoki, Parc National de nat. park Congo
176 A2 Nouâdhibou Mauritania
176 A3 Nouakchott Mauritania
176 A3 Nouâmghâr Mauritania
154 C2 Nouei Vietnam
125 G4 Nouméa New Caledonia
176 B3 Nouna Burkina
180 F5 Noupoort S. Africa
158 O2 Nousu Fin.
139 G4 Nov Tajik.
214 C1 Nova América Brazil
214 B3 Nova Esperança Brazil
214 D3 Nova Friburgo Brazil
170 G2 Nova Gradiška Croatia
214 C3 Nova Granada Brazil
214 D3 Nova Iguaçu Brazil
173 E6 Nova Kakhovka Ukr.
214 D2 Nova Lima Brazil
172 D4 Novalukoml' Belarus
173 D6 Nova Odesa Ukr.
214 C2 Nova Ponte Brazil
214 C2 Nova Ponte, Represa resr Brazil
170 C2 Novara Italy
214 C1 Nova Roma Brazil
189 H5 Nova Scotia prov. Canada
196 A2 Novato U.S.A.
214 E2 Nova Venécia Brazil
214 B1 Nova Xavantino Brazil
133 Q2 Novaya Sibir', Ostrov i. Rus. Fed.
132 G2 Novaya Zemlya is Rus. Fed.
171 L3 Nova Zagora Bulg.
167 F3 Novelda Spain
168 I7 Nové Zámky Slovakia
172 E3 Novgorodskaya Oblast' admin. div. Rus. Fed.
173 E5 Novhorod-Sivers'kyy Ukr.
139 J1 Novichikha Rus. Fed.
171 J3 Novi Iskŭr Bulg.
150 H1 Novikovo Rus. Fed.
170 C2 Novi Ligure Italy
171 L3 Novi Pazar Bulg.
171 I3 Novi Pazar Serbia
171 H2 Novi Sad Serbia
173 G6 Novoaleksandrovsk Rus. Fed.
173 G5 Novoanninskiy Rus. Fed.
210 F5 Novo Aripuanã Brazil
173 F6 Novoazovs'k Ukr.
139 G5 Novobod Tajik.
172 H3 Novocheboksarsk Rus. Fed.
173 G6 Novocherkassk Rus. Fed.
139 H2 Novodolinka Kazakh.
172 G1 Novodvinsk Rus. Fed.
212 F3 Novo Hamburgo Brazil
214 C3 Novo Horizonte Brazil
168 G6 Novohradské hory mts Czech Rep.
173 C5 Novohrad-Volyns'kyy Ukr.
138 F1 Novoishimskiy Kazakh.
138 D1 Novokaolinovyy Rus. Fed.
173 G6 Novokubansk Rus. Fed.
138 B1 Novokuybyshevsk Rus. Fed.
146 A1 Novokuznetsk Rus. Fed.
129 D3 Novolazarevskaya research stn Antarctica
139 H2 Novomarkovka Kazakh.
170 F2 Novo Mesto Slovenia
172 F4 Novomichurinsk Rus. Fed.
173 F6 Novomikhaylovskiy Rus. Fed.
172 F4 Novomoskovsk Rus. Fed.
173 E5 Novomoskovs'k Ukr.
173 D5 Novomyrhorod Ukr.
173 G5 Novonikolayevskiy Rus. Fed.
139 G1 Novonikol'skoye Kazakh.
173 E6 Novooleksiyivka Ukr.
138 D2 Novoorsk Rus. Fed.
138 F1 Novopokrovka Kustanayskaya Oblast' Kazakh.
139 F1 Novopokrovka Severnyy Kazakhstan Kazakh.
139 J2 Novopokrovka Vostochnyy Kazakhstan Kazakh.
150 D2 Novopokrovka Rus. Fed.
173 G6 Novopokrovskaya Rus. Fed.
173 I5 Novorepnoye Rus. Fed.
173 F6 Novorossiysk Rus. Fed.
133 L2 Novorybnaya Rus. Fed.
169 N2 Novorzhev Rus. Fed.
173 E6 Novoselivs'ke Ukr.
169 N1 Novosel'ye Rus. Fed.
138 C1 Novosergiyevka Rus. Fed.
173 F6 Novoshakhtinsk Rus. Fed.
150 C2 Novoshakhtinskiy Rus. Fed.
132 J4 Novosibirsk Rus. Fed.
Novosibirskiye Ostrova is Rus. Fed. see New Siberia Islands
172 D3 Novosokol'niki Rus. Fed.
172 H4 Novospasskoye Rus. Fed.
138 D2 Novotroitsk Rus. Fed.
173 E6 Novotroyits'ke Ukr.
173 D5 Novoukrayinka Ukr.
138 D2 Novoural'sk Rus. Fed.
173 I5 Novouzensk Rus. Fed.
139 H1 Novovarshavka Rus. Fed.
173 C5 Novovolyns'k Ukr.
173 F5 Novovoronezh Rus. Fed.
139 J2 Novoyegor'yevskoye Rus. Fed.
173 D4 Novozybkov Rus. Fed.
168 H6 Nový Jičín Czech Rep.
173 F5 Novyy Oskol Rus. Fed.
132 I3 Novyy Port Rus. Fed.
172 I3 Novyy Tor'yal Rus. Fed.
132 I3 Novyy Urengoy Rus. Fed.
146 F1 Novyy Urgal Rus. Fed.
140 D4 Now Iran
199 E4 Nowata U.S.A.
140 C3 Nowbarān Iran
141 E3 Now Deh Iran
137 L3 Nowdī Iran
144 D4 Nowgong India
187 I2 Nowleye Lake Canada
168 G4 Nowogard Poland
127 I5 Nowra Australia
140 C2 Nowshahr Iran
140 C2 Nowshahr Iran
144 C2 Nowshera Pak.
169 J6 Nowy Sącz Poland
169 J6 Nowy Targ Poland
203 E4 Noxen U.S.A.
154 C1 Noy, Xé r. Laos
154 C1 Noy, Xé r. Laos
132 I3 Noyabr'sk Rus. Fed.
186 C3 Noyes Island U.S.A.
166 F2 Noyon France
146 C2 Noyon Mongolia
181 F5 Nozizwe S. Africa
181 G6 Nqamakwe S. Africa
181 I4 Nqutu S. Africa
179 D5 Nsanje Malawi
178 B4 Ntandembele Dem. Rep. Congo
181 G3 Ntha S. Africa
171 K5 Ntoro, Kavo pt Greece
178 D4 Ntungamo Uganda
177 F3 Nuba Mountains Sudan
137 J1 Nubarashen Armenia
177 F2 Nubian Desert Sudan
215 B3 Ñuble r. Chile
Nüden Mongolia see Ulaanbadrah
210 D7 Nudo Coropuna mt. Peru
199 D6 Nueces r. U.S.A.
187 J2 Nueltin Lake Canada
206 G5 Nueva Arcadia Hond.
206 H5 Nueva Armenia Hond.
213 C2 Nueva Florida Venez.
215 F2 Nueva Helvecia Uruguay
215 B3 Nueva Imperial Chile
213 A4 Nueva Loja Ecuador
212 B6 Nueva Lubecka Arg.
207 G5 Nueva Ocotepeque Hond.
206 D2 Nueva Rosita Mex.
207 G5 Nueva San Salvador El Salvador
205 I4 Nuevitas Cuba
215 D4 Nuevo, Golfo g. Arg.
204 C2 Nuevo Casas Grandes Mex.
206 C2 Nuevo Ideal Mex.
207 E2 Nuevo Laredo Mex.
207 E2 Nuevo León state Mex.
178 E3 Nugaal watercourse Somalia
128 B7 Nugget Point N.Z.
125 F2 Nuguria Islands P.N.G.
128 F3 Nuhaka N.Z.
125 H2 Nui atoll Tuvalu
154 C2 Nui Ti On mt. Vietnam
146 B4 Nu Jiang r. China/Myanmar conv. Salween
126 A4 Nukey Bluff h. Australia
140 D3 Nūklok, Chāh-e well Iran
Nuku'alofa Tonga see Nuku'alofa
125 I4 Nuku'alofa Tonga
125 H2 Nukufetau atoll Tuvalu
123 J5 Nuku Hiva i. Fr. Polynesia
125 H2 Nukulaelae atoll Tuvalu
125 F2 Nukumanu Islands P.N.G.
Nukunono i. Pacific Ocean see Nukunonu
125 I2 Nukunonu atoll Pacific Ocean
138 D4 Nukus Uzbek.
124 C4 Nullagine Australia
124 C5 Nullarbor Plain Australia
148 F1 Nulu'erhu Shan mts China
126 F2 Numalla, Lake salt flat Australia
176 D4 Numan Nigeria
151 F6 Numata Japan
151 F7 Numazu Japan
159 J3 Numedal val. Norway
147 F7 Numfoor i. Indon.
127 F6 Numurkah Australia
189 H2 Nunaksaluk Island Canada
185 N3 Nunakuluut i. Greenland
Nunap Isua c. Greenland see Farewell, Cape
188 E2 Nunavik reg. U.S.A.
185 H3 Nunavut admin. div. Canada
202 E3 Nunda U.S.A.
127 I3 Nundle Australia
161 F5 Nuneaton U.K.
188 B3 Nungesser Lake Canada
184 B4 Nunivak Island U.S.A.

Nunkun

Sable, Île de

214 B4 **Reserva** Brazil
212 E3 **Resistencia** Arg.
171 I2 **Reșița** Romania
185 I2 **Resolute** Canada
185 L3 **Resolution Island** Canada
128 A6 **Resolution Island** N.Z.
207 G5 **Retalhuleu** Guat.
154 □ **Retan Laut, Pulau** *i.* Sing.
161 G4 **Retford** U.K.
166 G2 **Rethel** France
165 H2 **Rethem (Aller)** Germany
171 K7 **Rethymno** Greece
150 C2 **Rettikhovka** Rus. Fed.
165 K2 **Reuden** Germany
175 I6 **Réunion** *terr.* Indian Ocean
167 G2 **Reus** Spain
168 D6 **Reutlingen** Germany
196 D3 **Reveille Peak** U.S.A.
166 F5 **Revel** France
186 F4 **Revelstoke** Canada
204 B5 **Revillagigedo, Islas** *is* Mex.
186 C3 **Revillagigedo Island** U.S.A.
164 C5 **Revin** France
136 E6 **Revivim** Israel
144 E4 **Rewa** India
144 D3 **Rewari** India
194 E3 **Rexburg** U.S.A.
189 H4 **Rexton** Canada
196 A2 **Reyes, Point** U.S.A.
136 F3 **Reyhanlı** Turkey
158 B2 **Reykir** Iceland
219 G2 **Reykjanes Ridge** *sea feature* N. Atlantic Ocean
158 B3 **Reykjanestá** *pt* Iceland
158 B2 **Reykjavík** Iceland
207 E2 **Reynosa** Mex.
159 N4 **Rēzekne** Latvia
137 L3 **Rezvānshahr** Iran
161 D5 **Rhayader** U.K.
165 G3 **Rheda-Wiedenbrück** Germany
164 E3 **Rhede** Germany
164 E3 **Rhein** *r.* Germany *alt.* **Rhin (France),** *conv.* **Rhine**
164 F2 **Rheine** Germany
164 E4 **Rheinisches Schiefergebirge** *hills* Germany
164 F5 **Rheinland-Pfalz** *land* Germany
165 K1 **Rheinsberg** Germany
165 G6 **Rheinstetten** Germany
166 H2 **Rhin** *r.* France *alt.* **Rhein (Germany),** *conv.* **Rhine**
168 C5 **Rhine** *r.* Europe *alt.* **Rhein (Germany),** *alt.* **Rhin (France)**
203 G4 **Rhinebeck** U.S.A.
190 C3 **Rhinelander** U.S.A.
165 K2 **Rhinkanal** *canal* Germany
165 K2 **Rhinluch** *marsh* Germany
165 K2 **Rhinow** Germany
170 C2 **Rho** Italy
203 H4 **Rhode Island** *state* U.S.A.
171 M6 **Rhodes** Greece
171 M6 **Rhodes** *i.* Greece
194 D2 **Rhodes Peak** U.S.A.
171 K4 **Rhodope Mountains** Bulg./Greece
166 G4 **Rhône** *r.* France/Switz.
161 D4 **Rhyl** U.K.
214 E2 **Riacho** Brazil
214 D1 **Riacho de Santana** Brazil
215 D4 **Riachos, Islas de los** *is* Arg.
214 C1 **Rialma** Brazil
214 C1 **Rianópolis** Brazil
144 C2 **Riasi** India
155 B2 **Riau, Kepulauan** *is* Indon.
167 C1 **Ribadeo** Spain
167 D1 **Ribadesella** Spain
214 B3 **Ribas do Rio Pardo** Brazil
141 F4 **Ribat Qila** Pak.
179 D5 **Ribáuè** Moz.
160 E4 **Ribble** *r.* U.K.
159 J5 **Ribe** Denmark
164 A5 **Ribécourt-Dreslincourt** France
214 C4 **Ribeira** *r.* Brazil
214 C3 **Ribeirão Preto** Brazil
164 B5 **Ribemont** France
166 E4 **Ribérac** France
210 E6 **Riberalta** Bol.
173 D6 **Rîbnița** Moldova
168 F3 **Ribnitz-Damgarten** Germany
168 G6 **Říčany** Czech Rep.
197 E4 **Rice** U.S.A.
191 F2 **Rice Lake** *l.* Canada
190 B3 **Rice Lake** U.S.A.
190 A4 **Riceville** *IA* U.S.A.
202 D4 **Riceville** *PA* U.S.A.
181 J4 **Richards Bay** S. Africa
187 G3 **Richardson** *r.* Canada
199 D5 **Richardson** U.S.A.
203 H2 **Richardson Lakes** U.S.A.
184 E3 **Richardson Mountains** Canada
128 B6 **Richardson Mountains** N.Z.
197 F2 **Richfield** U.S.A.
203 F3 **Richfield Springs** U.S.A.
203 E3 **Richford** *NY* U.S.A.
203 G2 **Richford** *VT* U.S.A.
190 B5 **Richland** *IA* U.S.A.
192 C2 **Richland** *WA* U.S.A.
190 B4 **Richland Center** U.S.A.
202 C6 **Richlands** U.S.A.
127 I4 **Richmond** *N.S.W.* Australia
124 E4 **Richmond** *Qld* Australia
191 J3 **Richmond** Canada
128 D4 **Richmond** N.Z.
181 I4 **Richmond** *Kwazulu-Natal* S. Africa
180 E5 **Richmond** *N. Cape* S. Africa
160 F3 **Richmond** U.K.
190 E6 **Richmond** *IN* U.S.A.
202 A6 **Richmond** *KY* U.S.A.
203 I2 **Richmond** *ME* U.S.A.
191 F4 **Richmond** *MI* U.S.A.
202 E6 **Richmond** *VA* U.S.A.
203 G2 **Richmond** *VT* U.S.A.
128 D4 **Richmond, Mount** N.Z.
191 H4 **Richmond Hill** Canada
127 J2 **Richmond Range** *hills* Australia
180 B4 **Richtersveld Cultural and Botanical Landscape** *tourist site* S. Africa
180 B4 **Richtersveld National Park** S. Africa
202 B4 **Richwood** *OH* U.S.A.
202 C5 **Richwood** *WV* U.S.A.
139 J2 **Ridder** Kazakh.
191 J3 **Rideau** *r.* Canada
191 I3 **Rideau Lakes** Canada
196 D4 **Ridgecrest** U.S.A.
202 D4 **Ridgway** U.S.A.
187 I4 **Riding Mountain National Park** Canada
168 D6 **Riedlingen** Germany
164 D4 **Riemst** Belgium
165 L3 **Riesa** Germany
212 B8 **Riesco, Isla** *i.* Chile
180 D5 **Riet** *watercourse* S. Africa
159 M5 **Rietavas** Lith.
180 E6 **Rietbron** S. Africa
180 D3 **Rietfontein** S. Africa
170 E3 **Rieti** Italy
195 F4 **Rifle** U.S.A.
158 C1 **Rifstangi** *pt* Iceland
145 H3 **Riga** India
159 N4 **Rīga** Latvia
159 M4 **Riga, Gulf of** Estonia/Latvia
141 E4 **Rīgān** Iran
203 F2 **Rigaud** Canada
194 C2 **Riggins** U.S.A.
189 I3 **Rigolet** Canada
162 E5 **Rigside** U.K.
140 E3 **Rigū** Iran
145 E4 **Rihand** *r.* India
145 E4 **Rihand Dam** India
159 N3 **Riihimäki** Fin.
129 D3 **Riiser-Larsen Ice Shelf** *ice feature* Antarctica
129 E3 **Riiser-Larsen Sea** *sea* Southern Ocean
195 D5 **Riito** Mex.
170 F2 **Rijeka** Croatia
150 G5 **Rikuzen-takata** Japan
171 J3 **Rila** *mts* Bulg.
194 C3 **Riley** U.S.A.
166 G4 **Rillieux-la-Pape** France
169 J6 **Rimavská Sobota** Slovakia
186 G4 **Rimbey** Canada
170 E2 **Rimini** Italy
144 D2 **Rimo Glacier** India
189 G4 **Rimouski** Canada
165 H5 **Rimpar** Germany
162 D2 **Rimsdale, Loch** *l.* U.K.
145 G3 **Rinbung** China
206 D3 **Rincón de Romos** Mex.
144 E4 **Rind** *r.* India
158 J3 **Rindal** Norway
127 G8 **Ringarooma Bay** Australia
144 C4 **Ringas** India
145 G2 **Ring Co** *salt l.* China
164 E2 **Ringe** Germany
159 J3 **Ringebu** Norway
159 J4 **Ringkøbing** Denmark
163 E2 **Ringsend** U.K.
159 J5 **Ringsted** Denmark
Ringvassøy *i.* Norway *see* **Ringvassøya**
158 L1 **Ringvassøya** *i.* Norway
161 F7 **Ringwood** U.K.
215 B3 **Riñihue** Chile
215 B3 **Riñihue, Lago** *l.* Chile
155 E4 **Rinjani, Gunung** *vol.* Indon.
165 H2 **Rinteln** Germany
190 C4 **Rio** U.S.A.
210 C5 **Río Abiseo, Parque Nacional** *nat. park* Peru
214 A2 **Rio Alegre** Brazil
210 C4 **Riobamba** Ecuador
197 H2 **Rio Blanco** U.S.A.
210 E6 **Rio Branco** Brazil
213 E4 **Rio Branco, Parque Nacional do** *nat. park* Brazil
214 C4 **Rio Branco do Sul** Brazil
214 A3 **Rio Brilhante** Brazil
215 B4 **Río Bueno** Chile
213 E2 **Río Caribe** Venez.
215 D1 **Río Ceballos** Arg.
214 C3 **Rio Claro** Brazil
213 E2 **Rio Claro** Trin. and Tob.
215 D3 **Río Colorado** Arg.
215 D2 **Río Cuarto** Arg.
214 D3 **Rio de Janeiro** Brazil (City Plan **116**)
214 D3 **Rio de Janeiro** *state* Brazil
206 I7 **Río de Jesús** Panama
212 G3 **Rio do Sul** Brazil
206 I6 **Río Frío** Costa Rica
212 C8 **Río Gallegos** Arg.
212 C8 **Río Grande** Arg.
215 G2 **Rio Grande** Brazil
206 D3 **Río Grande** Mex.
207 E2 **Rio Grande** *r.* Mex./U.S.A. *alt.* **Bravo del Norte, Río**
199 D7 **Rio Grande City** U.S.A.
219 G8 **Rio Grande Rise** *sea feature* S. Atlantic Ocean
213 B2 **Ríohacha** Col.
210 C5 **Rioja** Peru
207 G3 **Río Lagartos** Mex.
211 K5 **Rio Largo** Brazil
166 F4 **Riom** France
214 A1 **Rio Manso, Represa do** *resr* Brazil
210 E7 **Río Mulatos** Bol.
215 C4 **Río Negro** *prov.* Arg.
214 C4 **Rio Negro** Brazil
215 F2 **Río Negro, Embalse del** *resr* Uruguay
173 G7 **Rioni** *r.* Georgia
215 G1 **Rio Pardo** Brazil
214 D1 **Rio Pardo de Minas** Brazil
215 D1 **Río Primero** Arg.
195 F5 **Rio Rancho** U.S.A.
197 G6 **Rio Rico** U.S.A.
215 D1 **Río Segundo** Arg.
213 A3 **Ríosucio** Col.
215 D2 **Río Tercero** Arg.
210 C4 **Río Tigre** Ecuador
153 A4 **Rio Tuba** Phil.
214 E2 **Rio Verde** Brazil
207 E3 **Río Verde** Mex.
207 C4 **Río Verde** Mex.
214 A2 **Rio Verde de Mato Grosso** Brazil
196 E2 **Rio Vista** U.S.A.
214 A2 **Riozinho** *r.* Brazil
169 O5 **Ripky** Ukr.
160 F3 **Ripley** *England* U.K.
161 F4 **Ripley** *England* U.K.
202 B5 **Ripley** *OH* U.S.A.
201 B5 **Ripley** *TN* U.S.A.
202 C5 **Ripley** *WV* U.S.A.
167 H1 **Ripoll** Spain
160 F3 **Ripon** U.K.
196 B3 **Ripon** *CA* U.S.A.
190 C4 **Ripon** *WI* U.S.A.
161 D6 **Risca** U.K.
150 G2 **Rishiri-tō** *i.* Japan
136 E5 **Rishon LeẔiyyon** Israel
141 F5 **Rish Pish** Iran
159 J4 **Risør** Norway
158 J3 **Rissa** Norway
159 N3 **Ristiina** Fin.
158 O2 **Ristijärvi** Fin.
158 O1 **Ristikent** Rus. Fed.
180 F4 **Ritchie** S. Africa
129 D3 **Ritscher Upland** *mts* Antarctica
158 L2 **Ritsem** Sweden
196 C3 **Ritter, Mount** U.S.A.
165 G1 **Ritterhude** Germany
167 E3 **Rituerto** *r.* Spain
194 C2 **Ritzville** U.S.A.
215 D2 **Rivadavia** *Buenos Aires* Arg.
215 C2 **Rivadavia** *Mendoza* Arg.
212 D2 **Rivadavia** Arg.
215 B1 **Rivadavia** Chile
170 D2 **Riva del Garda** Italy
206 C1 **Riva Palacio** Mex.
206 H5 **Rivas** Nicaragua
141 E3 **Rīvash** Iran
215 D3 **Rivera** Arg.
215 F1 **Rivera** Uruguay
176 B4 **River Cess** Liberia
203 G4 **Riverhead** U.S.A.
127 F5 **Riverina** *reg.* Australia
180 D7 **Riversdale** S. Africa
181 H5 **Riverside** S. Africa
196 D5 **Riverside** U.S.A.
126 C5 **Riverton** Australia
187 J4 **Riverton** Canada
128 B7 **Riverton** N.Z.
194 E3 **Riverton** U.S.A.
189 H4 **Riverview** Canada
166 F5 **Rivesaltes** France
201 D7 **Riviera Beach** U.S.A.
203 I1 **Rivière-Bleue** Canada
189 G4 **Rivière-du-Loup** Canada
173 C5 **Rivne** Ukr.
128 D4 **Riwaka** N.Z.
142 C5 **Riyadh** Saudi Arabia
140 D3 **Riza** *well* Iran
137 H2 **Rize** Turkey
Rizhao China *see* **Donggang**
136 E4 **Rizokarpason** Cyprus
140 E4 **Rīzū'īyeh** Iran
159 J4 **Rjukan** Norway
159 I4 **Rjuvbrokkene** *mt.* Norway
176 A3 **Rkîz** Mauritania
159 J3 **Roa** Norway
161 G5 **Roade** U.K.
158 J2 **Roan** Norway
197 H2 **Roan Cliffs** *ridge* U.S.A.
166 G3 **Roanne** France
201 C5 **Roanoke** *AL* U.S.A.
190 C5 **Roanoke** *IL* U.S.A.
202 D6 **Roanoke** *VA* U.S.A.
200 E4 **Roanoke** *r.* U.S.A.
201 E4 **Roanoke Rapids** U.S.A.
197 H2 **Roan Plateau** U.S.A.
163 B6 **Roaringwater Bay** Ireland
206 H4 **Roatán** Hond.
158 M3 **Röbäck** Sweden
141 F4 **Robat** *r.* Afgh.
140 E4 **Robāţ** Iran
140 E3 **Robāţ-e Khān** Iran
127 F8 **Robbins Island** Australia
126 C6 **Robe** Australia
163 B4 **Robe** *r.* Ireland
126 D3 **Robe, Mount** *h.* Australia
165 K1 **Röbel** Germany
188 E3 **Robert-Bourassa, Réservoir** *resr* Canada
199 C6 **Robert Lee** U.S.A.
194 D3 **Roberts** U.S.A.
127 J2 **Roberts, Mount** Australia
196 D2 **Roberts Creek Mountain** U.S.A.
158 M2 **Robertsfors** Sweden
145 E4 **Robertsganj** India
199 E5 **Robert S. Kerr Reservoir** U.S.A.
180 C6 **Robertson** S. Africa
175 A4 **Robertsport** Liberia
125 C4 **Robertstown** Australia
189 F4 **Roberval** Canada
185 L1 **Robeson Channel** Canada/Greenland
160 G3 **Robin Hood's Bay** U.K.
149 □ **Robin's Nest** *h. Hong Kong* China
200 C4 **Robinson** U.S.A.
124 B4 **Robinson Ranges** *hills* Australia
125 E5 **Robinvale** Australia
197 G5 **Robles Junction** U.S.A.
197 G5 **Robles Pass** U.S.A.
187 I4 **Roblin** Canada
186 F4 **Robson, Mount** Canada
199 D7 **Robstown** U.S.A.
207 F4 **Roca Partida, Punta** *pt* Mex.
170 E6 **Rocca Busambra** *mt. Sicily* Italy
215 F2 **Rocha** Uruguay
160 E4 **Rochdale** U.K.
214 A2 **Rochedo** Brazil
164 D4 **Rochefort** Belgium
166 D4 **Rochefort** France
188 F2 **Rochefort, Lac** *l.* Canada
172 G2 **Rochegda** Rus. Fed.
190 C5 **Rochelle** U.S.A.
126 F6 **Rochester** Australia
161 H6 **Rochester** U.K.
190 D5 **Rochester** *IN* U.S.A.
190 A3 **Rochester** *MN* U.S.A.
203 H3 **Rochester** *NH* U.S.A.
202 E3 **Rochester** *NY* U.S.A.
161 H6 **Rochford** U.K.
165 K3 **Rochlitz** Germany
166 C2 **Roc'h Trévezel** *h.* France
186 D2 **Rock** *r.* Canada
190 B5 **Rock** *r.* U.S.A.
156 C4 **Rockall** *i.* N. Atlantic Ocean
219 H2 **Rockall Bank** *sea feature* N. Atlantic Ocean
129 B4 **Rockefeller Plateau** *plat.* Antarctica
190 C4 **Rockford** U.S.A.
187 H5 **Rockglen** Canada
124 F4 **Rockhampton** Australia
190 C1 **Rock Harbor** U.S.A.
201 D5 **Rock Hill** U.S.A.
124 B5 **Rockingham** Australia
201 E5 **Rockingham** U.S.A.
203 G2 **Rock Island** Canada
190 B5 **Rock Island** U.S.A.
198 D1 **Rocklake** U.S.A.
203 F2 **Rockland** Canada
203 H3 **Rockland** *MA* U.S.A.
203 I2 **Rockland** *ME* U.S.A.
190 C2 **Rockland** *MI* U.S.A.
126 E6 **Rocklands Reservoir** Australia
197 H3 **Rock Point** U.S.A.
203 H3 **Rockport** U.S.A.
198 D3 **Rock Rapids** U.S.A.
194 F2 **Rock Springs** *MT* U.S.A.
194 E3 **Rock Springs** *WY* U.S.A.
199 C6 **Rocksprings** U.S.A.
202 D3 **Rockton** Canada
190 D6 **Rockville** *IN* U.S.A.
202 E5 **Rockville** *MD* U.S.A.
203 I2 **Rockwood** U.S.A.
195 G4 **Rocky Ford** U.S.A.
202 B5 **Rocky Fork Lake** U.S.A.
191 F2 **Rocky Island Lake** Canada
201 E5 **Rocky Mount** *NC* U.S.A.
202 D6 **Rocky Mount** *VA* U.S.A.
186 G4 **Rocky Mountain House** Canada
194 F3 **Rocky Mountain National Park** U.S.A.
192 D2 **Rocky Mountains** Canada/U.S.A.
186 F4 **Rocky Mountains Forest Reserve** *nature res.* Canada
164 B5 **Rocourt-St-Martin** France
164 C5 **Rocroi** France
159 J3 **Rodberg** Norway
159 J5 **Rødbyhavn** Denmark
189 I3 **Roddickton** Canada
162 B3 **Rodel** U.K.
164 E1 **Roden** Neth.
165 J4 **Rödental** Germany
215 C1 **Rodeo** Arg.
206 C2 **Rodeo** Mex.
197 H6 **Rodeo** U.S.A.
166 F4 **Rodez** France
165 K5 **Roding** Germany
139 I1 **Rodino** Rus. Fed.
141 G5 **Rodkhan** Pak.
172 G3 **Rodniki** Rus. Fed.
138 D2 **Rodnikovka** Kazakh.
Rodos Greece *see* **Rhodes**
Rodos *i.* Greece *see* **Rhodes**
218 I6 **Rodrigues Island** Mauritius
124 B4 **Roebourne** Australia
124 C3 **Roebuck Bay** Australia
181 H2 **Roedtan** S. Africa
164 D3 **Roermond** Neth.
164 B4 **Roeselare** Belgium
185 J3 **Roes Welcome Sound** *sea chan.* Canada
165 J2 **Rogätz** Germany
199 E4 **Rogers** U.S.A.
191 F3 **Rogers City** U.S.A.
196 D4 **Rogers Lake** U.S.A.
194 D3 **Rogerson** U.S.A.
202 B6 **Rogersville** U.S.A.
188 E3 **Roggan** *r.* Canada
217 N8 **Roggeveen Basin** *sea feature* S. Pacific Ocean
180 D6 **Roggeveld** *plat.* S. Africa
180 D6 **Roggeveldberge** *esc.* S. Africa
158 K2 **Rognan** Norway
194 A3 **Rogue** *r.* U.S.A.
196 A2 **Rohnert Park** U.S.A.
168 F6 **Rohrbach in Oberösterreich** Austria
164 F5 **Rohrbach-lès-Bitche** France
144 B4 **Rohri Sangar** Pak.
144 D3 **Rohtak** India
154 B1 **Roi Et** Thai.
123 J5 **Roi-Georges, Îles du** *is* Fr. Polynesia
164 E5 **Roisel** France
159 M4 **Roja** Latvia
215 E2 **Rojas** Arg.
144 E3 **Rojhan** Pak.
207 E3 **Rojo, Cabo** Mex.
155 E2 **Rokan** *r.* Indon.
159 N5 **Rokiškis** Lith.
158 M2 **Roknäs** Sweden
173 C5 **Rokytne** Ukr.
214 B3 **Rolândia** Brazil
198 F4 **Rolla** U.S.A.
159 J3 **Rollag** Norway
128 D5 **Rolleston** N.Z.
191 H2 **Rollet** Canada
201 F7 **Rolleville** Bahamas
191 I2 **Rolphton** Canada
124 E4 **Roma** Australia
Roma Italy *see* **Rome**
181 G4 **Roma** Lesotho
159 L4 **Roma** Sweden
201 E5 **Romain, Cape** U.S.A.
189 H3 **Romaine** *r.* Canada
169 M7 **Roman** Romania
219 H6 **Romanche Gap** *sea feature* S. Atlantic Ocean
147 E7 **Romang, Pulau** *i.* Indon.
171 K1 **Romania** *country* Europe
146 D1 **Romanovka** Rus. Fed.
173 G5 **Romanovka** *Saratovskaya Oblast'* Rus. Fed.
139 J1 **Romanovo** Rus. Fed.
166 G4 **Romans-sur-Isère** France
184 B3 **Romanzof, Cape** U.S.A.
166 H2 **Rombas** France
153 B3 **Romblon** Phil.
153 B3 **Romblon** *i.* Phil.
Rombo, Ilhéus Secos ou do *i.* Cape Verde *see* **Secos, Ilhéus**
170 E4 **Rome** Italy (City Plan **112**)
201 C5 **Rome** *GA* U.S.A.
203 I2 **Rome** *ME* U.S.A.
203 F3 **Rome** *NY* U.S.A.
191 F4 **Romeo** U.S.A.
161 H6 **Romford** U.K.
166 F2 **Romilly-sur-Seine** France
Romitan Uzbek. *see* **Romiton**
138 F5 **Romiton** Uzbek.
202 D5 **Romney** U.S.A.
161 H6 **Romney Marsh** *reg.* U.K.
173 E5 **Romny** Ukr.
159 J5 **Rømø** *i.* Denmark
166 E3 **Romorantin-Lanthenay** France
154 B5 **Rompin** *r.* Malaysia
161 F7 **Romsey** U.K.
143 A3 **Ron** India
154 C1 **Ron** Vietnam
162 C1 **Rona** *i. Scotland* U.K.
162 C3 **Rona** *i. Scotland* U.K.
162 □ **Ronas Hill** U.K.
211 H6 **Roncador, Serra do** *hills* Brazil
125 F2 **Roncador Reef** Solomon Is
167 D4 **Ronda** Spain
159 J3 **Rondane Nasjonalpark** *nat. park* Norway
213 C3 **Rondón** Col.
213 E4 **Rondon, Pico** *h.* Brazil
214 A2 **Rondonópolis** Brazil
135 F3 **Rondu** Pak.
149 C5 **Rong'an** China
149 B4 **Rongchang** China
152 B5 **Rongcheng** China
152 B5 **Rongcheng Wan** *b.* China
145 G3 **Rong Chu** *r.* China
216 G6 **Rongelap** *atoll* Marshall Is
149 C5 **Rongjiang** China
149 C6 **Rong Jiang** *r.* China
145 H5 **Rongklang Range** *mts* Myanmar
128 E4 **Rongotea** N.Z.
149 C5 **Rongshui** China
149 D6 **Rongxian** *Guangxi* China
149 B4 **Rongxian** *Sichuan* China
159 K5 **Rønne** Denmark
159 K4 **Ronneby** Sweden
129 B3 **Ronne Entrance** *str.* Antarctica
129 B3 **Ronne Ice Shelf** *ice feature* Antarctica
165 H2 **Ronnenberg** Germany
164 B4 **Ronse** Belgium
181 G3 **Roodepoort** S. Africa
164 E1 **Roodeschool** Neth.
Roordahuizum Neth. *see* **Reduzum**
144 D3 **Roorkee** India
164 C3 **Roosendaal** Neth.
197 G5 **Roosevelt** *AZ* U.S.A.
197 G1 **Roosevelt** *UT* U.S.A.
186 D3 **Roosevelt, Mount** Canada
129 B5 **Roosevelt Island** *i.* Antarctica
186 E2 **Root** *r.* Canada
190 B4 **Root** *r.* U.S.A.
172 J2 **Ropcha** Rus. Fed.
166 D4 **Roquefort** France
213 E4 **Roraima** *state* Brazil
210 F2 **Roraima, Mount** Guyana
158 J3 **Røros** Norway
158 J2 **Rørvik** Norway
169 O6 **Ros'** *r.* Ukr.
210 □ **Rosa, Cabo** *c. Galapagos Is* Ecuador
201 D2 **Rosa, Lake** Bahamas
206 B2 **Rosa, Punta** *pt* Mex.
196 C4 **Rosamond** U.S.A.
196 C4 **Rosamond Lake** U.S.A.
215 E2 **Rosario** Arg.
204 A2 **Rosario** *Baja California* Mex.
206 D2 **Rosario** *Coahuila* Mex.
206 C3 **Rosario** *Sinaloa* Mex.
206 B2 **Rosario** *Sonora* Mex.
153 B2 **Rosario** Phil.
153 B3 **Rosario** Phil.
213 B2 **Rosario** Venez.
215 E2 **Rosario del Tala** Arg.
215 F1 **Rosário do Sul** Brazil
214 A1 **Rosário Oeste** Brazil
206 A1 **Rosarito** *Baja California* Mex.
206 B2 **Rosarito** *Baja California Sur* Mex.
170 F5 **Rosarno** Italy
203 F4 **Roscoe** U.S.A.
166 C2 **Roscoff** France
163 C4 **Roscommon** Ireland
190 E3 **Roscommon** U.S.A.
163 D5 **Roscrea** Ireland
196 C2 **Rose, Mount** U.S.A.
205 L5 **Roseau** Dominica
187 J5 **Roseau** U.S.A.
127 F8 **Rosebery** Australia
189 I4 **Rose Blanche** Canada
194 B3 **Roseburg** U.S.A.
191 E3 **Rose City** U.S.A.
160 G3 **Rosedale Abbey** U.K.
177 F3 **Roseires Reservoir** Sudan
123 H4 **Rose Island** *atoll* American Samoa
199 E6 **Rosenberg** U.S.A.
159 I4 **Rosendal** Norway
181 G4 **Rosendal** S. Africa
168 F7 **Rosenheim** Germany
186 C4 **Rose Point** Canada
170 F3 **Roseto degli Abruzzi** Italy
187 H4 **Rosetown** Canada
187 I4 **Rose Valley** Canada
196 B2 **Roseville** *CA* U.S.A.
190 B5 **Roseville** *IL* U.S.A.
127 J1 **Rosewood** Australia
172 D2 **Roshchino** *Leningradskaya Oblast'* Rus. Fed.
150 D2 **Roshchino** *Primorskiy Kray* Rus. Fed.
141 E3 **Roshkhvār** Iran
180 B3 **Rosh Pinah** Namibia
139 G5 **Roshtqal'a** Tajik.
170 D3 **Rosignano Marittimo** Italy
171 K2 **Roșiori de Vede** Romania
159 K5 **Roskilde** Denmark
172 E4 **Roslavl'** Rus. Fed.
158 P1 **Roslyakovo** Rus. Fed.
127 G9 **Ross** Australia
186 C2 **Ross** *r.* Canada
128 C5 **Ross** N.Z.
128 E4 **Ross, Mount** *h.* N.Z.
170 G5 **Rossano** Italy
163 C3 **Rossan Point** Ireland
199 F5 **Ross Barnett Reservoir** U.S.A.
189 G3 **Ross Bay Junction** Canada
163 B6 **Rosscarbery** Ireland
129 A5 **Ross Dependency** *reg.* Antarctica
125 F3 **Rossel Island** P.N.G.
129 B5 **Ross Ice Shelf** *ice feature* Antarctica
189 H5 **Rossignol, Lake** Canada
129 B5 **Ross Island** *i.* Antarctica
163 E5 **Rosslare** Ireland
161 A5 **Rosslare Harbour** Ireland
165 K3 **Roßlau** Germany
191 I3 **Rossmore** Canada
176 A3 **Rosso** Mauritania
170 C3 **Rosso, Capo** *c. Corsica* France
161 E6 **Ross-on-Wye** U.K.
173 F5 **Rossosh'** Rus. Fed.
190 D1 **Rossport** Canada
186 C2 **Ross River** Canada
129 B5 **Ross Sea** *sea* Antarctica
165 I5 **Roßtal** Germany
158 K2 **Røssvatnet** *l.* Norway
190 D5 **Rossville** U.S.A.
165 L3 **Roßwein** Germany
186 D3 **Rosswood** Canada
137 J3 **Ŗōst** Iraq
141 H2 **Rostāq** Afgh.
140 D5 **Rostāq** Iran
187 H4 **Rosthern** Canada
168 F3 **Rostock** Germany
172 F3 **Rostov** Rus. Fed.
173 F6 **Rostov-na-Donu** Rus. Fed.
173 G6 **Rostovskaya Oblast'** *admin. div.* Rus. Fed.
158 M2 **Rosvik** Sweden
201 C5 **Roswell** *GA* U.S.A.
195 F5 **Roswell** *NM* U.S.A.
147 G5 **Rota** *i.* N. Mariana Is
165 I5 **Rot am See** Germany
147 E8 **Rote** *i.* Indon.
165 H1 **Rotenburg (Wümme)** Germany
165 J4 **Roter Main** *r.* Germany
165 J5 **Roth** Germany
165 G4 **Rothaargebirge** *hills* Germany
160 F2 **Rothbury** U.K.
160 F2 **Rothbury Forest** U.K.
165 I5 **Rothenburg ob der Tauber** Germany
161 G7 **Rother** *r.* U.K.
129 E2 **Rothera** *research stn* Antarctica
128 D5 **Rotherham** N.Z.
161 F4 **Rotherham** U.K.
162 E3 **Rothes** U.K.
162 C5 **Rothesay** U.K.
190 C3 **Rothschild** U.S.A.
161 G5 **Rothwell** U.K.
127 F4 **Roto** Australia
128 C5 **Rotomanu** N.Z.
170 C3 **Rotondo, Monte** *mt. Corsica* France
128 D4 **Rotoroa, Lake** N.Z.
128 F3 **Rotorua** N.Z.
128 F3 **Rotorua, Lake** N.Z.
168 F6 **Rott** *r.* Germany
165 J5 **Röttenbach** Germany
165 I5 **Rottendorf** Germany
168 G7 **Rottenmann** Austria
164 C3 **Rotterdam** Neth.
165 I3 **Rottleberode** Germany
164 E1 **Rottumeroog** *i.* Neth.
164 E1 **Rottumerplaat** *i.* Neth.
168 D6 **Rottweil** Germany
125 H3 **Rotuma** *i.* Fiji
158 K3 **Rötviken** Sweden
165 K5 **Rötz** Germany
166 F1 **Roubaix** France
166 E2 **Rouen** France
128 B6 **Rough Ridge** N.Z.
Roulers Belgium *see* **Roeselare**
189 F3 **Roundeyed Lake** Canada
160 F3 **Round Hill** U.K.
127 J3 **Round Mountain** Australia
196 D2 **Round Mountain** U.S.A.
197 H3 **Round Rock** U.S.A.
194 E2 **Roundup** U.S.A.
162 E1 **Rousay** *i.* U.K.
203 G2 **Rouses Point** U.S.A.
166 F5 **Roussillon** *reg.* France
181 G5 **Rouxville** S. Africa
Rouyn Canada *see* **Rouyn-Noranda**
191 H1 **Rouyn-Noranda** Canada
158 N2 **Rovaniemi** Fin.
173 F5 **Roven'ki** Rus. Fed.
170 D2 **Rovereto** Italy
154 C2 **Rôviĕng Tbong** Cambodia
170 D2 **Rovigo** Italy
170 E2 **Rovinj** Croatia
173 H5 **Rovnoye** Rus. Fed.
127 H2 **Rowena** Australia
153 A4 **Roxas** Phil.
153 B2 **Roxas** Phil.
153 B3 **Roxas** Phil.
153 B4 **Roxas** Phil.
201 E4 **Roxboro** U.S.A.
128 B6 **Roxburgh** N.Z.
126 B3 **Roxby Downs** Australia
195 F4 **Roy** U.S.A.
163 E4 **Royal Canal** Ireland
145 F4 **Royal Chitwan National Park** Nepal
190 C1 **Royale, Isle** *i.* U.S.A.
181 H4 **Royal Natal National Park** S. Africa
191 F4 **Royal Oak** U.S.A.
166 D4 **Royan** France
164 A5 **Roye** France
161 G5 **Royston** U.K.
173 D6 **Rozdil'na** Ukr.
173 E6 **Rozdol'ne** Ukr.
173 F6 **Rozivka** Ukr.
140 C3 **Rozveh** Iran
173 G4 **Rtishchevo** Rus. Fed.
161 D5 **Ruabon** U.K.
179 B5 **Ruacana** Namibia
178 D4 **Ruaha National Park** Tanz.
128 F3 **Ruahine Range** *mts* N.Z.
128 E3 **Ruapehu, Mount** *vol.* N.Z.
128 B7 **Ruapuke Island** N.Z.
128 G2 **Ruatoria** N.Z.
172 D4 **Ruba** Belarus
142 C6 **Rub' al Khālī** *des.* Saudi Arabia
150 H3 **Rubeshibe** Japan
196 B2 **Rubicon** *r.* U.S.A.
173 F5 **Rubizhne** Ukr.
139 J2 **Rubtsovsk** Rus. Fed.
184 C3 **Ruby** U.S.A.
197 E1 **Ruby Lake** U.S.A.
197 E1 **Ruby Mountains** U.S.A.
149 D5 **Rucheng** China
202 D5 **Ruckersville** U.S.A.
145 E4 **Rudauli** India
141 F4 **Rudbar** Afgh.
137 L3 **Rūdbār** Iran
Rūd-i-Shur *watercourse* Iran *see* **Shūr, Rūd-e**
159 J5 **Rudkøbing** Denmark
146 F2 **Rudnaya Pristan'** Rus. Fed.
172 J3 **Rudnichnyy** Rus. Fed.
172 D4 **Rudnya** Rus. Fed.
138 E1 **Rudnyy** Kazakh.
150 D2 **Rudnyy** Rus. Fed.
132 G1 **Rudol'fa, Ostrov** *i.* Rus. Fed.
165 J4 **Rudolstadt** Germany
148 F3 **Rudong** China
140 C2 **Rūdsar** Iran
190 E2 **Rudyard** U.S.A.
179 D4 **Rufiji** *r.* Tanz.
215 D2 **Rufino** Arg.
176 A3 **Rufisque** Senegal
179 C5 **Rufunsa** Zambia
148 F3 **Rugao** China
161 F5 **Rugby** U.K.
198 C1 **Rugby** U.S.A.
161 F5 **Rugeley** U.K.
168 F3 **Rügen** *i.* Germany
202 B4 **Ruggles** U.S.A.
165 I5 **Rügland** Germany
140 B5 **Ruḩayyat al Ḩamr'ā'** *waterhole* Saudi Arabia
178 C4 **Ruhengeri** Rwanda
159 M4 **Ruhnu** *i.* Estonia
164 E4 **Ruhr** *r.* Germany
149 F5 **Rui'an** China
195 F5 **Ruidoso** U.S.A.
149 E5 **Ruijin** China
187 M2 **Ruin Point** Canada
179 D4 **Ruipa** Tanz.
206 C3 **Ruiz** Mex.
213 B3 **Ruiz, Nevado del** *vol.* Col.
136 F5 **Rujaylah, Ḩarrat ar** *lava field* Jordan
159 N4 **Rūjiena** Latvia
140 C5 **Rukbah** *well* Saudi Arabia
145 E3 **Rukumkot** Nepal
178 D4 **Rukwa, Lake** Tanz.
140 E5 **Rūl Ḑadnah** U.A.E.
141 E3 **Rūm** Iran
162 B4 **Rum** *i.* U.K.
171 H2 **Ruma** Serbia
140 B5 **Rumāh** Saudi Arabia
177 E4 **Rumbek** Sudan
201 F7 **Rum Cay** *i.* Bahamas
203 H2 **Rumford** U.S.A.
166 G4 **Rumilly** France
124 D3 **Rum Jungle** Australia
150 G3 **Rumoi** Japan
148 E3 **Runan** China
128 C5 **Runanga** N.Z.
128 F2 **Runaway, Cape** N.Z.
161 E4 **Runcorn** U.K.
179 B5 **Rundu** Namibia
158 L3 **Rundvik** Sweden
154 B3 **Rŭng, Kaôh** *i.* Cambodia
154 B3 **Rŭng Sânlŏem, Kaôh** *i.* Cambodia
148 E3 **Runhe** China
159 O3 **Ruokolahti** Fin.
146 A3 **Ruoqiang** China
145 H4 **Rupa** India
215 B4 **Rupanco, Lago** *l.* Chile
126 E6 **Rupanyup** Australia
155 B2 **Rupat** *i.* Indon.
188 E3 **Rupert** *r.* Canada
194 D3 **Rupert** U.S.A.
188 E3 **Rupert Bay** Canada
179 D5 **Rusape** Zimbabwe
171 K3 **Ruse** Bulg.
152 A5 **Rushan** China
161 G5 **Rushden** U.K.
190 B4 **Rushford** U.S.A.
190 C4 **Rush Lake** U.S.A.
145 H3 **Rushon** India
139 G5 **Rushon** Tajik.
190 B5 **Rushville** *IL* U.S.A.
198 C3 **Rushville** *NE* U.S.A.
126 F6 **Rushworth** Australia
199 E6 **Rusk** U.S.A.
201 D7 **Ruskin** U.S.A.
187 I4 **Russell** *Man.* Canada
203 F2 **Russell** *Ont.* Canada
128 E1 **Russell** N.Z.
198 D4 **Russell** U.S.A.
185 I2 **Russell Island** Canada
125 F2 **Russell Islands** Solomon Is
186 F2 **Russell Lake** Canada
201 C4 **Russellville** *AL* U.S.A.
199 E5 **Russellville** *AR* U.S.A.
200 C4 **Russellville** *KY* U.S.A.
165 G4 **Rüsselsheim** Germany
132 G3 **Russian Federation** *country* Asia/Europe
139 H1 **Russkaya-Polyana** Rus. Fed.
150 C3 **Russkiy, Ostrov** *i.* Rus. Fed.
137 J1 **Rust'avi** Georgia
181 G2 **Rustenburg** S. Africa
199 E5 **Ruston** U.S.A.
147 E7 **Ruteng** Indon.
197 E2 **Ruth** U.S.A.
165 G3 **Rüthen** Germany
191 H2 **Rutherglen** Canada
161 D4 **Ruthin** U.K.
172 H3 **Rutka** *r.* Rus. Fed.
203 G3 **Rutland** U.S.A.
161 G5 **Rutland Water** *resr* U.K.
187 G2 **Rutledge Lake** Canada
Rutög China *see* **Dêrub**
191 G2 **Rutter** Canada
158 N2 **Ruukki** Fin.
140 E5 **Rū'us al Jibāl** *hills* Oman
179 D5 **Ruvuma** *r.* Moz./Tanz.
136 F5 **Ruwayshid, Wādī** *watercourse* Jordan
140 D5 **Ruweis** U.A.E.
149 D5 **Ruyuan** China
139 F1 **Ruzayevka** Kazakh.
172 H4 **Ruzayevka** Rus. Fed.
148 D3 **Ruzhou** China
169 I6 **Ružomberok** Slovakia
178 C4 **Rwanda** *country* Africa
140 D2 **Ryābād** Iran
172 H2 **Ryadovo** Rus. Fed.
162 C5 **Ryan, Loch** *b.* U.K.
172 F4 **Ryazan'** Rus. Fed.
172 G4 **Ryazanskaya Oblast'** *admin. div.* Rus. Fed.
172 G4 **Ryazhsk** Rus. Fed.
132 E2 **Rybachiy, Poluostrov** *pen.* Rus. Fed.
138 D3 **Rybachiy Poselok** Uzbek.
139 J3 **Rybach'ye** Kazakh.
172 F3 **Rybinsk** Rus. Fed.
172 F3 **Rybinskoye Vodokhranilishche** *resr* Rus. Fed.
172 I4 **Rybnaya Sloboda** Rus. Fed.
169 I5 **Rybnik** Poland
172 F4 **Rybnoye** Rus. Fed.
186 F3 **Rycroft** Canada
159 K4 **Ryd** Sweden
129 B3 **Rydberg Peninsula** *pen.* Antarctica
161 F7 **Ryde** U.K.
161 H7 **Rye** U.K.
160 G3 **Rye** *r.* U.K.
173 E5 **Ryl'sk** Rus. Fed.
127 H4 **Rylstone** Australia
138 B3 **Ryn-Peski** *des.* Kazakh.
151 F5 **Ryōtsu** Japan
146 E4 **Ryukyu Islands** Japan
218 M3 **Ryukyu Trench** *sea feature* N. Pacific Ocean
169 K5 **Rzeszów** Poland
173 G4 **Rzhaksa** Rus. Fed.
172 E3 **Rzhev** Rus. Fed.

S

140 E3 **Sa'ābād** Iran
140 D4 **Sa'ādatābād** Iran
165 J6 **Saal an der Donau** Germany
165 J3 **Saale** *r.* Germany
165 J4 **Saalfeld** Germany
164 E5 **Saar** *r.* Germany
164 E5 **Saarbrücken** Germany
159 M4 **Saaremaa** *i.* Estonia
158 N2 **Saarenkylä** Fin.
164 E5 **Saargau** *reg.* Germany
158 N3 **Saarijärvi** Fin.
158 N2 **Saari-Kämä** Fin.
158 M1 **Saarikoski** Fin.
164 E5 **Saarland** *land* Germany
164 E5 **Saarlouis** Germany
137 L2 **Saatlı** Azer.
215 D3 **Saavedra** Arg.
136 F5 **Sab' Ābār** Syria
171 H2 **Šabac** Serbia
167 H2 **Sabadell** Spain
151 E7 **Sabae** Japan
155 E1 **Sabah** *state* Malaysia
154 B5 **Sabak** Malaysia
137 K2 **Sabalan, Kūhhā-ye** *mts* Iran
155 E4 **Sabalana, Kepulauan** *is* Indon.
144 D4 **Sabalgarh** India
Sabamagrande Hond. *see* **Sabanagrande**
205 H4 **Sabana, Archipiélago de** *is* Cuba
206 H5 **Sabanagrande** Hond.
213 B2 **Sabanalarga** Col.
136 D1 **Şabanözü** Turkey
214 D2 **Sabará** Brazil
143 C2 **Sabari** *r.* India
144 C5 **Sabarmati** *r.* India
170 E4 **Sabaudia** Italy
141 E3 **Sabeh** Iran
180 E5 **Sabelo** S. Africa
141 F4 **Sāberi, Hāmuūn-e** *marsh* Afgh./Iran
177 D2 **Sabhā** Libya
140 B6 **Şabḩā'** Saudi Arabia
144 D3 **Sabi** *r.* India
181 J2 **Sabie** Moz.
181 J2 **Sabie** *r.* Moz./S. Africa
181 I2 **Sabie** S. Africa
206 D2 **Sabinas** Mex.
207 D2 **Sabinas Hidalgo** Mex.
199 E6 **Sabine Lake** U.S.A.
137 L1 **Sabirabad** Azer.
153 B3 **Sablayan** Phil.
185 L5 **Sable, Cape** Canada
201 D7 **Sable, Cape** U.S.A.
125 F3 **Sable, Île de** *i.* New Caledonia

Sekoma

153 B4 **San Jose de Buenavista** Phil.
210 F7 **San José de Chiquitos** Bol.
206 B2 **San José de Comondú** Mex.
215 E1 **San José de Feliciano** Arg.
206 A2 **San José de Gracia** Mex.
206 C2 **San José de Gracia** Mex.
213 D2 **San José de Guanipa** Venez.
215 C1 **San José de Jáchal** Arg.
206 B2 **San José de la Brecha** Mex.
215 D1 **San José de la Dormida** Arg.
215 B3 **San José de la Mariquina** Chile
206 B3 **San José del Cabo** Mex.
213 B4 **San José del Guaviare** Col.
215 F2 **San José de Mayo** Uruguay
213 C3 **San José de Ocuné** Col.
206 B1 **San José de Primas** Mex.
207 D2 **San José de Raíces** Mex.
215 C1 **San Juan** Arg.
215 C1 **San Juan** *prov.* Arg.
213 A3 **San Juan** *r.* Col.
206 D2 **San Juan** Mex.
206 H6 **San Juan** *r.* Nicaragua/Panama
153 C4 **San Juan** Phil.
205 K5 **San Juan** Puerto Rico
196 B4 **San Juan** *r. CA* U.S.A.
197 H3 **San Juan** *r. UT* U.S.A.
213 D3 **San Juan** Venez.
213 E2 **San Juan** *r.* Venez.
206 G5 **San Juan, Punta** *pt* El Salvador
212 E3 **San Juan Bautista** Para.
207 E4 **San Juan Bautista Tuxtepec** Mex.
206 H5 **San Juancito** Hond.
215 B4 **San Juan dela Costa** Chile
206 I6 **San Juan del Norte** Nicaragua
206 I6 **San Juan del Norte, Bahía de** *b.* Nicaragua
213 C2 **San Juan de los Cayos** Venez.
213 D2 **San Juan de los Morros** Venez.
206 C2 **San Juan del Río** Mex.
207 E3 **San Juan del Río** Mex.
206 H6 **San Juan del Sur** Nicaragua
206 A2 **San Juanico, Punta** *pt* Mex.
197 F4 **San Juan Mountains** U.S.A.
144 D1 **Sanju He** *watercourse* China
212 C7 **San Julián** Arg.
215 E1 **San Justo** Arg.
143 A2 **Sankeshwar** India
143 D1 **Sankh** *r.* India
164 F4 **Sankt Augustin** Germany
168 D7 **Sankt Gallen** Switz.
168 D7 **Sankt Moritz** Switz.
Sankt-Peterburg Rus. Fed. *see* **St Petersburg**
168 G6 **Sankt Pölten** Austria
164 F5 **Sankt Wendel** Germany
144 D2 **Sanku** India
206 B3 **San Lázaro, Sierra de** *mts* Mex.
136 G3 **Şanlıurfa** Turkey
215 E2 **San Lorenzo** Arg.
210 F8 **San Lorenzo** Bol.
210 C3 **San Lorenzo** Ecuador
195 F6 **San Lorenzo** Mex.
167 E1 **San Lorenzo** *mt.* Spain
206 A1 **San Lorenzo, Isla** *i.* Mex.
210 C6 **San Lorenzo, Isla** *i.* Peru
212 B7 **San Lorenzo, Monte** *mt.* Arg./Chile
167 C4 **Sanlúcar de Barrameda** Spain
206 A2 **San Lucas** *Baja California Sur* Mex.
206 B3 **San Lucas** *Baja California Sur* Mex.
206 B3 **San Lucas, Cabo** *c.* Mex.
215 C2 **San Luis** Arg.
215 C2 **San Luis** *prov.* Arg.
207 G4 **San Luis** Guat.
197 E5 **San Luis** *AZ* U.S.A.
197 G5 **San Luis** *AZ* U.S.A.
215 C2 **San Luis, Sierra de** *mts* Arg.
207 D3 **San Luis de la Paz** Mex.
196 B4 **San Luis Obispo** U.S.A.
196 B4 **San Luis Obispo Bay** U.S.A.
206 D3 **San Luis Potosí** Mex.
206 D3 **San Luis Potosí** *state* Mex.
196 B3 **San Luis Reservoir** U.S.A.
204 B2 **San Luis Río Colorado** Mex.
170 E6 **San Marco, Capo** *c. Sicily* Italy
207 G5 **San Marcos** Guat.
207 E4 **San Marcos** Mex.
199 D6 **San Marcos** U.S.A.
170 E3 **San Marino** *country* Europe
170 E3 **San Marino** San Marino
129 E2 **San Martín** *research stn* Antarctica
212 C3 **San Martín** Arg.
215 C2 **San Martín** Arg.
210 F6 **San Martín** *r.* Bol.
213 B4 **San Martín** Col.
212 B7 **San Martín, Lago** *l.* Arg./Chile
206 D3 **San Martín de Bolaños** Mex.
215 B4 **San Martín de los Andes** Arg.
196 A3 **San Mateo** U.S.A.
215 D4 **San Matías, Golfo** *g.* Arg.
213 D2 **San Mauricio** Venez.
149 F4 **Sanmen** China
149 F4 **Sanmen Wan** *b.* China
148 D3 **Sanmenxia** China
210 F6 **San Miguel** *r.* Bol.
213 B4 **San Miguel** *r.* Col.
206 G5 **San Miguel** El Salvador
206 J6 **San Miguel** Panama
197 G6 **San Miguel** *AZ* U.S.A.
196 B4 **San Miguel** *CA* U.S.A.
197 H2 **San Miguel** *r.* U.S.A.
153 B3 **San Miguel Bay** Phil.
206 D3 **San Miguel de Allende** Mex.
215 E2 **San Miguel del Monte** Arg.
212 C3 **San Miguel de Tucumán** Arg.
196 B4 **San Miguel Island** U.S.A.
153 A5 **San Miguel Islands** Phil.
206 J6 **San Miguelito** Panama
207 E4 **San Miguel Sola de Vega** Mex.
149 E5 **Sanming** China
153 B3 **San Narciso** Phil.
170 F4 **Sannicandro Garganico** Italy
215 E2 **San Nicolás de los Arroyos** Arg.
196 C5 **San Nicolas Island** U.S.A.
181 F3 **Sannieshof** S. Africa
176 B4 **Sanniquellie** Liberia
169 K6 **Sanok** Poland
207 E3 **San Pablo** Mex.
153 B3 **San Pablo** Phil.
153 B3 **San Pascual** Phil.
215 E2 **San Pedro** *Buenos Aires* Arg.
212 D2 **San Pedro** Arg.
207 H4 **San Pedro** Belize
210 F7 **San Pedro** Bol.
176 B4 **San-Pédro** Côte d'Ivoire
206 B3 **San Pedro** Mex.
197 G5 **San Pedro** *watercourse* U.S.A.
167 C3 **San Pedro, Sierra de** *mts* Spain
196 C5 **San Pedro Channel** U.S.A.
213 C3 **San Pedro de Arimena** Col.
206 D2 **San Pedro de las Colonias** Mex.
212 E2 **San Pedro de Ycuamandyyú** Para.
206 G5 **San Pedro Sula** Hond.
170 C5 **San Pietro, Isola di** *i. Sardinia* Italy
162 E5 **Sanquhar** U.K.
210 C3 **Sanquianga, Parque Nacional** *nat. park* Col.
204 A2 **San Quintín, Cabo** *c.* Mex.
215 C2 **San Rafael** Arg.
196 A3 **San Rafael** U.S.A.
197 G2 **San Rafael** *r.* U.S.A.
213 C2 **San Rafael** Venez.
197 G2 **San Rafael Knob** *mt.* U.S.A.
196 C4 **San Rafael Mountains** U.S.A.
210 F6 **San Ramón** Bol.
170 B3 **San Remo** Italy
213 C1 **San Román, Cabo** *c.* Venez.
167 B1 **San Roque** Spain
199 D6 **San Saba** U.S.A.
215 E1 **San Salvador** Arg.
205 J4 **San Salvador** *i.* Bahamas
207 G5 **San Salvador** El Salvador
212 C2 **San Salvador de Jujuy** Arg.
167 E2 **San Sebastián de los Reyes** Spain
170 E3 **Sansepolcro** Italy
170 F4 **San Severo** Italy
149 F5 **Sansha** China
149 D6 **Sanshui** China
170 G2 **Sanski Most** Bos.-Herz.
149 C5 **Sansui** China
210 E7 **Santa Ana** Bol.
207 G5 **Santa Ana** El Salvador
125 G3 **Santa Ana** *i.* Solomon Is
196 D5 **Santa Ana** U.S.A.
196 D6 **Santa Anna** U.S.A.
213 B3 **Santa Bárbara** Col.
206 G5 **Santa Bárbara** Hond.
206 C2 **Santa Bárbara** Mex.
196 C4 **Santa Barbara** U.S.A.
196 B4 **Santa Barbara Channel** U.S.A.
196 C5 **Santa Barbara Island** U.S.A.
212 C3 **Santa Catalina** Chile
206 I6 **Santa Catalina** Panama
196 D5 **Santa Catalina, Gulf of** U.S.A.
Santa Catalina de Armada Spain *see* **Santa Catalina de Armada**
167 B1 **Santa Cataliña de Armada** Spain
196 C5 **Santa Catalina Island** U.S.A.
207 D2 **Santa Catarina** Mex.
210 E4 **Santa Clara** Col.
205 I4 **Santa Clara** Cuba
196 B3 **Santa Clara** *CA* U.S.A.
197 F3 **Santa Clara** *UT* U.S.A.
215 F2 **Santa Clara de Olimar** Uruguay
196 C4 **Santa Clarita** U.S.A.
170 F6 **Santa Croce, Capo** *c. Sicily* Italy
212 C8 **Santa Cruz** *r.* Arg.
210 F7 **Santa Cruz** Bol.
215 B2 **Santa Cruz** Chile
153 B2 **Santa Cruz** Phil.
153 B3 **Santa Cruz** Phil.
153 A3 **Santa Cruz** Phil.
196 A3 **Santa Cruz** U.S.A.
195 E5 **Santa Cruz** *watercourse* U.S.A.
210 □ **Santa Cruz, Isla** *i. Galapagos Is* Ecuador
207 G5 **Santa Cruz Barillas** Guat.
214 E2 **Santa Cruz Cabrália** Brazil
167 F3 **Santa Cruz de Moya** Spain
176 A2 **Santa Cruz de Tenerife** Canary Is
212 F3 **Santa Cruz do Sul** Brazil
196 C4 **Santa Cruz Island** U.S.A.
125 G3 **Santa Cruz Islands** Solomon Is
215 E1 **Santa Elena** Arg.
210 B4 **Santa Elena, Bahía de** *b.* Ecuador
206 H6 **Santa Elena, Cabo** *c.* Costa Rica
170 G5 **Santa Eufemia, Golfo di** *g.* Italy
215 E1 **Santa Fe** Arg.
215 E1 **Santa Fe** *prov.* Arg.
206 I6 **Santa Fe** Panama
195 F5 **Santa Fe** U.S.A.
214 B2 **Santa Helena de Goiás** Brazil
148 B4 **Santai** *Sichuan* China
139 J3 **Santai** *Xinjiang* China
212 B8 **Santa Inés, Isla** *i.* Chile
215 C3 **Santa Isabel** Arg.
125 F2 **Santa Isabel** *i.* Solomon Is
207 G5 **Santa Lucia** Guat.
215 F2 **Santa Lucia** *r.* Uruguay
195 B4 **Santa Lucia Range** *mts* U.S.A.
214 A2 **Santa Luisa, Serra de** *hills* Brazil
176 □ **Santa Luzia** *i.* Cape Verde
206 B2 **Santa Margarita, Isla** *i.* Mex.
212 C3 **Santa María** Arg.
156 A7 **Santa Maria** *i.* Azores
211 G4 **Santa Maria** *Amazonas* Brazil
212 F3 **Santa Maria** Brazil
215 F1 **Santa Maria** *r.* Brazil
176 □ **Santa Maria** Cape Verde
204 C2 **Santa María** *r.* Mex.
210 D4 **Santa María** Peru
196 B4 **Santa Maria** U.S.A.
181 I3 **Santa Maria, Cabo de** *c.* Moz.
167 C4 **Santa Maria, Cabo de** *c.* Port.
201 F7 **Santa Maria, Cape** Bahamas
215 B3 **Santa María, Isla** *i.* Chile
211 I5 **Santa Maria das Barreiras** Brazil
214 D1 **Santa Maria da Vitória** Brazil
213 D2 **Santa María de Ipire** Venez.
206 C2 **Santa María del Oro** Mex.
206 D3 **Santa María del Río** Mex.
171 H5 **Santa Maria di Leuca, Capo** *c.* Italy
125 G3 **Santa Maria Island** Vanuatu
213 B2 **Santa Marta** Col.
213 B2 **Santa Marta, Sierra Nevada de** *mts* Col.
196 C4 **Santa Monica** U.S.A.
196 C5 **Santa Monica Bay** U.S.A.
211 I6 **Santana** Brazil
214 B2 **Santana** *r.* Brazil
215 G1 **Santana da Boa Vista** Brazil
215 F1 **Santana do Livramento** Brazil
213 A4 **Santander** Col.
167 E1 **Santander** Spain
197 G5 **Santan Mountain** *h.* U.S.A.
170 C5 **Sant'Antioco** *Sardinia* Italy
170 C5 **Sant'Antioco, Isola di** *i. Sardinia* Italy
167 G3 **Sant Antoni de Portmany** Spain
196 C4 **Santa Paula** U.S.A.
211 J4 **Santa Quitéria** Brazil
211 H4 **Santarém** Brazil
167 B3 **Santarém** Port.
206 D2 **Santa Rita** Mex.
213 C2 **Santa Rita** Venez.
214 B2 **Santa Rita do Araguaia** Brazil
196 E3 **Santa Rita Park** U.S.A.
215 D3 **Santa Rosa** *La Pampa* Arg.
215 C4 **Santa Rosa** *Río Negro* Arg.
212 F3 **Santa Rosa** Brazil
207 G4 **Santa Rosa** Mex.
196 A2 **Santa Rosa** *CA* U.S.A.
195 F5 **Santa Rosa** *NM* U.S.A.
206 H6 **Santa Rosa, Parque Nacional** *nat. park* Costa Rica
206 C5 **Santa Rosa de Copán** Hond.
215 D1 **Santa Rosa del Río Primero** Arg.
210 D5 **Santa Rosa do Purus** Brazil
196 E5 **Santa Rosa Island** U.S.A.
206 A2 **Santa Rosalía** Mex.
194 C3 **Santa Rosa Range** *mts* U.S.A.
197 C5 **Santa Rosa Wash** *watercourse* U.S.A.
215 C2 **Santa Vitória do Palmar** Brazil
196 C5 **Santee** U.S.A.
201 E5 **Santee** *r.* U.S.A.
167 G3 **Sant Francesc de Formentera** Spain
212 F3 **Santiago** Brazil
176 □ **Santiago** *i.* Cape Verde
215 B2 **Santiago** Chile
215 B2 **Santiago** *admin. reg.* Chile
205 J5 **Santiago** Dom. Rep.
206 B3 **Santiago** Mex.
206 I6 **Santiago** Panama
153 B2 **Santiago** Phil.
210 □ **Santiago, Isla** *i. Galapagos Is* Ecuador
206 C3 **Santiago, Río Grande de** *r.* Mex.
207 F4 **Santiago Astata** Mex.
167 B1 **Santiago de Compostela** Spain
205 I4 **Santiago de Cuba** Cuba
212 D3 **Santiago del Estero** Arg.
206 C3 **Santiago Ixcuintla** Mex.
206 C2 **Santiago Papasquiaro** Mex.
215 F2 **Santiago Vazquez** Uruguay
206 C2 **Santiaguillo, Laguna de** *l.* Mex.
187 M2 **Santianna Point** Canada
167 G3 **Sant Joan de Labritja** Spain
167 G2 **Sant Jordi, Golf de** *g.* Spain
214 E **Santo Amaro** Brazil
214 C4 **Santo Amaro, Ilha de** *i.* Brazil
214 E **Santo Amaro de Campos** Brazil
214 C **Santo André** Brazil
212 F **Santo Angelo** Brazil
176 □ **Santo Antão** *i.* Cape Verde
214 D **Santo Antônio** *r.* Brazil
214 E **Santo Antônio, Cabo** *c.* Brazil
214 B **Santo Antônio da Platina** Brazil
214 E **Santo Antônio de Jesus** Brazil
214 A **Santo Antônio de Leverger** Brazil
210 E **Santo Antônio do Içá** Brazil
214 D **Santo Antônio do Monte** Brazil
211 G **Santo Corazón** Bol.
205 K5 **Santo Domingo** Dom. Rep.
207 G5 **Santo Domingo** Guat.
206 A1 **Santo Domingo** *Baja California* Mex.
206 B2 **Santo Domingo** *Baja California Sur* Mex.
206 D3 **Santo Domingo** *San Luis Potosí* Mex.
206 H5 **Santo Domingo** Nicaragua
213 C2 **Santo Domingo** *r.* Venez.
192 E4 **Santo Domingo Pueblo** U.S.A.
167 E1 **Santoña** Spain
152 D2 **Santong He** *r.* China
214 D1 **Santo Onofre** *r.* Brazil
171 K6 **Santorini** *i.* Greece
214 C3 **Santos** Brazil
214 D3 **Santos Dumont** Brazil
213 D3 **Santos Luzardo, Parque Nacional** *nat. park* Venez.
219 F7 **Santos Plateau** *sea feature* S. Atlantic Ocean
206 C1 **Santo Tomás** Mex.
206 H5 **Santo Tomás** Nicaragua
210 D6 **Santo Tomás** Peru
212 E3 **Santo Tomé** Arg.
197 F3 **Sanup Plateau** U.S.A.
212 B7 **San Valentín, Cerro** *mt.* Chile
206 G5 **San Vicente** El Salvador
153 B2 **San Vicente** Phil.
210 C6 **San Vicente de Cañete** Peru
213 B4 **San Vicente del Caguán** Col.
170 D3 **San Vincenzo** Italy
170 E5 **San Vito, Capo** *c. Sicily* Italy
149 C4 **Sanxia Shuiku** *resr* China
149 C7 **Sanya** China
148 C3 **Sanyuan** China
152 C2 **Sanyuanpu** China
141 F2 **S. A. Nyýazow Adyndaky** Turkm.
178 B4 **Sanza Pombo** Angola
214 C3 **São Bernardo do Campo** Brazil
212 E3 **São Borja** Brazil
214 C3 **São Carlos** Brazil
214 C1 **São Domingos** Brazil
214 B2 **São Domingos** *r.* Brazil
211 H6 **São Félix** Brazil
211 H5 **São Félix** Brazil
214 E3 **São Fidélis** Brazil
176 □ **São Filipe** Cape Verde
214 D1 **São Francisco** Brazil
211 K5 **São Francisco** *r.* Brazil
212 G3 **São Francisco do Sul** Brazil
215 F1 **São Gabriel** Brazil
214 D3 **São Gonçalo** Brazil
214 D2 **São Gotardo** Brazil
214 C1 **São João da Aliança** Brazil
214 E3 **São João da Barra** Brazil
214 C3 **São João da Boa Vista** Brazil
167 B2 **São João da Madeira** Port.
214 D1 **São João do Paraíso** Brazil
214 D3 **São João Nepomuceno** Brazil
214 C3 **São Joaquim da Barra** Brazil
155 A6 **São Jorge** *i.* Azores
213 D5 **São José** Brazil
214 E3 **São José do Calçado** Brazil
215 G2 **São José do Norte** Brazil
214 C3 **São José do Rio Preto** Brazil
214 D3 **São José dos Campos** Brazil
214 C4 **São José dos Pinhais** Brazil
214 D3 **São Lourenço** Brazil
214 A2 **São Lourenço** *r.* Brazil
214 A2 **São Lourenço, Pantanal de** *marsh* Brazil
215 G1 **São Lourenço do Sul** Brazil
211 J4 **São Luís** Brazil
214 C3 **São Manuel** Brazil
214 C2 **São Marcos** *r.* Brazil
211 J4 **São Marcos, Baía de** *b.* Brazil
214 E2 **São Mateus** Brazil
214 E2 **São Mateus** *r.* Brazil
156 A6 **São Miguel** *i.* Azores
214 C2 **São Miguel** *r.* Brazil
166 G3 **Saône** *r.* France
176 □ **São Nicolau** *i.* Cape Verde
214 C3 **São Paulo** Brazil (City Plan **116**)
214 C3 **São Paulo** *state* Brazil
219 H5 **São Pedro e São Paulo** *is* N. Atlantic Ocean
211 J5 **São Raimundo Nonato** Brazil
214 D2 **São Romão** Brazil
211 K5 **São Roque, Cabo de** *c.* Brazil
214 D3 **São Sebastião** Brazil
214 D3 **São Sebastião, Ilha do** *i.* Brazil
214 C3 **São Sebastião do Paraíso** Brazil
215 G1 **São Sepé** Brazil
214 B2 **São Simão** Brazil
214 B2 **São Simão, Barragem de** *resr* Brazil
147 E6 **Sao-Siu** Indon.
São Tiago *i.* Cape Verde *see* **Santiago**
176 C4 **São Tomé** *i.* São Tomé and Príncipe
214 E3 **São Tomé, Cabo de** *c.* Brazil
176 C4 **São Tomé and Príncipe** *country* Africa
214 C3 **São Vicente** Brazil
176 □ **São Vicente** *i.* Cape Verde
167 B4 **São Vicente, Cabo de** *c.* Port.
136 C1 **Sapanca** Turkey
138 D4 **Saparmyrat Türkmenbaşy** Turkm.
124 C2 **Saparua** Indon.
136 C2 **Şaphane Dağı** *mt.* Turkey
206 J7 **Sapo, Serranía del** *mts* Panama
176 B4 **Sapo National Park** Liberia
150 G3 **Sapporo** Japan
170 F4 **Sapri** Italy
155 D4 **Sapudi** *i.* Indon.
199 D4 **Sapulpa** U.S.A.
141 E3 **Sāqī** Iran
140 B2 **Saqqez** Iran
137 K3 **Sarā** Iran
140 B2 **Sarāb** Iran
137 K5 **Sarābe Meymeh** Iran
154 B2 **Sara Buri** Thai.
141 F2 **Saragt** Turkm.
210 C4 **Saraguro** Ecuador
171 H3 **Sarajevo** Bos.-Herz.
141 F2 **Sarakhs** Iran
138 D2 **Saraktash** Rus. Fed.
138 C2 **Saralzhin** Kazakh.
145 H4 **Saramati** *mt.* India
139 H2 **Saran'** Kazakh.
155 D3 **Saran, Gunung** *mt.* Indon.
203 G2 **Saranac** *r.* U.S.A.
203 F2 **Saranac Lake** U.S.A.
171 I5 **Sarandë** Albania
215 F2 **Sarandí del Yí** Uruguay
215 F2 **Sarandí Grande** Uruguay
153 C5 **Sarangani** *i.* Phil.
153 C5 **Sarangani Bay** Phil.
153 C5 **Sarangani Islands** Phil.
153 C5 **Sarangani Strait** Phil.
172 H4 **Saransk** Rus. Fed.
154 A1 **Saraphi** Thai.
132 G4 **Sarapul** Rus. Fed.
213 C3 **Sarare** *r.* Venez.
201 D7 **Sarasota** U.S.A.
144 B5 **Saraswati** *r.* India
173 D6 **Sarata** Ukr.
194 F3 **Saratoga** U.S.A.
203 G3 **Saratoga Springs** U.S.A.
155 D2 **Saratok** *Sarawak* Malaysia
173 H5 **Saratov** Rus. Fed.
173 H5 **Saratovskaya Oblast'** *admin. div.* Rus. Fed.
172 I4 **Saratovskoye Vodokhranilishche** *resr* Rus. Fed.
171 K5 **Saratsina, Akrotirio** *pt* Greece
141 F5 **Saravan** Iran
Saravan Laos *see* **Salavan**
154 A2 **Sarawa** *r.* Myanmar
155 D2 **Sarawak** *state* Malaysia
136 A1 **Saray** Turkey
136 B3 **Sarayköy** Turkey
136 D2 **Sarayönü** Turkey
141 F5 **Sarbāz** Iran
141 F5 **Sarbāz** *r.* Iran
141 E3 **Sarbīsheh** Iran
170 D2 **Sarca** *r.* Italy
137 L3 **Sarcham** Iran
144 E3 **Sarda** *r.* India/Nepal
145 E3 **Sarda** *r.* Nepal
144 C3 **Sardarshahr** India
137 L5 **Sardasht** Iran
140 B2 **Sar Dasht** Iran
Sardegna *i. Sardinia* Italy *see* **Sardinia**
213 B2 **Sardinata** Col.
170 C4 **Sardinia** *i. Sardinia* Italy
137 K3 **Sardrūd** Iran
140 C5 **Sareb, Rās as** *pt* U.A.E.
158 L2 **Sareks nationalpark** *nat. park* Sweden
158 L2 **Sarektjåkkå** *mt.* Sweden
141 G2 **Sar-e Pol** Afgh.
141 G3 **Sar-e Pol** Afgh.
140 B3 **Sar-e Pol-e Z̄ahāb** Iran
140 D4 **Sare Yazd** Iran
139 H5 **Sarez, Kŭli** *l.* Tajik.
219 E4 **Sargasso Sea** Atlantic Ocean
144 C2 **Sargodha** Pak.
177 D4 **Sarh** Chad
140 D2 **Sārī** Iran
171 L7 **Saria** *i.* Greece
144 D2 **Sarigh Jilganang Kol** *salt l.* Aksai Chin
136 B2 **Sarıgöl** Turkey
137 I1 **Sarıkamış** Turkey
136 D3 **Sarıkavak** Turkey
144 D4 **Sarila** India
154 □ **Sarimbun Reservoir** Sing.
124 E4 **Sarina** Australia
136 E2 **Sarıoğlan** Turkey
Sar-i-Pul Afgh. *see* **Sar-e Pol**
177 D2 **Sarīr Tibesti** *des.* Libya
137 I2 **Sarısu** Turkey
152 C4 **Sariwŏn** N. Korea
136 C2 **Sarıyar Barajı** *resr* Turkey
136 B1 **Sarıyer** Turkey
136 F2 **Sarız** Turkey
144 E4 **Sarju** *r.* India
139 I3 **Sarkand** Kazakh.
144 B4 **Sarkari Tala** India
136 C2 **Şarkikaraağaç** Turkey
136 F2 **Şarkışla** Turkey
173 C7 **Şarköy** Turkey
141 G4 **Sarlath Range** *mts* Afgh./Pak.
166 E4 **Sarlat-la-Canéda** France
138 C5 **Şarlawuk** Turkm.
147 F7 **Sarmi** Indon.
159 K3 **Särna** Sweden
137 K5 **Sarneh** Iran
170 C1 **Sarnen** Switz.
191 F4 **Sarnia** Canada
173 C5 **Sarny** Ukr.
155 B3 **Sarolangun** Indon.
150 H2 **Saroma-ko** *l.* Japan
171 J6 **Saronikos Kolpos** *g.* Greece
173 C7 **Saros Körfezi** *b.* Turkey
144 C4 **Sarotra** India
172 G4 **Sarova** Rus. Fed.
141 H3 **Sarowbī** Afgh.
173 H6 **Sarpa, Ozero** *l. Respublika Kalmykiya - Khalm'g-Tangch* Rus. Fed.
173 H5 **Sarpa, Ozero** *l. Volgogradskaya Oblast'* Rus. Fed.
159 J4 **Sarpsborg** Norway
166 H2 **Sarrebourg** France
164 F5 **Sarreguemines** France
167 C1 **Sarria** Spain
167 F2 **Sarrión** Spain
164 C6 **Sarry** France
170 C4 **Sartène** *Corsica* France
140 C3 **Sarud, Rūdkhāneh-ye** *r.* Iran
141 G5 **Saruna** Pak.
137 L4 **Sārūq** Iran
137 J2 **Şärur** Azer.
141 F4 **Saru Tara** *tourist site* Afgh.
137 K4 **Sarvābād** Iran
168 H7 **Sárvár** Hungary
140 D4 **Sarvestān** Iran
139 G4 **Saryagash** Kazakh.
139 G2 **Saryarka** *plain* Kazakh.
138 E3 **Sarybasat** Kazakh.
139 H4 **Sary-Bulak** Kyrg.
139 I4 **Sary-Jaz** *r.* Kyrg.
139 F2 **Sarykamys** *Karagandinskaya Oblast'* Kazakh.
138 C3 **Sarykamys** *Mangistauskaya Oblast'* Kazakh.
138 D4 **Sarykamyshskoye Ozero** *salt l.* Turkm./Uzbek.
139 G4 **Sarykemer** Kazakh.
139 H2 **Sarykiyak** Kazakh.
138 F1 **Sarykol'** Kazakh.
139 H5 **Sarykol Range** *mts* China/Tajik.
139 H3 **Sarykomey** Kazakh.
139 I3 **Saryozek** Kazakh.
138 B2 **Saryozen** *r.* Kazakh./Rus. Fed.
139 H3 **Saryshagan** Kazakh.
139 F3 **Sarysu** *watercourse* Kazakh.
138 B3 **Sarytash** Kazakh.
139 H5 **Sary-Tash** Kyrg.
Sary Yazikskoye Vodokhranilishche *resr* Turkm. *see* **Saryýazy Suw Howdany**
141 F2 **Saryýazy Suw Howdany** *resr* Turkm.
139 H3 **Saryyesik-Atyrau, Peski** *des.* Kazakh.
139 I4 **Saryzhaz** Kazakh.
138 B4 **Sarzha** Kazakh.
197 G6 **Sasabe** U.S.A.
145 F4 **Sasaram** India
151 A8 **Sasebo** Japan
187 H4 **Saskatchewan** *prov.* Canada
187 I4 **Saskatchewan** *r.* Canada
187 H4 **Saskatoon** Canada
133 M2 **Saskylakh** Rus. Fed.
206 H5 **Saslaya** *mt.* Nicaragua
206 H5 **Saslaya, Parque Nacional** *nat. park* Nicaragua
181 G3 **Sasolburg** S. Africa
172 G4 **Sasovo** Rus. Fed.
203 F5 **Sassafras** U.S.A.
176 B4 **Sassandra** Côte d'Ivoire
170 C4 **Sassari** *Sardinia* Italy
165 G3 **Sassenberg** Germany
168 F3 **Sassnitz** Germany
139 J3 **Sasykkol', Ozero** *l.* Kazakh.
173 H6 **Sasykoli** Rus. Fed.
176 A3 **Satadougou** Mali
151 B9 **Sata-misaki** *c.* Japan
144 C5 **Satana** India
143 A2 **Satara** India
181 I2 **Satara** S. Africa
173 G4 **Satinka** Rus. Fed.
145 G5 **Satkhira** Bangl.
143 B2 **Satmala Range** *hills* India
144 E4 **Satna** India
139 F2 **Satpayev** Kazakh.
144 C5 **Satpura Range** *mts* India
151 B9 **Satsuma-hantō** *pen.* Japan
154 B2 **Sattahip** Thai.
165 I5 **Satteldorf** Germany
144 D2 **Satti** India
169 K7 **Satu Mare** Romania
154 B4 **Satun** Thai.
215 E1 **Sauce** Arg.
206 D2 **Sauceda** Mex.
197 F5 **Sauceda Mountains** U.S.A.
206 C1 **Saucillo** Mex.
159 I4 **Sauda** Norway
139 G4 **Saudakent** Kazakh.
158 C2 **Sauðárkrókur** Iceland
142 B4 **Saudi Arabia** *country* Asia
165 F3 **Sauerland** *reg.* Germany
190 D4 **Saugatuck** U.S.A.
203 G3 **Saugerties** U.S.A.
198 E2 **Sauk Center** U.S.A.
190 C4 **Sauk City** U.S.A.
166 G3 **Saulieu** France
191 E2 **Sault Sainte Marie** Canada
190 E2 **Sault Sainte Marie** U.S.A.
139 G1 **Saumalkol'** Kazakh.
147 F7 **Saumlakki** Indon.
166 D3 **Saumur** France
209 G7 **Saunders Island** S. Sandwich Is
145 F4 **Saura** *r.* India
138 B3 **Saura** Kazakh.
179 C4 **Saurimo** Angola
144 D5 **Sausar** India
171 I2 **Sava** *r.* Europe
206 H5 **Savá** Hond.
127 F8 **Savage River** Australia
125 I3 **Savai'i** *i.* Samoa
173 G5 **Savala** *r.* Rus. Fed.
176 C4 **Savalou** Benin
190 B4 **Savanna** U.S.A.
201 D6 **Savannah** *GA* U.S.A.
201 B5 **Savannah** *TN* U.S.A.
201 D5 **Savannah** *r.* U.S.A.
201 E7 **Savannah Sound** Bahamas
154 C1 **Savannakhét** Laos
205 I5 **Savanna-la-Mar** Jamaica
188 B3 **Savant Lake** Canada
143 A3 **Savanur** India
158 M3 **Sävar** Sweden
171 L5 **Savaştepe** Turkey
176 C4 **Savè** Benin
179 D6 **Save** *r.* Moz.
140 C3 **Sāveh** Iran
158 O3 **Saviaho** Fin.
132 F3 **Savinskiy** Rus. Fed.
170 C2 **Savona** Italy
159 O3 **Savonlinna** Fin.
158 O3 **Savonranta** Fin.
166 H4 **Savoy** *reg.* France
137 I1 **Şavşat** Turkey
159 K4 **Sävsjö** Sweden
158 O2 **Savukoski** Fin.
137 H3 **Savur** Turkey
Savu Sea *sea* Indon. *see* **Sawu, Laut**
144 D4 **Sawai Madhopur** India
154 A1 **Sawankhalok** Thai.
195 F4 **Sawatch Range** *mts* U.S.A.
162 A6 **Sawel Mountain** *h.* U.K.
154 A3 **Sawi, Ao** *b.* Thai.
127 J3 **Sawtell** Australia
190 B2 **Sawtooth Mountains** *hills* U.S.A.
147 E7 **Sawu, Laut** *sea* Indon.
161 G4 **Saxilby** U.K.
161 I5 **Saxmundham** U.K.
158 K2 **Saxnäs** Sweden
139 I3 **Sayak** Kazakh.
146 B1 **Sayano-Shushenskoye Vodokhranilishche** *resr* Rus. Fed.
Sayat Turkm. *see* **Saýat**
138 E5 **Saýat** Turkm.
207 G4 **Sayaxché** Guat.
140 D4 **Sāyen** Iran
142 D6 **Sayḩūt** Yemen
138 A2 **Saykyn** Kazakh.
Säylac Somalia *see* **Saylac**
178 E2 **Saylac** Somalia
146 D2 **Saynshand** Mongolia
139 J3 **Sayram Hu** *salt l.* China
199 D5 **Sayre** *OK* U.S.A.
202 E4 **Sayre** *PA* U.S.A.
206 D4 **Sayula** *Jalisco* Mex.
207 F4 **Sayula** *Veracruz* Mex.
186 D4 **Sayward** Canada
138 B3 **Sazdy** Kazakh.
144 C2 **Sazin** Pak.
172 E3 **Sazonovo** Rus. Fed.
176 B2 **Sbaa** Alg.
176 C1 **Sbeitla** Tunisia
160 D3 **Scafell Pike** *h.* U.K.
162 B4 **Scalasaig** U.K.
170 F5 **Scalea** Italy
162 □ **Scalloway** U.K.
163 C5 **Scalp** *h.* Ireland
162 B3 **Scalpay** *i. Scotland* U.K.
162 C3 **Scalpay** *i. Scotland* U.K.
127 H8 **Scamander** Australia
162 E2 **Scapa Flow** *inlet* U.K.
162 C4 **Scarba** *i.* U.K.
191 H4 **Scarborough** Canada
213 E2 **Scarborough** Trin. and Tob.
160 G3 **Scarborough** U.K.
153 A3 **Scarborough Shoal** *sea feature* Phil.
162 A2 **Scarp** *i.* U.K.
165 I1 **Schaale** *r.* Germany
165 I1 **Schaalsee** *l.* Germany
164 C4 **Schaerbeek** Belgium
168 D7 **Schaffhausen** Switz.
165 J3 **Schafstädt** Germany
164 C2 **Schagen** Neth.
164 C2 **Schagerbrug** Neth.
180 B3 **Schakalskuppe** Namibia
141 F4 **Schao** *watercourse* Afgh./Iran
168 F6 **Schärding** Austria
164 B3 **Scharendijke** Neth.
165 I5 **Schebheim** Germany
165 H1 **Scheeßel** Germany
189 G3 **Schefferville** Canada
Schelde *r.* Belgium *see* **Scheldt**
164 C3 **Scheldt** *r.* Belgium
197 E2 **Schell Creek Range** *mts* U.S.A.
165 I2 **Schellerten** Germany
203 G3 **Schenectady** U.S.A.
165 H1 **Schenefeld** Germany
164 C2 **Schermerhorn** Neth.
162 D4 **Schiehallion** *mt.* U.K.
165 K6 **Schierling** Germany
164 E1 **Schiermonnikoog** Neth.
164 E1 **Schiermonnikoog** *i.* Neth.
164 E1 **Schiermonnikoog Nationaal Park** *nat. park* Neth.
165 G1 **Schiffdorf** Germany
165 J1 **Schilde** *r.* Germany
164 D4 **Schinnen** Neth.
170 D2 **Schio** Italy
165 K3 **Schkeuditz** Germany
164 E4 **Schleiden** Germany
165 J4 **Schleiz** Germany
168 D3 **Schleswig** Germany
165 H1 **Schleswig-Holstein** *land* Germany
165 I4 **Schleusingen** Germany
165 H4 **Schlitz** Germany
165 G3 **Schloss Holte-Stukenbrock** Germany
165 H4 **Schlüchtern** Germany
165 I5 **Schlüsselfeld** Germany
165 I4 **Schmalkalden, Kurort** Germany
165 G3 **Schmallenberg** Germany
165 K4 **Schneeberg** Germany
165 J3 **Schneidlingen** Germany
165 H1 **Schneverdingen** Germany
203 G3 **Schodack Center** U.S.A.
190 C3 **Schofield** U.S.A.
196 □1 **Schofield Barracks** *military base* U.S.A.
164 D2 **Schokland** *tourist site* Neth.
165 K1 **Schönebeck** Germany
165 J2 **Schönebeck (Elbe)** Germany
165 I2 **Schöningen** Germany
165 H5 **Schöntal** Germany
203 I2 **Schoodic Lake** U.S.A.
190 E4 **Schoolcraft** U.S.A.
164 C3 **Schoonhoven** Neth.
165 I5 **Schopfloch** Germany
165 I2 **Schöppenstedt** Germany
165 F1 **Schortens** Germany
127 H9 **Schouten Island** Australia
124 E2 **Schouten Islands** P.N.G.
190 D1 **Schreiber** Canada
203 G3 **Schroon Lake** U.S.A.
197 F5 **Schuchuli** U.S.A.
163 B6 **Schull** Ireland
187 J2 **Schultz Lake** Canada
196 C2 **Schurz** U.S.A.
164 F2 **Schüttorf** Germany
203 G3 **Schuylerville** U.S.A.
165 J5 **Schwabach** Germany
168 D6 **Schwäbische Alb** *mts* Germany
165 H5 **Schwäbisch Hall** Germany
168 E6 **Schwabmünchen** Germany
165 G2 **Schwaförden** Germany
165 H4 **Schwalmstadt-Ziegenhain** Germany
165 K5 **Schwandorf** Germany
155 D3 **Schwaner, Pegunungan** *mts* Indon.
165 G1 **Schwanewede** Germany
165 H2 **Schwarmstedt** Germany
165 L3 **Schwarze Elster** *r.* Germany
165 I1 **Schwarzenbek** Germany
165 K4 **Schwarzenberg** Germany
164 E4 **Schwarzer Mann** *h.* Germany
180 B2 **Schwarzrand** *mts* Namibia
Schwarzwald *mts* Germany *see* **Black Forest**
168 E7 **Schwaz** Austria
168 G4 **Schwedt an der Oder** Germany
165 G5 **Schwegenheim** Germany
164 E5 **Schweich** Germany
165 I4 **Schweinfurt** Germany
165 L3 **Schweinitz** Germany
165 K1 **Schweinrich** Germany
181 F3 **Schweizer-Reneke** S. Africa
164 F3 **Schwelm** Germany
168 D6 **Schwenningen** Germany
165 J1 **Schwerin** Germany
165 J1 **Schweriner See** *l.* Germany
165 G5 **Schwetzingen** Germany
168 D7 **Schwyz** Switz.
170 E6 **Sciacca** *Sicily* Italy
170 F6 **Scicli** Italy
161 A8 **Scilly, Isles of** U.K.
202 B5 **Scioto** *r.* U.S.A.
197 F2 **Scipio** U.S.A.
194 F1 **Scobey** U.S.A.
161 I5 **Scole** U.K.
127 I4 **Scone** Australia
162 E4 **Scone** U.K.
185 P2 **Scoresby Land** *reg.* Greenland
219 G9 **Scotia Ridge** *sea feature* S. Atlantic Ocean
219 G9 **Scotia Sea** S. Atlantic Ocean
191 G4 **Scotland** Canada
162 D4 **Scotland** *admin. div.* U.K.
186 D4 **Scott, Cape** Canada
129 B5 **Scott Base** *research stn* Antarctica
181 I5 **Scottburgh** S. Africa
198 C4 **Scott City** U.S.A.
129 B5 **Scott Coast** *coastal area* Antarctica
202 D4 **Scottdale** U.S.A.
185 K2 **Scott Inlet** Canada
129 A6 **Scott Island** *i.* Antarctica
187 H3 **Scott Lake** Canada
129 E4 **Scott Mountains** *mts* Antarctica
198 C3 **Scottsbluff** U.S.A.
201 C5 **Scottsboro** U.S.A.
200 C4 **Scottsburg** U.S.A.
127 G8 **Scottsdale** Australia
195 E5 **Scottsdale** U.S.A.
196 A3 **Scotts Valley** U.S.A.
190 D4 **Scottville** U.S.A.
196 D3 **Scotty's Junction** U.S.A.
162 C2 **Scourie** U.K.
162 □ **Scousburgh** U.K.
162 E2 **Scrabster** U.K.
203 F4 **Scranton** U.S.A.
162 B4 **Scridain, Loch** *inlet* U.K.
160 G4 **Scunthorpe** U.K.
161 H7 **Seaford** U.K.
203 F5 **Seaford** U.S.A.
191 G4 **Seaforth** Canada
153 A4 **Seahorse Shoal** *sea feature* Phil.
187 J3 **Seal** *r.* Canada
180 E7 **Seal, Cape** S. Africa
126 E5 **Sea Lake** Australia
203 I3 **Seal Island** U.S.A.
189 H3 **Seal Lake** Canada
180 F7 **Seal Point** S. Africa
197 E3 **Seaman Range** *mts* U.S.A.
160 G3 **Seamer** U.K.
197 E4 **Searchlight** U.S.A.
199 F5 **Searcy** U.S.A.
196 D4 **Searles Lake** U.S.A.
190 E4 **Sears** U.S.A.
203 I2 **Searsport** U.S.A.
196 B3 **Seaside** *CA* U.S.A.
194 B2 **Seaside** *OR* U.S.A.
162 E6 **Seaton** U.K.
194 B2 **Seattle** U.S.A.
186 B2 **Seattle, Mount** Canada/U.S.A.
203 F5 **Seaville** U.S.A.
203 H3 **Sebago Lake** U.S.A.
206 A1 **Sebastián Vizcaíno, Bahía** *b.* Mex.
203 I2 **Sebasticook** *r.* U.S.A.
155 E2 **Sebatik** *i.* Indon.
136 C1 **Seben** Turkey
171 J2 **Sebeş** Romania
155 C4 **Sebesi** *i.* Indon.
191 F4 **Sebewaing** U.S.A.
172 D3 **Sebezh** Rus. Fed.
136 E3 **Sebil** Turkey
136 G1 **Şebinkarahisar** Turkey
203 I2 **Seboeis Lake** U.S.A.
203 I2 **Seboomook** U.S.A.
203 I2 **Seboomook Lake** U.S.A.
201 D7 **Sebring** U.S.A.
173 G5 **Sebrovo** Rus. Fed.
210 B5 **Sechura** Peru
210 B5 **Sechura, Bahía de** *b.* Peru
165 H5 **Seckach** Germany
203 H2 **Second Lake** U.S.A.
176 □ **Secos, Ilhéus** *is* Cape Verde
128 A6 **Secretary Island** N.Z.
181 H3 **Secunda** S. Africa
143 B2 **Secunderabad** India
198 E4 **Sedalia** U.S.A.
143 B2 **Sedam** India
126 C5 **Sedan** Australia
166 G2 **Sedan** France
128 E4 **Seddon** N.Z.
128 C4 **Seddonville** N.Z.
141 E3 **Sedeh** Iran
203 I2 **Sedgwick** U.S.A.
176 A3 **Sédhiou** Senegal
168 G6 **Sedlčany** Czech Rep.
136 E6 **Sedom** Israel
197 G4 **Sedona** U.S.A.
170 B6 **Sédrata** Alg.
159 M5 **Šeduva** Lith.
165 I1 **Seedorf** Germany
163 D5 **Seefin** *h.* Ireland
165 J2 **Seehausen** Germany
165 J2 **Seehausen (Altmark)** Germany
179 B6 **Seeheim** Namibia
165 G5 **Seeheim-Jugenheim** Germany
180 E6 **Seekoegat** S. Africa
197 E5 **Seeley** U.S.A.
129 B4 **Seelig, Mount** *mt.* Antarctica
165 H2 **Seelze** Germany
166 E2 **Sées** France
165 I3 **Seesen** Germany
165 I1 **Seevetal** Germany
176 A4 **Sefadu** Sierra Leone
181 G1 **Sefare** Botswana
171 L5 **Seferihisar** Turkey
181 G1 **Sefophe** Botswana
159 J3 **Segalstad** Norway
153 A5 **Segama** *r. Sabah* Malaysia
155 B2 **Segamat** Malaysia
165 K2 **Segeletz** Germany
172 E2 **Segezha** Rus. Fed.
139 F3 **Segiz** *salt l.* Kazakh.
139 F3 **Segiz, Ozera** *lakes* Kazakh.
167 F3 **Segorbe** Spain
176 B3 **Ségou** Mali
213 B3 **Segovia** Col.
167 D2 **Segovia** Spain
172 E2 **Segozerskoye, Ozero** *resr* Rus. Fed.
167 G1 **Segre** *r.* Spain
177 D2 **Séguédine** Niger
176 B4 **Séguéla** Côte d'Ivoire
199 D6 **Seguin** U.S.A.
215 D1 **Segundo** *r.* Arg.
167 F3 **Segura** *r.* Spain
179 C6 **Sehithwa** Botswana
181 H4 **Sehlabathebe National Park** Lesotho
144 D5 **Sehore** India
158 M1 **Seiland** *i.* Norway
199 D4 **Seiling** U.S.A.
158 M3 **Seinäjoki** Fin.
188 B4 **Seine** *r.* Canada
166 E2 **Seine** *r.* France
166 D2 **Seine, Baie de** *b.* France
166 F2 **Seine, Val de** *val.* France
169 K3 **Sejny** Poland
155 B3 **Sekayu** Indon.
180 E2 **Sekoma** Botswana

South Fox Island

167 H5 Sidi Aïssa Alg.
167 G4 Sidi Ali Alg.
176 B1 Sidi Bel Abbès Alg.
170 C7 Sidi Bouzid Tunisia
170 D7 Sidi El Hani, Sebkhet de salt pan Tunisia
176 A2 Sidi Ifni Morocco
176 B1 Sidi Kacem Morocco
154 A5 Sidikalang Indon.
162 E4 Sidlaw Hills U.K.
129 B4 Sidley, Mount mt. Antarctica
161 D7 Sidmouth U.K.
186 E5 Sidney Canada
194 F2 Sidney MT U.S.A.
198 C3 Sidney NE U.S.A.
203 F3 Sidney NY U.S.A.
202 A4 Sidney OH U.S.A.
201 D5 Sidney Lanier, Lake U.S.A.
145 H5 Sidoktaya Myanmar
136 E5 Sidon Lebanon
172 G3 Sidorovo Rus. Fed.
214 A3 Sidrolândia Brazil
181 I3 Sidvokodvo Swaziland
Sidzhak Uzbek. see Sijjaq
166 F5 Sié, Col de pass France
169 K4 Siedlce Poland
165 G4 Siegen Germany
154 B2 Siĕmréab Cambodia
170 D3 Siena Italy
169 I5 Sieradz Poland
145 H3 Si'erdingka China
215 D4 Sierra, Punta pt Arg.
199 B6 Sierra Blanca U.S.A.
215 C4 Sierra Colorada Arg.
206 H5 Sierra de Agalta, Parque Nacional nat. park Hond.
215 D4 Sierra Grande Arg.
176 A4 Sierra Leone country Africa
219 H5 Sierra Leone Basin sea feature N. Atlantic Ocean
219 H5 Sierra Leone Rise sea feature N. Atlantic Ocean
196 C4 Sierra Madre Mountains U.S.A.
206 D2 Sierra Mojada Mex.
213 C2 Sierra Nevada, Parque Nacional nat. park Venez.
213 B2 Sierra Nevada de Santa Marta, Parque Nacional nat. park Col.
196 B2 Sierraville U.S.A.
197 G6 Sierra Vista U.S.A.
168 C7 Sierre Switz.
158 N3 Sievi Fin.
149 C6 Sifang Ling mts China
171 K6 Sifnos i. Greece
167 F5 Sig Alg.
185 M2 Sigguup Nunaa pen. Greenland
169 K7 Sighetu Marmației Romania
169 L7 Sighișoara Romania
143 C5 Sigiriya Sri Lanka
154 □ Siglap Sing.
155 A1 Sigli Indon.
158 C1 Siglufjörður Iceland
153 B4 Sigma Phil.
168 D6 Sigmaringen Germany
164 E4 Signal de Botrange h. Belgium
197 E5 Signal Peak U.S.A.
164 C5 Signy-l'Abbaye France
190 A5 Sigourney U.S.A.
219 C4 Sigsbee Deep sea feature G. of Mexico
207 H5 Siguatepeque Hond.
167 E2 Sigüenza Spain
176 B3 Siguiri Guinea
159 N4 Sigulda Latvia
154 B3 Sihanoukville Cambodia
148 F3 Sihong China
144 E5 Sihora India
149 D6 Sihui China
158 N2 Siikajoki Fin.
158 N3 Siilinjärvi Fin.
137 H3 Siirt Turkey
139 G4 Sijjaq Uzbek.
155 B3 Sijunjung Indon.
186 E3 Sikanni Chief Canada
186 E3 Sikanni Chief r. Canada
144 C4 Sikar India
141 H3 Sikaram mt. Afgh.
176 B3 Sikasso Mali
139 K3 Sikeshu China
199 F4 Sikeston U.S.A.
146 F2 Sikhote-Alin' mts Rus. Fed.
171 K6 Sikinos i. Greece
144 B5 Sikka India
145 G4 Sikkim state India
158 L2 Siksjö Sweden
152 B4 Sikuaishi China
155 E1 Sikuati Sabah Malaysia
167 C1 Sil r. Spain
153 C4 Silago Phil.
159 M5 Šilalė Lith.
206 D3 Silao Mex.
153 B4 Silay Phil.
165 H1 Silberberg h. Germany
145 H4 Silchar India
136 B1 Şile Turkey
143 C2 Sileru r. India
139 H2 Silety Kazakh.
139 H1 Silety r. Kazakh.
139 H1 Siletyteniz, Ozero salt l. Kazakh.
170 C6 Siliana Tunisia
136 D3 Silifke Turkey
145 G3 Siling Co salt l. China
171 L2 Silistra Bulg.
136 B1 Silivri Turkey
159 K3 Siljan l. Sweden
159 J4 Silkeborg Denmark
159 N4 Sillamäe Estonia
144 C5 Sillod India
181 I3 Silobela S. Africa
145 G3 Silong China
199 E6 Silsbee U.S.A.
158 N2 Siltaharju Fin.
141 F5 Sīlūp r. Iran
159 M5 Šilutė Lith.
137 H2 Silvan Turkey
144 C5 Silvassa India
190 B2 Silver Bay U.S.A.
195 E5 Silver City U.S.A.
190 C1 Silver Islet Canada
194 B3 Silver Lake U.S.A.
196 D4 Silver Lake l. CA U.S.A.
190 D2 Silver Lake l. MI U.S.A.
163 C5 Silvermine Mountains hills Ireland
196 D3 Silver Peak Range mts U.S.A.
202 E5 Silver Spring U.S.A.
196 C2 Silver Springs U.S.A.
126 D3 Silverton Australia
161 D7 Silverton U.K.
191 F3 Silver Water Canada
207 G4 Silvituc Mex.
153 B3 Simara i. Phil.
191 H2 Simard, Lac l. Canada
137 K5 Şīmareh, Rūdkhāneh-ye r. Iran
145 F4 Simaria India
136 B2 Simav Turkey
136 B2 Simav Dağları mts Turkey
178 C3 Simba Dem. Rep. Congo
191 G4 Simcoe Canada
191 H3 Simcoe, Lake Canada
143 D1 Simdega India
178 D2 Simēn mts Eth.
Simēn Mountains Eth. see Simēn
Simeuluë i. Indon. see Simeulue
155 A2 Simeulue i. Indon.
173 E6 Simferopol' Ukr.
145 E3 Simikot Nepal
213 B3 Simití Col.
196 C4 Simi Valley U.S.A.
195 F4 Simla U.S.A.
169 K7 Şimleu Silvaniei Romania
164 E4 Simmerath Germany
164 F5 Simmern (Hunsrück) Germany
196 C4 Simmler U.S.A.
197 F4 Simmons U.S.A.
201 F7 Simm's Bahamas
158 N2 Simojärvi l. Fin.
206 D2 Simon Mex.
186 F4 Simonette r. Canada
187 I4 Simonhouse Canada
168 D7 Simplon Pass Switz.
124 D4 Simpson Desert Australia
190 D1 Simpson Island Canada
196 D2 Simpson Park Mountains U.S.A.
159 K5 Simrishamn Sweden
153 A5 Simunul i. Phil.
146 H2 Simushir, Ostrov i. Rus. Fed.
143 A2 Sina r. India
177 F2 Sīnā', Shibh Jazīrat pen. Egypt
155 A2 Sinabang Indon.
154 A5 Sinabung vol. Indon.
164 C5 Sinai, Mont h. France
206 B2 Sinaloa state Mex.
170 D3 Sinalunga Italy
149 C5 Sinan China
152 C4 Sinanju N. Korea
145 H5 Sinbyugyun Myanmar
136 F2 Sincan Turkey
213 B2 Sincé Col.
213 B2 Sincelejo Col.
201 D5 Sinclair, Lake U.S.A.
186 E4 Sinclair Mills Canada
180 B2 Sinclair Mine Namibia
162 E2 Sinclair's Bay U.K.
144 D4 Sind r. India
153 B4 Sindañgan Phil.
155 C4 Sindangbarang Indon.
144 B4 Sindari India
168 D6 Sindelfingen Germany
143 B2 Sindgi India
144 B4 Sindh prov. Pak.
143 B3 Sindhnur India
136 B2 Sındırgı Turkey
144 D6 Sindkhed India
144 C5 Sindkheda India
152 C4 Sin-do i. China
172 I2 Sindor Rus. Fed.
145 F5 Sindri India
144 B3 Sind Sagar Doab lowland Pak.
172 I3 Sinegor'ye Rus. Fed.
171 L4 Sinekçi Turkey
167 B4 Sines Port.
167 B4 Sines, Cabo de c. Port.
158 N2 Sinettä Fin.
176 B4 Sinfra Côte d'Ivoire
177 F3 Singa Sudan
144 E3 Singahi India
144 D2 Singa Pass India
154 □ Singapore country Asia
154 B5 Singapore Sing. (City Plan 102)
154 □ Singapore, Strait of Indon./Sing.
155 E4 Singaraja Indon.
154 B2 Sing Buri Thai.
191 G3 Singhampton Canada
178 D4 Singida Tanz.
124 C2 Singkang Indon.
155 C2 Singkawang Indon.
154 A5 Singkil Indon.
127 I4 Singleton Australia
152 D4 Sin'gye N. Korea
143 C5 Sinharaja Forest Reserve nature res. Sri Lanka
152 D3 Sinhŭng N. Korea
170 C4 Siniscola Sardinia Italy
138 E2 Siniy-Shikhan Rus. Fed.
170 G3 Sinj Croatia
124 C2 Sinjai Indon.
137 H3 Sinjār Iraq
137 H3 Sinjār, Jabal mt. Iraq
137 I3 Sīnji Iran
177 F3 Sinkat Sudan
Sinkiang aut. reg. China see Xinjiang Uygur Zizhiqu
152 C4 Sinmi-do i. N. Korea
165 G4 Sinn Germany
211 H2 Sinnamary Fr. Guiana
171 M2 Sinoie, Lacul lag. Romania
173 E7 Sinop Turkey
152 D3 Sinp'a N. Korea
152 E3 Sinp'o N. Korea
152 D4 Sinp'yŏng N. Korea
152 D4 Sinsang N. Korea
165 G5 Sinsheim Germany
155 D2 Sintang Indon.
205 L5 Sint Eustatius i. Neth. Antilles
164 B3 Sint-Laureins Belgium
205 L5 Sint Maarten i. Neth. Antilles
164 C3 Sint-Niklaas Belgium
199 D6 Sinton U.S.A.
164 D4 Sint-Truiden Belgium
213 A2 Sinú r. Col.
152 C3 Sinŭiju N. Korea
164 F4 Sinzig Germany
153 B5 Siocon Phil.
168 I7 Siófok Hungary
168 C7 Sion Switz.
163 D3 Sion Mills U.K.
198 D3 Sioux Center U.S.A.
198 D3 Sioux City U.S.A.
198 D3 Sioux Falls U.S.A.
188 E3 Sioux Lookout Canada
207 G5 Sipacate Guat.
153 E4 Sipalay Phil.
152 C2 Siping China
187 J3 Sipiwesk Canada
187 J3 Sipiwesk Lake Canada
144 C5 Sipra r. India
201 C5 Sipsey r. U.S.A.
155 A3 Sipura i. Indon.
206 H5 Siquia r. Nicaragua
153 B4 Siquijor Phil.
153 B4 Siquijor i. Phil.
144 B5 Sir r. Pak.
143 B3 Sira India
159 I4 Sira r. Norway
140 D5 Şīr Abū Nu'āyr i. U.A.E.
154 B2 Si Racha Thai.
Siracusa Sicily Italy see Syracuse
186 E4 Sir Alexander, Mount Canada
137 G1 Şiran Turkey
141 G5 Siranda Lake Pak.
140 D5 Şīr Banī Yās i. U.A.E.
137 L3 Sīrdān Iran
139 G4 Sirdaryo Uzbek.
124 C3 Sir Edward Pellew Group is Australia
190 A3 Siren U.S.A.
140 E5 Sīrīk Iran
154 B1 Siri Kit, Khuan Thai.
140 D4 Sīrīz Iran
186 D2 Sir James MacBrien, Mount Canada
140 D4 Sīrjān Iran
140 D4 Sīrjān salt flat Iran
126 B5 Sir Joseph Banks Group is Australia
144 E4 Sirmour India
137 I3 Şırnak Turkey
143 C2 Sironcha India
144 D4 Sironj India
143 B2 Sirpur India
196 C3 Sirretta Peak U.S.A.
140 D5 Sīrrī, Jazīreh-ye i. Iran
144 C3 Sirsa Haryana India
145 E4 Sirsa Uttar Pradesh India
186 F4 Sir Sandford, Mount Canada
143 A3 Sirsi Karnataka India
144 D3 Sirsi Uttar Pradesh India
143 B2 Sirsilla India
177 D1 Sirte Libya
177 D1 Sirte, Gulf of Libya
143 A2 Sirur India
137 I2 Şirvan Turkey
159 N5 Širvintos Lith.
137 J4 Sīrwān r. Iraq
186 F4 Sir Wilfrid Laurier, Mount Canada
170 G2 Sisak Croatia
154 C2 Sisaket Thai.
207 G3 Sisal Mex.
180 E3 Sishen S. Africa
137 K2 Sisian Armenia
190 C1 Siskiwit Bay U.S.A.
154 B2 Sisŏphŏn Cambodia
196 B4 Sisquoc r. U.S.A.
198 D2 Sisseton U.S.A.
203 J1 Sisson Branch Reservoir Canada
141 F4 Sistān reg. Iran
127 F8 Sisters Beach Australia
141 E5 Sīt Iran
144 C5 Sitamau India
153 A5 Sitangkai Phil.
144 E4 Sitapur India
171 L7 Siteia Greece
181 I3 Siteki Swaziland
171 J4 Sithonias, Chersonisos pen. Greece
214 C1 Sítio da Abadia Brazil
214 D1 Sítio do Mato Brazil
186 B3 Sitka U.S.A.
144 B3 Sitpur Pak.
164 D4 Sittard Neth.
145 H4 Sittaung Myanmar
165 H1 Sittensen Germany
161 H6 Sittingbourne U.K.
145 H5 Sittwe Myanmar
149 □ Siu A Chau i. Hong Kong China
206 H5 Siuna Nicaragua
145 F5 Siuri India
143 B4 Sivaganga India
143 B4 Sivakasi India
140 D4 Sivand Iran
136 F2 Sivas Turkey
136 B2 Sivaslı Turkey
137 G3 Siverek Turkey
137 G2 Sivrice Turkey
136 C2 Sivrihisar Turkey
181 H3 Sivukile S. Africa
177 E2 Sīwah Egypt
144 D3 Siwalik Range mts India/Nepal
145 F4 Siwan India
144 C4 Siwana India
166 G5 Six-Fours-les-Plages France
148 E3 Sixian China
190 E4 Six Lakes U.S.A.
163 D3 Sixmilecross U.K.
181 H2 Siyabuswa S. Africa
148 F3 Siyang China
138 A4 Siyäzän Azer.
148 C1 Siyitang China
140 D3 Sīyunī Iran
Sjælland i. Denmark see Zealand
171 I3 Sjenica Serbia
159 K5 Sjöbo Sweden
158 L1 Sjøvegan Norway
173 E6 Skadovs'k Ukr.
158 C3 Skaftárós r. mouth Iceland
158 C1 Skagafjörður inlet Iceland
159 J4 Skagen Denmark
159 J4 Skagerrak str. Denmark/Norway
194 B1 Skagit r. Canada/U.S.A.
186 B3 Skagway U.S.A.
158 N1 Skaidi Norway
158 L1 Skaland Norway
158 K2 Skalmodal Sweden
159 J4 Skanderborg Denmark
203 E3 Skaneateles Lake U.S.A.
190 C2 Skanee U.S.A.
171 K5 Skantzoura i. Greece
159 K4 Skara Sweden
159 M4 Skärgårdshavets nationalpark nat. park Fin.
159 J3 Skarnes Norway
169 J5 Skarżysko-Kamienna Poland
158 M2 Skaulo Sweden
169 I6 Skawina Poland
186 D3 Skeena r. Canada
186 D3 Skeena Mountains Canada
161 H4 Skegness U.K.
158 M2 Skellefteå Sweden
158 M2 Skellefteälven r. Sweden
158 M2 Skelleftehamn Sweden
160 E4 Skelmersdale U.K.
163 E4 Skerries Ireland
159 J4 Ski Norway
171 J5 Skiathos i. Greece
163 B6 Skibbereen Ireland
158 M1 Skibotn Norway
160 D3 Skiddaw h. U.K.
159 J4 Skien Norway
169 J5 Skierniewice Poland
176 C1 Skikda Alg.
160 G4 Skipsea U.K.
126 E6 Skipton Australia
160 E4 Skipton U.K.
160 G4 Skirlaugh U.K.
159 J4 Skive Denmark
158 C2 Skjálfandafljót r. Iceland
159 J5 Skjern Denmark
159 I3 Skjolden Norway
139 H5 Skobeleva, Pik mt. Kyrg.
158 I3 Skodje Norway
Skoganvarre Norway see Skoganvarri
158 N1 Skoganvarri Norway
163 F6 Skokholm Island U.K.
190 D4 Skokie U.S.A.
138 E2 Skol' Kazakh.
161 B6 Skomer Island U.K.
171 J5 Skopelos i. Greece
172 F4 Skopin Rus. Fed.
171 I4 Skopje Macedonia
173 F5 Skorodnoye Rus. Fed.
159 K4 Skövde Sweden
203 I2 Skowhegan U.S.A.
159 M4 Skrunda Latvia
186 B2 Skukum, Mount Canada
181 I2 Skukuza S. Africa
196 D3 Skull Peak U.S.A.
190 B5 Skunk r. U.S.A.
159 M4 Skuodas Lith.
159 K5 Skurup Sweden
159 L3 Skutskär Sweden
173 D5 Skvyra Ukr.
162 B3 Skye i. U.K.
171 K5 Skyros Greece
171 K5 Skyros i. Greece
159 J5 Slagelse Denmark
158 L2 Slagnäs Sweden
155 C4 Slamet, Gunung vol. Indon.
163 E4 Slane Ireland
163 E5 Slaney r. Ireland
172 D3 Slantsy Rus. Fed.
173 G5 Slashchevskaya Rus. Fed.
190 D1 Slate Islands Canada
170 G2 Slatina Croatia
171 K2 Slatina Romania
187 G2 Slave r. Canada
176 C4 Slave Coast Africa
186 G3 Slave Lake Canada
139 I1 Slavgorod Rus. Fed.
169 N2 Slavkovichi Rus. Fed.
Slavonija reg. Croatia see Slavonija
171 H2 Slavonija reg. Croatia
171 H2 Slavonski Brod Croatia
173 C5 Slavuta Ukr.
173 D5 Slavutych Ukr.
150 B3 Slavyanka Rus. Fed.
173 F6 Slavyansk-na-Kubani Rus. Fed.
172 D4 Slawharad Belarus
168 H3 Sławno Poland
161 G4 Sleaford U.K.
126 A5 Sleaford Bay Australia
163 A5 Slea Head Ireland
162 C3 Sleat pen. U.K.
162 C3 Sleat, Sound of sea chan. U.K.
188 E2 Sleeper Islands Canada
190 D3 Sleeping Bear Dunes National Lakeshore nature res. U.S.A.
190 D3 Sleeping Bear Point U.S.A.
173 H7 Sleptsovskaya Rus. Fed.
129 C3 Slessor Glacier glacier Antarctica
199 F6 Slidell U.S.A.
163 A5 Slievanea h. Ireland
163 D3 Slieve Anierin h. Ireland
163 D5 Slieveardagh Hills Ireland
163 C4 Slieve Aughty Mountains hills Ireland
163 D3 Slieve Beagh h. Ireland/U.K.
163 C5 Slieve Bernagh hills Ireland
163 D4 Slieve Bloom Mountains hills Ireland
163 B5 Slievecallan h. Ireland
163 B3 Slieve Car h. Ireland
163 F3 Slieve Donard h. U.K.
163 B4 Slieve Elva h. Ireland
Slieve Gamph hills Ireland see Ox Mountains
163 C3 Slieve League h. Ireland
163 B5 Slieve Mish Mountains hills Ireland
163 B5 Slieve Miskish Mountains hills Ireland
163 A3 Slievemore h. Ireland
163 D4 Slieve na Calliagh h. Ireland
163 D5 Slievenamon h. Ireland
163 D2 Slieve Snaght h. Ireland
162 B3 Sligachan U.K.
163 C3 Sligo Ireland
163 C3 Sligo Bay Ireland
159 L4 Slite Sweden
171 L3 Sliven Bulg.
172 H2 Sloboda Rus. Fed.
172 I2 Slobodchikovo Rus. Fed.
171 L2 Slobozia Romania
186 F5 Slocan Canada
164 E1 Slochteren Neth.
172 C4 Slonim Belarus
164 C2 Slootdorp Neth.
164 D2 Sloten Neth.
164 D2 Slotermeer l. Neth.
161 G6 Slough U.K.
169 I6 Slovakia country Europe
170 F1 Slovenia country Europe
170 F1 Slovenj Gradec Slovenia
173 F5 Slov"yans'k Ukr.
154 C3 S'Lung, B'Nom mt. Vietnam
168 H3 Słupsk Poland
158 L2 Slussfors Sweden
172 C4 Slutsk Belarus
163 A4 Slyne Head Ireland
133 L4 Slyudyanka Rus. Fed.
203 I3 Small Point U.S.A.
189 H3 Smallwood Reservoir Canada
172 D4 Smalyavichy Belarus
169 M3 Smarhon' Belarus
180 E5 Smartt Syndicate Dam resr S. Africa
187 I4 Smeaton Canada
171 I2 Smederevo Serbia
171 I2 Smederevska Palanka Serbia
202 D4 Smethport U.S.A.
173 D5 Smila Ukr.
164 E2 Smilde Neth.
159 N4 Smiltene Latvia
139 G1 Smirnovo Kazakh.
186 G3 Smith Canada
196 C2 Smith U.S.A.
202 C6 Smith r. U.S.A.
184 C2 Smith Bay U.S.A.
186 D4 Smithers Canada
181 G5 Smithfield S. Africa
201 E5 Smithfield NC U.S.A.
194 E3 Smithfield UT U.S.A.
203 E5 Smith Island MD U.S.A.
203 F6 Smith Island VA U.S.A.
202 D6 Smith Mountain Lake U.S.A.
186 D3 Smith River Canada
191 I3 Smiths Falls Canada
185 K2 Smith Sound sea chan. Canada/Greenland
127 F8 Smithton Australia
196 C1 Smoke Creek Desert U.S.A.
186 F4 Smoky r. Canada
127 J3 Smoky Cape Australia
188 D3 Smoky Falls Canada
198 C4 Smoky Hill r. U.S.A.
198 D4 Smoky Hills U.S.A.
186 G4 Smoky Lake Canada
158 I3 Smøla i. Norway
138 B2 Smolenka Rus. Fed.
172 E4 Smolensk Rus. Fed.
172 E4 Smolenskaya Oblast' admin. div. Rus. Fed.
139 K1 Smolenskoye Rus. Fed.
171 K4 Smolyan Bulg.
150 C3 Smolyaninovo Rus. Fed.
188 D4 Smooth Rock Falls Canada
188 C3 Smoothrock Lake Canada
187 H4 Smoothstone Lake Canada
158 N1 Smørfjord Norway
203 F5 Smyrna DE U.S.A.
201 C5 Smyrna GA U.S.A.
202 C4 Smyrna OH U.S.A.
203 I1 Smyrna Mills U.S.A.
158 D2 Snæfell mt. Iceland
160 C3 Snaefell h. U.K.
186 A2 Snag (abandoned) Canada
194 D3 Snake r. U.S.A.
197 E2 Snake Range mts U.S.A.
194 D3 Snake River Plain U.S.A.
Snare Lakes Canada see Wekweêtì
125 G6 Snares Islands N.Z.
158 K2 Snåsa Norway
164 D1 Sneek Neth.
163 B6 Sneem Ireland
180 F6 Sneeuberge mts S. Africa
189 H3 Snegamook Lake Canada
161 H5 Snettisham U.K.
132 J3 Snezhnogorsk Rus. Fed.
170 F2 Snežnik mt. Slovenia
169 J4 Śniardwy, Jezioro l. Poland
173 E6 Snihurivka Ukr.
162 B3 Snizort, Loch b. U.K.
194 B2 Snohomish U.S.A.
194 B2 Snoqualmie Pass U.S.A.
158 K2 Snøtinden mt. Norway
187 I2 Snowbird Lake Canada
161 C4 Snowdon mt. U.K.
161 D5 Snowdonia National Park U.K.
197 G4 Snowflake U.S.A.
203 F5 Snow Hill MD U.S.A.
201 E5 Snow Hill NC U.S.A.
187 I4 Snow Lake Canada
126 C4 Snowtown Australia
194 D3 Snowville U.S.A.
127 H6 Snowy r. Australia
127 H6 Snowy Mountains Australia
127 G9 Snug Australia
189 I3 Snug Harbour Nfld Canada
191 G3 Snug Harbour Ont. Canada
154 C2 Snuŏl Cambodia
199 D5 Snyder OK U.S.A.
199 C5 Snyder TX U.S.A.
179 E5 Soalala Madag.
144 B2 Soan r. Pak.
179 E5 Soanierana-Ivongo Madag.
152 D6 Soan-kundo is S. Korea
213 B3 Soata Col.
162 B3 Soay i. U.K.
152 D6 Sobaek-sanmaek mts S. Korea
177 F4 Sobat r. Sudan
147 G7 Sobger r. Indon.
151 B8 Sobo-san mt. Japan
211 J6 Sobradinho, Barragem de resr Brazil
211 J4 Sobral Brazil
173 F7 Sochi Rus. Fed.
152 D5 Sŏch'ŏn S. Korea
123 I5 Society Islands Fr. Polynesia
214 C3 Socorro Brazil
213 B3 Socorro Col.
195 F5 Socorro U.S.A.
204 B5 Socorro, Isla i. Mex.
142 D7 Socotra i. Yemen
154 C3 Soc Trăng Vietnam
167 E3 Socuéllamos Spain
196 D4 Soda Lake U.S.A.
158 N2 Sodankylä Fin.
144 D2 Soda Plains Aksai Chin
194 E3 Soda Springs U.S.A.
159 L3 Söderhamn Sweden
159 L4 Söderköping Sweden
159 L4 Södertälje Sweden
177 E3 Sodiri Sudan
178 D3 Sodo Eth.
159 L3 Södra Kvarken str. Fin./Sweden
181 H1 Soekmekaar S. Africa
164 D3 Soerendonk Neth.
165 G3 Soest Germany
164 D2 Soest Neth.
127 H4 Sofala Australia
171 J3 Sofiya Bulg.
Sofiya Bulg. see Sofiya
158 O2 Sofporog Rus. Fed.
171 L6 Sofrana i. Greece
151 G10 Sōfu-gan i. Japan
145 H3 Sog China
213 B3 Sogamoso Col.
137 G1 Soğanlı Dağları mts Turkey
164 F2 Sögel Germany
159 I4 Søgne Norway
159 I3 Sognefjorden inlet Norway
153 C4 Sogod Phil.
172 H2 Sogra Rus. Fed.
148 A3 Sogruma China
136 C1 Söğüt Turkey
152 D7 Sŏgwip'o S. Korea
Sohâg Egypt see Sūhāj
144 D5 Sohagpur India
161 H5 Soham U.K.
125 F2 Sohano P.N.G.
143 C1 Sohela India
144 D3 Sohna India
152 E3 Sŏho-ri N. Korea
152 C6 Sohŭksan S. Korea
164 C4 Soignes, Forêt de for. Belgium
164 C4 Soignies Belgium
158 N3 Soini Fin.
166 F2 Soissons France
144 C4 Sojat India
153 B4 Sojoton Point Phil.
173 C5 Sokal' Ukr.
152 E4 Sokch'o S. Korea
171 L6 Söke Turkey
173 G7 Sokhumi Georgia
176 C4 Sokodé Togo
149 □ Soko Islands Hong Kong China
172 G3 Sokol Rus. Fed.
169 K4 Sokółka Poland
176 B3 Sokolo Mali
165 K4 Sokolov Czech Rep.
150 C3 Sokolovka Rus. Fed.
169 K4 Sokołow Podlaski Poland
176 C3 Sokoto Nigeria
176 C3 Sokoto r. Nigeria
173 C5 Sokyryany Ukr.
144 D3 Solan India
128 A7 Solander Island N.Z.
143 A2 Solapur India
213 B2 Soledad Col.
196 B3 Soledad U.S.A.
213 E2 Soledad Venez.
207 E4 Soledad de Doblado Mex.
173 G6 Solenoye Rus. Fed.
158 K2 Solfjellsjøen Norway
137 H2 Solhan Turkey
172 G3 Soligalich Rus. Fed.
161 F5 Solihull U.K.
132 G4 Solikamsk Rus. Fed.
138 C2 Sol'-Iletsk Rus. Fed.
207 H4 Solimón, Punta pt Mex.
164 F3 Solingen Germany
180 A1 Solitaire Namibia
137 L1 Şollar Azer.
158 L3 Sollefteå Sweden
165 K3 Söllichau Germany
165 H3 Solling hills Germany
165 I3 Sollstedt Germany
165 G4 Solms Germany
172 F3 Solnechnogorsk Rus. Fed.
155 B3 Solok Indon.
207 G5 Sololá Guat.
125 G2 Solomon Islands country Pacific Ocean
124 F2 Solomon Sea P.N.G./Solomon Is
139 K2 Soloneshnoye Rus. Fed.
190 B2 Solon Springs U.S.A.
147 E7 Solor, Kepulauan is Indon.
168 C7 Solothurn Switz.
172 E1 Solovetskiye Ostrova is Rus. Fed.
172 H3 Solovetskoye Rus. Fed.
170 G3 Šolta i. Croatia
141 E2 Solṭānābād Iran
141 E3 Solṭānābād Iran
165 H2 Soltau Germany
172 D3 Sol'tsy Rus. Fed.
203 E3 Solvay U.S.A.
159 K4 Sölvesborg Sweden
162 E6 Solway Firth est. U.K.
179 C5 Solwezi Zambia
151 G6 Sōma Japan
136 A2 Soma Turkey
164 B4 Somain France
178 E3 Somalia country Africa
218 H5 Somali Basin sea feature Indian Ocean
134 C6 Somaliland reg. Somalia
179 C4 Sombo Angola
171 H2 Sombor Serbia
206 D3 Sombrerete Mex.
144 C4 Somdari India
203 I2 Somerest Junction U.S.A.
159 M3 Somero Fin.
200 C4 Somerset KY U.S.A.
190 E4 Somerset MI U.S.A.
202 D5 Somerset PA U.S.A.
181 F6 Somerset East S. Africa
185 I2 Somerset Island Canada
203 G3 Somerset Reservoir U.S.A.
180 C7 Somerset West S. Africa
203 H3 Somersworth U.S.A.
199 D6 Somerville Reservoir U.S.A.
140 B3 Someydeh Iran
159 K4 Sommen l. Sweden
165 J3 Sömmerda Germany
189 G3 Sommet, Lac du l. Canada
144 B5 Somnath India
190 C5 Somonauk U.S.A.
206 H5 Somotillo Nicaragua
206 H5 Somoto Nicaragua
215 C4 Somuncurá, Mesa Volcánica de plat. Arg.
145 F4 Son r. India
206 I7 Soná Panama
139 G2 Sonaly Karagandinskaya Oblast' Kazakh.
139 G2 Sonaly Karagandinskaya Oblast' Kazakh.
145 F5 Sonamukhi India
145 G5 Sonamura India
143 C1 Sonapur India
144 D4 Sonar r. India
145 H4 Sonari India
152 F2 Sŏnbong N. Korea
152 C4 Sŏnch'ŏn N. Korea
172 E2 Sondaly Rus. Fed.
159 J5 Sønderborg Denmark
165 I3 Sondershausen Germany
170 C1 Sondrio Italy
143 B2 Sonepet India
144 B5 Songadh India
148 E4 Songbu China
Sông Cau Vietnam see Sông Câu
154 D2 Sông Câu Vietnam
179 D5 Songea Tanz.
152 D3 Sŏnggan N. Korea
152 D2 Songhua Hu resr China
152 C1 Songhuajiang China
150 B1 Songhua Jiang r. China
152 C1 Songhua Jiang r. China
152 E2 Songjiang Jilin China
148 F4 Songjiang Shanghai China
152 D2 Songjianghe China
149 C4 Songkan China
154 B4 Songkhla Thai.
139 H4 Songköl l. Kyrg.
148 F1 Song Ling mts China
152 D5 Sŏngnam S. Korea
152 C4 Songnim N. Korea
178 B4 Songo Angola
179 D5 Songo Moz.
148 B3 Songpan China
145 G4 Songsak India
152 D7 Sŏngsan S. Korea
148 D3 Song Shan mt. China
152 D3 Songshu China
149 C4 Songtao China
149 F5 Songxi China
148 D3 Songxian China
152 C1 Songyuan China
149 D4 Songzi China
154 D2 Sơn Ha Vietnam
154 D3 Sơn Hai Vietnam
144 D3 Sonipat India
158 N3 Sonkajärvi Fin.
149 B6 Sơn La Vietnam
141 G5 Sonmiani Pak.
141 G5 Sonmiani Bay Pak.
165 J4 Sonneberg Germany
214 D2 Sono r. Minas Gerais Brazil
211 I6 Sono r. Brazil
197 F6 Sonoita Mex.
197 F6 Sonoita watercourse Mex.
197 G6 Sonoita U.S.A.
206 B1 Sonora r. Mex.
206 B2 Sonora state Mex.
196 B3 Sonora CA U.S.A.
199 C6 Sonora TX U.S.A.
140 B3 Sonqor Iran
213 B3 Sonsón Col.
207 G5 Sonsonate El Salvador
149 B6 Sơn Tây Vietnam
181 H5 Sonwabile S. Africa
215 F1 Sopas r. Uruguay
177 E4 Sopo watercourse Sudan
171 K3 Sopot Bulg.
169 I3 Sopot Poland
168 H7 Sopron Hungary
139 H4 Sopu-Korgon Kyrg.
170 E4 Sora Italy
143 D2 Sorada India
159 L3 Söråker Sweden
152 E4 Sŏrak-san mt. S. Korea
188 F4 Sorel Canada
127 G9 Sorell Australia
127 G9 Sorell Lake Australia
136 E2 Sorgun Turkey
167 E2 Soria Spain
132 C2 Sørkappøya i. Svalbard
140 D3 Sorkh, Kūh-e mts Iran
140 D3 Sorkheh Iran
158 K2 Sørli Norway
159 J5 Sorø Denmark
145 F5 Soro India
173 D5 Soroca Moldova
214 C3 Sorocaba Brazil
138 C1 Sorochinsk Rus. Fed.
139 K1 Sorokino Rus. Fed.
147 G6 Sorol atoll Micronesia
147 F7 Sorong Indon.
178 D3 Soroti Uganda
158 M1 Sørøya i. Norway
167 B3 Sorraia r. Port.
158 L1 Sørreisa Norway
126 F7 Sorrento Australia
170 F4 Sorrento Italy
179 B6 Sorris Sorris Namibia
158 L2 Sorsele Sweden
153 C3 Sorsogon Phil.
172 D2 Sortavala Rus. Fed.
158 K1 Sortland Norway
172 I2 Sortopolovskaya Rus. Fed.
172 I3 Sorvizhi Rus. Fed.
152 D5 Sŏsan S. Korea
181 H2 Soshanguve S. Africa
173 F4 Sosna r. Rus. Fed.
215 C2 Sosneado mt. Arg.
172 J2 Sosnogorsk Rus. Fed.
139 I2 Sosnovka Kazakh.
172 H2 Sosnovka Arkhangel'skaya Oblast' Rus. Fed.
132 F3 Sosnovka Murmanskaya Oblast' Rus. Fed.
172 G4 Sosnovka Tambovskaya Oblast' Rus. Fed.
158 P2 Sosnovyy Rus. Fed.
159 O4 Sosnovyy Bor Rus. Fed.
169 I5 Sosnowiec Poland
173 F6 Sosyka r. Rus. Fed.
213 A4 Sotara, Volcán vol. Col.
158 O2 Sotkamo Fin.
215 D1 Soto Arg.
207 E3 Soto la Marina Mex.
207 G3 Sotuta Mex.
178 B3 Souanké Congo
176 B4 Soubré Côte d'Ivoire
203 F4 Souderton U.S.A.
171 L4 Soufli Greece
166 E4 Souillac France
164 D5 Souilly France
176 C1 Souk Ahras Alg.
Sŏul S. Korea see Seoul
166 D5 Soulom France
Soûr Lebanon see Tyre
167 H4 Sour el Ghozlane Alg.
187 I5 Souris Man. Canada
189 H4 Souris P.E.I. Canada
187 I5 Souris r. Canada/U.S.A.
211 K5 Sousa Brazil
176 D1 Sousse Tunisia
166 D5 Soustons France
180 E4 South Africa, Republic of country Africa
208 South America
191 G3 Southampton Canada
161 F7 Southampton U.K.
203 G4 Southampton U.S.A.
187 L2 Southampton Island Canada
202 E6 South Anna r. U.S.A.
161 F4 South Anston U.K.
189 H2 South Aulatsivik Island Canada
124 D5 South Australia state Australia
218 L7 South Australian Basin sea feature Indian Ocean
199 F5 Southaven U.S.A.
195 F5 South Baldy mt. U.S.A.
160 F3 South Bank U.K.
202 B4 South Bass Island U.S.A.
191 F3 South Baymouth Canada
190 D5 South Bend IN U.S.A.
194 B2 South Bend WA U.S.A.
201 E7 South Bight sea chan. Bahamas
202 D6 South Boston U.S.A.
128 D5 Southbridge N.Z.
203 G3 Southbridge U.S.A.
South Cape pt U.S.A. see Ka Lae
201 D5 South Carolina state U.S.A.
203 I2 South China U.S.A.
155 C1 South China Sea Pacific Ocean
198 C2 South Dakota state U.S.A.
203 G3 South Deerfield U.S.A.
161 G7 South Downs hills U.K.
161 G7 South Downs National Park nat. park U.K.
181 F2 South-East admin. dist. Botswana
127 G9 South East Cape Australia
127 H6 South East Forests National Park Australia
218 J7 Southeast Indian Ridge sea feature Indian Ocean
217 L10 Southeast Pacific Basin sea feature S. Pacific Ocean
187 I3 Southend Canada
162 C5 Southend U.K.
161 H6 Southend-on-Sea U.K.
190 A5 South English U.S.A.
180 E2 Southern admin. dist. Botswana
128 C5 Southern Alps mts N.Z.
124 B5 Southern Cross Australia
187 J3 Southern Indian Lake Canada
177 E4 Southern National Park Sudan
216 E10 Southern Ocean World
201 E5 Southern Pines U.S.A.
209 G7 Southern Thule S. Sandwich Is
162 D5 Southern Uplands hills U.K.
162 F4 South Esk r. U.K.
190 B6 South Fabius r. U.S.A.
216 G7 South Fiji Basin sea feature S. Pacific Ocean
195 F4 South Fork U.S.A.
South Fork South Branch r. U.S.A. see Potomac, South Fork South Branch
190 E3 South Fox Island U.S.A.

South Geomagnetic Pole

Tecolutla

201 D6 **Suwannee** *r.* U.S.A.
123 I4 **Suwarrow** *atoll* Cook Is
137 J5 **Suwayqīyah, Hawr as** *imp. l.* Iraq
137 H6 **Suwayr** *well* Saudi Arabia
177 F2 **Suways, Khalīj as** Egypt
177 F1 **Suways, Qanāt as** Egypt
152 D5 **Suwŏn** S. Korea
138 C4 **Suz, Mys** *pt* Kazakh.
151 F6 **Suzaka** Japan
172 G3 **Suzdal'** Rus. Fed.
148 E3 **Suzhou** *Anhui* China
148 F4 **Suzhou** *Jiangsu* China
152 C3 **Suzi He** *r.* China
151 E6 **Suzu** Japan
151 E7 **Suzuka** Japan
151 E6 **Suzu-misaki** *pt* Japan
158 N1 **Sværholthalvøya** *pen.* Norway
132 C2 **Svalbard** *terr.* Arctic Ocean
173 F5 **Svatove** Ukr.
154 C3 **Svay Riĕng** Cambodia
159 K3 **Sveg** Sweden
159 N4 **Sveķi** Latvia
159 I3 **Svelgen** Norway
158 J3 **Svellingen** Norway
159 N5 **Švenčionėliai** Lith.
159 N5 **Švenčionys** Lith.
159 J5 **Svendborg** Denmark
158 L1 **Svensbu** Norway
Svensby Norway *see* **Svensbu**
158 K3 **Svenstavik** Sweden
Sverdlovsk Rus. Fed. *see* **Yekaterinburg**
173 F5 **Sverdlovs'k** Ukr.
185 I1 **Sverdrup Channel** Canada
171 I4 **Sveti Nikole** Macedonia
146 F2 **Svetlaya** Rus. Fed.
172 B4 **Svetlogorsk** *Kaliningradskaya Oblast'* Rus. Fed.
132 J3 **Svetlogorsk** Rus. Fed.
173 G6 **Svetlograd** Rus. Fed.
172 B4 **Svetlyy** *Kaliningradskaya Oblast'* Rus. Fed.
138 E2 **Svetlyy** *Orenburgskaya Oblast'* Rus. Fed.
173 H5 **Svetlyy Yar** Rus. Fed.
172 D2 **Svetogorsk** Rus. Fed.
158 C2 **Svíahnúkar** *vol.* Iceland
171 L4 **Svilengrad** Bulg.
171 J2 **Svinecea Mare, Vârful** *mt.* Romania
172 C4 **Svir** Belarus
172 E2 **Svir'** *r.* Rus. Fed.
171 K3 **Svishtov** Bulg.
168 H6 **Svitava** *r.* Czech Rep.
168 H6 **Svitavy** Czech Rep.
173 E5 **Svitlovods'k** Ukr.
172 I4 **Sviyaga** *r.* Rus. Fed.
146 E1 **Svobodnyy** Rus. Fed.
158 K1 **Svolvær** Norway
171 J3 **Svrljiške Planine** *mts* Serbia
173 D4 **Svyetlahorsk** Belarus
161 F5 **Swadlincote** U.K.
161 H5 **Swaffham** U.K.
124 F4 **Swain Reefs** Australia
201 D5 **Swainsboro** U.S.A.
123 H4 **Swains Island** *atoll* American Samoa
179 B6 **Swakopmund** Namibia
160 F3 **Swale** *r.* U.K.
125 G3 **Swallow Islands** Solomon Is
187 I4 **Swan** *r.* Canada
161 F7 **Swanage** U.K.
126 E5 **Swan Hill** Australia
186 F4 **Swan Hills** Canada
Swan Islands *is* Hond. *see* **Cisne, Islas del**
187 I4 **Swan Lake** Canada
161 H6 **Swanley** U.K.
126 C5 **Swan Reach** Australia
187 I4 **Swan River** Canada
127 I4 **Swansea** *N.S.W.* Australia
127 H9 **Swansea** *Tas.* Australia
161 D6 **Swansea** U.K.
161 D6 **Swansea Bay** U.K.
203 I2 **Swans Island** U.S.A.
203 G2 **Swanton** U.S.A.
181 G2 **Swartruggens** S. Africa
197 F2 **Swasey Peak** U.S.A.
191 G1 **Swastika** Canada
144 B2 **Swat** *r.* Pak.
Swatow China *see* **Shantou**
181 I3 **Swaziland** *country* Africa
159 K3 **Sweden** *country* Europe
194 B2 **Sweet Home** U.S.A.
201 C5 **Sweetwater** *TN* U.S.A.
199 C5 **Sweetwater** *TX* U.S.A.
194 E3 **Sweetwater** *r.* U.S.A.
180 D7 **Swellendam** S. Africa
168 H5 **Świdnica** Poland
168 G4 **Świdwin** Poland
168 G4 **Świebodzin** Poland
169 I4 **Świecie** Poland
203 H2 **Swift** *r.* U.S.A.
187 H4 **Swift Current** Canada
187 H5 **Swiftcurrent Creek** *r.* Canada
186 C2 **Swift River** Canada
163 D2 **Swilly, Lough** *inlet* Ireland
161 F6 **Swindon** U.K.
163 C4 **Swinford** Ireland
168 G4 **Świnoujście** Poland
162 F5 **Swinton** U.K.
166 I3 **Swiss Tectonic Area Sardona** *tourist site* Switz.
166 H3 **Switzerland** *country* Europe
163 E4 **Swords** Ireland
172 E2 **Syamozero, Ozero** *l.* Rus. Fed.
172 G2 **Syamzha** Rus. Fed.
169 N3 **Syanno** Belarus
172 E2 **Syas'troy** Rus. Fed.
172 H3 **Syava** Rus. Fed.
190 C5 **Sycamore** U.S.A.
127 I4 **Sydney** Australia (City Plan **102**)
189 H4 **Sydney** Canada
187 K4 **Sydney Lake** Canada
189 H4 **Sydney Mines** Canada
173 F5 **Syeverodonets'k** Ukr.
165 G2 **Syke** Germany
172 I2 **Syktyvkar** Rus. Fed.
201 C5 **Sylacauga** U.S.A.
158 K3 **Sylarna** *mt.* Norway/Sweden
145 G4 **Sylhet** Bangl.
172 G2 **Syloga** Rus. Fed.
168 D3 **Sylt** *i.* Germany
201 D5 **Sylvania** *GA* U.S.A.
202 B4 **Sylvania** *OH* U.S.A.
186 G4 **Sylvan Lake** Canada
201 D6 **Sylvester** U.S.A.
186 E3 **Sylvia, Mount** Canada
171 L6 **Symi** *i.* Greece
173 E5 **Synel'nykove** Ukr.
138 C4 **Syngyrli, Mys** *pt* Kazakh.
139 J2 **Syntas** Kazakh.
129 E4 **Syowa** *research stn* Antarctica
170 F6 **Syracuse** *Sicily* Italy
198 C4 **Syracuse** *KS* U.S.A.
203 E3 **Syracuse** *NY* U.S.A.
139 F4 **Syrdar'ya** *r.* Kazakh.
Syrdar'ya Uzbek. *see* **Sirdaryo**
136 G4 **Syria** *country* Asia
137 G5 **Syrian Desert** Asia
171 L6 **Syrna** *i.* Greece
171 K6 **Syros** *i.* Greece
159 N3 **Sysmä** Fin.
172 I2 **Sysola** *r.* Rus. Fed.
172 I4 **Syzran'** Rus. Fed.
168 G4 **Szczecin** Poland
168 H4 **Szczecinek** Poland
169 J4 **Szczytno** Poland
169 J7 **Szeged** Hungary
169 I7 **Székesfehérvár** Hungary
169 I7 **Szekszárd** Hungary
169 J7 **Szentes** Hungary
168 H7 **Szentgotthárd** Hungary
170 G1 **Szigetvár** Hungary
169 J7 **Szolnok** Hungary
168 H7 **Szombathely** Hungary

T

153 B3 **Taal, Lake** Phil.
153 B3 **Tabaco** Phil.
181 H5 **Tabankulu** S. Africa
136 G4 **Ţabaqah** Syria
125 F2 **Tabar Islands** P.N.G.
136 E4 **Tabarja** Lebanon
170 C6 **Tabarka** Tunisia
140 E3 **Ţabas** Iran
207 F4 **Tabasco** *state* Mex.
141 E4 **Tabāsin** Iran
140 C4 **Tābask, Kūh-e** *mt.* Iran
210 E4 **Tabatinga** Col.
153 B2 **Tabayoc, Mount** Phil.
127 F5 **Tabbita** Australia
176 B2 **Tabelbala** Alg.
187 G5 **Taber** Canada
145 F3 **Tabia Tsaka** *salt l.* China
125 H2 **Tabiteuea** *atoll* Kiribati
159 N4 **Tabivere** Estonia
153 B3 **Tablas** *i.* Phil.
153 B3 **Tablas Strait** Phil.
128 F3 **Table Cape** N.Z.
180 C6 **Table Mountain** *h.* S. Africa
198 E4 **Table Rock Reservoir** U.S.A.
214 A2 **Tabocó** *r.* Brazil
168 G6 **Tábor** Czech Rep.
178 D4 **Tabora** Tanz.
139 G4 **Taboshar** Tajik.
176 B4 **Tabou** Côte d'Ivoire
140 B2 **Tabrīz** Iran
123 I3 **Tabuaeran** *atoll* Kiribati
142 A4 **Tabūk** Saudi Arabia
127 J2 **Tabulam** Australia
139 I1 **Tabuny** Rus. Fed.
125 G3 **Tabwémasana, Mount** Vanuatu
159 L4 **Täby** Sweden
206 D4 **Tacámbaro** Mex.
207 F5 **Tacaná, Volcán de** *vol.* Mex.
206 J6 **Tacarcuna, Cerro** *mt.* Panama
139 J3 **Tacheng** China
168 F6 **Tachov** Czech Rep.
153 C4 **Tacloban** Phil.
210 D7 **Tacna** Peru
194 B2 **Tacoma** U.S.A.
215 F1 **Tacuarembó** Uruguay
215 G2 **Tacuarí** *r.* Uruguay
206 B1 **Tacupeto** Mex.
213 E4 **Tacutu** *r.* Brazil
160 F4 **Tadcaster** U.K.
176 C2 **Tademaït, Plateau du** Alg.
125 G4 **Tadin** New Caledonia
178 E2 **Tadjourah** Djibouti
136 G4 **Tadmur** Syria
187 J3 **Tadoule Lake** Canada
189 G4 **Tadoussac** Canada
152 D4 **T'aebaek-sanmaek** *mts* N. Korea/S. Korea
152 C5 **Taech'ŏng-do** *i.* N. Korea
152 C4 **Taedasa-do** N. Korea
152 D4 **Taedong-gang** *r.* N. Korea
152 C5 **Taedong-man** *b.* N. Korea
152 E6 **Taegu** S. Korea
152 C6 **Taehŭksan-kundo** *is* S. Korea
152 D5 **Taejŏn** S. Korea
152 D7 **Taejŏng** S. Korea
152 E5 **T'aepaek** S. Korea
161 C6 **Taf** *r.* U.K.
125 I3 **Tafahi** *i.* Tonga
167 F1 **Tafalla** Spain
176 B4 **Tafiré** Côte d'Ivoire
212 C3 **Tafi Viejo** Arg.
140 C3 **Tafresh** Iran
140 D4 **Taft** Iran
196 C4 **Taft** U.S.A.
141 F4 **Taftān, Kūh-e** *mt.* Iran
173 F6 **Taganrog** Rus. Fed.
173 F6 **Taganrog, Gulf of** Rus. Fed./Ukr.
153 C3 **Tagapula** *i.* Phil.
153 B3 **Tagaytay City** Phil.
153 B4 **Tagbilaran** Phil.
145 E2 **Tagchagpu Ri** *mt.* China
163 E5 **Taghmon** Ireland
186 C2 **Tagish** Canada
170 E1 **Tagliamento** *r.* Italy
167 I4 **Tagma, Col de** *pass* Alg.
153 C4 **Tagoloan** *r.* Phil.
153 B4 **Tagolo Point** Phil.
138 D4 **Tagta** Turkm.
141 F3 **Tagtabazar** Turkm.
125 F3 **Tagula Island** P.N.G.
153 C5 **Tagum** Phil.
167 B3 **Tagus** *r.* Port./Spain *alt.* **Tajo (Spain),** *alt.* **Tejo (Portugal)**
186 F4 **Tahaetkun Mountain** Canada
154 B4 **Tahan, Gunung** *mt.* Malaysia
176 C2 **Tahat, Mont** *mt.* Alg.
146 E1 **Tahe** China
128 D1 **Taheke** N.Z.
123 I5 **Tahiti** *i.* Fr. Polynesia
141 F4 **Tahlab, Dasht-i-** *plain* Pak.
199 E5 **Tahlequah** U.S.A.
196 B2 **Tahoe, Lake** U.S.A.
196 B2 **Tahoe City** U.S.A.
184 H3 **Tahoe Lake** Canada
199 C5 **Tahoka** U.S.A.
176 C3 **Tahoua** Niger
186 D4 **Tahtsa Peak** Canada
153 C6 **Tahuna** Indon.
176 B4 **Taï, Parc National de** *nat. park* Côte d'Ivoire
149 □ **Tai A Chau** *Hong Kong* China
152 B3 **Tai'an** China
148 E2 **Tai'an** China
148 C3 **Taibai Shan** *mt.* China
152 A5 **Taibo Ding** *h.* China
149 F5 **T'aichung** Taiwan
128 C6 **Taieri** *r.* N.Z.
148 D2 **Taigu** China
148 D2 **Taihang Shan** *mts* China
128 E3 **Taihape** N.Z.
148 E3 **Taihe** *Anhui* China
149 E5 **Taihe** *Jiangxi* China
149 E4 **Taihu** China
146 E3 **Tai Hu** *l.* China
149 C5 **Taijiang** China
148 E3 **Taikang** China
Tai Lam Chung Res. China *see* **Tai Lam Chung Shui Tong**
149 □ **Tai Lam Chung Shui Tong** *resr* *Hong Kong* China
126 C5 **Tailem Bend** Australia
Tai Long Bay China *see* **Tai Long Wan**
149 □ **Tai Long Wan** *b.* *Hong Kong* China
149 F5 **T'ailuko** Taiwan
141 F3 **Taimani** *reg.* Afgh.
149 □ **Tai Mo Shan** *h.* *Hong Kong* China
149 F6 **T'ainan** Taiwan
171 J5 **Tainaro, Akrotirio** *pt* Greece
149 E5 **Taining** China
149 □ **Tai O** *Hong Kong* China
214 D1 **Taiobeiras** Brazil
149 F5 **T'aipei** Taiwan
149 C6 **Taiping** China
155 E2 **Taiping** Malaysia
152 E1 **Taipingchuan** China
148 A2 **Taipingpu** China
149 □ **Tai Po** *Hong Kong* China
Tai Po Hoi *b.* *Hong Kong* China *see* **Tolo Harbour**
151 C7 **Taisha** Japan
149 C6 **Taishan** China
149 F5 **Taishun** China
Tai Siu Mo To *is* *Hong Kong* China *see* **Brothers, The**
164 C3 **Taissy** France
212 B7 **Taitao, Península de** *pen.* Chile
128 D5 **Tai Tapu** N.Z.
149 F5 **T'aitung** Taiwan
158 C2 **Taivalkoski** Fin.
158 N1 **Taivaskero** *h.* Fin.
149 F4 **Taiwan** *country* Asia
149 F5 **Taiwan Strait** China/Taiwan
148 F3 **Taixing** China
139 G1 **Taiynsha** Kazakh.
148 D2 **Taiyuan** China
148 D2 **Taiyue Shan** *mts* China
148 F3 **Taizhou** *Jiangsu* China
149 F4 **Taizhou** *Zhejiang* China
149 F4 **Taizhou Wan** *b.* China
152 C3 **Taizi He** *r.* China
142 B7 **Ta'izz** Yemen
207 G5 **Tajamulco, Volcán de** *vol.* Guat.
170 C7 **Tajerouine** Tunisia
139 G5 **Tajikistan** *country* Asia
144 B4 **Tajjal** Pak.
144 D4 **Taj Mahal** *tourist site* India
157 C3 **Tajo** *r.* Spain *alt.* **Tejo (Portugal),** *conv.* **Tagus**
154 A1 **Tak** Thai.
140 B2 **Takāb** Iran
151 C7 **Takahashi** Japan
128 D4 **Takaka** N.Z.
144 D5 **Takal** India
151 D7 **Takamatsu** Japan
151 E6 **Takaoka** Japan
128 F4 **Takapau** N.Z.
128 E2 **Takapuna** N.Z.
151 F6 **Takasaki** Japan
180 F2 **Takatokwane** Botswana
180 D1 **Takatshwaane** Botswana
151 C8 **Takatsuki-yama** *mt.* Japan
151 E6 **Takayama** Japan
154 B4 **Tak Bai** Thai.
151 E7 **Takefu** Japan
139 G4 **Takelí** Tajik.
151 B8 **Takeo** Japan
151 B9 **Take-shima** *i.* N. Pacific Ocean *see* **Lancourt Rocks**
140 C2 **Takestān** Iran
151 B8 **Taketa** Japan
154 C3 **Takêv** Cambodia
137 J7 **Takhādīd** *well* Iraq
154 C3 **Ta Khmau** Cambodia
139 F1 **Takhtabrod** Kazakh.
Takhtakupyr Uzbek. *see* **Taxtako'pir**
137 L5 **Takht Apān, Kūh-e** *mt.* Iran
141 G4 **Takhteh Pol** Afgh.
140 C2 **Takht-e Soleymān** *mt.* Iran
137 K3 **Takht-e Soleymān** *tourist site* Iran
144 B2 **Takht-i-Bahi** *tourist site* Pak.
144 B3 **Takht-i-Sulaiman** *mt.* Pak.
Takht-i-Suleiman *mt.* Iran *see* **Takht-e Soleymān**
150 G3 **Takikawa** Japan
150 H2 **Takinoue** Japan
128 A6 **Takitimu Mountains** N.Z.
186 D3 **Takla Lake** Canada
186 D3 **Takla Landing** Canada
135 G3 **Taklimakan Desert** China
Taklimakan Shamo *des.* China *see* **Taklimakan Desert**
139 G5 **Takob** Tajik.
145 H3 **Takpa Shiri** *mt.* China
186 C3 **Taku** *r.* Canada
154 A3 **Takua Pa** Thai.
176 C4 **Takum** Nigeria
125 F2 **Takuu Islands** *atoll* P.N.G.
215 F2 **Tala** Uruguay
141 F4 **Talab** *r.* Iran/Pak.
172 D4 **Talachyn** Belarus
143 B4 **Talaimannar** Sri Lanka
144 C5 **Talaja** India
215 C1 **Talampaya, Parque Nacional** *nat. park* Arg.
145 H4 **Talap** India
210 B4 **Talara** Peru
Talar-i-Band *mts* Pak. *see* **Makran Coast Range**
139 G4 **Talas** *r.* Kazakh./Kyrg.
139 H4 **Talas** Kyrg.
139 G4 **Talas Ala-Too** *mts* Kyrg.
147 E6 **Talaud, Kepulauan** *is* Indon.
167 D3 **Talavera de la Reina** Spain
133 Q3 **Talaya** Rus. Fed.
153 C5 **Talayan** Phil.
185 K2 **Talbot Inlet** Canada
127 H4 **Talbragar** *r.* Australia
215 B2 **Talca** Chile
215 B3 **Talcahuano** Chile
143 D1 **Talcher** India
138 D4 **Taldyk** Uzbek.
139 I3 **Taldykorgan** Kazakh.
139 I3 **Taldykorganskaya Oblast'** Kazakh.
139 G2 **Taldysay** Kazakh.
139 I4 **Taldy-Suu** Kyrg.
137 L5 **Taleh Zang** Iran
140 C2 **Tālesh** Iran
139 I4 **Talgar** Kazakh.
161 D6 **Talgarth** U.K.
126 A4 **Talia** Australia
147 E7 **Taliabu** *i.* Indon.
153 C4 **Talibon** Phil.
143 B2 **Talikota** India
137 I1 **T'alin** Armenia
143 A3 **Taliparamba** India
153 B4 **Talisay** Phil.
153 C4 **Talisayan** Phil.
137 L2 **Talış Dağları** *mts* Azer./Iran
172 H3 **Talitsa** Rus. Fed.
155 E4 **Taliwang** Indon.
201 C5 **Talladega** U.S.A.
137 I3 **Tall 'Afar** Iraq
201 C6 **Tallahassee** U.S.A.
127 G6 **Tallangatta** Australia
201 C5 **Tallassee** U.S.A.
137 H3 **Tall Baydar** Syria
Tallimarjon Uzbek. *see* **Tallimarjon**
138 F5 **Tallimarjon** Uzbek.
159 N4 **Tallinn** Estonia
136 F4 **Tall Kalakh** Syria
137 I3 **Tall Kayf** Iraq
137 I3 **Tall Kūjik** Syria
163 C5 **Tallow** Ireland
199 F5 **Tallulah** U.S.A.
137 I3 **Tall 'Uwaynāt** Iraq
166 D3 **Talmont-St-Hilaire** France
173 D5 **Tal'ne** Ukr.
177 F3 **Talodi** Sudan
189 G2 **Talon, Lac** *l.* Canada
141 H2 **Tāloqān** Afgh.
138 B2 **Talovaya** Kazakh.
173 G5 **Talovaya** Rus. Fed.
185 I3 **Taloyoak** Canada
144 C2 **Tal Pass** Pak.
159 M4 **Talsi** Latvia
141 F4 **Tal Sīyāh** Iran
212 B3 **Taltal** Chile
187 G2 **Taltson** *r.* Canada
140 B3 **Talvār, Rūdkhāneh-ye** *r.* Iran
158 M1 **Talvik** Norway
127 H2 **Talwood** Australia
173 G5 **Taly** Rus. Fed.
126 E4 **Talyawalka** *r.* Australia
190 A4 **Tama** U.S.A.
213 B2 **Tamalameque** Col.
176 B4 **Tamale** Ghana
213 A3 **Tamana** *mt.* Col.
125 H2 **Tamana** *i.* Kiribati
151 C7 **Tamano** Japan
176 C2 **Tamanrasset** Alg.
145 H4 **Tamanthi** Myanmar
203 F4 **Tamaqua** U.S.A.
161 C7 **Tamar** *r.* U.K.
181 G1 **Tamasane** Botswana
207 E2 **Tamaulipas** *state* Mex.
206 C2 **Tamazula** Mex.
207 E4 **Tamazulápam** Mex.
207 E3 **Tamazunchale** Mex.
178 D3 **Tambach** Kenya
176 A3 **Tambacounda** Senegal
145 F4 **Tamba Kosi** *r.* Nepal
155 C2 **Tambelan, Kepulauan** *is* Indon.
153 A5 **Tambisan** *Sabah* Malaysia
127 G6 **Tambo** *r.* Australia
155 E4 **Tambora, Gunung** *vol.* Indon.
127 G6 **Tamboritha** *mt.* Australia
172 G4 **Tambov** Rus. Fed.
172 G4 **Tambovskaya Oblast'** *admin. div.* Rus. Fed.
167 B1 **Tambre** *r.* Spain
153 A5 **Tambulanan, Bukit** *h.* *Sabah* Malaysia
177 E4 **Tambura** Sudan
153 A5 **Tambuyukon, Gunung** *mt.* *Sabah* Malaysia
176 A3 **Tâmchekkeţ** Mauritania
138 D2 **Tamdy** Kazakh.
Tamdybulak Uzbek. *see* **Tomdibuloq**
213 C3 **Tame** Col.
167 C2 **Tâmega** *r.* Port.
145 H4 **Tamenglong** India
170 B7 **Tamerza** Tunisia
144 D5 **Tamia** India
207 E3 **Tamiahua** Mex.
207 E3 **Tamiahua, Laguna de** *lag.* Mex.
154 A4 **Tamiang, Ujung** *pt* Indon.
143 B4 **Tamil Nadu** *state* India
172 F1 **Tamitsa** Rus. Fed.
136 C7 **Ţāmīyah** Egypt
138 E2 **Tamkamys** Kazakh.
145 F4 **Tamkuhi** India
154 D2 **Tam Ky** Vietnam
201 D7 **Tampa** U.S.A.
201 D7 **Tampa Bay** U.S.A.
159 M3 **Tampere** Fin.
207 E3 **Tampico** Mex.
154 □ **Tampines** Sing.
146 D2 **Tamsagbulag** Mongolia
148 B1 **Tamsag Muchang** China
168 F7 **Tamsweg** Austria
145 H4 **Tamu** Myanmar
207 E3 **Tamuín** Mex.
145 F4 **Tamur** *r.* Nepal
127 I3 **Tamworth** Australia
161 F5 **Tamworth** U.K.
139 I2 **Tan** Kazakh.
178 D4 **Tana** *r.* Kenya
151 D8 **Tanabe** Japan
158 O1 **Tana Bru** Norway
158 O1 **Tanafjorden** *inlet* Norway
155 C4 **Tanah, Tanjung** *pt* Indon.
178 D2 **T'ana Hāyk'** Eth.
155 E3 **Tanahgrogot** Indon.
147 E7 **Tanahjampea** *i.* Indon.
155 A3 **Tanahmasa** *i.* Indon.
153 A6 **Tanahmerah** Indon.
154 B4 **Tanah Merah** Malaysia
154 B4 **Tanah Rata** Malaysia
124 D3 **Tanami Desert** Australia
154 C3 **Tân An** Vietnam
184 C3 **Tanana** U.S.A.
170 C2 **Tanaro** *r.* Italy
153 C4 **Tanauan** Phil.
148 F3 **Tancheng** China
152 E3 **Tanch'ŏn** N. Korea
206 D4 **Tancítaro, Cerro de** *mt.* Mex.
176 B4 **Tanda** Côte d'Ivoire
145 E4 **Tanda** India
153 C4 **Tandag** Phil.
171 L2 **Ţăndărei** Romania
153 A5 **Tandek** *Sabah* Malaysia
144 D2 **Tandi** India
215 E3 **Tandil** Arg.
215 E3 **Tandil, Sierra del** *hills* Arg.
144 B4 **Tando Adam** Pak.
144 B4 **Tando Bago** Pak.
126 E4 **Tandou Lake** *imp. l.* Australia
163 E3 **Tandragee** U.K.
143 B2 **Tandur** India
128 F3 **Taneatua** N.Z.
151 B9 **Tanega-shima** *i.* Japan
154 A1 **Tanen Taunggyi** *mts* Thai.
202 E5 **Taneytown** U.S.A.
176 B2 **Tanezrouft** *reg.* Alg./Mali
178 D4 **Tanga** Tanz.
128 E2 **Tangaehe** N.Z.
145 G4 **Tangail** Bangl.
125 F2 **Tanga Islands** P.N.G.
143 C5 **Tangalla** Sri Lanka
178 C4 **Tanganyika, Lake** Africa
140 D2 **Tangar** Iran
143 B4 **Tangasseri** India
149 B5 **Tangdan** China
140 D4 **Tang-e Kalleh** Iran
138 C5 **Tangeli** Iran
Tanger Morocco *see* **Tangier**
155 C4 **Tangerang** Indon.
165 J2 **Tangerhütte** Germany
165 J2 **Tangermünde** Germany
141 F5 **Tang-e Sarkheh** Iran
148 B3 **Tanggor** China
145 H2 **Tanggulashan** China
145 G2 **Tanggula Shan** *mts* China
145 G2 **Tanggula Shankou** *pass* China
148 D3 **Tanghe** China
144 B2 **Tangi** Pak.
176 B1 **Tangier** Morocco
203 E6 **Tangier Island** U.S.A.
145 G4 **Tangla** India
154 □ **Tanglin** Sing.
145 H3 **Tangmai** China
145 F3 **Tangra Yumco** *salt l.* China
148 F2 **Tangshan** China
153 B4 **Tangub** Phil.
176 C3 **Tanguiéta** Benin
150 A1 **Tangwang He** *r.* China
149 C4 **Tangyan He** *r.* China
148 E3 **Tangyin** China
150 A1 **Tangyuan** China
158 N2 **Tanhua** Fin.
154 C3 **Tani** Cambodia
145 H3 **Taniantaweng Shan** *mts* China
147 F7 **Tanimbar, Kepulauan** *is* Indon.
153 B4 **Tanjay** Phil.
154 A5 **Tanjungbalai** Indon.
Tanjungkarang-Telukbetung Indon. *see* **Bandar Lampung**
155 C3 **Tanjungpandan** Indon.
155 B2 **Tanjungpinang** Indon.
154 A5 **Tanjungpura** Indon.
155 E2 **Tanjungredeb** Indon.
155 E2 **Tanjungselor** Indon.
144 B2 **Tank** Pak.
144 D2 **Tanktse** India
Tankuhi India *see* **Tamkuhi**
180 C6 **Tankwa-Karoo National Park** S. Africa
125 G3 **Tanna** *i.* Vanuatu
162 F4 **Tannadice** U.K.
159 K3 **Tännäs** Sweden
146 B1 **Tannu-Ola, Khrebet** *mts* Rus. Fed.
153 B4 **Tañon Strait** Phil.
144 B4 **Tanot** India
176 C3 **Tanout** Niger
145 E4 **Tansen** Nepal
149 F5 **Tanshui** Taiwan
139 I3 **Tansyk** Kazakh.
177 F1 **Ţanţā** Egypt
176 A2 **Tan-Tan** Morocco
126 D6 **Tantanoola** Australia
207 E3 **Tantoyuca** Mex.
143 C2 **Tanuku** India
159 J4 **Tanumshede** Sweden
126 C5 **Tanunda** Australia
152 E5 **Tanyang** S. Korea
178 D4 **Tanzania** *country* Africa
154 A3 **Tao, Ko** *i.* Thai.
152 B1 **Tao'er He** *r.* China
148 B3 **Tao He** *r.* China
149 D4 **Taojiang** China
148 C2 **Taole** China
152 B1 **Taonan** China
216 G6 **Taongi** *atoll* Marshall Is
170 F6 **Taormina** *Sicily* Italy
195 F4 **Taos** U.S.A.
176 B2 **Taoudenni** Mali
176 B1 **Taourirt** Morocco
149 E5 **Taoxi** China
149 D4 **Taoyuan** China
149 F5 **T'aoyüan** Taiwan
159 N4 **Tapa** Estonia
153 B5 **Tapaan Passage** Phil.
207 F5 **Tapachula** Mex.
211 G4 **Tapajós** *r.* Brazil
155 A2 **Tapaktuan** Indon.
215 E3 **Tapalqué** Arg.
207 F4 **Tapanatepec** Mex.
154 A5 **Tapanuli, Teluk** *b.* Indon.
210 F5 **Tapauá** Brazil
210 E5 **Tapauá** *r.* Brazil
176 B4 **Tapeta** Liberia
154 A3 **Ta Pi, Mae Nam** *r.* Thai.
153 B5 **Tapiantana** *i.* Phil.
190 C2 **Tapiola** U.S.A.
154 B4 **Tapis, Gunung** *mt.* Malaysia
145 F4 **Taplejung** Nepal
149 □ **Tap Mun Chau** *i.* *Hong Kong* China
202 E6 **Tappahannock** U.S.A.
202 C4 **Tappan Lake** U.S.A.
140 C3 **Tappeh, Kūh-e** *h.* Iran
144 C5 **Tapti** *r.* India
128 D4 **Tapuaenuku** *mt.* N.Z.
153 B5 **Tapul** Phil.
153 B5 **Tapul Group** *is* Phil.
213 D5 **Tapurucuara** Brazil
137 J4 **Ţaqţaq** Iraq
214 B1 **Taquaral, Serra do** *hills* Brazil
Taquarí Brazil *see* **Alto Taquarí**
211 G7 **Taquarí** *r.* Brazil
214 A2 **Taquarí, Pantanal do** *marsh* Brazil
214 A2 **Taquari, Serra do** *hills* Brazil
214 C3 **Taquaritinga** Brazil
214 B3 **Taquaruçu** *r.* Brazil
163 D5 **Tar** *r.* Ireland
127 I1 **Tara** Australia
163 E4 **Tara, Hill of** Ireland
176 D4 **Taraba** *r.* Nigeria
210 E7 **Tarabuco** Bol.
Ţarābulus Libya *see* **Tripoli**
213 C4 **Taracua** Brazil
144 E4 **Tarahuwan** India
145 G4 **Tarai** *reg.* India
155 E2 **Tarakan** Indon.
153 A6 **Tarakan** *i.* Indon.
136 C1 **Taraklı** Turkey
127 H5 **Taralga** Australia
169 I3 **Taran, Mys** *pt* Rus. Fed.
127 H4 **Tarana** Australia
144 D5 **Tarana** India
144 C3 **Taranagar** India
128 E3 **Taranaki, Mount** *vol.* N.Z.
167 E2 **Tarancón** Spain
162 A3 **Taransay** *i.* U.K.
170 G4 **Taranto** Italy
170 G4 **Taranto, Golfo di** *g.* Italy
210 C5 **Tarapoto** Peru
128 E4 **Tararua Range** *mts* N.Z.
169 O6 **Tarashcha** Ukr.
210 D5 **Tarauacá** Brazil
210 E5 **Tarauacá** *r.* Brazil
123 G3 **Tarawa** *atoll* Kiribati
128 F3 **Tarawera** N.Z.
128 F3 **Tarawera, Mount** *vol.* N.Z.
139 G4 **Taraz** Kazakh.
167 F2 **Tarazona** Spain
167 F3 **Tarazona de la Mancha** Spain
139 J3 **Tarbagatay** Kazakh.
139 J3 **Tarbagatay, Khrebet** *mts* Kazakh.
162 E3 **Tarbat Ness** *pt* U.K.
144 C2 **Tarbela Dam** Pak.
163 B5 **Tarbert** Ireland
162 B3 **Tarbert** *Scotland* U.K.
162 C5 **Tarbert** *Scotland* U.K.
166 E5 **Tarbes** France
201 E5 **Tarboro** U.S.A.
126 A3 **Tarcoola** Australia
127 G3 **Tarcoon** Australia
127 G5 **Tarcutta** Australia
146 F2 **Tardoki-Yangi, Gora** *mt.* Rus. Fed.
127 J3 **Taree** Australia
126 E3 **Tarella** Australia
132 K2 **Tareya** Rus. Fed.
140 C6 **Ţarfā', Baţn aţ** *depr.* Saudi Arabia
194 E2 **Targhee Pass** U.S.A.
171 K2 **Târgovişte** Romania
171 J2 **Târgu Jiu** Romania
169 L7 **Târgu Mureş** Romania
169 M7 **Târgu Neamţ** Romania
169 M7 **Târgu Secuiesc** Romania
140 B3 **Tarhān** Iran
177 D1 **Tarhūnah** Libya
148 C2 **Tarian Gol** China
140 D5 **Tarif** U.A.E.
167 D4 **Tarifa** Spain
167 D4 **Tarifa, Punta de** *pt* Spain
210 F8 **Tarija** Bol.
147 F7 **Tariku** *r.* Indon.
142 C6 **Tarīm** Yemen
135 G3 **Tarim Basin** China
139 J4 **Tarim He** *r.* China
Tarim Pendi *basin* China *see* **Tarim Basin**
141 G3 **Tarīn Kowt** Afgh.
147 F7 **Taritatu** *r.* Indon.
137 J4 **Tarjīl** Iraq
181 F6 **Tarka** *r.* S. Africa
181 G6 **Tarkastad** S. Africa
198 E3 **Tarkio** U.S.A.
132 I3 **Tarko-Sale** Rus. Fed.
176 B4 **Tarkwa** Ghana
153 B3 **Tarlac** Phil.
165 H1 **Tarmstedt** Germany
166 F4 **Tarn** *r.* France
158 K2 **Tärnaby** Sweden
141 G3 **Tarnak** *r.* Afgh.
169 L7 **Târnăveni** Romania
169 J5 **Tarnobrzeg** Poland
172 G2 **Tarnogskiy Gorodok** Rus. Fed.
169 J5 **Tarnów** Poland
145 E3 **Taro Co** *salt l.* China
140 D4 **Ţārom** Iran
176 B1 **Taroudannt** Morocco
126 D6 **Tarpeena** Australia
201 E7 **Tarpum Bay** Bahamas
140 C3 **Tarq** Iran
146 E2 **Tarqi** China
170 D3 **Tarquinia** Italy
176 □ **Tarrafal** Cape Verde
167 G2 **Tarragona** Spain
158 L2 **Tårrajaur** Sweden
127 G9 **Tarraleah** Australia
127 G4 **Tarran Hills** Australia
128 B6 **Tarras** N.Z.
167 G2 **Tàrrega** Spain
136 E3 **Tarsus** Turkey
138 C4 **Tarta** Turkm.
212 D2 **Tartagal** Arg.
137 K1 **Tärtär** Azer.
137 K1 **Tärtär** *r.* Azer.
166 D5 **Tartas** France
159 N4 **Tartu** Estonia
136 E4 **Ţarţūs** Syria
214 E2 **Tarumirim** Brazil
173 H6 **Tarumovka** Rus. Fed.
154 A4 **Tarutao, Ko** *i.* Thai.
154 A5 **Tarutung** Indon.
170 E1 **Tarvisio** Italy
140 E4 **Tarz** Iran
139 H3 **Tasaral** Kazakh.
138 F3 **Tasbuget** Kazakh.
188 E4 **Taschereau** Canada
143 A2 **Tasgaon** India
152 A3 **Tashan** China
145 G4 **Tashigang** Bhutan
137 J1 **Tashir** Armenia
140 D4 **Tashk, Daryācheh-ye** *l.* Iran
Tashkepri Turkm. *see* **Daşköpri**
139 H4 **Tash-Kömür** Kyrg.
138 C2 **Tashla** Rus. Fed.
189 F2 **Tasialujjuaq, Lac** *l.* Canada
188 F2 **Tasiat, Lac** *l.* Canada
155 C4 **Tasikmalaya** Indon.
189 G2 **Tasiujaq** Canada
138 B2 **Taskala** Kazakh.
139 J3 **Taskesken** Kazakh.
136 E1 **Taşköprü** Turkey
137 I2 **Taşlıçay** Turkey
216 F8 **Tasman Abyssal Plain** *sea feature* Australia
218 O7 **Tasman Basin** *sea feature* Tasman Sea
128 D4 **Tasman Bay** N.Z.
127 G9 **Tasman Head** Australia
127 F9 **Tasmania** *state* Australia
128 D4 **Tasman Mountains** N.Z.
127 H9 **Tasman Peninsula** Australia
125 F5 **Tasman Sea** Pacific Ocean
136 F1 **Taşova** Turkey
176 C2 **Tassili n' Ajjer, Parc National de** *nat. park* Alg.
196 B3 **Tassajara Hot Springs** U.S.A.
139 G3 **Tasty** Kazakh.
138 F2 **Tasty-Taldy** Kazakh.
137 J2 **Tasūj** Iran
133 M3 **Tas-Yuryakh** Rus. Fed.
169 I7 **Tatabánya** Hungary
147 E7 **Tatamailau, Foho** *mt.* East Timor
173 D6 **Tatarbunary** Ukr.
132 I4 **Tatarsk** Rus. Fed.
146 G1 **Tatarskiy Proliv** *str.* Rus. Fed.
172 I4 **Tatarstan, Respublika** *aut. rep.* Rus. Fed.
140 B2 **Tatavi** *r.* Iran
151 F7 **Tateyama** Japan
151 E6 **Tate-yama** *vol.* Japan
186 F2 **Tathlina Lake** Canada
142 B5 **Tathlīth, Wādī** *watercourse* Saudi Arabia
127 H6 **Tathra** Australia
187 J2 **Tatinnai Lake** Canada
173 H5 **Tatishchevo** Rus. Fed.
194 A1 **Tatla Lake** Canada
186 D3 **Tatlatui Provincial Park** Canada
Tat Mailau, Gunung *mt.* East Timor *see* **Tatamailau, Foho**
127 G6 **Tatong** Australia
Tatra Mountains *reg.* Poland *see* **Tatry**
169 I6 **Tatry** *reg.* Poland
186 B3 **Tatshenshini** *r.* Canada
173 G5 **Tatsinskiy** Rus. Fed.
151 D7 **Tatsuno** Japan
139 H4 **Tatty** Kazakh.
214 C3 **Tatuí** Brazil
186 E4 **Tatuk Mountain** Canada
199 C5 **Tatum** U.S.A.
126 F6 **Tatura** Australia
137 I2 **Tatvan** Turkey
159 I4 **Tau** Norway
211 J5 **Taua** Brazil
214 D3 **Taubaté** Brazil
165 H5 **Tauber** *r.* Germany
165 H5 **Tauberbischofsheim** Germany
165 K3 **Taucha** Germany
165 H4 **Taufstein** *h.* Germany
139 H3 **Taukum, Peski** *des.* Kazakh.
180 F3 **Taung** S. Africa
147 B4 **Taunggyi** Myanmar
147 B5 **Taung-ngu** Myanmar
154 A2 **Taungnyo Range** *mts* Myanmar
161 D6 **Taunton** U.K.
203 H4 **Taunton** U.S.A.
165 F4 **Taunus** *hills* Germany
128 F3 **Taupo** N.Z.
128 E3 **Taupo, Lake** N.Z.
159 M5 **Tauragė** Lith.
128 F2 **Tauranga** N.Z.
170 G5 **Taurianova** Italy
128 D1 **Tauroa Point** N.Z.
136 D3 **Taurus Mountains** Turkey
138 B3 **Taushyk** Kazakh.
Tauu Islands P.N.G. *see* **Takuu Islands**
136 B3 **Tavas** Turkey
161 I5 **Taverham** U.K.
167 C4 **Tavira** Port.
161 C7 **Tavistock** U.K.
154 A2 **Tavoy** Myanmar
154 A2 **Tavoy Point** Myanmar
150 B3 **Tavrichanka** Rus. Fed.
136 B2 **Tavşanlı** Turkey
161 C6 **Taw** *r.* U.K.
153 A5 **Tawai, Bukit** *mt.* *Sabah* Malaysia
191 F3 **Tawas Bay** U.S.A.
191 F3 **Tawas City** U.S.A.
155 E2 **Tawau** *Sabah* Malaysia
161 D6 **Tawe** *r.* U.K.
144 C2 **Tawi** *r.* India
153 A5 **Tawi-Tawi** *i.* Phil.
149 F6 **Tawu** Taiwan
207 E4 **Taxco** Mex.
138 D4 **Taxiatosh** Uzbek.
144 C2 **Taxila** *tourist site* Pak.
139 H5 **Taxkorgan** China
138 E4 **Taxtako'pir** Uzbek.
186 C2 **Tay** *r.* Canada
162 E4 **Tay** *r.* U.K.
162 E4 **Tay, Firth of** *est.* U.K.
162 D4 **Tay, Loch** *l.* U.K.
153 B3 **Tayabas Bay** Phil.
158 P1 **Taybola** Rus. Fed.
162 C5 **Tayinloan** U.K.
186 E3 **Taylor** Canada
197 G4 **Taylor** *AZ* U.S.A.
191 F4 **Taylor** *MI* U.S.A.
190 B6 **Taylor** *MO* U.S.A.
198 D3 **Taylor** *NE* U.S.A.
199 D6 **Taylor** *TX* U.S.A.
203 E5 **Taylors Island** U.S.A.
200 B4 **Taylorville** U.S.A.
142 A4 **Taymā'** Saudi Arabia
133 K3 **Taymura** *r.* Rus. Fed.
133 L2 **Taymyr, Ozero** *l.* Rus. Fed.
133 K2 **Taymyr Peninsula** Rus. Fed.
Taymyr, Poluostrov *pen.* Rus. Fed. *see* **Taymyr Peninsula**
139 G1 **Tayncha** Kazakh.
154 C3 **Tây Ninh** Vietnam
206 C2 **Tayoltita** Mex.
138 B2 **Taypak** Kazakh.
138 C2 **Taysoygan, Peski** *des.* Kazakh.
153 A4 **Taytay** Phil.
153 B3 **Taytay** Phil.
153 A4 **Taytay Bay** Phil.
141 F3 **Tayyebād** Iran
133 J2 **Taz** *r.* Rus. Fed.
176 B1 **Taza** Morocco
137 J4 **Tāza Khurmātū** Iraq
137 K2 **Tazeh Kand** Azer.
202 B6 **Tazewell** *TN* U.S.A.
202 C6 **Tazewell** *VA* U.S.A.
187 H2 **Tazin** *r.* Canada
187 H3 **Tazin Lake** Canada
177 E2 **Tāzirbū** Libya
167 I4 **Tazmalt** Alg.
132 I3 **Tazovskaya Guba** *sea chan.* Rus. Fed.
Tbilisi Georgia *see* **T'bilisi**
173 H7 **T'bilisi** Georgia
173 G6 **Tbilisskaya** Rus. Fed.
178 B4 **Tchibanga** Gabon
177 D2 **Tchigaï, Plateau du** Niger
177 D4 **Tcholliré** Cameroon
169 I3 **Tczew** Poland
206 C3 **Teacapán** Mex.
128 A6 **Te Anau** N.Z.
128 A6 **Te Anau, Lake** N.Z.
207 F4 **Teapa** Mex.
128 G2 **Te Araroa** N.Z.
128 E2 **Te Aroha** N.Z.
128 E3 **Te Awamutu** N.Z.
160 E3 **Tebay** U.K.
187 J2 **Tebesjuak Lake** Canada
176 C1 **Tébessa** Alg.
170 B7 **Tébessa, Monts de** *mts* Alg.
212 E3 **Tebicuary** *r.* Para.
155 A2 **Tebingtinggi** Indon.
155 B3 **Tebingtinggi** Indon.
170 C6 **Tébourba** Tunisia
170 C6 **Téboursouk** Tunisia
173 H7 **Tebulos Mt'a** Georgia/Rus. Fed.
176 B4 **Techiman** Ghana
212 B6 **Tecka** Arg.
164 F2 **Tecklenburger Land** *reg.* Germany
207 E3 **Tecolutla** Mex.

V

W

ACKNOWLEDGEMENTS

Maps, design and origination by Collins Geo, HarperCollins Publishers, Glasgow

Population statistics: UN Department of Economic and Social Affairs Population Division

Earthquake data: United States Geological Survey (USGS) National Earthquakes Information Center, Denver, USA

Köppen classification map: Kottek, M., J. Grieser, C. Beck, B. Rudolf, and F. Rubel, 2006: World Map of the Köppen-Geigerclimate classification updated. *Meteorol. Z.*, 15, 259–263. http://koeppen-geiger.vu-wien.ac.at

Climate Change 2007: Impacts, Adaptation and Vulnerability, summary for Policymakers, Intergovernmental Panel on Climate Change

Population map: Center for International Earth Science Information Network (CIESIN), Columbia University

International Food Policy Research Institute (IFPRI); and World Resources Institute (WRI). 2000 Gridded Population of the World (GPW), Version 3. Palisades, NY: CIESIN, Columbia University Available at http://sedac.ciesin.columbia.edu/plue/gpw

IMAGE CREDITS

Pages 4–17
Blue Marble: Next Generation. NASA's Earth Observatory

Pages 18–19
NRSC Ltd/Science Photo Library and Blue Marble: Next Generation. NASA's Earth Observatory

Pages 20–21
NASA/GSFC/METI/ERSDAC/JAROS and U.S./Japan ASTER Science Team

Pages 22–23
MODIS/NASA

Pages 24–25
NASA/GSFC/METI/ERSDAC/JAROS, and U.S./Japan ASTER Science Team

Pages 26–27
USGS EROS DATA CENTER

Pages 28–33
IKONOS images courtesy of GeoEye. Copyright 2008

Pages 34–35
IKONOS image courtesy of GeoEye/Telespazio. Copyright 2009

Pages 90–91
Haiti earthquake: © United Nations Development Programme. Licensed under the Creative Commons 2.0 Attribution License. Mount Bromo: ©Michael Pitts/naturepl.com

Pages 92–93
Cyclone Nargis: MODIS/NASA

Pages 94–95
McCarty Glacier: NSIDC/U. S. Grant. (top); NSIDC/Bruce F. Molina (bottom)

Pages 96–97
Hong Kong, China: ©Justin Guariglia/Getty Images

Pages 100–101
Lake Eyre: MODIS/NASA
Mississippi: ASTER/NASA
Caspian Sea: MODIS/NASA
Madagascar: MODIS/NASA

Cover
Amazon River, Brazil: Image courtesy of the Image Science and Analysis Laboratory, NASA Johnson Space Center

http://eol.jsc.nasa.gov

GeoEye   www.geoeye.com
NASA   earthobservatory.nasa.gov
NASA   rapidfire.sci.gsfc.nasa.gov
NASA   asterweb.jpl.nasa.gov/index.asp
TeleGeography   www.telegeography.com
United States Geological Survey   www.usgs.gov

KEY TO THE MAP PAGES

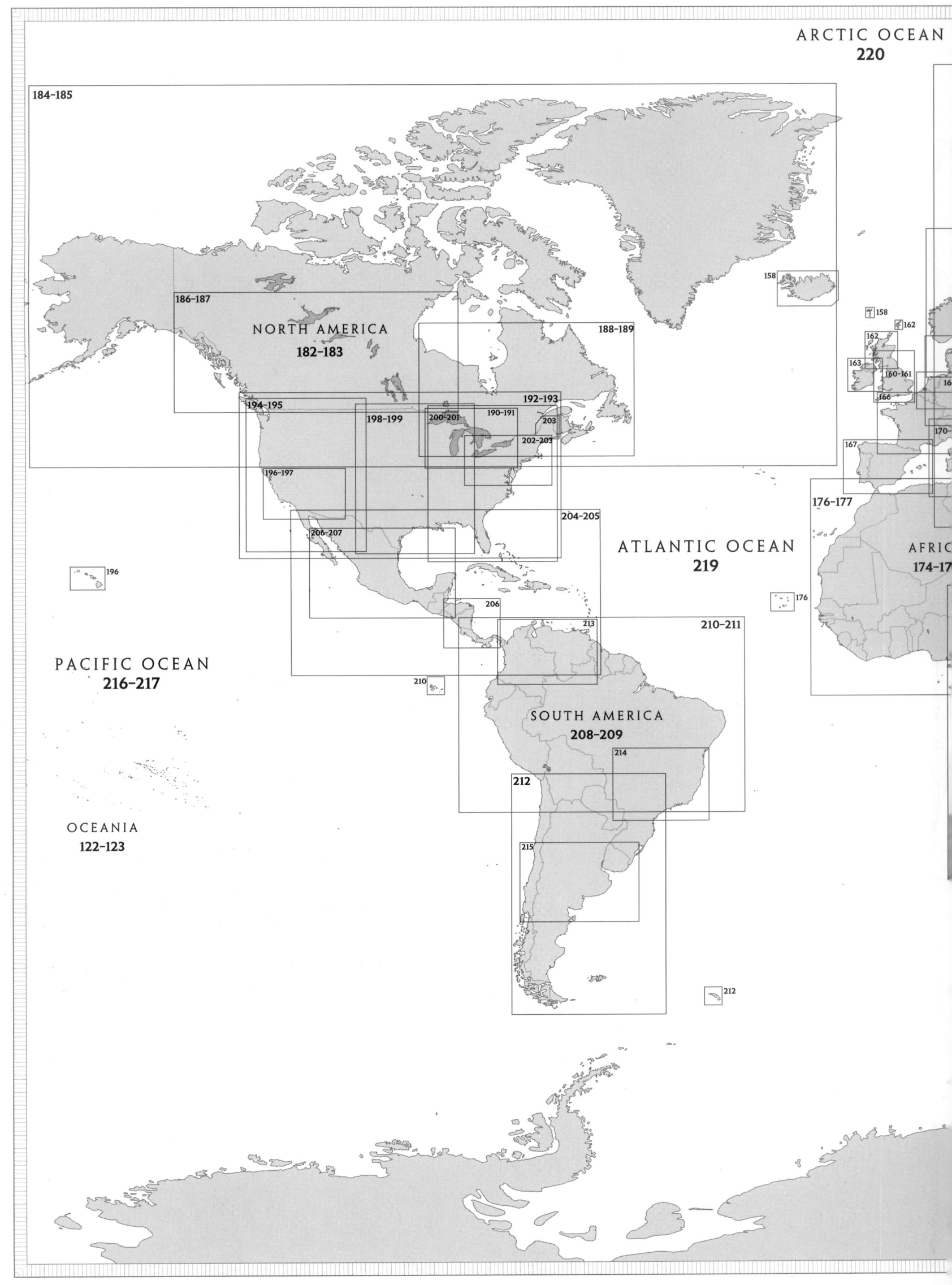

ARCTIC OCEAN
220

184–185

186–187

NORTH AMERICA
182–183

188–189

192–193

194–195

198–199

200–201

190–191

203

202–203

196–197

204–205

206–207

196

ATLANTIC OCEAN
219

206

213

210–211

210

212

SOUTH AMERICA
208–209

214

215

PACIFIC OCEAN
216–217

OCEANIA
122–123

158

158

162

162

163

160–161

166

16

170–

167

176–177

AFRIC
174–17

176